Semiconductor Radiation Detectors

Series in Sensors

Semiconductor Radiation Detectors

Alan Owens

*Institute of Experimental and Applied Physics,
Prague, The Czech Republic*

CRC Press
Taylor & Francis Group
Boca Raton London New York

CRC Press is an imprint of the
Taylor & Francis Group, an **informa** business

CRC Press
Taylor & Francis Group
6000 Broken Sound Parkway NW, Suite 300
Boca Raton, FL 33487-2742

First issued in paperback 2021

ISBN 13: 978-0-367-77968-9 (pbk)
ISBN 13: 978-1-138-07074-5 (hbk)

Library of Congress Cataloging-in-Publication Data

Names: Owens, Alan, author.
Title: Semiconductor radiation detectors / Alan Owens.
Other titles: Sensors series.
Description: Boca Raton, FL : CRC Press, Taylor & Francis Group, [2019] | Series: Series in sensors | Includes bibliographical references and index.
Identifiers: LCCN 2019001092| ISBN 9781138070745 (hbk ; alk. paper) | ISBN 1138070742 (hbk ; alk. paper) | ISBN 9781315114934 (ebk) | ISBN 1315114933 (ebk)
Subjects: LCSH: X-rays–Measurement–Materials. | Gamma ray detectors–Materials. | X-ray diffractometer–Materials. | Semiconductors. | Compound semiconductors.
Classification: LCC QC481.5 .O943 2019 | DDC 539.7/7–dc23
LC record available at https://lccn.loc.gov/2019001092

Visit the Taylor & Francis Web site at
www.taylorandfrancis.com

and the CRC Press Web site at
www.crcpress.com

For Cecilia, Katie, Andrea and Thelma

Contents

List of Acronyms

0D	Zero Dimensional
1D	One Dimensional
2D	Two Dimensional
3D	Three Dimensional
ADC	Analogue to Digital Converter
ADR	Adiabatic Demagnetization Refrigerator
AES	Auger Electron Spectroscopy
ALD	Atomic Layer Deposition
APD	Avalanche Photo Diode
ASIC	Application Specific Integrated Circuit
BCC	Body Centered Cubic
BESSY	Berliner Elektronenspeicherring-Gesellschaft für Synchrotronstrahlung
C-V	Capacitance-Voltage
CB	Conduction Band
CCD	Charged Coupled Device
CCE	Charged Collection Efficiency
CD	Circular Dichroism
CMOS	Complementary Metal Oxide Semiconductor
CTLM	Circular Transfer Length Method
CVD	Chemical Vapor Deposition
Cz	Czochralski
DAC	Digital to Analogue Conversion
DEPFET	DEpleted P-Channel Field Effect Transistor
DESY	Deutsches Elektronen-SYnchrotron
DEZ	Diethylzinc
DLTS	Deep Level Trap Spectroscopy
DMS	Dilute Magnetic Semiconductor
DNA	Deoxyribonucleic Acid
DOS	Density of States
DQE	Detective Quantum Efficiency
DSP	Digital Signal Processing
ENC	Equivalent Noise Charge
EPD	Etch Pit Density
ESRF	European Synchrotron Research Facility
EXAFS	Extended X-Ray Absorption Fine Structure
FCC	Face Centered Cubic
FE	Field Emission
FET	Field Effect Transistor
FWHM	Full-Width at Half Maximum
FZ	Float Zone
GCMS	Gas-Chromatography Mass Spectrometry
GCR	Galactic Cosmic Rays
GDMS	Glow-Discharge Mass Spectrometry
GFET	Graphene Field Effect Transistor
GMR	Giant Magneto-Resistance
HASYLAB	HAmburger SYnchrotronstrahlungsLABor
HEMT	High Electron Mobility Transistor
HBT	Heterojunction Bipolar Transistor
HCP	Hexagonal Closed Packed
HPB	High Pressure Bridgman
HVPE	Hydride Vapor Phase Epitaxy

I-V	Current-Voltage
ICP-MS	Inductively Coupled Plasma, Mass Spectroscopy
ICP-OES	Inductively Coupled Plasma Optical Emission Spectroscopy
LBL	Laurence Berkeley National Laboratory
LEC	Liquid Encapsulated Czochralski
LED	Light-Emitting Diode
LD	Laser Diode
LN2	Liquid Nitrogen
LA	Longitudinal Acoustic
LO	Longitudinal Optical
LPE	Liquid Phase Epitaxy
LPI	Lines Per Inch
LSB	Least Significant Bit
MBE	Molecular Beam Epitaxy
MCA	Multi Channel Analyser
MEMS	Micro-Electro-Mechanical System
MIGS	Metal-Induced Gap States
MIP	Minimum Ionizing Particle
MIR	Mid Infrared
MOCVD	Metal Organic Chemical Vapor Deposition
MSB	Most Significant Bit
MSM	Metal-Semiconductor-Metal
MTF	Modulation Transfer Function
MTPVT	Multi-Tube Physical Vapor Transport
NC	Nano Crystalline
NEP	Noise Equivalent Power
NIEL	Non Ionizing Energy Loss
NIR	Near Infrared
NIST	National Institute of Standards and Technology
NPL	National Physical laboratory
PECVD	Plasma-Enhanced Chemical Vapor Deposition
PICTS	Photo Induced Current Transient Spectroscopy
PIXE	Particle Induced X-Ray Emission
PKO	Primary Knock-On
PL	Photoluminescence
PMT	Photo Multiplier Tube
PTB	Physikalisch-Technische Bundesanstalt
PUR	Pile-Up Rejection
PVD	Physical Vapor Deposition
QD	Quantum Dot
QE	Quantum Efficiency
QW	Quantum Well
RAS	Reflection Anisotropy Spectroscopy
RBS	Rutherford Back Scattering
RF	Radio-Frequency
RHEED	Reflection High Energy Electron Diffraction
ROST	Rapid Optical Surface Treatment
RT	Room Temperature
RTD	Rise Time Discrimination
SAA	South Atlantic Anomaly
SBD	Surface Barrier Detector
SBH	Schottky Barrier Height
SE	Spectroscopic Ellipsometry
SEM	Scanning Electron Microscopy
SEP	Solar Energetic Particles
SNM	Special Nuclear Material

SPE	Solid Phase Epitaxy
SRIM	Stopping and Range of Ions in Matter
STJ	Superconducting Tunnel Junction
STM	Scanning Tunneling Microscopy
SRH	Shockley Read Hall
SSL	Solid-State Lighting
TA	Transverse Acoustic
TAC	Time Amplitude Converter
TE	Thermionic Emission
TEC	Thermo Electric Cooler
TEM	Transmission Electron Microscopy
TES	Transition Edge Sensor
TFE	Thermally Assisted Field Emission
THM	Travelling Heater Method
TLM	Transfer Length Method
TMA	Trimethylaluminum
TMG	Trimethylgallium
TMI	Trimethylindium
TMZ	Travelling Molten Zone
TO	Transverse Optical
TOF	Time-of-Flight
TRIM	Transport of Ions in Matter
TSC	Thermally Stimulated Current
UHV	Ultra High Vacuum
UV	Ultraviolet
VAM	Vacuum Ampoule Method
VB	Valence Band
VCz	Vapor Pressure Controlled Czochralski
VDWE	Van Der Waals Epitaxy
VGF	Vertical Gradient Freeze
VPE	Vapor Phase Epitaxy
VPG	Vapor Phase Growth
WBXRT	White Beam X-Ray Tomography
ZB	Zinc blende
XAFS	X-Ray Absorption Fine Structure
XANES	X-Ray Absorption Near Edge Structure
XPS	X-Ray Photoelectron Spectroscopy
XRD	X-Ray Diffraction
XRT	X-Ray Tomography

Preface

The study of semiconductors has grown substantially since the first investigations of conductivity in natural crystals in the early 1920s and the first conceptualization and rudimentary understanding of the solid state in the 1930s. It has been one of the main drivers of modern physics and is now ingrained in virtually all areas of physics. In fact, by exploiting the microscopic properties it deals with it has enabled the majority of modern technology – from material science and crystal growth, through to electronics. The application of semiconductors to radiation detection and measurement through the understanding of the electronic properties of matter has come to dominate radiation physics – a field that did not exist 150 years ago. Of special interest in radiation detection and measurement has been the rise of the compound semiconductors. The term "compound semiconductor" encompasses a wide range of materials, most of which crystallize in the zincblende, wurtzite or rock salt crystal structures. These and are so useful because of the shear range of compounds available and the wide range of physical properties they encompass compared to the elemental semiconductors, Sn-a, C, Si and Ge. They are generally derived from elements in Groups II through VI of the periodic table. Most elements in these groups are soluble within each other, forming homogeneous solid solutions over large ranges of miscibility. In addition to binary materials (such as GaAs or InP), most compounds are also soluble within each other, making it possible to synthesize ternary (*e.g.*, AlGaAs, HgCdTe) quaternary (*e.g.*, In- GaAsP, InGaAlP) and higher-order solutions, simply by alloying binary compounds together. In terms of radiation detection, compound semi-conductors were among the first direct detection media to be investigated over 70 years ago. However, material problems caused by impurities, high-defect densities and stoichiometric imbalances have limited their usefulness. Only recently has substantial progress been made in the technological aspects of material growth and detector fabrication, allowing compound semiconductor radiation detectors to become serious competitors to laboratory standards such as Ge and Si.

In this book, we shall examine the properties of semiconductors, their growth and characterization and the fabrication of radiation sensors with emphasis on the X- and gamma-ray regimes. We explore their promise and limitations and discuss where the future may lie, for example in material development by lattice hybridization and dimensional reduction or in radiation detection by exploiting DNA-origami, reduced dimensionality or other obscure degrees of freedom of the electron. The purpose of the book is two-fold (a) to serve as a textbook for those new to the field of compound semiconductor development and its application to radiation detection and measurement and (b) as a general reference book for the established researcher. As with all enterprises, the production of this book has greatly benefited from the efforts and inputs of a large number of people, most notably colleagues in the Science and Robotic Exploration Directorate of the European Space Agency. I am particularly indebted to Dr. Tone Peacock who as Head of the Science Payload and Advanced Concepts Office instigated and nurtured much of the Agency's research efforts in compound semiconductors. In addition, I would like to thank the following people: Dr. V. Gostilo of Baltic Scientific Instruments, Latvia, who elucidated much of the practical information presented and Dr. C. Hansson of Redlen Technology, Canada, for inputs and reviewing the text.

Alan Owens
Noordwijk, The Netherlands
November 2018.

About the Author

Dr. Alan Owens holds an undergraduate honours degree in physics and physical electronics and earned his doctorate in astrophysics from the University of Durham, United Kingdom. He spent over 35 years engaged in the design and construction of novel detection systems for X- and gamma-ray astronomy. For the majority of that time he was employed as a staff physicist at the European Space Agency's European Space Research and Technology Centre (ESTEC) in the Netherlands, involved in the development and exploitation of new technologies for space applications. Much of this work revolved around compound semiconductors for radiation detection and measurement, which by their very nature involve materials and systems at a low maturity level. Consequently, he was involved in all aspects of a systematic and long-term program on material assessment, production, processing, detector fabrication and characterization for a large number of compound semiconductors. He is currently a Senior Advisor at the Institute of Experimental and Applied Physics in Prague, the Czech Republic.

1

Introduction to Radiation and Its Detection

An Historical Perspective

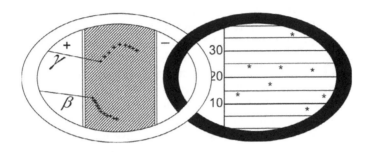

Reproduction of the cartoon on the front cover of P.J. van Heerden's seminal Doctoral thesis, entitled *"The Crystal Counter: A New Instrument in Nuclear Physics"* (Rijksuniversiteit Utrecht, July 1945). The left side shows a stylistic depiction of the "crystal counter", which was essentially a solid state ionization chamber – the precursor of all modern semiconductor energy resolving radiation detectors. The right side of the image shows the "deflections" (pulse heights) of individual events when the crystal was exposed to an external γ-ray source.

CONTENTS

1.1 The Discovery of Radiation

The study of radiation and radiation detection really begins with the discovery of X-rays by Röntgen in 1895 [1], who while investigating cathode rays using a Hittorf-Crookes tube, observed that when the rays hit the glass wall, a mysterious radiation was given off which could fog photographic plates and cause various materials to fluoresce. The rays became known as Röntgen rays. The first corroborative reports of radiation detection took place almost immediately afterwards – which is curious since, apart from photographic plates, radiation detectors had not yet been developed. In 1896, Brandes [2] reported seeing an *"effect"* which he described as a faint *"blue-gray"* glow that seemed to originate from the eye itself when standing close to an X-ray tube. In his first communication, Röntgen [1], had stated that the eye was insensitive to the new rays, but later in his third communication [3] reported seeing "*a feeble sensation of light that spread over the whole field of vision*". Whilst the mechanism was not understood, the observed effects (known by the grandiose but vacuous title of "radiation phosphenes"[1]), were assumed to be due to the direct action of the X-rays on the photoreceptors of the retina [4]. Much later, when radiotherapy became a standard medical modality, the visual effects of X-rays were immediately apparent and were attributed to Cherenkov radiation generated in the ocular media by secondary electrons [5].

While investigating Röntgen's work on X-rays, Becquerel [6] decided to test Poincare's hypothesis [7] that the emission of X-rays could be related to phosphorescence, essentially the delayed emission of light by a substance after its exposure to light. To do this, he placed crystals of potassium uranyl sulfate ($K_2UO_2(SO_4)_2.2H_2O$) on top of a copper Maltese cross and a photographic plate wrapped in black paper. He had originally planned to expose the uranium salts to sunlight before placing them on the cross and plate, believing that the uranium would absorb the sun's energy and then emit it as X-rays. However, the sky was overcast, so he placed the entire assembly into a darkened bureau draw and waited for the weather to improve. It did not, and so after several days he decided to develop the plate anyway and was surprised to see a distinct image of the cross (see Fig. 1.1). Since the plate had not been exposed to light and the crystals were non-luminous, the only conclusion that could be drawn was that the crystals were emitting a previously unknown energetic radiation which became known as Becquerel rays or uranium rays. At first the relationship between uranium rays and Röntgen rays was not clear. Becquerel rays seemed to have intermediate properties between light and Röntgen rays [8]. Of significance, he also observed that an electroscope loses its charge under the effect of the radiation, meaning that the radiation produces charges in the air. In 1898, Marie Curie [9] discovered that thorium minerals also behaved like uranium and suggested that a new radioactive element may be found in pitchblende based on the fact that it was more active than metallic uranium itself. In the same year, Pierre and Marie Currie announced the discovery of two new elements, polonium and radium [10,11] and concluded that uranium, thorium, polonium and radium all emit uranium rays. They coined the word "radioactivity", although at the time the meaning of the word was subtly different from today's. Today, we understand radioactivity to be the property exhibited by certain types of matter to emit energy and subatomic particles spontaneously. At the time, it was more an expression of how active these elements were with respect to metallic uranium. Thus, radium was more "active" than polonium which was more "active" than thorium which was more "active" than uranium. The true nature of the radiation was not revealed until 1899 when Rutherford [12] showed by absorption and conduction measurements that the emanations from uranium consisted of two components. He called the less penetrating component "alpha radiation" and the more penetrating one "beta radiation". Magnetic deflection measurements showed both to be particular in nature and to be of high energy. In the same year, Becquerel measured the mass-to-charge ratio (*e/m*) for beta particles by the method J. J. Thomson used to study cathode rays, which led to the identification of the electron [13]. He found that *e/m* for a beta particle is the same as for Thomson's electron and therefore suggested that the beta particle is, in fact, an electron. Later work by Rutherford showed that alpha particles were bare helium nuclei [14]. Also in 1900, Villard [15] demonstrated the presence of an even more penetrating ray emitted by radium. Later experiments showed that they frequently accompanied alpha and beta emission. In 1903, Rutherford renamed Villard's rays "gamma-rays" following the prosaic naming convention he had used for the hard and soft components of Becquerel's uranium rays.

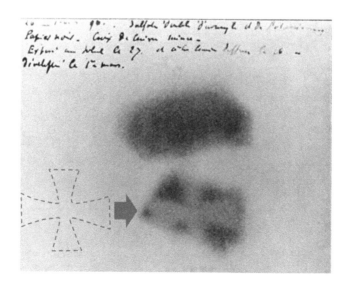

FIGURE 1.1 An exposed photographic plate made by Henri Becquerel [6] showing the fogging effects of "*invisible rays*" emitted by two crystals of uranium salt. The shadow of a metal Maltese cross sandwiched between the plate and the lower crystal is visible.

---

[1] A phosphene is defined as a visual response caused by stimuli other that the normal entry of light into the pupil.

The discovery of radium and polonium filled two empty places in the Periodic Table. However later studies showed that some radioactive elements had the same chemical properties as known stable elements but differed in the amount of radioactivity. Since this appeared to contract the Daltonian model of the elements, (*i.e*, that two elements could not occupy the same place in the Periodic Table) these new "elements" (now known as isotopes) were referred to as radioelements, identified by adding letters to the original parent element (for example UrX, ThA, ThB, ThC, ThX, RaA, RaB, RaC, *etc.*). In 1903, Rutherford and Soddy concluded that radioactive elements were undergoing a spontaneous transformation from one radioelement to another and that the emanations they were detecting were the signature of that transition [16].

1.1.1 Understanding the Atom and Its Structure

In 1911, Rutherford and co-workers observed that while a beam of alpha particles passed through a thin gold foil undeflected, a few were elastically scattered through very large angles [17]. This was completely unexpected since theoretical models of the atom at the time, assumed that atoms consisted of spheres of positive charge in which the electrons were uniformly embedded – the so-called "plum-pudding" model. As such, impinging alpha particles should pass through attenuated but with minimal scattering. Rutherford concluded that the bulk of the mass contained in the gold atoms must be concentrated in a tiny, central region, which we now know as the nucleus [17] and led directly to the more familiar sun and planet type model in which the atom is mostly empty space with the positive charge confined in a tiny compact core, surrounded by an orbiting cloud of electrons. As Rutherford described it at the time "*The mobile electrons constitute, so to speak, the bricks of the atomic structure, while the positive electricity acts as the necessary mortar to bind them together*". The vexed problem of why the electrons did not radiate energy according to classical electromagnetic theory and as a consequence spiral into the nucleus was explained by Niels Bohr in 1913, who assumed that the orbits of the electrons were quantized [18]. An atomic system, he claimed, "*can only exist in certain stationary states in which revolving electrons do not emit energy. Only when the system changes abruptly from a higher state E2 to a lower state E1 will the energy difference appear as radiation*". By 1916, the nuclear atom, which now had become the Bohr-Rutherford model of the atom, was generally accepted in the physics community. Later experiments in which light materials were bombarded by alpha particles were able to show that elements could be transmuted into other elements and by studying the reaction products, the existence of the proton and later the neutron [19] were inferred. Thus by 1932, all the "standard" types of radiation had been discovered and quantified and as a result a workable theoretical model of the atom and nuclear structure realized. The historical timeline of the key events in the discovery of radioactive emanations and the subsequent unravelling of its various components is given in Fig. 1.2.

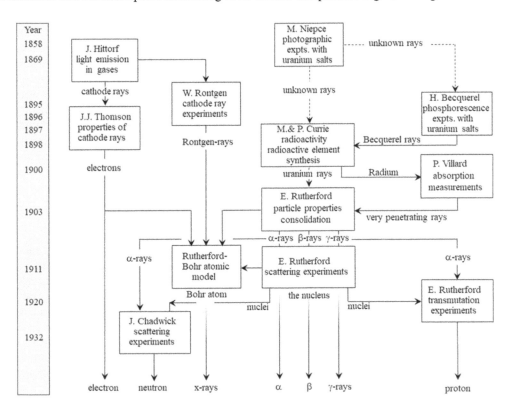

FIGURE 1.2 Flow diagram illustration of the key events following the discovery of radioactive emanations, their effects and the subsequent unravelling of their various components. Most of the initial discoveries took place within a relatively short period of 15 years and can be attributed to a handful of people. By 1932, all of the "standard" types of radiation as we understand them today had been discovered and quantified and a workable theoretical model of the atom and nuclear structure realized.

1.2 Radiation Detection

Although not appreciated at the time, the earliest description of a radiation detector can be attributed to Niépce de Saint-Victor who in 1858 observed [20] that "*a design traced on a sheet of cardboard with a solution of uranyl nitrate[2] and applied on a sheet of sensitive paper prepared with silver bromide, imprints its image*". More importantly, in 1861 Niépce [21] stated that uranium salts emitted some sort of radiation that was invisible to the human eye,

> *... this persistent activity ... cannot be due to phosphorescence, for it would not last so long, according to the experiments of Mr. Edmond Becquerel[3]; it is thus more likely that it is a radiation that is invisible to our eyes.* [21]

This was some 35 years before Henri Becquerel [6] was credited with the discovery of radioactivity in 1896. However, it was not until the work of Röntgen [1], that an urgent need for a radiation detector materialized. Early detection systems were based on photographic emulsions, although Röntgen had initially used a sheet of barium platinocyanide[4] to view the fluorescence produced by the X-ray photons. These observations were visual and could not be recorded, and so it was shortly after this that photographic plates were adapted for radiographic purposes. Glass plates, flexible films and sensitized papers were already available, and indeed Röntgen's original communication [1] indicated the importance of the photographic plate as a means of permanently recording an image. Consequently, photographic techniques evolved rapidly over the next two decades. Firstly, with the introduction safety film (cellulose triacetate) in 1908 and then with the introduction of an X-ray specific film in 1914, spurred on by World War I. This was a direct consequence of the lack of glass to use in the production of photographic plates, coupled to the war-related rapid increase in demand for medical X-rays.

Fluorescent intensifying screens used in combination with film (that would later evolve into scintillation detectors) were introduced as early as 1896 but failed to gain acceptance at the time, primarily due to the extreme amount of afterglow, excessive "graininess" caused by the size of the fluorescent crystals and the non-uniformity of the coating. Screen quality improved steadily over the next few decades, and the utility of using intensifying screens greatly improved with the introduction of "*duplized*" film by Kodak in 1918, in which photographic emulsion was coated on both sides of the film base instead of one surface only. Now two screens could be used to "sandwich" the film in a single cassette, increasing sensitivity of the film. This would eventually lead to the modern "cardboard" or rigid cassette versions routinely used in dental and medical applications.

1.2.1 Early Monitoring Devices

During his many experiments on X-rays, Röntgen had also observed that X-rays were able to discharge electrified bodies in air [22] and postulated that the effect was due to some change in the air which then acted on the electrified body. Thomson [23] showed that that air would conduct electricity when traversed by X-rays and Thomson and Rutherford [24] went on to demonstrate that this conduction was due to ionization (*i.e.,* the stripping of electrons from air molecules by the radiation). By placing two conducting plates with opposite charges at opposite ends of an air filled chamber, they were able to quantify the intensity of X-rays by measuring current flowing through the chamber. This led the development of the ionization chamber (for a review, see Frame [25]). Conventionally, the term "ionization chamber" is used to describe detectors that collect charge created by direct ionization of a gas through the application of an electric field. Specifically, it uses only the charge created by interaction between the incident radiation and the gas and does not exploit gas multiplication mechanisms used in Geiger-Müller or proportional counters. As such, ionization chambers tend to be mostly used for radiation intensity monitoring.

1.2.2 Early Recording Devices

Emulsions, screens and ionization chambers are essentially integrating devices and as such provide no information on individual particles or events. However, in 1903 Crookes [26] introduced the spinthariscope[5] which consisted of a small screen coated with zinc sulfide fixed to the end of a tube, with a tiny amount of radium salt suspended on the tip of a moveable pointer a short distance from the screen (see Fig. 1.3 left). When the screen was viewed through a lens on the other end of the tube, faint random scintillations could be observed which became more numerous and brighter, the nearer the source was brought to the screen (see Fig. 1.3 right). When very close, "*the flashes followed each other so quickly that the surface looks like a turbulent, luminous sea*". The device was initially regarded as a curiosity rather than a scientific instrument until Rutherford and Geiger [27] demonstrated that, on a uniform screen, each scintillation corresponded to the impact of an individual alpha particle. They did this by incorporating a zinc sulphide screen at the end of an ionization chamber and comparing the number of scintillation

---

[2] $UO_2(NO_3)_2$
[3] The father of Henri Becquerel.
[4] A fluorescent complex salt formed by the union of a compound of platinum and cyanide with another cyanide.
[5] from the Greek σπινθήρ (*spinthēr*) meaning "spark".

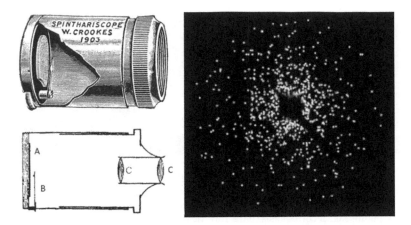

FIGURE 1.3 Left: cross-sectional image of a Crookes Spinthariscope (top) showing its essential components (bottom): A = ZnS screen; B = radium nitrate source painted onto the tip of a moveable pointer; C = eyepiece lens. Right: a snap-shot of scintillation "flashes" produced by alpha particle impacts when viewed through the eyepiece.

flashes from the screen with electrical signals from the chamber. This work was greatly aided by Regener [28] who, while using the scintillation method to record α-particles from polonium, outlined a methodology by which the scintillation method could routinely be used to quantify the flux of alpha rays. This method was widely used in nuclear physics for about two decades, although it had major disadvantages resulting from the fact that it imposed considerable strain and fatigue on the observer – the light flashes were faint, and the human eye could only count up to a maximum of about 50 events per minute for a short period of time. Marginal improvements in the accuracy of the technique were made by employing a dual ocular microscope and using two observers to independently register counts by marking a moving tape – in fact, the first application of the coincidence counting technique in nuclear science.

Count rate and human limitations were overcome in 1908 by Rutherford and Geiger [27] who developed an electronic counting system based on a gas-filled detector that utilized a Townsend avalanche discharge to amplify the initial ionization products. By using gas multiplication, a measurable electronic pulse could be obtained even for a single ionizing event. The design was later refined by Geiger and Müller [29] in 1928, culminating in the Geiger-Müller tube – a device still in extensive use today. Its main limitations are that it is unable to distinguish different radiation types or determine the amount of energy deposited. It should be noted that the only real difference between the various types of gas counter is the strength of the electric field applied across the sensitive volume.

1.2.3 Electro-Optical Approaches

The next significant advance in the state of art occurred in 1930, with the invention of the Kubetsky tube [30]. This represented the fusion of two technologies: the photoelectric tube developed around 1913 based on the photoelectric effect discovered by Hertz in 1887 [31] and an electron amplification stage patented by Slepian in 1923 [32], which utilized the secondary electron emission process first reported by Austin and Starke in 1902 [33]. Kubetsky [30] successfully integrated the two (specifically, a photoelectric-effect cathode and single electron amplification stage) into a single vacuum glass envelope. This later evolved into the more familiar photomultiplier tube (PMT) in early 1934 [34]. The first device had an electron amplification gain of only 8 but could operate at events rates >10 kHz. Within a year, electron gains of over one million were achieved [35]. The first application-to-radiation measurement was reported by Morgan [36] in 1942 who constructed an X-ray exposure meter by coating the input window of an RCA[6] PMT with calcium tungstate crystals. The measured output current of the PMT was found to be a function of X-ray intensity. This was closely followed by Curran and Baker [37], who in 1944 constructed what can be considered the first electronic scintillation counter, by coupling a ZnS screen to an RCA PMT. The first gamma-ray spectrometer can be attributed to Kallmann [38] who in 1947 coupled a PMT[7] to a block of Naphthalene and observed scintillation signals from both beta and gamma-ray radiation. Although Kallmann reported a proportionality between the energy of radiation source and height of the scintillation signals, the spectral acuity was poor. Modern-day high-resolution spectroscopy [39] can be traced back to the work of Hoftstader [40] who in 1948 coupled a PMT to a newly developed thallium activated alkali halide scintillation crystal. For the first time, not only could single event scintillation signals be efficiently discriminated and registered, but the energy they deposited simultaneously recorded with (up till that time) unprecedented spectral precision [41]. This is illustrated in Fig. 1.4 in which we show

---

[6] Radio Corporation of America, company founded in 1919.
[7] Reportedly, bartered for 10,000 Lucky Strike cigarettes in war-torn Berlin.

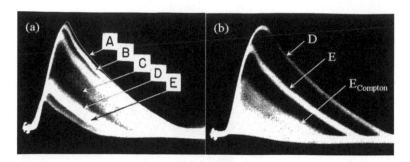

FIGURE 1.4 Photograph of an oscilloscope screen showing the collective pulse heights obtained from a ^{124}Sb radioactive source (from Hofstadter and McIntyre [41]) illustrating the high spectral resolution that can be obtained by coupling the newly developed doped alkali halide scintillator, NaI(Tl), with a photomultiplier tube. Each of the bright bands represents a unique gamma-ray energy. In (*a*) we show total spectrum. *A* represents the weak gamma-ray line from the 2.09 MeV photopeak, *B* corresponds to the 1.67 MeV photopeak, *C* represents the weak escape peak from the 2.09 MeV line, *D* is the weak 0.72 MeV line and *E* is the strong line from the 0.60 MeV photopeak. In (*b*) we show a gain expanded representation of (*a*). Here the bright band in the middle is the 0.60 MeV photopeak, under it lies the Compton distribution for this line, above the 0.60 MeV line is the weak 0.72 MeV line.

a photograph of the oscilloscope traces of pulse heights taken from the paper of Hofstadter and McIntyre [41], who measured pulse height distributions from a ^{124}Sb radioactive source using a 0.5 inch NaI(Tl) crystal coupled to an EMI PMT. ^{124}Sb decays by β emission to ^{124}Te with the emission of multiple gamma-rays, the strongest of which are at energies of 0.60 MeV and 1.69 MeV and weaker lines at 0.72 MeV and 2.09 MeV. The energy resolution and therefore spectral acuity is so good that the principal gamma-ray lines appear as separate bands in the trace. In (*a*) we show total spectrum. *A* represents the weak gamma-ray line from the 2.09 MeV photopeak, *B* corresponds to the 1.69 MeV photopeak, *C* represents the weak escape peak from the 2.09 MeV line, *D* is the weak 0.72 MeV line and *E* is the strong line from the 0.60 MeV photopeak. In (*b*) we show a gain expanded representation of (*a*). Here the bright band in the middle is the 0.60 MeV photopeak, under it lies the Compton distribution for this line, above the 0.60 MeV line is the weak 0.72 MeV line. The introduction of scintillation detectors not only led to a revolution in spectroscopy but also in electronics[8] since the conventional way of recording events was impractical for the large amount of simultaneous temporal and spectral data generated by these systems.

1.3 Early Work with Semiconductors

The beginning of semiconductor research can be traced to a report by Faraday in 1833 on the negative temperature coefficient of resistance of suphuret of silver (silver sulphide, Ag_2S) [43]. Faraday described experiments on a new compound that had a poor conductivity at room temperature but became comparable to metallic conductors at elevated temperatures (~175°C). This was the first observation of any semiconductor property and was significant at the time as it was completely at variance with the observed properties of metals and electrolytes[9] whose conductivity decreases with temperature (or alternately, whose resistance increases with temperature). Subsequent work on semiconducting materials focused largely on the conduction properties of crystals and the rectification properties of crystal/metal interfaces. The latter being driven by the advent of radio in the early 1900s and in particular the development of point-contact rectifiers for detecting millimetre electromagnetic waves (radio signals).

During this period some other surprising discoveries were made (although not recognized at the time). For example, in 1907 Round [44] recorded that during an investigation of the *"unsymmetrical passage of current through a contact of carborundum, a curious phenomenon was noted"*. He went on to describe how the crystal gave out a yellow glow when a potential of 10 volts was applied between two points on the crystal. In other crystals a light green, orange or blue glow was observed. Lossev [45] investigated these effects in detail and was able to demonstrate the presence of a thin (~10 μm) surface layer separated from the bulk by some kind of interface. He noted that the interface showed a sudden change in conductivity when bias was applied and correctly postulated that light was then emitted by a phenomenon that was essentially the inverse of the photoelectric effect recently proposed by Einstein [46]. In actuality, what Round had reported was the first observation of electroluminescence from SiC (carborundum), and what he had constructed was a primitive light-emitting diode (based on a Schottky, rather than a p-n junction). The interface postulated by Lossev was essentially the potential barrier formed at the junction and the bias-induced emission driven by minority carrier injection (tunneling) through the barrier into the

---

[8] For example the rapid development of the "kick-sorter" [42] – an early mechanical form of a Pulse Height Analyzer (PHA).

[9] A substance that produces an electrically conducting solution when dissolved in a polar solvent, such as water.

semiconductor. Light is then emitted following recombination. The observation of different colours from different crystals was presumably the effects of various impurities on the bandgap (*i.e.*, unintentional bandgap engineering).

1.3.1 Photoconduction Detectors

In terms of radiation detection, early work was carried out by exploiting the photoconduction properties of semiconductors. In fact, all semiconductors and insulators are photoconductors to some extent. The photoconductive effect was first reported by Willoughby Smith in 1873 who, while working on the transatlantic submarine cable, observed that crystalline selenium offers considerably less electrical resistance when exposed to light than when it is kept in the dark [47]. The effect was exploited several years later when Fritts [48] reported constructing the first solar cell based on selenium. The basic detection mechanism is as follows. A voltage is applied across a highly doped semiconducting material *via* two contacts. Light incident on the crystal generates electron–hole pairs which increase the carrier concentration and in turn, the conductivity of the material. By applying a small bias, a measurable current will flow through the crystal. Photoconductors differ from crystal counters[10] in several important ways; most notably they are operated as integrating devices since their large intrinsic conductivity means they are insensitive to individual particles or quanta. Radiation detectors based on photoconduction are still widely used in both analog and digital applications across a very broad wavelength range, from γ-rays to microwaves.

Early work on photoconductors concentrated on diamond. For example, photo-conductive diamond ultraviolet detectors were studied in the 1920s by Gudden and Pohl [49] who investigated their absorption characteristics. In diamond both the electrons with energies in the conduction band and holes in the valence band are free to move under the influence of an applied electric field. The movement of electrons and holes creates a current which can be amplified. Unfortunately, not all the electrons and holes are removed by the electric field because of trapping by crystal imperfections. This creates a space charge which generates an internal electric field opposed to the external field (the so-called polarization effect). The net result is a decreasing effective electric field which reduces the amplitude of charge pulses to a level too small to be registered by the electronics. Hence, the count rate decreases with time. In addition, research using diamond was necessarily limited due to its high intrinsic value and the difficulty of obtaining large geological diamonds with predictable properties. Because of this, a number of other materials were also studied during this period – most notably lead salts (*e.g.*, PbS) and the thallous halides. The "*photoelectric primary current phenomena*" in TlBr crystals was first described by Lehfeldt [50] in 1933 who later, in a study of the photoconductivities of thallium chloride, thallium bromide, silver chloride and silver bromide, concluded that the photoconducting properties of the thallium salts and the silver salts showed many similarities [51]. Ionizing radiation was also observed to induce noticeable conductivity in crystals although the effects were small and not reproducible. For example, Frerich and Warminsky [52] demonstrated a response to alpha and beta radiation in crystal phosphors. A complete bibliography of this early work up to 1932 may be found in Jaffe [53].

1.3.2 Do Semiconductors Exist?

At this juncture, it is worthwhile to emphasize that work carried out during this period gave highly variable and inconsistent results. With hindsight, this was almost entirely due to purity issues. Worse still, since materials in the 1930s were so impure, it was not possible to link theory with experiment. Consequently, many researchers were unsure whether semiconductors existed at all. In fact, in 1930 Gudden [54] proposed that no chemically pure material would be a semiconductor and that conductivity in semiconductors was purely due to minute impurities present, rather than being a property of the bulk material itself. He stated that "*semiconductors in the scientific sense of the word – if they exist at all – are by far scarcer than originally assumed*".[11] Following on, in 1931 Wolfgang Pauli famously reflected the views of many researchers in a letter to Rudolf Peierls, "*Über Halbleiter sollte man nicht arbeiten, das ist eine Schweinerei, wer weiss, ob es überhaupt Halbleiter gibt*". ("*One should not work on semiconductors, that's a mess, who knows whether there are semiconductors at all*".)

1.3.3 Theoretical Stagnation and Salvation

Shortly after the discovery of the electron by J.J. Thomson in 1897 [13], several scientists proposed theories of electron-based conduction in metals which were in line with Hall effect measurements. In 1878 Hall [55] showed that when a current-carrying conductor is placed in a magnetic field, the Lorentz force deflects the charge carriers and "*presses*" them against one side of the conductor. This generates a small transverse voltage which varies as the applied field. The constant of proportionality (the Hall coefficient) is defined as the ratio of the induced electric field to the product of the current density and the applied magnetic field and is dependent on the type, number and properties of the charge carriers that constitute the current. Its sign reflects the charge of the carrier. While this was found to be nearly always negative, sometimes the Hall

---

[10] An early name for a semiconductor radiation detector operated as a "solid state ionization chamber".
[11] A view he held until 1939.

coefficient was measured to be positive. This was not understood and remained as an anomaly until the concept of vacant states or holes in semiconductors was established in the 1930s. Until that time, there was no general consensus regarding semiconductors and whether their observed properties (negative temperature coefficient of resistance, photoconductivity, rectification, photovoltaic effect and electroluminescence) had any common origin. Indeed, theory had developed primarily to explain the conduction properties of metals and insulators and was largely based on the classical approach of Drude [56] from 1900. In 1908, Koenigsberger [57] proposed a dissociation theory of conduction in which he considered two populations of mobile carriers, one associated with the dissociated electrons and the other with the remaining positive ions. Each population could be assigned a temperature dependence and separate mobility. The dissociation energy, Q, associated with a given material could be used to define a particular conductive state [58]. For example, for insulators, Q is infinite, whereas for metals it tends to zero at high temperatures, meaning that the electrical resistivity should rise with increasing temperature, but more importantly the resistivity should become infinitely large at zero absolute temperature. Materials that have a finite Q value, he classified as "*variable conductors*", a term that was later relabelled as "*semiconductors*" by his student J. Weiss [59]. For these materials the resistivity should decrease exponentially with increasing temperature which was found to be the case. Keonigsberger [57] also observed that derived dissociation energies and mobilities of variable conductors were strongly correlated with material purity and structural perfection – which goes a long way to explain the poor reproducibly of results at that time. A physical explanation would become apparent in the late 1920s, when Frenkel [60] and later Wagner and Schottky [61] developed models of point defects in lattices to explain ionic conductivity in ionic crystals.

Semiconductor theory, thus far, had followed an empirically driven classical approach that had led to predictions broadly in line with observation but not based on any solid theoretical foundation. Indeed, the microphysics was not at all understood. The first comprehensive semiconductor theory was published in 1931 and can be attributed to Wilson [62,63] who adapted the quantum theory of solids being developed by Bloch [64], Peierls [65] and others to create the first realistic model of semiconductor behaviour. Bloch [64] had worked out the quantum mechanical solution for a periodic crystalline system where the fields are subject to the regular placement of its component nuclei. This was based on earlier attempts to apply and solve Schrödinger's equation for a periodic lattice. Within a year, Kronig and Penney [66] had derived a simple analytically soluble one-dimensional model for a periodic square-well potential. In the same year, Peierls [65] introduced the concept of the forbidden gap. In 1931, Wilson merged these ideas into the now familiar band theory of semiconductors based on a conduction band and valence band separated by the forbidden gap (see Fig. 1.5). In two classic papers, Wilson [62,63] was able to not only demonstrate quantitatively the difference between metals and insulators, but also distinguish between extrinsic and intrinsic semiconductors, by taking into account impurities (later to evolve into the concept of donors and acceptors). He suggested that conduction in a semiconductor could be enhanced if impurities were used to introduce energy states into the forbidden gap close to the conduction band [63] as shown in Fig. 1.5. Thermal energies could then be sufficient to lift electrons from this level into the conduction band where they can move freely and transport current. These became known as donor electrons but would soon be complemented by that of acceptors with energy levels just above the valence band. The intriguing inference was that electrons from the valence band could become trapped in the acceptor levels leaving a hole behind in the valence band. However, Wilson considered only electrons. The empty states (the holes[12]) in a nearly full valence band of a semiconductor were later shown by Heisenberg [68] to act globally in the exact same manner as equivalent "positive charges", explaining for the first time why positive Hall coefficients were observed in some semiconductors. Once quantum theory had been applied to semiconductors, progress became rapid. For example, in 1938 Mott [69] and Schottky [70] independently developed models of the potential barrier and current flow through a metal-semiconductor junction. Davydov [71] developed a theory of the copper-oxide rectifier, identifying the effect of the p–n junction and the importance of minority carriers and surface states.

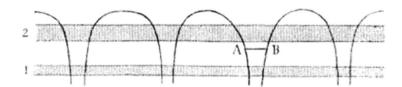

FIGURE 1.5 The first energy band diagram of an extrinsic semiconductor sketched by Wilson (taken from [62]). The periodic curved lines show the potential of an electron upon which is superimposed a simple 1-D representation of the valence (1) and conduction bands (2), respectively. A and B show a representation of an intermediate discrete energy level introduced into the bandgap by an impurity. Thermal energies suffice to lift electrons from this level into band 2 where they can move freely and transport current.

---

[12] It was William Shockely who first adopted the term "holes" in a monograph published in 1950 [67]. However, the term and hence the concept were not accepted by the community at the time – most researchers considered electrons only. While this is possible quantum mechanically, it is far more complicated than assuming the existence of holes *a-priori*.

1.3.4 Crystal Counters

A major leap forward in radiation detection occurred with the development of the "crystal counter" in the mid-1940s. In essence, a crystal counter is a particle detector in which the sensitive material is a dielectric (non-conducting) crystal mounted between two metallic electrodes. It is in effect a solid state ionization chamber, where the ionization released by the particle is a measure of its energy. The use of a solid as a counter was quickly recognized to provide several significant advantages over gaseous detectors; namely, high stopping power, small size (since the density of solids are ~1,000 times greater than gases) and high-speed operation. Most of the initial work was done with diamond and alkali halide crystals because of their insulating properties and relatively high purity. It was considered necessary to use an insulator so that an electric field could be applied for the purpose of charge collection without prohibitive electronic noise arising from leakage current through the device. The work was supported by the theoretical work of Hecht [72] who, in 1932, had already demonstrated a relationship between an ionization-induced current in a crystal and the electric field strength.

The first practical implementation of a crystal counter can be attributed to van Heerden [73], who in 1945 demonstrated that silver chloride crystals when cooled to low temperatures were capable of detecting individual γ-rays, alpha particles and beta particles. This was not an integrating device but was fully spectroscopic. Previous to this work, the energy spectra of charged particles had been determined *a priori* using a magnetic spectrograph – the detector then logged events either as a time-averaged photocurrent or as individual non-spectroscopic pulses on an oscilloscope or galvanometer. Van Heerden had initially reasoned that it must be possible to find substances in which a beta particle could be stopped over a short distance while releasing sufficient ionization to be measured. This led to the question: *what type of ionization is released in solids, liquids and gases?* Long collection times coupled with high ion recombination rates precluded liquids and gasses, and van Heerden [74] concluded that the solids most likely to fit the detection criteria would be those crystals that show photoelectric conductivity. Such crystals show a response to light and therefore would presumably do the same to ionizing particles if made thick enough. Crystals of silver chloride were grown from the melt and diced into 1 cm diameter cylinders of thickness 1.7 mm. Silver electrodes were deposited on each side of the crystal. At room temperature the crystal showed ionic conductivity but when cooled to liquid air temperature the crystal became a perfect insulator, allowing a large bias of up to 2.5 kV to be applied across the crystal. In Fig. 1.6 we show the crystal counter response to magnetically selected 400 keV beta particles which is spectroscopic with a measured FWHM of ~20% (from ref [75].). In addition, the response of the device was found to be linear over the energy range 0 to 1 MeV. Subsequent measurements showed that the device was also responsive to alpha particles and gamma-rays.

Van Heerden's work was later extended to the thallous halides by Hofstadter [76] in 1947, who reasoned that if silver chloride behaved as a crystal counter when cooled, then it would be likely that the thallium salts would do the same, based on their similarities in photoconductivity observed by Lehfeldt [50]. Hofstadter [77] fabricated sample counters from a mixture of TlBr and TlI. The samples were one inch in diameter and 2 mm thick. Platinum electrodes, half an inch in diameter were sputtered onto each side of the disk. A schematic diagram of the detector and preamplifier front-end is shown in Fig. 1.7 (from [77]). Like silver chloride, the sample was a conductor at room temperature. When connected to an amplifier and a power supply, the detector showed no pulses on an oscilloscope when exposed to a radium gamma-ray source but it did show photoconductivity and when cooled to liquid nitrogen temperature became insulating. When a 500 V bias was applied, it displayed a counting response to the gamma-ray source with individual pulses observed on an oscilloscope. Experiments at different temperatures

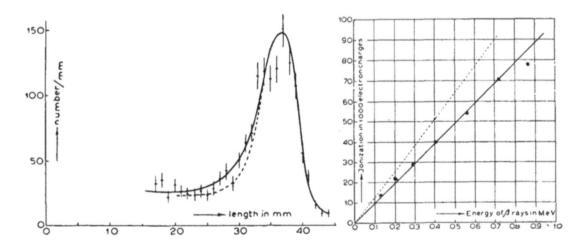

FIGURE 1.6 Left: the measured energy loss spectrum of a silver chloride (AgCl) crystal counter to 400 keV β-particles (from van Heerden and Milatz [75]). The applied bias was 200 V. The dashed curve is the theoretically predicted distribution. Right: the relation between the ionization and the beta particle energy (from [75]). The dotted line is the so-called "saturation curve" which in effect is the measured curve corrected for charge collection efficiency.

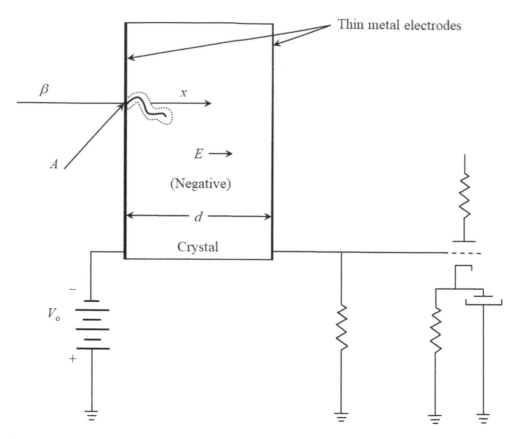

FIGURE 1.7 A schematic diagram of the key components of the crystal conduction counter of Hofstadter (reproduced from [77], courtesy Platts, a division of the McGraw-Hill Companies, Inc.).

showed that the device became conductive at temperatures above $-70°C$. Hofstadter concludes, "*the question of what properties determine which crystal will count is not yet answered*". Frerich [78] addressed this question by showing that all crystals which give pulses belong to the class of crystal phosphors, at least under the temperature conditions at which they were studied. Specifically, they show the same optical phenomena as a crystal phosphor, and the pulses induced by nuclear particles are related to the primary and secondary photo-currents,[13] characteristic of crystal phosphors.

By 1949, the number of materials found to exhibit "crystal counter" behaviour had increased significantly [79] (CdS, ZnS, AgCl, AgBr, TlCl, TlBr, TlJ,[14] type-2 diamonds,[15] solid Ar). At this time, the elemental semiconductors were not anticipated to be responsive to radiation, since silicon and germanium are not insulators and the application of an electric field to sweep out the charge produced by an incident particle would result in unacceptably high leakage currents. However, several months after Hofstadter's work, McKay [80] successfully measured the polonium alpha-ray spectrum using a reverse biased, point contact Ge detector. McKay circumvented low resistivity and charge trapping issues by constructing a very thin detector with an evaporated metal surface layer which acts as a rectifying contact whilst simultaneously creating a very high electric field across this layer using a point contact.

McKay and McAfee [81] subsequently extended this work producing both germanium and silicon p-n junction detectors. Schottky barrier Ge detectors were introduced in 1955, followed by Au-Ge surface barrier detectors in 1959. The actual surface barrier is a thin interfacial layer in which a high density of electron traps has been introduced (usually by etching), which acts like p-type material, essentially "hardening" the rectifying properties of the metal contact. However, apart from alpha spectroscopy applications, much of the work in semiconductors went unnoticed for almost a decade because of their very small counting efficiencies coupled with the disappointing performance of crystal counter detectors in general [82]. However, more significantly, crystal counter work was completely overshadowed by the introduction of large volume, efficient, scintillation counters introduced by Kallmann [36] in 1947 and in particular, the high-resolution alkali-halide scintillation spectrometers introduced by Hofstadter [40] in 1948.

---

[13] The primary photocurrent is the direct result of the absorbed quanta. The secondary photocurrent is the result of an increase in conductivity arising from the passage of the primary photocurrent (for a review, see Hughes [79]).

[14] Thallium(I) iodide.

[15] Very rare natural diamonds that are nitrogen free, clear, and do not UV absorb light.

1.4 Post-1960 Evolution

Renewed interest in "crystal counters" was sparked in the 1960s by two parallel events – the first the rapid development of elemental semiconductors and the second the emergence of compound semiconductors. The development of Ge and Si radiation detectors [83] arose as a consequence of the explosive rise of the semiconductor industry. Diffused junction and surface barrier detectors had already found widespread application for alpha particle spectroscopy and as ΔE detectors for particle identification [84]. Figure 1.8 shows early commercial Si surface barrier detector technology; (a) a simplified diagram of the manufacturing technique and (b) the practical implementation. For more energetic radiation, it was not until the introduction of the ion-drifting process, first demonstrated by Pell in 1960 [85], that a practical method was found by which large volume (and therefore efficient) semiconductor detectors could be made. In this process, the imperfect and low resistivity germanium could be closely compensated after crystal growth by drifting interstitial lithium donors into the bulk, neutralizing excess acceptors in p-type material. This allowed semi-insulating material to be produced, allowing higher biases to be applied, leading to better charge collection and larger active volumes. About the same time, Li drifting was also applied to Si. While this was a significant step forward, Li drifting has the major disadvantage that detectors must be kept cold, even in storage; otherwise the Li drifts out of the detector. Fortunately, by the late seventies, purification techniques had advanced sufficiently that high purity material could now be routinely produced and compensation was no longer needed. Indeed, at present, Ge(HP) detectors with volumes in excess of 200 cm^3 volume are readily available with FWHM energy resolutions of 2% at 1,333 keV.

The second major event to renew interest in "crystal counters" was the successful synthesis of large crystals of CdTe by de Nobel in 1959 [86], which led to the growth of a large number of other compound semiconductors, most notably a range of II-VI compounds. The potential of such materials was quickly recognized, especially their ability to operate without cryogenics [87]. Meanwhile, the synthesis of III-V materials was slower to take off, primarily because they did not occur naturally and therefore the range of potential compounds available was not fully appreciated at the time. Initial research concentrated on GaAs and InP – and then almost exclusively on LED and laser applications driven by their optoelectronic properties. The application to radiation detection did not begin until 1960 when the first bulk semi-insulating GaAs device was used to detect ^{60}Co gamma-rays. In fact, of the compound semiconductor groups, the first high-resolution spectroscopy results at gamma-ray wavelengths were actually obtained with GaAs rather than CdTe detectors in 1970 [88]. This was largely due to the development of Cr doped semi-insulating substrates in 1960, coupled with the introduction of epitaxial growth techniques in 1965.

1.4.1 The Current Situation

Since the 1960s, there has been a steady increase in the number and quality of semiconductors available. However, corresponding progress in the development and performance of radiation detection systems has been incremental but

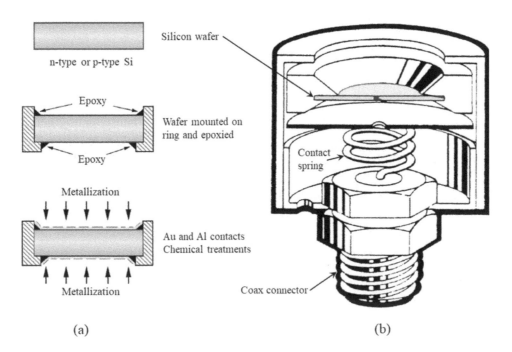

(a)   (b)

FIGURE 1.8 (a) Simplified diagram of the technological steps involved in the fabrication of a Si surface barrier detector. (b) A cross-sectional view of an early device mounted on a coaxial readout connector (reprinted with permission by Advanced Measurement Technology, Inc. – ORTEC).

hardly spectacular, and the reader is referred to the reviews of McGregor and Harmon [89] and Owens and Peacock [90]. Currently, almost any compound can be grown, but in general, detector performances are still plagued by material problems caused by severe micro-crystallinity, high defect and impurity densities and stoichiometric imbalances. However, some significant advances have been made – particularly in material development and the manipulation of materials at the microscopic level. For example, defect engineering [91], in which the mechanical and electrical properties of materials can be tailored at the microscopic level by the precise manipulation of the type, concentration and spatial distribution of defects within a crystalline solid. Recent work now targets structures with nm dimensions. Examples where the selective introduction of specific point, line and plane defects can make significant improvements in performance include compensation in semi-insulating materials, improving radiation hardness, tailoring mechanical properties, tuned optoelectronic properties and "gettering" of unwanted impurities.

A second area in material advancement has been to apply the concepts of "lattice hybridization" [92] and "dimensional reduction" [93] to synthesize whole new classes of compounds whose properties can be essentially tailored. So, for example, in "lattice hybridization", semiconductors belonging to different structural families can be hybridized to form a child compound which retains the best properties of the precursor semiconductors. Similarly, "dimensional reduction" combines the chemical concept of dimensional reduction [94] (*i.e.*, performing specific structural transformations to a compound using a reduction agent) with precise theoretical electronic structure calculations to craft a variant with pre-determined properties.

Another area where significant progress has been made is in the understanding of carrier transport. Recently, much effort has focused paradoxically on the qualities that degrade detector performance. Specifically, we refer to the pivotal work of Luke [95] and others in developing single-carrier sensing techniques. These techniques not only improve performance by mitigating the effects of the poorest carrier, but also point the way to future detector development – as, for example, the controlled movement or channelling of charge, as dramatically demonstrated in silicon drift detectors [96]. Although these developments are still ongoing, at the present time compound semiconductor radiation detectors have evolved sufficiently to fill certain niche applications and are slowly becoming viable competitors to laboratory standards like Si or Ge. The historical timeline of the key events in the development of radiation detectors up till the present day is shown in Fig. 1.9.

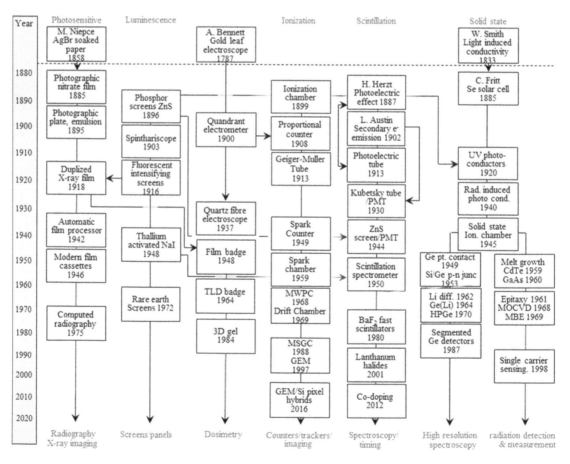

FIGURE 1.9 Flow diagram illustration of the development and evolution of radiation detector systems following the initial discovery of X-rays and radioactive emanations in 1895. The figure should be viewed in combination with Fig. 1.2, which illustrates the key events following the discovery of radioactive emanations, their effects and the subsequent unravelling of various components following the evolution of increasingly sophisticated instrumentation.

1.4.1.1 Other Technologies

In other areas of detector technology, for example scintillation and gas detectors, progress has also been made. In the case of scintillators, spectroscopy had stagnated after the introduction of the thallium doped alkali halides in 1948 [40] until a breakthrough was made in 2000 with the introduction of the cerium doped lanthanum halides and related compounds [97,98]. Previous to this date, the best achievable gamma-ray energy resolutions at 662 keV for 3-inch standard crystals were typically around 7% FWHM [99]. By comparison, the energy resolution of an equivalently sized $LaBr_3:Ce^{3+}$ crystal is 3% [100], and recent experiments with co-doping suggest that a further improvement to 2% is possible [101]. These results are comparable with the best results obtained with a compound semiconductor such as CdZnTe [102]. At present, the outstanding issue in scintillation physics is understanding and mitigating the effects of non-proportionality [99], which currently limits the attainable energy resolution in all scintillators (for a summary see [103]).

Gas detection systems have also seen a significant evolution since the late 1960s, driven mainly by high-energy particle physics. By the 1970s, Multiwire Proportional Chambers (MWPCs) and subsequently drift chambers, had become standards at particle accelerations, outperforming by orders of magnitude the rate capabilities of the previous generation of systems based on spark and bubble chambers [104]. Despite their successful use in particle physics experiments, MWPCs have several intrinsic limitations, namely, a rate limitation induced by slow moving ions created during the multiplication process, limited multi-track resolution dictated by the wire spacing and aging issues caused by the polymerization of the gasses used. A lot of these limitations were solved by the introduction of Micro-Strip Gas Counters (MSGCs) in the late 1980s, which provided a two orders of magnitude rate improvement over MWPCs and a ten-fold improvement in the multi-track resolution. However, because of the high internal electric fields employed, they were also susceptible to destructive discharges. These problems were solved with the introduction of Micro-Pattern Gas Detectors in the late 1990s [105]. The most notable derivative of this is the Gas Electron Multiplier (GEM) offering large sensitive areas, sub-mm positional accuracy, single electron sensitivity and long-term operational stability and reliability (for a review see Sauli [106]). Currently, GEM and solid state technologies are being merged by coupling the micro gas cells to pixelated silicon readouts enabling a positional and timing accuracy that is unmatched by any other gas detector.

1.5 Future Directions

Solid state semiconducting detectors now dominate virtually all areas of radiation detection. However, it is becoming increasingly apparent that conventional detection techniques based on the generation and separation of charge carriers using electric fields will reach an impasse in terms of sensitivity, primarily because the signal-to-noise ratio in such a system is dependent on not only the number of (information) carriers generated, but also on the level of thermally generated noise. Both are intimately dependent on the bandgap – decreasing the bandgap significantly will increase the number of information carriers but also cause a disproportionate increase in thermal noise. Even if the latter is reduced by cooling, the maximum number of carriers that can be generated following the initial ionization event is ultimately limited by the Fano factor. That is, roughly 90% of the energy deposited following the initial ionization cascade is lost in the form of lattice vibrations, phonons and plasmons. As a consequence, only ~10% of the initial energy will ever be available as information carriers. A major step forward would be to look beyond simply "collecting electrons" and exploit other obscure internal degrees of freedom of the electron in addition to its charge for non-volatile information processing, for example, utilizing its spin or alternately the valley degree of freedom implicit in the band structure of some semiconductors. Alternatively, one could exploit quantum effects on the nano-scale [107]. These three avenues are currently being explored, although research is still in its infancy.

1.5.1 Exploring the Nano-Scale Properties of Materials

One approach is to exploit the unique electronic and optical material properties that can be achieved when matter is manipulated on nanometre scale lengths in the form of nano-materials and composites. In these materials, the physical and electronic properties can be significantly modified by reducing the physical extent of at least one of the dimensions to the order of the exciton Bohr radius – typically around 5 nm. A semiconductor confined in one dimension is termed a quantum well and described as a two-dimensional system. Reducing the dimensionality further and confining in two dimensions results in 1-D quantum wires which, amongst other effects, exhibit quantized conductance. Finally, further reduction of the dimensions leads to a 0-D system known as a quantum dot (QD). There are a number of ways that reduced dimensionality can be exploited for radiation detection. For example, dispersing boron-rich 0-D nanoparticles into a counter gas can produce a thermal neutron sensitivity in standard ionization proportional counters, without the inefficiencies normally associated with self-absorption of the $(n,^{10}B)$ reaction products [108]. The exploitation of nano-composite materials could potentially lead to a new class of scintillators that would rely not on conventional crystal growth techniques, but on the assembly of high-Z scintillating QDs in a transparent matrix. The advantages of this approach are numerous: cost, simplicity, scalability and the ability to match the emitted scintillation light to a particular readout device (*e.g.*, APD, PMT *etc.,*) simply by tuning the size of the nanoparticles. Currently most work in nanotechnology is being carried out on 1-D structures. For example, it is found that 1-D freestanding nanowire structures can endure significantly more strain both radially and axially than films. Heterostructures based on semiconductor nanowires can therefore be formed without dislocations even in the case of materials that have a large lattice mismatch. This opens up substantial opportunities for various mismatched material combinations, such as InAs/InP and GaAs/GaP [109]. To fully exploit reduced dimensionality, current research is focusing on device fabrication based on nano-building blocks [110] and in particular the integration of nanowire elements into complex functional architectures. 2-D systems based on graphene are already being investigated for indirect radiation detection by exploiting the fact that in graphene, electrons move according to the laws of relativistic quantum physics described by the mass-free Dirac equation rather than the Schrödinger effective-mass equation – in essence, the electrons and holes behave as massless Dirac fermions and can be made to respond almost instantly to any perturbing influence, in, say, an absorber, induced by interacting radiation [111].

1.5.2 Exploiting New Degrees of Freedom

The second promising avenue is to exploit other degrees of freedom which are at present mainly regarded as academic curiosities. Conventional radiation detectors work by manipulating the electron's charge by the means of electric fields. This charge is generated through a process of ionization by the radiation and represents the basic information carriers of the system (*i.e.*, the signal). The signal-to-noise ratio in such a system is dependent not only on the number of carriers generated, but also on the level of thermally generated noise. Both are intimately dependent on the bandgap. For high-resolution systems, it is necessary to resort to low bandgap materials necessitating cryogenic cooling to suppress thermally generated carriers, which will seriously degrade the signal. A major step forward in detector performance may be possible, by exploiting other obscure internal degrees of freedom of the electron instead of solely relying on the electron's charge for non-volatile information extraction, for example, utilizing the electron's spin [112]. In an electron, spin behaves like angular momentum but is not related to any real rotational motion. Consequently, the spin of an electron can be switched more quickly than charge can be moved round, and as a consequence, spintronic[16] devices should operate at far higher speeds than their electronic counterparts [113]. For radiation detection, a spintronic detector would utilize the interaction between the electromagnetic field associated with incident radiation and the spin polarized electrons and not rely on energy deposition of the incoming radiation or the controlled manipulation of charge carriers for its signal. This means that the signal is essentially decoupled from its environment compared to the charge degree of freedom, potentially offering a higher degree of noise immunity and non-volatility. The actual implementation for a radiation detector is discussed in Chapter 13.

Similarly, it should be possible to exploit the valley degree of freedom using the valleytronic properties [114] of the band structure of certain semiconductors. In these materials (known as multivalley semiconductors), the local maxima/minima of the valence/conduction bands present can form multiple valleys inside the first Brillouin zone whose minima can occur at equal energies but at different positions in momentum space. By utilizing a suitable "valley-filter" (see Chapter 13), it is possible to control the number of electrons that occupy a particular valley. If this is now applied to the ionization generated in a detector by incident radiation, the resultant valley "polarization" can be used to transmit/process information [115]. As in spintronics, this new degree of freedom behaves mathematically in a similar way to the electron spin and also acts like an additional intrinsic angular momentum of the electron.

1.5.3 Biological Based Detection Systems

The third avenue of future development is biological based detection systems and so-called intelligent photonics. In 1977, the first highly conducting polymer (chemically doped polyacetylene) [116] was discovered, which demonstrated that organic molecules could be used as electrically active materials. Subsequently, it has been found that semiconductor systems can be directly interfaced with living cells [117] as in the case of nano-structured silicon, or by "*semiconductorizing*" living systems using DNA scaffolding [118] recently proposed to produce photonic arrays and energy-harvesting assemblies [119].

REFERENCES

[1] W.C. Röntgen, "Eine neue art von strahlen", in *Sitzungsberichte der Würsburger Physik-medic. Gesellschaft*, eds. Würzburg, Verlag und Druck Der Stahel'Schen K. Hof- und Universitäts-Buch- und Kunsthandlung, Wurzburg (1896), pp. 137–147. http://www.deutschestextarchiv.de/book/show/roentgen_strahlen02_1896

[2] G. Brandes, "Über die Sichtbarkeit der Röntgenstrahlen", *S. B. Preuss. Akad. Wiss.*, Vol. 1 (1896), pp. 547–550.

[3] W.C. Röntgen, "Dritte Mitt., Weitere Beobachtungen über die Eigenschaften der XStrahlen", *S. B. Preuss. Akad. Wiss.*, Vol. 26 (1897), pp. 576–592.

[4] L.E. Lipetz, "The X ray and radium phosphenes", *Br. J. Ophthalmol.*, Vol. 39 (1955), pp. 577–598.

---

[16] A contraction of spin transport electronics or spin-based electronics.

[5] K.D. Steidley, R.M. Eastman, R.J. Stabile, "Observations of visual sensations produced by cerenkov radiation from high energy electrons", *Int. J. Radiat. Oncol. Biol. Phys.*, Vol. **17** (1989), pp. 685–690.

[6] H. Becquerel, "Sur les radiations émises par phosphorescence", *Comptes rendus hebdomadaires des séances de l'Académie des sciences*, Vol. **122** (1896), pp. 420–421.

[7] H. Poincaré, "Les rayons cathodiques et les rayons Röntgen", *Revue générale des sciences pures et appliquées*, Vol. **7** (1896), pp. 52–59.

[8] J.J. Thomson, "The Röntgen rays", *Nature*, Vol. **53** (1896), pp. 581–583.

[9] M. Curie, "Rayons émis par les composés de l'uranium et du thorium", *Comptes rendus hebdomadaires des seances de l'Academie des sciences*, Vol. **126** (1898), pp. 1101–1103.

[10] P. Curie, M. Curie, "Sur une substance nouvelle radioactive contenue dans la pechblende", *Comptes rendus hebdomadaires des seances de l'Academie des sciences*, Vol. **127** (1898), pp. 175–178.

[11] P. Curie, M. Curie, M.G. Bémont, "Sur une nouvelle substance fortement radio-active, contenue dans pechblende", *Comptes rendus hebdomadaires des seances de l'Academie des sciences*, Vol. **127** (1898), pp. 1215–1217.

[12] E. Rutherford, "Uranium radiation and the electrical conduction produced by it", *Philos. Mag.*, series 5, Vol. **47**, no. 284 (1899) pp. 109–163.

[13] J.J. Thomson, "Cathode rays", *Philos. Mag.*, Vol. **44**, no. 269 (1897) pp. 293–316.

[14] E. Rutherford, T. Royds, "The nature of the α particle from radioactive substances", *Philosophical Magazine*, Vol. **17** (1909) pp. 281–286.

[15] P. Villard, "Sur la réflexion et la réfraction des rayons cathodiques et des rayons déviables du radium", *Comptes rendus hebdomadaires des seances de l'Academie des sciences*, Vol. **130** (1900), pp. 1010–1012.

[16] E. Rutherford, F. Soddy, "Radioactive change", *Philos. Mag.*, Series 6, Vol. **5** (1903) pp. 576–591.

[17] E. Rutherford, "The scattering of α and β particles by matter and the structure of the atom", *Philos. Mag.*, Series 6, Vol. **21** (1911) pp. 669–688.

[18] N. Bohr, "On the constitution of atoms and molecules, part II systems containing only a single nucleus", *Philos. Mag.*, Vol. **26**, no. 153 (1913) pp. 476–502.

[19] J. Chadwick, "Possible existence of a neutron", *Nature*, Vol. **129**, no. 3252 (1932), p. 312.

[20] M. Niépce de Saint-Victor, "Deuxième mémoire sur une nouvelle action de la lumière", *Comptes rendus hebdomadaires des seances de l'Academie des sciences*, Vol. **46** (1858), pp. 448–452.

[21] M. Niépce de Saint-Victorc, "Cinquième mémoire sur une nouvelle action de la lumière", *Comptes rendus hebdomadaires des seances de l'Academie des sciences*, Vol. **53** (1861), pp. 33–35.

[22] W.C. Röntgen, "Eine neue Art von Strahlen. 2. Mitteilung", in *Aus den Sitzungsberichten der Würzburger Physik.-medic. Gesellschaft Würzburg*, eds. Würzburg, Verlag und Druck Der Stahel'Schen K. Hof- und Universitäts-Buch- und Kunsthandlung, Wurzburg (1896), pp. 11–17.

[23] J.J. Thomson, "On the discharge of electricity produced by the Rontgen rays, and the effects produced by these rays on dielectrics through which they pass", *Proc. R. Soc. Lond.*, Vol. **59** (1895), pp. 274–276.

[24] J.J. Thomson, E. Rutherford, "On the passage of electricity through gases exposed to Röntgen rays", *Philos. Mag.*, series 5, Vol. **42**, no. 258 (1896), pp. 392–407.

[25] P.W. Frame, "A history of radiation detection instrumentation", *Health Phys.*, Vol. **88**, no. 6 (2005), pp. 612–637.

[26] W. Crookes, "Certain properties of the emanations of radium", *Chem. News*, Vol. **87** (1903), p. 241.

[27] E. Rutherford, H. Geiger, "An electrical method of counting the number of α-particles from radio-active substances", *Proc. R. Soc. A*, Vol. **81** (1908), pp. 141–162.

[28] E. Regener, "Über Zählung der a-Teilchen durch die Szintillation und die Größe des elektrischen Elementarquantums", *Verhandlungen der Deutschen Physikalischen Gesellschaft*, Vol. **10** (1908), pp. 78–83.

[29] H. Geiger, W. Müller, "Elektronenzählrohr zur Messung schwächster Aktivitäten", *Naturwissenschaften*, Vol. **16**, no. 31 (1928), pp. 617–618.

[30] L.A. Kubetsky, "Multiple amplifier", *Proc. Inst. Radio Eng.*, Vol. **25** (1937), pp. 421–433.

[31] H. Hertz, "Ueber einen Einfluss des ultravioletten Lichtes auf die electrische Entladung", *Ann. Physik*, Vol. **31** (1887), pp. 983–1000.

[32] J. Slepian, Westinghouse Electric, "Hot Cathode Tube" U.S. Patent 1,450,265, Issued April 3 (1923).

[33] L. Austin, H. Starke, "Ueber die Reflexion der Kathodenstrahlen und eine damit verbundene neue Erscheinung secundaerer Emission", *Ann. Physik*, Vol. **314** (1902), pp. 271–292.

[34] H. Iams, B. Salzberg, "The secondary emission phototube", *Proceedings of the IRE*, Vol. **23** (1935), pp. 55–64.

[35] V.K. Zworykin, G.A. Morton, L. Malter, "The secondary-emission multiplier – A new electronic device", *Proc. Inst. Radio Eng.*, Vol. **24** (1936), pp. 351–375.

[36] R.H. Morgan, "An exposure meter for roentgenography", *Am. J. Roentgenol Radium Ther.*, Vol. **547** (1942), pp. 777–784.

[37] S.C. Curran, W.R. Baker, "A photoelectric alpha-particle detector", *Rev. Sci. Instr.*, Vol. **19** (1948), p. 116.

[38] H. Kallmann, "Lesson in a colloquium in Berlin–Dahlem, reported by W. Bloch, "Kann man Elektronen sehen", *Natur und Technik*, Vol. **13** (1947), pp. 15–17.

[39] R. Hofstadter, M.H. Stein, "Twenty five years of scintillation counting", *IEEE Trans. Nucl. Sci.*, Vol. **22** (1975), pp. 13–18.

[40] R. Hofstadter, "Alkali halide scintillation counters", *Phys. Rev.*, Vol. **74** (1948), pp. 100–101.

[41] R. Hofstadter, J.A. McIntyre, "Measurement of gamma-ray energies with single crystals of NaI(T1)", *Phys. Rev.*, Vol. **80** (1950), pp. 631–637.

[42] S.G.F. Frank, O.R. Frisch, G.G. Scarrott, "A mechanical kick-sorter (pulse size analyser)", *Philos. Mag.*, Series 7, Vol. **42**, no. 329 (1951), pp. 603–611.

[43] M. Faraday, "On a new law of electric conduction", *Phil. Trans. R. Soc.*, Vol. **123** (1833), pp. 507–515.

[44] H.J. Round, "A note on carborundum", *Elect. World*, Vol. **49** (1907), p. 309.

[45] O.V. Lossev, "Luminous carborundum detector and detection effect and oscillations with crystals", *Philos. Mag.*, Vol. **5** (1928), pp. 1024–1044.

[46] A. Einstein, "Über einen die Erzeugung und Verwandlung des Lichtes betreffenden heuristischen Gesichtspunkt", *Annalen der Physik*, Vol. **322**, no. 6 (1905), pp. 132–148.

[47] W. Smith, "Effect of light on selenium during the passage of an electric current", *Nature*, Vol. **7**, no. 173 (1873), p. 303.

[48] C.E. Fritts, "On a new form of selenium photocell", *Am. J. Sci.*, Vol. **26** (1883), p. 465.

[49] B. Gudden, R. Pohl, "Das Quantenäquivalent bei der lichtelektrischen Leitung", *Z. Physik*, Vol. **17** (1923), pp. 331–346.

[50] W. Lehfeldt, "Ober die elektrische. Leitfahigkeit von Einkristallen", *Z. Physik*, Vol. **85** (1933), pp. 717–726.

[51] W. Lehfeldt, "Zur Elektronenleitung in Silber- und Thalliumhalogenid-kristallen", *Nachr. Ges. Wiss. Gottingen Math. Physik. Klasse, Fachgruppe II*, Vol. **1** (1935), p. 171.

[52] R. Frerichs, R. Warminsky, "Die Messung von β- und γ-Strahlen durch inneren Photoeffekt in Kristallphosphoren", *Naturwissenschaften*, Vol. **33**, no. 8 (1946), p. 251.

[53] G. Jaffe, "Über den Einfuß von α-Strahlen auf den Elektrizitätsdurchgang durch Kristalle", *Physikalische Zeitschrift*, Vol. **33** (1932), pp. 393–399.

[54] B. Gudden, "Über die Elektrizitätsleitung in Halbleitern", *Sitzungbarickle der Physmediz*, Vol. **62** (1930), pp. 389–302.

[55] E.H. Hall, "On a new action of the magnet on electric currents", *Am. J. Math.*, Vol. **II** (1879), pp. 287–292.

[56] P. Drude, "Zur Elektronentheorie der Metalle", *Ann. Phys. Leipzig*, Vol. **1** (1900), pp. 566–613.

[57] J. Koenigsberger, K. Schilling, "Über die Leitfähigkeit einiger fester Substanzen", *Phys. Z.*, Vol. **9** (1908), pp. 347–352.

[58] J. Koenigsberger, "Das elektrische Verhalten der variablen Leiter und deren Beitrage zur Elektronentheorie", *Jahrb. Radioakt. Elecktron*, Vol. **11** (1914), pp. 84–142.

[59] J. Weiss, "*Experimentelle Beiträge zur Elektronentheorie aus dem Gebiet der Thermoelektrizität*", Inaugural Dissertation Albert-Ludwigs Universität, Freiburg im Breisgau (1910).

[60] J. Frenkel, "Über die Wärmebewegung in festen und flüssigen Köpern", *Z. Phys.*, Vol. **35** (1926), pp. 652–669.

[61] C. Wagner, W. Schottky, "Theorie der geordneten Mischphasen", *Z. Phys. Chem.*, Vol. **11** (1930), pp. 163–210.

[62] A.H. Wilson, "The theory of electronic semiconductors I", *Proc. Roy. Soc. A*, Vol. **133** (1931), pp. 458–491.

[63] A.H. Wilson, "The theory of electronic semiconductors II", *Proc. Roy. Soc. A*, Vol. **134** (1931), pp. 277–287.

[64] F. Bloch, "Über die Quantenmechanik der Elektronen in kristallgittern", *Z. Phys.*, Vol. **52** (1928), pp. 555–560.

[65] R.E. Peierls, "Zur Theoric der elektrischen und thermischeri Leitfahigkeit von Metallen", *Ann. Phys. Leipzig*, Vol. **4** (1930), pp. 121–148.

[66] R. Kronig, W.G. Penney, "Quantum mechanics of electrons in crystals lattices", *Proc. Roy. Soc. A*, Vol. **130** (1931), pp. 499–513.

[67] W. Schockley, "*Electrons and Holes in Semiconductors*", D. van Nostrand Co. Inc., Princeton, NJ (1950), ISBN-10: 0882753827.

[68] W. Heisenberg, "Zum Paulischen Ausschliessungsprinzip", *Ann. Phys. Leipzig*, Vol. **10**, no. 5 (1931), pp. 888–904.

[69] N.F. Mott, "Note on the contact between a metal and an insulator or semiconductor", *Proc. Camb. Philos. Soc.*, Vol. **34** (1938), pp. 568–572.

[70] W. Schottky, "Halbleitertheorie der Sperrsschicht", *Naturwissenschaften*, Vol. **26** (1938), p. 843.

[71] B. Davydov, "On the rectification of current at the boundary between two semiconductors", *Compt. Rend. Doklady Acad. Sci. USSR*, Vol. **20** (1938), pp. 279–282.

[72] K. Hecht, "Zum Mechanismus des lichtelektrischen Primärstromes in isolierenden Kristallen", *Z. Phys.*, Vol. **77** (1932), pp. 235–245.

[73] P.J. Van Heerden, "The crystal counter: A new instrument in nuclear physics", PhD Dissertation, Rijksuniversiteit Utrecht, July (1945).

[74] P.J. Van Heerden, "The crystal counter: A new apparatus in nuclear physics for the investigation of β and γ-rays. Part I", *Physica*, Vol. **16**, no. 6 (1950), pp. 505–516.

[75] P.J. Van Heerden, J.M.W. Milatz, "The crystal counter: A new apparatus in nuclear physics for the investigation of β and γ-rays. Part II", *Physica*, Vol. **16**, no. 6 (1950), pp. 517–527.

[76] R. Hofstadter, "Thallium halide crystal counter", *Phys. Rev.*, Vol. **72** (1947), pp. 1120–1121.

[77] R. Hofstadter, "Crystal counters-I", *Nucleonics*, Vol. **4**, no. 2 (1949), pp. 20–27.

[78] R. Frerichs, "On the relations between crystal counters and crystal phosphors", *J. Opt. Soc. Am.*, Vol. **40** (1950), pp. 219–221.

[79] A.L. Hughes, "Photoconductivity in crystals", *Rev. Mod. Phys.*, Vol. **8** (1936), pp. 294–315.

[80] K.G. McKay, "A germanium counter", *Phys. Rev.*, Vol. **76** (1949), p. 1537.

[81] K.G. McKay, K.B. McAfee, "Electron multiplication in silicon and germanium", *Phys. Rev.*, Vol. **91** (1953), pp. 1079–1084.

[82] A.G. Chynoweth, "Conductivity crystal counters", *Am. J. Phys.*, Vol. **20** (1952), pp. 218–226.

[83] G.L. Miller, W.M. Gibson, P.F. Donovan, "Semiconductor particle detectors", *Ann. Rev. Nucl. Sci.*, Vol. **12** (1962), pp. 189–220.

[84] J.W. Mayer, "The development of the junction detector", *IRE Trans. Nucl. Sci.*, Vol. **NS-7** (1960), pp. 178–180.

[85] E.M. Pell, "Ion drift in an n-p junction", *J. Appl. Phys.*, Vol. **31** (1960), pp. 291–302.

[86] D. de Nobel, "Phase equilibria and semiconducting properties of Cadmium Telluride", *Phillips Res. Rept.*, Vol. **14** (1959), pp. 361–399.

[87] M.B. Prince, P. Polishuk, "Survey of materials for radiation detection at elevated temperatures", *Trans. Nucl. Sci.*, Vol. **NS-14** (1967), pp. 537–543.

[88] J.E. Eberhardt, R.D. Ryan, A.J. Tavendale, "High resolution radiation detectors from epitaxial n-GaAs", *Appl. Phys. Lett.*, Vol. **17** (1970), pp. 427–429.

[89] D.S. McGregor, H. Harmon, "Room-temperature compound semiconductor radiation detectors", *Nucl. Instr. Meth.*, Vol. **A395** (1997), pp. 101–124.

[90] A. Owens, A. Peacock, "Compound semiconductor radiation detectors", *Nucl. Instr. Meth.*, Vol. **A531** (2004), pp. 18–37.

[91] S. Ashok, J. Chevallier, P. Kiesel, T. Ogino, eds., *"Semiconductor defect engineering – materials, synthetic structures and devices II"*, Materials Research Society Symposium Proceedings, Warrendale, PA, Vol. **994** (2007).

[92] J.D. Martin, A.M. Dattelbaum, T.A. Thornton, R.M. Sullivan, J. Yang, M.T. Peachey, "Metal halide analogues of chalcogenides: A building block approach to the rational synthesis of solid-state materials", *Chem. Mater.*, Vol. **10** (1998), pp. 2699–2713.

[93] J. Androulakis, S.C. Peter, H. Li, C.D. Malliakas, J.A. Peters, Z. Liu, B.W. Wessels, J.-H. Song, H. Jin, A.J. Freeman, M. G. Kanatzidis, "Dimensional reduction: A design tool for new radiation detection materials", *Adv. Mater.*, Vol. **23** (2011), pp. 4163–4167.

[94] E.G. Tulsky, J.R. Long, "Dimensional reduction: A practical formalism for manipulating solid structures", *Chem. Mater.*, Vol. **13**, no. 4 (2001), pp. 1149–1166.

[95] P.N. Luke, "Single-polarity charge sensing in ionization detectors using coplanar electrodes", *Appl. Phys. Lett.*, Vol. **65** (1994), pp. 2884–2886.

[96] E. Gatti, P. Rehak, "Semiconductor drift chamber – An application of a novel charge transport scheme", *Nucl. Instr. Meth.*, Vol. **A225** (1984), pp. 608–614.

[97] E.V.D. van Loef, P. Dorenbos, C.W.E. van Eijk, H.U. Gudel, K.W. Kraemer, "High-energy-resolution scintillator: Ce^{3+} activated $LaCl_3$", *Appl. Phys. Lett.*, Vol. **77** (2000), pp. 1467–1469.

[98] E.V.D. van Loef, P. Dorenbos, C.W.E. van Eijk, H.U. Gudel, K.W. Kraemer, "High-energy-resolution scintillator: Ce^{3+} activated $LaBr_3$", *Appl. Phys. Lett.*, Vol. **79** (2001), pp. 1573–1575.

[99] P. Dorenbos, J.T.M. de Haas, C.W.E. van Eijk, "Non-proportionality in the scintillation and the energy resolution with scintillation crystals", *IEEE Trans. Nucl. Sci.*, Vol. **42**, no. 6 (1995), pp. 2190–2202.

[100] F.G.A. Quarati, P. Alan Owens, J.T.M. Dorenbos, G. de Haas, B.N. Blasi, C. Boiano, S. Brambilla, F. Camera, R. Alba, G. Bellia, C. Maiolino, D. Santonocito, M. Ahmed, N. Brown, S. Stave, H.R. Weller, Y.K. Wu., "High energy gamma-ray spectroscopy with $LaBr_3$ scintillation detectors", *Nucl. Instr. Meth.*, Vol. **A629** (2011), pp. 157–169.

[101] M.S. Alekhin, J.T.M. de Haas, I.V. Khodyuk, K.W. Krämer, P.R. Menge, V. Ouspenski, P. Dorenbos, "Improvement of γ-ray energy resolution of $LaBr_3$: Ce^{3+} scintillation detectors by Sr^{2+} and Ca^{2+} co-doping", *Appl. Phys. Lett.*, Vol. **102**, no. 16 (2013), pp. 161915-1–161915-4.

[102] A. Owens, T. Buslaps, V. Gostilo, H. Graafsma, R. Hijmering, A. Kozorezov, A. Loupilov, D. Lumb, E. Welter, "Hard X- and gamma-ray measurements with a large volume coplanar grid CdZnTe detector", *Nucl. Instr. Meth.*, Vol. **A563** (2006), pp. 242–248.

[103] I.V. Khodyuk, P. Dorenbos, "Trends and patterns of scintillator nonproportionality", *IEEE Trans. Nucl. Sci.*, Vol. **59**, no. 6 (2012), pp. 3320–3331.

[104] G. Charpak, F. Sauli, "Multiwire proportional chambers and drift chambers", *Nucl. Instr. Meth.*, Vol. **162** (1979), pp. 405–428.

[105] A. Oed, "Position sensitive detector with microstrip anode for electron multiplication with gases", *Nucl. Instr. Meth.*, Vol. **A263** (1988), pp. 351–359.

[106] F. Sauli, "The gas electron multiplier (GEM): Operating principles and applications", *Nucl. Instr. Meth.*, Vol. **A805** (2016), pp. 2–24.

[107] J.T. Lue, "Physical properties of nanomaterials", in *Encyclopedia of Nanoscience and Nanotechnology*, ed. H.S. Nalwa, vol. **5**, American Scientific Publishers, Valencia, CA (2007), pp. 1–46.

[108] F.D. Amaro, C.M.B. Monteiro, J.M.F. dos Santos, A. Antognini, "Novel concept for neutron detection: Proportional counter filled with ^{10}B nanoparticle aerosol", *Sci. Rep.*, Vol. **7**, Article number 41699 (2017), pp. 1–6.

[109] G. Zhang, K. Tateno, H. Gotoh, T. Sogawa, "Towards new low-dimensional semiconductor nanostructures and new possibilities", *NTT Tech. Rev.*, Vol. **8**, no. 8 (2010), pp. 1–8.

[110] S.J. Koh, "Controlled placement of nanoscale building blocks: Toward large-scale fabrication of nanoscale devices", *JOM*, Vol. **59**, no. 3 (2007), pp. 22–28.

[111] M. Foxe, G. Lopez, I. Childres, R. Jalilian, C. Roecker, J. Boguski, I. Jovanovic, Y.P. Chen, "Graphene field-effect transistors on undoped semiconductor substrates for radiation detection", *IEEE Trans. Nanotech.*, Vol. **11**, no. 3 (2012), pp. 581–587.

[112] N. Gary, S. Teng, A. Tiwari, H. Yang, "*Room temperature radiation detection based on spintronics*", IEEE/MIC/RTSD, 19th International Workshop on Room Temperature Semiconductor Detectors, Anaheim, CA, 27 October–3 November, Conference record (2012), pp. 4152–4155.

[113] D.D. Awschalom, M.E. Flatté, N. Samarth, "Spintronics", *Sci. Am.*, Vol. **286** (2002), pp. 66–73.

[114] V.T. Renard, B.A. Piot, X. Waintal, G. Fleury, D. Cooper, Y. Niida, D. Tregurtha, A. Fujiwara, Y. Hirayama, K. Takashina, "Valley polarization assisted spin polarization in two dimensions", *Nat. Commun.*, Vol. **6**, no. 7230 (2015), pp. 1–8.

[115] J. Isberg, M. Gabrysch, J. Hammersberg, S. Majdi, K.K. Kovi, D.J. Twitchen, "Generation, transport and detection of valley-polarized electrons in diamond", *Nat. Mater.*, Vol. **12** (2013), pp. 760–764.

[116] C.K. Chiang, C.R. Fincher, J.Y.W. Parker, A.J. Heeger, H. Shirakawa, E.J. Louis, S.C. Gau, A.G. MacDiarmid, "Electrical conductivity in doped polyacetylene", *Phys. Rev. Lett.*, Vol. **39**, no. 17 (1977), pp. 1098–1101.

[117] S.C. Bayliss, R. Heald, D.I. Fletcher, L.D. Buckberry, "The culture of mammalian cells on nanostructured silicon", *Adv. Mater.*, Vol. **11**, no. 4 (1999), pp. 318–321.

[118] K. Matczyszy, J. Olesiak-Banska, "DNA as scaffolding for nano-photonic structures", *J. Nanophotonics*, Vol. **6**, no. 1 (2012), pp. 064505-1–064505-15.

[119] W. Su, V. Bonnard, G.A. Burley, "DNA-templated photonic arrays and assemblies: Design principles and future opportunities", *Chem.-Eur. J.*, Vol. **17** (2011), pp. 7982–7991.

2

Semiconductors

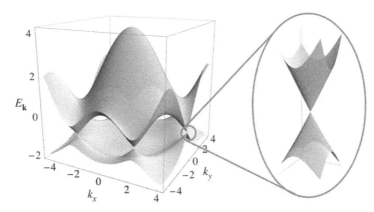

Frontispiece. The band structure of graphene (from Castro Neto *et. al. Rev. Mod. Phys.*, Vol. **81** (2009), pp. 109–162). The conduction band and the valence band touch at six discrete points, the so-called K points, each of which can be divided into two in-equivalent sets of three points each. The points within each set are all equivalent because they can reach each other by reciprocal lattice vectors. The two in-equivalent points are called K and K`, which form the valley pseudo-spin degree of freedom in graphene. The term "valley" stems from the similarity of the shape of k space in the vicinity of these points with a valley.

CONTENTS

2.1 Metals, Semiconductors and Insulators

Solids are usually divided into metals and non-metals with the non-metals comprised of insulators and semiconductors. They may be grouped by their resistivity, ρ, which can range between 10^{-8} Ω-cm (metals) and 10^{18} Ω-cm (insulators) as illustrated in Fig. 2.1. Semiconductors[1] are intermediate between the conductors and insulators and typically have resistivities between 10^{-2} Ω-cm and 10^{8} Ω-cm. Virtually all inorganic and many organic materials belong to this group. The metals comprise those materials with very low resistivities ($\sim 10^{-6}$ Ω-cm). However, within this group, some have resistivities that are quite high – in fact $\sim 10^{2}$ to 10^{3} times that of canonical metals, such as Cu. For this reason, metals such as As, Bi and Sb are known as semi-metals. At the other end of the resistivity scale, the distinction between insulators[2] and semiconductors can be somewhat vague, since some wide-gap materials that were previously regarded as insulators are now considered semiconductors (*e.g.*, diamond). Since conductivity is a strong function of temperature, the difference between an insulator and a semiconductor really depends on the temperature at which the material's properties are being evaluated. We can qualitatively define a semiconductor as a material that displays a noticeable electrical conductivity at, or above, room temperature. A true insulator, on the other hand, will either melt or sublime before noticeable conductivity can be measured (*e.g.*, quartz).

2.2 Energy Band Formation

Whether a material is classified as a metal, semiconductor or insulator depends on its band structure, which is a natural consequence of its crystal structure, and specifically the periodic arrangement of its atoms. The potential energy associated with these atoms varies significantly over inter-atomic distances, and so the macroscopic properties of the crystal can only be

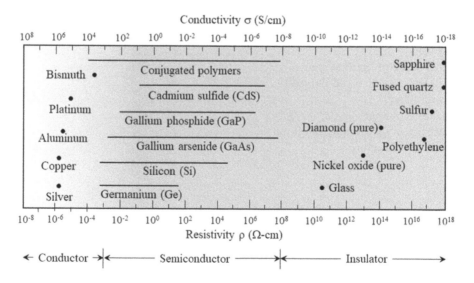

FIGURE 2.1 Typical range of resistivities/conductivities for insulators, semiconductors and conductors. Semiconductors exist in the shaded region. At resistivities above 10^8 Ω-cm, the distinction between insulators and semiconductors is blurred and ultimately depends on temperature, since an "insulating" material can only become semiconducting (in the sense it can pass a current) if a sufficient number of electrons can be excited into the conduction band.

---

[1] Because semiconductors display such a wide range of conductivities, they were initially classified as "variable conductors" by the German Physicist Johann Koenigsberger in 1914. It was his student, Josef Weiss, who first coined the word *Halbleiter* (semiconductor) in its modern sense in his doctoral thesis of 1910.

[2] From the late Latin "*insulatus*" meaning "made like an island".

adequately described using quantum mechanics. A comprehensive treatment of quantum mechanics is beyond the scope of this chapter; the reader is referred to standard texts, such as Kittel [1], Wenckebach [2] and Shockley [3]. We will approach the formation of energy bands qualitatively using the Bohr theory of the atom as a starting point.

Bohr [4] combined a sun and planet model of the atom with quantum theory to explain why negatively charged electrons can form stable orbits around a small positively charged nucleus. Electrostatic attraction had long been thought to provide the force to hold the electron in orbit, but by postulating that the electrons could only occupy discrete "shells" or energy levels, Bohr was able to explain why the electrons did not lose energy continuously and spiral catastrophically into the nucleus. According to the Bohr model for a hydrogen atom, these energy levels are given by

$$E_{\mathrm{H}} = -\frac{m_o q^4}{8\varepsilon_o^2 h^2 n^2} = -\frac{13.6}{n^2} \text{ eV}, \tag{1}$$

where m_o is the free electron mass, q is the electronic charge, ε_o is the permittivity of free space, h is Plank's constant and n is the principal quantum number that defines each quantized energy level. For $n = 1$, that is the ground state, $E_{\mathrm{H}} = -13.6$ eV[3] and for $n = 2$, the first excited state, $E_{\mathrm{H}} = -3.4$ eV. In the limit, when $n \to \infty$, $E_{\mathrm{H}} \to 0$ and the electron is no longer bound, the atom is ionized.

When two atoms approach one another, the electron wave functions begin to overlap and these energy levels split into two, as a consequence of Pauli's exclusion principle [5]. By the same argument, when N atoms come together to form a solid, each energy level splits into N separate but closely spaced levels, thereby resulting in a continuous band of energy levels for each quantum number, n. It should be noted that the Bohr model is approximate and only applicable for the simplest atom – hydrogen. For crystalline solids, the detailed energy band structures can only be calculated using quantum mechanics. Figure 2.2 shows the effects on the energy levels when isolated C atoms are brought together to form a diamond lattice structure. As the average interatomic distance between the atoms is reduced, the discrete $2s$ and $2p$ levels begin to split, into N closely spaced energy levels, where N is the number of atoms. As $N \to \infty$, these levels merge to form a continuous band – partly because the separation between energy levels becomes comparable with the lattice vibrational energy to which the electrons are strongly coupled, and partly because the separation in energy levels also becomes comparable with the time averaged energy uncertainty due to the Heisenberg uncertainty principle [6]. As the atoms approach their equilibrium interatomic spacing, this band separates into two, becoming the conduction band and the valence[4] band. At the equilibrium position, where the forces of attraction and repulsion between the atoms balance, the total energy of the electrons and the lattice is minimized. The region separating the conduction and valence bands is termed the forbidden gap, or bandgap, and is characterized by a bandgap energy, E_g. This gap uniquely determines the conduction properties of materials and represents the minimum energy to generate an electron-hole pair, the essential information carriers in a semiconductor. From Fig. 2.2, we see that a reduction in lattice spacing from its equilibrium value will increase the

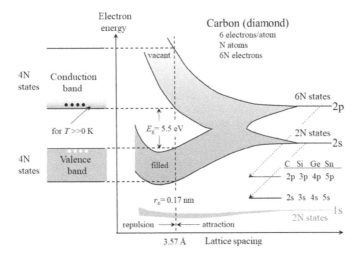

FIGURE 2.2 Formation of energy bands as a diamond crystal is formed by bringing together isolated carbon atoms – the energy bands essentially reflect the hybridization of the s and p levels. At the nominal lattice spacing, the forces of attraction and repulsion between atoms balance. The equilibrium position can be adjusted by pressure and temperature. The conventional one-dimensional representation of band structure at the nominal lattice spacing is given by the left diagram.

---

[3] By convention, an unbound electron has zero energy, which means that at the level $n = \infty$, $E\infty = 0$eV. Hence, any bound state (*i.e.*, $n < \infty$) will have energy < 0, since this is the energy to the system you need to add to the system to make the bound state free (*i.e.*, raising its energy to zero).

[4] Named the valence band by analogy to the valence electrons of individual atoms.

size of the bandgap while an increase in lattice spacing will decrease the bandgap. In reality, many other bands are formed out of the lower atomic levels, but only the valence and conduction bands play an active role in the conduction process, since all bands below the valence band are filled, and all bands above the conduction band are empty. It should be noted that electrons must move between states to conduct electric current, and so full bands do not contribute to the electrical conductivity. Only when electrons are excited into the conduction band are they free to move between states and thus contribute to current flow.

2.3 General Properties of the Bandgap

Fig. 2.3 shows schematically energy band diagrams for the three classes of solids: insulators, semiconductors and metals. These bands represent the outermost energy levels that an electron can occupy. Electrons in the valence band cannot move between atoms, whereas electrons in the conduction band can freely move from atom to atom and act as mobile charge carriers with which to conduct current. This is because the strength of the valence bond is 10 times that of the free electron bond, meaning that valence electrons are essentially fixed. The strength of the valence bond also explains why the movement of free carriers in the conduction band does not compromise the integrity of the lattice structure. In insulators, the bandgap is relatively large, and thermal energy or an applied electric field cannot raise the uppermost electron in the valence band to the conduction band. In metals, the conduction band is either partially filled, as in the case of a classical metal, such as Ag and Cu, or overlaps the valence band, as in the case of a semi-metal such as Sn or graphite. In both cases, there is no bandgap and current can readily flow in these materials. In semiconductors, the bandgap is smaller than that of insulators to such an extent that thermal energy can excite electrons to the conduction band. However, if the temperature (and therefore thermal energy) is lowered sufficiently, all semiconductors become insulators. Metals, on the other hand, are conductors at any temperature. Table 2.1 classifies materials according to bandgap energy, E_g., and consequently, the free carrier density, n, since its magnitude in pure materials is a direct consequence of the bandgap for a set temperature.

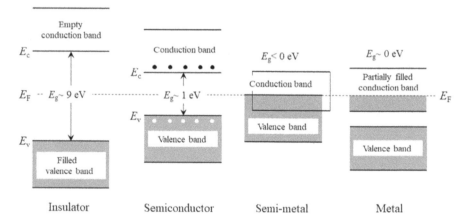

FIGURE 2.3 Schematic energy band diagrams for an insulator, a semiconductor and a metal. Two cases are given for the later: (1) semimetals with overlapping valence and conduction bands. Examples are Sn, Zn, Pb and graphite and (2) classical metals with partially filled conduction bands (*e.g.*, Cu, Au and Ag). The Fermi level, E_F, is the energy level at which an *average* of 50% of the available quantum states are filled by an electron and relates the probable location of electrons within a band. For metals, the Fermi level lies in the conduction band while for insulators and semiconductors the Fermi level lies in the bandgap.

TABLE 2.1

Classification of materials according to their energy gap, E_g, and carrier density, n, at room temperature. Values are given for pure, that is, intrinsic, materials. We arbitrarily chose a bandgap of 4 eV as the dividing line between insulators and semiconductors, since the actual distinction between the two is somewhat blurred and depends on the temperature in practice, as that determines the number of free carriers in the conduction band.

Material	E_g(eV)	n (cm^{-3})
Metal	no energy gap	10^{22}
Semi-metal	$E_g \leq 0$	10^{17}–10^{21}
Semiconductor	$0 < E_g < 4$	$< 10^{17}$
Insulator	$E_g \geq 4$	$\ll 1$

2.3.1 Hole Concept

In semiconductors, holes play an extremely important role in the electronic properties of the material. Although conceptually simple (usually defined as the absence of an electron in the valence band that is assumed to behave as a positive electron), it is actually difficult to describe, since the hole as a particle does not physically exist. It is a quasi-particle whose very "existence" is inextricably linked to the presence of the periodic potential provided by the presence of the crystal lattice. In 1931, Heisenberg [7] had already shown that empty states (holes) in a nearly full valence band of a semiconductor act globally in the exact same manner as equivalent "positive charges". The term "hole" was originally adopted by William Shockley in a monograph published in 1950 [3] but was not accepted by the community at the time – most researchers considered electrons only. Whilst this is possible quantum mechanically, it is far more complicated than assuming the existence of a few hole quasi-particles *a-priori* in a normally full band. A hole is created when an electron in the normally full valence band has sufficient energy to transit into the conduction band. The electron leaves behind an empty, discrete energy state in a band that is normally fully occupied by electrons. This "hole" perturbs the charge neutrality of the band, and a positive charge, whose absolute value is the same as the charge of the missing electron, prevails. Because the valence band is essentially full of electrons (in other words all energy states are occupied) the electrons are not free to move, since the next quantum states are almost always occupied and according to the Pauli exclusion principle, electrons cannot occupy multiple states. Thus, a field cannot produce acceleration in a valence electron. However, the holes *can* exchange their quantum states with the surrounding valence electrons and are thus free to move. Holes have a positive effective mass and normally occupy states at the top of the valence band where their energy is a minimum. The application of a field will cause an acceleration and the holes will transit through the valence band, carrying with them a charge and consequently a current.

2.3.2 Carrier Generation and Recombination

Free electrons may be generated by exciting electrons from the valence band into the conduction band leaving behind a free hole in the valence band. Thus, the carrier generation process simultaneously creates equal numbers of electrons and holes. There are several mechanisms by which free carriers can be generated, and these may be individually targeted for specific applications. For example, free carriers may be generated directly by thermal excitation (as in the case of thermal imaging), optical excitation (as in the case of optoelectronic devices) or direct ionization by charged particles (as in the case of radiation detectors). However, it should be appreciated that even in the absence of any external energy source, electron-hole pairs are constantly being generated by the thermal energy stored in the crystal lattice. However, at any temperature an equilibrium will be reached when the mean creation rate of electron hole pairs is equal to the recombination rate. The mean lifetime, τ_e, of a free electron is the average time that the electron exists in a free state before recombination. Similarly, the mean lifetime, τ_h, of a free hole is the average time that it exists in a free state before recombination. In an intrinsically pure semiconductor, $\tau_e = \tau_h$. In Si and Ge, these lifetimes are typically of the order of 10^{-3} s at room temperature. However, the lifetimes in compound semiconductors are considerably shorter, by some three to five orders of magnitude. Carrier recombination itself can take place by several mechanisms, the most important of which are illustrated in Fig. 2.4. These are (a) direct band-to-band radiative recombination, (b) deep level defect (trap) mediated or (c) non-radiative band-to-band Auger recombination. In the recombination process, electrons drop from the conduction band to the valence band – the energy difference being released as photons, phonons, photons and phonons or the transmission of the energy to a third particle. As shown in Fig. 2.4, band-to-band recombination occurs when an electron falls from its conduction band state into the empty valence band state associated with a hole, releasing a photon within a 1–10 ns time scale. The probability of band-to-band recombination is proportional to the product of the free electron and vacant hole densities. For indirect bandgaps,[5] this probability is small. Auger recombination involves three particles – an electron and a hole, which recombine in a band-to-band transition transferring the resulting energy to another electron or hole. The Auger recombination rate is proportional to the product of the concentrations of the respective reaction partners and is only relevant at high carrier concentrations. Carriers most effectively recombine *via* traps, which are energy states created in the forbidden energy gap by impurities or dislocations. These can hold an electron or hole long enough for a pair to be completed. The recombination process for deep level traps is usually non-radiative, so the energy must be dissipated as phonons. The probability of recombination is dependent on the product of the free electron and hole densities and on the number of carrier traps and is described by the Shockley, Read and Hall model [8,9].

In most materials, one of the above processes will have a much higher probability of occurrence than the others and will largely determine the effective lifetime, τ, of the excited carriers. This is given by

$$\frac{1}{\tau} = \frac{1}{\tau_{\text{rad}}} + \frac{1}{\tau_{\text{trap}}} + \frac{1}{\tau_{\text{Aug}}},\tag{2}$$

---

[5] See Section 2.3.5.

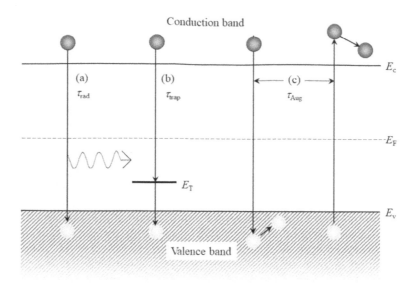

FIGURE 2.4 Schematic of carrier recombination mechanisms in semiconductors illustrating (a) radiative emission, (b) deep level trap mediated, or (c) non-radiative band-to-band Auger recombination.

where τ_{rad}, τ_{trap}, τ_{Aug} are the radiative, trap and Auger lifetimes, respectively. In high-quality direct bandgap semiconductors, radiative emission is likely to be the dominant mechanism. In heavily doped semiconductors ($>10^{17}$ atoms cm^{-3}), Auger recombination is usually more likely, since the Auger rate is proportional to the square of the doping density. For most semiconductors, however, especially those with indirect bandgaps, short lifetimes are mainly due to impurities and defects, especially for those traps and defects that introduce energy levels near the middle of the gap.

2.3.2.1 Excitons

When an electron hole pair is created following excitation in an insulator or semiconductor, they may remain bound together by their mutual Coulomb attraction, even though the electron resides in the conduction band and the hole in the valence band. This attraction provides a stabilizing force, and consequently, the system has slightly less energy than the unbound electron and hole. The strength of the coupling, or the binding energy, is weak and is typically around 10–100 meV in bulk materials. The bound electron-hole pair is called an *exciton* [10] and is essentially a quantum of excitation that behaves as an electrically neutral quasi-particle that can move through the crystal transporting energy but not electric charge. The bound pair is analogous to a neutral hydrogen atom in which an electron is bound electrostatically to a proton, except that the radius of the exciton is far larger, as the mass of the hole is much smaller than the proton mass. In fact, the radius can be much larger than the lattice constant, since a screening effect reduces the Coulomb attraction between the electron and hole further, depending on the dielectric constant of the material. The excitonic states are at energies just below the conduction band and can exist in a series of bound states in the gap. Since excitonic binding energies are so small, all excitons are unstable at room temperatures leading to recombination in which the electron drops into the hole and the exciton dissociates back into a free electron and free hole. Typical exciton lifetimes are in the range ps to ns. The energy of the exciton may be converted into light (photoluminescence), or it may be transferred to an electron of a neighbouring atom in the solid.

2.3.3 Pressure Dependence of the Bandgap

The size of the bandgap is found to vary with both pressure and temperature largely due to distortion of the lattice. Applying a high compressive stress reduces the lattice spacing that increases the bandgap. An approximate expression relating pressure to bandgap energy is

$$E_{\text{g}}(P) = E_{\text{g}}(0) + aP + bP^2,$$
(3)

where $E_{\text{g}}(0)$ is the bandgap energy at zero pressure, P is the hydrostatic pressure and a and b are empirical constants. To first order, it is found that the pressure dependences of many III-V direct gap materials are similar at $\sim+50$ meV per GPa (*e.g.*, [11,12]). Since 1 GPa corresponds to 65 tons per square inch, it can be seen that the effect is very small and can be neglected for almost all applications. However, for optoelectronic applications, measuring the variation of photoluminescence with pressure can be very useful in understanding the underlying band structure of a material and its structural properties. At the

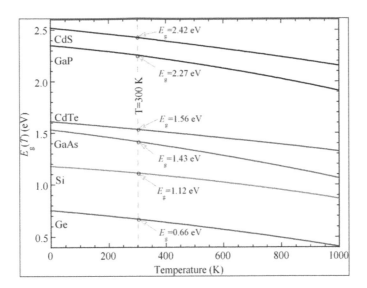

FIGURE 2.5 Temperature dependence of the bandgap energy for common semiconductors from Groups IV, III-V and II-VI compounds. Bandgap energies decrease by ~0.4 meV per degree K for most semiconductors.

highest hydrostatic pressures (usually, tens of GPa), some semiconductors can undergo a phase transition and lose their semiconducting properties completely, becoming metals.

2.3.4 Temperature Dependence of the Bandgap

As the temperature of a semiconductor increases, the amplitude of atomic vibrations increases, which in turn increases interatomic spacings. The lattice therefore expands, as witnessed by the material's linear expansion coefficient. The motion of the atoms about their equilibrium lattice points affects the width of the energy levels – at higher temperatures, they become broader. In addition, increasing the interatomic spacing decreases the potential seen by the electrons in the material, which in turn reduces the size of the energy bandgap. For example, the energy gap in silicon at 0 K is 1.17 eV. At room temperature, it is 1.12 eV. The relationship between bandgap energy, E_g, and temperature, T, can be described by the semi-empirical formulation of Varshni [13],

$$E_g(T) = E_g(0) - \frac{\alpha T^2}{T + \beta},$$

(4)

where $E_g(0)$ is the bandgap at absolute zero degrees K and α and β are material constants. In Table 2.2 we list these parameters for a range of materials from Groups IV, III-V and II-VI semiconductors and in Fig. 2.5 we plot the functional relationship for some of the more common materials. For all semiconductors, like Si, GaAs, and CdS, the bandgap changes by ~0.3–0.5 meV per degree K; so for the temperatures normally encountered by detection systems this effect can be neglected. In reality, the dominant effect of temperature on semiconductor properties is on carrier densities that vary exponentially with temperature (see Section 2.8.1.1).

2.3.5 Bandgap Morphology

The one-dimensional bandgap picture introduced in Sections 2.2 and 2.3 is over-simplified in that it does not adequately describe the true morphology of band structure and does not distinguish, for instance, between direct and indirect bandgap semiconductors. In order to do so, one has to consider the relationship between momentum and energy as a function of crystal direction (*e.g.*, see Fig. 2.6). The reason that the energies of the states can be broadened into a band without violating any exclusion law is that the energy now also depends on the magnitude of the wave vector, *k*, which in quantum mechanics, is used to represent the momentum *hk* of a particle. This relationship is known as energy-momentum dispersion. In essence, the periodic potential imposed by the crystal lattice modulates the band structure at a basic scale length of the smallest unit cell. The *E-k* space encompassed by this cell is known as the first Brillouin zone and is a particularly important concept in understanding the electronic and thermal properties of crystalline solids, because within this zone the band structure is unique – outside it is replicated. Although difficult to visualize, Brillouin zones can be explained as follows. Every periodic structure has two related lattices associated with it. The first is the real space lattice; this describes the periodic structure. The second is the reciprocal lattice (see Chapter 3), which determines how the periodic structure interacts with waves. Simply stated, Brillouin zones are polyhedra in reciprocal space that are the geometrical equivalent of Wigner-Seitz (primitive) cells in real space. Physically,

TABLE 2.2

Temperature dependence of bandgap constants for a number of semiconductors from Groups IV, III-V and II-VI (from ref [14]).

Parameter **Material**	$E_g(T = 0°K)$ (eV)	$\alpha \times 10^4$ (eV/K)	β (K)
Group IV			
Ge	0.744	4.07	230
Si	1.170	4.18	406
4H-SiC	3.26	6.5	1300
C (diamond)	5.49	5.0	1067
Group III-V			
InSb	0.234	2.50	136
InAs	0.420	2.50	75
InP	1.421	3.63	162
InN	1.994	2.1	453
GaSb	0.811	3.75	176
GaAs	1.519	5.41	204
GaP	2.338	5.771	372
GaN	3.47	5.99	504
AlSb	1.686	3.43	226
AlAs	2.239	6.0	408
AlP	2.49	3.5	130
AlN	6.20	8.3	575
Group II-VI			
CdTe	1.606	3.10	108
CdSe	1.846	4.05	168
CdS	2.583	4.02	147
ZnTe	2.394	4.54	145
ZnSe	2.825	4.90	190
ZnS	3.841	5.32	240

Brillouin zone boundaries represent Bragg planes that reflect (diffract) waves having particular wave vectors so that they cause constructive interference. The closer an electron is to the boundary of a Brillouin zone in k-space, the closer it is to being reflected by the actual crystal lattice. In particle terms, the reflection occurs by virtue of the interaction of the electron with the periodic array of positive ions that occupy the lattice points. The stronger the interaction, the more the electron's energy is affected. A number of zones can be defined. The first Brillouin zone is a uniquely defined primitive cell in reciprocal space. Planes related to points on the reciprocal lattice give the boundaries of this cell. The second Brillouin zone is the set of points that can be reached from the first zone by crossing only one Bragg plane, the third by crossing two planes and so on. However, these higher order zones are rarely considered, and as a result, the first Brillouin zone is often simply referred to as *the* Brillouin zone. At its boundary, energy has two values, the lower belonging to the first Brillouin zone and the upper to the second zone. Thus, there is a definite energy gap between the first and second Brillouin zones; this gap corresponds to the forbidden band.

2.3.5.1 Electrons in Solids

An electron in a vacuum can be described by a de Broglie plane wave, in which the electron momentum can be expressed as

$$\boldsymbol{p} = \hbar \boldsymbol{k}, \tag{5}$$

where $\hbar = h/2\pi$ and $\boldsymbol{k}$ is the wave vector of the de Broglie wave. The electron's velocity is then given by

$$\boldsymbol{v} = \hbar \boldsymbol{k}/\boldsymbol{m} \tag{6}$$

and its energy by

$$E = E_o + \frac{\hbar^2 (k)^2}{2m}, \tag{7}$$

which is also known as the dispersion relationship. The first term, E_o, represents the electron potential energy, and the second term its kinetic energy. In a similar fashion, a loosely bound electron in a crystalline solid can be represented by a modified plane wave. The corresponding wave function is a solution of the Schrödinger equation [15] for a periodic potential and is called the Bloch function. It has the form

$$\psi_k(\mathbf{r}) = \exp\,(i\boldsymbol{k} \cdot \boldsymbol{r})\varphi_k(\boldsymbol{r}), \tag{8}$$

where $\boldsymbol{r}$ is the position vector and $\varphi_k(\boldsymbol{r})$ is a periodic function with the period of the lattice potential field. Eq. (8) may be regarded as a plane wave $exp(i\boldsymbol{k}.\boldsymbol{r})$ with an amplitude $\varphi_k(\boldsymbol{r})$ modulated by the period of the crystal lattice. The band structure of the electron energy spectrum in a crystalline solid arises as a consequence of the interference of the Bloch waves [16].

From Eq. (7) we note that the electron energy is a function of $\boldsymbol{k}$ and also all parameters describing the motion of electrons are related to $\boldsymbol{k}$ and the function $E(\boldsymbol{k})$. In particular, whenever the Wulff-Bragg condition between the de Broglie wavelength of an electron and the period of the crystal lattice is fulfilled, the electron wave will be totally reflected from the lattice. Owing to the interference of the incident and reflected waves, a standing wave arises. An electron with the corresponding wave vector $\boldsymbol{k}_o$ cannot propagate and discontinuities in the electron energy spectrum occur corresponding to the energy levels at the lower and upper limits of both the conduction and valence bands, as shown in Fig. 2.3.

2.3.5.2 Electrons in the Conduction Band

In $\boldsymbol{k}$ space a discontinuity in the electron energy spectrum corresponds to an extreme point of the function $E(\boldsymbol{k})$. At this point, $dE/dk = 0$. Around this point the electron energy can be approximated by the quadratic function,

$$E(k) = E(k_o) + \frac{1}{2}\sum_{i=1}^{3} \frac{\partial^2 E}{\partial k_i^2}\,(k_i - k_{oi})^2, \tag{9}$$

where k_i denotes the appropriate coordinate in $\boldsymbol{k}$ space, and $\boldsymbol{k}_o$ is the position vector of the extreme point. The term "$E(k_o)$" may be either a minimum or maximum in a permitted band. At the minimum of the conduction band, $E(k_o) = E_c$, while at the maximum of the valence band, $E(k_o) = E_v$. This is illustrated in Fig. 2.6. The electrons in the conduction band occupy the energy states slightly above the bottom of the conduction band, their distribution in $\boldsymbol{k}$ space being accurately described by Eq. (9). Note that the number of electrons in the conduction band is much less than the number of available states. Hence, the electrons feel practically no restriction due to the exclusion principle to occupy any of the energy states. As such, within a crystal they can move and be accelerated; in many respects they behave like electrons in a vacuum.

In Fig. 2.6 we show the detailed band structure for Si and GaAs in which the energy, E, is plotted against the crystal momentum, $\boldsymbol{k}$, for two crystal directions. For Si, (see Fig. 2.6(a)) we note that the momentum of holes at the top of the valence band is different from the momentum of electrons at the bottom of the conduction band. Therefore, in order to satisfy energy and momentum conservation laws, a transition between the two states requires a momentum transfer to the crystal lattice *via* its vibrational modes. Since a simple direct transition is not possible, Si is known as an indirect bandgap semiconductor. The equivalent diagram for a direct bandgap semiconductor (GaAs) is shown in Fig. 2.6(b). Here we see that the minimum of the conduction band lies directly above the maximum of the valence band. Transitions can take place with the absorption or release of a photon of wavelength

$$\lambda_o(\mu) = \frac{1.239}{E_g(\text{eV})} \tag{10}$$

without the mediation of a third body to conserve momentum. This is a highly efficient process and underlies the basic principle of operation behind the light-emitting diode, semiconductor laser and solar cell. In contrast, photon absorption or emission in indirect bandgap materials is very inefficient, since the process requires the simultaneous emission or absorption of a phonon. Such a correlated transition has a very low probability. Hence, optoelectronic devices are nearly always fabricated from direct bandgap materials.

Note in Fig. 2.6 that Si also has a direct bandgap (a) and GaAs an indirect bandgap (b) – albeit with larger bandgap energies. The classification into indirect and direct bandgap materials is based on which gap has the smallest energy and

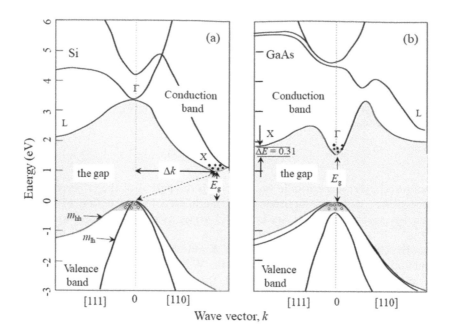

FIGURE 2.6 Energy band structures of (a) Si and (b) GaAs. Circles (o) indicate free holes in the valence bands, and filled circles (•) indicate free electrons in the conduction bands. Γ, X and L refer to different conduction band minima along the main crystallographic directions in the crystal. Si is an indirect bandgap material while GaAs is a direct bandgap material. The important difference is that for the direct bandgap material, an electron can transit between the lowest potential in the conduction band to the highest potential in the valence band without a change in momentum, Δk, whereas for an indirect bandgap material it cannot do so without the mediation of a third body (*e.g.*, phonon) to conserve momentum. Here m_{hh} and m_{lh} show the valence band maxima that contain heavy and light holes (see Section 2.3.5.3).

therefore will be the dominant source of transitions between the conduction and valence bands. The valence band maximum always occurs at $k = 0$, so we can define a direct-band semiconductor as one where the conduction-band minimum also occurs at $k = 0$ and an indirect-band semiconductor where it occurs at $k \neq 0$. Semi-metals are an interesting variant of indirect materials in that the bottom of the conduction band is usually lower than the top of the valence band; hence, the bandgap energy can actually be negative.

2.3.5.3 Band Sub-Structure

In Fig. 2.7 we show an *E-k* diagram for GaAs in which we have simplified the band structure shown in Fig. 2.6(b) to accentuate the conduction band minima. We see there are actually three valleys, which occur along different crystallographic directions in the crystal: the (000), (100) and (111) directions. These are known as the Γ (at $k = 0$), L, and X valleys (which both occur at non-zero values of k). The Γ valley[6] has the minimum energy and corresponds to the direct bandgap with $E_g = 1.43$ eV. Except under high field excitation, the L and X valleys contain very few electrons and therefore do not contribute to the conduction process. In common with Group IV and other III-V semiconductors, GaAs has three valence bands with maxima at $k = 0$. These are the light-hole, heavy-hole and split-off bands. The split-off band is due to spin-orbit interactions. Its maximum occurs at a lower energy than the other two valence bands and consequently contains virtually no holes. Because the heavy hole band is wider than the light hole band, holes tend to concentrate in it. The relative numbers of heavy holes and light holes (n_{hh} and n_{lh}) can be derived from a density of states calculation and is found to be a function only of their effective masses (see Sections 2.4 and 2.8.1.1). For Si, $n_{hh} = 6\, n_{lh}$, while for virtually all Group III-V and II-VI materials, $n_{hh} > 10\, n_{lh}$.

2.4 Effective Mass

Effective mass is a useful concept when describing the transport of electrons through a crystal and is essentially a heuristic approach that relates the quantum behaviour of electrons moving in the periodic potential of a crystal lattice to a semi-classical model. The effective mass is an important concept as it affects many of the electrical properties of the semiconductor, such as

---

[6] The Γ valley is always positioned at the center of the first Brillouin zone.

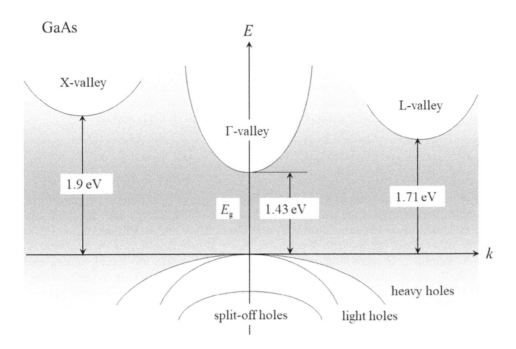

FIGURE 2.7 A simplified E-k diagram of GaAs showing the three valleys (L, Γ and X) in the conduction band. For direct gap materials, the height of the Γ valley represents the normally quoted bandgap. The valence band is comprised of three bands with different curvatures. Thus, there are three effective hole masses – heavy holes, light holes and split-off holes (see Section 2.4).

electron and hole mobilities, which in turn determine conductivity. From Fig. 2.6 we see that a particle's energy is a function of its crystal momentum. For an electron close to the conduction band minimum, the energy versus momentum relationship given by Eq. (7) can be approximated by a parabolic distribution

$$E(\boldsymbol{k}) \approx E_{\mathrm{o}} + a_2(\boldsymbol{k} - \boldsymbol{k}_{\mathrm{o}})^2, \tag{11}$$

where a_2 is constant that describes the curvature of the relationship. Here, $E_o = E_c$ is the "ground state energy" corresponding to a free electron at rest, and $\boldsymbol{k}_{\mathrm{o}}$ is the wave vector at which the conduction band minimum occurs. The above approximation holds for sufficiently small increases in E above Ec. Combining Eqs. (7) and (11) and taking $\boldsymbol{k}_{\mathrm{o}} = 0$, we can define the curvature coefficient a_2 associated with the parabolic minimum of the conduction band in terms of an *effective mass* m_{e}^{*}, thus

$$a_2 = \frac{\hbar^2}{2m_e^*}. \tag{12}$$

Substituting into Eq. (11) and taking the second derivative of the functional form of $E(\boldsymbol{k})$ with respect to $\boldsymbol{k}$ at the band minimum, $\boldsymbol{k}_{\mathrm{o}}$, we can derive an analytical expression for the effective mass

$$m_{\mathrm{e}}^* = \frac{\hbar^2}{\left(\frac{\partial^2 E_{\mathrm{c}}(k)}{\partial k^2}\right)}, \tag{13}$$

Note: this expression can be easily derived from Newton's 2nd law of motion, taking into account the electron's interaction with the lattice and the effect on its inertia. The quantity $(\partial^2 E_c / \partial k^2)$ is in fact the curvature of the band. Thus, the effective mass is an inverse function of the curvature of the E-k diagram: weak curvature gives a large effective mass and strong curvature a small effective mass. Depending on the nature of the periodic lattice potential, effective masses may be lighter or heavier than the free electron mass, m_{o}. From Fig. 2.6 we see that the conduction band actually has three minima or valleys and thus three effective masses. In GaAs these are 0.067 m_{o}, 0.85 m_{o} and 0.85 m_{o}, for the Γ, L and X valleys, respectively. In this case, carrier dynamics are dominated by the effective mass of the Γ valley since the other two valleys are essentially unpopulated under normal conditions.

For holes we have a dispersive relationship similar to that of electrons, with the effective mass m_{h}^{*} now determined from the curvature of the valence band maximum at $\boldsymbol{k} = 0$, that is,

$$m_h^* = \frac{\hbar^2}{\left(\frac{\partial^2 E_v(k)}{\partial k^2}\right)}. \tag{14}$$

However, the situation is somewhat more complicated as there are three valence bands centered on $k = 0$. Thus, there are three effective masses, corresponding to heavy holes, light holes and split-off holes. Since holes typically occupy the light-hole and heavy-hole bands only, the effective scalar mass of the valence band can be expressed as

$$m_h^* = \left(m_{hh}^{*\frac{3}{2}} + m_{lh}^{*\frac{3}{2}}\right)^{\frac{2}{3}}, \tag{15}$$

where m_{hh}^* is the effective mass of heavy holes and m_{lh}^* is the effective mass of light holes. Theoretically [17], it is found that the electron and heavy-hole effective masses are strongly related to the bandgap energy and, for the Γ conduction band, are approximately equal to each other for most cubic Group VI and Group III-V and II-VI materials (*i.e.*, $m_e^* \sim m_{hh}^*$). The mass of the heavy hole, on the other hand, shows little dependence, being largely determined by interactions with other bands more distant in energy. As a result, it changes little between materials and typically has a value of $m_{hh}^* \sim 0.4\ m_o$. Holes in the split-off band have roughly the same effective mass as those in the light-hole band, owing to their similar band curvatures.

Equations (13) and (14) also show that for a parabolic dispersion relationship, the second derivative of E with respect to k is a constant. Consequently, the effective mass is a constant as well and is independent of energy. Since the energy bands depend on temperature and pressure, the effective masses can also be expected to have such dependences. Note: the above discussions have assumed a simplified single dimension E-k space for the conduction band as depicted in Fig. 2.7. However, the wave vector k is a tensor; consequently, there are separate masses for each vector component. For a direct gap cubic material, like GaAs, the Γ valley has a spherical constant energy surface near its center, and the effective masses are essentially the same in each crystallographic direction; we need only specify a single value for the effective mass. This is illustrated in Fig. 2.8. For indirect gap materials (*e.g.*, Si and Ge), this is not the case; the relevant constant energy surfaces form six ellipsoids in k-space in the case of Si (X valley) and eight in the case of Ge (L valley). Fortunately, the ellipsoids frequently have longitudinal and rotational symmetry; thus, we need only specify two components for the effective mass: a longitudinal component m_l^* (defined along the $\langle 100 \rangle$ axis for the X-valley and the $\langle 111 \rangle$ axis for the L valley) and a transverse component m_t^* (taken as the minor or transverse axis of the ellipsoid). Generally, m_l^* is much larger than m_t^*.

Table 2.3 lists the average weighted effective electron and hole masses for a range of semiconducting materials. Values are given in units of the electron rest mass, m_o. As the table shows, Group III-V compounds, such as GaAs and InSb, tend to have far smaller electron effective masses than tetrahedrally coordinated Group IV elements like Si and Ge. Group II-VI materials tend to have effective electron masses intermediate between Groups IV and III-V. For all three groups, the hole effective masses are similar.

2.4.1 Polarons

In an ionic lattice, a free electron polarizes the ions, causing a change in their equilibrium position. As the electron moves through the crystal, it drags this ion displacement with it, becoming essentially surrounded by a region of lattice polarization and deformation. This can be visualized as a cloud of phonons that accompanies the electron on its passage through the crystal. The entire ensemble is known as a polaron, which can be considered a quasi-particle composed of a charge and its accompanying polarization field. The process is called the polaronic effect. The drag by necessity leads to an increase in the

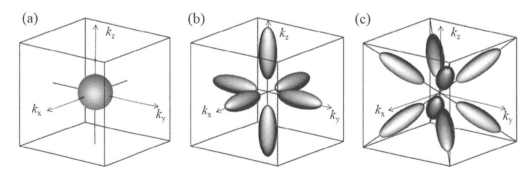

FIGURE 2.8 Ellipsoids of constant energy in the vicinity of the conduction-band minima for (a) GaAs with isotropic minimum at the Γ point, (b) silicon with six equivalent anisotropic minima ($m_l/m_t = 5$ not to scale) along the $\langle 100 \rangle$ axis and (c) germanium with eight equivalent anisotropic minima along the $\langle 111 \rangle$ axis. The cube indicates the $\langle 100 \rangle$ direction for cubic materials.

TABLE 2.3

Effective electron and hole masses for a number of semiconductors [18–23]. Depending on the nature of the periodic lattice potential, the effective masses may be lighter or heavier than the free electron mass, m_o. Note that the effective masses for the elemental semiconductors are larger than for compounds.

Material	Electron effective mass $m_e^* = m/m_o$	Hole effective mass $m_h^* = m/m_o$
Group IV		
Ge	0.55	0.37
Si	1.08	0.56
α-SiC	0.62	1.00
Group III-V		
InSb	0.013	0.60
InAs	0.023	0.40
InP	0.077	0.64
InN	0.055	1.70
GaSb	0.042	0.39
GaAs	0.067	0.50
GaP	0.82	0.60
GaN	0.20	1.25
AlSb	0.12	0.98
AlAs	0.15	0.51
AlP	0.13	0.64
AlN	0.40	3.53
Group II-VI		
CdS	0.21	0.80
CdSe	0.13	0.45
CdTe	0.11	0.35
HgTe	0.03	0.40
ZnO	0.28	0.59
ZnS	0.30	0.57
ZnSe	0.17	1.44
ZnTe	0.12	0.60
Group IV-VI		
PbS	0.25	0.25
PbSe	0.30	0.34
PbTe	0.17	0.20
Group I-VII		
AgCl	0.36	1.25
AgBr	0.29	1.11
AgI	0.53	1.41
CuCl	0.43	4.20
CuBr	0.21	1.40
CuI	0.33	0.38

electron mass and a corresponding decrease in mobility by an amount $(a\,\alpha + b\,\alpha^2 + \ldots)$, where a and b are constants and α is the Fröhlich coupling constant given by

$$\alpha = \frac{1}{2}\frac{e^2}{\hbar}\sqrt{\frac{2m^*}{\hbar\omega_{LO}}}\left(\frac{1}{\varepsilon_\infty} - \frac{1}{\varepsilon_o}\right),\qquad(16)$$

which is a measure of the strength of the electron-phonon interaction. Here, ε_o is the static dielectric constant, ε_∞ is the high-frequency dielectric constant, m^* is the effective mass given by the band structure (*i.e.*, not adjusted for polaronic effects), and ω_{LO} is the angular frequency of longitudinal optical phonons. The "polaron mass", m_p, is given by

$$m_p = m^* \left(1 + \frac{\alpha}{6} + 0.025\alpha^2 + \ldots \right), \tag{17}$$

for $\alpha \leq 1$, and

$$m_p = m^* \frac{16}{81\,\pi^4} \alpha^4, \tag{18}$$

when $\alpha \gg 1$. The energy of the electron is also lowered by its interaction with the lattice. Relative to the uncoupled case, the energy E_o for $k = 0$ is given by

$$E_o = -\left(\alpha + 0.0098\ \alpha^2 + \ldots \right)\hbar\omega_o, \quad \alpha \leq 1 \quad \text{and}$$
$$E_o = -\left(2.83 + 0.106\ \alpha^2 \right)\hbar\omega_o, \quad \alpha \gg 1. \tag{19}$$

When the coupling constant is small (≤ 1), the radius of a polaron is much larger than the lattice constant of the material, and the polaron is known as a large or Fröhlich-type polaron. In this case, the ion displacement is only a perturbative effect, and the transport proceeds as usual, but with a reduced mobility. When α becomes large (say, in the range 3–6), as is the case for strongly ionic crystals such as the alkali halides, the polaron can be confined to a volume of one unit cell or less by the electron–phonon interaction. In this case, transport proceeds *via* thermally activated hopping.

Summarizing, we see from the equations that polaronic effects only become significant when α becomes of order unity or larger. In real crystals, the measured values of α are essentially zero for Group IV semiconductors, lie in the range 0.02–0.2 for common III-V materials, increasing to 0.3–0.5 for II-VI materials and ~2 for I-VII materials. Therefore, for compound semiconductors, polaronic effects should be only significant when considering Group I-VII materials or some of the heavier ionic compounds, such as TlBr($\alpha = 2.6$).

2.5 Carrier Mobility

Mobility is a fundamental parameter in semiconductor physics because it essentially describes the conduction process. Free electrons in a solid have a velocity that depends on their energy. In the absence of an electric or magnetic field, the electron energy approaches that of the Maxwell thermal agitation energy given by

$$\frac{1}{2} m_e^* v_{th}^2 = \frac{3}{2} kT, \tag{20}$$

where m^*_e is the electron effective mass and v_{th} the average thermal velocity. Simply stated, a particle's effective mass is the mass it appears to have when calculated using a classical transport model and is described in detail in the previous section. The thermal motion of an individual electron can be visualized as a succession of random scattering from collisions with lattice atoms, impurity atoms and other scattering centers with no net direction. The average distance between collisions is called the mean free path, and the average time between collisions is termed the mean free time, τ_c. Upon applying an electric field, ξ, the electrons will be accelerated by a force equal to $-q\xi$ in an opposite direction to the electric field with a drift velocity, v_n. The net result is that the electrons are transported through the crystal at this velocity. The momentum of the electron is equal to $-q\,\xi\,\tau_c$. Therefore, $m_e^* v_e = -q\,\xi\,\tau_c$, or

$$v_n = -\left(\frac{q\tau_c}{m_e^*}\right)\xi. \tag{21}$$

Eq. (21) shows that the drift velocity is proportional to the applied electric field. The constant of proportionality (shown in brackets) is known as the electron mobility, μ_e. Hence,

$$v_n = -\mu_e \xi, \tag{22}$$

where $\mu_e = q\tau_c/m_e^*$. A similar expression can be written for holes

TABLE 2.4

Carrier scattering mechanisms in semiconductors. The individual temperature and material dependences are given. Here, N_{ion} is the number of ionized impurities, N_{def} is the number of defects, N_{disl} is the area dislocation density and d is the piezoelectric coefficient. Note: in this context $f(T)$ means that there is not a simple relationship between the mobility and temperature.

Mechanism	Scattering center	Dependences
Lattice scattering	Acoustic phonon	$\mu_{acoustic}(T) \sim T^{-3/2}$
	Deformation potential scattering	$\mu_{def}(T) \sim T^{-3/2}$
	Polar acoustic scattering (piezoelectric scattering)	$\mu_{piez}(T) \sim d^{-2}\, T^{-1/2}$
	Optical phonon	$\mu_{optical}(T) \sim T^{-1/2}$
	Polar optical scattering	$\mu_{po}(T) \sim f(T)$
	Non-polar optical scattering	$\mu_{npo}(T) \sim T^{-3/2}$
Impurity scattering	Impurity scattering	$\mu_{impurity}(T) \sim T^{+3/2}$
	Ionized impurity scattering	$\mu_{ionized}(T) \sim 1/N_{ion}\, T^{+3/2}$
	Neutral impurity scattering	$\mu_{neutral}(T) \sim$ const
	Defect scattering	$\mu_{defect}(T) \sim 1/N_{def}\, T^{+3/2}$
	Displacement scattering	$\mu_{disl}(T) \sim 1/N_{disl}\, T^{-1/2}$

$$v_p = \mu_p \xi. \tag{23}$$

The difference in signs between Eqs. (22) and (23) arises because electrons drift in the opposite direction to the electric field whilst holes drift in the same direction. It should be noted that because the electrons and holes have different effective masses, they will have different drift velocities for the same electric field and therefore different mobilities.

In reality, carrier velocities and therefore mobilities are not as high as predicted by Eqs. (22) and (23) due to various scattering mechanisms that impede the motion of carriers in real materials. These are summarized in Table 2.4. They can be grouped into two main categories: lattice scattering and impurity and/or defect scattering.

2.5.1 Lattice Scattering

In a perfect crystal, all atoms are identical and arranged periodically in space with perfect, long-range translation order. This is illustrated in Fig. 2.9(a) in which we show a perspective view though the internal structure of a perfect Si lattice projected onto the [110] plane; in (b) we show an STM topographic image of the surface of an actual high-quality crystal revealing a high degree of symmetry as well as its intrinsic 7×7 structure. In Fig. 2.9(c) we show a one-dimensional cross-section through one of the top lattice planes showing the periodic potential experienced by an electron travelling through the lattice, as described by the Bloch function given by Eq. (8). The potential seen by an electron is completely periodic and does not impede its movement through the lattice. In this case, the crystal has no resistance to current flow and behaves like a superconductor. However, due to random lattice vibrations, the perfect periodicity of the lattice is lost and the motion of the electron disrupted. It is for this reason we introduce the concept of effective mass – the mass a carrier would have in free space or a perfect crystal to achieve its actual velocity in a non-perfect crystal in response to a given electric field.

Atomic vibrations occur, even at zero temperature (a quantum mechanical effect) and increase in amplitude with temperature. In fact, the temperature of a solid is just a measure of the average vibrational activity of its constituent atoms and molecules. At room temperature, atoms typically vibrate at a frequency of the order of 10^{13} vibrations per second, with amplitudes of a few thousandths of a nanometre. These vibrations displace atoms transiently from their regular lattice sites, destroying the lattice's perfect periodicity. In a sense, these vibrations may be thought of as imperfections or defects that act as scattering centers for transiting carriers, creating a resistance to current flow. The magnitude of the vibrations is strongly temperature dependent: as the temperature is increased, lattice vibrations increase and consequently, the carrier velocity and therefore mobility decrease. In fact, at high temperature, lattice vibrations dominate the mobility function, and its contribution to the overall mobility begins to decrease as $\mu_{lattice} \propto T^{-3/2}$. In addition, for ionic materials, dipoles can be formed when the lattice is deformed. The resulting electric fields can also strongly scatter carriers.

Quantitatively, the scattering process (and indeed most solid-state phenomena) cannot be adequately described using static lattice theory as it is necessary to take into account the motion of the constituent atoms around their (ideal) equilibrium

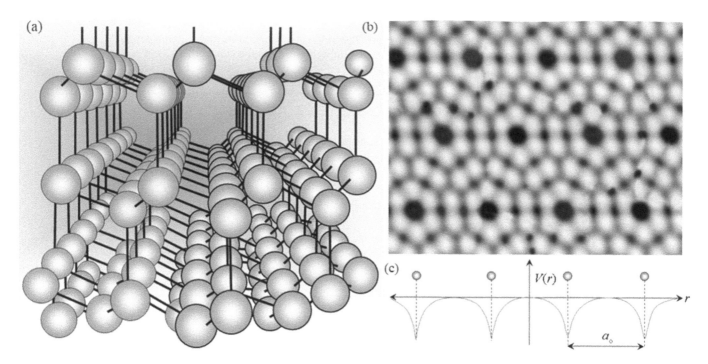

FIGURE 2.9 Illustration of a perfect Si crystal lattice. (a) A perspective view though the intrinsic structure projected on the [110] plane. The [111] direction is vertical. The [111] surface is obtained by cutting the vertical bonds in a horizontal plane (adapted from [24]). (b) An STM topographic image of the [111] surface revealing its intrinsic 7×7 structure. The large black circles are atoms in the first atomic layer. The smaller black circles the second layer, the smaller grey circles (difficult to discern) the third layer and so on. The size of the image is roughly 10×8 nm. (c) A one-dimensional cross-section through one of the top lattice planes showing the periodic potential experienced by an electron travelling through the lattice, as described by the Bloch function given by Eq. (8). a_o is the lattice parameter, $V(r)$ is the potential.

position. Since the lattice is a periodic and "fixed" structure, these atoms can only vibrate in certain modes; they essentially can be considered a set of coupled harmonic oscillators. As such, the vibration modes are essentially quantized. The coupling of these oscillators can be then be considered an exchange of particles, more commonly known as phonons.[7] Solids with more than one type of atom, such as compound semiconductors, exhibit two types of lattice vibration – an acoustic mode characterized by very low frequencies and an optical mode characterized by high frequencies. Each mode has its own associated phonon, which is absorbed or emitted when carriers are scattered by lattice waves.

2.5.1.1 Acoustic Modes and Acoustic Phonon Scattering

Acoustic phonons are coherent movements of lattice atoms out of their equilibrium positions which follow a cos (ωx) type of relationship, similar to a spring. If the displacement is in the longitudinal direction of propagation, some atoms will be closer together in some regions and further apart in others – a situation similar to the propagation of sound waves in air, namely, a compressive wave with zero rotation. Displacement in the transverse direction (*i.e.,* perpendicular to the propagation direction) is comparable to waves in water, in this case, a shear wave with zero divergence. The associated phonons are often abbreviated as LA (Longitudinal Acoustic) and TA (Transverse Acoustic) phonons, respectively.

Non-polar acoustic scattering is also known as deformation potential scattering because the deformation of the lattice waves changes the band edge, which gives rise to a deformation potential that becomes the origin of scattering. The carrier mobility due this mechanism is given by

$$\mu_{\text{def}} = \frac{2\sqrt{2\pi}e\hbar c_1}{3m^{*5/2}E^2}\left(\frac{1}{kT}\right)^{\frac{3}{2}}, \tag{24}$$

---

[7] The name *phonon* comes from the Greek word φωνή (phonē), which translates as *sound* or *voice* because long-wavelength phonons give rise to sound.

where E is the hydrostatic deformation potential and c_l is related to the sound velocity in that material. Deformation potential scattering is important at high temperatures and is the dominating mechanism in non-polar semiconductors, such as silicon and germanium, at and above room temperature.

Polar acoustic scattering (also known as piezoelectric scattering) occurs in compound semiconductors, due to their polar nature.[8] In these materials, a polarization field arises from the distortion of the basic unit cell when strain is applied along certain lattice directions. These fields may then scatter carriers by deflecting them. The effect is similar to deformation potential scattering, except the direction of the phonon wave vector is important. In strongly ionic crystals, such as II-VI or I-VII materials, piezoelectric scattering can be stronger than deformation potential scattering. The mobility is given by

$$\mu_{pz} = \frac{16\sqrt{2\pi}}{3} \frac{\hbar\varepsilon}{m^{3/2}eK^2} \left(\frac{1}{kT}\right)^{\frac{1}{2}}, \tag{25}$$

where K is related to the piezoelectric coefficient. In most other semiconductors, the effect is only important at low temperatures where other scattering mechanisms are weak.

2.5.1.2 Optical Modes and Optical Phonon Scattering

Optical phonons are out-of-phase movements of the atoms in the lattice, which occur if the lattice is made of atoms of different charge or mass. In this case, one atom will move to the left, while its neighbour moves to the right. This type of movement is called optical because in ionic crystals, such as sodium chloride, they are excited by infrared radiation. As with acoustic phonons, longitudinal and transverse modes exist and are abbreviated as LO and TO phonons, respectively.

Optical phonon scattering is the predominant scattering mechanism in compound semiconductors at high temperatures or high electric fields. Mobilities due to *optical* phonon scattering only tend to vary as $T^{-1/2}$. Depending on the crystal ionicity, there are two forms of optical scattering – polar optical scattering and non-polar optical phonon scattering. For the elemental semiconductors, such as Si and Ge, which only bond covalently, there is no polar optical scattering and scattering proceeds mainly by acoustic scattering and non-polar optical scattering at higher temperatures. In compound semiconductors, like GaAs, polar scattering is a very strong mechanism near room temperature. The mobility is given by

$$\mu_{pos} = \left(\frac{m^*}{m_o}\right)^{-\frac{3}{2}} T^{\frac{1}{2}} \left[\exp\left(\frac{\theta_D}{T}\right) - 1\right], \tag{26}$$

where θ_D is the Debye temperature.

At higher temperatures, non-polar optical scattering becomes an important scattering mechanism. Mathematically, it can be treated as a type of deformation potential scattering and consequently the mobility function has a similar form, *i.e.*,

$$\mu_{npo} = c \left(\frac{1}{kT}\right)^{-\frac{3}{2}}. \tag{27}$$

where c is a constant.

2.5.2 Impurity Scattering

In real semiconductors, the presence of impurities, interstitials, substitutionals and dislocations also upsets the periodicity of the lattice potential by creating donors and accepter states, which are typically ionized and thus charged. Impurity scattering results when a charge carrier travels past an ionized donor or acceptor and is perturbed by its Coulomb potential field. Unlike lattice scattering, ionized impurity scattering becomes less significant at high temperatures as the carriers move faster and are less effectively scattered. Consequently, the mobility increases. In addition to temperature, the mobility is inversely proportional to the number of ionized impurities, $N_{ionized}$,

$$\mu_{ionized} = \frac{c}{Z^2\sqrt{m^*}N_{ionized}} (kT)^{\frac{3}{2}}, \tag{28}$$

---

[8] Clearly not applicable for the elemental semiconductors.

where c is a constant. In compound (alloy) semiconductors, this type of scattering is always present since one of the species of atoms can always be considered an impurity. Therefore, for these materials, impurity scattering is sometimes referred to as alloy scattering. Note that carriers can also be scattered by structural defects, such as dislocations. If these defects are ionized, the scattering effect is similar to the case of impurities, although mobilities are proportional to T.

$$\mu_{\mathrm{disl}} = \frac{c}{N_{\mathrm{disl}}} T. \tag{29}$$

Here c is a constant and N_{disl} is the area dislocation density. Note that this form of scattering is only important for dislocation densities $> 10^7$ cm^{-2}.

When the impurities are neutral, carriers are only scattered when they impinge on impurities. So unless very high impurity concentrations are involved ($>10^{19}$/cm^3), the effect is negligible, with little or no temperature dependence. The mobility due to neutral impurity scattering is given by

$$\mu_{\mathrm{neutral}} = \frac{\pi^2 m^* q^3}{10\,\varepsilon_o \varepsilon_r N_{\mathrm{neut}} h}, \tag{30}$$

where N_{neut} is the density of neutral impurities. Note that in hyper-pure spectrometer grade Ge, neutral scattering is an important contribution to the total charge drift mobility.

An approximation to the overall mobility function can be written as a combination of the contributions from lattice vibrations (phonons) and impurities using Mattheisen's rule.[9]

$$\frac{1}{\mu} = \frac{1}{\mu_{\mathrm{lattice}}} + \frac{1}{\mu_{\mathrm{impurities}}} = \frac{1}{\mu_{\mathrm{acoustic}}} + \frac{1}{\mu_{\mathrm{optical}}} + \frac{1}{\mu_{\mathrm{ionized}}} + \frac{1}{\mu_{\mathrm{neutral}}} + \frac{1}{\mu_{\mathrm{defect}}} + \text{any others.} \tag{31}$$

Hence, the smallest mobility component dominates.

Mobilities are influenced by a number of factors. For example, because of differences in the properties of the valence and conduction bands, there is a significant difference in the mobilities of n and p type materials. Mobility is also strongly affected by temperature (see Fig. 2.10). It is lower at very low and high temperatures; it has its maximum value in the middle of the temperature range around 100K and then falls off again. Numerically, electron mobilities are generally in the thousands of cm^{2}V^{-1}s^{-1} range, while hole mobilities are in the hundreds of cm^{2}V^{-1}s^{-1} range. They tend to be larger in direct bandgap materials than in indirect bandgap materials. Note if the majority carriers are holes, the scattering mechanisms by heavy and light holes should be considered separately.

Figure 2.10 shows the temperature dependence of electron and hole mobilities for a number of Group III-V materials along with the elemental semiconductors Si and Ge. For comparison, the inset of Fig. 2.10(a) also shows the mobility temperature dependence of the two main scattering processes. Because both Si and Ge can be grown with impurity levels less than one part in 10^{12}, there is no impurity scattering, even at low temperatures. Consequently, their electron mobilities continue to rise with decreasing temperature. Group III-V materials, on the other hand, cannot be grown to the same purity levels and show a peak in mobility at ~100 K. At lower temperatures, impurity and defect scattering begins to dominate the overall mobility function (Eq. 31) and electron mobilities fall off with decreasing temperature. The same trends are seen in hole mobilities (see Fig. 2.10(b)). As expected for extrinsic semiconductors, the mobility due to lattice scattering shows a strong dependence on doping levels at temperatures for which scattering is dominant (see Section 2.8.2.1). For example, the electron and hole mobilities in Si at room temperature are $\mu_e = 1400$cm^{-2}V^{-1}s^{-1} and $\mu_h = 400$cm^{-2}V^{-1}s^{-1}, respectively, for dopant concentrations below 10^{15} cm^{-3}. These fall to $\mu_e \sim 300$cm^{-2}V^{-1}s^{-1} and $\mu_h \sim 100$cm^{-2}V^{-1}s^{-1} for concentrations above 10^{18} cm^{-3}.

2.6 Carrier Velocity

The ultimate speed of semiconductor materials depends on carrier velocity, which in turn can be related to the effective mass through the carrier's mobility. Assuming a simple kinetic model of electronic transport, in which the electrons undergo elastic billiard-ball like collisions with fixed and immobile crystal ions, the maximum obtainable charge carrier velocity is given by

---

[9] Developed from the study of electrical conduction in metals and alloys by Augustus Matthiessen in 1864.

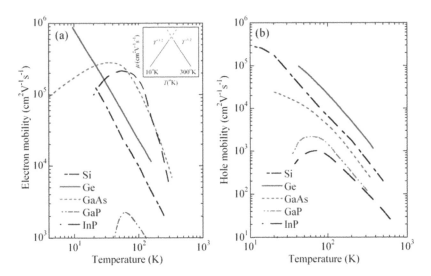

FIGURE 2.10 Temperature dependence of (a) electron mobilities and (b) hole mobilities for a number of Group IV and III-V semiconductors [25–31]. The rollover in mobilities for some Group III-V materials below ~100K is due to impurity scattering. For comparison, the inset in (a) illustrates the mobility temperature dependence of the two main scattering processes.

$$v_\mathrm{e} = \mu_\mathrm{e}\xi = \left(e\tau/m_\mathrm{e}^*\right)\xi, \tag{32}$$

where ξ is the applied external electric field, e is the electronic charge and τ is the carrier collision lifetime. A similar expression exists for holes. Thus, a low effective mass is a necessary property for high-bandwidth applications like communications and explains why power amplifiers in cell phones are increasingly based on GaAs rather than on Si technology. For radiation detectors, a low effective mass means the detector can operate effectively in high count-rate environments, such as those encountered at synchrotron radiation facilities.

2.7 High Field Phenomena

The drift velocity derived above is based on the so-called low-field mobility; it applies only to sufficiently small electric fields. The low field mobility is driven by scattering from distortions of the perfect lattice, with the electron gaining energy from the electric field between collisions at such distortion sites. The drift velocity, in turn, displays a linear dependence and the electric field.

2.7.1 Saturated Carrier Velocities

At sufficiently high electric fields (> ~3 kV cm^{-1}) the electron gains sufficient energy to interact inelastically with the host atoms of the crystal. Since the density of such atoms is high, compared to the density of scattering sites, this new mechanism determines carrier velocities at sufficiently high fields, causing them to become independent of the field. In this case, the semiconductor is said to be in a state of velocity saturation. For semiconductors that show saturation, such as Group IV semiconductors, it can be approximated by

$$v(\xi) = \frac{\mu_\mathrm{e}\xi}{1 + \mu_\mathrm{e}\xi/v_\mathrm{sat}}, \tag{33}$$

where ξ is the electric field and v_sat is the saturation velocity. The electric field at which velocities become saturated is referred to as the critical field, ξ_cr, and ranges from 0.1 MV cm^{-1} for Ge to 6 MV cm^{-1} for diamond. Values for the saturation velocities in common semiconductors at 300K are listed in Table 2.5.

2.7.2 Hot Electrons

In Fig. 2.11 we show the electron drift velocity as a function of electric field for a number of semiconductors. The linear dependence at small fields illustrates the low field mobility, with GaAs having a larger low-field mobility than silicon and

3

Crystal Structure

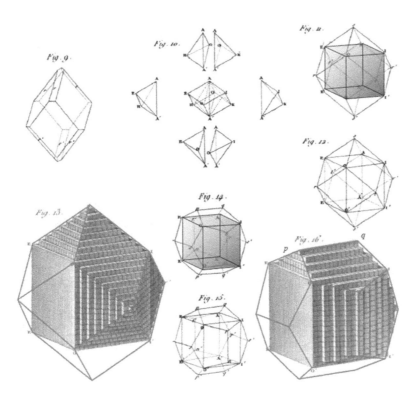

Illustrations from R.-J. Haüy's 1822 seminal work on crystal structure, *Traité de Cristallograhie*. Based on the law of rational indices, deduced from the stacking laws required to build the natural faces of crystals from elementary blocks, the figures show how piling up the so-called cubic "integrant molecules" can form the rhombic-dodecahedron {110} faces of garnet, shown in Fig. 13 or the pentagonal dodecahedron ½{210} faces of pyrite, shown in Fig. 16 (the canonical shapes for these crystals have been outlined in red for clarity). Notice how the decrescence of the molecular lamellae in the proportion of 2:1 leads to an interfacial angle at plane *pq* of 126° 52' 12" (the expected value is 127° 56' 08").

CONTENTS

To make a radiation detector, we first need to grow a crystal of the particular semiconductor material. The structure, atomic constituents and defects contained in the grown crystal dictate the physical characteristics of that crystal and its usefulness as a detector sensor. As such, a good understanding of crystals and crystal structure is highly important in the development of new types of detectors.

3.1 Introduction

Probably the first historical references to crystals and their uses date back to the 6th millennium BC and describe how the ancient Sumerians included crystals in magic formulas (*e.g.*, [1]), and later on the ancient Egyptians documented the healing properties of crystals in 1500 BC [2]. The term "crystal" is derived from the Greek word κρύσταλλος (krystallos) meaning *"clear ice or ice cold"* and was originally used to refer to materials that looked similar to ice (such as "rock crystal", the colourless form of quartz), since ice was the most obvious manifestation of crystal structure in the ancient world. Indeed, as most quartz crystals the Greeks were familiar with originated from the Alps, they concluded that rock crystal must be a new type of ice. The first plausible explanation was put forward by Theophrastus [3], a student of Aristotle who in 300 BC proposed that crystals were stones whose atoms had tiny hooks and eyes that grasped each other in regular geometric patterns. He speculated that crystals were, in fact, melted stone that had frozen so solid that it could not be melted again. Plato [4] came up with a more metaphysical explanation (428–348 BC). He proposed that nature was made up of powers or qualities of air, earth, fire and water. These powers were associated with and contained within four fundamental geometric shapes of nature that manifest themselves in the shape of crystals. In a later variation on a theme, Bauer[1] (1494–1555) assigned four alchemical qualities of salinity, ignitability, earthiness and metallicity to explain the properties of crystals and metals [5]. In his writings, Paracelsus[2] (1493–1541), the Swiss-German alchemist and philosopher, proposed that crystals grew from seed in a fashion analogous to that of fruit-bearing plants [6]. By the middle of the 17th century the old metaphysical-based and frequently dogmatic perspective applied to natural science was being replaced by a new rational and evidence-based approach. In 1665, Hooke [7] proposed a "corpuscular" explanation of crystal structure and stated that structure arises *"only from three or four several positions of Globular particles"*. Along the same lines, Huygens [8] deduced that the rhombohedral cleavage planes of calcite resulted from the stacking of flat ellipsoids, stating that crystal structure *"comes from the arrangement of the small invisible equal particles of which they are composed"*. In 1669, the Swiss philosopher Nicholas Steno [9] advanced crystallography significantly by precisely measuring the various angles at which crystals intersect and observed that although crystals of quartz and hematite appear in a great variety of shapes and sizes, the same interfacial angles persisted in every specimen. Although he failed to make the connection to an atomic lattice structure, the discovery drew attention to the significance of crystal form, leading to the law of rational crystallographic indices of Haüy [10] in 1801 and later formalized by Miller [11] in 1839. By the 18th century, crystallography was becoming a science with an over-riding emphasis on the mathematical description of crystalline symmetry. The culmination of this work was the seminal treatise of Bravais [12] in 1849, who worked out a mathematical theory of crystal symmetry based on the concept of a crystal lattice. The experimental verification and determination of structure was made possible by von Laue, Friedrich and Knipping's discovery of the diffraction of X-rays by crystals in 1912 [13], which is often said to mark the start of modern crystallography. That discovery, which came 17 years after the discovery of X-rays by Röntgen [14], allowed W.H. Bragg and W.L. Bragg to experimentally determine the basic crystal structures of diamond, sodium chloride, potassium chloride and potassium bromide by X-ray diffraction [15].

---

[1] He later adopted the Latin version of his name Gregorius Agricola.
[2] Aureolus Philippus Theophrastus Bombast named himself Paracelsus, meaning "like Celsus", after the Roman encyclopaedist.

3.2 Crystal Lattices

The structure of a crystal is usually defined in terms of lattice points, which mark the positions of the atoms forming the basic unit cell of the crystal. Cullity [16] defines a lattice point as "*an array of points in space so arranged that each point has (statistically) identical surroundings*". The word "statistically" is introduced to allow for solid solutions, where fractional atoms would otherwise be required. The defining property of a crystal is thus its inherent symmetry, by which we mean that under certain operations the crystal remains unchanged. For example, rotating the crystal 180 degrees about a certain axis may result in an atomic configuration that looks identical to the original configuration. The crystal is then said to have a two-fold rotational symmetry about this axis. In addition to rotational symmetries, a crystal may have translational symmetries and symmetries in the form of mirror planes.

3.2.1 The Unit Cell

In chemistry and mineralogy, a crystal is generally considered a solid in which the constituent atoms, molecules or ions are packed in a regularly ordered, repeating pattern of unit cells extending in all three spatial dimensions, globally forming the lattice. The unit cell is the smallest building block of a crystal, consisting of atoms, ions or molecules whose geometric arrangement defines a crystal's characteristic symmetry and whose repetition in space not only produces the crystal lattice, but also reproduces the full symmetry of the crystal. Thus, for a given crystal, the maximum number of unit cells that can be assembled is physically limited by the crystal face, and its shape is related to the shape of the unit cell. Each unit cell is defined in terms of its lattice points – the points in space about which the particles are free to vibrate in a crystal. Unit cells can be divided into two main types, primitive and non-primitive. Primitive is a simple cubic structure whereas non-primitive includes body-centered and face-centered cell structures. With reference to Fig. 3.1(b) the three simplest unit cell structures are as follows:

- The **primitive or simple cubic** system (abbreviated cP) consists of one lattice point on each corner of a cube and therefore contains a total of eight lattice points. Each atom at a lattice point is then shared equally among eight adjacent cubes. Therefore, the unit cell contains a total of one atom ($^1/_8 \times 8$).
- The **body-centered cubic** system (abbreviated bcc or cI) has one lattice point at each corner of the unit cell and one lattice point in the center (total nine). It therefore contains a total of two atoms per unit cell ($^1/_8 \times 8 + 1$).
- The **face-centered cubic** system (abbreviated FCC or cF) has a lattice point on each face of the unit cell in addition to the eight corner points (total 14 lattice points). In terms of the number of atoms the unit cell contains, we note that the corners each contribute 1/8 of an atom and the sides, one half of an atom, giving a total of four atoms per unit cell ($^1/_8 \times 8$ from the corners plus $^1/_2 \times 6$ from the faces).

Unit cells are defined by their lattice parameters, which are the lengths of the cell edges (denoted by **a**, **b** and **c**) and the angles between them (α, β and γ). The positions of the atoms inside the unit cell are in turn described by a set of atomic positions (x_i, y_i, z_i) measured from a lattice point.

3.2.2 Relationship between Lattices, Unit Cells and Atomic Arrangement

Figure 3.1 illustrates the relationship between the basic unit cell, its atomic arrangement and the corresponding lattice for the most common cubic crystal structures. In (a) we show macroscopic views of three common cubic lattice structures – a primitive or simple cubic lattice, a BCC lattice and a FCC lattice. The figures show a 2×2 replication of the basic unit cell shown in (b). In both (a) and (b), ball and stick representations are shown for clarity. While these are convenient for visualization, the actual structures adopted by crystalline solids are those that bring particles in closest contact to maximize the attractive forces between them. In many cases, the particles that make up the solids are spherical or approximately so, so the actual atomic configuration may be considered as tightly packed collections of "kissing" spheres. This is illustrated in Fig. 3.1(c), in which we show the close-packed versions of the unit cells shown in Fig. 3.1(b).

The coordination number of a particular crystal system is a particularly useful quantity because it essentially describes the number of bonds each atom can have in a given structure and how the atoms are packed together. It is defined as the number of number of particles immediately surrounding a central atom in a crystal structure (*i.e.,* its nearest neighbours). This is most easily seen by drawing out the lattice by replicating a number of unit cells in all directions, picking an atom near the middle and counting the number of nearest neighbour atoms. The larger the coordination number (more bonds), the greater the packing efficiency and therefore stability of the lattice. It also suggests that interstitial compounds (see Section 3.3) are less likely to be formed by materials that have unit cells with high coordination numbers, since there is less interstitial volume. In Fig. 3.1(d) we illustrate this by showing the corresponding space-filling view for each unit cell. Only the portion of each atom that belongs to the unit cell is indicated. For FCC close-packed structures, each sphere has 12 equidistant

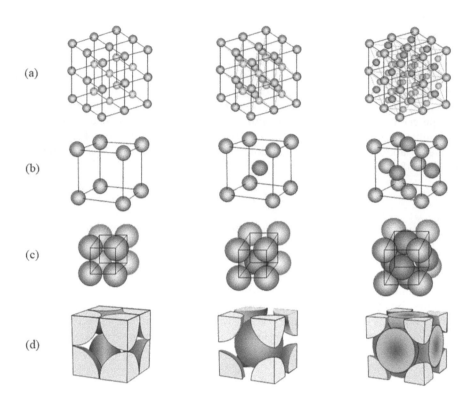

FIGURE 3.1 Figure illustrating the relationship among the basic unit cell, its atomic arrangement and the corresponding lattice for the most common cubic crystal structures. In (a) we show macroscopic views of three common cubic lattice structures – a primitive or simple cubic lattice, a BCC lattice and an FCC lattice. All lattices are based on a 2 × 2 replication of the basic unit cell shown in (b). In both (a) and (b), ball and stick representations are shown for clarity. Similarly, we identify body- and face-centered atoms by using a different colour (orange). In (c) we show the close-packed versions of the unit cells and in d) the corresponding space-filling view of each unit cell. Only the portion of each atom that belongs to the unit cell is indicated.

nearest neighbours. The coordination number is thus 12. In this case 74% of the total volume of the structure is occupied by spheres, or alternately there is 26% empty space. In comparison, each sphere in a BCC structure has a coordination number of 8, and 68% of the space is occupied (32% is empty). In the simple cubic structure, the coordination number is 6 and only 52% of the space is occupied (48% is empty).

3.2.3 Crystal Systems and the Bravais Lattice

Crystal systems are a grouping of crystal structures according to an axial system used to describe their lattice. Each system consists of a set of three axes in a particular geometrical arrangement. There are seven unique crystal systems. The simplest and most symmetric is the cubic (or isometric) system, having the symmetry of a cube. It is defined as having four three-fold rotational axes oriented at 109.5 degrees (the tetrahedral angle) with respect to each other, forming the diagonals of the cube. The other six systems, in order of decreasing symmetry, are hexagonal, tetragonal, rhombohedral (also known as trigonal), orthorhombic, monoclinic and triclinic.

Combining the central crystal systems with the various possible lattice centerings leads to the Bravais lattices (sometimes known as space lattices). Bravais [12] showed that crystals could be divided into 14 unit cells for which (a) the unit cell is the simplest repeating unit in the crystal; (b) opposite faces of a unit cell are parallel and (c) the edge of the unit cell connects equivalent points. There are 14 unique Bravais lattices, based on the seven basic crystal systems, which are distinct from one another in the translational symmetry they contain. All crystalline materials fit in to one of these arrangements. The 14 three-dimensional lattices, classified by crystal system, are shown in Fig. 3.2.

3.2.4 The Pearson Notation

While the Bravais lattice designations identify crystal types, they cannot uniquely identify particular crystals. There are several systems for classifying crystal structure, most of which are based on assigning a specific letter to each of the Bravais lattices. However, with the exception of the Pearson classification, few are self-defining. The Pearson notation [18] is a simple and convenient scheme and is based on the so-called Pearson symbols, of which there are three. The first symbol is a lower case letter designating the crystal type (*i.e.,* a = triclinic, m = monoclinic, o = orthorhombic, t = tetragonal, h = hexagonal and rhombohedral, c=cubic). The second symbol is a capital letter that designates the lattice centerings (*i.e.,* P = primitive,

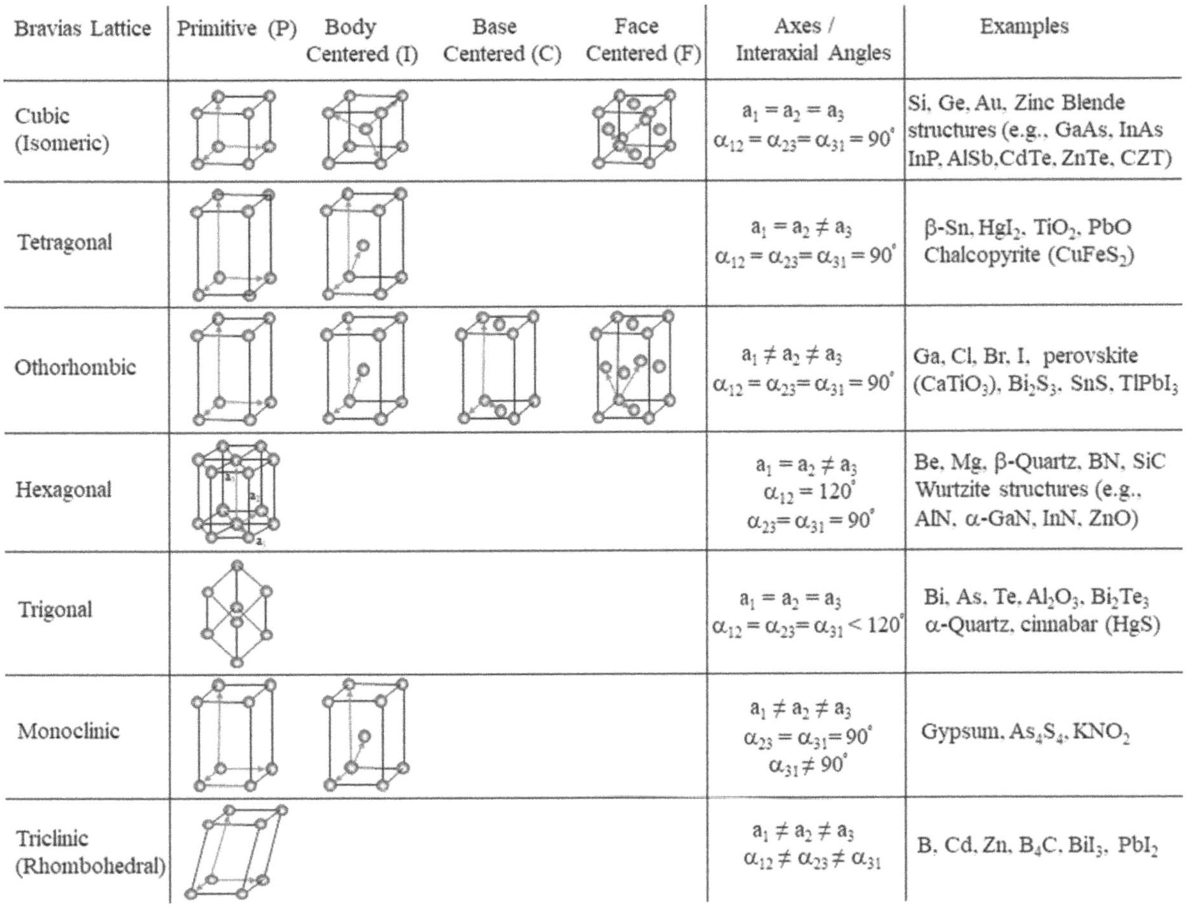

Bravias Lattice	Primitive (P)	Body Centered (I)	Base Centered (C)	Face Centered (F)	Axes / Interaxial Angles	Examples
Cubic (Isomeric)					$a_1 = a_2 = a_3$ $\alpha_{12} = \alpha_{23} = \alpha_{31} = 90°$	Si, Ge, Au, Zinc Blende structures (e.g., GaAs, InAs InP, AlSb, CdTe, ZnTe, CZT)
Tetragonal					$a_1 = a_2 \neq a_3$ $\alpha_{12} = \alpha_{23} = \alpha_{31} = 90°$	β-Sn, HgI_2, TiO_2, PbO Chalcopyrite ($CuFeS_2$)
Othorhombic					$a_1 \neq a_2 \neq a_3$ $\alpha_{12} = \alpha_{23} = \alpha_{31} = 90°$	Ga, Cl, Br, I, perovskite ($CaTiO_3$), Bi_2S_3, SnS, $TlPbI_3$
Hexagonal					$a_1 = a_2 \neq a_3$ $\alpha_{12} = 120°$ $\alpha_{23} = \alpha_{31} = 90°$	Be, Mg, β-Quartz, BN, SiC Wurtzite structures (e.g., AlN, α-GaN, InN, ZnO)
Trigonal					$a_1 = a_2 = a_3$ $\alpha_{12} = \alpha_{23} = \alpha_{31} < 120°$	Bi, As, Te, Al_2O_3, Bi_2Te_3 α-Quartz, cinnabar (HgS)
Monoclinic					$a_1 \neq a_2 \neq a_3$ $\alpha_{23} = \alpha_{31} = 90°$ $\alpha_{31} \neq 90°$	Gypsum, As_4S_4, KNO_2
Triclinic (Rhombohedral)					$a_1 \neq a_2 \neq a_3$ $\alpha_{12} \neq \alpha_{23} \neq \alpha_{31}$	B, Cd, Zn, B_4C, BiI_3, PbI_2

FIGURE 3.2 Schematic of the seven basic crystal systems and 14 conventional Bravais lattices (for a review see [17]). The lattice centerings are: **P** = primitive centering, **I** = body centered, **F** = face centered, **C** = base centered and **R** = rhombohedral (hexagonal class only).

C = side-face centered, F = all-face centered, I^3 = body centered, R = rhombohedral). Thus, the 14 unique Bravais lattices can be characterized by two-letter mnemonics as summarized in Table 3.1.

The third Pearson symbol is a number, which designates the number of atoms in the conventional unit cell. Therefore, a diamond structure, which is cubic, face centered and has eight atoms in its unit cell is represented by $cF8$. To use the Pearson system effectively, however, we need to know a structure, or prototype, that is the classic example of that particular structure. For example, both GaAs and MgSe have a Pearson designation cF8, but GaAs has a classic zincblende (*i.e.*, ZnS) structure and MgSe a "NaCl" type structure. Both structures are formed by two interpenetrating FCC lattices but differ in how the two lattices are positioned relative to one another.

3.2.5 Space Groups

While Pearson symbols categorize crystal structure into particular patterns and are easy to use and conceptually simple, not every crystal structure is uniquely defined. The space group designation, also known as the International or Hermann-Mauguin [19] system, is a mathematical description of the symmetry inherent in a crystal's structure and is also represented by a set of numbers and symbols. The space groups in three dimensions are made from combinations of the 32 crystallographic point groups with the 14 Bravais lattices, belonging to one of the basic crystal systems. This results in a space group being some combination of the translational symmetry of a unit cell including lattice centering and the point group symmetry operations of reflection and rotation improper rotation. The combination of all these symmetry operations results in a total of 230 unique space groups describing all possible crystal symmetries. The International Union of Crystallography publishes comprehensive tables [20] of all space groups and assigns each a unique number. For example, rock salt is given the number "225" (in Hermann-Mauguin notation it is designated "Fm3m"). The relationships among the basic crystal systems, the Bravais lattices and the point and space groups are shown in Fig. 3.3.

---

[3] From the German *innenzentrierte* meaning body centered.

TABLE 3.1

The 14 space (Bravais) lattices and their Pearson symbols.

Crystal system	Bravais lattice	Pearson symbol
Cubic	Primitive	*cP*
	Face-centered	*cF*
	Body-centered	*cI*
Tetragonal	Primitive	*tP*
	Body-centered	*tI*
Orthorhombic	Primitive	*oP*
	Base-centered	*oS*
	Face-centered	*oF*
	Body-centered	*oI*
Monoclinic	Primitive	*mP*
	Base-centered	*mS*
Triclinic (anorthic)	Primitive	*aP*
Rhombohedral	Primitive	*hR*
Hexagonal	Primitive	*hP*

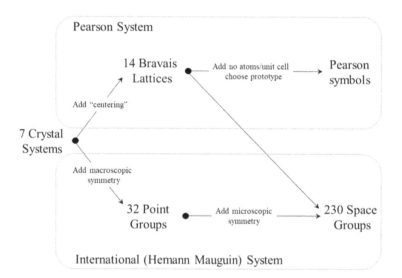

FIGURE 3.3 The inter-relationship among the basic crystal systems, the Bravais lattices, Point groups and the Pearson Symbols and Space groups. The macroscopic symmetry elements are those operations (*e.g.,* reflection and translation) that take place over unit cell dimensions, whereas the microscopic symmetry elements add small translations (less than a unit cell vector) to the macroscopic symmetry operations. A point group is a representation of the ways the macroscopic symmetry elements (operations) can be self-consistently arranged around a single geometric point. There are 32 unique ways this can be achieved.

3.2.6 Miller Indices

The classification of crystallographic directions and planes is important, since the physical and electrical properties of materials can vary significantly, depending on the plane (*e.g.,* optical and thermal properties, reactivity, dislocations, conductivity and transport to name a few). This can have far-reaching consequences, particularly when processing semiconductor materials. Since crystalline structures are repetitive, there exist families of equivalent directions and lattice planes centered on the unit cell. The orientation of a crystal plane may be defined by considering how the plane intersects the main crystallographic axes of the solid. The Miller Indices[4] [11] are a set of numbers that quantify the reciprocals of the intercepts and thus may be used to uniquely identify the plane. The number of indices will match with the dimension of the crystal or lattice. For example, in 1-D there will be one index and in 2-D there will be two, *etc.* A direction is expressed in terms of its ratio of unit vectors in the form $[u, v, w]$ where u, v and w are integers.[5] A family of crystallographically equivalent

---

[4] Named after the British mineralogist W.H. Miller (b-1801, d-1880).

[5] Note that in crystallography, as well as the nomenclature, the shape of the brackets is also convention.

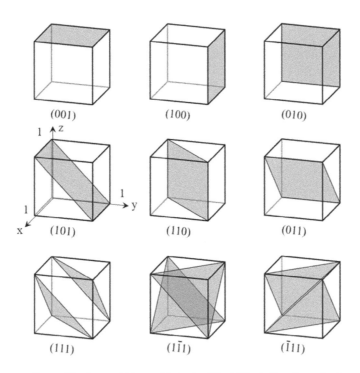

FIGURE 3.4 Examples of different crystallographic planes within a cubic lattice. The Miller indices for each plane are indicated in parenthesis (image courtesy Christophe Dang Ngoc Chan [21]).

directions is expressed as $<uvw>$. A plane is expressed as (lmn), where l, m and n are integers and a family of crystal-lographically equivalent planes is expressed as $\{lmn\}$. As an example, Fig. 3.4 illustrates how Miller indices are used to define different planes in a cubic crystal. The (100), (010) and (001) planes correspond to the faces of the cube. The (111) plane is tilted with respect to the cube faces, intercepting the x, y and z axes at 1, 1 and 1, respectively. In the case of a negative axis intercept, the corresponding Miller index is given as an integer with a bar over it, thus ($\bar{1}00$), which is similar to the (100) plane but intersecting the x axis at -1.

3.3 Underlying Crystal Structure of Compound Semiconductors

Compound semiconductors are derived from elements in Groups I to VII of the periodic table. They are very useful because of the sheer range of compounds available, compared to the elemental semiconductors, Sn-α, C, Si and Ge. Most elements in these groups are soluble within each other, forming homogeneous solid solutions. By solid solutions we mean a mixture of two crystalline solids that co-exist as a new crystalline solid that remains in a single homogeneous phase. The mixing is usually accomplished by melting the two solids together at high temperatures and cooling the combination to form the new solid. The phase in this context is understood to be a homogeneous, physically distinct and mechanically separable portion of the material with a given chemical composition and structure. Phases may be substitutional or interstitial solid solutions, ordered alloys or compounds, amorphous substances or even pure elements. The nature of the interface separating various phases is very much like a grain boundary.

Solid solutions[6] occur when atoms of a particular element can be incorporated into the lattice of another element without altering its crystal structure. The solute may incorporate into the solvent crystal lattice *substitutionally*, by replacing a solvent particle in the lattice, or *interstitially*, by fitting into the space between solvent particles as illustrated in Fig. 3.5. In general, compound semiconductors fall into the *substitutional* category while ceramics and clays fall in to the *interstitial* category. In both substitutional and interstitial solid solutions, the overall atomic structure is preserved, although there will be a slight distortion of the lattice. This distortion can disrupt the physical and electrical homogeneity of the solvent material.

In order for the atoms to form solid solutions over large ranges of miscibility, they should satisfy the Hume-Rothery[7] rules [22], namely that,

---

[6] When foreign atoms are incorporated into a crystal structure, whether in substitutional or interstitial sites, we say that the resulting phase is a solid solution of (a) the matrix material (the solvent) and (b) the foreign atoms (the solute).

[7] After the British metallurgist William Hume-Rothery (b. 1899, d. 1968).

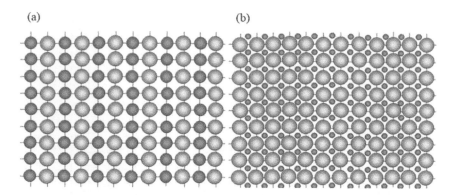

FIGURE 3.5 (a) schematic of an ordered substitutional cubic lattice in which an atom of one element replaces an atom of the host element in an alternating sequence. The ability to form a stable lattice depends on whether the two species can satisfy the Hume-Rothery rules. (b) example of an interstitial lattice in which the atoms of one element fit interstitially into the spaces in the lattice of the host element.

a) they have comparable atomic radii[8] allowing substitution without large mechanical distortion,

b) individually their crystal structures are the same,

c) the two species have similar valencies, and

d) their electronegativities are similar to avoid the creation of inter-metallic compounds.

The rationale behind the Hume-Rothery rules is as follows. The atomic radii (a) must be similar, because if the substituted atom is too large, considerable strain will develop in crystal lattice and the components must have similar crystal structure (b) if solubility is to occur over all proportions. However, this is less important if small proportions of solute are being added – such as in the doping of semiconductors. Similar valences (c) and electronegativities (d) indicate that components have similar bonding properties. In fact, components should have the same valence and an electronegativity difference close to zero to archive maximum solubility. The more electropositive one element and electronegative the other, the greater the likelihood of forming an intermetallic compound instead of a substitutional solid solution. It should be appreciated that the Hume-Rothery rules are a semi-empirical guide and by themselves not sufficient to ensure extensive solid solutions for all qualifying systems. For example, binary alloys of silver or copper satisfy all four rules but do not form a continuous series of solid solutions; instead they form a eutectic system[9] with limited primary solubility.

Mooser and Pearson [23,24] expanded on the work of Hume-Rothery [22] and elaborated a set of simple rules to test whether an element or compound will show semiconducting properties. For materials that contain at least one atom per molecule from Groups IV to VII of the periodic table, these are

1) The bonds are predominantly covalent,

2) The bonds are formed by a process of electron sharing which leads to completely filled *s* and *p* subshells on all atoms in elemental semiconductors. For compound semiconductors, the condition has to apply for each constituent element,

3) The presence of vacant "metallic" orbitals on some atoms in a compound will not prevent semiconductivity provided that these atoms are not bonded together,

4) The bonds form a continuous array in one, two or three dimensions throughout the crystal.

Elements from Group IV to Group VII will satisfy conditions (1) and (2).[10] Compounds that follow the second condition can be represented by the following equation [24].

$$n_e/n_a + b = 8 \tag{1}$$

Here, n_e is the number of valence electrons <u>per molecule</u>, n_a is the number of Group IV to Group VII atoms <u>per molecule</u> and b is the average number of covalent bonds formed by one of these atoms with other atoms of Groups IV to VII. Thus

---

[8] Experimentally it is found that they should differ by no more than 15%.

[9] A eutectic system describes a homogeneous solid mix of substances that are able to dissolve in one another as liquids but upon cooling begin to separate simultaneously into an intimate mixture of solids. The phrase most commonly refers to a mixture of alloys.

[10] Completed octets, that is closed *s* and *p* sub-shells, can only occur in atoms from Groups IV to VII of the periodic table (excluding the transition metals).

for GaAs, $n_e = 8$, $n_a = 1$ and $b = 0$, which satisfies the above condition. Note, for the elemental semiconductors, Eq. (1) is equivalent to the 8-*N* rule, which states that each atom in a covalent crystal should have 8-N nearest neighbours – *N* being ordinal number of the atomic group to which the atom belongs.

In addition to binary materials (such as GaAs or InP), most binary systems are also soluble within each other, making it possible to synthesize *ternary* (*e.g.,* AlGaAs, HgCdTe) *quaternary* (*e.g.,* InGaAsP, InGaAlP) and higher order solutions, simply by alloying binary compounds together. Thus semiconductor compounds can be built up hierarchically as illustrated in Fig. 3.6.

3.3.1 Lattice Constant and Bandgap Energy of Alloy Semiconductors

Consider two semiconductors *A* and *B* and the semiconductor alloy $A_x B_{1-x}$ where *x* is the alloy composition or alloy mole fraction. If we assume that *A* has a lattice constant a^A and *B* has a lattice constant a^B, then the lattice constant of the alloy, $a^{AB}(x)$, is given by Vegard's law [26], which at constant temperature empirically relates the crystal lattice constant of an alloy to the concentrations of its constituent elements, thus

$$a^{AB}(x) = ax^A + (1-x)a^B \qquad (2)$$

Similarly, the effective bandgap energy $E_g(x)$ is given by

$$E_g(x) = xE_g^A + (1-x)E_g^B + x(1-x)E_b \qquad (3)$$

where E_g^A and E_g^B are the corresponding bandgap energies of elements *A* and *B,* and E_b is a constant known as the bowing parameter which is a consequence of the microscopic bonding and structural properties of the alloys [27]. Tabulated values of E_b for an extensive range of binary, ternary and quaternary alloys may be found in [28,29]. In effect, Eqs. (2) and (3) mean that most material properties (such as lattice constant, bandgap, refractive index, thermal properties and mechanical constants) can be adjusted smoothly by the composition of these alloys. This dramatically increases the freedom of choosing the best compound for a particular application. However, it should be noted that for higher order alloys (quaternaries and higher), Eq. (3) does not hold, due the presence of multiple minima in the conduction band which, in theory, should be taken into account. In fact, for the higher order alloys, the bandgap can change type (direct to indirect and *vice-versa*), depending on composition and whether the semiconductor is strained or not. Generally, however, a so-called "one valley" bandgap model can give reasonable agreement over a limited range of compositions.

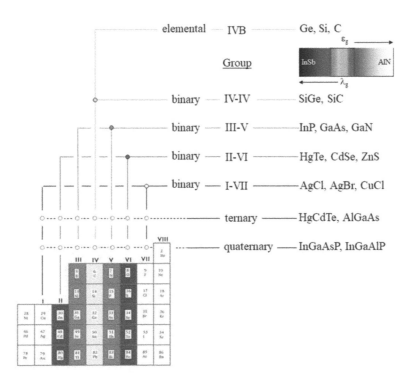

FIGURE 3.6 Diagram illustrating the relationship of the elemental and compound semiconductors (from [25]). Examples of the compound type are given and are listed by increasing bandgap energy or alternatively decreasing wavelength, from the infrared to the ultraviolet. InSb and AlN delineate the extremes of the range in which compound semiconductors lie (0.17 eV – 6.2 eV).

3.3.2 Bonding

Structurally, compound semiconductors are crystals formed by combinations of elements in which each element tries to attain a closed valence shell of 8 electrons. It achieves this by sharing (donating or accepting) electrons with its nearest neighbours. The simplest semiconducting materials are composed of a single atomic element, with the basic atom having four electrons in its valence band, supplemented by covalent bonds to four neighbouring atoms to complete the valence shell (see Fig. 3.7a). These elemental semiconductors use atoms from groups IV of the periodic table. Binary semiconductors are composed of two atoms – one from group N (where $N < 4$) and the other from group M (where $M > 4$) such that $N + M = 8$, thus filling the valence band. Fig. 3.6 shows the relevant section of the periodic table from which the bulk of the semiconductor groups are derived.

3.3.3 Common Semiconductor Structures

Most semiconductors used in radiation detection assume a cubic lattice structure and fall into few basic structures, which are described below. Note that some compounds can also crystallize into both cubic and hexagonal forms. For historical reasons, actual crystal structures are generally named after the minerals that most commonly assume that shape.

Diamond. The "diamond structure" is the arrangement of the Group IV atoms in diamond, silicon and germanium. The structure is illustrated in Fig. 3.8. The configuration belongs to cubic crystal family and can be envisioned as two interpenetrating face centered cubic sub-lattices with one sub-lattice staggered from the other by one quarter of the distance along a diagonal of the cube. Each atom has four equidistant, tetrahedrally coordinated, nearest neighbours. The atom-atom bond direction is <111> and the interbond angle is 109 degrees 28 min.

Zincblende. Named after the mineral zinc blende or "sphalerite", which is the principal ore of zinc. In fact it consists largely of crystalline ZnS. The zincblende structure is the stable form of many III-V and II-VI compounds and is structurally identical to that of diamond described previously, except that alternate atoms are respectively from Group III and of Group V for III-V materials and Groups II and VI for II-VI materials (see Fig. 3.9). Among the materials that classify into this structure are virtually all of the Group III-V compounds (*e.g.,* GaAs, InAs, InSb and InP).

Wurtzite is the less frequently encountered mineral form (a polymorph) of zinc sulfide and is named after French chemist Charles-Adolphe Wurtz.[11] In some semiconductors, such as GaN, although a cubic structure is meta-stable, the stable form is an atomic arrangement (the "wurtzite" structure) in which the atoms are slightly displaced from their cubic-structure positions and form a crystal of hexagonal symmetry. Thus, this structure is a member of the hexagonal crystal system.

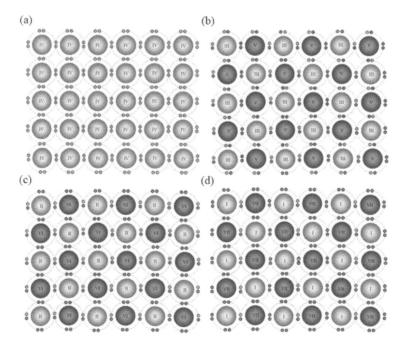

FIGURE 3.7 Illustration showing how semiconductors bond to form closed valence shells. Examples are given for the four most common semiconductor groups; (a) group IV elemental semiconductors, such as Si and Ge (b) Group III-V semiconductors, such as GaAs and InP, (c) Group II-VI compounds, such as ZnS and CdTe, and (d) Group I-VII compounds such as AgCl and AgBr.

---

[11] (b. 1817, d. 1884).

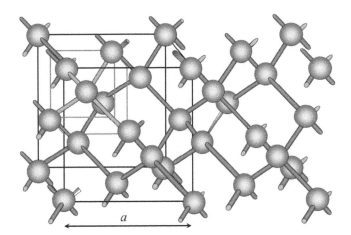

FIGURE 3.8 The diamond lattice structure. Each atom has four equidistant, tetrahedrally coordinated, nearest neighbours. The unit cell is outlined by the cube of dimension (lattice parameter), *a*. The primitive cell containing one lattice point is shown by the lighter coloured bonds.

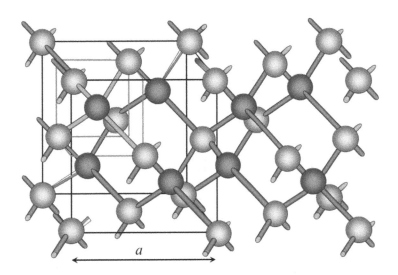

FIGURE 3.9 The Zincblende lattice structure which is most common structure for binary compound semiconductors. Here, the light and dark spheres denote the atoms of the binary elements. The unit cell is outlined by the cube of dimension (lattice parameter), *a*. The primitive cell is shown by the lighter coloured bonds.

Specifically in the case of ZnS, the structure consists of tetrahedrally coordinated zinc and sulfur atoms that are stacked in an alternating pattern (see Fig. 3.10). The principal difference with the zincblende lattice structure is that wurtzite contains four atoms per primitive unit cell instead of two. Among the compounds that can assume a wurtzite configuration are AgI, ZnO, CdS, CdSe, α-SiC, GaN and AlN. In most of these compounds, wurtzite is not the favoured form of the bulk crystal, but the structure can be observed in some nanocrystal forms.

Hexagonal. The hexagonal structure is depicted in Fig. 3.11. It has the same symmetry as a right prism with a hexagonal base. The hexagonal system is uniaxial, meaning it is based on one major axis, in this case a six-fold rotational axis that is perpendicular to the other axes. Examples of materials that crystallize into a hexagonal structure include silicon carbide and boron nitride and of course all wurtzite forms.

3.3.4 Polycrystalline and Amorphous Structures

The majority of semiconductors solidify into regular periodic crystallographic patterns (Fig. 3.12(a)). However, they can also form amorphous[12,13] solid solutions in which the arrangement of the atoms exhibits no periodicity or long-range order at all

---

[12] From the ancient Greek, "άμορφος", meaning without shape or form.
[13] In fact, almost any material can be made amorphous by rapid solidification from the melt (molten state). This condition is unstable, and the solid will crystallize in time.

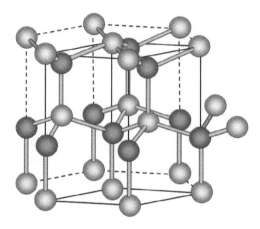

FIGURE 3.10 The wurtzite lattice structure, which is the second most common structure for binary compound semiconductors. Here, the light and dark spheres denote the atoms of individual elements.

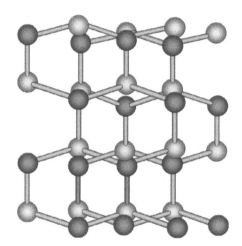

FIGURE 3.11 A hexagonal lattice structure. The light and dark spheres denote the atoms of individual elements.

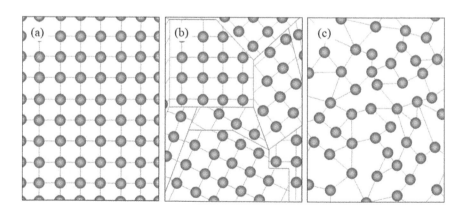

FIGURE 3.12 Illustration of macroscopic crystal structures in a semiconductor. While the majority of semiconductors solidify into regular periodic patterns shown in (a), they can also form polycrystals, shown in (b), a collection of individual grains of crystalline material separated by grain boundaries, or (c) amorphous solid solutions in which there is little long-range order.

(Fig. 3.12(c)). The presence or absence of this long-range order has a profound influence on the electronic and photonic properties of a material and in fact, carrier mobilities and diffusion lengths degrade as the order is reduced. Even amongst crystalline materials, we can distinguish between single-crystal and polycrystalline materials. Polycrystalline materials are solids that are comprised of a collection of crystallites (grains) connected to each other with random orientations and separated from one another by areas of relative disorder known as grain boundaries (see Fig. 3.12(b)). In this case, order can only be maintained within the grains.

3.4 Crystal Formation

Crystal growth can occur naturally or artificially and is a consequence of two processes – nucleation followed by growth. It occurs because it is energetically favourable, in that the potential energy of the atoms or molecules is lowered when the atoms or molecules form bonds with each other. The process starts with the nucleation stage. Several atoms or molecules in a supersaturated vapor or liquid start forming clusters, since the bulk free energy of the cluster is less than that of the vapor or liquid. If the crystallization of material occurs from a single nucleation point, a single crystal will be produced. However, unless the growth is strictly controlled, growth can occur from multiple nucleation points. In this case, growth from a particular nucleation site will continue until it impinges on the growth from an adjacent nucleation site. The net result is that many crystals begin to grow in the liquid and a polycrystalline solid forms. The final sizes of the individual crystals depend on the number of nucleation points. Each crystal is known as a grain and the interface between grains is known as a grain boundary. The interfacial atoms at the boundary between the grains have no crystalline structure and are said to be disordered.

Both nucleation and crystal growth are very dependent upon temperature, pressure and the composition of the surrounding fluid/vapor (including ionic concentrations necessary for crystal growth and the effects of other ions that may act as poisons to inhibit growth). Rapid cooling generally results in more nucleation points and smaller grains, whilst slow cooling results in larger grains with lower strength, hardness and ductility. Growth is also dependent on the availability of surface area (in the case of a crystal that has already nucleated). To effect growth, the ions must be transported to the surface of the crystal and react with the surface. The reaction products, in turn, must be removed to allow the nucleation process to continue. Crystal morphology is also dependent upon the manner (that is, rate and direction) in which the crystal grows. Many compounds have the ability to crystallize with different crystal structures – a phenomenon called polymorphism. Each polymorph is a different thermodynamic solid state, and crystal polymorphs of the same compound exhibit different physical properties, such as dissolution rate, shape (angles between facets and facet growth rates) and melting point.

3.5 Crystal Defects

Until now, we have considered perfectly regular crystal structures, which do not exist in practice. All real crystals contain defects that can distort a lattice to the point where it is far from being perfectly regular. This affects not only its structural and thermal properties, but also its electronic properties. Natural crystals always contain a myriad of defects (impurities, mechanical defects, *etc.*) due to the uncontrolled nature by which they were formed. While laboratory prepared crystals also contain defects, they are usually at a level orders of magnitude less, since considerable control can be exercised over defect type, concentration and distribution. However, even neglecting structural imperfections, a material that is 6N pure (99.9999%) still contains $\sim 10^{17}$ impurity atoms cm^{-3}. The importance of defects and their effects depends upon the material, type of defect and the properties that are being considered. For example, properties such as stiffness and density are not overtly affected by the presence of defects whilst others, such as mechanical strength, ductility, conductivity and crystal growth are very much structure dependent and as such particularly sensitive to the presence of defects. In fact, high precision controlled doping with small concentrations of chemical impurities is the key technology by which the electronics industry tailors the electronic properties of silicon and germanium to specific applications.

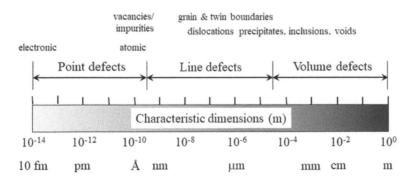

FIGURE 3.13 Figure illustrating the characteristic scale length of the various crystal defects listed in Table 3.2. Note we have included electronic defects, which while not structural in the same sense as vacancies or impurity atoms, reflect errors in the charge distribution in solids caused by inexact stoichiometry. They are particularly prevalent in ionic crystals.

3.5.1 Defect Classification and Morphology

The word "defect" means a native defect (vacancy, interstitial, antisite, *etc.*, see next section), an impurity (meaning an atom of a different kind from the host atoms) or any combination of isolated defects, starting with small clusters and aggregates and leading to larger defect structures such as precipitates, grain boundaries, surfaces and voids, *etc.* The characteristic scales of these range from Ångstroms to metres as illustrated in Fig. 3.13. However, it is important to note that all defect structures are built out of Ångstrom-sized building blocks and therefore understanding their macroscopic properties begins at the atomic scale. Crystalline defects can be conveniently grouped into four main categories, based on their geometry:

1) <u>point</u> or zero-dimensional defects that occur at a single lattice point,

2) <u>linear</u> or one-dimensional defects occurring along a row of atoms. These are known as line defects,

3) <u>area</u> defects occurring over a two-dimensional surface of the crystal. These are known as plane defects, and

4) <u>volume</u> or three-dimensional defects, generally known as bulk defects.

These are described in detail in the following sections and summarized in Table 3.2. It is important to note that unlike the elemental semiconductors, any deviation from perfect stoichiometry[14] in a compound semiconductor must necessarily result in defects – a surplus of atoms of whatever kind will be incorporated into the lattice as point defects, agglomerates or precipitates. The large thermal gradients encountered during the growth process invariably lead to mechanical stress, which gives rise to dislocations. In addition, most compounds are mechanically soft, which means that relatively small stresses are sufficient to cause plastic deformation, even at low temperatures, again giving rise to dislocations.

TABLE 3.2

List of crystal imperfections. The dimensions of a point defect are close to those of an interatomic space. With linear defects, their length is several orders of magnitude greater than their width. Surface defects have a small depth, while their width and length may be several orders larger. Volume defects (voids, pores and cracks) may have substantial dimensions in all measurements (*i.e.* at least a few tens of Å).

Imperfection		
Type class		**Description**
Point defects		Point or zero-dimensional defects
	Electronic	Errors in the charge distribution in solids caused by inexact stoichiometry
	Interstitial	An extra atom in an interstitial site
	Schottky	An atom missing from correct site
	Frenkel	An atom displaced to an interstitial site, creating a nearby vacancy
	Anti-site	One component of a binary component occupying a lattice site of the other component
Line defects		Linear or one-dimensional defects
	Edge dislocation	A row of atoms marking the edge of a crystallographic plane extending only part way into the crystal
	Screw dislocation	A row of atoms that appear to spiral about a normal crystallographic plane
Plane defects		Area or two-dimensional defects
	Lineage boundary	A boundary between two adjacent perfect regions in the same crystal that are slightly tilted with respect to each other
	Grain boundary	A boundary between two crystals in a polycrystalline solid
	Twin boundary	A grain boundary whose lattice structures are mirror images of each other in the plane of the boundary
	Stacking fault	A boundary between closed-packed regions of a crystal having alternate stacking sequences
Bulk defects		Volume or three-dimensional defects
	Precipitates	A conglomeration of impurity atoms
	Inclusions	Melt solution droplets, conglomerations of precipitates
	Voids	Macroscopic holes in the lattice
	Cracks	Linear clusters of voids connected to a "stress riser". Separation of crystal planes

---

[14] For a crystal containing more than one kind of atom to be stoichiometric, the ratio of atoms in the unit cell must be *exactly* the same as in the entire crystal.

3.5.2 Point Defects

Point defects are localized defects of atomic dimensions. They are the smallest structural elements, or imperfections, to cause a departure from a perfect lattice structure. They include self-interstitial atoms, interstitial impurity atoms, substitutional atoms and vacancies. In addition, to these "physical" structural point defects, electronic defects are simultaneously formed, which reflect errors in the charge distribution caused by inexact stoichiometry. They are mainly prevalent in ionic crystals and are free to move under the influence of an electric field.

The various types of structural point defects are listed in Table 3.2 and illustrated in Fig. 3.14. A self-interstitial atom is an extra atom that has found its way into an interstitial void in the crystal structure. The creation of a self-interstitial causes substantial distortions in the surrounding lattice and costs more energy than the creation of a vacancy ($Q_i > Q_v$). Consequently, under equilibrium conditions, self-interstitials are present in much lower concentrations than vacancies are. Interstitial impurity atoms, on the other hand, are much smaller than the bulk matrix atoms and fit into the open spaces between the host atoms of the lattice structure.

A substitutional atom is an atom of a different type from the bulk atoms that has replaced a host atom in the lattice. Substitutional atoms are usually close in size (within approximately 15%) to the bulk atoms; otherwise, a solid solution will not form. If the substitutional atoms are much smaller the host atoms, they will be incorporated into the lattice interstitially. If much larger, physically they will mechanically disrupt the lattice and a new phase will form. It should also be noted that for appreciable solid solubility to take place, the crystal structures of both types of atoms must be the same (see Section 3.3). In general, substitutional impurities are electronically active, meaning they will alter the electronic properties of the material, whereas interstitial contaminants are generally electronically inactive.

Vacancies[15] are empty spaces where an atom should be but is missing. In fact, in many crystals, diffusion (mass transport by atomic motion) can only take place because of vacancies. They are particularly common at elevated temperatures when atoms frequently and randomly change their positions, leaving behind empty lattice sites. The equilibrium number of vacancies, N_v, increases exponentially with temperature, T, and is described by a Boltzmann distribution,

$$N_v = N \exp\left(-Q_v/kT\right), \tag{4}$$

where N is the total number of atoms (lattice sites) in the material, k is Boltzmann's constant and Q_v is the energy needed to form a vacancy in a perfect crystal. In intermediate mass materials, Q_v is typically around 1 eV per atom, which means that at room temperature $N_v/N = {\sim}10^{-16}$, whereas at 1,000°C it is ${\sim}10^{-4}$. When a solid is heated a new higher equilibrium concentration of vacancies is established, usually first at crystal surfaces and then in the vicinity of dislocations and grain boundaries. These provide sites for the atoms that have left their normal lattice site. Vacancies gradually spread throughout the crystal (from the surfaces into the bulk). On cooling, the vacancy concentration is lowered by a process of diffusion to grain boundaries or dislocations, which act as sinks.

If the missing atom is no longer in the vicinity of the hole or has migrated to the surface, it is known as a Schottky defect. Those that migrate to the surface generally become incorporated into the lattice at the surface. If the atom has moved into an adjacent interstitial site and outside the recombination volume, it is known as a Frenkel defect. A Frenkel defect is in

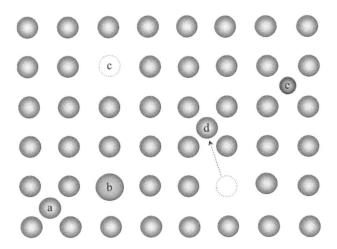

FIGURE 3.14 Schematic illustrating the different types of point defect in a crystalline material. These are (a) self-interstitial atom, (b) substitutional impurity atom, (c) Schottky defect, (d) Frenkel defect and (e) interstitial impurity atom.

---

[15] For historic reasons, vacancies in many ionic crystals are also called centers (*e.g.,* luminescence centers and colour centers).

reality a pair of defects – an empty lattice site and an extra interstitially positioned atom. In compound semiconductors, the interstitial ion will always be the positively charged one (*i.e.*, a cation,[16] because it is generally smaller than a negatively charged ion and thus fits better into interstitial sites. In other words, its formation enthalpy is lower. Schottky and Frenkel defects tend to form during the growth process and are frozen into the lattice as the crystal crystallizes. They can also be created by energetic particle interactions leading to a phenomenon called displacement damage in radiation physics. The energy of formation is closely related to the crystal binding energy and is typical around 10 eV. Although in both cases the crystal remains neutral (since the total number of positive and negative ions is the same), these defects can affect semiconductor properties as they allow ionic conduction. However in tetrahedrally coordinated materials (*e.g.*, IV, or III-V semiconductors), bonding is predominantly covalent rather than ionic. Consequently, activations energies are high and ionic conduction can be neglected, which may not be the case in ionic materials (*e.g.*, Group I-VII semiconductors).

Another form of point defect is the anti-site defect, which is neither a vacancy, an interstitial nor an impurity. It occurs when an anion[17] (cation) replaces a cation (anion) on a regular cation (anion) lattice site.

A defect is known as "native" if it is a vacancy, a self-interstitial, an anti-site or a complex formed of these.

3.5.3 Line Defects (Dislocations)

A dislocation, or line or linear defect, is a line through the crystal along which crystallographic registry is lost. It is formed in a crystal when an atomic plane is misaligned and terminates within the crystal instead of passing all the way through it producing lattice distortions centered on a line. This generally occurs as a consequence of the solidification process, plastic deformation, vacancy condensation or atomic mismatch in solid solutions. The resulting irregularity in spacing is most severe along a perpendicular line called the line of dislocation. Consequently, line defects are commonly referred to as dislocations. Physically, they may be pictured as simply the edge of an extra inserted fractional plane of atoms. Dislocations are of importance in determining the strength of ductile metals.

The concept of the dislocation was first proposed by Orowan [30], Polanyi [31] and Taylor [32] in 1934 to explain two key observations about the plastic deformation of crystalline materials. The first was that the stress required to plastically deform a crystal is much less (by a factor of 10^2 to 10^3) than the stress calculated for a defect-free crystal structure. The second observation was that a material *work hardens* when it has been plastically deformed and subsequently requires a greater stress to deform it further. However, it was not until 1947 that the existence of dislocations was experimentally verified [33]. Later advances in electron microscopy techniques showed that dislocations were mobile in a material – this being the origin of plastic (ductile versus brittle) behaviour. Only in the last few years have electron microscopy techniques advanced sufficiently to allow the atomic structure around a dislocation to be resolved.

There are two extreme types of dislocations, the edge type and the screw type – both of which are described in sections 3.5.2.1 and 3.5.2.2. Any particular dislocation is usually a mixture of these two types.

A line defect can be thought of as a one-dimensional array of point defects, which are out of position in the crystal structure and occur when the crystal is subjected to stresses beyond the elastic limit of the material. These stresses generally arise from thermal and mechanical processing during the growth and detector fabrication process. The energy to create a dislocation is of the order of ~100 eV per mm of dislocation line, whereas it takes only a few eV to form a point defect, which is a few nm in extent. Thus, forming a number of point defects is energetically more favourable than forming a

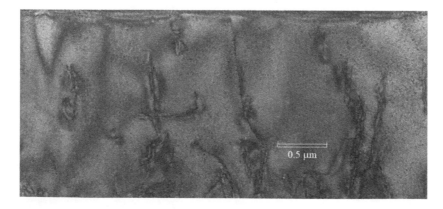

FIGURE 3.15 Transmission Electron Microscopy image of a large number of dislocations in a GaN crystal grown by metalorganic chemical vapor deposition.

---

[16] The electropositive component in a compound (*e.g.*, Ga in GaAs, Cd in CdTe).
[17] The electronegative component in a compound (*e.g.*, As in GaAs, Te in CdTe).

dislocation. Dislocations interact with chemical and other point defects. The presence of a dislocation is usually associated with an enhanced rate of impurity diffusion leading to the formation of diffusion pipes. This effect gives rise to the introduction of trapping states in the bandgap, altering the electrical properties of the devices. In addition, dislocations move when a stress is applied, leading to slip and plastic deformation of the lattice. Note that a line of dislocation cannot end within a crystal unless it forms a complete loop. Generally, it ends at the sample surface, in which case it can be visualized by etching (acids preferentially etch the intersections of dislocations with the crystal surface), infrared, X-ray or electron transmission techniques (see Fig. 3.15). In fact, etching is routinely used to quickly assess material quality in terms of number of dislocations at a surface[18] – the result is usually expressed in terms of the Etch Pit Density. Dislocations can also be visualized by decoration (metallic impurities will tend to precipitate on dislocations forming so-called decorated boundaries). Depending on the size of the dislocation, these can be easily observed using optical microscopy.

3.5.3.1 Edge Dislocations

An edge dislocation can easily be visualized as an extra half-plane of atoms inserted into the lattice orthogonal to the growth direction, as illustrated in Fig. 3.16(a). The dislocation is called a line defect because the locus of defective points produced in the lattice by the dislocation lie along a line. This line runs along the top of the extra half-plane. The inter-atomic bonds are significantly distorted only in the immediate vicinity of the dislocation line (denoted by the symbol, ⊥, which runs into the paper), forming a strain field on either side of the line with associated impacts on the mechanical strength of the crystal. In addition, the disruption of the lattice in this region leaves behind a string of dangling bonds,[19] which act as trapping centers with subsequent impacts on electronic properties.

In all dislocations, the distortion is very intense near the dislocation line where the atoms do not have the correct number of neighbours. Just above the line, the atoms are in compression; just below they are in tension. The local expansion around the ⊥ symbol is the actual dislocation. The region is called the core of dislocation. At the core, the local strain is extremely high, whereas a few atomic distances away from the center the distortion is so small that the crystal is almost perfect locally. The dotted line shown in Fig. 3.16(a) is known as the slip plane and represents the plane in which the dislocation can move (left or right) by the application of a suitable force resulting in plastic (permanent) deformation.

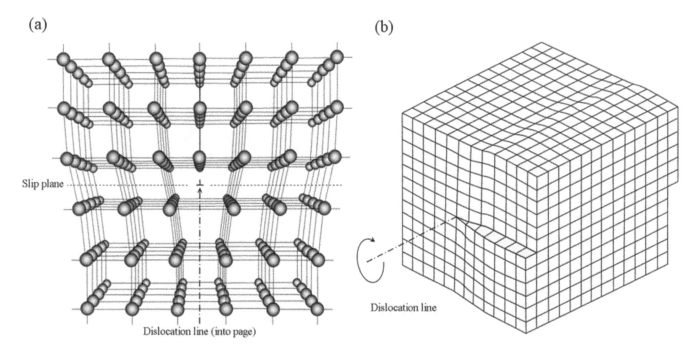

FIGURE 3.16 Illustrations of line defects (a) edge dislocation in a cubic primitive lattice (modified from [35]). Normally the symbol ⊥ is used to represent a positive dislocation (*i.e.,* an extra fractional plane) and T is used to represent a negative dislocation (a missing fractional plane. The dotted line shown in Fig. 3.16(a) is known as the slip plane and represents the plane in which the dislocation can move (left or right) by the application of a suitable force resulting in plastic (permanent) deformation (b) screw dislocation (modified from [36]).

---

[18] For example, the average density of dislocations in detector grade CdZnTe is 10^4–10^5 cm^{-2} [34].
[19] Defined as an unsatisfied valence on an immobilized atom.

3.5.3.2 Screw Dislocations

The screw dislocation was introduced by Burger in 1939 [37]. It is also called Burger's dislocation. If the dislocation is such that a step or ramp is formed by the displacement of atoms in a plane in the crystal, then it is referred to as a screw dislocation and is essentially a shearing of one portion of the crystal with respect to another by one atomic distance. The displacement occurs on either side of a "screw dislocation line" (see Fig. 3.16(b)), which forms the boundary between the slipped and unslipped atoms in the crystal and forms a spiral ramp winding around the line of the dislocation. The pitch of the screw may be left-handed or right-handed and one or more atom distances per rotation. The distortion is very small in regions away from the screw dislocation, while atoms near the center are in regions of high distortion so much so that the local symmetry in the crystal is completely destroyed. In this case, the atoms near the center of the screw dislocation are not stretched linearly on the lattice as in an edge dislocation but are on a twisted or sheared lattice. In some materials, such as SiC, growth can commonly proceed *via* spiral growth. It was suggested by Frank [38], that a presence of a screw dislocation in the crystal provides a step onto which atoms can be adsorbed under vapor growth. The whole spiral can then rotate around its emergence point with uniform angular velocity and stationary shape giving rise to the growth morphology illustrated in Fig. 3.17.

3.5.4 Plane Defects

A plane defect is a disruption of the long-range stacking sequence of atomic planes. Plane defects are generally manifested as stacking faults or twin regions. This type of error is only applicable for structures that can be close packed, for example, HCP and FCC structures. BCC structures have no close packed planes and therefore do not have a stacking sequence and stacking faults. A stacking fault is a change in the stacking sequence over a few atomic spacings. As such, it is a simple two-dimensional defect. The normal stacking sequence in an HCP crystal is AB AB AB, etc., whilst in an FCC crystal it is ABC ABC ABC, etc. Here, A B and C represent the three distinct crystallographic planes that allow the crystal to be most efficiently packed or stacked together, whilst replicating and preserving the basic unit cell structure. This is illustrated in Fig. 3.18 for both HCP and FCC structures. Whilst the closed packed structure for the hexagonal lattice visually resembles that of its Bravais lattice (*i.e.,* the sequence BAB), that of the FCC is not so obvious and is in fact formed of hexagonal planes along the diagonal of the corresponding Bravais lattice. For HCP crystals, the normal stacking sequence is AB AB AB ... etc. If the structure now switches to AB AB AB C AB AB, there is a stacking fault present at C. For FCC crystals, two main kinds of stacking fault can occur. The first type is where one plane is missing, that is, the stacking sequence is ABC A_C ABC .... and is known as an intrinsic stacking fault (the fault is indicted by _). Near the fault, the packing sequence is similar to an HCP lattice. Alternately, if an additional plane is present (*i.e.,* ABC AB*A*C ABC), the defect is called an extrinsic stacking fault. An extended stacking fault in which the order of stacking is reversed is called a twin lamella (*e.g.,* ABC ABC *B*AC *B*ABC ABC).

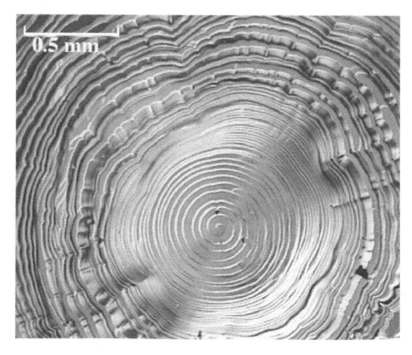

FIGURE 3.17 A screw dislocation growth spiral in SiC (taken from [39]). It is formed by a shear ripple extending from side to side. The atomic planes perpendicular to it constitute a helicoid, like the thread of a screw.

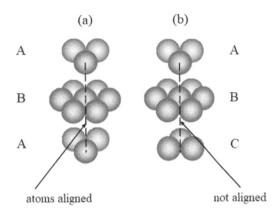

FIGURE 3.18 Illustration showing how planes A, B, C are comprised and stacked to form the basic hexagonal (HCP) and FCC lattice structures (Fig. 3.18(a) and (b), respectively). As can be seen, the HCP structure contains only two types of planes in an alternating AB AB arrangement, whereas the FCC structure contains three types of planes in an ABC ABC sequence. Notice for the HCP structure, how the atoms of the third plane are in exactly the same position as the atoms in the first plane, whereas for the FCC structure the atoms in rows A and C are no longer aligned.

3.5.4.1 Twinning

If a stacking fault continues over a large number of atomic spacings, it can produce a second stacking fault that is the twin of the first one, for example, a stacking sequence ABC ABC AB*C*BA CBA CBA. Their joint interface is called twin boundary and can be thought of as a special type of grain boundary across which there is a mirror image mis-orientation of the lattice structure. Twin boundaries generally occur in pairs such that the change in orientation introduced by the first boundary is restored by the second (see Fig. 3.19(a)). The region in between is known as a twin region. These are the most frequently encountered boundaries in most FCC crystals. These imperfections are not thermodynamically stable; rather, they are meta-stable imperfections. They arise primarily from the clustering of line defects into a plane. Twins can also be produced when a shear force, acting along the twin boundary, causes the atoms to shift out of position.

3.5.4.2 Grain Boundaries

Another type of plane defect is a grain boundary, which is the interface or boundary between two crystallites or grains in a polycrystalline material. Note a twin boundary is also a grain boundary with a high degree of symmetry. A number of grain boundaries are illustrated in Fig. 3.19(b). These are structural defects that decrease the electrical and thermal conductivity of the material. The high interfacial energy and relatively weak boundaries often make them preferred sites for the onset of corrosion or the precipitation of new phases from the solid during growth. Grains can range in size from nanometres to millimetres, and their orientations are usually rotated with respect to neighbouring grains. The interface between grains is typically two to five atomic diameters wide. The manipulation of grain sizes can be used to limit the lengths and motions of dislocations since they essentially act as obstacles to dislocation motion. This is a well-known strengthening technique in metallurgy known as "grain-boundary" or "Hall-Petch" strengthening (named after E.O. Hall [40] and N.S. Petch [41] who showed that the strength of mild steel varies reciprocally as the root of its grain size). For any crystalline material, the relation between yield stress and grain size is given by

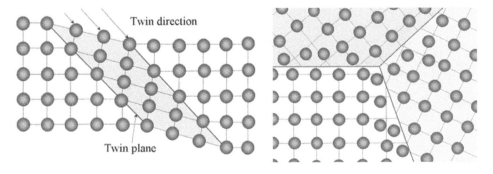

FIGURE 3.19 (a) Illustration of a twin region bounded by twin planes or boundaries. Twin boundaries are the most frequently encountered boundaries in many FCC crystals and are in fact a special case of a grain boundary with a high degree of symmetry (b) Grain boundaries among three crystals (grains) of the same phase. The grains differ in mutual orientations.

$$\sigma_y = \sigma_o + \frac{k_y}{\sqrt{d}} \qquad (5)$$

where σ_y is the yield stress, σ_o is a materials constant for the starting stress for dislocation movement (or the resistance of the lattice to dislocation motion), k_y is the strengthening coefficient (a constant specific to each material) and d is the average grain diameter. The smaller the grain size, the smaller the repulsion stress felt by a grain boundary dislocation and the higher the applied stress needed to propagate dislocations through the material. Theoretically, a material could be made infinitely strong if the grains are made infinitely small. However, this is limited because of the finite size of the unit cell of the material. Even then, if the grains of a material were of the size of a single unit cell, the material would actually be amorphous, not crystalline, since there is no long-range order and dislocations cannot be defined in an amorphous material. In any case, recent studies [42] on solids composed of nanoscale grains suggest that the Hall-Petch relation breaks down at a critical grain size of around 10 nm, below which a transition from dislocation-mediated yielding to grain boundary sliding occurs and the material progressively softens with decreasing grain size.

Twins and grain boundaries can be considered arrays of linear dislocations that can affect charge transport in several ways. For example, they can (1) form potential barriers that guide the carriers, (2) accumulate secondary phases that block carrier transport, (3) attract impurities, which increase charge trapping and (4) form conduction bands, which make the boundaries highly conductive. The effect on charge transport is illustrated in Fig. 3.20, which shows a comparison of spatially resolved optical and X-ray response images of two 50×50 mm^2, 3 mm thick slices of a CdZnTe crystal grown by the High Pressure Bridgman method. In the optical images shown in Fig. 3.20(a), numerous grain boundaries and twins are apparent. In Fig. 3.20(b) we show the corresponding X-ray response, in terms of count rate, to a finely collimated ^{57}Co radioactive source, illustrating poor charge collection at the grain boundaries (indicated by the white areas). Interestingly, no correlation was found with the numerous twin boundaries observed inside the grains, indicating that twins have a negligible effect on the electric field and charge collection of semi-insulating CdZnTe devices.

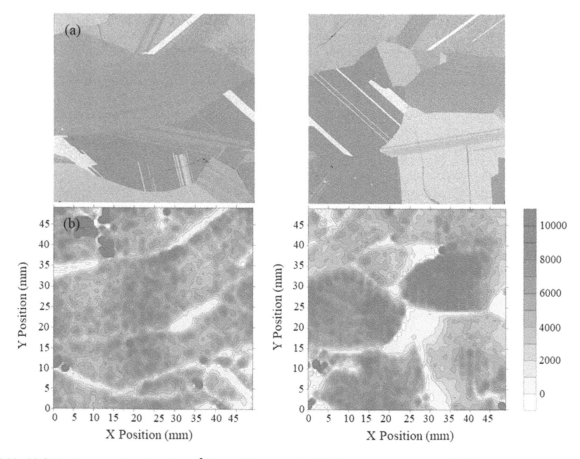

FIGURE 3.20 (a) Optical images of two 50×50 mm^2, 3 mm thick slices of a CdZnTe crystal grown by the High Pressure Bridgman method (from [43]). Numerous grain boundaries and twins are apparent in the image. (b) The crystal count rate response, measured with a ^{57}Co radioactive source is shown in the lower images, illustrating poor charge collection at the grain boundaries. Interestingly, no correlation was found with the numerous twin boundaries observed inside the grains, indicating that twins have a negligible effect on the electric field and charge collection of semi-insulating CdZnTe devices.

3.5.5 Volume or Bulk Defects

Volume or bulk defects, as the name suggests, are defects in three dimensions and occur on a macroscopic scale compared to point defects. These include voids, pores, cracks, foreign inclusions, precipitates and other phases. All these defects are normally introduced during the growth and processing phases and can act as stress raisers. In metals, foreign particles are sometimes deliberately added to strengthen the parent material in a process known as dispersion hardening. These particles act as obstacles, preventing the movement of dislocations that facilitate plastic deformation.

Volume defects in crystals are three-dimensional aggregates of atoms or vacancies. It is common to divide them into four classes in an imprecise classification that is based on a combination of the size and effect of the particle. The four categories are precipitates, which are a fraction of a micron in size and decorate the crystal; second phase particles or dispersants, which vary in size from a fraction of a micron to the normal grain size (10–100 μm), but are intentionally introduced into the microstructure; inclusions, which vary in size from a few microns to macroscopic dimensions, and are relatively large, undesirable particles that entered the system as dirt or were formed by precipitation; and voids, which are holes in the solid formed by trapped gases or by the accumulation of vacancies.

In compound semiconductors used for radiation detection, perhaps, the most common forms of bulk defects are precipitates and inclusions. They are distinguishable primarily by size – precipitates tend to have characteristic dimensions in the 10–100 nm range and inclusions in the 01–10 microns range. Until recently, they were considered the same defect, primarily because a large precipitate looks pretty much like a small inclusion. They are now better defined in terms of their respective generation mechanisms. Precipitates are clusters of impurity atoms introduced into the matrix, which form small regions of a different phase – meaning a region of space occupied by a physically homogeneous material. In essence, precipitates are small particles. They tend to be more common in bulk growth materials, such as CdTe. In metallurgy, precipitates are commonly used to increase the strength of structural alloys by acting as obstacles to the motion of dislocations. Their efficiency in doing this depends on their size, their internal properties and their distribution through the lattice. However, their role in the microstructure is to modify the behaviour of the matrix rather than to act as separate phases in their own right. In radiation detectors, precipitates act as efficient trapping centers. In CdZnTe, the cumulative (statistical) effect of high concentrations of Te precipitates leads to a progressive degradation of energy resolution and efficiency loss, depending on their number density and size distribution. For example, large Te precipitates (or inclusions) of >10 μm already are detrimental to a device's responses at concentrations of 10^4 cm^{-3}, whilst smaller ones, say <2 μm, can be tolerated up to 10^6 cm^{-3}, depending on the device's thickness [44]. Examples of precipitates and their effects in CdZnTe are shown in Fig. 3.21, which compares spatially resolved IR and X-ray images measured with a 1-mm thick CdZnTe crystal [45]. Te precipitates are clearly identified in the IR micrograph images (Fig. 3.21(a)). The corresponding X-ray response of

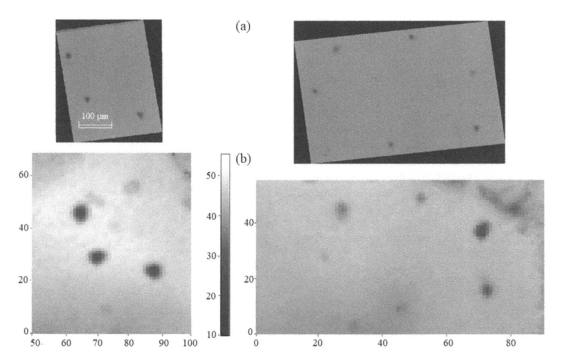

FIGURE 3.21 (a) IR micrograph images of Te precipitates measured in a 1-mm thick CZT crystal (from [45]) (b) The lower images for the corresponding X-ray response of the crystal when operated as a simple planar detector. The dark spots in this case correspond to a drop in the detector response, demonstrating the link between precipitates and poor device performance. The scans were performed by using a 10 × 10 μm^2, 85 keV X-ray beam. In some cases, the typical triangular shapes of precipitates are recognizable in the X-ray maps.

the crystal when operated as a simple planar detector is shown in Fig. 3.21(b). In this case, the dark spots correspond to a drop in the detector response, demonstrating a clear link between precipitates and poor charge collection. Inclusions are melt-solution droplets,[20] foreign particles or large precipitate particles incorporated into the growing interface. They are generally undesirable constituents of crystal microstructure, since they act as preferential sites for failure, weakening the useful strength of structural alloys. In II-VI compounds they can be reduced by post-growth annealing.

A gross type of bulk defect is a void (or pore), a region in which there are a large number of missing atoms (that is holes) in the lattice. They are usually caused by gases that are trapped during solidification (*e.g.*, air bubbles – commonly known as porosity), shrinkage when a material solidifies (known as cavitation) or vacancy condensation in the solid state. Mechanically, their principal effect is to decrease material strength and promote fracture under small loads. Electrically, charge collection is zero in these regions.

3.6 Defect Engineering

While the presence of defects is generally considered to have adverse effects on material properties, in many situations the controlled introduction of specific imperfections can be beneficial. For example, defects play an important role in various technological processes such as annealing, precipitation, diffusion sintering, oxidation and others [46]. In fact, a perfect crystal has very limited uses and most of the useful mechanical and electrical properties of materials are the direct results of defects. This has led to new branch of material science, described as defect engineering with numerous applications in the fabrication of semiconductors and devices with improved performance and/or new properties. The fact that such small amounts of impurity atoms can significantly alter the electrical properties of semiconductors was largely the driving force behind the development of the transistor and literally opened up the entire field of solid-state device technology. Defect engineering involves the manipulation of the type, concentration, spatial distribution or mobility of defects within a crystalline solid. Recent work now targets structures with nm dimensions. In order to control the behaviour of these defects and optimize performance in specific applications, various techniques have been developed and implemented (*e.g.*, see [46]). The defects themselves can be introduced by a number of means. Some can be grown in (*e.g.*, dislocations, vacancies, interstitials, swirls and voids[21]). Others are process-induced (*e.g.*, dislocations, precipitates, metals and twinning).

Defect formation affects semiconductor properties in a variety of ways, depending on the type of defect. Point defects typically affect electronic properties such as carrier type, concentration or mobility [47,48]. Extended defects affect physical properties, such as strength or toughness [49] and at elevated temperatures can serve as sources or sinks of point defects. Surfaces do the same, interacting through both bond-exchange [50,51] and electrostatic [50] mechanisms. Most defects can act as sites where electrons and holes can efficiently recombine [52–54], although typically degrading the performance of the host material in applications ranging from optoelectronics to photocatalysis. We discuss the effects of each class of defect in the following sections.

3.6.1 Point Defect Engineering

The addition of small quantities of impurities to metals to form alloys is a well-known way of improving mechanical strength and corrosion resistance – as in the addition of 0.1% to 2% carbon to iron to form steel. Similarly, the electrical behaviour of semiconductors is largely controlled by crystal imperfections. Indeed, the conductivity of silicon can be altered by over eight orders of magnitude through the addition of minute amounts of electrically active dopant elements that create donor or acceptor states close to conduction or valence band edges. In this case, each atom of dopant, substitutionally incorporated, represents a point defect in the silicon lattice. Point defects are inherent to the equilibrium state and are thus determined by temperature, pressure and composition of a given system.

Point defect engineering has also been used to improve the radiation hardness of silicon particle detectors [55], using a process known as the Diffusion Oxygenated Float Zone technique. Oxygen impurities are introduced into the bulk using high temperature diffusion, which has a high affinity for capturing radiation-induced vacancies and interstitials. It is estimated that for charged hadron irradiation, silicon detectors produced using this technique are three times harder than conventional silicon detectors [54].

3.6.2 Line Defect Engineering

The existence of dislocations (line defects) in crystals provides a mechanism by which permanent change of shape or mechanical deformation can occur. While the existence of dislocations in crystals insures ductility (ability to deform), the

---

[20] In binary compounds, AB, where A and B represent the constituent elemental atoms, droplets form in A- or B-rich melts due to morphological instabilities at the crystallization interface. The size and morphology of the resultant inclusion depends on the growth conditions.

[21] Also known as crystal orientated pits for historic reasons.

theoretical strength of crystalline solids is, however, drastically reduced by their presence. Note that a crystalline solid free of dislocations is brittle and practically useless as an engineering material and in fact, virtually all mechanical properties of crystalline solids are to a significant extent controlled by the behaviour of line imperfections.

In semiconductors, however, the presence of dislocations is generally undesirable because they serve as sinks for metallic impurities and disrupt diffusion profiles. However, the ability of dislocations to sink impurities may be engineered into a wafer fabrication advantage (*i.e.*, it may be used in the removal of impurities from the wafer, a technique known as "gettering") [56]. It uses large densities of surface defects as sinks, generated, for example, by abrading the back of a slice of a semiconductor or by ion bombardment. The slice is then heated so the impurities diffuse from the sub-surface active regions of the bulk to the surface where they, precipitated by the defects, become electrically inactive.

From an optoelectronic point of view, a line defect introduced into an otherwise perfect 3D photonic crystal will act as a light guide [57], confining photons, not by internal reflection as in conventional waveguides, but by a mechanism more analogous to quantum confinement. This is because the guided mode for those photons whose energy lies within the photonic bandgap can only exist along the path of the defect. The great advantage of this mode of propagation is that it is lossless and ultra-compact – highly desirable qualities when considering the potential fabrication of photonic integrated circuits.

3.6.3 Plane Defect Engineering

The presence of two-dimensional imperfections known as *Bloch walls*, allow (to a large part) a ferromagnetic material (such as iron, nickel or iron oxide) to be magnetized and demagnetized. These interfaces are boundaries between two regions of the crystal that have different magnetic states. As magnetization occurs, these defects migrate and by their motion provide the material with a net magnetic moment. Without the existence of Bloch walls, all ferromagnetic materials would be permanent magnets. In fact, electromagnets would not exist if it were not for this type of defect.

In conclusion, almost none of the properties that led to the above engineering accomplishments are found in, or could be achieved by, a "perfect" crystal. The benefits of using imperfect crystals have led to the concept of "defect engineering".

REFERENCES

[1] C. Walke and M. Dick, "*The Induction of the Cult Image in Ancient Mesopotamia -The Mesopotamian Mis Pi Ritual*", State Archives of Assyria Literary Texts (SAALT1), Helsinki (2001).

[2] B. Ebbell, "*The Papyrus Ebers; the Greatest Egyptian Medical Document*", Trans. from the Ebers papyrus, Levin and Munksgaard, Ejnar Munksgaard, Copenhagen (1937).

[3] Theophrastus, "*De Lapidibus*", in *Eis Organon Aristotelous [Opera Graece]*, Aldus Manutius, Venice, Vol. **2** (1495–1498).

[4] "*Plato, the Collected Dialogues Including the Letters*", ed. by E. Hamilton, H. Cairns, with Introduction and Prefatory Notes, Bollingen Series LXXI, Princeton University Press, Princeton, NJ (1961).

[5] G. Agricolae, "*De Ortu et Causis Subterraneorum [And Other Works]*", H. Frobenius and N. Episcopius, Basel (1546).

[6] A. E. Waite, "*The Hermetic and Alchemical Writings of Aureolus Philippus Theophrastus Bombast of Hohenheim, Called Paracelsus the Great*", Vol. **1**, Hermetic Chemistry, James Elliott and Co., London (1894).

[7] R. Hooke, "*Micrographia, or, Some Physiological Descriptions of Minute Bodies Made by Magnifying Glasses, with Observations and Inquiries Thereupon*", J. Martyn, J. Allestry, London (1665).

[8] C. Huygens, "*Treatise on Light in Which are Explained the Causes of that Which Occurs in Reflexion, & in Refraction; and Particularly in the Strange Refraction of Iceland Crystals*", Trans. S. Thompson; Dover, New York (1962).

[9] N. Stenonis, "*De Solido Intra Solidum*", naturaliter contento dissertationis prodromus (1669).

[10] R.-J. Haüy, "*Traité de mineralogy*", Louis, Paris, Vol. **1** (1801).

[11] W. H. Miller, "*A Treatise on Crystallography*", Cambridge [Eng.]: Printed for J. & J.J. Deighton; J.W. Parker, London (1839).

[12] A. Bravais, "*Sur les polyédres de forme symétrique*", *Compt. Rendus Séances Acad. Sci.*, Vol. **14** (1849), pp. 1–40.

[13] M. von Laue, "*Eine quantitative prüfung der theorie für die interferenz-erscheinungen bei Röntgenstrahlen*", *Sitzungsberichte Der Kgl. Bayer. Akad. Der Wiss*, Munich (1912), pp. 363–373.

[14] W. C. Rontgen, "*Eine neue art von strahlen*", in *Sitzungsberichten Der Würzburger Physik-Medic. Gesellschaft*, Würzburg: Verlag und Druck der Stahel'schen K.B. Hof- und Universitätsbuch- und Kunsthandlung, no 9 (1895), pp. 1–12. https://archive.org/details/eineneueartvonst1896rntg2/page/n3

[15] W. H. Bragg and W. L. Bragg, "*The structure of some crystals as indicated by their diffraction of X-rays*", *Proc. R. Soc. Lond.*, Vol. **A89** (1913), pp. 277–291.

[16] B. D. Cullity, "*Elements of X-Ray Diffraction*", Addison-Wesley Pub. Co., Reading, MA (1956). Revised edition: B.D. Cullity, S.R. Stock, "*Elements of X-Ray Diffraction*", 2nd Ed., Prentice Hall, Upper Saddle River, NJ (2001) **ISBN**: 978-0201610918.

[17] N. W. Ashcroft and N. D. Mermin, "*Solid State Physics*", 1st ed. Brooks Cole Pub. Co., California (1976), pp. 112–129, ISBN-13: 978-0030839931.

[18] W. B. Pearson, "*A Handbook of Lattice Spacings and Structures of Metals and Alloys*", Vol. **2**, Pergamon Press, Oxford (1967).

[19] "*Internationale Tabellen Zur Bestimmung Von Kristallstrukturen*", ed. C. Hermann, Gebruder Borntraeger, Berlin, Vol. **I** and **II** (1935).

[20] T. Hahn, "*International Tables for Crystallography, Volume A: Space Group Symmetry*", 5th ed. Springer-Verlag, Berlin, New York (2002) ISBN 978-0-7923-6590-7.

[21] http://commons.wikimedia.org/wiki/File:Indices_miller_plan_exemple_cube.png

[22] W. Hume-Rothery, R. W. Smallman, and W. Haworth, "*The Structure of Metals and Alloys*", The Institute of Metals, London (1969).

[23] E. Mooser and W. B. Pearson, "The chemical bond in semiconductors", *J. Electron.*, Vol. 1 (1956), pp. 629–645.

[24] E. Mooser and W. B. Pearson, "Recognition and classification of semiconducting compounds with tetrahedral sp^3 bonds", *J. Chem. Phys.*, Vol. 26 (1957), pp. 893–899.

[25] A. Owens and A. Peacock, "Compound semiconductor radiation detectors", *Nucl. Instr. And Meth.*, Vol. A531 (2004), pp. 18–37.

[26] L. Vegard, "Die Konstitution der Mischkristalle und die Raumfüllung der Atome", *Z. Phys.*, Vol. 5 (1921), pp. 17–26.

[27] J. E. Bernard and A. Zunger, "Optical bowing in zinc chalcogenide semiconductor alloys", *Phys. Rev.*, Vol. B34 (1986), pp. 5992–5995.

[28] Landolt-Börnstein, "*Semiconductors–New Data and Updates for II-VI Compounds*", Vol. 44A, Springer-Verlag, Berlin (2009) ISSN 978-3-540-74391-0 (Print) 978-3-540-74392-7 (Online).

[29] Landolt-Börnstein, "*Semiconductors – New Data and Updates for I-VII, III-V, III-VI and IV-VI Compounds*", Vol. 44B, Springer-Verlag, Berlin (2009) ISBN 978-3-540-48528-5.

[30] E. Orowan, "Zür Kristallplastizität I-III", *Z. Physik*, Vol. 89 (1934), Teil I pp. 605–613; Teil II pp. 614-633, Teil II, pp. 634–659.

[31] M. Polanyi, "Über eine Art Gitterstörung, die einem Kristall plastisch machen könnte", *Z. Physik*, Vol. 89 (1934), pp. 660–664.

[32] G. I. Taylor, "The mechanism of plastic deformation of crystals I-II", *Proc. Roy. Soc.*, Vol. A145 (1934), Part I pp. 362–387; Part II pp. 388–404.

[33] P. Lacombe, "Report of a conference on strength of solids", University of Bristol, England, *The Physical Society, London* (1948), pp. 91–94.

[34] A. Koyama, A. Hichiwa, and R. Hirano, "Recent progress in CdZnTe crystals", *J. Electron. Mater.*, Vol. 28 (1999), pp. 683–687.

[35] P. J. McNally, "Techniques: 3D imaging of crystal defects", *Nature*, Vol. 496, no. 7443 (2013), pp. 37–38.

[36] Y. Ran, Y. Zhang, and A. Vishwanath, "One-dimensional topologically protected modes in topological insulators with lattice dislocations", *Nature Physics*, Vol. 42 (2009), pp. 298–303.

[37] J. M. Burgers, "Some considerations of the field of stress connected with dislocations in a regular crystal lattice", *Proc. K. Ned. Akad. Wet.*, Vol. 42 (1939), part I pp. 293–325; part II pp. 378–399.

[38] F. C. Frank and J. H. van der Merwe, "One-dimensional dislocations. I. Static theory", *Proc. Roy. Soc. Lond. Series A, Mathematical and Physical Sciences*, Vol. A198 (1949), pp. 205–216.

[39] M. Syväjärvi, "*Epitaxial Growth and Characterization of SiC*", PhD thesis (1999), Linköping University, LITH-Tek-Lic-694 (1998).

[40] E. O. Hall, "The deformation and ageing of mild steel: III Discussion and results", *Proc. Phys. Soc. London*, Vol. B64 (1950), pp. 747–753.

[41] N. S. Petch, "The cleavage strength of polycrystals", *J. Iron Steel Inst.*, Vol. 174 (1953), pp. 25–28.

[42] P. Cavaliere, "Mechanical properties of nanocrystalline materials", in *Handbook of Mechanical Nanostructuring*, ed. M. Aliofkhazraei, Wiley-VCH Verlag GmbH, Weinheim (2016), pp. 3–16, ISBN–13:978-3527335060.

[43] C. Szeles and M. C. Driver, "Growth and properties of semi-insulating CdZnTe for radiation detector applications", *Proc. SPIE*, Vol. 3446 (1998), pp. 2–9.

[44] A. E. Bolotnikov, N. M. Abdul-Jabber, O. S. Babalola, G. S. Camarda, Y. Cui, A. M. Hossain, E. M. Jackson, H. C. Jackson, J. A. James, K. T. Kohman, A. L. Luryi, and R. B. James, "Effects of Te inclusions on the performance of CdZnTe radiation detectors", *IEEE, Trans. Nucl. Sci.*, Vol. 55, no. 5 (2008), pp. 2757–2764.

[45] G. A. Carini, A. E. Bolotnikov, G. S. Camarda, G. W. Wright, L. Li, and R. B. James, "Effect of Te precipitates on the performance of CdZnTe (CZT) detectors", *Appl. Phys. Lett.*, Vol. 88, no. 14 (2006), pp. 143515–1–143515–3.

[46] "*Semiconductor defect engineering - materials, synthetic structures and devices II*", Materials Research Society Symposium Proceedings, Editors S. Ashok, J. Chevallier, P. Kiesel and T. Ogino, Warrendale, PA, Vol. 994 (2007).

[47] E. G. Seebauer and M. C. Kratzer, "Charged point defects in semiconductors", *Mater. Sci. Eng. R.*, Vol. 55 (2006), pp. 57–149.

[48] E. G. Seebauer and M. C. Kratzer, "*Charged Semiconductor Defects: Structure, Thermodynamics and Diffusion*", Springer, New York (2008).

[49] D. Hull and D. J. Bacon, "*Introduction to Dislocations*", Butterworth-Heinemann, Oxford (2001).

[50] K. Dev, M. Y. L. Jung, R. Gunawan, R. D. Braatz, and E. G. Seebauer, "Mechanism for coupling between properties of interfaces and bulk semiconductors", *Phys. Rev. B.*, Vol. 68 (2003), pp. 195311–1–195311–6.

[51] E. G. Seebauer, K. Dev, M. Y. L. Jung, R. Vaidyanathan, C. T. M. Kwok, J. W. Ager, E. E. Haller, and R. D. Braatz, "Control of defect concentrations within a semiconductor through adsorption", *Phys. Rev. Lett.*, Vol. 97 (2006), pp. 055503–1–055503–4.

[52] W. Shockley and W. T. Read, "Statistics of the recombinations of holes and electrons", *Phys. Rev.*, Vol. 87 (1952), pp. 835–842.

[53] S. M. Sze, "*Semiconductor Devices, Physics and Technology*", Wiley, New York (2002).

[54] K. Vanheusden, W. L. Warren, C. H. Seager, D. R. Tallant, J. A. Voigt, and B. E. Gnade, "Mechanisms behind green photoluminescence in ZnO phosphor powders", *J. Appl. Phys.*, Vol. 79 (1996), pp. 7983–7990.

[55] G. Lindstrom, Z. Kohout, V. Linhart, B. Sopko, *et. al.*, "Developments for radiation hard silicon detectors by defect engineering results by the CERN RD48 (ROSE) collaboration", *Nucl. Instr. And Meth.*, Vol. **A465** (2001), pp. 60–69.

[56] "Gettering and defect engineering in semiconductor technology XV", in *15th Gettering and Defect Engineering in Semiconductor* Technology *(GADEST 2013)*, ed. J.D. Murphy, September 22–27, 2013, Oxford, UK (2014) ISBN-13: 978-3-03785-824-0.

[57] M. Tokushima and H. Yamada, "Light propagation in a photonic-crystal-slab line-defect waveguide", *IEEE J. Quant. Electron.*, Vol. **38**, no. 7 (2002), pp. 753–759.

4

Growth Techniques

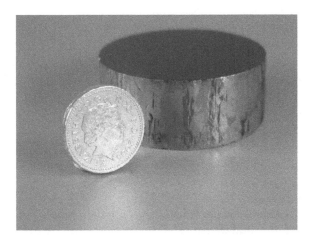

A 2-inch diameter, 0.8-inch thick single CdZnTe crystal grown by the Multi-Tube Physical Vapor Transport (MTPVT) technique (photo courtesy Kromek®).

CONTENTS

4.1 Introduction

Crystals are structures that are formed from a regular repeated pattern of connected atoms or molecules extending in all three spatial dimensions. Growth occurs as a consequence of a crystallization process, in which new atoms or ions condense into the characteristic basic shape of the Bravais lattice. The term "crystal" is derived from the ancient Greek word κρύσταλλος (krystallos) meaning *"clear ice or ice cold"* and was originally used to refer to ice crystals only. However, as the centuries progressed, the meaning of the word expanded, incorporating and replacing words like condensation and coagulation. Artificially induced crystal growth has been practiced for millennia. For example, there is evidence, in the form of broken briquetage,[1] that seawater was crystallized by evaporation at a number of sites in central and Eastern Europe from the 5th millennium BC. In fact, the earliest evidence for the deliberate recrystallization of salt can be traced back to eastern Romania in 6050 BC. In 79 AD, the Roman philosopher Gaius Plinius Secundus (better known as Pliny the Elder) reported the crystallization of a number of salts, for example vitriols[2] in his 37 volume encyclopaedia *"Naturalis Historia"* [1]. Over the next few centuries, knowledge of the crystallization phenomenon and the processes by which it can occur were driven by the medieval alchemists. For example, the Spanish alchemist Geber, whose writings date from the 14th century, describe the preparation and purification of various materials by recrystallization as well as by sublimation and distillation. In the writings of Paracelsus (1493–1541), the Swiss-German alchemist and philosopher proposed that crystals grew from seed, in a fashion analogous to that of fruit-bearing plants. The physical foundations for crystal growth were laid down in the 17th century with an increasing interest in crystal formation and especially attempts to formalize it. In fact, the subject of crystal growth was treated as part of crystallography and did not have an independent identity until the last quarter of the 19th century when a systematic understanding of the subject was achieved when Haüy's molecular crystal structure theory [2] was coupled with Gibbs' thermodynamically derived phase equilibrium concept [3]. In essence, the growth of single crystals can be regarded as a phase transformation into the solid state from the solid, liquid or vapor state. Of these, liquid to solid and vapor to solid transformations are the most important[3] for the production of semiconductor crystals and have resulted in a number of "standard" growth processes. Based on these ideas and parallel advances in chemistry, a new impetus in developing crystal growth techniques began, driven largely by the commercial prospects of producing large synthetic gemstones. By the late 1880s, Verneuil [4] had developed the flame-fusion growth method to produce rubies and sapphires. However, more importantly, for the first time Verneuil's method made it possible to carry out systematic investigations of high-temperature crystallization processes. His principles of nucleation and growth control were later adapted and utilized in most other melt growth techniques. For example, in 1917 Czochralski [5] suggested that crystals could be pulled directly from the melt and in 1926 Kyropoulos [6] replaced the crystal-pulling process from the melt with directed crystallization by decreasing the melt's temperature. At the same time, Bridgman [7] proposed the single crystal growth method using crucibles with conical bottoms. The method involved heating polycrystalline material above its melting point and slowly cooling it from one end of the crucible where a seed crystal is located. As the temperature drops, melt material crystallizes on the seed assuming the same crystallographic orientation and progressively forms a single crystal along the length of the crucible. The method was improved by Stockbarger [8] who discovered that the quality of the crystals could be significantly improved by the careful control of the temperature gradient at the melt/crystal interface. In the 1950s, research had begun on epitaxial

---

[1] A coarse ceramic material used to make evaporation vessels.

[2] Vitriol is an archaic name for a sulfate. The name is derived from the Latin word "vitreolum" for glassy, as crystals of several metallic sulfates resemble pieces of coloured glass.

[3] Solid–solid phase transformations are rarely employed, except to grow single crystals of fast ionic and superionic conductors and their alloys.

growth techniques, which became fully developed by the 1960s driven mainly by the semiconductor industry's need for very high quality single crystal films. In fact, in March 1961, Fairchild introduced the 2N914, one of its most successful transistors whose structures were fabricated entirely using epitaxy. The process was quickly adopted by its competitors. Since this time, ever more sophisticated techniques have been developed, such as MOCVD and MBE, which involve the precise manipulation of elements at the atomic level.

4.2 Base Material Production

To produce radiation detectors with good charge collection properties, high-quality single-crystal material that is both crystallographically and stoichiometrically as perfect as possible must to be grown. The overall process is illustrated in Fig. 4.1 and begins with the production of the starting material or precursors. Typically, these are synthesized from pure elements or purchased from commercial vendors with nominal purity of 99.9999%.[4] The precursors are then grown into the desired semiconductor, usually in a polycrystalline form, using a melt technique. This material is then purified using zone refining or distillation techniques before being used to produce thin seeds or substrates with as near prefect single crystal properties as possible. These will be used to grow ingots or wafers of similar or higher quality crystalline material. For some materials, these ingots can be grown directly during the last zone-refining stage. Wafers will subsequently be divided into detector crystals. If ingots are produced, they will be sawn into wafers and then diced into detectors. Substrates are usually grown by bulk techniques because of cost considerations, starting with the purified form of the elements that make up the element or compound.

4.2.1 Material Purification

The preparation of semiconductor material usually begins with the purification of the starting material down to impurity levels below 0.001% (10 parts per million or 5N) and in some cases 1 ppm (6N). Typical solution-based purification methods

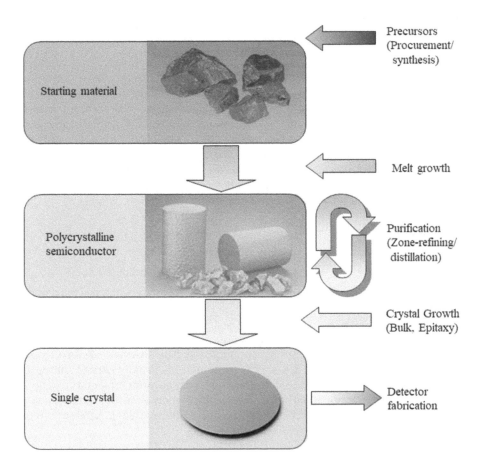

FIGURE 4.1 Steps required to manufacture single crystalline material for detector production.

---

[4] Commonly known as 6 N, where "6" refers to the total number of "9s" before and after the decimal point.

with which to achieve this goal are (1) recrystallization, (2) vacuum sublimation, both static and dynamic, (3) vacuum distillation and (4) zone refining.

4.2.1.1 Recrystallization

Recrystallization, also known as fractional crystallization, is a procedure for purifying an impure compound in a solvent and is commonly used to produce silicon stock material for the semiconductor industry. The method of purification is based on the principle that the solubility of most solids increases with increased temperature. This means that as temperature increases, the amount of solute that can be dissolved in a solvent increases. In practice an impure compound is dissolved in a solvent (the impurities must also be soluble in the solvent), to produce a highly concentrated solution at a high temperature. If the solution is then cooled, the solubility of the impurities in the solution and the substance being purified, decreases. Solubility curves can be used to predict the outcome of a recrystallization procedure, but ideally the substance to be purified crystallizes before the impurities, excluding their molecules from the growing crystal lattice and leaving them behind in the solution. The slower the rate of cooling, the larger the crystals that form. The result is a purer crystal in a soup of impurities. A filtration process must then be used to separate the more pure crystals. The procedure can be repeated. The advantage of this technique is that it is a very simple and effective way of purifying a sample. Its main disadvantage is that it takes a long time.

4.2.1.2 Vacuum Distillation/Sublimation

Vacuum distillation and sublimation are methods routinely applied to the purification of materials, and in fact both processes are similar. In essence, distillation is a method of separating mixtures based on differences in volatilities of components in a boiling liquid mixture. In the distillation of a liquid, the liquid is heated until it reaches the temperature at which its vapor pressure equals the total pressure of the system. If the vapors are then allowed to expand into a cooler zone, they will condense, leaving the impurities behind in the liquid. If a solid is distilled, it first melts, then vaporizes and then condenses and solidifies – so the distillation of a solid is just like distillation of a liquid except for the additional steps of melting and solidification. Some solids have a sufficiently high vapor pressure they will vaporize, or "distil" without melting; such behaviour is called sublimation, that is, the substance transits directly from the solid phase to a gas phase without passing through an intermediate liquid phase.

Sublimation is a useful procedure for purification if the impurities are essentially non-volatile and if the desired substance has a vapor pressure of a few Torr at its melting point. In vacuum sublimation the sample is placed under reduced pressure to permit sublimation at lower temperatures. Since elements and compounds condense at unique temperatures, the use of a controlled cold surface will recover the required solid leaving the non-volatile residue impurities behind. There are numerous variations of this method using either static or dynamic vacuum. Dynamic vacuum is more effective for less volatile samples. Vacuum sublimation is commonly used in the purification of organic semiconductors [9], $A^IB^{II}C^V$ semiconductors [10] and HgI_2 [11].

Distillation is most commonly used to purify volatile elemental precursors before compounding, for example, Cd, Hg and Te, used to produce HgCdTe. For these elements, the vapor pressures of the most common impurities (*e.g.*, Zn, Mg, Cu ...) are many orders of magnitude less at their melting points, allowing easy separation. The method is most effective if dynamically pumped; this has a number of advantages over conventional distillation, namely:

   i) the rate of evaporation increases with decrease in residual pressure,

   ii) better separation of volatile constituents can be achieved at lower pressure,

   iii) the possibility of interaction of distillate with ambient gas is reduced or eliminated and

   iv) working at a lower temperature reduces the possibility of metal/container interactions and reduces the running costs of the process.

Vacuum distillation is usually carried out in sealed retort containing an electrically heated crucible containing the charge to be purified. The material is then heated while the system is pumped and the distillate allowed to condense out on a cold plate or region close to the crucible. In the case of Cd and Te, distillation is usually carried out at a temperature of 400–500°C and vacuum pressure of 10^{-4}–10^{-5} Torr. Because it is already a liquid at room temperature, Hg is usually distilled at a lower temperature of ~200°C and a pressure of 100 Torr. One-stage distillation is usually sufficient to produce 5N material, provided commercial grade feedstock is used. Multiple distillation or additional zone refining is required to produce 6N material, which is a necessary starting point for detector grade material.

4.2.1.3 Zone Refining

Zone-refining was first developed in 1952 by Pfann [12] to purify germanium and belongs to a class of techniques known as fractional solidification, in which a separation is brought about by crystallization of a melt without solvent being added. It can be applied to the purification of almost every type of substance that can be melted and solidified (*e.g.*, elements, organic compounds and inorganic compounds) – in fact to any solute-solvent system having an appreciable concentration difference between the solid and liquid phases at equilibrium. Purities as high as 99.999% are often obtained by this technique. The technique works as follows. A moving heater traverses a long ingot of raw material and melts a small zone as it moves. This is illustrated schematically in Fig 4.2(a); the practical implementation is in Fig. 4.2(b). Depending on whether an impurity lowers or raises the melting point of a material, the passage of the molten zone through the charge will concentrate the impurity in either the molten or frozen phase, respectively. Specifically, if it lowers the melting point of the parent, it will be more soluble in the liquid than in the solid and if it raises the melting point, it will segregate more in the solid. The first case is by far the most common and forms the basis of zone refining. Let C_s be the solute concentration in the solid and C_L the solute concentration in the liquid zone. The ratio C_s/C_L, is known as the segregation or equilibrium distribution coefficient, k_o, of the impurity [13] and ranges typically between 10^{-5} and 10, depending on the impurity. Note that measured segregation coefficients are higher than the equilibrium values, since the impurity segregation during the growth is influenced by many factors, most importantly by incomplete mixing in the liquid. Therefore in practice, it is more realistic to define an effective segregation coefficient, k_{eff}, which takes account of these effects by describing the practical segregation behaviour of impurities during growth.[5]

If k_{eff} is less than unity (*i.e.*, $C_L > C_s$), the impurities will congregate in the liquid zone and be carried forward as the heater moves. The net effect is that the impurities accumulate at one end of an ingot (usually in the last-to-freeze region) leaving pure material in the central section – the impure section can be removed simply by cutting it off. It has been shown [14] that after one zone pass ($n = 1$), the relative solute concentration, C_s, in the solid at a distance x from the starting end, is a function of the initial concentration, C_o, the width of the molten zone, l, and the effective distribution coefficient, k_{eff}, and is given by,

$$C_S(x) = C_o[1 - (1 - k_{eff})\ \exp(-k_{eff}x/L)].$$ (1)

In practice, zone refining is repeated many times, and this is most efficiently carried out using multiple heaters as illustrated in Fig. 4.2(b). In Fig. 4.2(c), we show a plot of relative impurity content (C/C_o) along an ingot (expressed as L/l) for various numbers (n) of zone passes for an ingot 10 zones long and for an assumed distribution coefficient of 0.5. It can be seen that

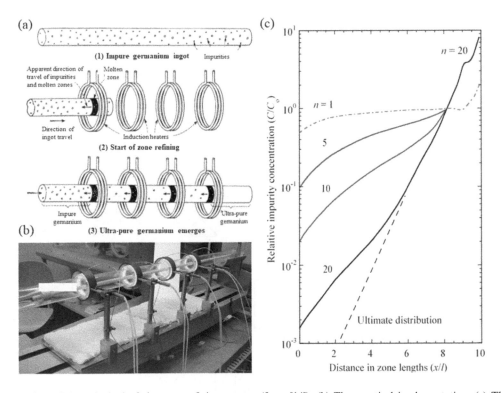

FIGURE 4.2 (a) Illustration of the principal of the zone refining process (from [14]). (b) The practical implementation. (c) The relative impurity distribution along a 10 zone lengths long ingot for various numbers (n) of zone passes for a distribution coefficient of 0.5 (adapted from [14]). Here C_o the initial uniform impurity concentration of the rod, l, is the length of the zone and L is the length of the ingot being refined. The distribution coefficient is defined as the ratio of the solute concentration in the solid to that in the liquid in equilibrium and should be less than unity for purification to occur.

---

[5] For slow solidification and thorough mixing $k_{eff} \approx k_o$, while for fast solidification with no mixing k_{eff} approaches unity.

as more zone passes are made, the impurity concentration at the beginning of the ingot drops lower and lower until it eventually reaches a limit called the ultimate distribution. Because the solid-liquid phase equilibria may not be favourable for all impurities, zone refining is often combined with other techniques to achieve ultrahigh purities. Note that besides impurity manipulation, zone melting is finding increasing use as a method of growing single crystals as exemplified by the float-zone technique (see Section 4.4.5), widely used for producing high quality Si.

4.3 Crystal Growth

Broadly speaking, semiconductor crystal growth can be reduced to two basic approaches – bulk growth, in which crystals are produced directly out of solution, and film growth, in which atomic layers are deposited onto a substrate[6] forming a film. In Fig. 4.3 we show a schematic of bulk and film growth technologies for compound semiconductors. In this chapter, following an initial introduction to the framework of crystal growth, each growth process will be discussed in detail.

4.3.1 Phases and Solidification

When materials change temperature they can change state; this is called a phase change. The obvious examples are melting, vaporization, condensing and freezing (or solidification), and whenever there is a phase change in a material, the atoms or molecules involved are rearranged. Most phase transformations begin with the formation of numerous small particles of the new phase that increase in size until the transformation is complete. The process evolves from an initially disordered phase in three steps. These are:

1) supersaturation or supercooling of a liquid or gas phase with respect to the component to be grown followed by

2) nucleation or crystallization, which is the process whereby nuclei (seeds) act as templates for crystal growth, and

3) the growth of the nuclei into single crystals of distinct phases.

The moment a crystal begins to grow is known as nucleation, and the point where it occurs is the nucleation point. Growth typically follows an initial stage of either homogeneous nucleation, in which nuclei form uniformly throughout the parent phase (*i.e.*, without the involvement of a "foreign" substance) or by heterogeneous nucleation (*i.e.*, surface catalyzation[7]). At the solidification temperature, atoms of a liquid, such as melted metal, begin to bond together at the nucleation points and start to form crystals. The final sizes of the individual crystals depend on the number of nucleation points. The crystals increase in size by the progressive addition of atoms and grow until they impinge upon adjacent growing

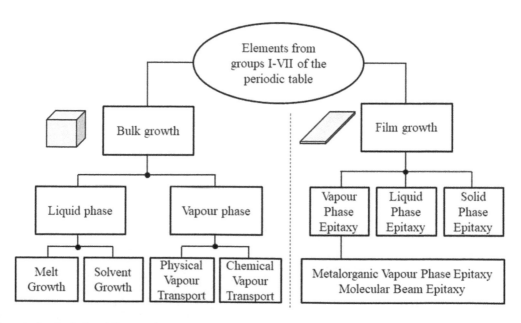

FIGURE 4.3 Schematic showing bulk and film growth methods for compound semiconductors. Each method is discussed in detail in sections of this chapter.

---

[6] A substrate is a solid, usually thin and planar material onto which a layer of another material is applied and to which the second material adheres.

[7] In which nuclei form at structural inhomogeneities, for example, container surfaces, impurities, grain boundaries, dislocations, etc.

crystals. The nucleation centers themselves may either occur spontaneously due to the conditions prevailing in the parent phase or be induced artificially, for example, by introducing a "seed" crystal to start the growth.

For chemically pure elements, the growth of crystalline material is relatively straightforward since the only phase changes are usually the melting of the starting material and its solidification around a seed or substrate. Generally a crystal will always be grown, the quality of which will depend mainly on the actual growing technique and the purity of the starting material. However, this is not the case for compounds. When elements are mixed together in their molten state and allowed to solidify, alloys are formed. The solution can crystallize into three principal forms – a solid solution, a eutectic[8] or a compound, depending on the relative strengths of the atomic bonds of its components. In Fig. 4.4 we show the atomic arrangement of each for a binary alloy composed of atoms A and B. We have assumed a binary alloy for simplicity, but the extension to ternary or quaternary alloys is straightforward and easily understood. Note that a number of pseudo binary compounds are used in radiation detection – a good example of which is cadmium zinc telluride ($CdZn_xTe_{(1-x)}$) in which a small percentage x of an additional element[9] is added (in this case Zn) to improve the electrical and mechanical properties of the basic compound (in this case CdTe).

In a solid solution (Fig. 4.4(a)), the atoms of elements A and B occupy random positions in the crystal, a classic example of which is brass, a Cu-Zn alloy. A solid solution is obtained because the strength of the A-B bond is intermediate between those of the A-A and the B-B bonds. A eutectic on the other hand (Fig. 4.4(b)), is formed when the A-B bond is weaker than the A-A bond and B-B bond, which means that each component will have a low solubility in the other. As a consequence, the liquid solution solidifies by separating into two distinct phases – one rich in element A and the other rich in element B – in essence a eutectic is an alloy of immiscible components, a good example of which is solder, which is an alloy of Pb and Sn. If now the A-B bond is stronger than either the A-A bond or B-B bond, a compound is formed (Fig. 4.4(c)) – for example, GaAs, InP, CdTe, *etc.* Each of these configurations is known as a phase and is defined as a homogeneous, physically distinct portion of matter that is present in a non-homogeneous system. It may be a single component or mixture. The component of a phase is the relative amount of the different atomic species it contains, while the amount of a phase is the fraction of the mixture that is in that particular phase. A solid may exist in single or multiple phases at the same time, depending on its temperature and thermodynamic state. This behaviour is most easily understood with reference to a phase diagram, which is a graphical representation of the combinations of temperature, pressure, composition, or other variables for which specific phases exist at equilibrium.[10] An example of a phase diagram is given in Fig. 4.5(a) for the CdTe binary system and (b) for the CdTe-ZnTe pseudo-binary system – both commonly used detection media for X-and gamma-rays.

Curves are given at constant pressure. For both, the relative compositions are plotted on the horizontal scale and the temperature on the vertical scale. Note: usually the mole fraction of the more volatile component is plotted on the horizontal axis. The overall composition represents the amounts of the components that are present in the alloy (*i.e.*, the amounts in the initial melt). In alloys, a two-phase region separates the liquid phase from the solid in the diagram. In this region, both solid and liquid coexist with varying compositions depending on temperature. The compositions of both the solid and the liquid may be determined at a particular temperature by taking suitable sections through the liquidus and solidus curves (so-called tie lines[11]). The relative fraction of solid and liquid at a particular temperature can be found using the "lever" rule. Briefly put, the lever rule allows us to calculate the concentration of phases present from a phase diagram, based on a mass balance assumption. For a more detailed explanation of phase diagrams, tie lines and the lever rule, see ref [15]. The amounts of

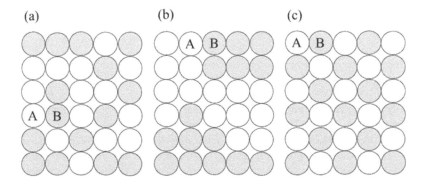

(a)   (b)   (c)

FIGURE 4.4 Schematic illustrating the possible configurations of two different elemental atoms in a binary alloy (denoted by A and B). (a) represents a solid solution, (b) segregation in a eutectic and (c) compound formation – in this case AB.

---

[8] An alloy of of immiscible components, from the Greek *eutektos* meaning "easily melted".
[9] The deliberate introduction of foreign atoms is called alloying.
[10] which is the state that matter aspires to, given sufficient time
[11] A tie line is an isothermal (constant temperature) line connecting the compositions of the two phases in a two-phase field. It is used to find the compositions of the phases in the two-phase field.

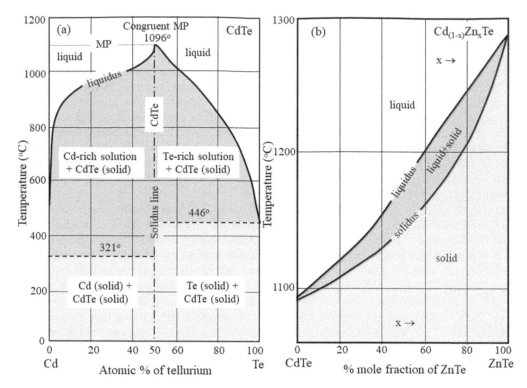

FIGURE 4.5 The phase diagrams for (a) the CdTe binary system and (b) the CdTe-ZnTe pseudo-binary system. Both curves are given at constant pressure.

phases can be calculated from the lever rule. The amount of solid is the ratio of the length of the tie line between the overall composition of the liquidus curve to the length of the tie line between the liquidus and solidus. The amount of liquid phase is the ratio of the line between the solidus and the overall composition divided by the length between the solidus and liquidus.

The liquidus curves shown in Figs 4.5(a) and (b) represent the lowest temperatures of a liquid as a function of composition. It forms the boundary between the liquid and the two-phase system, while the solidus line represents the highest temperature of a solid as a function of composition. Similarly, it represents the border between the solid phase and the two-phase region. Although not applicable to Fig. 4.5, for completeness we define the solvus line as the limit of solubility of one component in the other. It is generally the limit between a solid solution phase and a two-phase eutectic region. A dominant feature in Fig. 4.5(a) is the central compound CdTe containing equiatomic amounts of Cd and Te. This represents a single phase. Compounds generally have well-defined stoichiometries, which manifest themselves as vertical lines in the phase diagram. The strong bond between the Cd and Te atoms in CdTe is reflected in the melting temperature of the compound, which is higher than either of its components.

4.4 Bulk Growth Techniques

Bulk single crystals are usually grown from a liquid phase. The liquid may have approximately the same composition as the solid. It may be a solution consisting primarily of one component of the crystal or it may be a solution whose solvent constitutes a minor fraction of the crystal's composition. Generally melt growth techniques are used for the elemental semiconductors and Group II-VI materials. They are usually not preferred for III-V materials (e.g., GaAs, GaP, GaN) because the large difference in vapor pressures of the constituent components requires a high-pressure environment to keep the Group V element in the system. In addition, it is not feasible to grow bulk ternary materials, because of (a) the difficultly in controlling stoichiometry due to competing phases and (b) the greater difficulty in achieving equilibrium among the components during phase transitions of vapor-liquid-solid or vapor-solid without introducing crystal defects. Provided layers are not too thick, ternary and higher-order compounds are generally grown by MBE or MOCVD.

4.4.1 Hydrothermal Synthesis

Hydrothermal synthesis is a single crystal growth method that depends on the solubility of minerals in hot water under high pressure (for a review see [16]). The term "hydrothermal" is derived from geologic origins and usually refers to any heterogeneous reaction in the presence of aqueous solvents or mineralizers under high pressure and temperature conditions. It was first used in the 18th century to describe the action of water at elevated temperature and pressure in bringing about changes in the earth's crust leading to the formation of various rocks and minerals. The first report of the hydrothermal growth of crystals was by German geologist Karl Emil von Schafhäutl who in 1845 grew microscopic quartz crystals in a pressure cooker [17]. The technique is very similar to the processes by which natural quartz crystals are formed; indeed, hydrothermal synthesis is now commonly used worldwide to produce synthetic quartz and gemstones for industrial purposes – materials that are very difficult to synthesize by either melt or flux techniques. For semiconductors, the method is mainly used for growing single $A^{II}B^{VI}$ crystals, such as ZnO, for optoelectronic applications. Crystals produced by this method tend to have a higher degree of crystallinity, a lower defect density and high carrier mobilities when compared to comparable crystals produced by chemical vapor transport or melt techniques. Crystal growth is carried out in an autoclave, in which a nutrient is supplied along with water. A gradient of temperature is maintained across the growth chamber so that the hotter end dissolves the nutrient and the cooler end causes seed growth. Advantages of the hydrothermal method over other types of crystal growth include the ability to create crystalline phases that are not stable at the melting point. Also, materials that have a high vapor pressure near their melting points can be grown. The method is also particularly suitable for the growth of large good-quality crystals while maintaining good control over their composition. The main disadvantage of the method is the need for expensive autoclaves.

Hydrothermal synthesis can also be used for the purification of some compounds (for example, TlBr) by re-crystallization. In the case of TlBr [18] this is achieved by dissolving impure TlBr in solution. Impurity components that are more soluble than TlBr will be left in solution when the TlBr is re-crystallized, while less soluble components will not dissolve and will be left together with the residual nutrient as a by-product. In fact, some impurities, which are difficult to remove from the TlBr in the molten state, are soluble in water and are most easily removed by hydrothermal re-crystallization.

4.4.2 Czochralski (CZ)

Perhaps the most widely used bulk growth technique is the crystal-pulling or Czochralski (CZ) method[12] [5], in which the melt of the charge, usually high-quality polycrystalline material, is held in a vertical crucible, as illustrated in Fig. 4.6 (left). The top surface of the melt is held just above the melting temperature. A seed crystal[13] is then lowered into the melt and slowly withdrawn (see Fig. 4.7(a)). Both the seed crystal and the crucible are rotated during the pulling process, in opposite

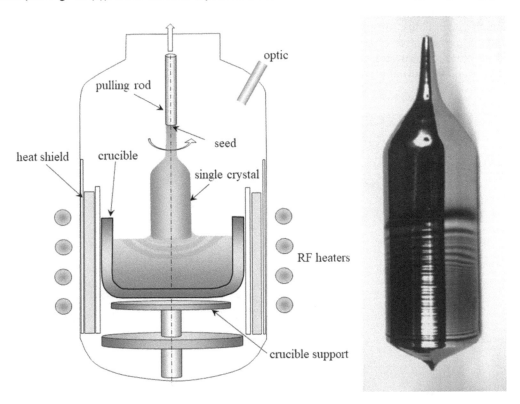

FIGURE 4.6 Left: schematic of a Czochralski crystal growth system used to produce Si, GaAs and InP substrate ingots. Right: a typical boule (ingot). The top of the crystal is called the seed end, or alternatively, the first to solidify end. The bottom is known as the "tail" or "tang" end.

---

[12] Named after the Polish chemist Jan Czochralski (b.1885, d. 1953).

[13] A seed crystal is a small piece of single crystal or polycrystal material from which a large crystal of typically the same material is to be grown.

directions. As the heat from the melt flows up the seed, the melt surface cools and the crystal begins to grow. The surface tension of the melt at the interface allows molten material to be pulled out of the crucible, where it cools and solidifies on the seed, thus forming a single crystal. The rotation of the seed about its axis ensures a roughly circular cross-section crystal and reduces radial temperature gradients. The rotation also has the added advantage that it inhibits the natural tendency of the crystal to nucleate along natural orientations to produce a faceted crystal. The initial pull rate is relatively rapid so that a thin neck is produced (Fig. 4.7(b)), since if the cross-section of the neck is less than that of the seed, thermal stress is reduced; this helps prevent the generation of dislocations in the pulled crystal. The melt temperature is now reduced and stabilized so that the desired ingot diameter can be formed (Fig. 4.7(c)). This diameter is generally maintained by controlling the pull rate. The pulling continues until the melt is nearly exhausted, at which time a tail is formed. The crystal diameter and length depend upon the temperature, pulling rate and the dimensions of the melt container; the crystal quality depends critically upon minimization of temperature gradients, which enhance the formation of dislocations. Pellets of dopant material can be added to the melt if extrinsic semiconductor material is required. The impurities are usually added and dissolved in the melt using solid impurities and are primarily used to increase the concentration of mobile carriers and subsequently the conductivity of the material.

The CZ technique is commonly used in Si, GaAs and InP production to produce substrate material and is not suited to detector grade material as too many impurities, originating mostly from the crucible, are left in the crystal. This reduces minority carrier lifetimes to unacceptably low levels making the production of high resistivity material impossible. The technique [19] yields long ingots (boules) with good circular cross-sections (see Fig. 4.6 right). The largest silicon ingots produced today are around 400 mm in diameter and 1 to 2 m in length and weigh up to 100 kg.

4.4.2.1 Liquid Encapsulated Czochralski (LEC)

Synthesizing compound semiconductor crystals is much more difficult than synthesizing elemental semiconductors, because the vapor pressures of the constituent materials are generally quite different. At the temperature required to melt the higher temperature material, the lower melting-point material has evaporated. For example, in the case of the compounds GaAs and InP, As and P have the lower melting points and tend to leave the melt and condense on the sidewalls. Metz *et al.* [20] circumvented this problem by using a liquid lid or encapsulate to seal the melt and thus prevent evaporation. The technique is generally referred to as liquid encapsulated Czochralski, or LEC. The encapsulate is made of an inert material that is (a) less dense than the melt – so it floats, forming the "lid" and (b) does not interact with or contaminate the melt (see Fig. 4.8). For most compounds, a molten layer of boron oxide (B_2O_3) is commonly used to prevent the evaporation of the volatile component as well as block oxygen and carbon contamination of the melt. The growth system is contained in a high-pressure atmosphere (up to 100 atm) of argon or nitrogen to prevent the volatile constituent from bubbling through the B_2O_3

FIGURE 4.7 Growth sequence in a Czochralski furnace (Images courtesy of Kinetic Systems Inc.). (a) Growth begins when a seed crystal is lowered into the melt. (b) As crystallization begins, the rod holding the seed crystal is slowly withdrawn. (c) By varying the pull rate, the diameter of crystal can be controlled, forming the basis of the ingot. (d) A view into an actual crucible during the drawing of an ingot.

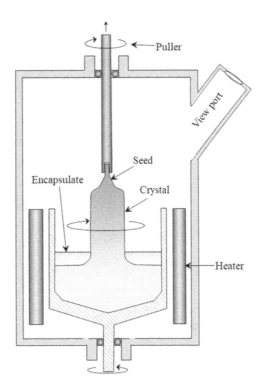

FIGURE 4.8 Schematic of a Liquid Encapsulated Czochralski furnace showing the position of the encapsulate. For compound materials, the encapsulate acts as a liquid lid, sealing the melt and preventing the evaporation of the more volatile component.

and escaping. Metz *et al.* [20] originally developed the method to grow Group IV-VI Pb compounds. It was further developed by Mullin *et al.* [21] for III-V compounds and is now the industry standard for the production of most semi-insulating Group III-V compound materials, such as GaAs, GaP, InP and InAs.

4.4.2.2 *Limitations of the Czochralski Method*

The CZ and LEC techniques have several limitations, which mainly relate to contamination issues and dislocation defect densities (usually expressed in terms of the Etch Pit Density or EPD[14]). Contamination usually arises from the encapsulate (usually boron when using B_2O_3) and/or the crucible (which can be fabricated from any of a number of materials, typically quartz, graphite, glassy carbon, BN and AlN). Crystal defects arise largely from temperature gradients across the melt. While these can be minimized using multi-zone heaters or careful heat-shield design, a reduction in gradients also raises the temperature at the crystal surface, which can diminish the compositional stability from the dissociation of volatile elements such as phosphorus or arsenic. Dissociation can be reduced by the introduction of an ambient of the volatile into the space above the melt. For LEC, growth dissociation can also occur from the crystal surface left exposed above the B_2O_3 encapsulate layer leading to twinning production. Crystal twinning emanating from the neck of the pulled crystal (see Figs. 4.7(c) and (d)) or the cone of the crucible can also be a problem. Twinning occurs when two separate crystals share some of the same crystal lattice points resulting in an intergrowth of two different crystals joined by a so-called twin boundary. Effective methods to reduce the formation of twinning are an optimization of the cone angle at the crystal shoulder and the application of a magnetic field to suppress temperature variations in the melt during growth.

4.4.2.3 *Vapor Pressure Controlled Czochralski (VCZ)*

In LEC, the main drawback of using low temperature gradients (<30 K cm^{-1}) to reduce the defect density is the high surface temperature of the crystal as it emerges from the encapsulate. For GaAs and InP, the resulting dissociation of the recently solidified material due to the high partial pressures of As and P adversely affects the growth kinetics and crystal perfection. The principle of VCZ is to seal the crystal growing system inside a pressurized container, which establishes an ambient of the most volatile component to suppress the dissociation of the crystals, resulting in mirrored rather than dull, pitted surfaces.

---

[14] The EPD is usually determined by etching the crystal surface with a molten alkali, such as KOH, and measuring the number of "pits" in the surface using an optical microscope. Each pit represents a single dislocation ending at the surface.

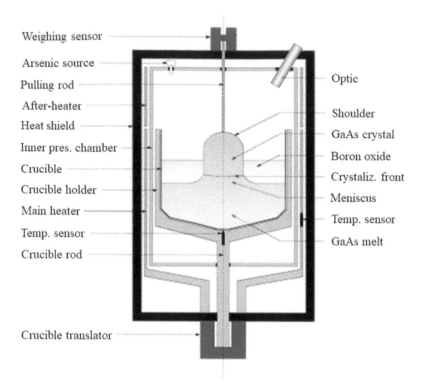

Weighing sensor

Arsenic source

Pulling rod

After-heater

Heat shield

Inner pres. chamber

Crucible

Crucible holder

Main heater

Temp. sensor

Crucible rod

Optic

Shoulder

GaAs crystal

Boron oxide

Crystaliz. front

Meniscus

Temp. sensor

GaAs melt

Crucible translator

FIGURE 4.9 An implementation of the vapor controlled Czochralski technique used for the production of GaAs (from [22]). The essential difference over the standard Czochralski method is the inner pressure chamber and an arsenic source to provide an ambient overpressure of the most volatile component.

The essential components of VCZ are shown in Fig. 4.9. In a variation on this theme, Rudolph and Kiessling [23] have proposed growing GaAs crystals by the VCZ method without a boron oxide encapsulate. Material thus grown shows markedly reduced boron content.

4.4.3 Bridgman-Stockbarger (B-S)

The second major bulk crystal growth technique is the Bridgman-Stockbarger[15] (B-S) method, which is essentially a controlled freezing process taking place under liquid-solid equilibrium conditions. The technique is an amalgam of two earlier methods. The Bridgman method [7] is characterized by the translation of a crucible containing a melt along a single axial temperature gradient in a furnace as illustrated in Fig. 4.10. The Stockbarger method [8] is a modification of Bridgman and employs a single-heat insulation buffer separating a vertical furnace into two zones, a high-temperature zone and an upper low-temperature zone. A B-S system consists of three major components: a growth chamber, temperature controllers and "boat" translation assembly. The "boat" is a crucible into which the starting material (commonly known as the "charge") is loaded with a seed crystal at one end. The boat is then heated until the charge melts and wets the seed crystal. The seed is used to crystallize the melt by slowly lowering the boat temperature starting from the seed end. The charge may be composed of either high quality polycrystalline material or carefully measured quantities of elements, which make up the desired compound. The boat is kept stationary during this process while the longitudinal furnace temperature is varied to form the crystal. Although the process can be carried out in a horizontal or vertical geometry, the easiest approach is to use a horizontal boat. However, in this case, the boule produced has a D-shaped cross-section. To produce circular cross-sections, vertical configurations have been developed for GaAs and InP.

4.4.3.1 High Pressure Bridgman (HPB)

High Pressure Bridgman (HPB) is commonly used for compounds with high melting points and disparate vapor pressures and thus is ideal for Group II-VI compounds such as CdTe and CdZnTe. In this implementation, the charge and seed are contained in a container, either a graphite crucible with tight lid or a sealed quartz ampoule. The container is also backfilled (as in the case of CdZnTe) with a high-pressure inert gas (~100 Atm) such as Ar, to suppress the loss of the most volatile component.

---

[15] Named after the American physicists P.W. Bridgman (b.1882, d.1961) and D.C. Stockbarger (b.1895, d.1952).

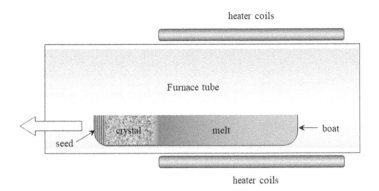

FIGURE 4.10 Schematic of the horizontal Bridgman method. The crystal is solidified by slowly withdrawing the charge from the heater.

Fig. 4.11 shows the horizontal and vertical versions of this technique. In both cases, the crystal grows from the melt by moving it along a region with a temperature gradient that extends from above to below the melting point. The growth may proceed by mechanically moving the ampoule or heating the furnace. In recent systems, the furnace consists of many heating zones with computer control of the temperature profile. The computer shifts the profile electronically so there are no moving parts within the furnace. However, the crucible, which is generally made of carbon-layered silica glass, gives rise to a number of problems: solid-liquid interface curvature, spurious nucleation of grains and twins and large thermal stresses during cooling of the crystal. Crystals grown by HPB are polycrystalline with large grains and twins. Nevertheless, the grains are large enough to obtain single crystals of volume of several cm^3. In addition, HPB material tends to suffer from macroscopic cracking, the formation of pipe defects and in the case of CdZnTe the formation of Te inclusions. For CdZnTe growth, precise control of stoichiometry can be achieved by controlling the partial pressure of its constituent elements. This is usually effected through Cd since it is has the highest vapor pressure and can thus be expected to have the greatest influence on the composition of the melt. Without such control, the melt will lose Cd and become Te rich. For commercially produced CdZnTe, a typical growth cycle lasts around 4 weeks, producing ingots up to 10 kg in weight [25].

4.4.3.2 Vertical Gradient Freeze (VGF)

The Vertical Gradient Freeze (VGF) method of Ramsperger and Melvin [26] is a derivative of vertical Bridgman. Raw polycrystalline material is molten in a preshaped crucible and directionally solidified from a single-crystal seed at the bottom of the crucible (Fig. 4.12). This is achieved by lowering the temperature while maintaining a positive temperature gradient in the melt. As the crystal growth is usually performed in multi-zone furnaces, there are many degrees of freedom and numerical modelling is required for the optimization of the furnace and the growth process. The main advantages of the

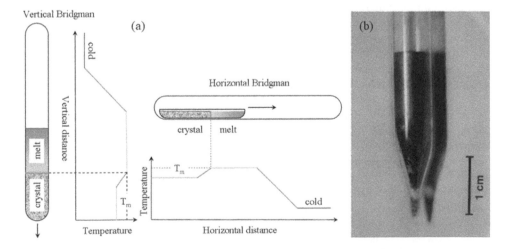

FIGURE 4.11 Left, crystal growth by high pressure vertical and horizontal Bridgman showing the temperature profiles across the charge (modified from [24]). T_m is the melting temperature. Right, a sealed vertical Bridgman charge.

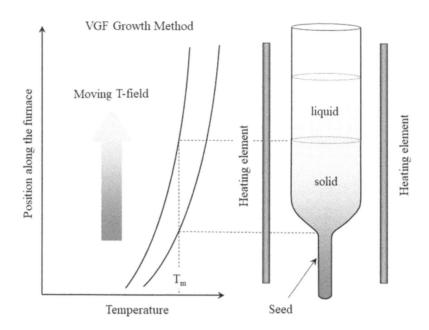

FIGURE 4.12 Essential elements of the Vertical Freeze Growth (VGF) method. Left: the furnace temperature profile. Right: the material state along the charge. T_m is the melting temperature.

VGF process include its scalability, low stress, high mechanical strength and a defect rate that is orders of magnitude lower than conventional compound semiconductor crystal growth techniques. For example, for high quality GaAs, EPD values of ~30 cm^{-2} can be achieved, as opposed to values of over 500 cm^{-2} for LEC growth.

In addition to producing high-purity bulk crystals, the techniques discussed previously are also used to produce crystals with specified electrical properties, such as high-resistivity materials or *n*- or *p*-type materials. For example, high resistivity or semi-insulating (SI) substrates are extremely useful in device isolation and for high-speed devices. However, it is difficult to produce high-resistivity Si substrates by bulk crystal growth, and resistivities are usually <10 Ω cm. Carrier trapping impurities such as chromium and iron can be added to the melt to produce material with resistivities of ~ 10^3 Ω cm.

4.4.4 Travelling Molten Zone (TMZ) or Heater Method (THM)

The conventional technique used for compounding Group II-VI semiconductors such as CdTe consists of unidirectional freezing of a molten charge within a Bridgman or modified Bridgman reactor. The Bridgman process, being a melt technique, has two inherent drawbacks: (i) the high compounding temperature can lead to material contamination from the crucible walls, and (ii) the potential exists for explosions associated with the highly exothermic synthesis reaction and the possibility of trapping uncompounded Cd within a "shell" of compounded material. This limits the size of the charges that can be safely processed in a single batch and thus increases production costs. Alternate growth techniques like zone refining are not used because of the volatility of Group II and Group VI elements as well as the compounds themselves. Hence, there has been little development of conventional zone-refining techniques for these compounds. However, a related technique called the THM or sometimes the travelling solvent method has attracted much interest. It is a low-temperature/pressure alternative technique based on solution growth and is the method of choice for the growth of CdTe [27].

In the THM synthesis of cadmium telluride, a saturated solution of cadmium in tellurium is held molten by a narrow heater and swept vertically through the length of the charge consisting of the precursor elements. The technique is illustrated in Fig. 4.13. The movement of the molten zone through the charge leads to progressive dissolution of the charge at the top liquid-to-solid interface, simultaneously leading to synthesis and growth of the compound at the bottom interface. This has two advantages. Firstly, it reduces the temperature of crystallization significantly below the melting point of the compound thus markedly reducing the vapor pressure of the components of the compound, effectively eliminating evaporation. Secondly, compared to other melt growth methods, impurity levels are low due to the refining effect of the travelling zone of liquid and because of the reduced growth temperature, it is possible to eliminate sub-grain boundaries. With this process, a constant, controlled but slow growth is achieved. However, the quality of grown crystals is very sensitive to the relative movement of the temperature profile that determines the growth rate. The THM has a number of advantages over melt techniques, namely, easy growth of binary and ternary alloys, less thermal stresses, uniform crystal composition [28] and a low temperature and pressure environment compared to HPB. In addition, whereas CdTe synthesis with conventional Bridgman is typically limited to around 1.5 kg, mainly because of crucible limitations, CdTe ingots produced by THM can

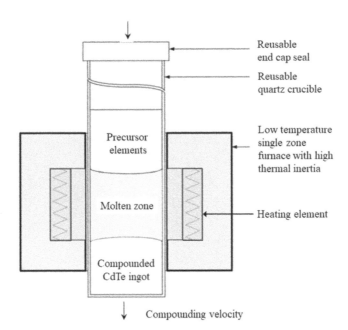

FIGURE 4.13 Schematic of the Travelling Heater Method (THM) (from Audet *et al.* [28], ©2007 IEEE), in which a zone of hot liquid travels slowly (a few millimetres per day) along a batch of previously synthesized polycrystalline material. A particular advantage of the method is that the refining effect of the travelling zone results in crystals of very high purity.

easily exceed 3.5 kg. This offers significant cost advantages, as well as benefits in terms of reduced material contamination (large ingots have a smaller surface-to-volume ratio, so less material is in contact with the crucible). Its main disadvantage is the very slow deposition rate of ~5 mm/day.

4.4.5 Float-Zone Growth Technique (FZ)

Since one of the major sources of crystal defects and contamination in conventional growth techniques (*e.g.*, CZ, LEC) is the contact of the melt with the crucible, a major improvement in crystal quality (impurity introduction, formation of dislocations and residual stress) can be achieved by dispensing with the crucible altogether. The float zone technique is one such method and is widely used for the production of high-purity silicon [29]. The technique is illustrated in Fig. 4.14 (left). A polysilicon rod with a seed crystal at the bottom is held in a vertical position. The rod is then lowered through an electromagnetic coil. An RF field is used to produce a local melted zone in the polycrystalline rod, which is dragged from the seed end to the holder, solidifying into crystalline material in its wake – as illustrated in Fig. 4.14 (right). The small molten segment remains fixed in place between the solid portions of the rod due to surface tension and has the added advantage of carrying impurities away with it, taking advantage of the low segregation coefficients of many impurities (*i.e.*, impurities are more soluble in the melt than the crystal). In fact, the method is a modification of the zone-refining technique originally developed by Pfann [12] for purifying germanium (see Section 4.2.2.3) but is now widely used to produce high-quality silicon for power devices and detector applications. The dimensions of ingots produced are generally less than 15 cm due to surface tension limitations during growth. The relative advantages and disadvantages of this technique are listed in Table 4.1.

4.4.6 Vapor Phase Growth (VPG)

Vapor growth of bulk single crystals is a method with great potential, but it has only been effectively exploited for a few materials, specifically those that decompose prior to melting or react with crucibles or solutions. Vapor phase growth (VPG) has several advantages over other techniques. For example, growth is usually carried out in closed ampoules, which is attractive in view of its experimental simplicity and minimal needs for complex process control. It addition, it can be performed at a lower temperature, which means that vapor-grown crystals can have considerably lower defect densities than those grown with conventional high-temperature methods. This has become particularly important in view of the ever-increasing demand for high-quality substrates for thin film applications. The main disadvantage of VPG compared to other techniques is that growth rates are low and the grown crystals small.

Crystal growth from the vapor phase can be carried out either using CVD or PVD techniques. CVD is based on chemical transport reactions occurring in a closed ampoule across two different temperature zones. A schematic of such a

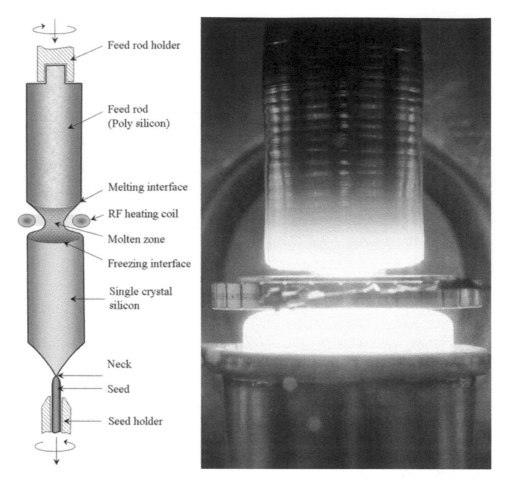

FIGURE 4.14 Left: Schematic of the Float-Zone growth technique (from [30]). Right: photograph of the growth area (courtesy: Topsil Semiconductor Materials A/S).

TABLE 4.1

Advantages and disadvantages of the Float-Zone crystal growth technique.

Advantages	Disadvantages
No crucible contamination	Expensive (driven by feedstock costs)
Shape control (*via* the liquid-solid interface)	Incongruent melting
In-built impurity control	Sensitivity of control parameters
Vapour pressure control	Relatively poor radial doping profiles
High growth rates (0.5–50 mm/min)	Boule diameter limited to ~15 cm

system is shown in Fig. 4.15(a). PVT on the other hand does not depend on a transport agent but relies on the dissociative sublimation of compounds. For binary materials, the PVT method takes advantage of the volatility of both compounds, especially for II-VI materials in which high volatility coupled with high melting points makes melt growth particularly difficult. In PVT, an ampoule containing polycrystalline material is heated to a temperature that causes the compound to sublime at a rate conducive to crystal growth. The ampoule is typically placed in a furnace having a temperature gradient over the length of the ampoule so that the polycrystalline source material sublimes at the end with the higher temperature. A schematic of such a system is shown in Fig. 4.15(b). In order to control deviations from stoichiometry a reservoir is sometimes used. Impurities with a higher vapor pressure will condense in the reservoir while those with a lower vapor pressure will remain in the source material. Thus, PVT also acts as a purification process. Typically, growth is carried out at temperatures lower than the melting point, which gives lower defect densities, in terms of voids and inclusions compared to melt-grown material. To date, the method is used to produce large crystals of HgI_2 and PbI_2. VPG has also been used to produce GaN, for which it is not practical to pull a crystal from the melt because the equilibrium N_2 pressure over congruent molten GaN is greater than 6 GP.

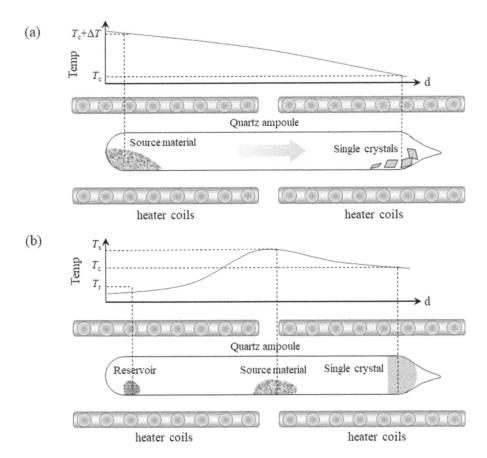

FIGURE 4.15 (a) Schematic of a sealed Chemical Vapor Deposition (CVD) system for the bulk growth of single crystals. (b) Schematic of a Physical Vapor Deposition (PVD) system for the bulk growth of single crystals.

TABLE 4.2

Summary of the properties of the two most common bulk crystal growth techniques – Czochralski and Bridgman-Stockbarger. For comparison, we also include details for the THM, since this has become the preferred growth method for some II-VI compounds.

Parameter	Czochralski	Bridgman-Stockbarger	THM
Commonly used for	Si, Ge, III-V semiconductors	III-V and II-VI semiconductors	Mainly II-VI semiconductors
Precursor	Uses seed crystal	Uses seed crystal	Solution growth
Crystal growth	Vertical	Vertical or horizontal	Vertical or horizontal
Crucible	Yes	Yes	Yes
Growth speed	1–2 mm/min	1 mm/hr	5 mm/day
Contamination	Oxygen contamination problems	Contamination from crucible	Low, due to the refining effect of the travelling zone of liquid.
Comments	Axial resistivity is poor. Long heat up/cool down times	Precise temperature gradient required	Low growth temperature Low defect and contamination
	For materials with k <1	Crystal perfection better than seed	Difficult to obtain near
	Impurities stay in liquid	Low k, impurities stay in liquid	Stoichiometric compounds

4.4.7 Discussion

In Table 4.2 we summarize the properties of growth techniques based on the two most common bulk crystal growth methods, Czochralski and Bridgman-Stockbarger. For completeness, we also include details for the THM, since this has become the preferred growth method for some II-VI compounds. In terms of specific growth systems, a number of comparative studies have been carried out with a view to maximizing ingot size while simultaneously minimizing dislocation defect densities. For example, in a comparison of the LEC, VCZ and VGF techniques for growing InP, Kawarabayashi *et al.* [31], concluded that overall, the VCZ technique was the most promising method for growing large diameter crystals

TABLE 4.3

Comparison of properties of large diameter InP crystal growth by conventional LEC, VCZ and the VGF methods (taken from Kawarabayashi *et al.* [31], ©1994 IEEE).

Parameter	Conventional LEC	VCZ	VGF
Size	O	O	Δ
EPD	×	O	O
Technology maturity	O	Δ	×

Key: O ≡ superior, Δ ≡ medium, × ≡ inferior

(>3 inches) with low dislocation densities. In fact, the EPDs were over an order of magnitude lower than those measured with conventional LEC. Their conclusions are summarized in Table 4.3.

4.5 Epitaxy

The word epitaxy is derived from the Greek words επι (*epi* meaning *"on"*) and ταξις (*taxis* meaning *"order"*). Thus, epitaxy refers to the ordered growth of one crystal upon another crystal. Epitaxial films may be grown from gaseous or liquid precursors. There are three main modes of epitaxial growth: (a) monolayer, (b) nucleated and (c) nucleation followed by monolayer. Monolayer growth occurs when the deposited atoms are more strongly bound to the substrate than they are to each other. The atoms aggregate to form monolayer islands of deposit, which enlarge, and eventually a complete monolayer coverage has taken place. The process is repeated for subsequent layer growth. In case of nucleated growth, the initial atoms deposited aggregate as small three-dimensional (3D) islands that increase in size as further deposition continues until they touch and intergrow to form a continuous film. This mode is favoured where the force of attraction between the deposited atoms is greater than that between them and the substrate. In the final mode, growth starts with the formation of a single or few monolayers on the substrate followed by subsequent nucleation of 3D islands on top of these monolayers.

4.5.1 Substrates

In order to structure the crystal correctly, the films must be grown on a substrate of material that is itself as nearly a structurally perfect crystal as possible. Epitaxy will then occur in such a way that the total energy at a substrate-film interface is minimal. Because the substrate acts as a seed crystal, the deposited film will attempt to replicate the lattice structure and orientation of the substrate, provided the shape and size of the lattices are not too dissimilar. Obviously, the ideal substrate is a highly crystallographic form of crystal of the material to be grown – the better the quality of the substrate, the better the quality of the resultant epitaxial film.

If a film is deposited on a substrate of the same composition, the process is called homoepitaxy; otherwise, it is called heteroepitaxy. For the elemental semiconductors Si and Ge, bulk crystal growth techniques are highly mature and high-quality substrates are readily available. For most compound semiconductors (with the possible exception of InP), they are not; thus it is necessary to grow layers on substrates of another crystalline material, hence heteroepitaxy. This is not ideal because, even if the substrate has the same crystal structure, the lattice spacing can be quite different making the formation of an interlinking boundary energetically unfavourable. The growing layer then acquires precisely the same structure as that of the substrate. This phenomenon is called pseudomorphism. As a result, a stressed film with its own distinctive structure grows. Unless the mismatch between crystals is elastically strained, the net result will be a bowing of the wafer, which can present complications during wafer processing. In Fig. 4.16 we show the effects of strain for two cases: (a) when the lattice parameter for the substrate is less than that of the film, then the film will be in compression and the wafer will bow downwards and (b) if the lattice parameter for the substrate is greater than that of the film, then the film is under tension and the film will bow upwards.

As film thickness increases in a mismatched system, strain will rise until it is greater than the energy required to create a dislocation, whereupon the crystal will partially or totally relax by the formation of dislocations. Further increases in film thickness will result in a film consisting of regions of relatively good fit separated by a series of misfit dislocations. In the worst cases, the epi-layer will delaminate from the substrate. In practice, structurally stable interfaces can only be formed if the change in lattice constant is kept to less than around 15% (a consequence of the Hume-Rothery rules). If this is the case, it is possible to grow highly crystalline films. For example, the nitrides GaN, AlN and InN are grown on SiC or sapphire (Al_2O_3) substrates and AlGaInP is grown on a GaAs substrate. Thus, it is clear that semiconductor technology is critically dependent on the availability of high-quality substrates with as large a diameter as possible. In summary, the most cost-effective way to produce large-area crystalline substrates is to use bulk crystal growth techniques because deposition rates are much higher than for epitaxial techniques and many substrate wafers can be cut from a single melt-grown ingot.

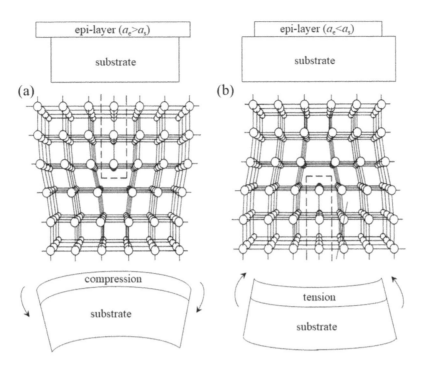

FIGURE 4.16 Examples of lattice mismatch between the substrate and the grown film showing the effects of bowing, which will occur if one or both crystals are inelastically strained. If the lattice parameter for the substrate a_s is less than that of the epi film case, a_e (a), then the film will be in compression and the wafer will bow downwards. If the converse is true (b), then the film is under tension and the film will bow upwards.

4.5.2 Strain and Electronic Properties

Strain in a crystalline semiconductor creates a proportional distortion in its key material properties. Specifically, the elastic strain energy stored in the material alters the band structure affecting its electronic properties. This includes the width and shape of the energy bands and the effective masses of the carriers. In practice, a lattice mismatch of less than 2% can modify the band structure by 100 meV or more. Strain can also have practical benefits. For example, the effective mass of an electron in a strained region is reduced, increasing its mobility. Inducing strain into the charge control region of, say, a transistor, can result in faster switching times. In this way, lattice-mismatched heterostructures containing strained films to enhance performance are rapidly growing in importance in semiconductor device technology.[16]

4.5.3 Lattice Matching

Unfortunately, it is not possible to produce high-quality substrates for most compound semiconductors, and although an iterative approach may work, it is usually not cost effective, since the original substrate is lost after each growth. In addition, the requirement that the lattice parameters differ by no more than a few percent imposes a significant constraint when forming higher-order alloys, since changing the composition of an alloy changes its average lattice constant because of the different atomic bonding radii of the constituent elements. This problem is best appreciated with reference to a graph of the energy gap versus lattice constant.

In Fig. 4.17 we plot bandgap energy versus the cubic lattice constant for the most common Group III-V binary compounds. For a possible range of ternary alloy systems, a solid line is generated between the starting binary materials. In the case of a quaternary compound, the boundary is laid out by four intersecting lines. Thus, we see for the GaAs-AlAs system, there is little variation in lattice constant,[17] which allows the continuous formation of solid solutions of Al in GaAs over the full range of Al substitution. However, for the GaAs-InAs and GaAs-GaSb systems, the incorporation of In to GaAs (to form the $Ga_{(1-x)}In_{(x)}As$ alloy) or Sb to GaAs (to form the $GaAs_{(1-y)}Sb_{(y)}$ ternary alloy) results in major shifts of the average lattice constant, and stable alloys cannot be produced across the entire range of In or Sb substitution.

---

[16] In the semiconductor industry, the straining of a silicon crystal is a well-known technique to increase charge carrier mobility, enhancing device performance. To be most effective, the strain in a silicon channel of a Complementary Metal Oxide Semiconductor (CMOS) device should be compressive to improve the hole conduction of p-type Metal Oxide Semiconductor (PMOS) transistors and tensile to improve the electron conduction of transistors n-type Metal Oxide Semiconductor (NMOS).

[17] The actual mismatch between GaAs and AlAs is 0.127%.

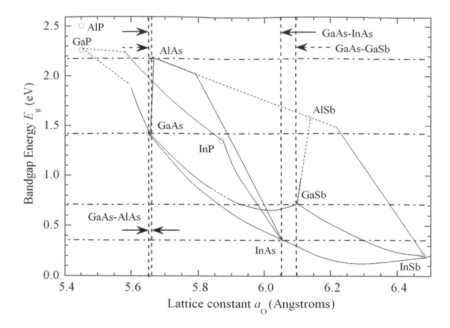

FIGURE 4.17 Bandgap energy, E_g, versus lattice constant, a_o, for the most common Group III-V ternary alloys at room temperature and their relationship to the participating binary (from [32]). The solid interconnect lines represent direct bandgap compounds while the dotted interconnect lines represent indirect bandgap compounds.

Fortunately, lattice-mismatched layers can be grown pseudomorphically (that is, defect-free) on a substrate with a different lattice constant if the epi-layer thickness is kept below a certain critical value, h_c, given by

$$h_c = a_e/14f \qquad (2)$$

where f is the lattice misfit[18] given by $(a_e-a_s)/a_e$ and a_e and a_s are the lattice constants of the epi-layer and substrate, respectively. Below the critical value, the lattice is elastically strained and its structure preserved. Above this thickness, plastic deformation occurs, leading to displacement of epi-layer atoms.

For completeness, in Fig. 4.18, we plot bandgap energy versus the cubic lattice constant for the most common Group II-VI binary compounds.

4.5.4 Van der Waals Epitaxy (VDWE)

For lattice mismatched materials, Koma *et al.* [33,34] have pointed out that the lattice-matching condition given by Eq. (2) can be significantly relaxed if the interlayer bonding at the interface proceeds *via* weak van der Waals interactions. The technique is known as Van der Waals epitaxy (VDWE) and is now being widely exploited to produce 2D materials and non-planar nanostructures by a process known as atomic scale "Lego" [35]. A van der Waal interface is formed when a layered material is grown on a cleaved face of another layered material. By layered, we mean materials whose crystal structure consists of planes of single-atom or polyhedral-thick layers of atoms with covalent or ionic bonding within the layer and van der Waals bonding between layers. Many layered materials are suitable for this purpose, for example GaSe, MoS_2 and $NbSe_2$ [36].

In the case of conventional three-dimensional epitaxy using grossly mismatched materials, some bonds will be unfulfilled, creating dangling bonds as illustrated in Fig. 4.18(a). After a few monolayers, the grown material separates into islands, disrupting uniform growth. Layered materials, on the other hand, are easily cleaved along the layers without producing dangling bonds on their cleaved surfaces. The epitaxial growth of another layered material onto these surfaces proceeds by van der Waals interactions, resulting in good heteroepitaxial growth – even when there is a large lattice mismatch between the grown and the substrate materials (Fig. 4.19(b)). In fact, Ohuchi *et al.* [37] have shown that very high-quality thin films can be produced with lattice mismatches of 10% – which should be compared to a few percent for conventional epitaxy. Initially, VDWE was limited to the growth of a layered material onto a layered substrate. Later, the technique was extended to the heteroepitaxial growth of layered materials onto three-dimensional substrates that have active dangling

---

[18] also known as the strain

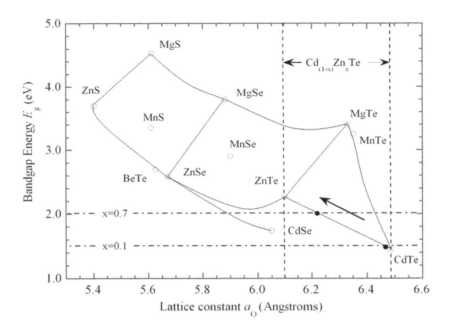

FIGURE 4.18 Bandgap energy, E_g, of Group II-VI compounds as a function of lattice constant a_o (from [32]). The dotted lines illustrate how altering the zinc fraction, x, in $Cd_{(1-x)}Zn_xTe$ alters the bandgap energy. Two cases are shown: $x = 0.1$, which provides optimum energy resolution at T = 243 K, and $x = 0.7$, which provides optimum energy resolution at room temperature.

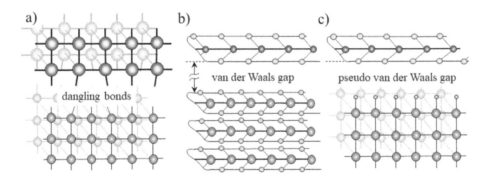

FIGURE 4.19 Van der Waals epitaxy (VDWE). (a) Conventional epitaxy of 3D materials with a larger lattice mismatch. (b) Van der Waals epitaxy between two mismatched layered materials. (c) The epitaxial growth of a mismatched layered material on a 3D substrate is achieved by regular termination of the surface dangling bonds by capping atoms.

bonds on their clean surfaces [38]. This is achieved by terminating the active dangling bonds with suitable "capping" atoms, thus rendering the surface inactive. The surface now behaves like the cleaved surface of a layered material, allowing a film of a layered material to be heteroepitaxially grown by van der Waals-like weak forces, as shown in Fig. 4.19(c). Similarly, Alaskar *et al.* [39] pointed out that VDWE could provide a potential route to heteroepitaxy between lattice mismatched 3D materials using a 2D buffer layered material, for example, the van der Waals epitaxial growth of GaAs on Si using a graphene buffer layer.

4.5.5 Bandgap Engineering

Unlike Si and Ge whose electronic and chemical properties are "fixed", those of compound semiconductors can be modified by bandgap or, alternatively, wavelength engineering [40]. In this case, the ability to tailor the bandgap to a particular wavelength has profound optoelectronic applications. A specific example of how this is useful for radiation detector applications is CdZnTe. In Fig. 4.18, we show the phase diagram for a range of Group II-VI materials, including CdTe and ZnTe. By altering the zinc fraction, x, in $Cd_{(1-x)}Zn_xTe$, the range of possible alloys moves along the line between CdTe and ZnTe, making it possible to optimize the noise performance of a CdZnTe radiation detector for a given operating temperature. Specifically, increasing the zinc fraction, x, increases the bandgap energy E_g, which can be described empirically [41] by

$$E_g(x) = 1.510 + 0.606x + 0.139x^2 \quad \text{eV.} \tag{3}$$

While an increase in E_g increases Fano noise due to carrier generation statistics, it simultaneously reduces shot noise due to thermal leakage currents. A trade-off can lead to a noise minimum at a given operating temperature and therefore an optimization of the spectral performance [42]. For example, a zinc fraction of 10% is optimum for operation at -30°C, whereas 70% is optimum at room temperature. Alloying of CdTe with Zn or ZnTe has additional benefits in that it increases the energy of defect formation [43] and mechanically strengthens the lattice resulting in a lowering of defect densities [44].

4.5.6 Semiconductor Structures

For highly crystalline material there are two basic structures considered in epitaxy when producing devices – homostructures and heterostructures. Both structures are formed by one or more interfaces or junctions and are essential elements in virtually all electronic devices. A homostructure is a semiconductor interface (known as a homojunction) that occurs between layers of the same semiconductor material (*i.e.*, materials of equal bandgap) but typically with different doping, for example a p-n junction. Such a junction provides control over the sign and density of charge carriers and when coupled with non-equilibrium charge injection forms the basis of modern electronics. A heterostructure on the other hand is built up of a combination of one or more heterojunctions of different semiconductors with different bandgaps and different electron affinities. This creates an alternating variation of the potential seen by electrons in the conduction band and holes in the valence band, allowing a high degree of control over the physical and transport properties of the junction, including the bandgap energy, refractive index, carrier mass and mobility. For device fabrication, double heterostructures are most often used, in which a semiconductor with small bandgap is sandwiched between two slices of a semiconductor with a larger bandgap. The smaller energy gap material forms energy discontinuities at the boundaries, confining the electrons and holes to the semiconductor with the smaller energy gap. If the width of this region is of the order of the de Broglie wavelength, quantum confinement occurs. The energies of particles in this layer are no longer continuous but are discrete and the double heterostructure now forms a quantum well bounded by potential barriers at the interfaces. Note that the notion of wells and barriers is interchanged for electrons and holes because they have opposite energies. Thus a material may act as a well for conduction electrons but as a barrier for holes. This kind of structure is routinely used for optoelectronic applications, for example, in solid state lasers, where recombination, light emission and population inversion coincide and occur entirely in the sandwiched layer.

A logical evolution of heterostructures is the superlattice (SL), which consists of a large number of periodically repeated heterostructures, grown through deposition of layers of different semiconductor materials on top of each other in the growth direction representing a repeated sequence of quantum barriers and wells (for a review see [45]). Note that a superlattice can also refer to a lower-dimensional structure such as an array of quantum dots or quantum wires. It is customary to classify SLs (or more generally multiple quantum well structures) according to the confinement energy schemes of their electrons and holes, which in turn are governed by the relative bandgaps of the semiconductors. These confinement schemes are usually labelled types I, II or III. In type I SLs the electrons and holes are both confined within the same layer forming the well. In type II the electrons and holes are confined in different layers. Type III SLs involve a semi-metal material, such as HgTe/CdTe and are similar to Type I in that the electrons and holes are confined to the same layer but are unique because the conduction and valence bands are inverted in the well material. In a true superlattice, the semiconductor layers separating the wells are also very thin so that the adjacent electron and hole wave functions overlap leading to strong inter-well coupling. The resultant tunneling currents can be exploited to produce various devices, such as tunnel diodes, resonant tunneling transistors and quantum cascade infrared lasers. Because of the thinness of the layers and the precision required to produce atomically well-defined structures, Metal Organic Chemical Vapor Deposition or Molecular Beam Epitaxy (see Sections 4.6.3.2 and 4.6.4.3) are usually used to fabricate devices.

4.6 Film Growth Techniques

The most commonly used epitaxial growth techniques are discussed in the following sections. These are Solid Phase Epitaxy (SPE), Liquid Phase Epitaxy (LPE), Vapor Phase Epitaxy (VPE), Metal Organic Chemical Vapor Deposition (MOCVD) also sometimes referred to as MOVPE (Metal Organic Vapor Phase Epitaxy) and Molecular Beam Epitaxy (MBE). Strictly, MOCVD and MPE are VPE derivatives. Whilst, epitaxial silicon is usually grown by VPE (a modification of CVD), LPE and MBE are used extensively for growing Group III-V compound semiconductors.

4.6.1 Solid Phase Epitaxy (SPE)

SPE is a growth technique in which a metastable amorphous layer is deposited onto a single crystal substrate that serves as a template for crystal growth [46]. The deposition of the amorphous layer may be carried out using a range of techniques, including evaporation, sputtering and CVD. The amorphous-to-crystal transformation occurs solely in the solid phase and may be induced by heating at ambient pressure. This encourages the reordering of atoms at the crystalline/amorphous interface with the resulting propagation of the interface towards the surface. Epitaxial growth then proceeds on a layer-by-layer basis. The amorphous layer and the substrate need not have the same elemental composition – provided that the lattices are well matched, a single crystal layer can be grown and in fact SPE is commonly used to grow buffer layers. However, it should be noted that SPE is a homoepitaxial technique not suitable for compounds because of the added complication that the cations and anions have to be located at specific lattice sites – a problem that does not exist for the elemental semiconductors. Although SPE can produce high-quality films, it is more commonly used as a technological step in device fabrication, for example: (1) producing complex planar Si structures, (2) repairing thin films amorphodized following ion implantation of electrically active impurities, (3) preparation of thin semiconductor layers with dopant concentrations in excess of solubility limits and (4) the growth of buffer layers on single crystalline substrates for improving heteroepitaxy of lattice mismatched heterostructures.

4.6.2 Liquid Phase Epitaxy (LPE)

LPE is a relatively simple epitaxial growth technique that evolved in the early 1960s and was widely used until the 1970s when it gradually gave way to more sophisticated techniques, such as MOCVD and MBE (for a review, see Mauk [47]). It is much less expensive than the latter techniques, and while it can produce very thin, uniform and high quality layers, it offers less control in interface abruptness when growing heterostructures. The method is commonly used for the growth of compound semiconductors, particularly ternary and quaternary III-V compounds on GaAs and InP substrates.

LPE refers to the growth of semiconductor crystals in which the growth constituents are transferred to the growing interface through a liquid medium. The difference between LPE and melt growth techniques like CZ is that in the latter growth occurs at the melting temperature of the solid, whereas in LPE growth occurs at a much lower temperature. This is made possible by the fact that a mixture of a semiconductor and a second element generally has a lower melting point than the pure semiconductor alone. For III-V materials, it is usually possible to use the Group III metal itself, as the solvent for the Group V element, which has the added value of avoiding the contamination inherent in a foreign solvent. Thus, taking GaAs as an example, the melting point of a mixture of GaAs and Ga is considerably lower than 1,238 °C, the melting point of pure GaAs. The actual melting point of the mixture is determined by the proportion of the constituents Ga and GaAs and can occur at several hundreds of °C below that of the binary solid. Growth begins by placing a GaAs seed crystal into the Ga-GaAs solution. As the solution cools, a single crystal GaAs begins to grow on the seed leaving an increasingly Ga-rich liquid mixture with an even lower melting point. Further cooling causes more GaAs to crystallize on the seed. The principal components of an LPE system are

- a furnace with high thermal mass and lateral temperature uniformity,
- a system to control the temperature of the furnace and hence that of the melts and substrate,
- a reservoir, or boat, to contain the melts and to separate them – from each other and the substrate,
- a means of bringing the melt solution and the substrate into contact at the start of the growth process and of separating the two at its conclusion and
- a reducing atmosphere to prevent the formation of oxides during growth.

4.6.2.1 Tipping Furnace Method

The earliest and simplest form of equipment to achieve LPE is the so-called tipping furnace method of Nelson [48] illustrated in Fig. 4.20(a). In this technique, a graphite boat contains a substrate wafer clamped at one end and a supersaturated melt solution at the other end. In the case of GaAs LPE, the substrate would be a GaAs wafer and the solution would consist of a Ga-GaAs mixture. The boat is inserted into a furnace in a tilted position with the melt solution at the lower end as shown in Fig. 4.20(a). As the system is heated above ~800°C, the GaAs source material begins to dissolve in the Ga solvent to saturate the liquid with As. The furnace is then tipped so that the liquid now flows under gravity over the substrate. Cooling the system causes the entire liquid to become supersaturated in As, causing epitaxial GaAs to grow on the substrate, which in turn allows the liquid near the interface to return to equilibrium. To terminate growth, the system is tilted back to its original position to allow the liquid to drain away. The advantages of the tipping technique are that it is relatively simple and can be used in closed systems to avoid vapor loss. This main disadvantages are that it lacks thickness control, de-wetting can occur at the solid-liquid interface and it is difficult to design a system that can grow more than one layer.

4.6.2.2 Vertical Dipping System

Another simple form of LPE is the dipping method first reported by Linares *et al.* [49] for the growth of garnet films. The apparatus is illustrated in Fig. 4.20(b) and consists of a static or rotating substrate immersed in a supersaturated melt

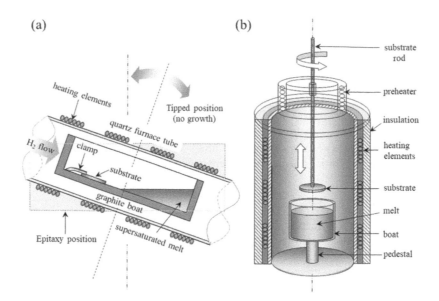

FIGURE 4.20 (a) The tipping furnace LPE system of Nelson [48]. A graphite boat contains a substrate clamped at one end and a supersaturated melt at the other end. The boat is tipped towards the left so that that the liquid now flows under gravity over the substrate, causing epitaxial growth on the substrate. The system is tilted back to its original position to terminate growth. (b) A dipping LPE system, in which a substrate is immersed into a bath of supersaturated melt solution until the desired film thickness is reached. The sample is then withdrawn from the solution and the remnants removed, by either gravity or spin-down rotation.

solution for a fixed time until the desired film thickness is reached. The sample is then withdrawn from the solution and the remnants removed, by either gravity or spin-down rotation. In this way, smooth films as thin as 1 µm can be grown. The advantages of dipping are low cost, high throughput and no necessary movement of the crucible. The disadvantages are that it is difficult to control the thickness of thin layers or grow double layers.

4.6.2.3 Sliding Boat Method

The dominant technique for LPE growth in use today is the sliding boat method developed by Panish *et al.* [50]; this has emerged as the preferred method for producing complex heterostructures. It is more complex than a simple dipping or tipping system but allows for multiple growth solutions and thus multilayer structures. The practical implementation of this technique is illustrated in Fig. 4.21. The device works as follows. The boat consists of an upper fixed portion with a number of reservoirs, which contain the melts and a movable slider located below containing a recess for the substrate. Both the boat and sliding system are usually manufactured from high purity graphite due to its ease of machining and its compatibility with corrosive liquids, such as Ga and As at elevated temperatures. The slider is pulled in sequence under several different melts allowing the growth of multiple layer structures. The liquid may also contain dopants that are to be introduced into the crystal. Growth is terminated by removing the substrate from beneath the final melt. Purified hydrogen is passed through the reactor to prevent ingress of water and air and to purge vapors from the melts. Even though the sliding boat method is generally used for multi-layer growth, for single melt regimes it has an advantage over a tipping system as the empty bins are useful in removing excess melt during wipe-off (termination of growth) after growth. Overall, its strengths are that it is a simple, low-cost technique with a high throughput and that because it is very close to being an equilibrium growth technique, it produces material with a very low native defect density. Additionally, it uses no toxic gasses, but uses easily handled solids. Its disadvantages are that it is a small-scale process with poor thickness uniformity and high surface roughness. Since the technique is close to equilibrium, careful management of the furnace temperature is essential since this controls the supersaturation of the melts. Too high a temperature results in the substrate or previously grown layers dissolving into the melt rather than the growth of further layers. In addition, the high growth rates (0.1 to 1 µm/minute) prevent the growth of multilayer structures with atomically abrupt interfaces. Nevertheless, heterostructures with interfaces graded over 10–20 Å can be grown, which is adequate for many applications. Its widest use has been in the growth of III-V materials for optoelectronic applications (*i.e.*, tuneable lasers, LEDs, photodetectors and Gunn diodes) and accounted for ~50% of the commercial epitaxy market in 2000.

The main limitations of the sliding boat LPE are actually imposed by geometry and specifically the gap between the sliding parts (typically, in the range 25 to 100 µm). If the gap between the boat and the slider is too wide, after-growth mixing with next solution can occur – this then limits the maximum thickness of the layer that can be grown (or alternately the maximum number of layers). On the other hand, if the gap is too narrow, scratching of LPE-grown layers will occur. This then limits

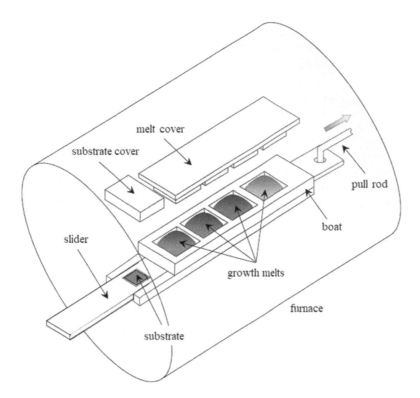

FIGURE 4.21 A schematic of a sliding "boat" LPE growth system, suitable for growing heterostructures. The substrate is placed in a recess pocket in the slider, such that slider, boat and top covers fit intimately. The melts are contained within wells in the boat. The slider (and therefore the substrate) is then pulled in sequence under several different melts to grow multiple layer structures.

the minimum layer thickness that can be grown. A slider-free growth system has been developed by Scheel [51], which allows LPE deposition of a large number of layers without the disadvantages arising from sliding container parts. In essence, solutions of different compositions are arranged in separate chambers of a helical screw-like inner graphite frame contained within a cylindrical outer chamber. A number of substrates are fixed to the inner wall of the outer chamber. By rotating the frame, the solutions can pass over the substrates without intermixing, and by controlled cooling, epitaxial layers are deposited onto the substrates. Scheel [51] reports that up to 15 layers of III-V compounds were produced in a double-screw device, with thicknesses between 0.1 and 10 μm and a thickness reproducibility of about 10%. In Table 4.4 we list strengths and weakness of the three main LPE growth techniques.

4.6.3 Vapor Phase Epitaxy (VPE)

A large class of epitaxial techniques relies on delivering the components that form a crystal from a gaseous environment. If the atoms or molecules emerging from the vapor can be deposited on the substrate in an ordered manner, epitaxial crystal growth can occur. There are two main delivery systems – CVD and PVD. The essential difference between the two is that

TABLE 4.4

Comparison of the three main LPE growth techniques; tipping, dipping and sliding boat.

Technique	Advantages	Disadvantages
Tipping	Simple Thickness control Closed system	Relatively high cost Small-scale production
Dipping	Simple Low cost High throughput Thick layer growth possible	Can't grow double layers Control of thin layers
Sliding boat	Thickness control Multiple layer growth possible Good crystal quality	High cost Thin layer thickness limited by "boat scratching" Difficult to grow thick layers (melt mixing)

CVD is based on chemical processes in gaseous phase and the substrate surface, whereas PVD is based on evaporation and condensation or the collisional impact of solid material on the substrate. Both are non-equilibrium[19] growth techniques requiring a high degree of temperature control and are described in the following sections. For a review of CVD see Jones and Hitchman [52] and for PVD, see Mattox [53].

4.6.3.1 Chemical Vapor Deposition (CVD)

CVD is a generic name for a suite of processes that include plasma enhanced CVD (PECVD), atomic layer deposition (ALD), metal-organic chemical vapor deposition (MOCVD) and a range of thermal processes. Typically in CVD, a heated substrate (wafer) is exposed to one or more volatile precursors, which then chemically react and/or decompose on the substrate surface to produce a uniform epitaxial film. This is what distinguishes CVD from other PVD processes, such as evaporation and reactive sputtering, in which vaporized material is deposited directly onto the substrate by adsorption. A large range of CVD systems have been developed, however they are all basically variations on a theme of a few common components. These are:

(1) a reactor chamber in which deposition takes place

(2) a gas delivery system that supplies the precursors to the reactor chamber. The precursors must be volatile but at the same time stable enough to be able to be delivered to the reactor.

(3) a mechanism for introducing and removing substrates

(4) an energy source to heat the precursors/substrate to a high enough temperature so they react/decompose

(5) an exhaust system for the removal of all other gaseous species other than those required for reaction/deposition, including the bi-products of the reaction

Generally, CVD reactors fall into two categories – hot wall and cold wall. A hot-wall reactor is essentially an isothermal furnace, which is often heated by external resistance elements. This means that the reactor walls and substrate are both at the same temperature, which results in a number of advantages. For example, because the reactor is isothermal, the substrate has a uniform temperature distribution, which in turn ensures uniform film growth. Other advantages include a tendency toward simpler operation, accommodation of several substrates and the ability to be operated at a range of temperatures and pressures. They usually do not require the low pressures that are necessary with PVD processes. Consequently, the vacuum system is simpler and less costly. The main disadvantage is that deposition also occurs on the reactor walls, which can become a source of contamination requiring frequent cleaning. In addition, homogeneous reactions affecting the deposition reactions and hence the structure of the films may take place in the vapor. Hot-wall reactors are generally used in those applications where high throughput is required, as reactor designs can more easily accommodate multiple wafer configurations.

In the cold-wall reactor design, the substrates are heated internally by resistive, inductive or IR radiation heating. The reactor walls thus remain cold, and usually no deposition occurs there, eliminating the risk of particles breaking loose from the walls. Furthermore, a low wall temperature reduces the risk of contaminating vapor/wall reactions. Other advantages are that (*a*) it has a higher deposition rate, (*b*) pressure and temperature can be tightly controlled and (*c*) in comparison to the hot-wall reactor, gas-phase reactions are suppressed. Its main disadvantage is that steep temperature gradients near the substrate surface can introduce natural convection, resulting in non-uniform film thickness and microstructure.

In both hot- and cold-wall reactors, precursor gasses (often diluted in carrier gasses, such as H_2, N_2 or Ar) are delivered into the reaction chamber at approximately ambient temperatures. The precursors have to be volatile (gaseous) and the chemical reactions need to be thermodynamically favourable in order to form a solid film (meaning that the Gibbs free energy has to decrease). The by-products also need to be volatile (gaseous) so that they can be easily removed from the system. Substrate erosion prior to introduction of the reactants can be a problem for some materials, as the substrate is at a very high temperature. As the precursors pass over, or come into contact with, a heated substrate, they react or decompose forming a solid phase on the substrate. The deposited films are usually only a few microns thick and are generally deposited at fairly high rates (of the order of a few hundred microns per hour). Volatile by-products are often produced, which are removed by gas flow through the reaction chamber.

For many optoelectronic applications, devices require layers with different carrier concentrations. This can be achieved by introducing the appropriate amounts of dopants (solid, liquid or gas) into the gas phase. Gas dopants are preferable because they provide a finer control on the concentration of the layer. The concentration is altered by controlling the partial pressure of the dopant gas as it enters the reactor chamber. However, impurities also change the deposition rate. Additionally, the

---

[19] Although many CVD processes can be considered near-equilibrium growth phenomena.

high temperatures at which deposition is performed may allow dopants to diffuse into the growing layer from other layers in the wafer ("autodoping"). Conversely, dopants in the source gas may diffuse into the substrate.

Three common forms of CVD reactor are illustrated in Fig. 4.22. These are the Horizontal reactor, the Vertical (or pancake) reactor and the Multi-barrel reactor. The Horizontal reactor is the simplest implementation and is shown in Fig. 4.22(a). Because the flow of gasses is parallel to the surface of the wafers (hence the term horizontal), growth is uniform laterally across a wafer (but non-uniform in the upstream and downstream directions. The Vertical reactor (Fig. 4.22(b)) improves uniformity since the gas flows at right angles to the surface of all of the substrates (hence the term vertical), which are also rotated during growth to further improve uniformity. However, one disadvantage is that turbulence can occur in the chamber due to the presence of stagnation points. The Multi-barrel reactor (Fig. 4.22(c)) utilizes parallel gas flow and hence it is actually a horizontal reactor). It is particularly suitable for high-volume production (for example, Si epitaxy).

In industry, CVD is widely used to apply solid thin-film protective coatings to surfaces that require wear, corrosion, high temperature or erosion protection. CVD is also used to produce high-purity bulk materials and powders, as well as artificial diamonds. The process is widely used in the semiconductor industry but mainly to deposit various films (*e.g.*, polycrystalline, amorphous and epitaxial Si, SiO_2, Si_2N_3) as part of the device fabrication process. The advantages of CVD are that it can produce high-purity conformal coatings at relatively high deposition rates. The disadvantages are that it is a high-temperature deposition technique (substrate temperatures can range from $\sim 600^\circ C$ to $1,000^\circ C$), it is hard to deposit alloys and it often uses and produces toxic and corrosive gasses, which must be removed and neutralized.

4.6.3.2 Metal Organic Chemical Vapor Deposition (MOCVD)

MOCVD is a CVD process using organic sources and, along with Molecular Beam Epitaxy (MBE), is widely used for precision heteroepitaxy. In contrast to liquid and conventional VPE, MOVPE is a highly non-equilibrium process due to the

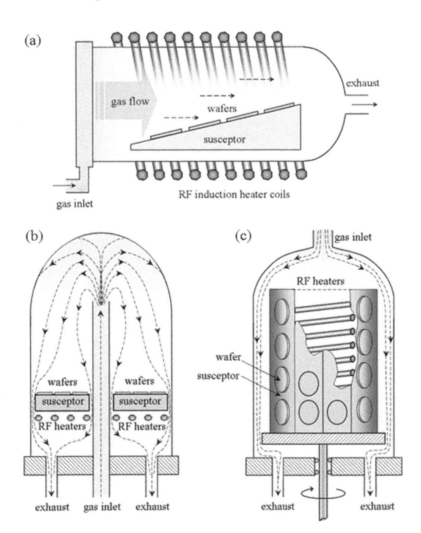

FIGURE 4.22 Reactors for CVD growth. (a) Horizontal reactor, (b) Vertical reactor and (c) Multi-barrel reactor. In all cases, the substrate temperature must be maintained uniformly over its area.

large thermodynamic driving force, in which metred amounts of metal organic reactants, along with metred amounts of hydrides are supplied to a heated reaction zone, whereby pyrolysis[20] of the reactants occurs and epitaxial growth takes place. The use of metal-organic compounds was first described by Manasevit in 1968 [54] and takes advantage of the fact that metal organic compounds are halide free and are often much more volatile and less thermally stable than metal halides, making it easier to get them to react at the deposition site. For a review, see Dapkus [55].

The growth process involves the forced convection of the metal organic vapor species over a heated substrate. The molecules striking the surface release the desired species, which then chemically react at the surface producing growth. For Group III-V semiconductors, the chemical processes involved are quite simple, in that an alkyl compound for the Group III element and a hydride for Group V element decompose in the 500 °C to 800 °C temperature range to form the III-V compound semiconductor. For example, gallium arsenide could be grown in a reactor on a substrate by introducing trimethylgallium (Ga(CH₃)₃, often abbreviated to TMG) and arsine (AsH₃), *via* the reaction,

$$Ga(CH_3)_3 + AsH_3 \rightarrow GaAs + CH_4 \uparrow . \tag{4}$$

This growth process is illustrated in Fig. 4.23. Similarly, indium phosphide is often synthesized using trimethylindium (In(CH₃)₃, often abbreviated to TMI) and phosphine (PH₃), via the reaction,

$$In(CH_3)_3 + PH \rightarrow InP + CH_4 \uparrow . \tag{5}$$

A typical MOCVD system is shown in Fig. 4.24. The reactor is a chamber made of a material that does not react with the chemicals being used, usually stainless steel or quartz. It must also withstand high temperatures. This chamber is composed of reactor walls, liner, a susceptor (which holds the substrate) gas injection units and temperature control units. To prevent overheating, cooling water must be flowing through the channels within the reactor walls. Special glasses, such as quartz or ceramic, are often used as the liner in the reactor chamber between the reactor wall and the susceptor. The susceptor is also made from a material resistant to the metalorganic compounds and is used to transfer heat from the RF heater to the substrate. Graphite is commonly used. For growing nitrides and related materials, a special coating on the graphite susceptor is necessary to prevent corrosion by ammonia (NH₃) gas. Gasses are

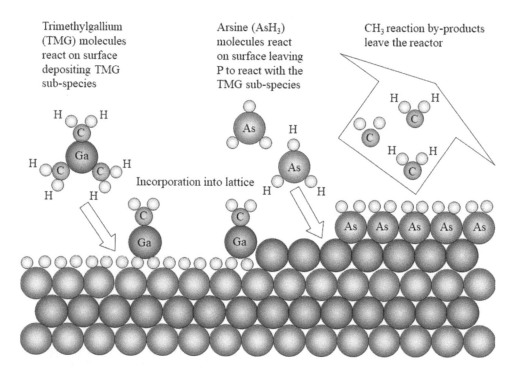

FIGURE 4.23 Figure illustrating the principle behind the MOCVD growth of GaAs. The precursor gasses, consisting of trimethylgallium (TMG) and arsine, are delivered to the substrate through a series of valves and bubblers, whereupon the TMG is adsorbed on the surface and then reacts with the arsine. This results in film growth and the release of volatile by-products, which are then removed by the gas flow and exhausted.

---

[20] The thermochemical decomposition of organic material at elevated temperatures in the absence of oxygen (or any halogen). The word is derived from the Greek-derived elements pyro "fire" and lysis "separating".

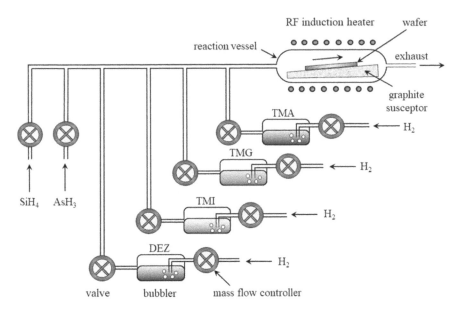

FIGURE 4.24 Schematic of the essential components of a typical MOCVD reactor system for the growth of a range of III-V compounds (modified from [56]). Here, the orgnometallic precursors are TMA = trimethylaluminum, TMG = trimethylgallium, TMI = trimethylindium, DEZ = diethylzinc, AsH_4 = arsine and SiH_4 = silane. The diagram shows the case using a horizontal reactor as shown in Fig. 4.22(a).

introduced via a complicated switching system using devices known as "bubblers". In a bubbler a carrier gas (usually nitrogen or hydrogen) is bubbled through the metal-organic liquid, which picks up some metal-organic vapor and transports it to the reactor. The amount of metal-organic vapor transported depends on the rate of carrier gas flow and the bubbler temperature.

Until recently, one of the main disadvantages of MOVPE when compared to MBE was the lack of real-time monitoring of the growth process. Reflection high-energy electron diffraction (RHEED) for example, cannot be used at the typical pressures (10–50 mbar), which characterize the MOVPE growth process, because the electron beam would be so strongly attenuated. However, optical techniques such as Reflectance Anisotropy Spectroscopy (RAS) can be applied for in-situ monitoring. RAS [57] is a non-destructive spectroscopic technique capable of sensing the stoichiometry and symmetry of the uppermost atomic monolayers of cubic semiconductors and metals. It works by measuring the reflectance of two beams of orthogonal linearly polarized light that are shone at normal incident onto a surface. The difference in signal is insensitive to the bulk response by symmetry but is very sensitive to surface anisotropy and as such can be used in a closed-loop feedback system to control single monolayer growth.

There are several varieties of MOCVD reactors – the most common being the atmospheric MOCVD and the low-pressure MOCVD (LPMOCVD) reactors. In atmospheric MOCVD, the growth chamber is essentially at atmospheric pressure, which alleviates the problems associated with vacuum generation but at the expense of a larger gas flow rate. In LPMOCVD, the growth chamber pressure is kept low. It uses less gas than does the atmospheric case, but growth rate is then lower – in fact even slower than in MBE. The principal advantages of MOCVD are that excellent uniformity in layer thickness, composition and carrier concentration can be achieved over a large area wafer. This technique also easily lends itself to the growth of abrupt heterointerfaces, allowing complex heterostructures to be grown layer by layer. The main disadvantage is that it uses large quantities of extremely toxic gasses.

4.6.3.3 The Multi-Tube PVT (MTPVT) Technique

In view of the difficulty of growing high vapor pressure compounds, such as CdTe, CdZnTe and HgTe, from the melt, Mullins *et al.* [58,59] have proposed the Multi-Tube PVT (MTVPT) technique, for growing II-VI materials directly from the vapor phase [60]. This has several advantages over traditional liquid phase techniques, namely improved structural perfection, purity and an inherent resistance to "self-poisoning". In this technique, control of vapor transport is critical to the growth, and this is achieved by separating the source and growth regions. A schematic cross-section through a MTPVT system used to grow CdZnTe is shown in Fig. 4.25. It is essentially a seeded vapor phase growth system, in which the source materials are sublimed and allowed to condense on a cooler seed crystal. It is based on the Markov [61] method of PVT, where the seed is separated from the container walls, ensuring wall-free growth. Cadmium telluride and zinc telluride source tubes, fabricated from quartz and heated by independent tubular furnaces, are connected to a third, similarly heated, quartz growth tube by a demountable quartz cross-member. The cross-member is heated to prevent condensation of sublimed source material prior to reaching the growth region. The transport passage incorporates a flow restrictor, which regulates the

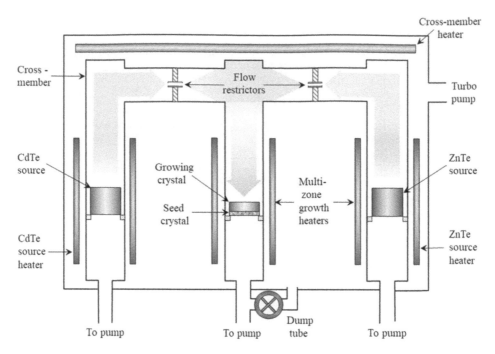

FIGURE 4.25 Schematic of the essential components of the "Multi-Tube" PVT (MTPVT) growth system used to grow CdZnTe (figure courtesy of Kromek®).

mass flow and decouples it from the source-sink temperature difference. In the growth tube, the seed wafer is located on a quartz pedestal separated from the inner wall of the tube by an annular gap in a Markov–Davydov configuration [61], ensuring wall-free growth. Accurate control of the temperature profile permits the vapor to condense and grow on a seed crystal in the center of the growth tube, whilst an annulus around the seed allows the removal of excess components and volatile impurities.

In contrast to liquid phase methods, vapor phase techniques offer lower growth temperatures and the potential to grow ternary and multinary alloys of uniform composition. Indeed, the flexibility of vapor phase growth has been widely exploited in the epitaxy of a wide range of optoelectronic materials and low-dimensional structures. In the context of the growth of single crystals of the relatively weak Group II-VI compounds, particularly CdTe, lower growth temperatures have shown improved structural quality of material, as well as a reduction in defects and in contamination by impurities from the growth system [59].

4.6.4 Physical Vapor Deposition (PVD)

PVD[21] or VPE encompasses a variety of vacuum deposition methods, such as sputtering, electron beam evaporation and MBE. In these systems, thin films are deposited by the condensation and adsorption of a vaporized form of a material onto the surface of a substrate, where film formation and growth proceed atomistically (for a review see [53]). The deposition usually takes place in high vacuum conditions at gas pressures, ranging from 10^{-3} Torr to 10^{-8} Torr. It involves purely physical processes such as high-temperature vacuum evaporation with subsequent condensation, or plasma sputter bombardment, rather than involving a chemical reaction at the surface, as in CVD. PVD is used mainly for homoepitaxy, since typical systems generally do not have the precise control systems needed for heteroepitaxy, as in MOCVD.

PVD is commonly preferred for thin film growth rather than CVD, since a wider range of films can be produced, the films produced tend to be more uniform, and the precursors and by-products are relatively benign compared to CVD. Another advantage is that the temperature of the substrate being coated is typically in the range of 200–500°C, considerably lower than temperatures associated with CVD (~500–1,000°C) leading to less severe temperature gradients. However, there are several disadvantages. For example, the deposition rates are relatively low, and it is a technologically more complex system since it is vacuum based. Also since PVD is a line-of-sight process it requires the substrate surface to be easily accessible, otherwise uneven deposition and/or shadowing effects can occur. Lastly, it is difficult to evaporate materials with low vapor pressures.

---

[21] Sometimes called vacuum deposition, because the process is usually carried out in an evacuated chamber.

4.6.4.1 Sputtering

A typical RF sputtering[22] system is shown schematically in Fig. 4.26. The substrate is placed in a vacuum chamber along with the source material and an inert gas (such as argon, neon, krypton or xenon) introduced at a pressure of $\sim 10^{-3}$ mbar. A noble gas is chosen since it will not react with either the target or the substrate. A gas plasma is struck, causing the gas to become ionized. DC plasmas are infrequently used because deposition is relatively slow compared to other techniques and the technique requires conductive electrodes to allow the flow of the current. Using an RF source has several advantages over a DC source, such as higher deposition rates at lower voltages and sputter pressures and insulating targets can be sputtered. Strong magnets, of strength ~ 200 Gauss, are usually placed under the target to confine the plasma to the region closest to the target plate, preventing the secondary electrons from leaving the plasma volume before they have had a chance to ionize the gas atoms. In this way, the ionization rate in the plasma volume is increased. The ions generated in the plasma are then accelerated into the target, causing atoms of the source material to be sputtered or "boiled–off" the surface by energy transfer. The resulting vapor condenses on all surfaces including the substrate. The average number of atoms ejected from the target per incident ion is called the sputter yield; it depends on the ion incident angle, the energy of the ion, the masses of the ion and target atoms and the surface binding energy of atoms in the target. The structure of the sputtered films is mainly amorphous – its physical and chemical properties being affected by target composition, ionized gas pressure and substrate temperature as well as the type of energetic particles impinging on the surface. The main advantages of sputtering are that it is simple, can be carried out at low temperature and most elements of the periodic table can be sputtered, including refractory materials, such as tungsten. Its main disadvantage is that semiconductor surfaces can be damaged by ion bombardment.

4.6.4.2 Evaporation

Evaporation is a high-temperature variant of sputtering in which source material is heated to the point where it starts to boil and evaporate. The evaporation is carried out in a vacuum chamber to ensure that the source molecules are evenly dispersed throughout the chamber, where they subsequently condense on all surfaces. Normal vacuum levels are in the medium to high range of 10^{-5} to 10^{-9} Torr. A wide variety of materials, including refractory metals such as tungsten and metal alloys, can be evaporated. Compared to sputtering, evaporation offers much higher deposition rates and is less damaging to the substrate, since the evaporated atoms have a Maxwellian energy distribution when they condense on the substrate. Sputtered atoms can have a hard velocity component, due to the plasma. On the negative side, adhesion and large area surface uniformity is generally poor compared to sputtering.

There are two popular evaporation technologies, namely resistive and e-beam evaporation. The choice of which method to use depends largely on the phase transition properties of the evaporant. In resistive evaporation, a tungsten filament wire or

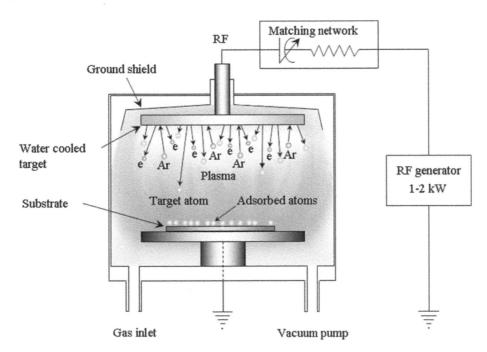

FIGURE 4.26 Schematic of a typical RF sputtering system. A plasma created by ionizing a sputtering gas (generally a chemically inert, heavy gas like Argon) bombards the target "sputtering" off target material, which can then impinge and adhere to the surface of a substrate, forming a film.

---

[22] The word sputtering is derived from the Latin word *sputare*, meaning "to emit saliva with noise".

foil is used to directly electrically heat the target (or alternately a ceramic boat containing the source material) to make the material evaporate. A typical system is shown schematically in Fig. 4.27(a). The advantage of this method is that is it simple and inexpensive, giving good coverage at low to moderate deposition rates. The disadvantages are that it is limited to low melting point metals and in the case of direct heating, the resistive heating element can react with the target to pollute the evaporant.

In e-beam evaporation, an electron beam of energy up to 15 keV is aimed at the source material causing local heating and evaporation. A schematic diagram of a typical system for e-beam evaporation is shown in Fig. 4.27(b). Since the electron beam can concentrate a large amount of energy into a very small area, high rates of deposition are possible. In fact, the energy level can exceed 10 million Watts per cm^2. Consequently, the evaporator must be water cooled to prevent it from melting. The advantages of e-beam evaporation are the high deposition rate and the fact that any material can be evaporated. Its main disadvantages are high capital costs and secondary X-ray emission from the target for beam energies above ~10 keV.

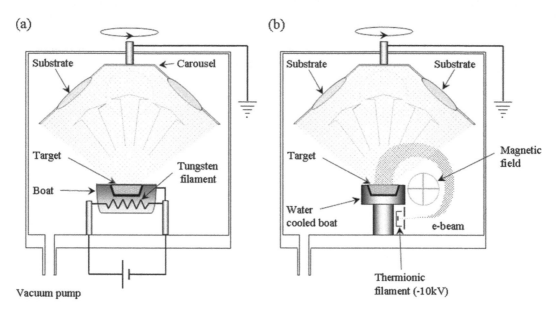

FIGURE 4.27 Typical systems for (a) resistive and (b) electron beam evaporation of materials. In both cases, the source material is heated to the boiling point and subsequently evaporates, whereupon it condenses on the substrate, forming a film.

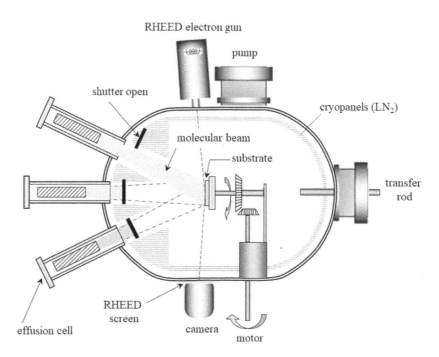

FIGURE 4.28 Schematic diagram of a MBE system, which in essence is a highly refined vacuum evaporation system. Consequently, the deposition is carried out in UHV conditions.

4.6.4.3 Molecular-Beam Epitaxy (MBE)

At its simplest, MBE is the most refined form of vacuum evaporation allowing precise epitaxy to be carried out. The technique was pioneered in the late 1960s at Bell Telephone Laboratories by Arthur [62] and Cho [63]. In MBE, a source material is heated to produce an evaporated beam of particles. These particles travel through a very high vacuum (10^{-8} Pa) to the substrate, where they condense. Thus, the method is ballistic rather than hydrodynamic with the incident beams derived from elemental sources. MBE has a lower throughput than other forms of epitaxy (~1–300 nm/min) but is a much more precise delivery system for the growth of mono-atomic layers. However, the slow deposition rates require proportionally better vacuum in order to achieve the same impurity levels as other deposition techniques. MBE has a number of advantages over CVD techniques. For example, because it is a low-temperature process, outdiffusion and autodoping are minimized, and unlike MOCVD, the gasses used to form the crystal are made up of single elements and not complex and often toxic molecules.

A schematic diagram of an MBE system is shown in Fig. 4.28. A critical feature is the extensive cryopaneling surrounding both the substrate station and the evaporation sources. The cryopanels are cooled to liquid nitrogen temperature (77K) and effectively cryopump the reactor volume, reducing the arrival rate of unwanted species as well as providing heat dissipation for both the evaporation sources and the substrate heater. The molecular beams are produced by evaporation or sublimation from heated liquids or solids contained in the crucibles, called effusion cells or Knudsen cells. At the pressures used in MBE equipment, collision-free beams emanate from the sources and interact chemically only with the substrate to form an epitaxial film. Growth proceeds according to the number of free atoms that stick to the growing surface. To improve growth

TABLE 4.5

Characteristics, advantages and disadvantages of the main epitaxial growth techniques.

Growth method	Features	Advantages	Disadvantages
Van der Waal Epitaxy (VDWE)	Strain-free growth taking advantage of weak van der Waal bonding at the growth interface	Capable of compensating for large lattices mismatches in layered materials and producing non-planar nanostructures. Can use a number of deposition techniques for actual growth.	Limited to 2D materials as a growth technique
Solid Phase Epitaxy (SPE)	Recrystallization of a metastable amorphous film placed on a crystalline substrate	Simple low cost. High growth rates (up to 0.1μm/s). High quality films. Low temperature process. Can create dopant concentrations in excess of solubility limits.	Suitable for a limited number of materials. Difficult to achieve thick layers. Quality of films very sensitive to substrate impurities.
Liquid Phase Epitaxy (LPE)	Growth from supersaturated solution onto a substrate.	Simple low-cost equipment and high throughput. Very high quality material. No toxic gasses and easily handled solids Good control of impurities. High growth rates (0.1–1μm/min). Suitable for selective growth.	Difficult to grow abrupt heterostructures. Limited substrate areas. Poor control over the growth of very thin layers. Re-dissolution of the grown material. Poor uniformity over large areas. High growth temperatures for some compounds (*e.g.*, 900°C for InP).
Chemical Vapor Deposition (CVD)	Thin film formation on a substrate by a chemical reaction of vapor-phase precursors	Produces high purity conformal coatings. Relatively simple reactors.	Limited number of substrates because of high temperature operation. Precursors and by-products can be toxic, pyrophoric or corrosive.
Vapor Phase Epitaxy (VPE)	Uses metal halides as transport agents to grow.	Relatively simple reactors. Extremely high purity material.	No Al compounds, thick layers. Toxic precursor gasses. Low growth rate.
Multi-Tube Physical Vapor Transport (MTPVT)	Controlled growth of high volatility II-VI compounds directly from the vapor phase.	Simple, low temperature process. In-situ compounding. High quality material. Reusable quartz ware. Low cost.	Lattice matched substrates difficult to find for some compounds
Metal Organic Chemical Vapor Deposition (MOCVD)	Uses metal-organic compounds as the sources.	High-quality material. Atomically abrupt interfaces. Low temp growth. High vapor pressure material growth possible. In-situ monitoring with (RAS).	Expensive and slow (~3 monolayer/sec = 3 μm/hr). Some sources very toxic. Tendency for C contamination. Several growth parameters to control.
Molecular Beam Epitaxy (MBE)	Deposit epi-layer in ultrahigh vacuum, in-situ characterization. Low temperature operation.	High quality material. Most precise epi-growth technique. Atomically abrupt interfaces. Less sensitive than other techniques to outdiffusion and autodoping. No toxic gasses. Easily handled. solids. Relatively simple chemistry. In-situ monitoring (RHEED).	Expensive and slower (~1 monolayer/sec = 1 μm/hr). UHV required. Hard to grow materials with high vapor pressure. Run-to-run reproducibility of layer thickness and composition. Surface "oval defects". Memory effect for P species.

uniformity, the substrate is also rotated. The composition of the grown epi-layer and its doping level depend on the relative arrival rates of the constituent elements and dopants, which in turn depend on the evaporation rates of the corresponding sources. Simple mechanical shutters in front of the beam sources are used to interrupt the beam fluxes, to start and to stop the deposition or doping. Typically, shutter speeds are of the order of ~ 0.1 to 0.3 seconds, which is less than the time taken to grow a monolayer (~1 second). This means that changes in composition and doping can be abrupt on an atomic scale.

Because MBE is carried out in ultra-high vacuum conditions, it has a unique advantage over other epitaxial growth techniques, in that the physical and chemical properties of the films can be monitored and controlled *in-situ* by surface sensitive diagnostic techniques, such as RHEED, Auger Electron Spectroscopy (AES), Spectroscopic Ellipsometry (SE) and RAS. In RHEED [64], which is the most commonly used technique, a collimated electron beam at, or near, 10 keV impacts the crystals surface at a grazing angle. The electrons diffracted by the surface lattice interfere and form beams, which result in streaks or lines on a phosphor screen. The spacing between these lines is proportional to the surface reciprocal lattice spacing. AES is a common non-destructive analytical technique [65] that makes use of the Auger effect to study the chemical and compositional nature of a surface. It relies on the spectroscopic analysis of secondary Auger electrons emitted from the surface following stimulation by an incident electron beam. It is useful for determining the composition of surface layers to a depth of about 2 nm for elements above He, with a spatial resolution greater than or equal to 100 nm. SE [66] is a non-destructive optical technique that measures the change of polarization upon reflection or transmission and compares it to a model to characterize composition, roughness, crystalline nature, electrical conductivity and other material properties of a surface. It is also a very useful tool in measuring optical functions of bulk materials. RAS is another non-destructive optical technique that measures the difference in reflectance from two normally incident polarized light beams to determine the symmetry and stoichiometry of the uppermost atomic layers of a surface [57]. However, while the technique is extremely sensitive, enabling growth monitoring and control on a sub-monolayer level, its use is limited to cubic material.

To summarize, the principal advantage of MBE is its precise control of film thickness, composition and doping, with atomically abrupt interfaces readily obtainable and consequently a high degree of low dimensional structure control. Compared to MOCVD, the process uses easily handled solids and no toxic gasses, and its chemistry is relatively straightforward. The main disadvantage of MBE is its high expense, as UHV and cryogenics are required and films are grown one layer at a time. Consequently, the technique does not lend itself to industrial-scale production.

In Table 4.5, we summarize the relative advantages and disadvantages of the main epitaxial growth techniques.

REFERENCES

[1] C. Plinii Secundi, *"Naturalis Historiæ"* (Pliny the Elder, *The Natural History*), available online at www.perseus.tufts.edu/hopper/text?doc=Plin.+Nat.+toc

[2] R.-J. Haüy, *"Traité de Cristallograhie"*, Bachelier and Huzard, Paris (1822).

[3] J.W. Gibbs, *"On the Equilibrium of Heterogeneous Substances, Collected Works"*, Longmans Green, New York (1928).

[4] A. Verneuil, "Memoire sur la Reproduction du Rubis par Fusion", *Ann. Chim. Phys.*, Vol. 3, no. 8 (1904), pp. 20–48.

[5] J. Czochralski, "Ein neues Verfahren zur Messung des Kristallisationsgeschwindigkeit der Metalle", *Z. Phys. Chem.*, Vol. 92 (1918), pp. 219–221.

[6] S. Kyropoulos, "Ein Verfahren zur Herstellung großer Kristalle", *Z. Anorg. U. Allg. Chem.*, Vol. 154, no. 1 (1926), pp. 308–313.

[7] P.W. Bridgman, "Certain physical properties of single crystals of tungsten, antimony, bismuth, tellurium, cadmium, zinc, and tin", *Proc. Am. Acad. Sci.*, Vol. 60, no. 9 (1925), pp. 303–383.

[8] D.C. Stockbarger, "The production of large single crystals of lithium fluoride", *Rev. Sci. Instrum.*, Vol. 7 (1936), pp. 133–136.

[9] H.-G. Jeon, Y. Kondo, S. Maki, E. Matsumoto, Y. Taniguchi, M. Ichikawa, "A highly efficient sublimation purification system using baffles with orifices", *Org. Electron.*, Vol. 11, no. 5 (2010), pp. 794–800.

[10] B.W. Montag, M.A. Reichenberger, N. Edwards, P. Ugorowski, M. Sunder, J. Weeks, D.S. McGregor, "Static sublimation purification process and characterization of LiZnP semiconductor material", *J. Cryst. Growth*, Vol. 419 (2015), pp. 133–137.

[11] V.M. Zaletin, N.V. Lyakh, I.N. Nozhkina, "Mass transport during growing mercuric iodide by static sublimation method", *Cryst. Res. Technol.*, Vol. 20 (1985), pp. 321–327.

[12] W.G. Pfann, "Principles of zone melting", in *Transactions of the American Institute of Mining and Metallurgical Engineers*, Metals & Materials Society, New York, NY, Vol. 194 (1952), pp. 747–753.

[13] J.K. Kennedy, G.H. Moates, "Continuous horizontal zone refining apparatus", *Rev. Sci. Instrum.*, Vol. 37, no. 11 (1966), pp. 1530–1533.

[14] W.G. Pfann, "Zone melting", *Science*, Vol. 135, no. 3509 (1962), pp. 1101–1109.

[15] W.D. Callister, D. Rethwisch, *"Materials Science and Engineering an Introduction"*, John Wiley & Sons, New York, 8th ed. (2009) ISBN 978-0-470-41997-7.

[16] K. Byrappa, M. Yoshimura, *"Handbook of hydrothermal technology"*, A Technology for Crystal Growth and Materials Processing, William Andrew Publishing, Noyes, Norwich, New York (2001) ISBN: 978-0-8155-1445-9.

[17] K.M. von Schafhäutl, "Die neuesten geologischen Hypothesen und ihr Verhältniß zur Naturwissenschaft überhaupt", *Gelehrte Anzeigen, Die Königliche Bayerische Akademie Der Wissenschaften*, Vol. 20 (1845) pp. 557–596.

[18] V. Kozlov, H. Andersson, V. Gostilo, M. Leskela, A. Owens, M. Shorohov, H. Sipila, "Improved process for the TlBr single-crystal detector", *Nucl. Instrum. Methods*, Vol. **A591** (2008), pp. 209–212.

[19] W. Lin, K.E. Benson, "The science and engineering of large-diameter Czochralski silicon crystal growth", *Ann. Rev. Mater. Sci.*, Vol. **17** (1987), pp. 273–298.

[20] E.P.A. Metz, R.C. Millen, R. Mazelsky, "A technique for pulling single crystals of volatile materials", *J. Appl. Phys.*, Vol. **33** (1962), pp. 2016–2017.

[21] J.B. Mullin, B.W. Straugham, W.S. Brickell, "Liquid encapsulation techniques", *J. Phys. Chem. Solids*, Vol. **26** (1965), pp. 782–784.

[22] J. Winkler, "*Beiträge zur Regelung des Czochralski-Kristallzüchtungsprozesses zur Herstellung von Verbindungshalbleitern*", PhD Thesis, Technische Universität Dresden, Germany, Shaker Verlag Aachen (2007).

[23] P. Rudolph, F.-M. Kiessling, "Growth and characterization of GaAs crystals produced by the VCz method without boric oxide encapsulation", *J. Cryst. Growth*, Vol. **292** (2006), pp. 532–537.

[24] U. Lachish, "*CdTe and CdZnTe Crystal Growth and Production of Gamma Radiation Detectors*", (2000). http://urila.tripod.com/crystal.htm

[25] R. Sudharsanan, K.B. Parham, N.H. Karam, "Cadmium zinc telluride detects gamma-rays", *Laser Focus World*, Vol. **32**, no. 6 (1999), pp. 199–204.

[26] H.C. Ramsperger, E.H. Melvin, "The preparation of large single crystals", *J. Opt. Soc. Am.*, Vol. **15**, no. 6 (1927), pp. 359–363.

[27] N. Audet, M. Cossette, "Synthesis of ultra-high purity CdTe ingots by the travelling heater method", *J. Electron. Mater.*, Vol. **34**, no. 6 (2002), pp. 683–686.

[28] N. Audet, B. Levicharsky, A. Zappettini, M. Zha, "Composition study of CdTe charges synthesized by the travelling heater method", *IEEE Trans. Nucl. Sci.*, Vol. **54**, no. 4 (2007), pp. 782–785.

[29] P.H. Keck, M.J.E. Golay, "Crystallization of silicon from a floating liquid zone", *Phys. Rev.*, Vol. **89**, no. 6 (1953), p. 1297.

[30] W. Zulehner, "Historical overview of silicon crystal pulling development", *Mater. Sci. Eng.*, Vol. **B-73** (2000), pp. 7–15.

[31] S. Kawarabayashi, M. Yokogawa, A. Kawasaki, R. Nakai, "Comparisons between conventional LEC, VCZ and VGF for the growth of InP crystals", *IEEE Proceedings of the Sixth International Conference on Indium Phosphide and Related Materials*, Santa Barbara, CA (1994), pp. 227–230.

[32] A. Owens, A. Peacock, "Compound semiconductor radiation detectors", *Nucl. Instrum. Methods*, Vol. **A531** (2004), pp. 18–37.

[33] A. Koma, K. Sunouchi, T. Miyajima, "Fabrication and characterization of heterostructures with subnanometer thickness", *Microelectron. Eng.*, Vol. **2** (1984), pp. 129–136.

[34] A. Koma, K. Sunouchi, T. Miyajima, "Fabrication of ultrathin heterostructures with van der Waals epitaxy", *J. Vac. Sci. Technol. B*, Vol. **3** (1985), p. 724.

[35] A.K. Geim, I.V. Grigorieva, "Van der Waals heterostructures", *Nature*, Vol. **499** (2013), pp. 419–425.

[36] A. Koma, "Van der Waals epitaxy – A new epitaxial growth method for a highly lattice-mismatched system", *Thin Solid Films*, Vol. **216**, no. 1 (1992), pp. 72–76.

[37] F.S. Ohuchi, B.A. Parkinson, K. Ueno, A. Koma, "Van der Waals epitaxial-growth and characterization of MoSe$_2$ thin-films on SnS$_2$", *J. Appl. Phys.*, Vol. **68** (1990), pp. 2168–2175.

[38] A. Koma, "Van der Waals epitaxy for highly lattice mismatched systems", *J. Cryst. Growth*, Vol. **201/202** (1999), pp. 236–241.

[39] Y. Alaskar, S. Arafi, D. Wickramaratne, M.A. Zurbuchen, L. He, J. McKay, Q. Lin, M.S. Goorsky, R.K. Lake, K.L. Wang, "Towards van der Waals epitaxial growth of GaAs on Si using a Graphene Buffer Layer", *Adv. Funct. Mater.*, Vol. **24**, no. 42 (2014), pp. 1–10.

[40] W. Faschinger, "Doping limits and bandgap engineering in wide gap II-VI compounds", in *Wide Bandgap Semiconductors*, ed. S.J. Pearton, William Andrew Publishing, New York (1999), pp. 1–37 ISBN 0-8155-1439-5.

[41] D. Olego, J. Faurie, S. Sivananthan, P. Raccah, "Optoelectronic properties of Cd$_{1-x}$Zn$_x$Te films grown by molecular beam epitaxy on GaAs substrates", *Appl. Phys. Lett.*, Vol. **47** (1985), pp. 1172–1174.

[42] J.E. Toney, T.E. Schlesinger, R.B. James, "Optimal bandgap variants of Cd$_{1-x}$Zn$_x$Te for high-resolution X-ray and gamma-ray spectroscopy", *Nucl. Instrum. Methods.*, Vol. **A428** (1999), pp. 14–24.

[43] A.W. Webb, S.B. Quadri, E.R. Carpenter, E.F. Skelton, "Effects of pressure on Cd$_{1-x}$Zn$_x$Te alloys (0 < x < 0.5)", *J. Appl. Phys.*, Vol. **61** (1987), pp. 2492–2494.

[44] J.F. Butler, C.L. Lingren, F.P. Doty, "Cd$_{1-x}$Zn$_x$Te gamma ray detectors", *IEEE Trans. Nucl. Sci.*, Vol. **39** (1992), pp. 605–609.

[45] H.T. Grahn, "*Semiconductor Superlattices: Growth and Electronic Properties*", World Scientific, Hackensack, NJ (1995) ISBN 9810220618, 9789810220617.

[46] G.L. Olson, J.A. Roth, "Solid phase epitaxy", in *Handbook of Crystal Growth*, ed. D.T.J. Hurle, Vol. **3**, Ch.7, Elsevier, Amsterdam (1994), pp. 255–312.

[47] M.G. Mauk, "Liquid phase epitaxy", in *Handbook of Crystal Growth: Thin Films and Epitaxy*, ed. T. Kuech, Elsevier, Amsterdam (2014), pp. 225–314, ISBN 9780444633057.

[48] H. Nelson, "Epitaxial growth from the liquid state and its application to the fabrication of tunnel and laser diodes", *RCA Rev.*, Vol. **24** (1963), pp. 603–615.

[49] R.C. Linares, R.B. McGraw, J.B. Schroeder, "Growth and properties of Yttrium iron garnet single-crystal films", *J. Appl. Phys.*, Vol. **36** (1965), pp. 2884–2886.

[50] M.B. Panish, S. Sumslki, I. Hayashi, "Preparation of multilayer LPE heterostructure with crystalline solid solutions of Al$_x$Ga$_{1-x}$As: Heterostructure lasers", *Metall. Trans. AIME*, Vol. **2** (1971), pp. 795–801.

[51] H.J. Scheel, "A new technique for multilayer LPE", *J. Cryst. Growth*, Vol. **42** (1977), pp. 301–308.

[52] A.C. Jones, M.L. Hitchman, "Overview of chemical vapour deposition", in *Chemical Vapour Deposition: Precursors, Processes and Applications*, eds. A.C. Jones, M.L. Hitchman, Royal Society of Chemistry, London (2009) pp. 1–36, ISBN: 978-0-85404-465-8.

[53] D.M. Mattox, *"Handbook of Physical Vapor Deposition (PVD) Processing: Film Formation, Adhesion, Surface Preparation and Contamination Control"*, Noyes Publications, Westwood, NJ (1998) ISBN 0-8155-1422-0.

[54] H.M. Manasevit, "Single-crystal gallium arsenide on insulting substrates", *Appl. Phys. Lett.*, Vol. **12**, no. 4 (1968), pp. 156–159.

[55] P.D. Dapkus, "Metalorganic chemical vapor deposition", *Ann. Rev. Mater. Sci.*, Vol. **12** (1982), pp.243–269.

[56] C.J. Hepburn, "Temperature dependent operation of vertical cavity surface emitting lasers (VCSELs)", M.Sc. Dissertation, University of Essex (2001).

[57] P. Weightman, D.S. Martin, R.J. Cole, T. Farrell, "Reflection anisotropy spectroscopy", *Rep. Progr. Phys.*, Vol. **68**, no. 6 (2005), pp. 1251–1341.

[58] J.T. Mullins, J. Carles, N.M. Aitken, A.W. Brinkman, "A novel 'Multi-tube' vapour growth system and its application to the growth of bulk crystals of cadmium telluride", *J. Cryst. Growth*, Vol. **208** (2000), pp. 211–218.

[59] J.T. Mullins, B.J. Cantwell, A. Basu, Q. Jiang, A. Choubey, A.W. Brinkman, B.K. Tanner, "Vapour phase growth of bulk crystals of cadmium telluride and cadmium zinc telluride on gallium arsenide", *J. Electron. Mater.*, Vol. **37**, no. 9 (2008), pp. 1460–1464.

[60] A.W. Brinkman, J. Carles, "The growth of crystals from the vapour", *Progr. Cryst. Growth Char. Mater.*, Vol. **37**, no. 4 (1998), pp. 169–209.

[61] E.V. Markov, A.A. Davydov, "Growing orientated single crystals of cadmium sulfide from the vapour phase", *Inorg. Mater.*, Vol. **11** (1975), pp. 1504–1506.

[62] J.R. Arthur, J.J. LePore, "GaAs, GaP, and GaAs$_x$P$_{1-x}$ epitaxial films grown by molecular beam deposition", *J. Vac. Sci. Technol.*, Vol. **6** (1969), pp. 545–548.

[63] A.Y. Cho, "Film deposition by molecular-beam techniques", *J. Vac. Sci. Technol.*, Vol. **8**, no. 5 (1971), pp. 31–38.

[64] A. Ichimiya, P.I. Cohen, *"Reflection High Energy Electron Diffraction"*, Cambridge University Press, Cambridge, UK (2004) ISBN 0-52-145373-9.

[65] D. Briggs, M.P. Seah, eds., *"Practical Surface Analysis, Vol I, Auger and X-Ray Photoelectron Spectroscopy"*, John Wiley & Sons, New York, 2nd ed. (1990) ISBN 0-47-192081-9.

[66] H.G. Tompkins, *"A Users's Guide to Ellipsometry"*, Academic Press Inc., London (1993) ISBN 0-12-693950-0.

5

Contacting Systems

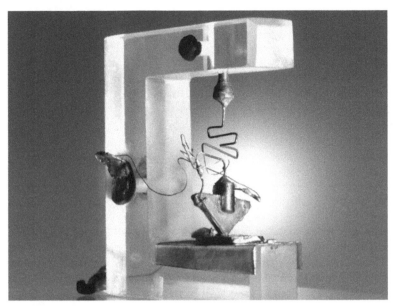

The first point contact transistor constructed at Bell Labs in 1947 (Bardeen and Brattain, "The Transistor, A Semiconductor Triode", *Phys. Rev.*, Vol. **74** (1948), pp. 230–231). The emitter and collector contacts are made of gold and are attached to the sides of a truncated wedge-shaped piece of insulating material whose apex is in mechanical contact with a block of germanium. The other face of the semiconductor is, in turn, in mechanical contact with a gold plate, which forms the base contact. Note that the "spring" providing the mechanical pressure is actually a bent paper clip (Photo credit: The Porticus Centre, Beatrice Consumer Products Inc., Subsidiary of Beatrice Companies Inc.).

CONTENTS

5.1 Introduction

Once charge is created in a semiconducting material, it must be extracted, and this is the function of the contacts. However, the formation of stable and laterally uniform contacts can be a major problem – the practical aspects of which will be elaborated in later sections. Depending on their current-voltage characteristics, contacting systems can be broadly classified into two groups (see Fig. 5.1). These are Ohmic and Schottky contacts.

We begin by summarizing some of the important aspects of contacting systems before addressing each in detail.

1) The contacts are usually metal, which means that the physics of metal semiconductor interfaces dictate how the contacts behave.

2) A metal-semiconductor interface results in the creation of a space charge region extending from the contact into the bulk of the semiconductor, which is depleted of carriers. It is known as the depletion or space charge layer.

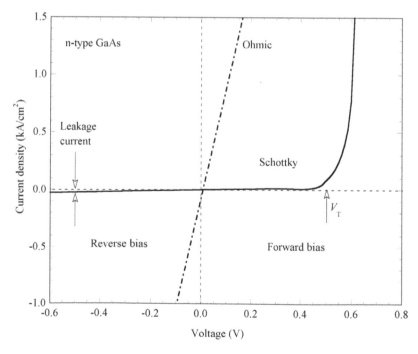

FIGURE 5.1 Examples of Ohmic and Schottky current-voltage characteristics obtained with metal-semiconductor contacts on *n*-type GaAs doped to 10^{15} cm^{-3}. The Schottky contact (solid line) shows strong rectifying behaviour (*i.e.,* passing current in the forward direction but blocking it in the reverse direction). Here V_T is the turn-on voltage above which the diode conducts. By contrast, the Ohmic contact (dashed line) passes current in both directions. A truly ohmic contact has a linear current-voltage (I-V) curve, obeying Ohm's law.

3) The presence of the space charge region generates a potential barrier.

4) The ability to conduct current across this barrier can changed by manipulating the height and width of this barrier.

5) If there is no hindrance to current flow by the barrier, the contact is said to be ohmic; if current is impeded by the barrier, the contact is Schottky and is rectifying (*i.e.*, behaves as a diode – it passes current in the forward direction but blocks it in the reverse direction as shown in Fig. 5.1).

6) Ohmic and Schottky junctions are two extreme cases of the same thing (the influence, or lack of it, of a potential barrier on the transport of current across the metal-semiconductor interface).

All metal-semiconductor junctions are Schottky in nature and rectifying – by which we mean that the current flow through the device will be asymmetric. However, depending on the actual metallization and semiconductor, such systems can display a wide range of behaviour, from essentially ohmic to strongly rectifying.

5.1.1 Low-Resistance or "Ohmic" Contacts

Ideally, low-resistance or "ohmic" contacts have linear current-voltage characteristics, that is, there is an unimpeded transfer of either carrier from one material to another, which does not depend on the polarity or direction of the bias (shown by the dashed line in Fig. 5.1). The advantage of a true Ohmic contact is that it does not affect or modify the detector signal. For practical reasons, Rhoderick [1] has proposed that this definition be extended to include all low resistance contacts – meaning that the resistance of the contact is negligible compared to the bulk resistance of the semiconductor and as such does not affect the I-V characteristics of the device. From an equivalent circuit point of view, the system behaves like a resistor.

5.1.2 Schottky or Blocking Contacts

In contrast to ohmic-like contacts, Schottky, or blocking, contacts have current-voltage characteristics that are non-linear and asymmetric as shown by the solid line in Fig. 5.1. The effect of a bias across the junction is to raise or lower the height of the potential barrier. Depending on the level of the bias, the barrier can be lowered sufficiently to allow current to pass unimpeded in one direction (the forward bias direction) or raised sufficiently, to prevent, or "block" the passage of current in the other direction (the reverse bias direction).

Practically, the "blocking" effect is exploited for rectification applications, which in its simplest form is a single metal-semiconductor junction – a Schottky diode. Schottky diodes are most often used in applications when a low voltage drop is desired, for example, in high-speed switching applications, mixers for heterodyne receivers and frequency multipliers for local oscillators. They are universally used to reduce leakage currents in low bandgap, low noise systems by blocking the replacement of free carriers in the bulk after bias is applied. In practical detection systems (see Chapters 6 and 8), they also have the advantage that by reverse biasing a detector, a higher bias can be applied across the bulk, thus improving the collection of charge following an ionization event in the semiconductor without significantly increasing the leakage current.

5.1.3 Contacting Technologies

Contacting can be achieved by several methods, such as applying metal-bearing paints and pastes, such as Aquadag,[1] melting metals directly on the semiconductor surface, evaporation, sputtering, molecular beam epitaxy, ion-implantation and others. These are described in detail in Chapters 4 and 6 and will not be repeated here. For a comprehensive review of contacting systems and semiconductor interfaces, see references [2–4].

5.2 Metal Semiconductor Interfaces

The properties of metal-semiconductor interfaces were first systematically investigated by Braun [5] in 1874, who observed that when metal sulfides were contacted by metal points, the electrical resistance varied with the magnitude and polarity of the applied voltage, which he described as "unilateral conduction". In essence, the current flow was impeded in one direction but not the other – a process labelled rectification. At the time, devices were called rectifiers but quickly became known as "*diodes*"[2] Later Pierce [6] showed that most semiconducting materials showed rectification properties when metal contacts were sputtered onto them. Braun's discovery remained little more than a scientific curiosity and found no practical

---

[1] "AQUueous Deflocculated Acheson Graphite" – a colloidal dispersion of pure graphite in a water carrier, manufactured by Acheson Industries, a subsidiary of ICI.

[2] The entomology of the word is derived from the Greek roots *di* (from δí), meaning "two", and *ode* (from ὁδóς), meaning "path".

application until the advent of radio in the early 1900s when Bose [7] demonstrated the detection of millimetre electromagnetic waves using galena (lead sulfide) crystals contacted by a metal point. This led directly to the introduction of point contact rectifiers (often referred to as "cat's whiskers") for radio wave signal detection [8,9]. Early theories of rectification were largely qualitative and assumed that the contact between the metal and the semiconductor was not perfect and as such introduced a large resistance that varied enormously with the direction of the current. For example, whilst not explicitly mentioning a potential barrier, Braun [5] attempted to explain rectification phenomena by a thin interface layer of extremely high resistance. Later attempts to explain the phenomenon by quantum mechanical tunneling through the layer proved unsatisfactory as the unimpeded current flowed in the wrong direction. In 1938, both Schottky [10] and independently Mott [11] pointed out that the observed direction of rectification could be explained by assuming that electrons have to pass over a potential barrier following the normal process of drift and diffusion through the semiconductor. This barrier became known as the Schottky barrier, which was quickly exploited for practical applications as the Schottky diode.

5.3 Schottky Barriers

When the surfaces of a metal and semiconductor come into intimate contact, the Fermi levels equalize. This is because electrons from the side with the higher Fermi level will move to the side on the junction with the lower Fermi level. As electrons fill the electronic states in the region of lower Fermi level, the Fermi level there rises. At the same time, as electrons leave the region of the higher Fermi level, the Fermi level there falls as states become available. Eventually the Fermi levels in both regions equalize and the overall flow of electrons stops. For a metal-metal interface, the charges reside on the surface because of the high carrier densities. However, for a metal-semiconductor interface, the lower free charge densities available in the semiconductor result in charge being depleted, not only from the interface but also from a finite depth into the semiconductor. For an n-type semiconductor, a positive charge builds up on the semiconductor side and due to the excess electrons a negative charge on the metal side. This in turn, results in an electric field being established and the creation of a potential barrier (the Schottky barrier) at the junction. This barrier prevents the further flow of electrons. Because the region over which charge is depleted extends a finite depth into the semiconductor, the energy bands on the semiconductor side must bend for the Fermi levels to equalize as is illustrated in Fig. 5.2. The height of the Schottky barrier is defined as the energy distance between the Fermi level, E_F, and the respective majority carrier band-edge at the interface – that is the valence-band maximum, E_v, for p-type semiconductors or the conduction-band minimum, E_c, for n-type semiconductors. In an n-type semiconductor, the barrier prevents the flow of electrons from the metal to the semiconductor but promotes the flow of holes. It is the opposite for a p-type semiconductor.

The height of the barrier, ϕ_b, is defined as the potential difference between the Fermi energy of the metal and the band edge where the majority carriers reside (the conduction band for an n-type semiconductor and the valence band for a p-type semiconductor). With reference to Fig. 5.2(b), the barrier height for an n-type semiconductor is given by

$$\phi_b = \phi_m - \chi_s, \tag{1}$$

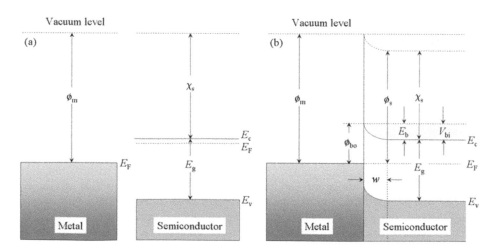

FIGURE 5.2 Energy band diagram of a metal and n-type semiconductor before (a) and after (b) contact is made. Here E_c is the conduction band energy, E_v is the valence band energy, E_F is the Fermi band energy, E_b is the amount of energy by which the band-edge is shifted at the interface to equalize the Fermi levels, E_g is the bandgap energy, χ_s is the electron affinity of the semiconductor, ϕ_b is the contact barrier height, ϕ_m is the metal work function, ϕ_s is the semiconductor work function, V_{bi} is the built in potential and w is the depletion width (*i.e.*, the width of the region depleted of carriers by the effect of the junction).

where φ_m is the work function of the metal and χ_s is the electron affinity of the semiconductor, given by

$$\chi_s = \phi_s - (E_c - E_F). \tag{2}$$

Here φ_s is the work function of the semiconductor defined as the minimum energy needed to remove an electron from the Fermi energy level, E_F, into vacuum. The electron affinity, in turn, is defined as the work required to remove an electron from an energy level at the bottom of the conduction band to an energy level corresponding to an electron at rest in vacuum outside the solid and beyond the range of the image force (see Section 5.3.1). Near the interface, the semiconductor will be depleted of electrons (the depletion layer) since E_F is further away from E_c (see Fig. 5.2(b)). Ionized donors remain, so a space charge will be built up in that region, resulting in a "built-in" electric potential near the interface. This built-in potential, V_{bi}, is the potential developed across the junction in order to equalize the metal-semiconductor Fermi levels and is in fact the potential barrier encountered by electrons in the semiconductor when approaching the metal in thermal equilibrium. It is given by

$$V_{bi} = \phi_m - \phi_s = \phi_b - (E_c - E_F). \tag{3}$$

The associated built-in electric field in the semiconductor causes electrons to move away from the interface. No electric field exists in the metal, since it is assumed a perfect conductor.

The height of the barrier from the metal side is fixed and determined only by the work function of the metal and the electron affinity of the semiconductor. However, the height of the barrier from the semiconductor side can be altered by applying a potential across the junction. It can lower the barrier height, increasing conduction, or raise it, reducing conduction. The effect is illustrated in Fig. 5.3 for an *n*-type semiconductor. Specifically, when a positive bias is applied to the metal, the Fermi level of the semiconductor is raised relative to the metal Fermi level reducing the barrier height (ϕ_b) and narrowing the depletion width (w). This is known as forward biasing. As the applied voltage is increased, the current flow increases (see Fig. 5.1). When a negative voltage is applied to the metal, the Fermi level of the semiconductor drops, increasing the band bending and the height of the barrier. An electron now requires more energy to overcome it and consequently less current flows. This is known as reverse biasing. For a *p*-type device the sign of the voltage is reversed (*i.e.,* it passes current when the metal is connected to the negative terminal and blocks current when it is connected to the positive terminal).

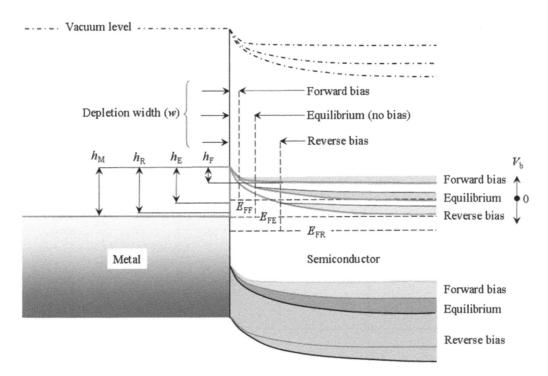

FIGURE 5.3 The energy band structure for a metal in contact with an *n*-type semiconductor showing the resulting band structure with (a) no bias applied (*i.e.,* equilibrium conditions), (b) under forward bias and (c) under reverse bias. In all three cases, the height of the barrier from the metal side, h_M, stays the same; however, from the semiconductor side the conduction band moves up and down with bias – increasing the height of the potential barrier h_R and the depletion width w with negative (reverse) bias and decreasing the potential barrier h_F and the depletion width w with positive (forward) bias. Here, E_{FE}, E_{FF} and E_{FR} refer to the Fermi levels under equilibrium, forward and reverse bias conditions.

Metal semiconductor junctions are primarily majority carrier[3] devices. Majority carriers flow from the semiconductor to the metal. Minority carrier injection into the semiconductor can usually be neglected. The current across a junction is given by

$$J = J_o \left(\exp \left(\frac{eV}{kT} \right) - 1 \right), \tag{4}$$

where J is the current density for an applied potential V, T is the absolute temperature, k is Boltzmann's constant and J_o is a constant that depends on the Schottky barrier, $\phi_{\beta o}$,

$$J_o = AT^2 \exp \left(-\frac{\phi_{\beta o}}{kT} \right), \tag{5}$$

where A is the Richardson constant for thermionic emission and is a material property. If under forward bias, the applied bias is greater than the turn on voltage, V_T (*i.e.*, greater than and opposite to the built-in potential) the current increases exponentially with applied potential. If the external potential is less than V_T (*i.e.*, adds to the built-in potential) the exponential term in Eq. (4) tends to zero and $J = -J_o$, which is known as the leakage or reverse bias saturation current. It arises from those thermally generated carriers that have enough energy to surmount the barrier and is thus a sensitive function of temperature (as can be seen in Eq. 5). In Fig. 5.1, we show the forward and reverse bias I/V curves. This characteristic shape leads to its unique rectifying behaviour.

5.3.1 Image Force

As we have seen once the barrier is formed, a space charge region depleted of mobile carriers is created in the semiconductor adjacent to the metal layer. Obviously, a thin layer of space charge with opposite polarity must also exist in the metal at the interface to complete the charge dipole and maintain charge neutrality. Consider a single electron close to the metal-semiconductor interface. Its charge attracts an opposite surface charge, which exactly balances it so that the electric field surrounding the electron does not penetrate beyond this surface charge, as shown in Fig. 5.4(a). This can be represented conceptually as the electron and another "electron" of positive charge, located at equal distance on the opposite side of the interface, as illustrated in Fig. 5.4(b). This charge is called the *image charge*. The difference between the actual surface charges and the image charge is that the fields in the metal are distinctly different. Image charges build up in the metal electrode of a metal-semiconductor junction as carriers approach the interface. Schottky [12] argued that the potential energy outside surface of the metal now consists of two components: an electrostatic part due to the dipole charge built up at the interface and an image part, both of which reduce the effective barrier height leading to a net increase in the current flow across the junction. The effect is illustrated in Fig. 5.5.

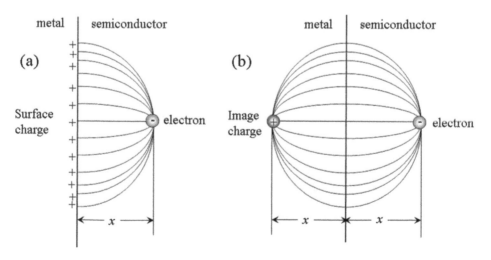

FIGURE 5.4   (a) Field lines and surface charges due to an electron in close proximity to a perfect conductor and (b) an equivalent representation of (a) showing the field lines and image charge of an electron.

---

[3] Simply, the charge carriers (electrons or holes) that are present in large quantity and consequently carry most of the electric charge or current.

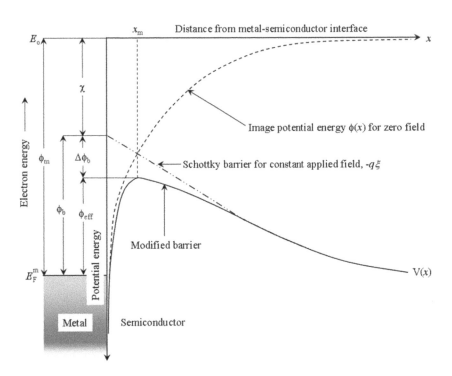

FIGURE 5.5 Modification of the Schottky barrier and subsequent lowering of the barrier height by the image potential. Here E^m_F is the Fermi level of the metal, E_o is the vacuum potential, ϕ_b is the barrier height at the metal semiconductor interface, $\Delta\phi_b$ is the reduction in barrier height, ϕ_{eff} is the effective barrier height, χ is the electron affinity and ϕ_m is the metal work function. The shaded area indicates the filled electronic states in the metal.

5.3.2 Image Force Reduction of the Schottky Barrier

To better understand the image charge and its effect on the barrier height, consider an electron in vacuum, at a distance x from a metal surface (see Fig. 5.4). From the Shockley-Ramo theorem [13,14], a positive charge will be induced on the metal at a distance $-x$ from its surface and will give rise to an attractive force between the two, known as the image force

$$F(x) = -\frac{q^2}{4\pi\varepsilon_o\varepsilon_s(2x)^2} = -\frac{q^2}{16\pi\varepsilon_o\varepsilon_s x^2}, \tag{6}$$

where q is the unit electronic charge, ε_o is the permittivity of free space and ε_s is the relative permittivity of the semiconductor. This force has associated with it an image potential energy that corresponds to the work done by the electron in moving from infinity to x, given by integrating Eq. (6),

$$\phi(x) = -\int_x^\infty F(x)\,dx = -\frac{q^2}{16\pi\varepsilon_o\varepsilon_d x}, \tag{7}$$

where ε_d is the image force dielectric constant, which for most semiconductors is nearly equal to the static dielectric constant.[4] When Eq. (7) is combined with the potential variation due to the electric field, we obtain the total potential energy, $V(x)$, at position x,

$$V(x) = -q\xi_m x - \frac{q^2}{16\pi\varepsilon_o\varepsilon_d x}, \tag{8}$$

where the field due to the charge in the depletion region is assumed to be constant and equal to the maximum electric field at the interface, ξ_m, which is given by

---

[4] The static dielectric constant primarily determines the depletion width through Poisson's equation. The image force dielectric constant is essentially the high-frequency dielectric constant by virtue of the potential sensed by moving charges from the image force.

$$\xi_m = \sqrt{\frac{2qN_d(V_{bi} - V_{bias})}{\varepsilon_o \varepsilon_d}}. \tag{9}$$

Here N_d is the semiconductor donor concentration, V_{bi} is the built-in potential, V_{bias} is the applied voltage and N_d is the ionized donor concentration.

The modification of the Schottky barrier and subsequent lowering of the barrier height by the image potential is illustrated in Fig. 5.5. The potential energy due to the distributed charge of the ionized donors in the depletion region reaches its maximum value at x_m when $\partial V(x)/\partial x = 0$.

$$x_m = \sqrt{\frac{q}{16\pi \varepsilon_o \varepsilon_s \xi_m}} \tag{10}$$

The corresponding maximum value of the potential energy is given by

$$V_{bi} = \phi_m - \phi_s = \phi_b - (E_c - E_F), \tag{11}$$

where $\Delta\varphi_b$ is the barrier height reduction given by [15]

$$\Delta\varphi_b = \sqrt{\frac{q\xi_m}{4\pi\varepsilon_o\varepsilon_d}} = 2\xi_{max}x_m, \tag{12}$$

which for a uniformly doped semiconductor is given by

$$\Delta\phi_b = \left[\frac{q^2 E_b N_d}{8\pi^2 \varepsilon_o^3 \varepsilon_s \varepsilon_d}\right]^{\frac{1}{4}} = \left[\frac{q^2 N_d}{8\pi^2 \varepsilon_o^3 \varepsilon_d^2 \varepsilon_s}\left(V_{bias} + V_d - \frac{kT}{q}\right)\right]^{\frac{1}{4}}, \tag{13}$$

Here E_b is the amount of energy by which the majority carrier band-edge is shifted at the interface to equalize the Fermi levels, V_{bias} is the externally applied bias across the depletion region and V_d is the diffusion potential associated with the barrier (that is ϕ_b minus the Fermi energy). Thus, the actual barrier height, ϕ_{eff}, is now $\phi_b - \Delta\phi$. We see from Eqs. (10) and (12) that with increasing applied electric field, the effective barrier height is continually reduced and the maximum of the potential barrier is moved closer to the metal surface.

Even though the barrier reduction tends to be small compared to the barrier height itself (typically, $\Delta\phi_b < 0.2\phi_b$), it is important since it depends on the applied voltage and leads to a voltage dependence of the reverse bias current (as witnessed by the finite slope in the reverse current in Fig. 5.1). Note that the image force is not zero at zero bias because the electric field at zero bias is not zero.

5.3.3 Barrier Width – Ideal Case

One consequence of the image force is that a depletion layer is created whose width is equal to that needed to support a potential change equal to the built-in potential. If we assume a uniform distribution of ionized impurities in the semiconductor, the shape of the potential energy barrier can be approximated by a one-dimensional solution of Poisson's equation

$$\phi(x) = qV(x) = \frac{q^2 N_d x^2}{2\varepsilon_s \varepsilon_o}, \tag{14}$$

for $0 \le x \le w$. Here, w is the width of the depletion layer, N_d is the ionized donor concentration in the semiconductor, ε_s the relative static dielectric constant and ε_o the permittivity of free space. The width of the depletion layer is related to the amount of energy, E_b, the band has been shifted in the depletion region (see Fig. 5.2) by

$$E_b = \phi_b - \phi_s - qV_{bias} = \frac{q^2 N_d w^2}{2\varepsilon_s \varepsilon_o}, \tag{15}$$

where, ϕ_b is the barrier height, ϕ_s is the position of the Fermi level relative to the conduction band edge and V_{bias} is the applied forward bias. Thus, the depletion width is given by

$$w = \sqrt{\frac{2\varepsilon_s\varepsilon_o(\phi_b - \phi_s - qV_{bias})}{qN_d}} \tag{16}$$

and varies as $V_{bias}^{1/2}$ and $N_d^{-1/2}$. Note that the dependences on impurity concentration and bias are stronger than for the corresponding reduction in barrier height by the image force, which varies as $V_{bias}^{1/4}$ and $N_d^{1/4}$ (see Section 5.3.2). Typically, depletion widths are of the order of tens of nm for moderately doped semiconductors. For example, the unbiased depletion depth for a Cr-Si junction doped to 10^{17}cm^{-3} is ~60 nm, which increases to 260 nm under 5V reverse bias.

5.3.3.1 Non-Ideal Case

In reality, the shape of a metal-semiconductor potential barrier is not truly parabolic because of the interaction of the image force potential, which modifies the energy distribution function as follows [16]

$$\phi(x) = \frac{q^2 N_d x^2}{2\varepsilon_s\varepsilon_o} - \frac{q^2}{16\pi\varepsilon_d\varepsilon_o(w - x)}, \tag{17}$$

where ε_d is the relative dynamic (high frequency) dielectric constant of the semiconductor [17]. Consolidating the barrier and material properties into one term, we obtain [16],

$$\frac{\phi(x)}{E_b} = \frac{V(x)}{E_b} = \frac{x^2}{w^2} - \frac{1}{8\pi(E_b/E_{11})^{3/2}(1 - x/w)}, \tag{18}$$

where $(E_b/E_{11})^{3/2} = N_d w^3 - \varepsilon_d/\varepsilon_s$. The numerator, E_{11}, is a constant of the material, given by

$$E_{11} = \frac{q^2}{2\varepsilon_o}\left[\frac{N_d}{\varepsilon_s\varepsilon_d^2}\right]^{\frac{1}{3}} = 9.05 \times 10^{-7}\left[\frac{N_d}{\varepsilon_s\varepsilon_d^2}\right]^{\frac{1}{3}} \text{ (eV)} \tag{19}$$

and is a measure of the modification of the barrier shape by the image force. In fact, the term $(E_b/E_{11})^{3/2}$ in Eq. (18) uniquely describes the deviation of the barrier shape from parabolic. The zero image force situation occurs when $(E_b/E_{11})^{3/2} \to \infty$.

In Fig. 5.6 we plot the barrier potential energy distribution for *n*-type GaAs, which includes the modification due to the image force for a range of donor carrier densities. We note that for a practical range of impurity donor concentrations[5] the barrier height is only reduced by ~20%. We also see that as the donor concentration increases, not only does the barrier height decrease, but the peak of the potential energy distribution shifts into the semiconductor. The distance, x_m, of the potential energy maximum from the metal-semiconductor interface is

$$x_m = \frac{\Delta\phi_b}{2\xi_i} = \left(\frac{q}{16\pi\varepsilon_d\varepsilon_o\xi_m}\right)^{\frac{1}{2}}, \tag{20}$$

where ξ_m is the maximum electric field that occurs at the metal-semiconductor interface. For Si, $\varepsilon_d = 12$ and x_m varies from 10–50 Å, depending on the applied field [17].

5.3.4 Junction Capacitance

The capacitance of a Schottky barrier is associated with its depletion region, which acts as a dielectric. In some respects, the barrier resembles a parallel plate capacitor in which the separation between the "plates" increases when reverse bias is applied and decreases when a forward bias is applied. In this context, the separation between the plates is equal to the width of the barrier (*i.e.*, the depletion width) given by

---

[5] The practical doping limit for *n*-type GaAs is ~2 × 10^{19} cm^{-3}, although higher concentrations are easier to achieve for *p*-type material.

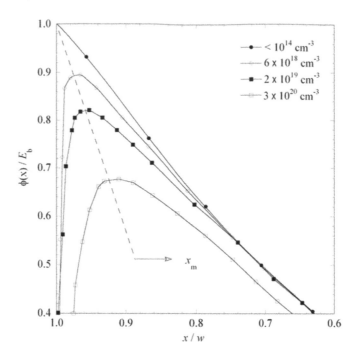

FIGURE 5.6 The effect of the image force on the shape of the potential barrier at a metal-semiconductor interface for *n*-type GaAs with a band-bending energy of 1 eV as a function of the ionized donor concentration, N_d. x_m is the position of the maximum height of the barrier from the metal-semiconductor interface.

$$w = \sqrt{\frac{2\varepsilon_s\varepsilon_0(V_{bi} - V_{bias})}{qN_d}}. \tag{21}$$

The total charge per unit junction area in the depletion region is then given by

$$Q = qN_dw = qN_d\sqrt{\frac{2\varepsilon_s\varepsilon_0(V_{bi} - V_{bias})}{qN_d}} = \sqrt{2\varepsilon_s\varepsilon_0 N_d(V_{bi} - V_{bias})} \ (C/cm^2). \tag{22}$$

The capacitance per unit area associated with this is given by

$$C = \frac{dQ}{dV_{bias}} = \frac{d}{dV_{bias}}\left[\sqrt{2\varepsilon_s\varepsilon_0 N_d(V_{bi} - V_{bias})}\right] = \sqrt{\frac{q\varepsilon_0\varepsilon_s N_d}{2(V_{bi} - V_{bias})}} = \frac{\varepsilon_0\varepsilon_s}{w} \ (F/cm^2). \tag{23}$$

Recall that for a parallel plate capacitor, $C \sim \varepsilon_0\varepsilon_s A/d$, where A is the area of the plate and d is the separation of the plates. However, while the analogy to a parallel plate capacitor implies that the capacitance should be constant, it is not in the metal-semiconductor junction case since the depletion layer width, w, varies with applied bias.

If we now square the reciprocal of Eq. (23), we obtain,

$$\left(\frac{1}{C_b}\right)^2 = \frac{2(V_{bi} - V_{bias})}{q\varepsilon_0\varepsilon_s N_d} = \frac{2V_{bi}}{q\varepsilon_0\varepsilon_s N_d} - \frac{2}{q\varepsilon_0\varepsilon_s N_d}V_{bias} \tag{24}$$

The utility of Eq. (24) is that a plot of $1/C^2$ versus V_{bias} is linear and can be used to extract both the semiconductor doping, N_d, and built-in potential barrier, V_{bi}, from the gradient and intercept respectively. Once V_{bi} is determined, the barrier height, φ_b, can be calculated from Eq. (3). Rearranging,

$$\phi_b = V_{bi} + (E_c - E_F). \tag{25}$$

The value of $(E_c - E_F)$ can be determined from the impurity concentration

$$(E_c - E_F) = \frac{kT}{q} \ln \left(\frac{N_c}{N_d} \right), \tag{26}$$

where N_c is the density of states in the conduction band given by

$$N_c = 2 \left(\frac{m_e^* kT}{2\pi \hbar^2} \right)^{\frac{3}{2}} = 2.51 \times 10^{19} \left(\frac{m_e^* T}{300} \right)^{\frac{3}{2}}. \tag{27}$$

5.3.5 Real Barriers

It was soon found experimentally (see for example, Schweikert [18]) that barrier heights in "real" metal-semiconductor systems vary proportionally with the metal work function, but with a slope much smaller than unity, typically between 0.1 and 0.3. The departure of experiment from theory was first explained in terms of localized electronic surface states [19] or "dangling bonds" resulting from immobilized atoms with unfulfilled valence [20,21] (*i.e.,* broken or missing bonds). The dangling bond may have an unpaired electron, two unpaired electrons or no unpaired electron and arises naturally at the surface of a solid because the atoms have neighbours on one side only. As a consequence, at a metal-semiconductor interface the wave functions of the metal electrons decay exponentially into the semiconductor forming a continuum of metal-induced gap states (MIGS) in the bandgap [22]. Physically, these states may extend up to one nm into the semiconductor and determine the barrier height in an ideal, abrupt, defect-free and laterally homogenous metal-semiconductor contact. However, in practice, there is limited agreement between experiment and theory, since surfaces are rarely ideal, because of defects introduced during interface formation [23], surface contamination and oxide layers introduced during chemical processing.

5.3.6 Metal-Induced Gap States (MIGS)

The MIGS are continuously distributed in energy within the forbidden gap and are characterized by a "neutral level", ϕ_o, such that the surface states are occupied up to ϕ_o and empty above it and hence the surface is electrically neutral. In general, the Fermi level and neutral level do not coincide. In this case, there is a net charge in the surface states. If, in addition there is a thin oxide layer between the metal and the semiconductor,[6] the charge in the surface states together with its image charge on the surface of the metal will form a dipole layer. This dipole layer will alter the potential difference between the semiconductor and the metal, which can be described by a simple modification [24] to the Schottky-Mott equation

$$\phi_b = \gamma(\varphi_m - \chi_s) + (1 - \gamma)(E_g - \phi_o) - \Delta\phi_b. \tag{28}$$

Here, E_g is the bandgap of the semiconductor in eV, ϕ_o is the position of neutral level (measured from the top of the valence band), $\Delta\phi_b$ is the reduction in barrier height due to image force lowering and γ is a weighting factor,[7] which depends mainly on the surface state density and the thickness of the interfacial layer and is given by

$$\gamma = \frac{\varepsilon_o \varepsilon_i}{\varepsilon_o \varepsilon_i + q\delta D_s}. \tag{29}$$

Here, D_s is the density of surface states per unit area per eV at the metal-semiconductor interface, ε_i is the relative permittivity of the interfacial layer and δ is its thickness. The interface states can be classified broadly into two groups – intrinsic and extrinsic. The intrinsic states arise because of the discontinuity in the crystal structure of the solids at the interface, while the extrinsic defects are the result of chemical reactions or damage to the surface of the semiconductor during the metal deposition process. If there are no surface states, $D_s = 0$. Therefore $\gamma = 1$ and Eq. (28) reduces to the classical Schottky-Mott equation given by Eq. (1), except for the image force lowering term, $\Delta\phi_b$. However, if the density of states is large, ε becomes small, $\gamma \to 0$ and ϕ_b approaches a limiting value of $(E_g - \phi_o)$. When this occurs, the Schottky barrier height (SBH) is no longer dependent on the metal work function and the Fermi level and is said to be pinned relative to the band edges by the surface states.

---

[6] This is invariably introduced during chemical processing of the semiconductor. Such an oxide film is referred to as an interfacial layer.
[7] Also known as the Schottky pinning factor.

5.3.7 Fermi Level Pinning

It is found experimentally that SBHs for a wide range of metals in contact with a particular semiconductor fall within a narrow range. In fact, this range is so narrow that it is described empirically by the "one-third" rule, that the SBH is roughly 1/3 of the bandgap in a p-type semiconductor and 2/3 of the bandgap in an n-type semiconductor. Thus, the Fermi level appears to be pinned at $E_g/3$ from the valence band maximum. Experimentally, Fermi level pinning is also found to be dependent largely on the type of bonding. For example, pinning tends to be more prevalent in covalently bonded compounds than in ionically bonded ones and therefore is related to the electro-negativity of a compound, since covalent substances tend to be less electronegative than ionic. Thus, we would expect pinned surfaces to be more common in Group IV and Group III-V materials than in II-VI or I-VII compounds. Experimentally, this is indeed found to be the case. In Fig. 5.7(a) we plot the quantity $E_c - E_F$ for Au contacts formed on various covalent semiconductors, as a function of bandgap energy. The data are well fitted by a straight line of slope 2/3. In contrast, Fig. 5.7(b) shows experimentally determined barrier heights for various metals on ZnS (a Group II-VI ionic semiconductor), which display a marked dependence on metal work function, varying as $\sim \phi_m^{0.7}$. For comparison, we show curves for a covalent compound (GaAs), which essentially show no variation, thus $\phi_b \sim \phi_m^{0.0}$. The difference has recently been attributed [26] to the observation that the penetration of MIGS in covalent semiconductors is deeper than in ionic semiconductors and hence changes in metal work function are more effectively screened out. Calculation has shown [27] that the penetration depth of MIGS is typically 0.1 nm in ionic semiconductors and ~0.3 nm in covalent semiconductors.

5.4 Current Transport across a Schottky Barrier

The current across a metal-semiconductor junction is mainly due to majority carriers and as we will see is strongly influenced by the doping concentration in the semiconductor, N_d, and the junction temperature, T. There are three principal components to the overall current through the device [28,29]. These are thermionic emission (TE), thermionic field emission (TFE) and field emission (FE). The first component gives rise to the current rectification in metal-semiconductor diodes and is primarily dependent on the temperature and the barrier height. The other two components involve quantum mechanical tunneling through the barrier and are critically dependent on the impurity concentration level. The effect of impurity concentration on the structure of the metal-semiconductor interface is illustrated in Fig. 5.8 for an n-type semiconductor.

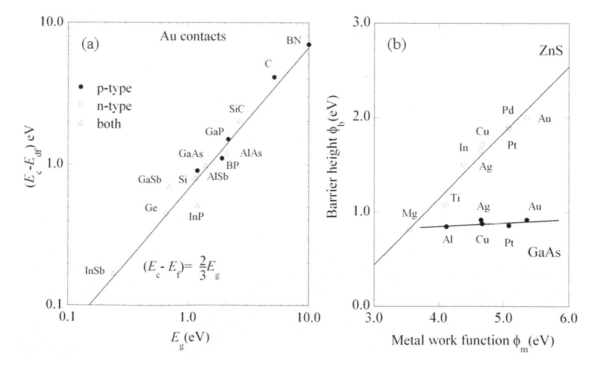

FIGURE 5.7 (a) The location of the Fermi level relative to the conduction band ($E_c - E_F$) barrier heights for Au contacts on various covalent semiconductors, plotted as a function of energy gap, illustrating the two-thirds rule for barrier height pinning for n-type semiconductors at the interface (adapted from ref [25]). (b) Experimentally determined barrier heights for various metals on an ionic semiconductor (ZnS) showing a clear dependence of the barrier height on work function. For contrast, we show data for a covalent semiconductor (GaAs), which has a much weaker dependence (from [25]).

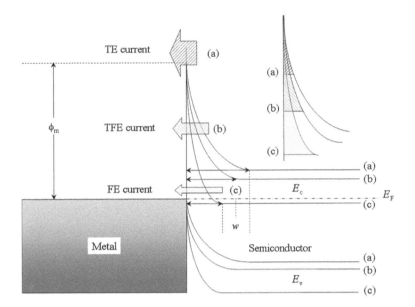

FIGURE 5.8 Conduction mechanisms through a metal/semiconductor interface with different semiconductor donor levels for an *n*-type semiconductor, corresponding to (a) a low residual doping level of < 10^{17}cm^{-3}, (b) an intermediate doping level of 10^{17}cm^{-3}–10^{18}cm^{-3} and (c) a high doping level of >10^{18}cm^{-3}. It can be seen that as the doping concentration increases, the barrier width decreases as the conduction band falls below the Fermi level. To the right of the barrier we show an additional schematic of the barrier in which the shaded regions give a visual indication of the fraction of carriers that can cross the barrier.

5.4.1 Thermionic Emission (TE)

Thermionic emission is the rate limiting process for electron emission over a barrier when the temperature is high and a non-degenerate semiconductor is used. In a Schottky diode, TE is the dominant current transport mechanism. The current through the device is a two-step process. Firstly, the electrons have to be transported through the depletion region by the processes of drift and diffusion, and secondly they must undergo emission over the barrier into the metal. Fig. 5.8(a) shows the interface band structure for an ideal Schottky barrier, which would be appropriate for a lightly doped semiconductor with a residual impurity doping concentration, $N_d < 10^{17}$cm^{-3}. In this case, the depletion width is too wide to allow quantum mechanical tunneling and therefore the only way for carriers to cross the barrier is by thermal excitation over it. The current through the barrier is dependent on the number of electrons that impinge on unit area of the metal per second and is given by

$$I_{TE} = AA^*T^2 \exp\left(\frac{-q\phi_b}{kT}\right)\left[\exp\left(\frac{-qV_{eff}}{nkT}\right) - 1\right], \tag{30}$$

where A is the cross-sectional area of the metal/semiconductor interface, A^* is the modified Richardson constant,[8] T is the temperature in K, k is Boltzmann's constant q is the electronic charge, V_{eff} is the effective bias across the interface and n is the ideality factor, which is a dimensionless quantity introduced to include contributions from other current-transport mechanisms. This mode of current transport is commonly referred to as thermionic emission current [29] and varies strongly with kT and barrier height. The ideality factor is a measure of the quality of the junction and is highly process dependent. For an ideal Schottky junction, $n = 1$; hence, the emission is purely thermionic. In practice, however, n lies in the range $1 < n < 2$ due to the presence of non-ideal effects, mainly image force reduction of the barrier [29] and edge leakage currents. In fact, in the absence of other effects, image force lowering of the barrier will result in an ideality factor given by

$$n = \frac{1}{\left(1 - \frac{\Delta\phi_b}{V}\right)}, \tag{31}$$

where $\Delta\phi_b$ is the image force lowering of the barrier given by Eq. (12) and V is the applied bias. If tunneling effects become significant, the ideality factor can reach values > 2. In this case, the junction performs better under reverse bias. In fact, as

---

[8] Modified form of the Richardson constant for thermionic emission from a metal where the free electron mass m, is substituted by the semiconductors effective electron mass m^*.

pointed out by Rideout [29], the deviation of the diode *n*-value from unity may be used as a measure of the relative contribution of thermionic emission and thermionic field tunneling to the conduction process.

5.4.2 Thermally Assisted Field Emission (TFE)

For a moderately doped semiconductor ($N_d \approx 10^{18}$ cm^{-3}), the depletion layer can become sufficiently thin that thermally assisted field emission (TFE) or alternately, thermally assisted tunneling, can take place in which hot carriers[9] begin to tunnel through the top of the barrier. This is illustrated in Fig. 5.8(b). The derivation of an analytical form describing the current through the barrier rapidly leads to intractable expressions, since it is generally derived by expanding the tunneling probability in a Taylor series about the Fermi energy of the source material. Padovani and Stratton [30] approached the problem by assuming a simple parabolic barrier shape by ignoring image force barrier lowering effects. It was found that the current due to TFE varies as

$$\exp \left(\frac{\phi_b}{E_{oo} \coth \left[\frac{E_o}{kT} \right]} \right), \tag{32}$$

where E_o is a constant given by,

$$E_o = E_{oo} \coth \left(\frac{E_{oo}}{kT} \right) \tag{33}$$

and E_{oo} is a tunneling parameter[10] related to the material properties of the semiconductor, given by the expression

$$E_{oo} = \left(\frac{qh}{4\pi} \right) \sqrt{\frac{N_d}{m^* \varepsilon_0 \varepsilon_r}} = 1.85 \times 10^{-11} \sqrt{\frac{N_d}{m_r \varepsilon_r}} \, (\text{meV}). \tag{34}$$

Here, h is Planck's constant, N_d is the impurity doping concentration and m^* is the effective mass of electron $= m_r m_o$.

Although TFE conduction has a complicated dependence on T and N_d, we can make some general statements. Since E_{oo} varies as $\sqrt{N_d}$, therefore E_{oo}/kT is proportional to $\sqrt{N_d}/T$ and as the temperature increases, the fraction of current transported due to thermionic emission also increases. However, if the doping is increased, the barrier width is reduced and the fraction due to quantum mechanical tunneling is enhanced.

For all cases apart from very low bias, the tunneling current, I_t, is given by

$$I_t = I_{to} \left[\exp \left(\frac{qV_{bias}}{E_o} \right) - 1 \right], \tag{35}$$

where I_{to} is the tunneling saturation current, which is a complicated function of temperature, barrier height and semiconductor parameters. If we treat the material properties empirically, then we can incorporate them into the diode equation and arrive at the following approximate form for the combined TE and TFE currents [28]

$$I_t = I_w \left[\exp \left(\frac{qV_{bias}}{nkT} \right) - \exp \left(\left(\frac{1}{n} - 1 \right) \frac{qV_{bias}}{kT} \right) \right], \tag{36}$$

which is applicable for both forward and reverse bias. When *n* equals unity, Eq. (36) reduces to the classical diode equation given by Eq. (30).

5.4.3 Field Emission (FE)

As the impurity concentration increases further, more and more thermal carriers begin to tunnel through increasingly wide regions of the barrier, until at some point ($N_d \geq 10^{19}$ cm^{-3}) significant numbers of carriers can even tunnel through the base as

---

[9] The term "hot carriers" refers to either holes or electrons (also referred to as "hot electrons") that have gained very high kinetic energy and are not in thermal equilibrium with the lattice. Typically, carrier energization or "heating" occurs in regions with very high electric fields.

[10] The quantity E_{oo} is directly related to the expression derived for the barrier transmission derived from a WKB quantum mechanical treatment.

illustrated in Fig. 5.8(c). Conduction now takes place through the entire barrier, and the carrier flow is referred to as field emission (FE), or carrier tunneling emission and is the preferred mode of current transport in metal-semiconductor ohmic contacts. The current flow in this case is found to depend mainly on the barrier height and the tunneling parameter E_{oo}, thus

$$I_{FE} \propto \exp\left(\frac{\phi_b}{E_{oo}}\right), \tag{37}$$

with little dependence on temperature. Since E_{oo} is proportional to $\sqrt{N_d}$, the forward bias characteristic for field-emission dominated conduction is strongly dependant on the doping concentration. The transition from FTE to FE is illustrated in Fig. 5.9. As with FTE, an analytic formulation of FE current transport is complicated by the necessity to use Taylor expansions. An additional complication is that the theories of FE and TFE are distinct and do not converge to one another as the fields and energies approach. However, Padovani and Stratton [30] showed that the transition between FTE and FE can be defined in terms of a constant c_1, given by

$$c_1 = \frac{1}{2E_{oo}} \log \left[\frac{4(E_b - E)}{\zeta}\right], \tag{38}$$

where E_b is the potential energy of the top of the barrier with respect to the Fermi level, E is the potential energy associated with an applied bias V and ζ is the energy of the Fermi level with respect to the bottom of the conduction band. FTE dominates when $1/c_1 < kT$ and FE when $1/c_1 > kT$.

5.4.4 Relative Contributions of TE, TFE and FE

A useful parameter indicative of the electron tunneling probability is the quantity kT/E_{oo}. When E_{oo} is high compared to the thermal energy kT, the probability of electron transport by tunneling increases. Therefore, the ratio kT/E_{oo} is a useful measure of the relative importance of the thermionic process to the tunneling process. For lightly doped semiconductors, E_{oo} will be small (Eq. 34) and $kT/E_{oo} \gg 1$. In this case, TE will dominate, and the contact is rectifying. For high levels of doping, $kT/E_{oo} \ll 1$ and FE now dominates and the contact is ohmic. For intermediate levels of doping ($kT/E_{oo} \approx 1$) neither field nor thermionic emission accurately describes the conduction process. Note, that both TE and TFE are temperature dependent while FE is not.

5.4.5 Estimated Contact Resistances for TE, FTE and FE Current Modes

The current-voltage relations and therefore the expected contact resistances for the three main current transport mechanisms can be calculated by applying the semi-classical, quantum mechanical approximation of Wentzel-Kramers-Brillouin (WKB) to the simple energy band model shown in Fig. 5.8. For thermionic emission, ρ_c is given by [26,31]

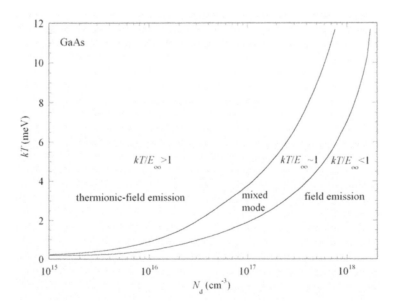

FIGURE 5.9 Ranges of temperature and donor concentration over which n-type GaAs Schottky diodes exhibit field and thermionic-field emission (adapted from [31], reproduced by permission of the Institution of Engineering & Technology).

$$\rho_c = C_1 \exp\left(\frac{q\phi_b}{kT}\right), \tag{39}$$

where $C_1 = (k/qA)T$. For contacts with heavy doping in which the tunneling process (FE) is the dominant current transport mechanism, ρ_c is given by

$$\rho_c = C_2 \exp\left(\frac{q\phi_b}{E_{oo}}\right) = C_2 \exp\left[\frac{4\pi\sqrt{\varepsilon m^*}}{h}\left(\frac{\phi_b}{\sqrt{N_d}}\right)\right], \tag{40}$$

where C_2 has a weak temperature dependence. For the contacts in which TFE is the dominant transport mechanism, ρ_c, is given by

$$\rho_c = C_3\left(\frac{\phi_b}{\sqrt{N_d}\coth(E_{oo}/kT)}\right), \tag{41}$$

where C_3 is a function of ϕ_b and T. These equations show that a reduction of ρ_c can be achieved by reducing the ϕ_b value and/or increasing the N_d value in the vicinity of the metal-semiconductor interface. In the case of TE and TFE, ρ_c can also be reduced by increasing the temperature.

5.4.6 Other Current Components

In addition to primary current mechanisms, image-force lowering of the potential barrier and surface-leakage currents are additional consequences of recombination in the depletion region.

5.4.6.1 Current Due to Image Force Lowering of the Potential Barrier

The current due to image force lowering of the potential barrier is difficult to calculate, since the image force reduces not only the barrier height but also the width leading to thermionic, field-assisted thermionic and field components. Rideout and Crowell [16] have modelled current transport across a Schottky barrier for three cases of image force lowering and show that the inclusion of the image force can lead to a significant increase in the magnitude of the current density.

5.4.6.2 Generation-Recombination Effects

The generation and recombination of carriers within the depletion region give rise to a parallel component of the thermionic emission current transport mechanism, which is equivalent to minority carrier injection [32]. Recombination is presumed to take place *via* Shockley-Read centers located near the middle of the bandgap. The current contribution, I_{gr}, due to this mechanism is given by [33]

$$I_{gr} = I_{gro}\left[\exp\left(\frac{qV_{bias}}{2kT}\right) - 1\right], \tag{42}$$

where I_{gro} is the generation-recombination saturation current, which in turn is given by

$$I_{gro} = \frac{qn_i wA}{2\tau_o}. \tag{43}$$

Here, n_i is the intrinsic carrier concentration of the semiconductor (which is proportional to $\exp(-qE_g/2kT)$), w is the width of the depletion layer and τ_o is the effective carrier lifetime within the depletion region. The relative importance of thermionic emission and of recombination in the depletion region depends on ϕ_b, E_g, T and τ. Both generation and recombination components tend to be important in metal-semiconductor interfaces with high barriers, materials with low lifetimes (such as GaAs), at low temperatures and/or junctions operated at low biases.

5.4.6.3 Surface Leakage Current

Surface leakage current, I_{leak}, is another parallel component of the total current, and even though it does not directly depend on the contact we include here for completeness. It is caused by leakage that flows near the edge of the contact instead of uniformly throughout the metal and across the metal-semiconductor interface. Its magnitude tends to be unpredictable but can usually be reduced significantly by careful design and fabrication techniques. Because it is primarily a surface phenomenon, it bypasses the metal/semiconductor interface altogether and is often thought of as a large leakage resistor, R_{leak}, in parallel with it. For a given detector bias, V_{bias}, the leakage current can be expressed as

$$I_{leak} = V_{bias}/R_{leak}. \tag{44}$$

5.4.7 Practical Application of Schottky Barriers

The rectifying properties of the Schottky barrier are exploited practically in the form of Schottky diodes,[11] which have a number of important advantages over p-n devices. For example, Schottky diodes are unipolar components; meaning only one type of charge, the majority carriers are responsible for current transmission. So, if the semiconductor body is doped *n*-type, only the *n*-type carriers (mobile electrons) play a significant role in the normal operation of the device. These carriers are quickly injected into the conduction band of the metal contact on the other side of the diode and as a consequence, the diode can conduct and cease conduction much faster than an ordinary p-n diode, in which device speed is dependent on the slower diffusion and recombination processes of minority carriers injected from both sides of the p-n junction. When coupled with a much lower conducting forward bias[12] (the so-called cut-in voltage), a Schottky diode is highly suitable for switching and high frequency applications. It also ensures a steeper I/V characteristic, which results in a much higher current for a given bias than for a p-n diode. As a result, Schottky diodes are preferred as low voltage, high current rectifiers. Another advantage of Schottky diodes is that there is essentially no recombination in the depletion region, and as a result ideality factors are very close to unity – in contrast to p-n diodes, where there is significant recombination in the depletion region and ideality factors range from 1.2 to 2.0.

As well as positive attributes, Schottky diodes have several disadvantages. For example, their blocking voltages are limited by the reverse current, which can be 3 to 4 orders of magnitude higher than in p-n devices and varies steeply with temperature. This can lead to thermal instability issues, which often limit the useful reverse voltage to well below the actual rating. Also, the electrical performance of a Schottky diode is critically dependent on the quality of the Schottky junction, which in turn is critically dependent on the material processing route. For some semiconductors this is poor, and rectification can more easily be achieved using p-n junctions.

5.5 Ohmic Contacts

All semiconductor devices generally need at least one ohmic contact; it is often the quality of this contact that significantly affects the performance of semiconductor devices. As discussed previously, the term ohmic refers in principle to a metal-semiconductor contact that (a) is non-injecting, (b) has a linear I-V characteristic in both directions and (c) has negligible contact resistance relative to the bulk or spreading resistance[13] of the semiconductor. In practice, the contact is usually acceptable if it does not perturb the performance of the device substantially and can supply the required current density with a voltage drop that is small compared to the drop across the active region, implying that the contact resistance should be small. A small contact resistance is important for other reasons. For example, the RC time constant associated with the contact resistance may limit the frequency response of devices.

Consider the metal/*n*-type semiconductor junction shown in Fig. 5.8. The conduction properties of the contact are determined by the current transport mechanisms discussed in the Schottky barrier section. However, a true Schottky contact is not the best system for extracting signals from a device since it is difficult to obtain (a) a low specific contact resistance and (b) a linear and symmetric current-voltage relationship within the limits of its intended use. Ohmic contacts are thus preferred. Consequently, considerable effort has been expended in making metal-semiconductor contacts look electrically like ohmic contacts. There are two main methods to achieve this. The first is to lower the barrier height by choosing a suitable contacting material. Whether a contact is ohmic or Schottky depends primarily on the work functions of the metal and the

---

[11] Also known as surface barrier or hot carrier diodes.

[12] A Schottky diode typically has a voltage drop in the range 0.15V – 0.45V, whereas a conventional p–n junction diode may have a voltage drop between 0.6–1.7 V.

[13] Due to the non-linearity of the electric field around the contact.

6.2.4 Etching

Even though finer polishing agents may reduce the visibility of defects, considerable damage can exist in the form of filled-in scratches, microcracks, pits and dislocations, particularly in softer materials. These defects will affect the electronic properties of the material, so prior to further processing, the polished surface is sometimes etched in order to remove the damaged layer, which for soft materials may be several hundred microns thick. Chemical etchants are commonly used, different for different crystals, although a bromine-methanol solution is commonly used for most materials. Etching is usually carried out in an etching cabinet equipped with services such as ventilation, de-ionized water supply and a compressed air system to clean and dry the crystals. Fig. 6.4 shows such an example, which also features an automatic titration system for preparing different etch solutions. A comprehensive list of etchants for over 50 different metals, semiconductors and clean room materials is given in reference [2] along with some etch rates. For Group III-V materials, a discussion of chemical etchants and their properties may be found in Faust [3].

Besides removing damaged layers, etchants are useful for revealing bulk defects present in the crystals, since they preferentially attack defect sites, revealing dislocations and other atomic arrangement faults. Wet chemical etching (or preferential etching) is one of the oldest methods used for the characterization of single crystals and today is the most commonly used technique for evaluating defects in semiconductor wafers due to its easy sample preparation and low cost. Other benefits are that it is sensitive and can be applied to large areas. The technique was originally employed in the 19th century for defining the crystallographic orientation of natural crystals, although at that time it was not understood that etch pits were formed due to the presence of crystallographic defects. The procedure for preferential etching is relatively simple. First, chemical mechanical polishing is applied to remove any specific layer and to flatten the surface. Next, etching is performed with a solution (usually a molten alkali, such as potassium hydroxide) that dissolves the material faster at the defects than at the perfect regions and makes the defects visible as etch pits. In the case of dislocations, when a dislocation line intersects the surface of a metallic material, the associated strain field locally increases the relative susceptibility of the material to acidic etching and an etch pit of regular geometrical format results. Not only can etching proceed preferentially along certain crystallographic axes, the shape of the etch pits can also be used for determining the orientation of the crystal surface. For example in Si, etch pits are elliptical in shape along the [000] crystallographic axis, whereas they are triangular or pyramidal in shape along the [111] axis (see Fig. 6.5). If the material is strained (deformed) and repeatedly re-etched, a series of etch pits can be produced that effectively trace defects. Dislocation densities in particular may be assessed in this manner by measuring the Etch Pit Density (EPD). The EPD is determined directly by measuring the number of "pits" in the surface optically using a variety of techniques depending on the spatial resolution required. Generally, phase contrast microscopy is used. The advantage of this technique is that it is simple, inexpensive and sensitive and requires no special knowledge, as in the case of TEM. If damage information is needed for different layers, step etching can be used. The specific drawbacks of the technique are that the results can be difficult to interpret and that they are not systematic for different materials and temperatures. In addition, detection limits are not specific. The EPDs for the elemental semiconductors tend to be low – typically < 100 cm^{-2} for Si wafers and $\sim 10^3$ cm^{-2} for the Ge used to make high purity radiation detectors, primarily due to impurities. The EPDs for compound semiconductors tend to be much higher (10^4–10^6 cm^{-2}) because of impurities and stoichiometric imbalances.

FIGURE 6.4 Etching cabinet with automatic titration system for preparing different etch solutions. The cabinet also contains services such as a de-ionized water supply and compressed air system to clean and dry the crystals (Image courtesy European Space Agency).

FIGURE 6.5 Dislocation etch pits in Si, showing the effects of preferential etching along different crystallographic directions. In the [100] orientation (a), the etch pits appear elliptical in shape, whereas along the [111] direction they can assume triangular or pyramidal shapes: (b) and (c).

6.2.5 Cleaning

During device processing, it is virtually impossible to screen the wafer/crystal from contaminants commonly encountered in process environments. These originate from the ambient air and storage ambient (also known as aging), as well as from materials introduced during the fabrication process (*e.g.*, gasses, chemicals, materials and water). Furthermore, manufacturing tools as well as the personnel operating them in the cleanroom are sources of contamination. Unless removed using dedicated cleaning operations, these contaminants will cause defect formation, performance degradation or even failure. In fact, in the semiconductor industry, functional devices cannot be produced with acceptable yields without thorough cleaning at all stages of the manufacturing process. Consequently, cleaning is the most applied operation in the semiconductor industry.

The most common contaminants encountered in the semiconductor-processing environment are (1) particles that adsorb on the sensitive surface, (2) metallic contaminants originating from liquid chemicals and water, (3) volatile organic compounds present in the ambient air, including clean-room air, (4) native or chemically induced oxides and (5) moisture, originating from the ambient air or residues of wet processes. An illustration of how these contaminants are adsorbed onto a hydrophilic silicon surface because of "aging" is shown in Fig. 6.6.

Cleaning involves the removal of this contamination using various cleaning steps and solutions, which may contain ammonia, methanol, hydrogen peroxide, hydrofluoric acid, hydrochloric acid and deionized water, depending on the semiconductor being processed. There are essentially two forms of cleaning – wet and dry. Wet cleaning is the most widely used technique in semiconductor device manufacturing, because of its overall efficiency coupled with its ability to remove all

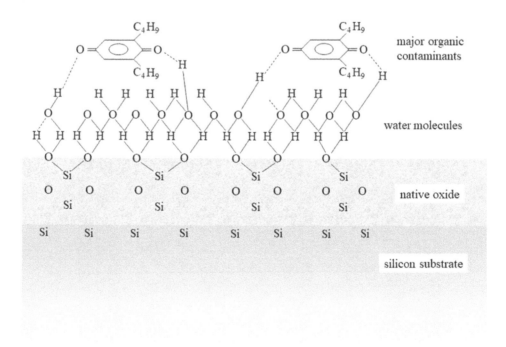

FIGURE 6.6 Surface contamination (moisture and organics) adsorbed onto a hydrophilic silicon surface because of "aging" (adapted from [4], (reproduced with permission from *J. Electrochem. Soc.* Copyright 1996, The Electrochemical Society).

kinds of contaminants. Dry cleaning generally supplements wet cleaning in selected applications, primarily related to surface conditioning; it is most often used to clean "aged" surfaces, prior to processing steps. The relative strengths and weakness of both techniques are summarized in Table 6.1.

6.2.5.1 Wet Cleaning

Wet cleaning is the most prevalent cleaning process used in the semiconductor industry and involves the removal of contaminants from a surface *via* selective chemical reaction in the liquid phase. Wet cleaning agents use combinations of acids, solvents, surfactants and deionized water to selectively dissolve, oxidize, etch and scrub contaminants from the wafer surface – the exact recipe depends on application and the extent of surface contamination; it is usually proprietary. The process is often enhanced by megasonic agitation.[4] An integral part of every wet cleaning sequence is rinsing in ultra-pure deionized (DI) water, often heated to increase rinse efficiency and often with ozone. The rinse sequence effectively stops chemical reactions on the surface and washes reactants and reaction products off the surface. A wet cleaning/rinse sequence is always completed with a wafer/crystal drying process.

Wet cleaning can be implemented by immersion, spray and spin cleaning. The simplest method is immersion in a cleaning bath. This method assures uniform exposure of both surfaces of the wafer/crystal to cleaning chemistries and is fully compatible with megasonic agitation. It also allows precise control of the bath temperature and assures uniform rinsing. Immersion cleaning is implemented in wet benches, which include several cleaning/rinsing tanks. For wafer processing, cassettes are moved in the desired sequences from tank to tank. In an alternative approach, known as centrifugal spray cleaning, cassettes with wafers are rotated in chemicals, or rinsing water, spray dispensed from the nozzles located in the in the center and on the outside walls of the process chamber. This technique offers savings on chemicals and DI water, but its uniformity in high-end applications may not be sufficient. The third approach is a spin cleaning, a single-piece process typically used in the case of large diameter wafers. In this technique, chemicals or rinsing water are dispensed onto the rapidly rotated wafer. To increase efficiency, a higher-pressure jet spray can be employed and the dispensing nozzle moved along the diameter of the rotating wafer. The relative performance of each technique may vary from application to application and selection will be based on the balance between cost of the chemicals and water and the required performance of processed devices. In some applications, wet cleaning methods can be supplemented with dry cleaning methods.

6.2.5.2 Dry Cleaning

Dry cleaning is the process of removing contaminants from the semiconductor surface via chemical reaction in the gas phase. This can be achieved, by (a) conversion of the contaminant into a volatile compound through chemical reaction, (b) "knocking" the contaminant off the surface *via* momentum transfer with a stream of inert gas or (c) as a result of surface irradiation (*i.e.*,

TABLE 6.1

A comparison between dry and wet cleaning techniques.

Technique	Wet	Dry	Dry
Method	Chemical solutions	Ion bombardment or chemical reactive	UV illumination/ozone
Environment	Atmosphere	Vacuum	ambient
Equipment	Bath	Vacuum chamber	UV/halogen lamp
Advantages	• Low cost, relatively easy to implement • High etching rate, high throughput • Good selectivity for most materials • Highly selective	• Highly selective • Capable of defining small feature sizes (<100 nm)	• Very low cost • Simple to implement • Most suitable for oxidative removal of adsorbed organics • Fast cleaning time 1 – 5 minutes
Disadvantages	• Inadequate for defining feature size • Potential of chemical handling hazards • Wafer contamination issues	• High cost, hard to implement • Low throughput • Poor selectivity • Potential radiation damage	• Only really developed for Si/GaAs wafers. Limitedto light organics – not effective for most inorganics or metals. Ozone extraction system required for UV lamp system
Directionality	Isotropic (except for etching crystalline materials)	Anisotropic (etching mainly normal to surface)	Anisotropic most effective when normal to the surface

---

[4] Megasonic is used in preference to ultrasonic agitation, since it does not cause the violent cavitation effects found at ultrasonic frequencies, thereby greatly reducing the likelihood of surface damage.

IR–heating, Halogen/UV–bond breaking/oxidation) sufficient to detach the contaminant from the surface. An example of the latter is lamp cleaning (also known as Rapid Optical Surface Treatment or ROST), which is an excellent example of a simple, yet efficient, dry cleaning process. In lamp cleaning, the semiconductor surface is heated by halogen lamps to a few hundred °C in ambient air. For Si wafers, an exposure of only ~30 s is sufficient to remove virtually all volatile contaminants accumulated during wafer processing and handling. The technique is also effective in removing organic compounds remaining after IPA (isopropyl alcohol) drying. The effectiveness of the technique is illustrated in Fig. 6.7, in which we show thermally desorbed gas chromatography spectra from a Si wafer surface that had been exposed to cleanroom air for 24 hours, allowing the buildup of organic combination. (a) was taken before lamp cleaning and (b) after. The peaks in the spectra *reflect* the composition and relative abundances of organic contaminants on the sample surface [5]. From the figure, we see a marked reduction in the peaks following lamp cleaning; in fact, Danel *et al.* [6] have shown that the method is even effective in not only removing the nm thick native oxide layer formed after brief exposure in air but can also be used as a metrological tool to measure its thickness.

6.2.5.3 Surface Conditioning

Both wet and dry cleaning operations remove chemical contaminants and particles from the semiconductor surface but can also leave it chemically active. In the case of wet chemistries, an important role is played by the aggressiveness of the DI water rinsing process employed. Thus, the specific chemical makeup of semiconductor surface is a by-product of the cleaning/rinsing process. Surface conditioning is used to alter the chemical state of the semiconductor surface in a controlled fashion. It is generally used prior to each processing step, and in each case the chemistry of the conditioning process is determined by the needs of the subsequent operation. For example, the requirements regarding chemical makeup of the surface prior to oxide etching are quite different to those required for contact deposition. Most often the goal is to stabilize the surface, preventing further chemical reactions before the next stage of processing. For example, a "clean" Si surface, will inevitably acquire a ~ 1 nm thick layer of spontaneously grown SiOx, containing Si-H and Si-OH groups resulting from the cleaning/rinsing processes as well as organic contaminants from the ambient. Such a surface is chemically active, which means that it will further undergo modifications of its chemical makeup depending on the ambient in which it is stored and the time of exposure. One way to condition such surface towards chemical passivity and

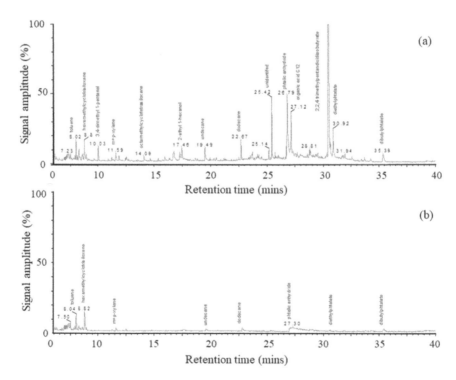

FIGURE 6.7 Gas Chromatography Mass Spectroscopy from the surface of an "aged" Si wafer. Each peak gives information on different organic species (contaminants)[5] (a) before (left) and (b) after (right) lamp cleaning at 400°C for 30s in ambient air (figure courtesy A. Danel). The retention time is the amount of time elapsed from the injection of a sample into the chromatographic system to the recording of the peak maximum of a particular chemical component in the spectrum.

---

[5] Strictly, what are being measured are ionized fragments. The different sets of fragments are compared to a library to identify possible molecules (for a review of TD-GC-MS see [5]).

ensure reproducibility of its characteristics is to saturate the Si bonds on the surface with hydrogen by careful emersion in $HF:H_2O$ solution. In this case, hydrogen atoms in the solution will terminate and neutralize the dangling surface Si atoms, preventing uncontrolled oxidation. This "hydrogen cap" can be easily removed immediately prior to subsequent processing steps, such as the epitaxial deposition of a Si or SiGe layer.

6.3 Electrode Deposition Methods

For any semiconductor detector, at least one pair of contacts is required to (a) supply bias, (b) collect the charge generated by the radiation the device is designed to detect and (c) control the interface properties such as Schottky barrier height and contact resistivity. By contacting, we mean a union, or junction, of surfaces. This can be achieved by several methods, such as applying conductive paints or pastes, melting metals directly on the semiconductor surface, metal sintering, physical vapor deposition (such as sputtering, evaporation and molecular beam epitaxy), ion-implantation and others. For simple geometries, contacts can be applied by transmission directly through a shadow mask, usually fabricated out of a thin metal foil such as brass. For more complex geometries, photolithographic techniques are used. Once deposited, the electrodes are usually thermally annealed using a conventional oven. This promotes the formation of an alloy between the metal and semiconductor, improving the bond; it is sometimes required to achieve the desired values of contact resistivity.

For a review of contacting systems and compound semiconductor interfaces, see references [7–14]. We review the more common techniques in the following sections. A summary of the relative pros and cons of each technique is given in Table 6.2.

TABLE 6.2

Comparison of commonly used contacting methods. Note that the tabulated values are somewhat subjective and only reliable when comparing methods within deposition techniques (*e.g.*, electrodeposition, PVD).

Parameter	Paint/pastes	Sputtering	Evaporation		Electrodeposition	
			Resistive	Electron beam	Electroplating	Electroless
Rate	N/A	One atomic layer/s	One atomic layer/s	1,000 atomic layers/s	1–10 µm/min	1- µm 10/hr
Choice of materials	Limited to a few metals and carbon	Almost unlimited	Limited to materials with low volatility	Limited, but more materials than resistive	Must be conductive	Almost unlimited
Purity	N/A composition fixed	Possibility of incorporating impurities	Possible contamination from heater wire	High	Not high	Not high
Substrate heating	N/A	Low using magnetron	Low	Can be a problem	N/A	N/A
Substrate damage	Can chemically react with substrate	Possible ionic bombardment damage	Very low	X-ray damage possible	Chemical compatibility with substrate?	Chemical compatibility with substrate?
In-situ cleaning	Not an option	Easily done with sputter etch	Not an option	Not an option	Not an option	Not an option
Alloy composition stoichiometry	N/A	Can be tightly controlled	Little or no control	Little or no control	Little or no control	Little or no control
Change in source material	Not really an option	Expensive	Easy	Easy	Relatively straight forward	Relatively straight forward
Decomposition of material	N/A	Low	High	Higher than resisitive evap.	High	Lower than electroplating
Uniformity	Poor over large areas	Good over large areas	Difficult to control	Difficult to control	Difficult to control	Good over large areas & complex shapes
Thickness control	Poor; limited to 100 microns	Good	Not easy to control	Not easy to control	Poor for complex shapes	Good
Adhesion	Often poor	Excellent	Poor compared to sputtering	Poor compared to sputtering	Often poor	Good
Capital equipment	Very inexpensive	Can be very expensive	Expensive	More expensive	Relatively inexpensive	Relatively inexpensive
Hazardous waste	No	No	No	No	Yes	Yes
Ease of use	Very easy	good	easy	Not so easy	Not so easy	Not so easy

6.3.1 Metal Paints and Pastes

A simple and convenient method for forming contacts (especially when prototyping or testing) is to apply a conductive paint or paste onto the surface to be contacted. Silver paint is commonly used as well as Aquadag[6] and derivatives such as Electrodag, which are colloidal dispersions of graphite and silver, respectively, in a carrier (water, in the case of Aquadag, and a fluoroelastomer resin in the case of Electrodag). All can be easily applied to most surfaces by conventional spray, brush or dip methods. However, while the applied films have good adhesion and low sheet resistance (<120 $\Omega/\square$), they are generally soft, making it difficult to attach wires without the use of a mechanical spring or conductive glue.

6.3.2 Electrodeposition

Electrochemical deposition, or electrodeposition [15] for short is a generic term for plating technologies including electroplating and electroless plating and is ideally suited for producing thin metallic films (*e.g.* copper, gold and nickel) with thicknesses ranging from ~1 µm to 100 µm. Electroplating [16] is an electrolytic process in which an electric current is used to reduce cations of a desired material, *M*, from a metal salt solution, *MA*, and coat the object to be plated with a thin layer of the material. In a sense, it is analogous to a galvanic cell acting in reverse; it was first demonstrated by the Italian chemist Luigi Brugnatelli in 1805. Plating is usually carried out in an electrolytic cell, consisting of two electrodes, an aqueous, organic or fused salt electrolyte and external source of current. The part to be plated forms the cathode of the circuit while the metal used to plate the cathode forms the anode. Both the anode and cathode are immersed in an electrolytic solution containing one or more dissolved metal salts as well as other ions that permit the flow of electricity. A power supply supplies a direct current to the anode, oxidizing the metal atoms that comprise it and allowing them to dissolve in the solution. They then are attracted to the cathode, where the dissolved metal ions, M^{n+} (where n is the number of electrons per ion taking part in the reduction) are reduced to metallic form, M. Specifically, the metal ions acquire electrons at the interface between the solution and the cathode and subsequently "plate out" on the cathode. In its simplest form, the reaction in an aqueous solution at the cathode is given by

$$M^{n+} + ne^- \Rightarrow M, \tag{1}$$

which is accompanied by a corresponding anodic reaction. The rate at which the anode is dissolved is equal to the rate at which the cathode is plated, which in turn is proportional to the current flowing through the circuit. In this manner, the ions in the electrolyte bath are continuously replenished by the anode. A schematic diagram of a typical electroplating system is shown in Fig. 6.8(a). The main advantage of electroplating is that it is relatively simple to carry out. Its main disadvantages are that it only works with conductive substrates and that because it is a low-energy electrochemical process, the ions can deposit on the cathode in an uncontrollable manner especially around edges. Additionally, the solutions used in the electrolytic bath are generally toxic and represent an environmental hazard.

In the electroless plating process [17,18], the substrate is immersed in a complex chemical solution that promotes spontaneous deposition on the substrate if the substrate's electrochemical potential is sufficiently high with respect to the

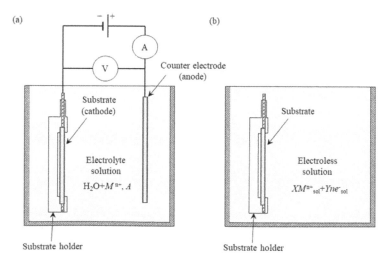

FIGURE 6.8 Simplified schematic of electrodeposition systems: (a) Electroplating configuration for plating metal *M* from an aqueous solution of a metal salt *MA* and (b) Electroless configuration for plating metal *M* from a solution of metal ions (XM^{n+}) and a reduction agent (*Yne*[-]).

[6] AQUueous Deflocculated Acheson Graphite manufactured by Acheson Industries, a subsidiary of ICI.

solution. Electroless deposition is in fact an autocatalytic, self-sustaining process, in which the source of metal ions M and the reducing electrons, e^-, are obtained from separate compounds in solution. The electrode reaction has the form

$$M^{n+} + Reducing_{solution} \Rightarrow M + Oxidation\,product_{solution}. \tag{2}$$

In principle, any water-based reducer can be used, although the reduction (redox) potential must be high enough to overcome the energy barriers inherent in liquid chemistry. For example, electroless nickel plating uses hypophosphite as the reducer, while other metals (like silver, gold and copper) typically use low-molecular weight aldehydes. A schematic diagram of a typical electroless plating system is shown in Fig. 6.8(b).

Electroless deposition offers a number of advantages over electroplating. For example, (a) only one electrode is required, (b) external power sources are not required and (c) complex shapes and a variety of surfaces types can be coated. Additionally, an important consequence of (b) is that plating will work with non-conductive substrates, since an electrolyte is not required to drive the process. Its disadvantages are that it is slower than electroplating, and it is more difficult to control film thickness, although uniformity and coverage are better. As with electroplating, the chemicals used are usually highly toxic and pose an environmental hazard.

6.3.3 Physical Vapor Deposition (PVD)

Physical Vapor Deposition (PVD) is an alternative process to electrodeposition. It is a vacuum deposition technique involving the transfer of material at the atomic level from a target to the surface to be coated. It is similar to Chemical Vapor Deposition (CVD), except the precursors are solid. However, PVD is more suited to contact formation, since higher deposition rates are possible at lower temperatures. In addition, the process does not produce hazardous by-products, which can be a problem with CVD. The most commonly used PVD coating processes are sputtering (using magnetic enhanced sources or "magnetrons", cylindrical or hollow cathode sources) and evaporation (typically using cathodic arc, resistive heating or electron beam sources). All of these processes occur in vacuum at a working pressure of 10^{-2} to 10^{-4} Torr and generally involve the generation of an atomic vapor of the material to be deposited. Additionally, reactive gasses such as nitrogen, acetylene or oxygen may be introduced into the vacuum chamber during metal deposition to create various compound coating compositions with tailored physical, structural and tribological properties. A comprehensive guide to the evaporation properties of common thin film materials used in metallization and contacting (including density, melting point, vapor pressures, preferred evaporation methods and crucible liners) can be found in ref [19].

6.3.3.1 Sputtering

Sputtering is a relatively quick and easy method for applying contacts and is commonly used for thin-film deposition, etching and surface composition studies. The technique was first described 150 years ago by Grove [20] in 1852 and later by Plücker [21] in 1858 who, whilst investigating glow discharges, reported the vaporization of a metal target and the subsequent re-deposition of material on a substrate in the form of a thin metal film.[7] Sputtering is a cold (low temperature) evaporation technique that utilizes a gas plasma to remove material from a target and deposit it onto the surface to be contacted, known as the substrate. For contacting purposes, this may take the form of a wafer or a piece of crystal. RF excitation is generally used to generate the plasma, although for conductive materials, DC can also be used. A typical RF sputtering system is described in Chapter 4 Section 4.6.4.1 and shown schematically in Fig. 4.26. The main advantages of sputtering are that the technique is simple, can be carried out at low temperature and most elements of the periodic table can be sputtered, including refractory materials, such as tungsten. Its main disadvantage is that semiconductor surfaces can be damaged by ion bombardment.

6.3.3.2 Evaporation

Evaporation is a high temperature variant of sputtering in which source material is heated to the point where it starts to boil and evaporate. The evaporation is carried out in a vacuum chamber to ensure that the source molecules are evenly dispersed throughout the chamber, where they subsequently condense on all surfaces. There are two popular evaporation technologies, namely resistive and e-beam evaporation. The two systems are described in detail in Chapter 4, Section 4.6.4.2 and shown schematically in Figs. 4.27(a) and (b). A wide variety of materials, including refractory metals such as tungsten and low vapor pressure metals such as platinum and alloys can be evaporated. Compared to sputtering, evaporation offers much higher deposition rates and is less damaging to the substrate, since the evaporated atoms have a Maxwellian energy

---

[7] In the current-voltage characteristic of a glow discharge, sputtering occurs in the abnormal discharge region after the normal discharge region and immediately before arcing.

distribution when they condense on the substrate. Sputtered atoms can have a hard velocity component, due to the plasma. On the negative side, adhesion and large area surface uniformity are generally poor compared to sputtering.

6.4 Lithography

For simple geometries, contacts can be applied by vapor transmission directly through a shadow mask, usually fabricated out of a thin metal foil such as brass. For more complex geometries, photolithographic techniques are used.[8] The technique is generally used to process complete wafers when multiple detectors are being produced. Photolithography is a process to remove selectively parts of a thin film or even the bulk of a substrate. It uses light to transfer a geometric pattern from a photo mask to a light-sensitive chemical, known as photo-resist, on the wafer. The photo-resist consists of three components: a resin base material, a photoconductive compound and a solvent to control the mechanical properties of the photo-resist. After exposure to light, a series of chemical treatments then etches the exposure pattern from the mask into the material beneath the photo-resist. The entire process is illustrated in Fig. 6.9.

The process begins by cleaning the metallized detector blank or wafer and baking it to remove any H_2O that may oxidize the sample. It is then covered with photo-resist by dipping or spin coating to produce a uniformly thick layer. The photo-resist is a viscous, liquid solution that is dispensed onto the wafer. For spin coating, the wafer is spun at 1,200 to 4,800 rpm for 30 to 60 seconds, which produces a layer between 0.5 and 2.5 µm thick. The photo-resist coated wafer is then prebaked to sensitize the photo-resist to UV light by driving off excess solvent. This is typically carried out at 100°C for 5 to 30 minutes. After prebaking, the photo-resist is exposed to a pattern of intense light, typically ultraviolet. Positive photo-resist, the most common type, becomes chemically less stable when exposed; negative photo-resist becomes more stable. This chemical change allows the unstable photo-resist to be removed by a special solution, called "developer" by analogy with photographic developer.

Exposure systems typically produce an image on the wafer using a photomask. The light shines through the photomask, which blocks it in some areas and lets it pass in others (an example of a photomask is shown in Fig. 6.10 (left); it was used to produce a series of GaAs diodes and pixel arrays. Generally, three forms are lithography are used depending on the application. In order of complexity these are: contact, proximity and projection. A contact printer is the simplest exposure system. It sets the photo mask in direct contact with the wafer and exposes it to a uniform light source. While offering simplicity and high resolution (approximately the wavelength of the radiation), the mask and wafer are particularly susceptible to damage precisely because they are in contact. However, this is particularly attractive for research and small-scale production processes because it uses inexpensive hardware and can still achieve high optical resolution.

Proximity printing reduces mask damage by setting the mask a fixed distance above the wafer (of the order of 20 mm). Unfortunately, the resolution limit is significantly increased, making proximity printing inappropriate for complex lithography. In addition, diffraction at the pattern edges causes light divergence.

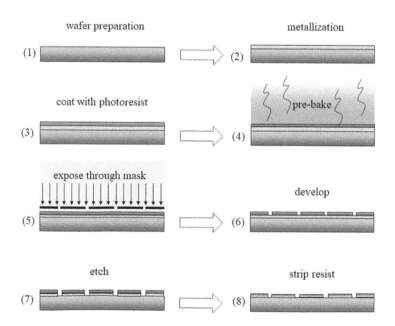

FIGURE 6.9 Example of a typical sequence of lithographic processing steps (positive resist) for forming the contacts to the array illustrated in Fig. 6.10.

---

[8] The word lithography derives from the Greek λίθος, meaning "stone", and γράφειν meaning "to write".

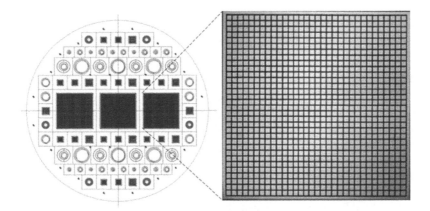

FIGURE 6.10 Left: a photolithographic mask used to produce a number of GaAs diodes and pixel arrays with guard rings (Image courtesy Oxford Instruments Analytical Oy). Right: a 32×32 pixel array processed using the mask (step 8, in Fig. 6.9).

For the needs of the semiconductor industry, projection printing is the most common method of exposure. Projection lithography derives its name from the fact that an image of the mask is projected on-to the wafer. Projection lithography has become a viable alternative to contact/proximity printing since the advent of computer-aided lens design and improved optical materials allowed the production of lens elements of sufficient quality to meet the exacting requirements of the semiconductor industry. Unlike contact or proximity masks, which cover an entire wafer, projection masks (also called "reticles") show only one die. Projection exposure systems (steppers) project the mask onto the wafer many times to create the complete pattern. The smaller imaging field simplifies the design and manufacture of the lens, but at the expense of a more complicated reticle and wafer stage. Also, the depth of focus (a few μm) restricts the thickness of the photo-resist and places strict demands on wafer flatness. In view of its complexity and significant capital costs, the technique is only suitable for large-scale production. Currently, the highest lithographic precision is achieved using X-rays from a synchrotron radiation source [22]. Having wavelengths below 1 nm, X-rays overcome the diffraction limits of optical systems, which allow smaller feature sizes to be achieved. The technique is largely being pursued by the semiconductor industry for Ultra Large Scale Integration (ULSI) with demonstrated structural fabrication precisions of 0.1 μm for lines and 0.5 μm for devices being achieved [23]. At present, X-ray lithography is only viable for large-scale microelectronics production, in view of the costs involved.

For both contact and proximity lithography, the mask covers the entire wafer; this requires that the light intensity be uniform across an entire wafer and the mask is precisely aligned to features already on the wafer. As modern processes use increasingly large wafers and multiple layering, these conditions become progressively more difficult to fulfil.

After exposure, the wafer is "hard-baked", typically at ~150 °C for ~25 mins. The hard bake solidifies the remaining photo-resist to make a more durable protecting layer for future ion implantation, wet chemical etching or plasma etching. In the etching step, a liquid ("wet") or plasma ("dry") chemical agent removes the layers in the areas that are not protected by photo-resist. These will become the contacts that are then either sputtered or evaporated onto the surface and adhere through the windows opened up during the etching procedure. In semiconductor fabrication, dry etching techniques are generally preferred, as they can be made anisotropic, in order to avoid significant undercutting of the photo-resist pattern. This is essential when the width of the features to be defined is similar to or less than the thickness of the material being etched, for example, when the aspect ratio approaches unity.

Dopants may be introduced by thermal diffusion, in a gaseous ambient or by ion implantation. In this process, ions are accelerated to a potential of 20 to 100 keV depending on the desired penetration depth and are initially distributed interstitially; however, in order to activate them electrically as donors or acceptors, they must be moved to substitutional lattice sites. This is achieved by heating (thermal annealing).

Once etching is completed, the photo-resist is no longer needed and must be removed from the substrate. This usually requires a liquid "resist stripper", which alters the resist chemically so that it no longer adheres to the substrate. Fig. 6.10 (right) shows a completed 32×32 GaAs pixel array after etching and photo-resist removal (Step 8 in Fig. 6.9). The mask it was produced from is shown the Fig. 6.10 (left).

6.5 Detector Assembly

For epitaxial produced material, the end product is a wafer of semiconducting material between 2 inches and 8 inches in diameter, although for most compound semiconductors, wafer sizes are limited to 2 inches. Once processed and patterned with particular device structures, on-wafer testing of each chip is carried out and the wafer is separated into individual chips, commonly referred to as wafer dicing. Depending on the wafer material and its thickness, dicing is achieved by (i) scribing along selected crystallographic planes and breaking, (ii) cutting with a high precision diamond blade or (iii) laser cutting. For melt grown crystals, the ingot or boule is first sliced into wafers using a diamond tip or wire saw and the surfaces lapped and polished prior to patterning and dicing. All other operations beyond this point are identical.

6.5.1 Detector Packaging

There are many ways to package a detector. Generally, the simpler the implementation, the noisier the result; so to achieve low-noise operation great care must be taken in the assembly and positioning of components. We will describe one commonly used implementation for small detectors (≤ 0.1 cm^3), which achieves low noise by minimizing stray capacitance and reducing thermally generated leakage currents and Johnson noise by cooling the detector crystal, FET and front-end feedback components. As an example, we will consider a simple HgI$_2$ planar detector, although the method is generally applicable to more complex geometries such as arrays. For simple detectors (*e.g.*, planar detectors), the diced chips are mounted on a ceramic header and the chip contacted by wire bonding (see Fig. 6.11). In the case of mechanically soft materials, such as HgI$_2$, TlBr and CdTe, contacting is usually achieved by gluing the wire with a silver epoxy. This prevents surface damage, which occurs easily close to the detector edges. The other end of the wire is wire bonded to one of the pins of the TO8 carrier. Irradiation of the detector is through the front contact. Since the mobility-lifetime product ($\mu\tau$) of electrons in most semiconductors is much greater than that for holes, negative bias is usually applied to the front electrode to obtain electron collection from the back electrode. This ensures that the trapped carrier (the holes) traverse the minimum amount of detector material and therefore have a reasonable chance of being collected at the anode. The signal from the detector is extracted directly into the gate of the pre-amplifier input FET also mounted on the ceramic holder along with the feedback components. This close proximity minimizes parasitic capacitance and reduces noise.

Cooling the detector and front-end components can lead to a substantial reduction in noise – up to a factor of 10 for Si detectors and 2 or more for GaAs and CdZnTe detectors. Cooling the detector reduces the number of thermally generated carriers and as a result the leakage current, while cooling the input FET increases its transconductance, which in turn reduces the electronic noise of the system. Moderate cooling is most easily achieved using a Peltier cooler (see Section 6.7.3). The ceramic substrate containing the detector and front-end components is coupled to the cold side of a Peltier cooler. A single-stage cooler can generate a maximum temperature differential of 60°C, leading to a typical substrate temperature of 240K at room temperature. A two-stage cooler can generate a temperature differential $\sim$ 80°C, leading to a substrate temperature of 215K at room temperature. An assembled detector head is illustrated in Fig. 6.11, showing the substrate, detector, wire bond and front-end FET. A temperature sensor is also mounted on the substrate to provide a direct reading of the detector temperature. The hot side of the Peltier cooler is glued to the TO8 carrier and heat removed to the external detector housing *via* the threaded screw. The entire assembly is capped with a thin steel hermetic vacuum enclosure, which is sometimes filled with a dry, heavy noble gas for low thermal conductivity. Water vapor condensation is prevented by careful sealing, and water absorbers can be employed inside the package. X-rays are incident on the detector through a thin (25–100 μm) beryllium window. The rest of the preamp is located outside the TO8 carrier on a PCB in which the TO8 pins are soldered or secured in a socket. For two-, three- or four-stage thermoelectric coolers, heat sinking *via* the mounting screw is usually not sufficient, and the entire base of the detector housing must be firmly attached to a suitably sized heat sink. As a rule of thumb, a heat sink thermal resistivity of ~2 K/W is typically required for most two-stage and three-stage Peltier coolers. Four-stage coolers require ~1 K/W. A thin layer of heat conductive epoxy or silicone grease should be applied to improve thermal contact between detector housing and heat sink. Holes drilled through the heat sink provide passage for the (insulated) TO8 pins to the PCB.

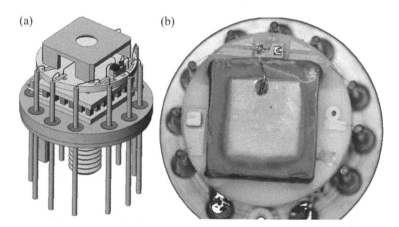

(a)   (b)

FIGURE 6.11   Left: detector packaging using a TO8 holder and Peltier cooler. The front-end FET and feedback resistor can be seen on the right side of the package. Right: a single HgI$_2$ crystal planar detector with Pd electrodes mounted on a ceramic disc on a TO8 holder (images courtesy of Oxford Instruments Analytical Oy).

6.5.1.1 Leakage Current

All detectors show finite conductivity – even when no signal is present. This leads to a steady-state current flowing through the detector, the magnitude of which depends on the bias applied to collect the signal charge carriers and whether ohmic or Schottky contacts are used. In most cases, the leakage current ultimately limits detector performance. For an ohmic detector, the leakage current should ideally be only dependent on the semiconductor resistivity and bias voltage, whereas for a Schottky detector it should be dependent on the nature of the contact. For junction detectors, minority carrier diffusion across the junction produces most of the bulk leakage current. In Fig. 6.12 we show the measured leakage currents obtained by Park *et al.* [24] for two similar-sized CdZnTe detectors, one fitted with ohmic contacts (Pt/CZT/Pt) and the other with Schottky contacts (In/CZT/Au). In the case of the ohmic detector, the dynamic resistance-voltage characteristics are nearly constant and correspond to the static resistance-voltage characteristics. The Schottky detector, on the other hand, shows non-linear current-voltage characteristics, that is, when forward biased a much larger current is obtained than when reverse biased.

Up to a point, the larger the bias, the better the charge collection efficiency, but also the larger the leakage current. The random fluctuations that occur in this current are a significant source of noise, since they are generally much greater in amplitude than the small transitory current pulse (signal) that appears following an ionizing event. For example, the resistivity of high purity float-zone grown silicon is ~10,000 Ω-cm. If a 1 mm thick slab of this silicon were cut with 1 cm^2 surface area and fitted with ohmic contacts, the electrical resistance between faces would be 1,000 Ω. An applied voltage of 100 V would therefore cause a leakage current through the silicon of 100 mA. By contrast, the peak current generated by a 100 keV X-ray (corresponding to the creation of a few times 10^4 radiation-induced charge carriers) would only be about 0.1 μA, a factor of 10^5 times smaller. If we consider a similar detector fabricated out of intrinsic GaAs with a resistivity of 10^8 Ω-cm, the leakage current would be 10 μA, whereas the signal would be only 0.5 μA – a factor of 20 lower. It is therefore essential to greatly reduce this current either by increasing the resistance of the detection medium or by the use of blocking contacts. As a rule of thumb, in critical applications the leakage current should not exceed about a nA to avoid significant resolution degradation and for Fano limited performance, the leakage current should be of the order of a 100 pA or less. For reference, typical bias voltages for Group III-V materials are in the range of 50–100 Volts and leakage current densities are typically a few nA cm^{-2}. For II-VI materials, applied voltages can be in the range 500–2,000V. In this case, leakage current densities of 20 nA cm^{-2} or more are quite common. For Fano limited operation, current densities need to be of the order of, or less than, 1 nA cm^{-2}. At these levels, leakage across the surface of the semiconductor can often become far more significant than bulk leakage.

An additional complication can arise in CdTe and related materials, which have a propensity to polarize – meaning the buildup of stored charge associated with the trapping of low mobility holes produced following an ionizing event [25]. In order to improve charge collection and minimize tailing effects caused by the hole trapping, it is usually necessary to increase the applied voltage to higher values than would otherwise be required. Unfortunately, the leakage current also increases, usually in a non-linear fashion, and additional measures may then be required to return the leakage to an acceptable level.

The leakage current itself is composed of two main components – a bulk component and a surface component, both of which can have a large effect on detector performance. These are described in later sections.

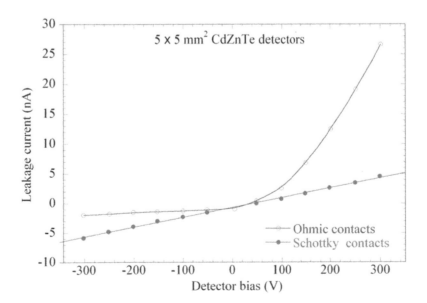

FIGURE 6.12 Current-voltage characteristics of a CdZnTe Ohmic detector and a CdZnTe Schottky detector measured by Park *et al.* [24]. Both measurements were carried out at room temperature (21°C).

6.5.1.2 Bulk Leakage Currents

Bulk leakage current, as the name implies, is a dark current that originates in the bulk or active detector volume as a consequence of the thermal generation of charge carriers and from the current generated by crystal imperfections and impurities. These disrupt the perfect periodicity of the crystal and introduce energy levels into the forbidden gap that act as generation and recombination centers for electrons and holes. Unlike other electrical properties of a semiconductor, the contribution of these centers is difficult to predict, in terms of magnitude and its dependence on the reverse bias.

In a critical examination of bulk leakage, Amman *et al.* [26] performed current-voltage and noise characterization measurements on 37 CdZnTe crystals from three different crystal suppliers. Simple 1×1 cm^2 planar detectors were fabricated with ohmic contacts and a guard ring structure to measure surface and bulk leakage separately. They found that there was a marked difference in the bulk leakage currents among materials supplied by different manufacturers and even among different samples from the same manufacturer. In some cases, the bulk leakage current shows no correlation with the bulk resistivity of the materials. In other cases, the bulk leakage currents tend to be lower for lower-resistivity materials, which is the opposite expected for ohmic contacts. In fact, they found that 35 samples actually exhibited non-ohmic behaviour and concluded that in most devices the leakage is dominated by surface current flowing along the sides of the device. They also found that this component often varies linearly with applied voltage, leading one to incorrectly conclude from the global I/V curve that the device is operating with ohmic contacts. Surprisingly, as Luke *et al.* [27] pointed out, this surface component does not seem to contribute significantly to detector noise. This can be explained if the surface conduction is resistive in nature generating a thermal noise contribution but not shot noise.

6.5.1.3 Suppressing Surface Leakage Currents

Surface leakage takes place at the surface of the detector and is particularly acute at the edges where potential gradients are greatest. Its magnitude is very dependent on material processing, surface preparation and passivation techniques as well as material type. The most obvious identifying characteristic of surface leakage is that it scales with the surface area of the device. If the device is fabricated in the geometry of a mesa, the relevant surface area is the area of the mesa sidewalls, which in turn is proportional to the periphery of the top of the mesa. Thus, for mesa devices, the dark current would have a component that is proportional to the mesa periphery, which can be identified as the surface leakage current. Components that are proportional to the device cross-sectional area can then be identified as bulk currents.

Surface leakage currents are usually attributed with electron states that develop on the surface of air-exposed semiconductor surfaces. Such states do not exist in the bulk of the semiconductor and provide another conduction path in parallel to bulk leakage currents. In addition, surface traps contribute to $1/f$ noise currents. Attempts to control surface components generally follow two approaches. The first is carried out as *ex-situ* processing steps, usually performed after growth and some device fabrication steps, in which a passivation material or treatment is applied to air-exposed surfaces. The disadvantage of such *ex-situ* steps is that they add complexity and cost and in general are only partially effective. However, in some materials passivation techniques can be quite effective as illustrated in Fig. 6.13, in which we show the effects of passivation carried out on one of two identical $10 \times 10 \times 1$ mm^3 CdZnTe detectors [28]. After fabrication, one detector had Au contacts applied, while the other had its surface passivated by immersion in a NH$_4$F/H$_2$O solution before electrode deposition. As can be seen, the effect of the passivation is to reduce the leakage current by a factor of ~ 3.

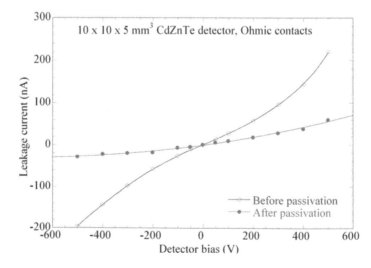

FIGURE 6.13 Leakage currents measured with two identically sized CdZnTe detectors. One detector was prepared with a surface passivation and the other prepared without (from [28], ©2008 IEEE).

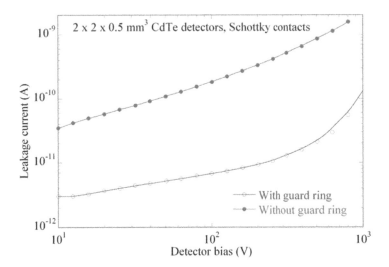

FIGURE 6.14 Leakage current of two CdTe diodes with an active area of 2×2 mm^2 and thickness 0.5 mm. Closed circles show the measured leakage current in the detector without a guard ring while open circles show the corresponding leakage current in the detector operated with a guard ring (from [29], ©2004 IEEE). In both cases, the operating temperature was 20°C.

A second technique routinely used to reduce surface leakage is using a guard ring fabricated on the sensitive surface. The guard ring operates in a fashion similar to guard rings in other detector types such as ionization chambers in that it prevents the flow of an undesirable charge component in preference to a desirable component. Physically, the guard ring is a conductive ring surrounding the signal contact but separated from it by a small gap. It is biased so that there is no potential difference across the surface between the contact and ring and no currents flows, ideally ensuring that only bulk leakage contributes to the signal. The reduction in surface leakage currents can be quite pronounced resulting in an overall reduction in leakage current of a factor of 2 or more [29,30], depending on type of electrodes. For example, for resistive electrodes the reduction tends to be a factor of 2 or 3, while for Schottky detectors, it can be an order of magnitude or more. The effect is nicely demonstrated in Fig. 6.14 (from Nakazawa *et al.* [29]), in which we show the leakage currents measured on two identical 2×2 mm^2, 0.5 mm thick CdTe Schottky detectors – one fitted with a guard ring and the other without.

Another advantage of a guard ring is that it reduces potential gradients at the edge of the detector's active area, thereby increasing the breakdown voltage rating. In fact, for large detector biases, a multiple guard ring system is commonly employed. These redistribute the electric field over a larger distance along the detector edge, thus preventing breakdown at the edge. A third advantage, which is important for spectroscopic applications, is that the guard ring collects most of the signal charge generated by those interactions close to or just outside the active area, which would otherwise be lost by diffusion and recombination, resulting in incomplete charge collection.

6.6 Processing Electronics

Radiation detectors generally require additional processing electronics to fully extract the energy and time information embedded in the signal. In essence, there are two types of signal pulses used in radiation measurement – these are linear and logic pulses. A linear pulse is a signal pulse carrying information in its amplitude and shape and is generally used to extract the energy information in the signal. Logic pulses, on the other hand, are signal pulses of standard size and shape that carry information only by their presence or absence and are usually used to extract timing information. Note that linear pulses can be and are readily converted into logic pulses, whenever sequenced or conditional logic is required for processing signals. However, the reverse process cannot be carried out, since there is no linear information in a logic pulse. A typical detector processing electronics chain is shown in Fig. 6.15. Each component will be discussed in the following sections.

6.6.1 Front End – Choice of Preamplifiers

The preamplifier is the first component in the signal processing chain. Its purpose is to collect the charge created within the detector and in spite of its name, does not act as an amplifier. Rather, it is the interface between the detector and the pulse processing electronics that follow. Its main function is to extract and integrate weak charge pulses and convert them into voltage pulses for amplification without significantly degrading the intrinsic signal-to-noise ratio. In order to achieve this, the preamplifier needs to be designed to match the characteristics of the detector – capacitance and impedance. To minimize loading on the detector and thus maximize signal, the preamp should present a high impedance load whilst providing a low

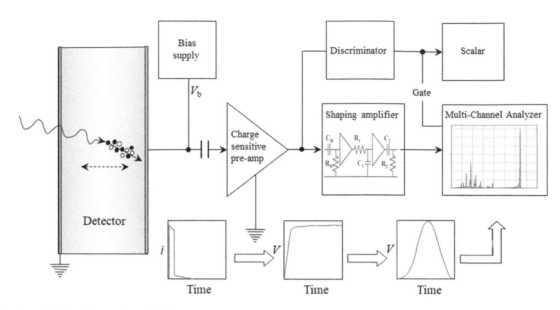

FIGURE 6.15 A typical signal processing chain for spectroscopy with a radiation detector.

impedance source for the shaping amplifier, especially since it may be necessary to drive several metres of cable. Correctly terminating the capacitance of the detector is another important consideration, since any mismatch can cause a substantial increase in noise. Consequently, the preamp should be located as close as possible to the detector to minimize stray capacitance. This has additional benefits of reducing microphonic noise, ground loops and radio frequency pickup. In practice, the detector is ac-coupled to the FET, which while the increased front-end capacitance increases noise, it allows signal extraction and high voltage biasing to be carried out by a single cable. This is normally preferred, since the other end (electrode) of the detector can be at ground potential making detector mounting relatively easy (*i.e.*, signal ground and bias ground can be the same). For dc-coupling, the detector signal electrode should be at virtual ground potential, meaning that the other end (the one usually used to mount the detector) is at a high potential.

With reference to Fig. 6.16, charge-sensitive preamplifiers work as follows. The charge from the detector is collected on C_i and C_f over a period of time, effectively integrating the detector current pulse. As the charge is collected, the voltage on the feedback capacitor rises, producing a step change in voltage. In the resistive feedback design, the output voltage is proportional to the total integrated charge as long as the time constant $R_f C_f$ is sufficiently longer than the duration of the

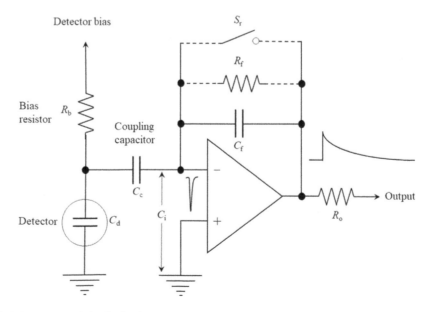

FIGURE 6.16 Conceptual designs for a resistive feedback and pulsed reset charge sensitive preamplifiers. In the resistive feedback design, the charge accumulated on the feedback capacitor, C_f, decays through the feedback resistor R_f with a characteristic time constant $R_f C_f$. In a pulsed reset design, a reset switch, S_r, is used instead of the resistor to drain the charge. In practice, a transistor switch is generally used.

input pulse. Here R_f is the feedback resistor whose primary function is to discharge the feedback capacitor, ready for the next input pulse. At the component level, the detector signal is connected to the gate of a low-noise junction FET whose open loop gain, A, is set sufficiently high so that the amplification is not affected by the input capacitance, such that

$$A >> (C_i + C_f)/C_f, \tag{4}$$

where C_i and C_f are the input and feedback capacitances, respectively. By definition, the output voltage V_{out} is related to the input voltage V_{in} at the inverting input by

$$V_{out} = -AV_{in}, \tag{5}$$

The voltage across C_f is given by

$$V_f = -(A+1)V_i. \tag{6}$$

The total charge, Q, stored on the input and feedback capacitors is given by

$$Q = C_i V_i + C_f V_f. \tag{7}$$

Substituting Eqs. (4) and (5) into Eq. (6) and rearranging,

$$V_{out} = -A \frac{Q}{C_i + (A+1)C_f} \cong -\frac{Q}{C_f}. \tag{8}$$

Thus, the output pulse height is only proportional to the charge deposited in the detector and more importantly, to the energy deposited by the interaction of radiation with the detector. The charge gain of the amplifier is then given by

$$G_c = \frac{V_{out}}{Q} \cong \frac{1}{C_f}, \tag{9}$$

which depends only on the feedback capacitance and most importantly, is insensitive to variations of detector input capacitance.

The temporal behaviour of the output pulse shape is illustrated in Fig. 6.17(a) and is characterized by a rise time, which is equal to the detector current pulse width, and a decay time, τ_f, given by $R_f C_f$, which is typically of the order of tens to hundreds of µs.

$$V_{out}(t) = \frac{Q}{C_f} \exp\left(\frac{-t}{R_f C_f}\right). \tag{10}$$

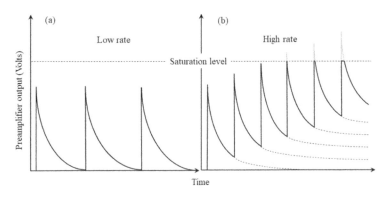

FIGURE 6.17 Output signal patterns from a resistive feedback charge sensitive preamplifier. The decay time of the output pulse is determined by the time constant of the circuit, given by the product $R_f C_f$. (a) Pulses that arrive at a time long compared to the time constant (b) Pile-up induced by pulses arriving too quickly. Those pulses that exceed the saturation level of the amplifier will be clipped and no longer accurately reproduce the charge deposited in the detector.

An electronic test pulse (or pulser) input is usually incorporated into the design and is used to calibrate radiation spectroscopy systems. A tail pulse generator with adjustable rise and decay times is commonly used for setting up and adjusting shaping and timing parameters. Using a constant output amplitude, the electronic noise present in the system can be directly determined from the width of the amplitude distribution, measured with a pulse height analysis system.

6.6.1.1 Limitations of Resistive Feedback Preamplifiers

The resistive feedback preamp illustrated in Fig. 6.16 suffers from two main limitations, both related to the feedback resistor. In normal operation at ordinary counting rates, the rising step caused by each detector event rides on the exponential tail of the previous event due to the long decay time, and the preamp output does not have a chance to return to the baseline. This does not create a serious problem, since the significant information in the output pulse is contained in its rising edge and a shaping amplifier is capable of extracting the pulse height from the rising edge of each pulse by careful shaping. However, as the counting rate increases, pulses begin to "pile up" on each other and the excursions of the preamp output move farther away from the baseline as illustrated in Fig. 6.17(b). The dc power supply eventually limits the voltage excursions, and this determines the maximum counting rate that can be accommodated without distortion of the output pulses.

The second limitation of resistive feedback designs is that the feedback resistor R_f is an intrinsic noise source because of the Johnson noise associated with it.

$$\langle en_J \rangle = \sqrt{4kTR_f \Delta f}, \tag{11}$$

Where $\langle en_J \rangle$ is the equivalent noise voltage that can be associated with R_f, k is Boltzmann's constant, T is the absolute temperature and Δf is the bandwidth. Clearly, Johnson noise can be minimized by selecting a higher R_f value. However, this approach is limited since a simple increase of R_f may lead to too long a time constant with subsequent impacts on count rate performance. Alternately, keeping the time constant low by reducing C_f is also limited, since if C_f becomes too low, the linearity of the preamp is affected.

6.6.1.2 Pile-Up and Baseline Restoration

These two shortcomings can be eliminated by removing the feedback resistor. However, without R_f, the charge pulses from the detector continue to be accumulated on the feedback capacitor in a staircase fashion[9] as shown in Fig. 6.18, until the output voltage saturates – being limited by the supply rail voltage. Some method must be provided to reset the preamp when the staircase approaches the maximum allowable voltage; this is usually achieved using a pulsed reset feedback mechanism, in which the output voltage level is used to trigger the threshold circuit, which then sends a pulse to a switch that subsequently discharges C_f to ground – in essence "resetting" the baseline. This is represented by the switch S_r in Fig. 6.16. Popular pulse reset methods include transistor reset [31], optoelectronic reset [32], drain feedback reset [33] and pentafet reset [34] configurations. All give much-improved noise performance compared to traditional resistive feedback.

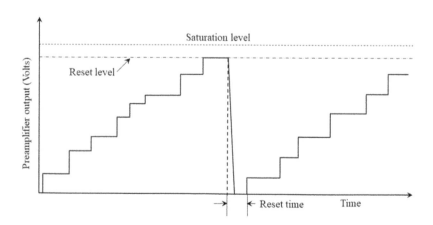

FIGURE 6.18 The output voltage of an active reset preamplifier. The reset is triggered when the voltage crosses the upper-level limit during the normal counting phase. Each upward step is an individual signal pulse. During the reset time, the amplifier cannot accept input pulses and thus the reset period represents dead time.

---

[9] Each upward step corresponds to an individual charge pulse at the input from particle interactions in the detector.

For the lowest noise operation, so-called "resistor-less" feedback circuits are employed [35]. In this implementation, the dc level of the gate of the front-end FET is kept at a constant voltage by forward biasing the gate-source junction of the input FET. As the leakage current and the signal charge from the detector accumulate on the input capacitance, the system will find equilibrium, where the total current of the detector equals the current through the forward-biased gate-source junction. This type of continuous reset mechanism of the preamplifier is particularly well suited for multi-element detectors where the crosstalk between neighbouring channels due to reset pulses can be a serious problem. The improvement in noise can be quite dramatic as illustrated in Fig. 6.19, in which we compare two ^{55}Fe spectra measured with a near Fano limited 250 μm×250 μm×40 μm GaAs pixel detector. One spectrum was taken using a matched resistive feedback preamplifier and the other with a matched resistor-less feedback preamplifier. The energy resolution measured using the conventional resistive preamplifier was 394 eV FWHM at 5.9 keV (pulser width = 378 eV) [36]. With the resistor-less feedback preamplifier, the FWHM energy resolution was 219 eV at 5.9 keV with a pulser resolution of 163 eV [37]. However, it should be pointed out that resistor-less feedback techniques can only be applied to detection systems with leakage currents in the pA range because of the inherently high open loop gain.

6.6.2 Shaping Amplifiers

Following the charge sensitive preamplifier, the signal is fed to a shaping amplifier, which essentially performs three functions. Firstly, it amplifies the preamplifier output signal from the mV level into the 0.1 to 10 V range, which facilitates accurate pulse amplitude determination using analog-to-digital converters and single-channel pulse-height analysers. Secondly, it enhances the signal-to-noise ratio by selectively filtering noise in the frequency domain of the preamplifier output signal. Thirdly, the shaping amplifier provides a shortened output pulse compared to the preamp allowing for a faster baseline restoration. This is especially important at high count rates, where pulses from consecutive events can "pile up" and spectroscopy is lost. Frequently, the requirement to handle high counting rates is in conflict with the need for optimum energy resolution – optimum energy resolution usually requires long pulse widths to prevent ballistic deficiency effects,[10] whereas high count rate recognition requires short pulse widths to prevent pile up and dead time. In such cases, a compromise pulse width must be selected to optimize the quality of information collected during the measurement. Many types of shaping amplifiers (semi-Gaussian, pseudo-Gaussian, quasi-triangular and others) use combinations of high-pass and low-pass filters, as described in the next section. A formal analysis of amplifier design is beyond the scope of this book, and the reader is referred to references [38–40]. Suffice it to say that circuit analysis is largely carried out by deriving the circuits transfer function using Laplace and Fourier transformations assuming a step function input pulse.

6.6.2.1 Pulse Shaping

For radiation spectroscopy, the simplest concept for pulse shaping is the use of a capacitor-resistor (*CR*) circuit followed by a resistor-capacitor (*RC*) circuit. The *CR* circuit is illustrated in Fig. 6.20(a) and acts as a high-pass filter, since the high

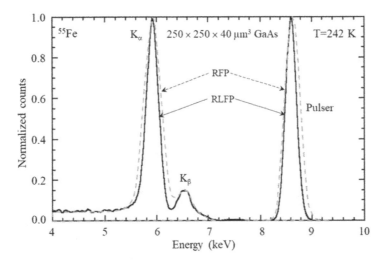

FIGURE 6.19 ^{55}Fe spectra measured with a near Fano limited 250 μm×250 μm×40 μm GaAs pixel detector. The energy resolution measured using a conventional resistive preamplifier was 394 eV FWHM at 5.9 keV (pulser width = 378 eV) [36]. With a resistor-less feedback preamplifier, the FWHM energy resolution was 219 eV at 5.9 keV with a pulser resolution of 163 eV [37].

---

[10] Ballistic deficiency is a measure of the inadequacy of signal integration. Specifically, it is the fractional deficit of the recorded output pulse height compared to that of an infinitely short input pulse of the same charge.

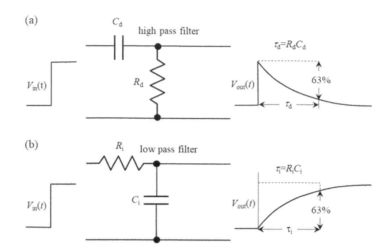

FIGURE 6.20 The time domain responses of *CR* high pass (a) and *RC* low pass (b) filters to a step function input pulse (a reasonable approximation to the long decay output pulses from charge sensitive preamplifiers). Historically, a *CR* high-pass filter is known as a *CR* differentiator, since for sufficiently small time constants, the signal looks "differentiated". Similarly, the *RC* low-pass filter is known as an RC integrator circuit, following similar arguments.

frequencies are preferentially passed through the capacitor and low frequency signals effectively blocked. Similarly, the *RC* circuit shown in Fig 6.20(b) acts as a low-pass filter. In this case, the low frequencies pass through the circuit unimpeded, whereas the high frequencies are effectively "shorted" to ground by the capacitor. Although these elementary filters are rarely used, they encompass the basic concepts essential for understanding the higher-performance, active filter techniques. The high-pass filter improves the signal-to-noise ratio by attenuating low frequencies that contain a lot of noise (*e.g.*, the shot noise due to the detector leakage current) but little signal. Similarly, the *RC* low-pass filter improves the signal-to-noise ratio by attenuating and removing high-frequency noise (*e.g.*, the noise due to the input FET of the preamplifier), which is similarly devoid of signal. Both circuits are characterized by a time constant $\tau = RC$, which is known as the shaping time. Historically, the *CR* high-pass filter is known as a *CR* differentiator since if the time constant of the circuit is made sufficiently small, the output voltage is almost proportional to the time derivative of the input wave form. Similarly, the *RC* low-pass filter is known as an *RC* integrator circuit, following similar arguments.

By combining *CR* and *RC* filters, it is possible to produce a "shaped" unipolar pulse as illustrated in Fig. 6.21. The input signal in this case is assumed to be a step function, which is a good approximation to an actual preamplifier output pulse, in view of its rapid rise time and very long decay time – in fact, much longer than the shaped output pulse width. This approximation is indeed fortunate, since a step input function greatly simplifies circuit analysis from a mathematics point of view.

Typically, the high-pass filter time constant, $\tau_d = C_d R_d$, is set equal to the integration time constant, $\tau_i = R_i C_i$, that is, $\tau = \tau_d = \tau_i$. For a step function input described by

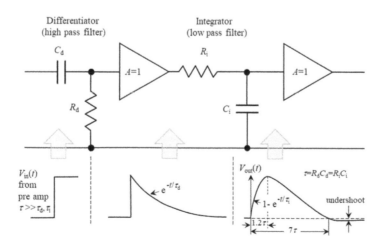

FIGURE 6.21 Basic shaping amplifier based on *CR-RC* filtering. The output pulse diagram shows the case when the time constant of the input *CR* differentiator stage is set equal to that of the *RC* integration stage. The input pulse shape is assumed to be a step function (a reasonable approximation to the long decay output pulses from charge sensitive preamplifiers).

$$V_{\text{in}}(t) = \begin{cases} V_{\text{o}} & t > 0 \\ 0 & t \leq 0 \end{cases}, \tag{12}$$

where V_{o} is the height of the step, the output pulse shape is given by

$$V_{\text{out}}(t) = \frac{V_{\text{o}}}{\tau} t \exp\left(-\frac{t}{\tau}\right). \tag{13}$$

The output pulse rises. slowly reaching its maximum at 1.2τ, known as the peaking time, before returning to the prepulse baseline level after 7τ.

If the *CR* high-pass filter is now followed by several stages of *RC* integration, the output pulse shape becomes more and more Gaussian depending on the number of integration stages. Consequently, such amplifiers are called semi-Gaussian shaping amplifiers. The output pulse shape can be described by

$$V_{\text{out}}(t) = \propto \left(\frac{t}{\tau}\right)^n \exp\left(-\frac{t}{\tau}\right), \tag{14}$$

where n is the number of integrations. In Fig. 6.22(a) we show the output waveform as a function of the number of integration stages, assuming equal time constants for all stages. In this case, the peaking time is equal to $n\tau$. Consequently, the pulse duration increases with the number of integration stages with subsequent impacts on the maximum count rate capability. In Fig. 6.22(b) we show the case in which the integration time constants are adjusted to give equal peaking times. This has the effect of resulting in a more symmetric pulse shape than in Fig. 6.22(a), a more rapid return to the baseline and an improvement in the signal-to-noise ratio. In fact, at the noise corner[11] time constant, semi-Gaussian shaping can reduce the output pulse width by 20–50% compared to a simple *CR-RC* filter, leading to a faster return to the baseline, improving high count rate capability.

In Fig. 6.23 we show the practical implementation of a semi-Gaussian shaping amplifier, incorporating pole-zero cancellation on the input signal and baseline restoration of the output signal (discussed in the following sections). The simple *CR-RC* networks have been replaced by active filters, which improves the SNR by 17% to 19% at the noise corner time constant.

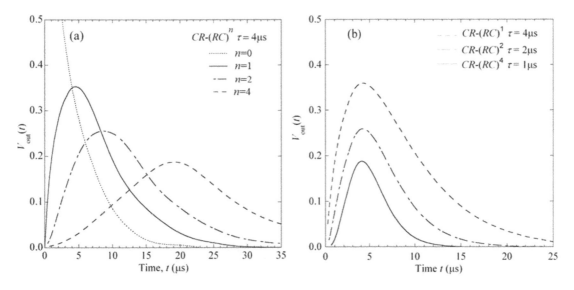

FIGURE 6.22 Illustration of semi-Gaussian shaping, in which we show the evolution of the output pulse shape with the number of integration stages. As can be seen, by increasing the number n of *RC* output stages (*i.e.*, $CR+RC_n$) the output pulse shape becomes more symmetrical, eventually approximating a Gaussian. In (a), we show the case when all time constants are equal (4μs). Note how the peaking time increases linearly with each additional integration stage. In (b), we show the case where the integration time constants are adjusted to give equal peaking times. This not only shortens the output pulse length for semi-Gaussian shaping but also results in a more symmetric shape, a quicker return to the baseline and an improvement in signal-to-noise ratio.

---

[11] Defined as the value of shaping time that minimizes the system noise (*i.e.*, when the series noise component is equal to the parallel noise component). For a more detailed discussion see Chapter. 8, Section 8.5.2.1.2.

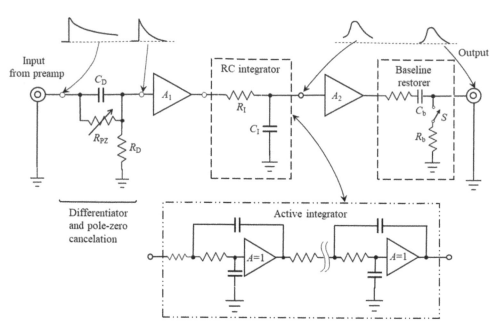

FIGURE 6.23 The practical implementation of a semi-Gaussian shaping amplifier, incorporating pole-zero cancellation on the input signal and gated baseline restoration of the output signal (from [41], reprinted with permission of Advanced Measurement Technology, Inc. – ORTEC).

6.6.2.2 Pole-Zero Cancellation (Tail Cancellation)

In a real preamp pulse, the falling tail follows a long exponential decay instead of being a constant as for a simple step function. Consequently, there is a small amplitude undershoot starting at about 7τ, which subsequently decays back to baseline with the much longer time constant of the preamplifier as illustrated in Fig. 6.23. At medium to high counting rates, a substantial fraction of the amplifier output pulses will ride on the undershoot of a previous pulse. The apparent pulse amplitudes measured for these pulses will be lower, which deteriorates the energy resolution. Most shaping amplifiers incorporate a pole-zero cancellation circuit to eliminate this undershoot. The benefit of pole-zero cancellation is improved peak shapes and energy resolution at high counting rates. Fig. 6.23 illustrates the pole-zero cancellation circuit and its effect. The preamplifier signal is applied to the input of the CR differentiator circuit. Depending on the relative preamp and differentiator time constants and the input signal rate, the output pulse from the differentiator exhibits an undershoot, which will be larger for longer shaping times. To cancel the undershoot, a variable resistor R_{pz} is added in parallel with the capacitor C_D, and adjusted. The result is an output pulse exhibiting a simple exponential decay to baseline with the desired differentiator time constant. This circuit is termed a pole-zero cancellation circuit because it uses a zero in the transfer function (expressed as a Laplace transformation) of the shaping circuit to cancel a pole present in the input pulse. Exact pole-zero adjustment is critical for good energy resolution and is achieved when $R_{PZ}C_D = \tau_{inp}$, where τ_{inp} is the time constant of the input components of the preamplifier.

6.6.2.3 Baseline Restoration

High performance spectroscopy amplifiers are dc-coupled throughout, except for the initial CR differentiator network located close to the amplifier input. This ensures peak position stability at high counting rates – for if intermediate stages were capacitively coupled, the baseline on which the output pulses sit would tend to shift downwards as the count rate increases. In fact, for all capacitively coupled circuits, the area of the signal above ground potential is equal to the area between ground potential and the shifted baseline, which is clearly rate dependent. Since the following pulse height measuring equipment usually references a pulse height to signal ground, the recorded pulse height will depend on rate. This is further reflected in a degradation of energy resolution. Whilst DC coupling prevents this, the DC offsets[12] in the earliest stages of the amplifier can be magnified by the amplifier gain into large and unstable offsets at the amplifier output – again resulting in incorrectly recorded pulse heights and a degradation in energy resolution. A baseline restorer (BLR) is used to solve both problems, by both AC coupling the output stage and simultaneously removing DC offsets by electronically tying the output signal baseline to ground in the absence of a signal. This is illustrated schematically in Fig. 6.23. An electronic circuit (not shown) senses when a pulse is about to arrive at the baseline restorer. In the absence of a signal, the switch, S, is closed, connecting the signal line to

---

[12] for example, the input offset of an operational amplifier stage

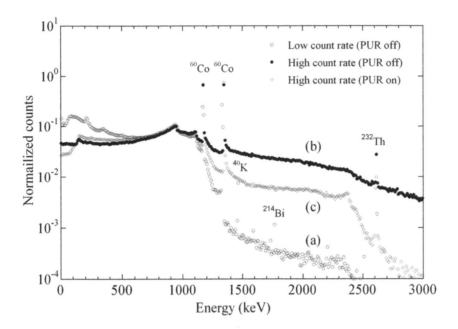

FIGURE 6.24 Demonstration of the effectiveness of pulse pile-up rejection in suppressing piled-up events measured with a 100 cm³ high purity germanium detector exposed to a ⁶⁰Co radioactive source. Note that several background radioactive lines are also apparent, particularly in the low count rate spectrum. Three curves are given normalized to the height of the 1173 keV peak. (a) is the spectrum measured at a low count rate (~500 ct s⁻¹), (b) measured at a high count rate (50k ct s⁻¹ and (c) is the same as (b) but with pulse pile-up rejection switched on.

ground through resistor, R_b, and thus establishing the prepulse baseline. When a signal pulse has been sensed, the switch, S, is opened immediately prior to its arrival. The signal then passes directly to the output through the capacitor, C_b, which also blocks the DC offset of the signal generated in previous stages. Since the signal baseline is always tied to ground immediately before and after a pulse arrives, frequency dependent baseline shifts at the output are greatly suppressed. Because the circuit is only active between pulses, it is known as a gated restorer. Needless to say, the $R_b C_b$ time constant should be much longer than the shaping time constant; otherwise, the BLR will act as in additional integration stage.

6.6.2.4 Pile-Up Rejection

Very high performance spectroscopy amplifiers also employ pile-up rejection to improve signal-to-noise ratio at very high counting rates. Pulse pile-up occurs when two incident particles arrive at the detector within the width of the shaping amplifier output pulse. In this case, their respective amplifier pulses pile up to form a composite output pulse. If the pulses are very close in time, the system will simply record the two pulses as a single event with combined pulse amplitude. This is also known as peak pile-up. If the pulses are spaced further apart, the system may accept both events and record them with incorrect pulse amplitude. This is known as tail pile-up. In either case, the events will end up in the wrong energy channels and the spectrum will be contaminated leading to incorrect results when the spectra are further analysed. The effect is illustrated in Fig. 6.24 in which we show a ⁶⁰Co spectrum acquired with a HPGe detector. At low count rates (~500 Hz), two peaks are apparent, one at 1173 keV and another at 1333 keV. A Compton continuum below the peak at 1173 keV is also apparent. Note that the number of spectral events is essentially zero above the peak at 1333 keV, however at high count rates (50 kHz). A continuum of events is apparent above the 1333 keV, on which are superimposed weak line features at 2 times 1171 keV and 2 times 1.333 keV.

A pile-up rejecter is used to prevent further processing of these distorted pulses. Practically, it is implemented by adding a "fast" pulse shaping amplifier with a very short shaping time constant in parallel with the main "slow" shaping amplifier. Pulses from the "slow" channel are then digitized by a discriminator to produce logic pulses whose duration is equal to the duration of a shaped output pulse. This duration is known as the inspection time. The fast shaping amplifier output, in turn, is digitized by a fast discriminator to produce short logic pulses. If two such pulses occur within the inspection time, an inhibit pulse is generated, which is used by the associated peak-sensing ADC or multichannel analyser to prevent further analysis of the piled-up event. The effect is demonstrated in Fig. 6.24, from which we can see that active pile-up rejection can substantially reduce the piled-up continuum at high counting rates.

6.6.3 Analog-to-Digital Conversion

A peak sensing analog-to-digital converter (ADC) measures the height of an analog pulse at its peak and converts that value to a digital number. The digital output is thus a proportional representation of the analog pulse height at the ADC input.

For sequentially arriving pulses, the digital outputs from the ADC are fed to a dedicated memory, or a computer, and sorted into a histogram (pulse height spectrum). This histogram represents the spectrum of input pulse heights. The dynamic range of the ADC is matched to the range of the spectroscopy amplifier output, usually 0 ~10 V. Although a peak-sensing ADC is mainly used for energy spectroscopy, it can also be used for time spectroscopy when a time-to-amplitude converter (TAC) is connected to the ADC input. The resulting histogram represents the time spectrum measured by the TAC. The combination of a peak sensing ADC, histogramming memory and a display of the histogram forms a multichannel analyser (MCA). Generally, ADCs are available in three architectures, which are described in the next three sections. A comparison of the properties of each are listed in Table 6.3.

6.6.3.1 Flash ADCs

Flash analog-to-digital converters (also known as parallel ADCs) provide the fastest way to convert an analog signal to a digital signal and are conceptually simplest to understand. A flash ADC is formed by a ladder chain of high-speed comparators,[13] each comparing the input signal to a reference voltage specific to that comparator, usually derived from a resistive divider. If the input voltage is greater than the reference voltage, the output voltage will be high corresponding to logic level one. If it is less, the output voltage will be zero corresponding to logic level zero. The architecture is illustrated in Fig. 6.25. For an N-bit converter, the circuit will employ 2^N-1 comparators. The resistive divider, in turn, will be formed by 2^N resistors, each providing a specific reference voltage to each comparator. For a linear conversion (*i.e.*, using equal value resistors), the voltage on a specific comparator will correspond to the binary equivalent of one least significant bit (LSB) more than the reference voltage of the comparator immediately below it and 1 LSB less than the comparator above it. Each comparator produces a 1 when its analog input voltage is higher than the reference voltage applied to it. Otherwise, the comparator output is 0. Thus, if the analog input is between V_5 and V_6, comparators c_1 through c_5 produce 1's. The remaining comparators will produce 0's. This is known as a thermometer code, since in a thermometer, there is mercury up to the point of the temperature under measurement but none above it. The comparator outputs then connect to the inputs of a priority encoder circuit, which then encodes the thermometer code as a binary number whose value is equal to the measured pulse height.

Flash ADCs are the most efficient of the ADC technologies for high-speed conversion, being limited only in comparator and gate propagation delay (typically < 20 ns). Consequently, they are ideal for applications requiring high speed and very large bandwidth. Unfortunately, flash technology is also the most component-intensive for any given number of output bits. A three-bit flash ADC requires seven comparators – a four-bit version would require 15 comparators. With each additional output bit, the number of required comparators doubles. As a consequence, they consume much more power than other ADC architectures and can be quite expensive. As such, flash ADCs are generally limited to 8-bit resolution (255 comparators). An additional advantage of the flash converter is that it is relatively easy to produce a non-linear mapping between the input and output using non-equal values of resistors in the divider network. This is particularly useful when digitizing signals that cover a wide dynamic range using, for example, a logarithmic or square root response.

TABLE 6.3

Summary of the Pros and Cons of commonly used ADC architectures. Nomenclature: DNL = Differential non-linearity, Msps = Mega samples per second.

Conversion type	Pros	Cons
Flash	Simple to implement Very fast conversion time (<10 ns) High speed (up to 500 Msps) Non-linear conversions easy to implement	Limited resolution (~ 8-bits) High power consumption (~W) Poor DNL Large die size – large input capacitance Prone to output glitches
Wilkinson	Good DNL (<0.1%) Medium die size	Slow conversion time (~100µs) Low speed (~0.1 Msps)
Successive approximation	Short conversion time (~ µs) Medium speed (~5 Msps) Mostly digital circuitry Small die size Low power consumption (<100mW)	Poor DNL, typically 10–20%

---

[13] Comparators are essentially open-loop-gain amplifiers used to test a logical condition. If the input to the non-inverting input is greater than that of the inverting input, the output is set to logical condition 1 (+5 volts). If the input to the non-inverting input is less than that of the inverting input, the output is set to logical condition 0 (0 volts). In essence, a comparator is a 1-bit analog-to digital-converter.

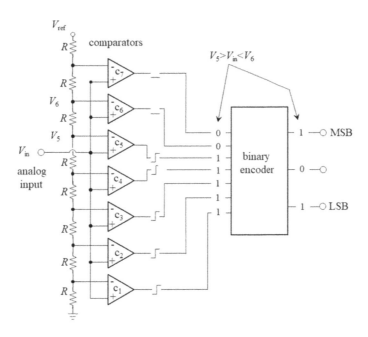

FIGURE 6.25 3-bit flash ADC architecture. If the analog input is between V_5 and V_6, comparators c_1 through c_5 produce 1's. The remaining comparators will produce 0's.

6.6.3.2 Wilkinson ADC

In a Wilkinson or linear ramp ADC, an input voltage is compared with that produced by a charging capacitor (see Fig. 6.26). The capacitor is then allowed to charge until its voltage is equal to the amplitude of the input voltage. This condition is tested by a comparator and when reached, the capacitor is allowed to discharge linearly, producing a ramp voltage. At the point when the capacitor begins to discharge, a gate pulse is generated, which remains on until the capacitor is completely discharged. Thus, the duration of the gate pulse is directly proportional to the amplitude of the input pulse. The gate pulse opens a linear gate, which is connected to a high-frequency oscillator. While the gate is open, a discrete number of clock pulses, N_c, pass through the linear gate and are counted by a counter. The time the linear gate is open is proportional to the amplitude of the

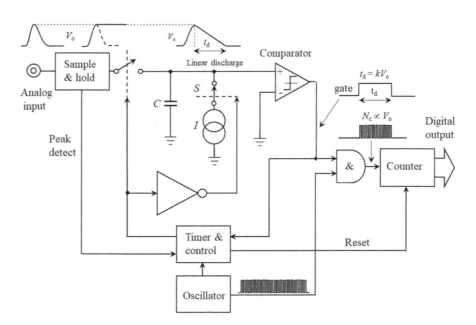

FIGURE 6.26 Principle of operation of a Wilkinson ADC showing critical waveforms. After the peak has been acquired by the sample and hold circuit (S/H), the input capacitor, C, is disconnected from the S/H allowed to discharge through a constant current source ensuring a linear discharge. At the same time, a pulse is generated that gates pulses from a high-frequency clock into a counter. When the capacitor is completely discharged, the comparator output goes low and the conversion is complete – the number of clock cycles recorded by the counter being proportional to the pulse amplitude. This number is then transferred into memory.

input pulse, so the number of clock pulses recorded by the counter is also proportional to the pulse amplitude. Consequently, the analog-to-digital conversion time becomes a function of the input pulse height. During the memory cycle, the contents of the counter are transferred to address N_c located in a histogramming memory and one count is added to the contents of that location. The counter is then reset in preparation for the next conversion cycle. The value N_c is usually referred to as the channel number. The total number of channels, in turn, is defined as the conversion gain, which generally ranges from 256 channels (8-bit) for low resolution applications to 16,384 channels (14-bit), for high-resolution applications. The main advantage of Wilkinson ADCs is its excellent linearity, with non-linearities being typically < 1%. The main disadvantage is the long conversion time for large pulse amplitudes, exacerbated by large analog-to-digital conversion gains.

6.6.3.3 Successive Approximation ADC

The successive approximation ADC is the most commonly used technique for analog-to-digital conversion. As the name suggests, it operates by using a binary search algorithm to converge on the amplitude of the input signal. The architecture is shown schematically in Fig. 6.27. Pulses from the shaping amplifier are first fed into a sample and hold, which detects the peak of the pulse and "holds" it at that level until the rest of the circuitry completes the conversion cycle. During the rise of the analog input pulse, switch S is closed and the voltage on capacitor C tracks the rise of the input signal. When the input signal reaches maximum height, S is opened, leaving C holding the maximum voltage of the input signal. After detection of the input pulse peak, the ADC begins its measurement process. First, the most significant bit of the digital-to-analog converter (DAC) is set to 1. If the comparator determines that the DAC output voltage is greater than the signal amplitude V_o, the most significant bit is reset; otherwise, it is left in the set condition. The test is then repeated by adding the next most significant bit, and so on. When all bits have been tested and set, the control logic then issues an EOC signal, which latches the last DAC digital word into memory. This bit pattern is thus a digital representation of the analog input pulse height, V_o, and is subsequently used as the address of the memory location to which one count is added to build the histogram representing the pulse-height spectrum.

If the ADC has n bits ($2n$ channels), n test cycles are required to complete the analysis, and this is the same for all pulse heights. Although successive-approximation ADCs are available for high-resolution applications (*i.e.*, having a large number of bits), their linearity is not good. The problem can be overcome by adding the sliding scale linearization, which works as follows. For each input signal, a random analog voltage is generated and added before pulse height analysis. If the generated random number is m, this results in the ADC reporting the analysis m channels higher than normal. By digitally subtracting the number m at the output of the ADC, the digital representation is brought back to its normal value. Due to its random nature, the added pulse averages the analysis of each input pulse height over adjacent channels (typically, 256 channels or 8-bit) in the successive approximation ADC. This improves the non-linearity significantly (< 1%). The advantages of the successive-approximation ADC with sliding scale linearization are low differential non-linearity and a fast conversion that is independent of the pulse height. An additional advantage is that the architecture consists of mostly digital circuitry, which results in improved stability compared to the Wilkinson design. In addition, successive approximation ADCs use fewer components that other architectures (*e.g.*, the system only uses one comparator), which results in low power.

6.6.4 Digital Signal Processing

With current analog pulse processing systems, the preamp signal from the detector is shaped, filtered and amplified by a shaping amplifier before being digitized by a peak-sensing ADC. The process can lead to significant dead time due to the sequential

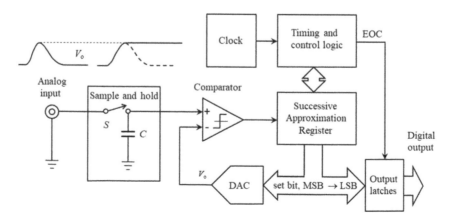

FIGURE 6.27 Operation of a successive-approximation ADC. Here the digital setting of DAC is sequentially adjusted by the control logic until its analog output is equal to V_o. The control logic then issues an end-of-conversion signal (EOC), which latches the last DAC digital word setting into memory.

delays introduced by the shaping time of the amplifier and the conversion time of the ADC. In digital signal processing (DSP) systems, the detector signal is digitized immediately after the preamplifier. The temporal evolution of the signal is preserved by repeated sampling using a flash ADC. Thus, the signal is no longer a continuous stream as in Fig. 6.28(a), but is instead a string of discrete values (representing the instantaneous voltages $V(t)$ at time t as shown in Fig. 6.28(b). Provided the sampling rate is high enough (typically 25–50 Msps) and the interval between samples small, the digital numbers will reproduce the pulse profile to reasonable accuracy. The digitized pulse can then be shaped digitally and the pulse height extracted. The key element controlling these operations is either a field-programmable gate array or a dedicated processor.

Digital pulse detection is generally based on linear trapezoidal filtering in which a trapezoidal function sequentially samples two data sets (windows) at a time, as illustrated in Fig. 6.28(b). Between the two data sets, an optional "gap" exists, represented by the flat top of the trapezoid. Generally, both windows have the same width. Every sample period, the filtering function averages the digitized values inside each window and subtracts the two sums before moving on in time to the next sample period.

Fig. 6.28(c) shows the response of the trapezoidal filter to a step-like input function, which illustrates not only why this type of filter is called "trapezoidal", but also how the various parameters of the filter can affect the shape of the output. For optimal performance, the gap should always be greater than the event rise time of the input. The presence of a pulse is recognized by a non-statistical difference between the two sums. The amplitude of the signal, V_o, at the "step" can then be determined from

$$V(t) = - \sum_{i \, (\text{before})} w_i v_i + \sum_{i \, (\text{after})} w_i v_i, \tag{15}$$

evaluated at t_o, where the first summation is the average over the measured voltage points, v_i, before the step (win1) and the second summation, the average over the voltage points taken after the step (win2) as shown in Fig. 6.28(b). The factors, w_i, in Eq. (15) are weighting constants that determine the type of average being computed. The primary differences between different digital filters lie in what set of weights $\{w_i\}$ are used and how the regions are selected for the computation of Eq. (15). Thus, for example, when the weighting values decrease with separation from the step, then the equation produces "cusp-like" filters. When the weighting values are constant, one obtains triangular (if the gap is zero) or trapezoidal filters. The reasoning behind using cusp-like filters is that since the points nearest the step carry more information about its height, they should be more strongly weighted in the averaging process. How one chooses the filter lengths results in time-variant (the lengths vary from pulse to pulse) or time-invariant (the lengths are the same for all pulses) filters. Traditional analog filters are time invariant. However, for X-ray detection applications, there is an advantage in using time-variant filters. Since the X-rays arrive randomly and the lengths between them vary accordingly, one can make maximum use of the available information by adjusting the window length on a pulse-by-pulse basis. In principal, the very best filtering is accomplished by using cusp-like weights and time-variant filter length selection.

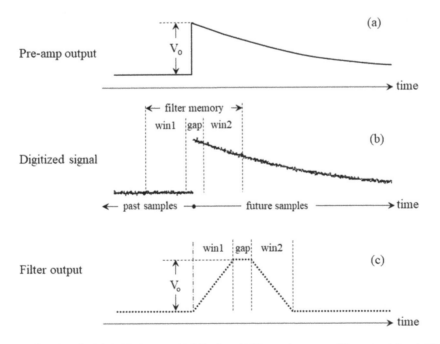

FIGURE 6.28 Schematic illustrating the principle behind trapezoidal filtering. (a) The analog preamplifier output signal, (b) The digitized preamplifier output signal and (c) The output of the trapezoidal filter. Note that the curve appears smooth since each point is an average over the data points in each window.

The benefit of using DSP is that it allows the implementation of signal filtering functions that are difficult to achieve using analog electronics (for example cusp filters). Digital filter algorithms also require considerably less overall processing time, so that the resolution remains fairly constant over a large range of count rates, whereas the resolution of analog systems typically degrades rapidly as the count rate increases. As a result, DSP provides a much higher throughput without significant resolution degradation. Improved system stability is another potential benefit since the detector signal is digitized much earlier in the signal processing chain, which minimizes the drift and instabilities associated with analog signal electronics. A final benefit is that since the signals are captured early in the acquisition process, off-line analysis can be applied later if a more complex data reduction is required. The main disadvantage of DSP is that it requires detailed knowledge of digital algorithms and information theory in general.

6.7 Cooling

Depending on the bandgap energy, it may be necessary to cool the detector in order to reduce thermally generated charge carriers to an acceptable level. This is particularly important when attempting to maximize the spectroscopic performance of a detection system. Otherwise, thermally induced leakage currents can easily dominate the achievable energy resolution of the detector or, if high enough, prevent detector operation. In Fig. 6.29 we show the intrinsic energy resolution that is potentially achievable as a function of bandgap energy for a range of compound semiconductors assuming an average value for the Fano factor of 0.14 [42]. For completeness, we also include the superconductors. This resolution is only achievable if the production of thermally generated carriers is reduced to insignificant levels, which can only be achieved by in most cases by cooling the detector. The type of cooling required for good spectroscopy is illustrated on the left side of the figure, from which we can see that room temperature operation can only be achieved for bandgaps above ~1.4 eV, while thermoelectric cooling can be used for Si down to Ge. We can thus define wide bandgap (WBG) semiconductors as having a bandgap energy conducive to room temperature operation (*i.e.*, > 1.4 eV) and narrow bandgap (NBG) compounds having a bandgap energy below this value.

For the highest precision measurements, cooling of the preamplifier front-end components is also necessary to reduce Johnson noise in the feedback resistor and series noise in the front-end FET. At and below the bandgap of Si, cryogenic cooling[14] is required and liquid nitrogen (LN$_2$) is generally used for temperatures down to 80K. This temperature is appropriate for the vast majority of narrow-gap compound semiconductors, including the indium based III-V compounds (InAs and InSb). For the

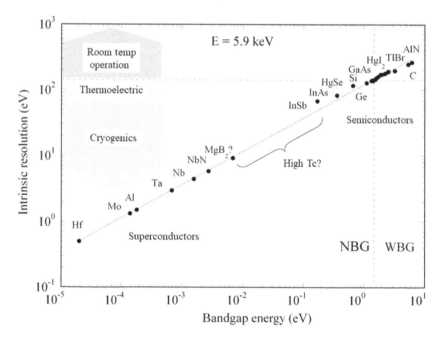

FIGURE 6.29 Left: the limiting energy resolution achievable for a range of compound semiconductors as a function of bandgap energy at 5.9 keV (from [42]). For completeness, we also include the superconductors. Curves are given for average values of the Fano factor (*i.e.*, 0.22 for superconductors and 0.14 for semiconductors). NBG and WBG show the regions in which the narrow bandgap and wide bandgap semiconductors lie.

---

[14] Cryogenics is the science that addresses the production and effects of temperatures less than 120K. The word stems from the Greek words *"kryos"* meaning "frost" and *"genic"* meaning "to produce."

smallest gap semiconductors (*e.g.*, the mercury based II-VI compounds) and superconductors, cooling to liquid helium temperatures (4K) or lower may be required. In the sections that follow, we will review the various cooling techniques.

6.7.1 Passive Cooling

As the name implies, passive cooling requires no active elements and no input power. Two types are commonly used: radiative cooling using radiators and cryogenic cooling using stored cryogens.

6.7.1.1 Radiators

Passive radiators are perhaps the simplest and most reliable cooling method and are most commonly used on spacecraft to reject heat from critical areas of the spacecraft to space. Consequently, radiators have surface finishes with high IR emittance to maximize heat rejection and low absorptance to limit heat input from external heat sources. The amount of heat radiated, Q, is described by the Stefan–Boltzmann law, which scales as

$$Q \sim T^4 - T^4_c, \tag{16}$$

where T is the temperature of the object to be cooled and T_c is the cold heat "sink" temperature. For far Earth orbit, T_c, can be a few K, while for near Earth orbit it is in the tens of K range. The main drawback of radiators is that they are limited in the amount heat they can lift, since they must have a colder space to radiate into and the process is relatively inefficient. Typically, the cooling capacity (which is the measure of the system's ability to remove heat) is in the milliwatt range for an object temperature of 70K; in practice, the lowest temperature that can be achieved with a single stage is around 50K when environmental and parasitic heat loads dominate heat flows. Additional factors in the performance are surface contamination and degradation, which effectively limit the operational lifetime of the radiator. However, even with these limitations, passive radiators can provide an elegant, low-tech, low mass and vibration-free solution for applications with modest cooling requirements.

6.7.1.2 Cryogenic Cooling

For detector operating temperatures below -40°C, passive cooling using a stored cryogen such as N_2 or He is commonly used over radiators, since cryogen cooling can achieve very high heat lifts at much lower temperatures with excellent temperature stability. In cryogenic cooling, heat is absorbed by boiling or sublimation, depending on whether a liquid or solid cryogen is used. In general, liquid nitrogen (LN2) is most commonly used and is the standard for HPGe and Si(Li) detectors, which

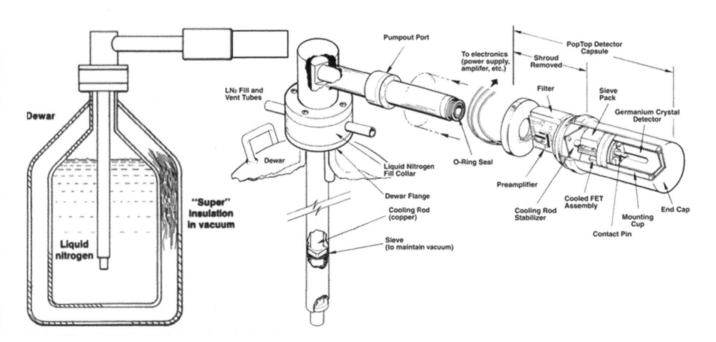

FIGURE 6.30 Left: a classical liquid nitrogen Dewar cooling systems for Ge detectors. Right: expanded detail of the cold-finger/detector assembly (from [43], reprinted with permission of Advanced Measurement Technology, Inc. – ORTEC).

require cooling to <120 K to operate as gamma- and hard X-ray detectors. At these temperatures, the detector must be encased in a vacuum tight container (cryostat) to reduce thermal conduction between the crystal and the air. A classical Ge cryogenic system is shown in Fig. 6.30. Since the preamp should be located as close as possible to the detector to reduce the overall capacitance, the front-end components of the preamp are integrated into the cold head. The entire cold assembly is kept under high vacuum for both thermal insulation and protection from contamination. All of the cryostat materials around the detector are usually constructed from low Z materials, such as, aluminum, magnesium, beryllium, teflon and mylar to reduce photon scatter. A variant of the classical system is used at the University of Leicester, UK, for the routine testing and calibration of a wide range of semiconductors and is shown schematic in Fig. 6.31. The device under test is held in a vacuum cryostat and is cooled by a cold finger connected at one end to a Dewar and the other end to a copper cold head, on which the detector and its front-end components are integrated. An ohmic heater is used to set the temperature of the cold head, which can be controlled to a precision of ±1°C over the temperature range -130°C to +30°C, using a platinum resistance thermometer in a feedback loop. Measurements take place by coupling the cryostat to a beamline or through a 50 μm thick Be window.

The use of LN$_2$ as a coolant is, at best, inconvenient. Maintenance, operating costs, the availability of LN$_2$ and the hazardous nature of the material limit the practicality of LN$_2$-cooled systems, no matter how desirable they might be from other standpoints. In fact, it is a consequence of these drawbacks that so much effort has been expended to develop mechanical coolers on the one hand and compound semiconductor materials that can operate at or near room temperature on the other hand.

6.7.2 Mechanical Cooling

Mechanical coolers, once considered a novelty, have undergone a rapid evolution over the last few decades and have now gained acceptance in applications requiring portability or detector operation in remote locations where LN$_2$ is not readily available. Examples include portable HPGe systems used for nuclear redemption and special nuclear materials monitoring. The useful amount of refrigeration they can achieve is thermodynamically limited by the Carnot coefficient of performance, which is defined as the ratio of the cooling provided at a given temperature to the energy consumed. For example, assuming the ideal case and

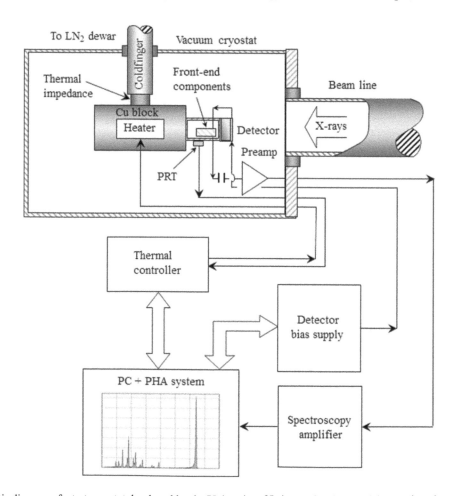

FIGURE 6.31 Schematic diagram of a test cryostat developed by the University of Leicester for characterizing semiconductor detectors.

a sink temperature of 300K, 2.75W of input power is required for each watt of cooling to maintain a cold temperature of 80K. This increases to 74W per watt of cooling to maintain a cold temperature of 4K. In reality, however, power inputs in practical refrigerators are at least a factor of ten times higher because of various inefficiencies.

Mechanical coolers generally consist of two distinct elements: a compressor and a cooler. In essence, the compressor supplies a pressure wave, which is used to drive a closed thermodynamic cycle, which takes place in the cooler. A detailed description of thermodynamic cycles is beyond the scope of this book, and the reader is referred to refs [44–46]. Generally, mechanical coolers operate in one of two thermodynamic modes, known as recuperative and regenerative cycles. Recuperative cryocoolers function through a continuous flow of refrigerant and employ only recuperative heat exchangers, where heat flows constantly from one location within the fluid flow loop to another. A regenerative cycle system operates with oscillating flows and oscillating pressures, analogous to an AC electrical system. In this case, the pressure is analogous to voltage and the mass or volume flow is analogous to current. They utilize a regenerator that stores thermal energy within its porous structure for one half of its cycle and then releases it. Heating occurs as the pressure is increasing, and cooling occurs as the pressure is decreasing. They almost always use high-pressure helium as the working fluid because of its ideal gas properties, its high thermal conductivity and high ratio of specific heats.

Five types of active cryocoolers are commonly used. Fig. 6.32 shows how they are related to each other, and each is described below. A survey of commercial cryocoolers with cooling powers less than several tens of watts can be found in Ter Brake *et al.* [47]. Data is provided on reliability, efficiency, size and mass.

1. <u>Stirling cycle.</u> These coolers consist of a compressor pump and a displacer unit with a regenerative heat exchanger, known as a regenerator. The system causes a working gas to undergo a Stirling cycle, which consists of two constant volume processes and two isothermal processes. The latest generation of Stirling cycle coolers have proved to be extremely reliable and efficient and are now routinely used for long-duration space flight applications. Two-stage devices can extend the lower temperature range from 60–80K to 15–30K for small thermal masses.

2. <u>Pulse tube.</u> Pulse tube coolers are similar to the Stirling cycle coolers although the thermodynamic processes are quite different. They consist of a compressor and a fixed regenerator. Since there are no moving parts at the cold end, they are more reliable and have lower vibration than Stirling cycle machines. A variant is the sorption cooler, which uses a thermochemical process to provide gas compression with no moving parts.

3. <u>Joule-Thompson (J-T).</u> These coolers work using the well-known Joule-Thomson (Joule-Kelvin), effect, which describes the thermodynamic process that occurs when a fluid expands from high pressure to low pressure at constant enthalpy (an isenthalpic process). In this case, a gas is forced through a thermally isolated porous plug, or throttle valve, by a mechanical compressor unit leading to isenthalpic[15] cooling. Although this is an irreversible process, with correspondingly low efficiency, J-T coolers are simple and reliable and have low electrical and mechanical noise levels.

4. <u>Gifford-McMahon (G-M).</u> These are closed-cycle coolers, based on the Gifford-McMahon thermodynamic cycle. Many different forms of this cycle exist (*e.g.*, the Solvay cycle). G-M cooler units rely on the controlled

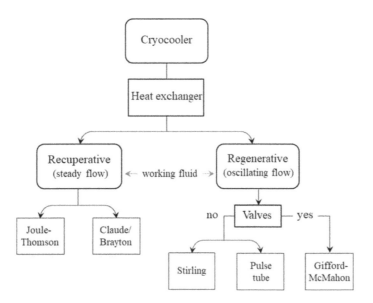

FIGURE 6.32 Top level classification of dynamic cryocoolers showing how different types are related to each other.

---

[15] *i.e.*, without a change in enthalpy

compression and expansion of the working gas between separate compression and cold head units and are similar to Stirling cryocoolers in operation, except that the compressor has valves. An advantage of the system is that the cold head can be placed at some distance from the compressor with flexible lines connecting the high- and low-pressure gas. While they are mechanically simple and reliable, they are relatively inefficient and suffer from low-frequency vibrations.

5. <u>Reverse Brayton</u>. Reverse Brayton coolers are complex machines consisting of a rotary compressor, a rotary turbo-alternator (expander) and a counter-flow heat exchanger (as opposed to the regenerator found in Stirling or Pulse Tube coolers). The compressor and expander use high-speed miniature turbines on gas bearings making these coolers difficult to miniaturize. They are characterized by high efficiency and heat-lifting capability and are practically vibration free. They are primarily used in applications where a large machine is inevitable, as in the case for temperatures less than 10K or when a large cooling capacity is required at higher temperatures.

The J-T and Brayton cycles are classified as recuperative cycle systems, while Stirling cycle, pulse tube and G-M coolers are classified as regenerative cycle systems (see Fig. 6.32).

6.7.2.1 *Application to Detectors*

For detector applications, early systems were based on the Solvay cycle [48], which may be regarded as a variant of the G-M thermodynamic cycle. Although they were effective in cooling to liquid nitrogen temperatures, they proved unreliable, consumed high power and gave poor performance due to mechanical vibrations. For example, the measured FWHM energy resolution at 5.9 keV for a 30 mm^2 planar Ge detector was 640 keV at 5.9 keV [48], whereas a resolution of 175 eV FWHM could be achieved using LN$_2$ cooling. In addition, the compressors required oil for lubrication. Consequently, they were not portable or even transportable.

The first commercial mechanically cooled HPGe systems appeared in the early 1980s. To reduce microphonics, the compression and expansion of the working substance (helium) occurred in physically separated units connected by flexible metal hoses. These systems were expensive to operate, had high power requirements, were bulky in size and were unreliable. In addition, they gave degraded system performance. In 1986, ORTEC® introduced a modified Solvay-cycle cooler. In this design, the detector was physically separated from the piston-driven compressor thereby reducing the effects of the microphonics shown in earlier devices. Stone *et al.* [49] reported a performance degradation of only 8% at 5.9 keV as typical for this type of system. In 2000, ORTEC introduced a modified J-T cooler known as the XCooler with much improved portability and reliability. In this system, the compressor was again separated from the cold head by a 10-ft gas hose. Compared to LN$_2$ cooling, the manufacturer claims that for coaxial detectors, there is no degradation in energy resolution above 500 keV and 10% for energies less than 500 keV (20% for planar detectors) [50].

Stirling closed cycle refrigerators are now the most commonly used, especially for space applications, since they offer a number of advantages over other cooling systems (for example, Rankine-cycle engines, J-T or Solvay cycle). Specifically,

1) The coolant is contained within the cooler itself and not in an external heat exchanger, thus ensuring small size (see Fig. 6.33),

2) They use a small amount of refrigerant and no phase change takes place, offering a constant high efficiency for low lifts,

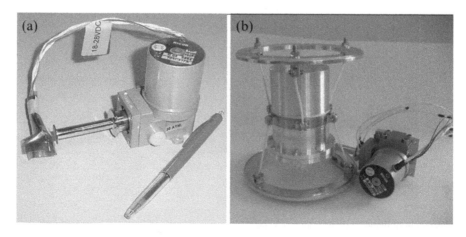

FIGURE 6.33   (a). A miniaturised Stirling cycle cooler produced by Ricor® Ltd [51]. At an ambient temperature of 23°C, the device is capable of providing a heat lift at the cold finger of 700 mW at 77 K. (b) coupled to an encapsulated 6 cm right circular HPGe detector (image courtesy Baltic Scientific Instruments®). The measured FWHM energy resolution was 5 keV at 1333 keV, limited by mechanical vibration.

3) unlike J-T or Solvay-cycle coolers, Stirling coolers use a piston and springs driven at very high rates to compress the refrigerant gas. This arrangement is oil free, thus eliminating contamination of the gas and thus the possibility of clogging,

4) Present designs are based on a hermetically sealed, free-piston linear motor assembly with gas bearings, achieving low noise and higher reliability by dispensing with performance-limiting mechanical interfaces.

Stirling cryocoolers were first developed in the early 1950s for the liquefaction of air in remote locations. By the 1960s, they were used in military night vision equipment for cooling the infrared detectors. These early devices were rotary types with crank drives and rubbing piston rings. Consequently, lifetimes were limited to a few hundred hours. Linearly driven Stirling cryocoolers were introduced in the 1970s that eliminated most of the rubbing contact and offered lifetimes of around one year with refrigeration powers in the range 0.15 to 1.75 W at 80K. Stirling coolers were first developed seriously for space applications after the introduction of flexure bearings to support the piston and displacer inside the cylinder walls with no rubbing contact. Lifetimes of at least 10 years can now be expected. Gas-bearing support of the piston was introduced in the 1990s as an alternative approach to achieve a long lifetime.

6.7.2.2 Reducing Microphonics

The major problem of all mechanical coolers is degraded performance due to microphonic noise arising from vibrations, created either by a moving piston in a compressor or even from the boiling of a refrigerant. These vibrations are conducted through the mechanical structure into the detector head causing small changes in capacitance, which in turn induce low frequency electrical oscillations at the preamplifier input. The resulting noise can severely degrade detector energy resolution as is demonstrated in Table 6.4, in which we list the measured FWHM energy resolutions at 1,332 keV for various HPGe gamma-ray spectrometers cooled by Stirling cycle engines [52–56]. Resolution values are given for two cases, one with the cooler switched on and with the other with the cooler switched off. In the latter case, it is assumed that the measured energy resolutions are indeed the limiting case; this has been verified experimentally by a number of groups. For example, in reference [49] the spectrometric performance was first measured by coupling the detector cryostat to a standard LN_2 cryostat, prior to system integration with a Stirling cryocooler. At an operating temperature of ~80 K, the energy resolutions measured at 122 keV and 1,332 keV using the LN_2 cryogen were found to be the same as when measured with the Stirling system when the cooler was momentarily switched off. From Table 6.4, we see that on average the energy resolution degrades by a factor of 1.4 when the cooler is active.

It is possible to limit microphonic effects by careful cryostat design. However, these mechanical options usually have adverse trade-offs for portable applications. For example, Sakai *et al.* [57] used a specially designed anti-microphonic crystal mount. A clear improvement in performance was achieved for a small planar HPGe detector but only at short and non-optimum shaping times. The most common method in use today is separating the detector from the moving compressor as much as possible. However, such a system is clearly non-portable. Broerman *et al.* [51] showed that a detector attached to a Ortec X-Cooler commercial system (with a separate head and compressor) had on average a 7% degradation in performance across the energy range 5.9 keV to 1,332 keV, versus the same detector cooled on liquid nitrogen. Lavietes *et al.* [58] used an active vibration control system in which software controlled a mechanical balance in a Stirling cycle engine

TABLE 6.4

Compilation of FWHM energy resolutions recoded at 1,332 keV for a number of Stirling engine cooled HPGe spectrometers. On average, the energy resolution degrades by a factor of 1.4 when the cooler is active. As expected, the dual opposed piston cooler has lower noise when active (as reflected in the measured energy resolution) than the single piston coolers.

Instrument	Detector volume (cm³)	Cooler model/ configuration	Energy resolution at 1,332 keV			Reference
			Cooler on	Cooler off	Ratio off/on	
R&D prototype	14	Sumitomo[16] SRS-210, single piston	3.3	2.4	1.4	[52]
MESSENGER GRS	98	Ricor[17] K508, single piston	3.5	2.4	1.5	[53]
KAGUYA GRS	252	Sumitomo[16] FS1ST, dual opposing piston	3.0	2.4	1.3	[54]
GeMini	98	Ricor[17] K508, single piston	3.8	2.4	1.6	[55]
Miniature GRS	170	Ricor[17] K508, single piston	4.0	3.0	1.3	[49]
Miniature GRS	158	Thales RM3, single piston	4.0	2.2	1.8	[56]

---

[16] www.shi.co.jp/quantum/eng/product/space/space.html

[17] www.ricor.com/

to compensate for the vibrations caused by the moving pistons. This method attempts to correct the microphonics, not at the detector but at the compressor. The drawback to this method is the additional electronics and thus power needed to control the balance, as well as the additional weight created by the balance itself. Today, nearly all cryocoolers used in space have active vibration suppression systems built into their drive electronics that reduce the peak unbalanced forces to less than 1% of their original levels. An alternate approach is to use two engines with dual opposed pistons coupled. In this case, the engines are connected at their hot ends and mounted in tandem to cancel vibration. Depending on the degree of dynamic balancing, vibration level amplitudes can be reduced by a factor of 100 or more at the expense of increased complexity and power. A large reduction can also be achieved using digital filtering in the signal processing chain. For example, Upp *et al.* [59] constructed a system centered on a 50 mm×30 mm HPGe crystal cryogenically cooled by a dual opposed piston Stirling cycle engine (see Fig. 6.34). The measured FWHM energy resolution at 1,333 keV was 15 keV. However, using low frequency filtering, the measured energy resolution improved dramatically to 2.1 keV FWHM at 1,333 keV [60]. Most recently, Kondratjev *et al.* [56] fabricated a spectrometer centered on a 6 cm diameter×5.6 cm high right circular HPGe cooled by a Thales RM3 single piston, miniature Sterling cycle cooler. With the mechanical cooler switched off, FWHM energy resolutions of 1.5 keV and 2.2 keV were obtained at 122 keV and 1,333 keV, respectively, at the nominal operating temperature of 90K. When the cooler was switched on the energy resolutions degraded to 2.5 keV and 4 keV, respectively. However, these improved significantly to 1.8 keV and 2.4 keV, when a low frequency rejection filter was incorporated into the signal chain.

6.7.3 Thermoelectric Cooling

For compound semiconductors, thermoelectric cooing provides a simple and relatively inexpensive method of cooling detectors for test and evaluation. Thermoelectric cooling makes use of the Peltier effect to create a heat flux at the junction of two dissimilar materials. A thermoelectric, or Peltier, cooler is a heat pump made up of many Peltier junctions, which converts electrical energy into a temperature gradient between two surfaces, namely the top and bottom of the device. Its main advantages over vapor-compression refrigerators are its lack of moving parts or circulating fluid, its small size and flexible shape (form factor). Its main disadvantage is that its relative efficiency is only 15–20% of that achieved with conventional compression cycle refrigeration systems. This can be a limitation in some applications, for example, in space, where spacecraft resources are generally very limited. In addition, the maximum temperature difference that can be achieved with a multi-stage device is less than 160K. Thus, from an ambient temperature of 300K, the minimum temperature achievable is only about 140K. At these temperatures, the coefficient of performance is very low ($\sim 10^{-4}$) as opposed to, say, a pulse tube cryocooler (up to $\sim 10^{-1}$). Consequently, thermoelectric coolers are not suitable for cryogenic cooling in their present form.

6.7.3.1 The Peltier Effect

Jean Charles Peltier [61] observed that when an electric current passed across the junction of two dissimilar conductors (a "thermocouple") there was a heating effect that could not be explained by Joule heating alone. In fact, depending on the direction of the current, the overall effect could be either heating or cooling, which can be explained as follows. When two conductors are placed in electric contact, the difference in the Fermi levels causes electrons to flow across the junction until

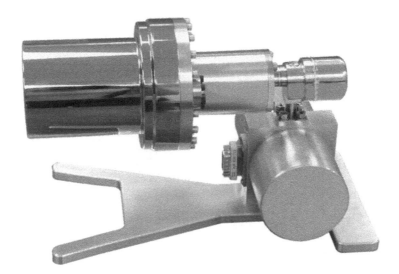

FIGURE 6.34 A mechanically cooled 50 mm×30 mm HPGe spectrometer (from [59]). The system uses a dual opposed piston Stirling cycle engine (SAX101-002B), manufactured by Hymatic Engineering, Ltd. Using low frequency filtering, the measured energy resolution is 2.1 keV FWHM at 1333 keV. The power consumption is still less than 16 Watts, making battery operation possible.

the change in electrostatic, or contact, potential brings the two Fermi levels into a common equilibrium. Current passing across the junction results in either a forward or reverse bias creating a temperature gradient as heat moves (or is "pumped") in the direction of charge carrier flow. Note that the charge carriers actually transfer the heat and release it on the opposite ("hot") side as the carriers move from a high- to a low-energy state. If the temperature of the hotter junction, T_h, is now kept low by removing the generated heat, then the temperature of the cold junction, T_c, will drop relative to the hot junction by an amount, $\Delta T = T_h - T_c$, which can be tens of degrees for a single stage cooler.

6.7.3.2 Quantifying the Effect

The amount of heat absorbed or released at the thermocouple junction is directly proportional to the current passing through it, i, and its duration, t,

$$Q = Pit, \tag{17}$$

where P is the Peltier coefficient, which is defined as the amount of heat emitted or absorbed at the junction of a thermocouple when a current of one ampere passes through it for one second. The Peltier coefficient depends upon the contact temperature of the two materials from which the thermocouple is formed and is usually defined in terms of the Seebeck coefficient, α, which is a measure of the thermoelectric "power" of the junction at temperature, T,

$$P = \alpha T. \tag{18}$$

Specifically, the Seebeck coefficient is a measure of the magnitude of an induced thermoelectric voltage in a material in response to a temperature difference across that material and the entropy per charge carrier in the material. α has units of V/K, though μV/K is more common. Values in the hundreds of μV/K, regardless of sign, are typical of good thermoelectric materials.

The conversion efficiency of a device (that is, the ratio of electrical power generated to the heat absorbed at the hot junction) can be expressed in terms of the Carnot efficiency and a specific material parameter, referred to as thermoelectric figure-of-merit, Z, defined as

$$Z = \frac{\alpha^2 \sigma}{\kappa}, \tag{19}$$

where σ and κ are the electrical and thermal conductivities of the thermoelectric material, respectively. Typical values of Z lie in the range $(2.5-3.2) \times 10^{-3}$ K^{-1}. To be efficient as a thermoelectric material, Z should be as large as possible over as wide a temperature range as possible. If Z is known, then the temperature difference, ΔT_{max} across the junction can be estimated from

$$\Delta T_{max} = \frac{1}{2} Z T_c^2, \tag{20}$$

where T_c is the cold side temperature. The typical temperature dependence ΔT_{max} versus Z is shown in Fig. 6.35 for four ambient temperatures, which in normal operation is the temperature to which the hot side is heat-sinked. The cooling capacity, or alternately the heat lift, is given by

$$Q = Q_{max} \left(1 - \frac{\Delta T}{\Delta T_{max}} \right), \tag{21}$$

where ΔT is the differential temperature to give the required cold junction temperature, (*i.e.* $T_c = T_h - \Delta T$). ΔT_{max} and Q_{max} are given by the manufacturer.

Since, α, σ and κ all have different temperature dependencies, for practical applications, the product ZT is more commonly used as a dimensionless metric with which to evaluate the effectiveness of a thermocouple, *viz*,

$$ZT = \frac{\alpha^2 \sigma T}{\kappa} = \frac{\alpha^2 T}{\kappa R}. \tag{22}$$

Here R is the electrical resistance of the thermocouple. A greater ZT indicates a greater thermodynamic efficiency, subject to certain provisions, particularly that the two materials in the couple have similar Z values. Typical values for the best commercial junctions are of the order ~1. However, to compete effectively with mechanical devices, values in

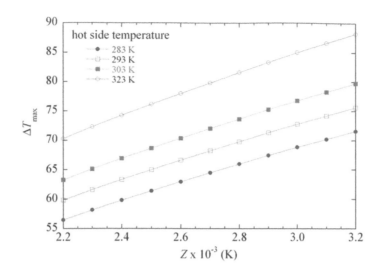

FIGURE 6.35 ΔT_{max} as a function of Z, for a single stage Peltier cooler. Data are given for four hot side temperatures, assumed to be "sinked" to the ambient temperature.

the range 3 – 4 are required. To date, the best reported ZT values are in the 3 – 4 range. Note that in bulk materials α, R and k are functions of the carrier concentration in the specific materials used and therefore correlate with each other – meaning they cannot be adjusted independently. From Eq. (22), we see that, at first glance, metals would appear to be a good choice for thermocouple materials, because of their low electrical resistance. However, they also have very high thermal conductivities, which would seriously reduce the heat gradient, thus lowering the overall ZT value. In practice, semiconductors are the materials of choice, because they have low conductivities and their resistivities can be controlled by doping. The bismuth chalcogenides, Bi_2Te_3 and Bi_2Se_3 comprise some of the best performing room temperature thermoelectric materials, with ZT values between 0.8 and 1.0. The material most often used is bismuth telluride (Bi_2Te_3), which can be doped to produce both n and p type material, which is then fabricated into individual elements, or pellets, to form the basic building blocks of a thermoelectric cooler (TEC) as described in section 6.7.3.3.

Unfortunately, while it is possible to make a simple thermoelectric device with a single semiconductor pellet, you can't pump an appreciable amount of heat through it. In order to achieve a greater thermoelectric heat-pumping capacity, multiple pellets are used together. Of course, the initial temptation would be simply to connect them in parallel – both electrically and thermally. While this is possible, it does not make for a very practical device. The problem is that typical TE semiconductor pellets are rated for only a very small voltage – as little as tens of millivolts – while it can draw a substantial amount of current. For example, a single pellet in an ordinary TE device might draw five amps or more with only 60 mV applied; if wired in parallel in a typical 254-pellet configuration, the device would draw over 1,270 amps with the application of that 60 mV (assuming that the power supply could deliver that much current). The only realistic solution is to wire the semiconductors in series, in a way that keeps them thermally in parallel (that is, pumping together in the same direction). Here, we might be tempted to simply zigzag the electrical connections from pellet to pellet to achieve a series circuit. This is theoretically workable; however, the interconnections between pellets introduce thermal shorting that significantly compromises the performance of the device. Fortunately, there is another option that gives us the desired electrical and thermal configuration while better optimizing the thermoelectric effect. By arranging n and p-type pellets in a "couple" (see Fig. 6.36) and forming a junction between them with a plated copper tab, it is possible to configure a series circuit that can keep all of the heat moving in the same direction. As shown in Fig. 6.36, with the free (bottom) end of the p-type pellet connected to the positive voltage potential and the free (bottom) end of the n-type pellet similarly connected to the negative side of the voltage, an interesting phenomenon takes place. The positive charge carriers (*i.e.*, "holes") in the p material are repelled by the positive voltage potential and attracted by the negative pole; the negative charge carriers (electrons) in the n material are likewise repelled by the negative potential and attracted by the positive pole of the voltage supply. In the copper tabs and wiring, electrons are the charge carriers; when these electrons reach the p material, they simply flow through the "holes" within the crystalline structure of the p-type pellet (remember, the charge carriers inherent in the material structure dictate the direction of heat flow). Thus, the electrons flow continuously from the negative pole of the voltage supply, through the n pellet, through the copper tab junction, through the p pellet and back to the positive pole of the supply – yet because we are using the two different types of semiconductor material, the charge carriers and heat all flow in the same direction through the pellets (bottom to top in the drawing). Thus, using these special properties of the TE "couple", it is possible to team many pellets together in rectangular arrays to create practical thermoelectric modules (see Fig. 6.37).

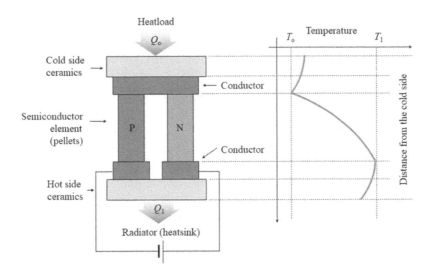

FIGURE 6.36 Simplified illustration of a single thermoelectric module showing the functional elements and the temperature differential across it (taken from [62], courtesy TEC Microsystems® GmbH).

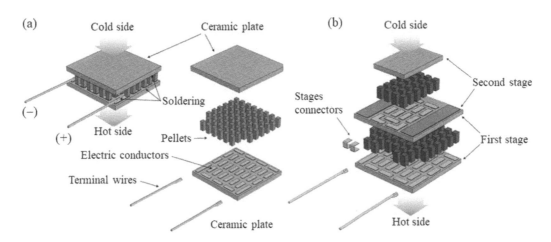

FIGURE 6.37 (a) A single stage TE module consisting of a matrix of pellets and a pair of hot and cold sides. (b) A multi-stage module can be viewed as two or more single stages stacked on top of each other. The construction of a multi-stage module is usually pyramidal – each lower stage is bigger than the upper stage. The topmost stage is the cold side (Image courtesy of TEC Microsystems GmbH).

These devices are able to not only pump appreciable amounts of heat; with their series electrical connection, they are matched to commonly available DC power supplies. The most common TE devices now in use are formed by connecting several alternating p- and n-type pellets, sufficient to be powered by a 12 to 16 V DC supply, drawing only 4 to 5 amps. This should be contrasted to a power supply demand of ~1,300 amps at 60 mV if all elements were connected in parallel.

6.7.3.3 Thermoelectric Cooler (TEC) Construction

The basic TEM unit is a thermocouple, which consists of a p-type and an n-type semiconductor element or pellets. Copper commutation tabs are used to interconnect pellets that are traditionally made of bismuth telluride. A typical TEM consists of thermocouples connected electrically in series and sandwiched between two alumina ceramic plates. When DC moves across the TEM, it causes temperature differential between thermoelectric modules (TEMs) sides. As a result, one TEM face, which is called cold, will be cooled while its opposite face, which is called hot, simultaneously is heated. If the heat generated on the TEM hot side is effectively dissipated into heat sinks and further into the surrounding environment, then the temperature on the TEM cold side will be much lower than that of the ambient, potentially by tens of degrees. The TEM's cooling capacity is proportional to the current passing through it. The TEM's cold side will consequently be heated and its hot side will be cooled once the TEM's polarity has been reversed. A TEC is comprised of many TEMs, each composed of thermoelectric (TE) couples, comprised of n- and p-type semiconductor legs that are connected electrically in series. The two semiconductors are positioned thermally in parallel and joined at one end by a conducting cooling plate (typically of copper or aluminum). They are then fixed by soldering and sandwiched between two ceramic plates (usually Al_2O_3), ceramics which serve as the hot and cold surfaces. The construction of a TE module is shown in Fig. 6.37. In practice, many TE couples (from several elements to hundreds of units), such as described above, are connected side-by-side in a single TEC unit. The heat-pumping capacity of a cooler is proportional to the current and the number of pairs in the unit and can have cooling capacities ranging from fractions of Watts to hundreds of Watts. The ceramic plates that form the hot and cold sides of the TE also supply the mechanical integrity of a TE module.

6.7.3.4 Performance

Thermoelectric junctions are generally only around 5 – 10% as efficient as an ideal refrigerator (Carnot cycle), compared with 40 – 60% achieved by conventional compression cycle systems (reverse Rankine systems using compression/expansion). Due to the relatively low efficiency, thermoelectric cooling is generally only used in environments where the solidstate nature (no moving parts, maintenance-free, compact size) outweighs pure efficiency. The TEC performance is a function of ambient

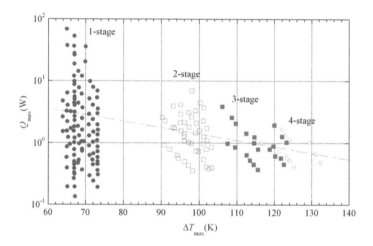

FIGURE 6.38 Maximum performance parameters of classical commercial thermoelectric modules (TEMs) (taken from [62], courtesy of RMT® Ltd.).

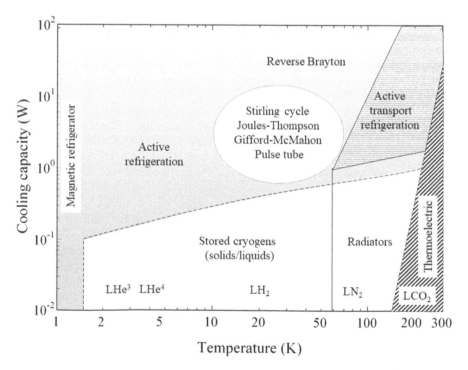

FIGURE 6.39 Graph illustrating the approximate range of cooling capacity for various passive and active coolers as a function of temperature. The cooling capacity is the measure of a cooling system's ability to remove heat.

temperature, hot and cold side heat exchanger (heat sink) performance, thermal load, Peltier module (thermopile) geometry and Peltier electrical parameters.

6.7.3.5 Sizing a TEC for a Detector

For a specific application, a TEC can be characterized by a set of operational parameters. These are:

1) ΔT – operating temperature difference (at known ambient temperature, T_a),

2) Q – operating cooling capacity,

3) I – applied or available current,

4) U – terminal voltage and dimensional restrictions.

The TEC cooling capacity, Q, depends on the number of pellets and their geometry. Low height pellets or/and larger pellet cross-sectional area provides higher cooling capacity, at the expense of increased operating current and total power consumption. Pellets with small cross-sectional area and tall pellets increase the maximum temperature difference and

TABLE 6.5

Summary of the cooling properties and relative pros and cons of the different cooling systems, approximately ordered by complexity.

Cooler	Typical temperature range (K)	Typical cooling capacity range (W)	Advantages	Disadvantages
Radiator				
1 stage (small)	80 – 100	0.001 – 0.1	Simple, low-cost, reliable, vibration free, no moving parts, long life.	Limited heat lift. Radiator view factor to cold sink may be limited.
2 stage (small)	50 – 80	0.001 – 0.05		
1 stage (large)	100 – 200	1 – 10		
Cryogen (liquid)			Simple, no moving parts. Constant temperature. Vibration-free. Large heat lift.	Cryogen availability, hazardous material. Limited operating lifetime.
CO_2	194.6	Unlimited		
N_2	77.5	-		
He^4	4.2	-		
He^3	3.2			
Cryogen (solid)			Reliable. No moving parts. Vibration-free. Higher heat content per kg.	Complex Dewar design. Complicated filling procedures.
NH_3	150.0–195.0	0.1–1		
CO_2	125.0–215.7			
N_2	43.4–63.4			
H_2	8.3–14.0			
Thermoelectric			Simple, reliable, low cost, no moving parts vibration-free, can be used to anneal detectors.	Limited heat lift. Power consumption can be high. Generates more heat than it removes on the hot side due to I^2R losses.
1 stage	230–300	0.1–100		
Multi-stage	145–230	0.01–1		
Stirling (1–2 stages)	15–80	0.001–15	High efficiency. Small compact. Moderate cost. Large heritage.	Dry, no lubrication. Vibration from displacer. Limited lifetime.
Pulse tube	40–80	0.8	High efficiency. Lower vibration, higher reliability, lower cost than Stirling and G-M.	Lower efficiency than Stirling. Gravity induced instability. Lower size limit for efficiency.
Jules-Thompson	3–90	Up to 20	No moving parts. Steady flow (no vibration). Long MTBF. Transport cold long distances.	Requires high pressures. Low efficiency below 90K. Small orifice susceptible to clogging.
Gifford-McMahon	3–77	0.001–15	Reliability and moderate cost. Split cycle allows remote cooling.	Large and heavy. Vibrations from displacer. Low efficiency.
Reverse Brayton	4.2–77	10–100	High cooling capacity. Good efficiency. Steady flow (low vibration). Long lifetime.	Requires large heat exchanger. Difficult to miniaturize. Expensive to fabricate.
Magnetic refrigeration (ADR)	0.05	0.01	Only way to reach these temperatures Low noise, compact.	Large magnetic field. Complex system because of the need to precool.

reduce TEC power consumption, at the expense of cooling capacity. A rough estimate of the operating temperature difference and cooling capacity can be obtained from,

$$Q = Q_{max}\left(1 - \frac{\Delta T}{\Delta T_{max}}\right) \text{ and } \Delta T = \Delta T_{max}\left(1 - \frac{Q}{Q_{max}}\right). \tag{23}$$

In general, if T across the thermoelectric is less than 55°C, then a single stage thermoelectric is sufficient. The theoretical maximum temperature difference for a single stage thermoelectric is between 65°C and 70°C, assuming operation in a vacuum and parasitic heat losses are minimized. Therefore, as a rule of thumb, if ΔT is required to be > than 55°C, then a multi-stage TEC should be considered. High ΔT can be achieved by stacking as many as 6 or 7 single-stage thermoelectrics on top of each other in a pyramid fashion. This is illustrated in Fig. 6.38, in which we show the maximum temperature difference as a function of maximum cooling capacity (corresponding to $\Delta T_{max} = 0$) for single-stage, two-stage, three-stage and four-stage TEMs (from [62]). From the figure, we see that the maximum cooling capacity decreases as the number of stages increases, while the maximum temperature differential increases. It is also clear why TECs are not used for Ge detectors. Assuming heat sinking at room temperature, the maximum temperature differential would need to be > 200 K to allow operation. Even with 7 stages, ΔT_{max} is only around 170K with a miniscule amount of heat pumping, insufficient to overcome the numerous parasitic heat losses.

6.7.4 Summary and Comparison of Cooling Systems

In this section, we summarize the cooling performances of the systems previously described. In Fig. 6.39 we show the approximate range of cooling capacity for various passive and active cooling technologies as a function of temperature, and in Table 6.5, we summarize of the relative pros and cons of these systems, approximately ordered by complexity.

REFERENCES

[1] "*PM5 Auto-Lap Precision Lapping & Polishing Machine Product Sheet*", Logitech Limited, Erskine Ferry Road, Glasgow G605EU, Scotland, U.K. (2005).
[2] Brigham Young University, Electrical and Computer Engineering, https://cleanroom.byu.edu/wet_etch
[3] J.W. Faust Jr., "Etching of the III–V semiconductor intermetallic compounds", in *Compound Semiconductors, Vol. 1, Preparation of III-V Compounds*, eds. R.K. Willardson, H.L. Goering, Reinhold, New York (1962), pp. 445–468.
[4] K. Saga, T. Hattori ""Identification and removal of trace organic contamination on silicon wafers stored in plastic boxes", *J. Electrochem. Soc.*, Vol. **143**, no. 10 (1996), pp. 3279–3284.
[5] E. Woolfenden, "Thermal desorption for gas chromatography", in *Gas Chromatography*, ed. C.F. Poole, Elsevier, Amsterdam, Chapter 10 (2012), pp. 235–289.
[6] A. Danel, C.L. Tsai, K. Shanmugasundaram, E. Kamieniecki, F. Tardif, J. Ruzyllo, "Cleaning of Si surfaces by lamp illumination", *Solid State Phenom., Ultra Clean Process. Silicon Surf. VI*, Vol. **92** (2003), pp. 195–198.
[7] E.H. Rhoderick, R.H. Williams, "*Metal-Semiconductor Contacts*", Clarendon Press, Oxford, 2nd ed. (1988).
[8] L.J. Brillson, "*Contacts to Semiconductors: Fundamentals and Technology*", Noyes Publ. Co., Park Ridge, NJ (1993).
[9] C. Wilmsen, "*Physics and Chemistry of III–V Compound Semiconductor Interfaces*", Plenum Press, New York (1985), p. 129.
[10] L.M. Porter, K. Das, Y. Dong, J.H. Melby, A.R. Virshup, "Contacts to wide-band-gap semiconductors", *Compr. Semicond. Sci. Technol.*, Vol. **4** (2011) pp. 44–85.
[11] A.G. Baca, F. Ren, J.C. Zolper, R.D. Briggs, S.J. Pearton, "A survey of ohmic contacts to III–V compound semiconductors", *Thin Solid Films*, Vol. **308** (1997), pp. 599–606.
[12] T.-J. Kim, P.H. Holloway, "Ohmic contacts to II–VI and III–V compound semiconductors", *Wide Bandgap Semiconductors*, ed. S. Pearton, William Andrew Publishing, New York (1999), pp. 80–150, ISBN 0-8155-1439-5.
[13] C.J. Palmstrøm, "Contacts for compound semiconductors: Ohmic type", in *Encyclopaedia of Materials: Science & Technology*, eds. K.H.J. Buschow, R. Cahn, M. Flemings, B. Ilschner, E. Kramer, S. Mahajan, P. Veyssiere, Pergamon Press, Elsevier, Oxford (2001), pp. 1581–1587.
[14] L.J. Brillson, "Contacts for compound semiconductors: Schottky barrier type", in *Encyclopaedia of Materials: Science & Technology*, eds. K.H.J. Buschow, R. Cahn, M. Flemings, B. Ilschner, E. Kramer, S. Mahajan, P. Veyssiere, Pergamon Press, Elsevier, Oxford (2001), pp. 1587–1595.
[15] M. Paunovic, M. Schlesinger, "*Fundamentals of Electrochemical Deposition*", Wiley, New York (1998).
[16] M. Schlesinger, M. Paunovic, "*Modern Electroplating*", Wiley, New York, 4th ed. (2000).
[17] S.S. Djokic, P. Cavallotti, "Electroless deposition: Theory and applications", in *Modern Aspects of Electrochemistry*, ed. S.S. Djokic, Springer, New York, No. 48, Chapter 6 (2010), pp. 251–289.
[18] S.S. Djokic, "Electroless deposition of metals and alloys", in *Modern Aspects of Electrochemistry*, eds. B.E. Conway, R. E. White, Kluwer Academic/Plenum Press, New York, No. 35, Chapter 2 (2002), pp. 51–133.
[19] T.F. Databook, (1990). Available by request from the Lebow Corporation, 5960 Mandarin Ave., Galeta CA 93117, USA.

[20] W.R. Grove, "On the electro-chemical polarity of gases", *Philos. Trans. R. Soc. London, A*, Vol. **142** (1852), pp. 87–101.

[21] J. Plücker, "Observations on the electrical discharge through rarefied gases", *Lond., Edinb. Dublin Philos. Mag.*, Vol. **16** (1858), pp. 409–418.

[22] A. Heuberger "X-ray lithography", *Microelectron. Eng.*, Vol. **5**, no. 1–4 (1986), pp. 3–38.

[23] U.S. Tandon, B.D. Pant, A. Kumar "An overview of X-ray lithography for use in semiconductor device preparation", *Vacuum*, Vol. **42**, no. 18 (1991), pp. 1219–1228.

[24] S.H. Park, J.H. Ha, J.H. Lee, H.S. Kim, Y.H. Cho, S.D. Cheon, D.G. Hong "Effect of temperature on the performance of a CZT radiation detector", *J. Korean Phys. Soc.*, Vol. **56**, no. 4 (2010), pp. 1079–1082.

[25] A.G. Kozorezov, V. Gostilo, A. Owens, F. Quarati, M. Shorohov, A. Webb, J.K. Wigmore, "Polarisation effects in thallium bromide X-ray detectors", *J. Appl. Phys.*, Vol. **108** (2010), pp. 1-1–1-10.

[26] M. Amman, P.N. Luke, J.S. Lee, E. Orlando, "Material dependence of bulk leakage current in CdZnTe detectors", *Proc. SPIE*, Vol. **6706** (2007), pp. 67060L-1–67060L-9.

[27] P.N. Luke, M. Amman, J.S. Lee, P.F. Manfredi, "Noise in CdZnTe detectors", *IEEE Trans. Nucl. Sci.*, Vol. **48** (2001), pp. 282–286.

[28] S.H. Park, J.H. Ha, Y.H. Cho, H.S. Kim, S.M. Kang, Y.H. Kim, J.K. Kim "Surface passivation effect on CZT-metal contact", *IEEE Trans. Nucl. Sci.*, Vol. **55**, no. 3 (2008), pp. 1547–1550.

[29] K. Nakazawa, K. Oonuki, T. Tanaka, Y. Kobayashi, K. Tamura, T. Mitani, G. Sato, S. Watanabe, T. Takahashi, R. Ohno, A. Kitajima, Y. Kuroda, M. Onishi "Improvement of the CdTe diode detector using a guard-ring electrode", *IEEE Trans. Nucl. Sci.*, Vol. **51**, no. 4 (2004), pp. 1881–1885.

[30] A. Owens, M. Bavdaz, H. Andersson, T. Gagliardi, M. Krumrey, S. Nenonen, A. Peacock, I. Taylor, L. Tröger, "The X-ray response of CdZnTe", *Nucl. Instrum. Methods.*, Vol. **A484** (2002), pp. 242–250.

[31] D.A. Landis, F.S. Goulding, R.H. Pehl, J.T. Walton, "Pulsed feedback techniques for detector radiation spectrometers", *IEEE Trans. Nucl. Sci.*, Vol. **NS-18** (1972), pp. 115–124.

[32] F.S. Goulding, J. Walton, D.F. Malone, "An optoelectronic coupling feedback preamplifier for high resolution nuclear spectroscopy", *Nucl. Instrum. Methods*, Vol. **71** (1969), pp. 273–279.

[33] E. Elad, "Drain feedback—A novel feedback technique for low-noise cryogenic preamplifiers", *IEEE Trans. Nucl. Sci.*, Vol. **19** (1972), pp. 403–411.

[34] T. Nashashibi, G. White, "A low noise FET with integrated charge restoration for radiation detectors", *IEEE Trans. Nucl. Sci.*, Vol. **37** (1990), pp. 452–456.

[35] G. Bertuccio, P. Rehak, D.M. Xi, "A novel charge sensitive preamplifier without feedback resistor", *Nucl. Instrum. Methods*, Vol. **A326** (1993), pp. 71–76.

[36] A. Owens, M. Bavdaz, A. Peacock, A. Poelaert, H. Andersson, S. Nenonen, L. Tröger, G. Bertuccio, "Hard X-ray spectroscopy using small format GaAs arrays", *Nucl. Instrum. Methods*, Vol. **A466** (2001), pp. 168–173.

[37] A. Owens, M. Bavdaz, A. Peacock, A. Poelaert, H. Andersson, S. Nenonen, H. Sipila, L. Tröger, G. Bertuccio, "High resolution X-ray spectroscopy using GaAs arrays", *J. Appl. Phys.*, Vol. **90** (2001), pp. 5376–5381.

[38] E. Kowalski, "*Nuclear Electronics*", Springer-Verlag, New York (1970).

[39] P.W. Nicholson, "*Nuclear Electronics*", Wiley, New York (1974) ISBN: 0471636975, 9780471636977.

[40] F.S. Goulding, D.A. Landis "Signal processing for semiconductor detectors", *IEEE Trans. Nucl. Sci.*, Vol. **29**, no. 3 (1982), pp. 1125–1141.

[41] Ortec data sheet, "*Introduction to Amplifiers*", www.ortec-online.com

[42] A. Owens, A. Peacock, "Compound semiconductor radiation detectors", *Nucl. Instrum. Methods.*, Vol. **A531** (2004), pp. 18–37.

[43] Ortec GMX Series Coaxial HPGe Detector product guide. Product Configuration Guide, www.ortec-online.com, Ortec, Oak Ridge, TN.

[44] G.K. White, P.J. Meeson, "*Experimental Techniques in Low-Temperature Physics*", Oxford University Press, Oxford, 4th ed. (2002).

[45] R. Radebaugh "Refrigeration for superconductors", *Proc. IEEE*, Vol. **92**, no. 10 (2004), pp. 1719–1734.

[46] R. Radebaugh, "Cryocoolers: State of the art and recent developments", *J. Phys. Cond. Mat.*, Vol. **21** (2009), pp. 1–9.

[47] H.J.M. Ter Brake, G.F.M. Wiegerinck "Low-power cryocooler survey", *Cryogenics*, Vol. **42**, no. 11 (2002), pp. 705–718.

[48] J. Marler, V. Gelezunas "Operational characteristics of a high purity germanium photon spectrometer cooled by a closed-cycle cryogenic refrigerator", *IEEE Trans. Nucl. Sci.*, Vol. **NS-20**, no. 1 (1973), pp. 522–527.

[49] R.E. Stone, V.A. Barkley, J.A. Fleming, "Performance of a gamma-ray and X-ray spectrometer using germanium and Si(Li) detectors cooled by a closed-cycle cryogenic mechanical refrigerator", *IEEE Trans. Nucl. Sci.*, Vol. **NS-33** (1986), pp. 299–302.

[50] E. Broerman, R. Keyser, T. Twomey, D. Upp, "A new cooler for HPGe detector systems", *Presented at the 23rd Brugge ESARDA meeting on Safeguards and Nuclear Materials Management*, May (2001).

[51] À. Pchelincev, À. Loupilov, R. Nurgaleev, O. Jakovlevs, À. Sokolov, V. Gostilo, A. Owens, "A miniature compact HPGe gamma-spectrometer for space applications", *J. Instrum*, Vol. **12** (2017), pp. P05017.

[52] M. Katagiri, H. Itoh, "Small electric-cooled germanium gamma-ray detector", in Radiation detectors and their uses. *Proceedings of the 8th Workshop on Radiation Detectors And Their Uses*, KEK Proceedings 94-7 (1994), pp. 174–179.

[53] M. Burks, C.P. Cork, D. Eckels, E. Hull, N.W. Madden, W. Miller, J. Goldsten, E. Rhodes, B. Williams, "Thermal design and performance of the gamma-ray spectrometer for the MESSENGER spacecraft", *IEEE Nucl. Sci. Symp. Conf. Rec.*, Vol. **1** (2004), pp. 390–394.

[54] N. Hasebe, E. Shibamura, T. Miyachi, T. Takashima, M. Kobayashi, O. Okudaira, N. Yamashita, S. Kobayashi, T. Ishizaki, K. Sakurai, M. Miyajima, M. Fujii, K. Narasaki, S. Takai, K. Tsurumi, H. Kaneko, M. Nakazawa, K. Mori, O. Gasnault, S. Maurice, C. d'Uston, R.C. Reedy, M. Grande, "Gamma-ray spectrometer (GRS) for lunar polar orbiter SELENE", *Earth Planet Space*, Vol. **60** (2008), pp. 299–312.

[55] L.E. Heffern, M.T. Burks, D.J. Lawrence, J.O. Goldsten, P.N. Peplowski, "Initial characterization of the gemini plus, a high-resolution gamma-ray spectrometer for planetary composition measurements", *48th Lunar and Planetary Science Conference*, 20–24 March (2017).

[56] A.V. Kondratjev, A. Pchelintsev, O. Jakovlevs, A. Sokolov, V. Gostilo, A. Owens, "Performance of a miniature mechanically cooled HPGe gamma-spectrometer for space applications", *J. Instrum.*, Vol. **13** (2018), pp. T01002.

[57] E. Sakai, Y. Murakami, H. Nakatani "Performance of a high-purity Ge gamma-ray spectrometer system using a closed-cycle cryogenic refrigerator", *IEEE Trans. Nucl. Sci.*, Vol. **NS-29**, no. 1 (1982), pp. 760–763.

[58] A.D. Lavietes, G.J. Mauger, E.H. Anderson "Electromechanically cooled germanium radiation detector system", *Nucl. Instrum. Methods*, Vol. **422**, no. 1–3 (1999), pp. 252–256.

[59] D.L. Upp, R.M. Keyser, T.R. Twomey "New cooling methods for HPGe detectors and associated electronics", *J. Radioanal. Nucl. Chem.*, Vol. **264**, no. 1 (2005), pp. 121–126.

[60] M.K. Schultz, R.M. Keyser, R.C. Trammell, D.L. Upp "Improvement of spectral resolution in the presence of periodic noise and microphonics for hyper-pure germanium detector gamma-ray spectrometry using a new digital filter", *J. Radioanal. Nucl. Chem.*, Vol. **271**, no. 1 (2007), pp. 101–106.

[61] J.C. Peltier, "Nouvelles expériences sur la caloricité des courants électrique", *Annales de Chimie et de Physique*, Vol. **56** (1834), pp. 371–386.

[62] G. Gromov *"Thermoelectric Cooling Modules"*, Business briefing: Global photonics applications & technology, World Markets Research Centre, London (2002), pp. 1–8.

7

Detector Characterization

Left: A selection of epitaxial GaAs diodes and arrays prior to testing and packaging. Right: Probe station used for resistivity, I-V and C-V measurements of semiconductor samples.

CONTENTS

7.1 Introduction

Following contacting, the detector is ready for characterization. This usually consists of several steps, and the results will determine the potential of the detector. Any characterization of a semiconductor should in principle include:

(a) chemical composition

(b) crystallographic structure

(c) electrical properties

(d) electronic properties

(e) characterization of defects

(f) performance testing

To carry out this analysis, many techniques have been developed, some of which are described in this chapter.

7.2 Chemical Analysis

The elemental composition and trace element impurities are known to be two important factors which greatly influence the physical and electronic properties of semiconductor materials. In fact, deviations from stoichiometry of the constituent elements by as little as 0.1% and trace impurities levels of as low as ppb may lead to serious performance issues. A number of physical and chemical analytical techniques have evolved for quantifying composition, stoichiometry and the level of trace contaminants; these are summarized Table 7.1. In general, analytical chemistry techniques are more accurate for the determination of the composition and stoichiometry, while physical techniques are more accurate for the determination of trace elements.

7.2.1 Compositional Analysis

For the stoichiometric analysis of semiconductors, high precision and accuracy with an error of less than or equal to 0.5% is a prerequisite. However, most instrumental or physical methods, such as optical spectroscopy, activation analysis, mass spectrometry *etc.*, are not suitable for high precision elemental analysis, with the exception of X-ray fluorescence, which is capable of achieving a precision of ± 0.1% when suitable standards are available. The main limitation of physical techniques is that the results are matrix dependent and rely on comparison with analysed standards or require standards for correction of instrumental parameters. For these reasons, the quantitative application of physical methods for compositional and stoichiometric analysis is limited [1].

Classical wet chemical techniques, based on gravimetry, titrimetry and electrochemistry can meet precision and accuracy requirements; these are commonly used for analysis of major elements and stoichiometry. Although tedious and time

TABLE 7.1

Summary of achievable precision and sensitivities of a number of physical and chemical compositional and trace analytical techniques (taken from [1]).

Technique	Applications	Sensitivity	Precision (%)
Wet chemistry gravimetry	Major and minor phase concentrations	100 mg to 1 g	0.003–0.03
		1–10 mg	0.1
Wet chemistry titrimetry	Major and minor phase concentrations: impurities	10^{-2} M in solution	0.01
		10^{-5} M in solution	0.1
		10^{-6}-10^{-7} in solution	0.2–1.0
Coulometry	Major phase concentrations	-	0.001–0.005
X-ray fluorescence spectrometry	Major and minor constituents	20–200 ppm generally;	0.1–0.5
		0.1 ppm with pre-concentration	2–10
ICPS-OES	Impurities	1–10 ppb in solution	1–3
-MS		1–10 ppt in solution	0.3–1
GDMS	Impurities	2–5 ppb	20

consuming, they are unaffected by the electrical properties of the element of interest as opposed to electrical methods in which information is restricted to elements of electro-active form while those of electro-inactive or electrically compensated species cannot be effectively determined. Conventional gravimetry, based on the weight of a reaction product, and titrimetry, based on the amount of a standard solution consumed in a reaction, can achieve precisions of ± 0.1%; with refinements in technique, this can be further extended to ± 0.01%. Coulometry is an electrochemical method based on the quantitative measurement of electric charge resulting from the quantitative electrochemical conversion of a constituent in the solution from one initial oxidation state to another well-defined oxidation state. Since electrons are essentially being used as the measured reagent, this method is capable of very high precision and accuracy. The most precise determination of stoichiometry is based on constant current coulometry, in which a reliability of 0.001%–0.01% can be obtained. Yang *et al.* [1], used constant potential and constant current coulometry in a stoichiometric analysis of several (III-V and II-VI) binary compounds, such as GaAs, CdTe and $Hg_xCd_{1-x}Te$ and ternary (1-III-VI) compounds such as $CuInS_2$, $CuInSe_2$ and $CuGa_x In_{1-x}Se_yTe_{2-y}$. They found that precisions of < 0.3% could be readily obtained.

7.2.2 Trace Analysis

Quite often, it is found that while as-grown material has good stoichiometry and crystallographic qualities, it has poor transport properties. This can usually be traced to an impurity in the precursors used to grow the crystal that acts as a majority carrier mu-tau "killer". The impurity levels needed to do this may only be a ppm or less, so it is not always listed in the manufacturer's assay of stock material. If an impurity is found, then the options are limited, either produce a purer material or change vendors. There are several physical techniques for detecting impurities to this level, the most sensitive of which are Inductively Coupled Plasma, Mass and Optical Emission Spectroscopy (ICP-MS and ICP-OES) and Glow-Discharge Mass Spectrometry (GDMS). A comparison of the two techniques is given in Table 7.2.

Inductively Coupled Plasma (ICP) Spectroscopy techniques [2] are so-called "wet" sampling methods whereby samples are introduced in liquid form for analysis. Plasma mass spectroscopy (MS) is highly sensitive and capable of the determination

TABLE 7.2

Comparison of the strengths and weaknesses of the GDMS and ICP techniques.

Technique	Pros	Cons
Inductively Coupled Plasma (ICP)	Can determine elemental trace compositions to very low detection limits (typically sub-ppb) with high accuracy and precision	The sample to be analysed must be digested and introduced to the plasma in the form of a liquid spray.
Spectroscopy	Many elements (up to 70 in theory) can be determined simultaneously in a single sample analysis.	Emission spectra are complex and inter-element interferences are possible.
	Instrumentation suitable for automation	
Glow	Full periodic table coverage (except H)	Sample un-homogeneity
Discharge	Sub-ppb detection	Samples must be vacuum compatible
Mass	Linear and simple calibration	Not suited for organic materials/polymers
Spectroscopy	Capability to analyse insulators	Smallest sample ~1 cm in diameter
(GDMS)	Depth profiling possible	

of an elemental range from 0–240 amu, to a sensitivity of below one part in 10^{12}. It utilizes an inductively coupled plasma to ionize the sample material and a mass spectrometer to separate and detect the ions. The technique is also capable of monitoring isotopic speciation for the ions of choice. The sample to be analysed is normally dissolved and then mixed with water before being sprayed into the plasma.

7.2.2.1 Inductively Coupled Plasma Spectroscopy (ICP-MS and ICP-OES)

Plasma optical emission spectroscopy (OES), also known as Atomic Emission Spectroscopy (AES), is a fast multi-element technique with a dynamic linear range and moderately low detection limits (~0.2–100 ppb). The instrument uses an inductively coupled plasma source to dissociate the sample into its constituent atoms or ions, exciting them to a level at which they emit light of a characteristic wavelength. A quantitative determination takes place based on the proportionality of the intensity of the radiation and the elemental concentrations in both calibration and analysis samples. Up to 60 elements can be screened per single sample run of less than one minute, and the samples can be analysed in a variety of aqueous or organic matrices.

7.2.2.2 Glow-Discharge Mass Spectrometry (GDMS)

Glow-Discharge Mass Spectrometry (GDMS) [3] is a technique that is capable of providing trace-level elemental quantification for a wide range of solid and powder materials. The sample to be analysed forms the cathode in a low pressure (~100 Pa) gas discharge or plasma. Argon is typically used as the discharge gas. Positive argon ions are accelerated towards the cathode (sample) surface with energies from hundreds to thousands of eV resulting in erosion and atomization of the upper atom layers of the sample. Only the sputtered neutral species are capable of escaping the cathode surface and diffusing into the plasma where they are subsequently ionized. The atomization and ionization processes are thus separated in space and time, which simplifies calibration and quantification as well as ensuring the near matrix independence of this technique. Although the analysis can be more time consuming than solution-based analytical methods, the sensitivity, the ease of calibration, flexibility and the capability to analyse a wide variety of sample forms and matrices is impressive. Additionally, besides bulk element compositions, it is possible to collect depth profiling information with very high sensitivity. The technique has a larger dynamic range and better detection limits than the ICP techniques and does not require sample dissolution.

7.3 Crystallographic Characterization

The physical and electronic properties of a particular semiconductor medium are intimately dependent on its crystal structure – the study of which is known as crystallography, deriving from the Greek words κρύσταλλος (*crystallon* meaning *frozen drop*) and γραφω (*grapho* meaning *write*). Crystallographic investigations not only give valuable information on the quality of the growth method, but also are key to understanding transport properties, particularly for asymmetric materials. Based on such information it is possible to pre-screen new materials or gauge the performance of existing materials for detector applications. Before the development of X-ray diffraction (XRD) techniques, crystallographic studies were based on geometry, which involved measuring the angles of crystal faces relative to theoretical reference axes (crystallographic axes) and establishing the symmetry of the crystal in question. Present-day crystallography is largely based on the interpretation and analysis of X-ray diffraction data, in which the crystal structure of the material acts as a diffraction grating, so-called "Bragg reflection".

7.3.1 Single Crystal X-Ray Diffraction

Single crystal X-ray diffraction, or Laue technique, is a non-destructive analytical technique commonly used for the study of crystal structures. It provides detailed information on the lattice of crystalline materials, including unit cell dimensions, bond lengths, bond angles and details on site ordering. XRD is based on the observation that when a crystal is illuminated with X-rays of a similar wavelength to the spacing of the atomic lattice planes, intense diffracted X-rays are produced for certain incident angles. The strongest diffracted beam will leave the crystal at the same angle as the incident beam when the scattered X-rays interfere constructively, and for this to happen the differences in the travel paths must be equal to integer multiples, n, of the wavelength, where n is known as the order of the reflection. The process is illustrated in Fig. 7.1. A general expression relating the wavelength of the incident X-rays, λ, angle of incidence, θ, and the spacing between the crystal lattice planes of atoms, d, is given by the Bragg relationship [4],

$$n\lambda = 2d \sin \theta \qquad (1)$$

In single-crystal XRD measurements, the X-rays are generated by an X-ray tube, conditioned to produce monochromatic radiation and collimated to a plane parallel beam at the sample. The sample is a single crystal of size 1 mm or larger that has

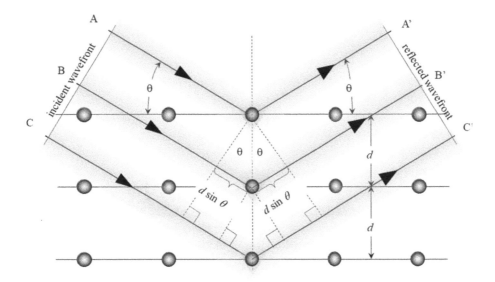

FIGURE 7.1 Schematic illustrating Bragg reflection. The diffracted x-rays exhibit constructive interference when the distances between paths AA', BB' and CC' differ by integer numbers of wavelengths (λ).

been polished to a high degree. It is mounted in an Eulerian cradle that can be orientated in three directions of space. The primary beam defines the frame of reference. All accessible reflex positions can be adjusted and measured with either a two-dimensional or a single monolithic detector. By changing the geometry of the incident rays, the orientation of the target crystal relative to the detector, all possible diffraction directions of the lattice can be attained. Generally, for most materials the scattering angles are known so that it is straightforward to position the detector approximately and then finely adjust the orientation of the crystal until a diffraction peak is observed. For crystalline material, the diffracted waves consist of sharp interference maxima (peaks) with the same symmetry as in the distribution of atoms. The widths of the peaks give information on the distribution of lattice plane spacing (*i.e.*, micro-strains) and on the grain size. In Fig. 7.2, we show a diffraction pattern measured from a ZnTe crystal that has a cubic structure. The angular position of each peak (2θ) can be used to calculate the spacing between planes within the structure using Eq. (1) ($d = 89.37/2\theta$).

X-ray diffraction methods are divided into single crystal diffraction and powder diffraction – the relative advantages and disadvantages of each are summarized in Table 7.3. Single-crystal XRD is generally used to determine large-scale crystalline structures ranging from simple inorganic solids to complex molecules, such as proteins. It is an extremely precise technique; however, it is difficult to set up and is time consuming. Powder XRD is used to characterize crystallographic structure, crystallite size and orientation in polycrystalline or powdered solid samples. It is widely used to identify unknown substances, by comparing diffraction data against a database maintained by the International Centre for Diffraction Data [5]. The main advantages of powder XRD over single-crystal XRD are that it is easy to use, simple to set up and gives rapid and accurate results. However, the data are more difficult to interpret.

TABLE 7.3

Summary of the pros and cons of single-crystal and powder XRD.

Single crystal XRD		Powder XRD	
Pros	**Cons**	**Pros**	**Cons**
Non-destructive	Must have single polished crystal	Can be rapid (<10 mins)	Requires tenths of a gram of material
No separate standards required	Sample setup time consuming	Minimal sample preparation on spectrometer	Sample must be ground into a powder
Provides very detailed structural information	Data collection takes ~4 hours	In most cases unambiguous material identification	Must have access to a standard reference file of inorganic compounds (d-spacings, *hkl*s)
		Data interpretation relatively simple	For unit cell determinations, indexing of patterns for non-isometric crystal systems is complicated
		XRD spectrometers widely available	

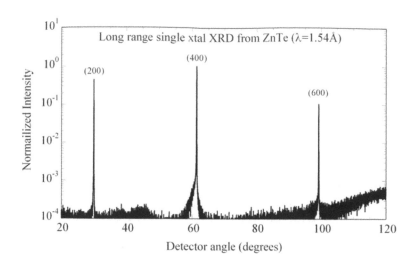

FIGURE 7.2 A single crystal X-ray diffraction pattern from a ZnTe crystal. The peaks in the pattern are due to diffraction of X-rays off the various crystallographic planes of atoms within the crystal. In this case, the (200), (400) and (600) reflections are clearly evident. The angular position of each peak (2θ) can be used to calculate the spacing between planes within the structure using Eq. (1). The derived lattice spacing was 6.18 ± 0.02 Å. The wavelength of the incident X-ray beam was 1.54 Å.

7.3.2 Powder Diffraction

Powder XRD is perhaps the most widely used XRD technique for characterizing materials since it is rapid, easy to use and gives accurate results. As the name suggests, the sample is a powder, consisting of fine grains of single crystalline material. The grain size used is typically around 8 μm and the sample volume is about 1 mm³ or more. For a random assortment of grains, enough grains will be orientated so that an adequate number of crystallites are always in a reflection position for every single Bragg reflex. A basic powder diffractometer consists of a source of monochromatic radiation, a sample holder and an X-ray detector situated on the circumference of a graduated circle centered on the specimen. Divergent slits located between the X-ray source and the specimen, and further divergent slits located between the specimen and the detector, limit scattered (non-diffracted) radiation, reduce background noise and collimate the radiation to produce a plane parallel beam at the sample. The detector and specimen holder are mechanically coupled with a goniometer so that a rotation of the detector through 2θ degrees occurs in conjunction with the rotation of the specimen through θ degrees, in a fixed 2:1 ratio. For typical powder patterns, data are collected at 2θ from ~5° to 70°, angles. The statistics can be improved by a suitable azimuthal rotation of the sample. Powder diffraction data can be collected using either transmission or reflection geometries.

Since the crystalline domains are randomly oriented in the sample, a 2-D diffraction pattern shows concentric rings of scattering peaks corresponding to the varying d spacing in the crystal lattice. The positions and the intensities of the peaks are used for identifying the underlying structure or phase of the material. Phase identification is particularly important. For example, the diffraction lines of graphite will be different from those of diamond, even though they both are made of carbon atoms. These differences give rise to their very different material properties. An example of a powder scan of $Cd_{1-x}Zn_xTe$ is given in Fig. 7.3.

7.3.3 Rocking Curve (RC) Measurements

Another type of scan that is closely related to a θ-2θ scan is a rocking curve (RC) scan. By passing over maxima, the intensity distribution displays a narrow peak whose angular width is a measure of the crystalline quality of the material. This peak is known as the rocking curve and is obtained in practice by keeping the scattering angle fixed and "rocking" the crystal from side to side. Rocking curves measure the amount of mosaicity in the crystal, an angular measure of the degree of long-range order of the unit cells. Lower mosaicity indicates better-ordered crystals and hence better X-ray diffraction. From a rocking curve measurement, it is possible to determine the mean spread in orientation of the different crystalline domains of a non-perfect crystal. If the crystalline particles are very small, it is possible to determine their size from an RC scan. RC scans also measure the underlying curvature in the lattice planes and give indirect information on strains that manifest themselves in the broadening of the peak. In order to obtain the RC, one need not perform a θ-2θ scan first, since the position for diffraction is usually well known. In this case, the detector is moved to the diffraction peak. Then, data are acquired by varying the orientation of the sample by an angle $\Delta\theta$ around its equilibrium position, whilst keeping the detector position fixed. For RC scans, the detector does not need to have a small angular acceptance, since we are not measuring a scattering angle or lattice parameter. RC scans can also be very useful for determining the efficacy of crystal-processing techniques. For example, in Fig. 7.4 we show double- and triple-axis XRD RCs for a $Cd_{1-x}Zn_xTe$ crystal before and after

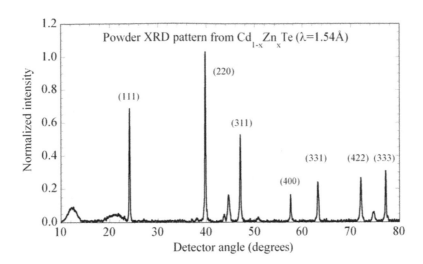

FIGURE 7.3 A powder XRF scan of a $Cd_{1-x}Zn_xTe$ crystal. The wavelength of the incident X-ray beam was 1.54 Å. Each of the crystal planes has been identified. The derived lattice spacing was 6.46 Å. By comparing this value with that expected from pure CdTe and using Vegard's law [6], the Zn fraction, x, was determined to be 4.4%.

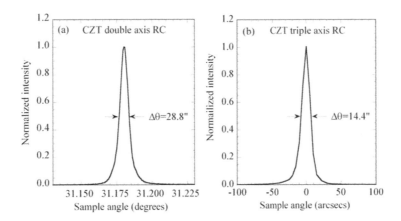

FIGURE 7.4 Examples of (a) double- and (b) triple-axis X-ray RC scans of a $Cd_{1-x}Zn_xTe$ crystal, taken before and after various post-processing treatments have been applied to remove surface damage. The wavelength of the incident X-ray beam was 1.54 Å. Triple-axis RC is generally only used on the highest quality crystals and can give quantitative information on mosaicity and strain in the crystal.

various post-processing treatments have been applied to remove surface damage. Triple-axis RC is generally only used on the highest quality crystals for which intrinsic width is very small (< 14 arc seconds) and can give quantitative information on mosaicity and strain in the crystal.

7.3.4 XRD and Detector Performance

Goorsky *et al.* [7] used triple-axis XRD to study the relationship between crystalline perfection and the spectral performance of radiation detectors for a number of compound semiconductors (GaAs, $Cd_{1-x}Zn_xTe$ and HgI_2). In all cases, mosaicity was found to be inversely related to detector performance – the lower the mosaicity, the better the spectral performance. In GaAs, low-angle grain boundaries were attributed to impaired detector performance, while in large HgI detectors, detector performance was more intimately related to deviations from stoichiometry. Interestingly, in HgI_2 detectors, no differences in crystallinity were found between detectors that displayed polarization effects and those that did not.

7.4 Electrical Characterization

Electrical measurements on semiconductors and devices are routinely performed to determine properties, such as resistivity (or conductivity), contact resistance and depletion depth.

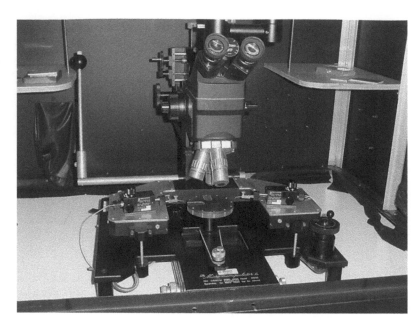

FIGURE 7.5 A Micromanipulator 6000 series Probe Station. The sample is held on the central 4-inch diameter chuck by vacuum suction. It can be raised to contact the micromanipulator needles (two shown) with a positional accuracy of a few microns using the microscope. The chuck can also rotate through 360° and can move ±2 inches in X and Y and ±0.7 inches in Z (the vertical direction).

7.4.1 Current-Voltage (I-V) Measurements

Current-voltage measurements are used to determine the electrical characteristics of semiconductor devices and test structures by measuring the current flowing across the device as a function of applied voltage. These measurements yield information on the barrier formation, bulk resistivities and contact performance in terms of specific resistivities. They are usually carried out on a probe station that utilizes manipulators to allow the precise positioning of thin needles on the surface of a semiconductor device. The essential functional elements are illustrated in Fig. 7.5. Depending on the accessory package, probe stations can measure currents as low as 1 FA over temperatures ranging from –70°C to +200°C or above. If the device is being electrically stimulated, the signal is acquired by the probe and displayed on an oscilloscope. Probe stations are extensively used in research and prototyping as it is often faster and more flexible to test a new electronic device or sample with a probe station than to wire bond and package the device before testing.

The detector response is ultimately dependent on the physics of the metal-semiconductor interfaces and the size of the barrier height, which will determine whether it is ohmic or Schottky. Ohmic contacts are formed when the work function of the metal contact matches the work function of the semiconductor leaving little or no potential barrier at the interface. In this case the metal-semiconductor contact has a very low resistance, which is independent of the applied voltage and may be expressed as $R = V/I$ = constant. Its current-voltage characteristic is therefore linear and symmetric about the origin. The slope of the I-V characteristic can be used to derive the resistivity of the bulk material. This quantity is important since it determines how much bias can be applied to the detector before breakdown is induced by excessive leakage current. Increased bias enhances charge collection efficiency and decreases response time. For radiation detectors, resistivities tend to be in the region of 10^6 to 10^{12} Ω cm.

Schottky contacts, on the other hand, display asymmetric current-voltage characteristics, *which allow high current to flow across when forward biased,* but block current flow under reverse bias. This behaviour is controlled by bias-dependent changes of the potential barrier height in the contact region, as described in Chapter 5. The current voltage characteristic is generally described by the empirical diode equation

$$I = I_\circ \left(\exp\left(\frac{qV}{nkT} \right) - 1 \right), \tag{2}$$

assuming that any series resistance in the device is negligible; otherwise, V should be replaced by $V\text{-}IR$, where R is the series resistance. Here, I_o is the reverse bias saturation or leakage current, T is the absolute temperature, k is Boltzmann's constant, q is the electronic charge, V is the bias across the detector and n is the ideality factor, a dimensionless quantity introduced to include contributions from other current-transport mechanisms (see Chapter 5). The saturation current is given by

$$I_o = \alpha A^{**} T^2 \exp\left(\frac{q\phi_{\beta o}}{nkT}\right), \tag{3}$$

where a is the anode area, A^{**} is the effective Richardson constant and $\phi_{\beta o}$ is the barrier height. The analysis of I-V data usually proceeds with a least-squares fit to Eq. (2), treating n and $\phi_{\beta o}$ as adjustable parameters. Most radiation detectors are operated under reverse bias to allow the application of a larger bias and therefore larger electric field. The slope of the reverse part of the characteristic gives the resistivity of the device. For successful operation as a radiation detector, this should be $> 10^6$ Ω cm.

Expression (2) is actually an approximation. Rhoderick [8] pointed out that such a relationship is physically implausible, since whatever mechanism is responsible for making n exceed unity must affect the reverse current, as well as the forward current. This means that the second term on the right-hand side should also contain n. Crowell and Rideout [9] have shown that the current/voltage relationship, which results from tunneling through the barrier, can be written in the form

$$I = I_o \exp\left(\frac{qV}{nkT}\right)\left\{1 - \exp\left(-\frac{qV}{kT}\right)\right\}, \tag{4}$$

which has the desired property that the reverse current depends on n. Rhoderick [8] noted that Eq. (4) is similar to current voltage relationships resulting from other transport mechanisms, such as the voltage dependence of the barrier height and electron-hole recombination in the depletion region. Based on this, Missous and Rhoderick [10] argued that Eq. (4) is, in fact, a generic form of current transport and that a plot of

$$\log\left[I/\left\{1 - \exp\left(-\frac{qV}{kT}\right)\right\}\right] \tag{5}$$

against V should be linear for all values of V, including reverse voltages. The effect is dramatically demonstrated in Fig. 7.6 in which we show current-voltage data plotted for an expitaxial Al/GaAs Schottky diode [10]. The left graph shows a conventional plot using Eq. (2), while the right side shows a plot using Eq. (5).

The success of this modified plot in linearizing the characteristic is immediately apparent, the graph being linear over the whole range from -1 V to +0.5 V. This allows a more accurate determination of the ideality factor, n, than Eq. (2), which in turn, facilitates a more accurate calculation of the saturation current I_o and barrier height. However, it should be pointed out that the ideality factor for this particular diode is very nearly unity. For this reason, we have also plotted data from a non-epitaxial Al/GaAs diode in Fig. 7.6 (b). For this diode, we estimate n to be ~1.18. The calculated curve using Eq. (5) overlies the curve for the epitaxial diode. The departure from ideal behaviour is obvious and is predominantly due to generation/recombination in the depletion region for reverse bias. As pointed out by Missous and Rhoderick [10], it is probably easier to interpret the reverse portion of the characteristic with the aid of a conventional plot.

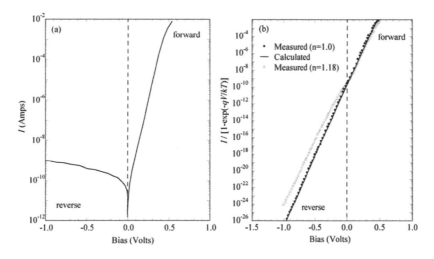

FIGURE 7.6 Current-voltage data for an epitaxial Al/GaAs Schottky diode (from Missous and Rhoderick [10]. Reproduced by permission of the Institution of Engineering & Technology). (a) Conventional logarithmic plot of I versus V. (b) Logarithmic plot of $I/\{1 - \exp(-qV/kT)\}$ versus V. For comparison, we over plot data for a non-ideal diode with an ideality factor of ~1.18. Note the calculated curve for this diode overlies the epitaxial diode calculated curve.

7.4.2 Contact Characterization

The formation of laterally stable and uniform contacts is of critical importance for the operation and reliability of compound semiconductor devices. In fact, the contacts are often the limiting factor affecting performance, as discussed in length in Chapter 5. In Fig. 7.7 we show a typical detector construction, which consists of a sandwich of a semiconductor between two metal contacts. R_c and R_{bulk} represent the resistances of the contact and the bulk semiconductor, so that the total impedance across the device is $R_{c1}+R_{bulk}+R_{c2}$. The contact resistance regions in which the contacts are formed may have a significant extent if a high degree of doping is used to reduce the Schottky barrier width (see Chapter 5) or if they are formed by alloying the metal with the semiconductor.

Contacts are usually characterized by their *contact resistance, R_c*, measured in ohms. This is the total resistance of the metal-semiconductor interface and is dependent on the area and the geometry of the contact. However, in reality the entire contact interface may not be available for conduction because of the physical properties of the interface, such as the barrier height of the metal, surface roughness, or preparation technique. All these factors are in principle independent of the contact geometry but in practice may strongly influence the value of the contact resistance. Hence, a more useful figure of merit for a contact is its *specific contact resistance ρ_c*, which is the contact resistance of a unit area contact and has units of Ω cm^2. It is particularly useful when comparing contacts of different sizes and is usually defined as the slope of the *I-V* curve at *V*=0,

$$\rho_c = \left(\frac{\partial J}{\partial V} \bigg|_{V=0} \right)^{-1} = \lim_{\Delta A \to 0} R_c \Delta A, \tag{6}$$

where J is the current density in units of Amps cm^{-2} and A is the area of the contact. Most semiconductor detectors have a sandwich (contact–semiconductor–contact) structure; the contact resistance, R_c, of a contact is then given by

$$R_c = \rho_c/A. \tag{7}$$

Depending on the semiconductor material and on the contact quality, ρ_c can vary anywhere between 10^{-3} Ω cm^2 to 10^{-8} Ω cm^2. A typical current density in a sandwich-type device can be as high as 10^4 Amps cm^{-2}. Hence, a specific contact resistance of, say, 10^{-5} Ω cm^2 (a value appropriate for LEDs and laser diodes [11]) would lead to a voltage drop of the order of 0.1 V. For the contact be ohmic in character, the voltage drop across the semiconductor must be much greater than this value. For Group III-V compounds, where biases can be of the order of volts, this can be a problem. However, for Group II-VI compounds where biases tend to be of the order of kV, it would not.

7.4.3 Measuring Contact Resistance

Measuring contact resistances directly is extremely difficult because of the resistive contributions from leads, probe contact resistance and bond wires. The transfer length method (TLM) is the classical approach for determining contact

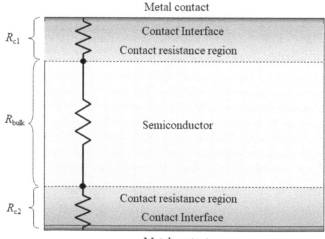

FIGURE 7.7 Typical detector construction, which consists of a sandwich of a metal contact, semiconductor and further metal contact. R_c and R_{bulk} represent the resistances of the contact and the bulk semiconductor. The contact resistance region is that region over which the contact is formed and may have a finite extent depending on how the contact was formed.

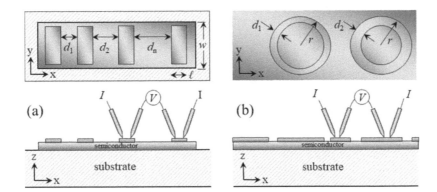

FIGURE 7.8 Contact resistance test patterns (from [12]) (a) Measurement configurations for the transfer length method (TLM) and (b) Circular transfer length method (CTLM). For the TLM measurement, the semiconductor has been etched away around the contacts to form a mesa in order to restrict current flow to adjacent contacts. For both TLM and CTLM, measurements are usually carried out using a four-probe technique as illustrated in the cross-sectional views.

and sheet[1] resistances in a single-layer material. The technique involves fabricating a series of identical contact pads of width w and length l, separated by varying distances: d_1, d_2, d_3, ... d_n (see Fig. 7.8 a) and applying probes to successive pairs of contacts. The contact and specific contact resistances are determined through the linear relationship between the resistance and the gap spacing between the contacts. For very small resistances, the TLM structure may require mesa etching to restrict current flow to adjacent contact pads and so negate so-called "current crowding". One way to eliminate this problem and the need for additional mesa etching is to use the circular transfer length method (CTLM). In this case, the use of concentric circular contacts ensures current confinement in the direction perpendicular to the contact; this is implicit in the design (see Fig. 7.8 b). In both cases, a four-point measurement[2] may be used to avoid the influence of external series resistances. One pair of probes carries the current and another pair senses the voltage. The measured resistance between two contacts separated by the distance d_i can be determined from the relationship

$$R_{\text{meas}} = R_c x + R_s y, \tag{8}$$

where R_c (Ω mm) is the contact resistance, R_s ($\Omega/\square$) is the sheet resistance[3] of the semiconductor and x and y depend on the geometry of the contact. For a TLM structure,

$$x = 2/w \tag{9}$$

and

$$y = d_i/w, \tag{10}$$

where w is the width of the pads. For a CTLM structure,

$$x = \frac{1}{2\pi} \ln \left(\frac{1}{r} + \frac{r}{r - d_i} \right) \tag{11}$$

and

$$y = \frac{1}{2\pi} \ln \left(\frac{r}{r - d_i} \right). \tag{12}$$

---

[1] The sheet resistance (R_s) is a measure of the resistance of thin films that have a uniform thickness and is commonly used to evaluate the outcome of semiconductor processes, such as doping or metallization. Its utility is that, unlike resistivity measurements, it can be directly measured using the four probe method and for thin films is simply related to the resistivity, $\rho = R_s$ / t, where t is the film thickness.

[2] Originally developed by Wenner [13] in 1915 to measure the earth's resistivity.

[3] Here given in units of ohms per square, which is used exclusively for sheet resistance precisely because it cannot be misinterpreted as a bulk resistance.

Fig. 7.9 illustrates graphically the measured resistance as a function of distance d_i. Fitting Eq. (8), we can derive the values of R_s and R_c from the slope and y-axis intercept of the curve, respectively. The specific contact resistance, ρ_c (Ω cm^2), is given by

$$\rho_c = -\frac{R_c^2}{R_s}. \tag{13}$$

The intercept of the fitted line[4] with the x-axis is equal to twice the transfer length l_t (also known as effective contact length), which is defined by

$$l_t = R_c/R_s \tag{14}$$

and is a measure for the "ohmic quality" of the metal contact. A shorter l_t, means a better ohmic contact. Note that the validity of the above analysis depends on the assumption that the transfer length l_t is much smaller than the contact length l [14].

7.4.4 Capacitance-Voltage (C-V) Measurements

This technique is used with detectors that utilize a metal-semiconductor junction (Schottky barrier) or a p-n junction in which a depletion region is created. In normal operation, this region is empty of conducting electrons and holes but may contain ionized donors and electrically active defects or traps. The depletion region with its enclosed ionized charges behaves like a capacitor. By varying the voltage across the junction, it is possible to vary the depletion width, w_d. The dependence of the depletion width upon the applied voltage provides information on the semiconductor doping profile and electrically active defect densities. Measurements are usually carried out on a probe station and may be done either using DC alone or DC together with a small amplitude AC signal.

Consider a Si p-n diode with $N_a = 10^{16}$ cm^{-3}, $N_d = 10^{17}$ cm^{-3} and area 10^{-4} mm^2. The space charge Q_{ac} per unit area of the depletion region under the influence of an applied bias V is given by

$$Q_{ac} = qN_dw_d = \sqrt{2q\varepsilon N_d(V_{bi} - V - kT/q)}, \tag{15}$$

where w_d is the depletion layer width, ε is the permittivity of semiconductor and V_{bi} is the built-in potential, defined in Chapter 5. The depletion layer capacitance per unit area is given by

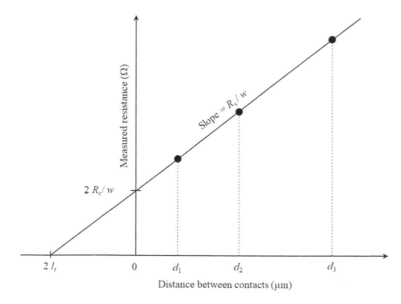

FIGURE 7.9 Evaluation of the contact and sheet resistances using TLM and CTLM measurements.

---

[4] Note: for CTLM structures, Marlow and Das [14] have pointed out that effects of finite contact radii and non-zero metal resistance can introduce large errors in the determination of the contact resistance if a linear fit is used to represent the voltage drop versus gap length data.

$$C = \frac{|\partial Q|}{\partial V} Q_{ac} = \sqrt{\frac{q\varepsilon_a N_d}{(V_{bi} - V - kT/q)}} = \frac{\varepsilon}{w}. \tag{16}$$

Thus, C is related simply to the number of carriers and depletion layer width.

In Fig. 7.10 we show a plot of capacitance measured with a capacitance metre versus applied bias, from which we see that the capacitance does indeed drop as the square root of $|V$-$V_{bi}|$ as the depletion region width increases.

Eq. (16) can be written in the form

$$\frac{1}{C^2} = \frac{2(V_{bi} - V - kT/q)}{q\varepsilon N_d}. \tag{17}$$

In the case of uniform doping concentration, N_d is constant within the depletion region and a straight line results from plotting $1/C^2$ against V (see Fig. 3.10, right ordinate). The slope of this line is obtained by differentiating Eq. (17) with respect to V,

$$\frac{d(1/C^2)}{dV} = -\frac{2}{q\varepsilon N_d}, \tag{18}$$

from which the doping concentration can be calculated: thus,

$$N_d = -\frac{2}{q\varepsilon \frac{d(1/C^2)}{dV}}. \tag{19}$$

The intercept of this line on the horizontal axis occurs at V_{bi}, the built-in potential (see Chapter 5). In this case, $V_{bi} = +0.7$ eV. If N_d is not constant, the doping profile can be deduced from iterating Eq. (19) with bias. The depletion-layer width at a particular bias is simply given by

$$w = \varepsilon/C_d. \tag{20}$$

Here C_d is the detector capacitance at the operational bias V.

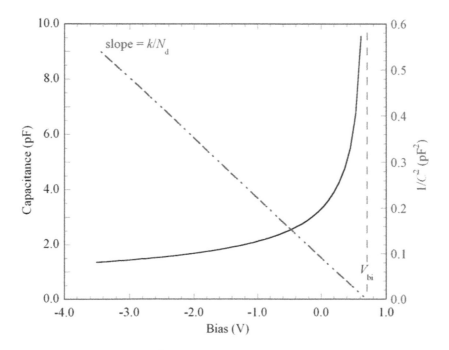

FIGURE 7.10 Capacitance (left ordinate solid line) and $1/C^2$ (right ordinate – dashed line) versus bias voltage of a p-n diode with $N_a = 10^{16}$ cm^{-3}, $N_d = 10^{17}$ cm^{-3}. The area of the diode is 10^{-4} cm^2.

7.5 Electronic Characterization

Electronic characterization of semiconductors is routinely performed for the analysis of a wide range of semiconductor and interface properties. Specific measurements typically carried out include (i) barrier heights, (ii) bandgap energy and the separation of an impurity level from the band edge, (iii) carrier concentrations, (iv) the mobility of electrons and holes, (v) lifetime and diffusion length of minority carriers, (vi) surface recombination velocity of carriers and (vii) the investigation of deep impurity levels. The carrier properties, particularly the transport properties, are probably the most important intrinsic parameters of a semiconducting material and dictate the type of detector and application it can reasonably fulfil. For example, materials with poor transport properties would not make planar detectors with good stopping power, since if the active depth is too thick the generated charge due to interacting particles would not be transported to the electrodes and collected.

7.5.1 Determining the Majority Carrier

In some materials, the majority carrier may not be known. This information is crucial when selecting a contact material. The hot or thermoelectric probe technique is a simple method for determining whether a semiconductor is n-type or p-type. Two probes make contact with the semiconductor surface as illustrated in Fig. 7.11. One is heated to 25 to 100°C greater than the other. At the hot probe, the thermal energy of the majority carrier is higher than at the cold probe so carriers will tend to diffuse away from the probe, driven by the temperature gradient. As they diffuse away from the hot probe, they leave behind the oppositely charged, immobile donor atoms, which results in a current flow towards the hot probe for p-type material and away from the hot probe for n-type material. Thus, a measurement of either the polarity of the short-circuit current or the open circuit voltage reveals the material type. The hot probe technique is most effective over the 10^{-3} to 10^3 Ω cm resistivity range.

7.5.2 Determining Effective Mass

The effective mass of carriers can be determined by cyclotron resonance measurements [15]. In this method, the frequency is measured at which there is strong attenuation of electromagnetic signals passing through a sample due to absorption between Landau levels created by a magnetic field. If the resonance frequency is ω, the effective mass is simply given by

$$m^* = qB/\omega,\tag{21}$$

where B is the applied magnetic field.

7.5.3 The Hall Effect

The Hall effect measurement is a widely used technique for determining basic carrier properties, such as majority carrier type, concentration and mobility. It is named after Edwin Hall who in 1879 placed a thin layer of gold in a strong magnetic field. He connected a battery to the opposite sides of this film and measured the current flowing through it. He discovered that a small voltage appeared across this film and that it was proportional to the strength of magnetic field multiplied by the

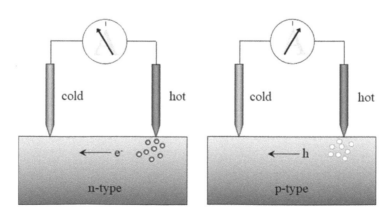

FIGURE 7.11 Illustration of the hot probe technique for determining the majority carrier type in semiconductors. Carriers diffuse more rapidly near the hot probe. This leads a flow of majority carriers away from the hot probe and a resultant electrical current towards (p-type) or away from (n-type) the hot probe.

current [16]. Hall effect measurements commonly use two sample geometries: (1) a long, narrow Hall bar and (2) the nearly square or circular van der Pauw configuration [17]. Each has advantages and disadvantages.

To understand the Hall effect, consider the rectangular sample geometry shown in Fig. 7.12. A voltage V_x is applied across the sample and a magnetic field B_x applied normal to the sample. Electrons and holes (shown by filled and open circles, respectively) flowing in the semiconductor will experience a force, bending their trajectories so that they build up on one side of the sample, creating a potential, V_H, across it, the so-called Hall voltage, as illustrated in Fig. 7.12. On the assumption that all the conduction electrons have the same drift velocity v_n and the same relaxation time τ_e, the resulting Lorentz force [18] acting on any electron is given by:

$$\vec{F} = -q(\vec{\varepsilon} + \vec{v}_n \times \vec{B}), \tag{22}$$

$$\vec{F} = m_e \frac{d\vec{v}_n}{dt} + m_e \frac{\vec{v}_n}{\tau_c}, \tag{23}$$

where $m_e v_n/\tau_e$ is a damping term. In the steady state, the derivative of v_n in Eq. (23) is zero. Combining Eqs. (22) and (23) and using the relation

$$\vec{J} = -qn_o\vec{v}_n \tag{24}$$

for electrons, we get

$$\vec{\varepsilon} = \frac{\vec{J}}{q\mu_n n_o} + \frac{\vec{J}}{qn_o} \times \vec{B}, \tag{25}$$

where $\mu_n = q\tau_e\, m_e$ is the mobility. Considering that $B_x = B_y = 0$ and $J_y = J_z = 0$, Eq. (25) can be written as two scalar equations:

$$\varepsilon_x = J_x/\sigma \tag{26}$$

and

$$\varepsilon_y = -\frac{J_x B_z}{qn_o} = -\mu_n B_z \varepsilon_x, \tag{27}$$

where $\sigma = q\mu_n n_o$ is the conductivity and relation (26) is Ohm's law. Eq. (27) expresses the fact that along the y direction the force on one electron due to the magnetic field ($q\mu_n B_z \varepsilon_y$) is balanced by a force ($-q\varepsilon_y$) due to the Hall field. Eq. (27) is generally written

$$\varepsilon_y = R_H J_x B_z, \tag{28}$$

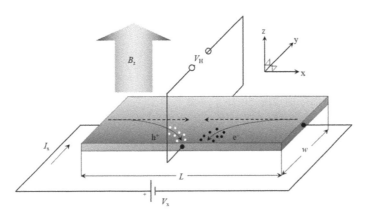

FIGURE 7.12 Schematic illustrating the sign convention and terminology for the Hall effect.

where $R_H = -1/q\,n_{eo}$ is the Hall constant. For a p-type semiconductor,

$$\varepsilon_y = R_H J_x B_z = \mu_p B_z \varepsilon_x\,, \tag{29}$$

with, $R_H = 1/qn_{ho}$ where n_{ho} is the equilibrium concentration of holes in the sample. It is thus seen that R_H has a negative sign for electrons and a positive sign for holes. For the rectangular bar geometry shown in Fig. 7.12, the Hall constant and the Hall mobility can be expressed by

$$R_H = \frac{V_H d}{I_x B} \tag{30}$$

and

$$\mu_H = \frac{V_H L}{V_x B w}\,, \tag{31}$$

where V_H is the measured Hall voltage, V_x is the applied voltage along the length, L, and I_x is the current. For a semiconductor with comparable concentrations of electrons and holes, R_H is given by

$$R_H = -\frac{(\mu_H^2 n_{ho} - \mu_{ne}^2 n_o)}{q(\mu_p n_{ho} - \mu_n n_{eo})^2} \tag{32}$$

and the interpretation of R_H and μ_H is ambiguous. However, if the material is strongly extrinsic, Eq. (31) reduces to

$$R_H = 1/qn_{eo} \; and \; \mu_H = \mu_n \tag{33}$$

for highly n-doped material and

$$R_H = 1/qn_{ho} \; and \; \mu_H = \mu_p \tag{34}$$

for highly p-doped semiconductors. Thus, in summary:

1. The results are only simply interpreted for strongly extrinsic material
2. The sign of R_H gives the majority carrier type and
3. The measured values of R_H and μ_H determine the majority carrier concentration and the mobility, respectively.

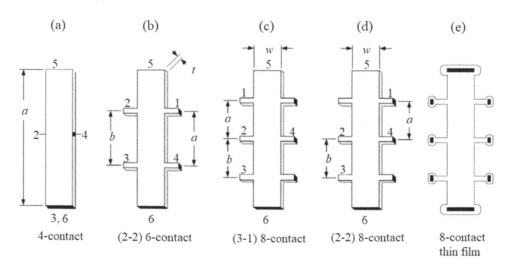

FIGURE 7.13 Hall measurement layouts and contact configurations (from ref [19].). For measurements, the magnetic field is applied alternately into and out of the page.

7.5.3.1 Hall Effect Measurements

Practical layouts and contact configurations for Hall measurements are shown in Fig. 7.13. The simplest arrangement for measuring the Hall voltage is the rectangular geometry shown in Fig. 7.13(a). However, a number of spurious voltages are included in this measurement due to the finite contact size. These spurious voltages are eliminated if four readings are taken by reversing the direction of the bias current I_x and the magnetic flux B. The true Hall voltage is then obtained by taking the average of the four readings. The contacts used for measuring the Hall voltage should be infinitesimally small, so that they do not perturb the current flow. In practice, the bridge shaped geometry shown in Fig. 7.13(b,c,d,e) is often used to reduce the distortion of the current. The projecting pads on the pattern allow a large area to be used for contacting the sample without a severe distortion of current flowing through it. The Hall voltage is measured between contacts 1 and 2 and then 3 and 4. The average is then taken as the final value. In some cases, it may not be convenient to cut the specimen in the form of a rectangular bar, in which case the van der Pauw geometry is often used.

7.5.3.2 Van Der Pauw Method

Van der Pauw [17] showed how to determine the resistivity, carrier concentration, and mobility of an arbitrary, flat sample as shown in Fig. 7.14, if the following conditions are met:

- The sample thickness must be much less than the width and length of the sample.
- The contacts are at the circumference of the sample (or as close to it as possible).
- The contacts are sufficiently small. Any errors given by their non-zero size will be of the order D/L, where D is the average diameter of the contact and L is the distance between the contacts.
- The sample is of uniform thickness, and
- the surface of the sample does not have isolated holes or islands.

The method involves passing current through two adjacent points on the perimeter of the shape and measuring the voltage across two other points on the perimeter of the sample. With reference to Fig. 7.14(a) the resistivity of the sample is given by

$$\rho = -\frac{\pi t}{\ln(2)} \frac{V_{43}}{I_{12}} + \frac{V_{14}}{I_{23}}, \tag{35}$$

where t is the sample thickness, V_{23} is V_2-V_3 and I_{12} refers to the current that enters the sample through contact 1 and leaves through contact 2. Two voltage readings are required with the van der Pauw sample, whereas the resistivity measurement on a Hall bar requires only one. This same requirement also applies to Hall coefficient measurements, so that equivalent measurements take twice as long with van der Pauw samples.

In the presence of a magnetic field normal to the sample surface, a current I_{ac} is established between two opposite contacts (*1* and *3*), and the voltage is measured between the contacts *2* and *4*, as shown in Fig. 7.14(b). The resistance defined as

$$R_{13,24} = V_{24}/I_{13} \tag{36}$$

is first measured with the magnetic field, -B, $R_{13,24}$ (+B) and then with the opposite direction of the field, -B, – $R_{13,24}$ (-B). If all the variations in the current and magnetic field polarity are considered, then the Hall constant is given by

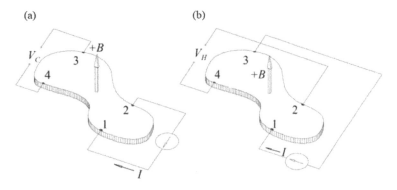

FIGURE 7.14 Van der Pauw arrangement [17] for (a) measuring resistivity and (b) the Hall coefficient in an arbitrary shaped sample (from ref [19].).

$$R_{\mathrm{H}} = (d/8B)\big(R_{13,24}(+B) - R_{13,24}(-B) + R_{24,13}(+B) - R_{24,13}(-B)\big)$$
$$+ R_{31,42}(+B) - R_{31,42}(-B) + R_{31,42}(+B) - R_{31,42}(-B)\big) \tag{37}$$

and the Hall mobility

$$\mu_{\mathrm{H}} = -\frac{R_{\mathrm{H}}}{\rho_{\mathrm{o}}}, \tag{38}$$

where ρ_{o} is the zero-field resistivity given by Eq. (35).

In practice, arbitrary shapes are not used since there are a number of "standard" geometries as shown in Fig. 7.15. Since they are all symmetrical about two axes (see Fig. 7.15), the van der Pauw relation simplifies to

$$R_{\mathrm{s}} = \pi\big(R_{34,12} + R_{13,24}\big)/2 \ln 2, \tag{39}$$

where

$$R_{34,12} = (V_3 - V_4)/I_{12} \tag{40}$$

and

$$R_{13,24} = (V_1 - V_3)/I_{24}. \tag{41}$$

Here the numbering of contacts goes from left to right and top to bottom. Compared to the conventional Hall measurements using bar geometries, the van der Pauw method has the advantage that only four contacts are required, simple geometries can be used and there is no need to measure sample widths or distances between contacts. However, its disadvantages are that measurements take twice as long and errors due to contact size and placement can be significant when using simple geometries.

7.6 Evaluating the Charge Transport Properties

Of the basic transport properties, carrier mobilities and mobility-lifetime or mu-tau ($\mu\tau$) products are the most useful and form the starting point for choosing a material for a particular energy range or application. They can be estimated as follows. Consider a simple planar detector of thickness, d. A bias, V, is applied across the detector giving a constant and uniform electric field, $E = V/d$ throughout the device. For the electric field strengths usually encountered in such detectors, the drift velocity v_{d} of the carriers is proportional to the electric field strength, E, with the constant of proportionality being the mobility μ, thus $v_{\mathrm{d}} = \mu E$. The instantaneous current is $I = qnv_{\mathrm{d}}$, where q is the charge on an electron and n is the number of carriers. The duration of a transient current pulse is determined by the distance the carriers must travel. Note that a consequence of the Shockley-Ramo theorem [20,21] is that charge induction on the contacts begins as soon as the carriers are created (or move) and not when the charges are deposited on the electrodes. There will clearly be two distinct current

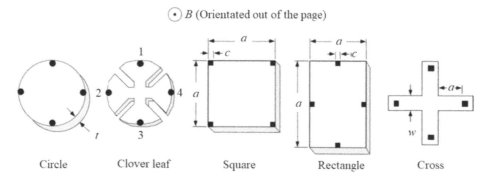

FIGURE 7.15 Common van der Pauw geometries (from ref [19]). The cross appears as a thin film pattern and the others are bulk samples. Note that the contacts are shown in black and the magnetic field is out of the page.

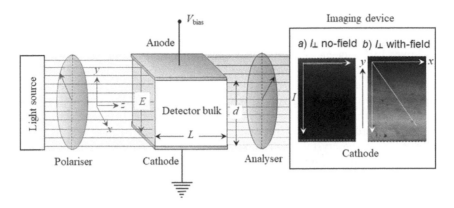

FIGURE 7.16 Pockels effect setup for internal electric field measurements in zincblend structured crystals. The polarizer is set at angle of -45° with respect to the direction of the applied electric field. IR transmission images of the detector crystal are then recorded using an analyser set at +45° to the applied electric field under both unbiased (electric field absent) and biased (electric field present) conditions. The insets (a) and (b) show the resulting Pockels images for an illuminated CdTe planar detector (the dark spots visible in the images are due to Te inclusions and precipitates). (Pockels images taken from A. Cola and I. Farella, "Electric Field and Current Transport Mechanisms in Schottky CdTe X-ray Detectors under Perturbing Optical Radiation", *Sensors*, Vol. 13 (2013), p. 9414, licenced under CC BY 3.0).

pulses, one from holes, I_h, and one from electrons, I_e. For most materials, the electrons will produce a much higher current for a shorter time, since they generally have much higher mobilities ($t \propto \mu^{-1}$ and $I \propto \mu$).

7.6.1 Probing the Electric Field

For conventional radiation detectors, for example planar devices, the distribution and stability of the internal electric field strongly affects charge collection, which in turn is heavily influenced by the uniformity of the detection medium, space charge effects and the nature of the electrical contacts. In some crystalline materials, these can be investigated using a simple non-invasive technique based on the Pockels electro-optic effect. It is named after the German physicist Friedrich Pockels, who in 1893 observed that a steady electric field applied to certain bifringent materials causes the refractive index to vary, approximately in proportion to the field strength. The effect is already widely exploited, mainly for laser applications (*e.g.,* example, electro-optical modulation in communication systems, Q switching, electronically controlled linear retarders and ultra-fast optical shutters). Its use in detector diagnostics and characterization is relatively new [22]. Fortunately, the Pockels effect is apparent in all zincblende structures to varying degrees and is particularly pronounced in II-VI compounds, such as CdTe[5] due to their large electro-optic coefficients.

Fig. 7.16 shows how the technique can be applied to map the internal electric field of a CdTe planar detector. A plane parallel beam of polarized light is shone through the detector bulk, parallel to the contact and perpendicular to the electric

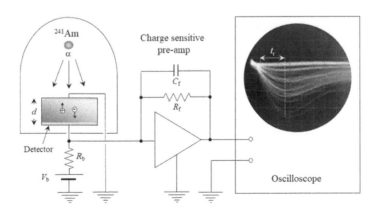

FIGURE 7.17 Experimental setup for determining carrier mobilities in a detector by the time-of-flight (TOF) method. The apparatus is setup for electron measurements. For hole measurements, the source should be placed under the detector or the charge signal on the cathode monitored and the bias reversed. For materials suitable for detector applications, rise times will typically be in the range 0.1–10 µs.

---

[5] In fact, CdTe has the highest linear electro-optical coefficient amongst II-VI compounds. For wavelengths around 1 µm, it has a value of 5.5×10^{-12} m/V (GaAs has a value of 1.1×10^{-12} m/V).

field. The wavelength of the light is chosen so that the crystal is transparent at that wavelength, which for CdTe occurs in the NIR around 800 nm, or equivalently ~1.5 eV. Under field-free conditions, the velocity of light will be the same in all directions. However, when an electric field is applied, the refractive index along the field direction changes, thus changing the velocity and the crystal becomes birefringent – meaning that two orthogonally polarized light waves will travel at different velocities through the crystal introducing a phase shift between the two waves. This phase difference varies linearly with the internal electric field intensity and the path length of light through the crystal. As described in Chapter 3 Section 3.3.3, a zincblende structure consists of two intersecting sub-lattices. In the case of CdTe, one will consist of Te atoms and the other Cd atoms. The application of an electric field causes the sub-lattices to be displaced with respect to each other by the inverse piezoelectric effect, which causes strain in the material proportional to the field strength. This in turn causes a displacement of charge throughout the crystal resulting in different binding energies along different planes. Light travels more slowly along planes where the binding force is strongest causing the shift in phase with respect to planes where the binding forces are weaker. This phase shift can be visualized and measured by illuminating the crystal with linearly polarized light whose E vector is orientated at -45° with respect to the applied electric field (see Fig. 7.16). After passing through the crystal, the phase shift will cause the beam to become elliptically[6] polarized. If upon exiting the crystal the light now interacts with an analyser polarizer orientated at +45°, only the components of the original polarized light that have been adjusted and are parallel to the analyser will be transmitted. In this case, the transmitted light $I(x,y)$ is given by [23],

$$I(x,y) = I_o(x,y) \sin^2[\alpha E(x,y)],$$ (42)

where

$$\alpha = \frac{\pi\sqrt{3}\, n_o^3\, r_{41} L}{2\lambda}.$$ (43)

Here $I_o(x,y)$ is the maximum light intensity transmitted through the unbiased detector with parallel polarizers, λ is the wavelength of the incident light, n_o is the field-free refractive index of the material, r_{41} is the electro-optic coefficient,[7] L is the optical path length through the crystal and $E(x,y)$ is the mean electric field intensity along the optical path. Since $\alpha E(x,y)$ is $\ll 1$, then

$$I(x,y) \approx I_o(x,y)\, [\alpha E(x,y)]^2$$ (44)

Thus $E \sim \sqrt{I}$. In a practical system, the transmitted radiation is captured by a suitably sensitized imaging device with good spatial and temporal resolution, for example a CCD. The insets (a) and (b) in Fig. 7.16 show two Pockels intensity images of a 10×10×1 mm thick CdTe planar detector taken with a IR sensitized CCD, averaged along the z direction [24]. The left panel (a) shows an image of the unbiased detector with crossed polarizers, from which we can see there is no variation in intensity across the detector depth, d. However, panel (b) shows an image taken with the detector under bias. In this case, the variation in light intensity shows the change of the internal electric field from low field (dark) to high field (light) as a function of detector depth.

As well as probing the electric field, this technique is particularly useful for characterizing base material in terms of uniformity and quality and evaluating the effects of new contact materials [24] or studying the time evolution of phenomena such as charge polarization due to space charge effects [25]. The Pockels effect has even been proposed as a method with which to detect high-energy radiation [26] in using semiconductors for their electro-optic properties rather than their semiconducting properties.

7.6.2 Estimating the Mobilities

The use of the Hall method for determining mobilities is not suitable for highly insulating materials because of the difficulty in measuring nA or smaller currents. Various other experimental techniques can be used. Most are based on TOF, such as the canonical Haynes-Shockley experiment [27], which measures the transit times of injected carriers across a bar of semiconducting material. However, while this technique has a high pedagogical value it requires complicated sample preparation and a dedicated measurement system. Consequently, it is little used outside undergraduate teaching laboratories.

---

[6] In fact, depending on the magnitude of the phase shift, the polarization changes from linear to elliptical to circular to elliptical and back to linear again with increasing shift.

[7] For zincblend crystals (point group 43m) the only non-vanishing electro-optical tensor elements are $r_{41} = r_{52} = r_{63}$).

For an existing detection system, a simpler procedure can be employed in which a detector is stimulated to produce transient ionization events near one contact and the response to the event measured on the other contact. Analysis of the transient pulse shape then provides information on the carrier transit time and on trapping times. The stimulation can be in the form of laser pulses, alpha particles or gamma-rays. An attractive feature of this technique is that the motion of both electrons and holes can be observed separately by reversing the polarity of the applied bias or directing the ionization to the other contact. For example, with reference to Fig. 7.17, electron mobilities can be determined by measuring the anode signal rise time, t_{re}, on a digital oscilloscope for near cathode events. For trapping lengths greater than the detector thickness, the rise time of signal reflects the carrier drift time, thus $t_r \cong t_d$, across the detector. Since $v_d = \mu E$, $v_d = d/t_d$ and $E = V/d$, then,

$$\mu_e = \frac{d^2}{V_b t_{de}} = \frac{d^2}{V_b t_{re}}.$$ (45)

In a sense, this method is a derivative of the Haynes-Shockley technique [27] but is less complicated. While inherently simple, it has several drawbacks, the most notable being the difficulty in determining the start and end times of the events for long transit times, particularly if thermal de-trapping becomes significant. The later can introduce a slow exponential component to the rise time signal. Generally, most experimenters use the 10% and 90% or 10% and 95% amplitude points when determining rise times. Even so, 5% accuracies in the electron mobility determination can be achieved.

By way of practical example, Fig. 7.18 shows the anode and cathode signals collected from a 1 cm^2, 1 cm thick HgI$_2$ pixel detector [28] following stimulation by gamma-radiation. The detector consists of four 1 mm^2 pixels patterned in a large anode plane – each pixel separated from the plane by 1 mm. The bottom of the detector consists of a single 1 cm^2 planar cathode. From Fig. 7.18, we derive an electron drift time of ~7 μs. The external electric field is 2500 Vcm^{-1} from which we estimate the electron mobility, μ_e to be 60 cm^{2}V^{-1}s^{-1}. In principle, hole mobilities can be estimated in a similar manner by measuring the rise time of the cathode signal for a near anode event. However, since hole mobilities are generally much poorer, this can only be achieved realistically using thin detectors or between test structures on thick detectors. For the thick detector, the hole signal from the cathode is indistinguishable from the noise, as illustrated by curve B in the inset of Fig. 7.18. Fortunately, for pixel detectors hole mobility measurements can usually be carried out between pixels and/or guard rings. This is illustrated by the waveform shown by curve A of the inset. This was obtained by applying a positive bias to the anode plane relative to a pixel and recording the pixel's response. From the figure, we deduce a risetime of 5 μs. In this case the external electric field was set to 2500 Vcm^{-1}, which for a 1 mm gap, yields a hole mobility of $\mu_h = 8$ cm^{2}V^{-1}s^{-1}.

It is important to note that the mobilities derived from drift or conductivity measurements, μ_d, can be quite different from those derived from Hall effect measurements, μ_H. The ratio μ_H/μ_d is usually close to unity for direct bandgap semiconductors but can be greater than unity for indirect bandgap materials.

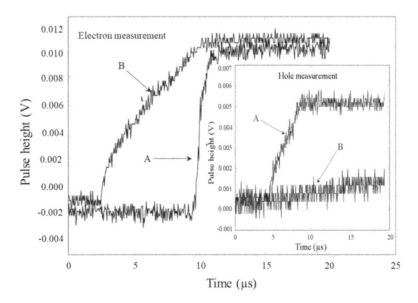

FIGURE 7.18 Measured drift time of electrons for a near cathode event in a 1 cm thick HgI$_2$ pixelated detector (adapted from [28], ©2003 IEEE). Waveform *A* is the anode (pixel) signal, while waveform *B* is the cathode signal. Note that the pixel signal is generated only when the electrons are close to the pixel. From the figure, the electron drift time is estimated to be 7 μs. The inset shows pulse waveforms from an event near the anode. In this case, holes generate the signal. However, since they cannot travel the full thickness of the detector, the signal from the cathode is indistinguishable from noise (curve B), while a significant charge can be induced between the anode plane and pixel (curve A).

7.6.3 Estimating the Mu-Tau (μτ) Products

The mu-tau products of a material are usually derived using the Hecht equation [29], which relates the electron and hole mean drift lengths to the amount of charge collected from the electrodes – usually expressed in terms of the Charge Collection Efficiency (CCE). The CCE is defined as the ratio of the induced charge at the contact, Q, divided by the total charge created as electron–hole pairs in the material, Q_o. Assuming a uniform field across the detector and negligible trapping, the CCE can be expressed as

$$CCE = \frac{Q}{Q_o} = \frac{\lambda_e}{d}\left[1 - \exp\left(-\frac{(d - x_o)}{\lambda_e}\right)\right] + \frac{\lambda_h}{L}\left[1 - \exp\left(-\frac{x_o}{\lambda_h}\right)\right], \tag{46}$$

where d is the detector thickness, x_o is the distance from the irradiated electrode to the point of charge creation and λ_e and λ_h are the carrier drift lengths in the applied electric field, E, given by, $\lambda_e = \mu_e\tau_e E$ and $\lambda_h = \mu_h\tau_h E$. Here, μ_e and μ_h are the electron and hole mobilities and τ_e and τ_h are the corresponding lifetimes. It follows from Eq. (46) that the CCE depends not only on λ_e and λ_h, but also on the location where the charge was created. In the case in which the interaction depth of the incident event is very close to one of the contacts, the induced signal will be due almost exclusively to the drift of one of the carriers. For example, in a material in which electrons are the majority carrier, the signal will be only sensitive to electrons for events close to the cathode ($x_o \sim 0$). Eq. (46) then reduces to

$$CCE = \frac{Q}{Q_o} = \frac{\mu_e\tau_e V}{d^2}\left[1 - \exp\left(-\frac{d^2}{\mu_e\tau_e V}\right)\right]. \tag{47}$$

Practically, the mu-tau product is then derived by recording the collected charge (generated by laser light, low energy X-ray or α-particles) as a function of applied bias and best fitting the results to Eq. (47) with $\mu_e\tau_e$ the fitting parameter.

In Fig. 7.19 we show a typical Hecht plot obtained from a 20 mm², 0.5 mm thick, TlBr planar detector measured using [241]Am alpha particles [30]. Pulse height spectra were obtained from −10 V to −350 V and the electron mobility-lifetime product derived by fitting Eq. (47). A value for $\mu_e\tau_e$ of $(2.8 \pm 0.2) \times 10^{-3}$ cm²V⁻¹ was obtained. Based on the measured electron mobility of (1040 ± 20) cm² V⁻¹s⁻¹, the effective electron lifetime[8] in this material is calculated to be $\tau_e = 6$ μs. From Fig. 7.19, we see that the curvature of the best-fitted Hecht curve is different from that defined by the data points as indicated by the arrows. Veale *et al.* [31] have pointed out that for data showing enhanced CCE values at low biases, a better fit to the data can be achieved by modifying the Hecht equation to include a voltage offset term, such that $V = V_b - V_o$, where V_b is the applied bias and V_o is a constant offset voltage. This is equivalent to the intercept of the X-axis at CCE = 0.

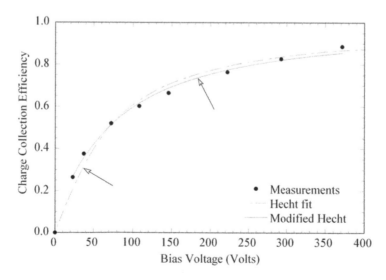

FIGURE 7.19 Charge collection efficiency (CCE) versus applied bias relationship for a 20 mm², 0.5 mm thick TlBr planar detector. From a best fit of Eq. (47), the mobility-lifetime product of electrons was estimated to be 2.8×10^{-3} cm²V⁻¹ (adapted from [25], ©2009 IEEE). The deviations from the fitted curve indicated by the arrows are most likely due to the assumptions implicit in the Hecht equation – no de-trapping or surface recombination. The dashed line shows the improvement in fit using a modified Hecht equation in which the fitted bias is offset by a delta amount.

---

[8] in effect the trapping time

Physically, it can be thought of as an internal voltage of opposite polarity, produced by a polarization field within the device. The dashed line in Fig. 7.19 shows a best-fit Hecht function with a voltage offset, which is clearly a better fit for the data. In this case, the derived $\mu_e\tau_e$ is $(2.34 \pm 0.03)\times10^{-3}$ cm^{2}V^{-1}.

In principle, the mu-tau product for holes can be determined by carrying out the same analysis as for the electrons. However, it is not always possible to isolate the hole signal because of its much poorer mu-tau product. This is illustrated in Fig. 7.20 in which we show measured energy-loss spectra in a 4 mm thick HgI$_2$ detector [28] from both the anode and cathode. It can be seen that the anode hole signal peak appears at a much lower channel number than the cathode electron signal due to significant hole trapping. This means that in practice there will be an insufficient range of peak-channel positions with bias to fit a Hecht function. In this situation, the hole mu-tau can be determined using the relative electron and hole peak pulse height distributions at the nominal bias. The hole mu-tau product can then be calculated from the ratio

$$\frac{Q_a}{Q_c} = \mu_e\tau_e \frac{\left[1 - \exp\left(-d^2/\mu_e\tau_e V\right)\right]}{\left[1 - \exp\left(-d^2/\mu_h\tau_h V\right)\right]}, \tag{48}$$

where the ratio of Q_a and Q_c is derived from the ratio of the relevant pulse height peak channels shown in Fig. 7.20.

Rearranging Eq. (48), we get

$$\mu_h\tau_h = \frac{-d^2}{V \ln\left[1 - \left(\frac{Q_a}{Q_c}\right)\mu_e\tau_e\right]}. \tag{49}$$

Alternately, when no hole peak can be discerned, it is still possible to evaluate the hole mobility-lifetime product using the average charge collection efficiency model of Ruzin and Nemirovski [32]. Consider the uniform illumination of the cathode by photons. For a high enough energy X-ray source, the electron and holes will be generated uniformly throughout the detector volume and both will contribute to the signal. The average charge collection efficiency is given by

$$\langle Q \rangle = \frac{N_0 q}{d\left(1 - \exp\left(-Md\right)\right)} \left\{(\lambda_e + \lambda_h)\left(1 - \exp\left(-Md\right)\right)\right.$$
$$\left. + \frac{\lambda_e M \exp\left(-d/\lambda_e\right)}{M - 1/\lambda_e}\left(\exp(-d(M - 1/\lambda_e)) - 1\right) + \frac{\lambda_h M}{M + 1/\lambda_h}\left(\exp\left(-d(M + 1/\lambda_h)\right) - 1\right)\right\}, \tag{50}$$

where M is the absorption coefficient of the photons of energy E, d the detector thickness, λ_e the mean free path for electrons ($=\mu_e\tau_e E$) and λ_h the mean free path for holes ($=\mu_h\tau_h E$). The average charge deposited on the electrode can be determined from the mean pulse heights of events as a function of bias. By fitting Eq. (50) to the data and fixing the mu-tau products for the electrons derived from a Hecht analysis, it is possible to extract the hole mobility-lifetime product, $\mu_h\tau_h$.

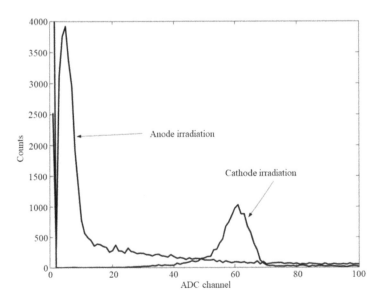

FIGURE 7.20 Anode and cathode pulse height distributions from a 4 mm thick HgI$_2$ pixel detector (from ref. [28], ©2003 IEEE). The effect of significant hole trapping is observed in the much-reduced hole pulse height distribution.

7.6.4 Limitations of the Hecht Equation

While the Hecht procedure is relatively straightforward, it suffers from several limitations in that it assumes (a) a uniform internal electric field, (b) carrier trapping is permanent (that is, no de-trapping) and (c) surface recombination effects are negligible. In polarizable materials, these may not be valid and the net effect of a Hecht analysis is to underestimate mu-tau products by factors of 3 or more. For Group II-VI materials, it is now widely recognized that highly localized defects on the surface are extremely efficient recombination centers and can be a dominant mechanism for controlling carrier lifetime. This can lead to a significant decrease in spectroscopic performance and affect a Hecht analysis by reducing the effective carrier lifetimes. The magnitude of the effect is also dependent on how the crystal was processed during detector fabrication (*i.e.,* contacting, mechanical polishing, chemical etching and passivation), and as a result, resistivities (of which recombination is a manifestation) can vary by many orders of magnitude. As a result, conditioning of the surfaces has remained little more than a "black art"; it is not generally considered in the data analysis. Thus, many reported mobility-lifetime values must be regarded with reservation, since both mu-tau products and surface recombination velocities are strongly dependent on surface treatment.

As an example of the importance of considering surface effects, Fig. 7.21 shows the electron charge collection efficiency as a function of bias voltage for a $10\times10\times2$ mm^3 CdZnTe detector [33]. The sample was polished and rinsed in methanol, treated in a standard 5% bromine-in-methanol etching solution for two minutes. Au contacts were then deposited by sputtering, and Pd leads were attached to contacts using a colloidal graphite suspension in water. Finally, the devices were covered with a protective coating. The detector was then irradiated on both 10×10 mm^2 surfaces, labelled A and B. The solid lines are theoretical fits with the Hecht equation. The fitted electron mobility-lifetime products for the two surfaces are listed in Table 7.4 from which we see that the electron $\mu\tau$ product for surface A is twice as large as that of surface B. Cui *et al.* [33] attribute the difference to surface recombination effects – in short, the two *surfaces* are different. While this cannot affect mobility, it will modify the carrier lifetime resulting in a measured mu-tau product, which is different from the bulk value. The bulk and surface contributions to the measured lifetime can be described by the equation

$$\frac{1}{\tau} = \frac{1}{\tau_b} + \frac{1}{\tau_s}. \tag{51}$$

Here τ_b and τ_s are electron lifetimes in the bulk and surface. Following the approach of Many [34], who considered the surface recombination and bulk trapping times in photoconducting CdS, Eq. (47) can be written as

$$CCE = \frac{\mu\tau_b V}{d^2}\left[1 - \exp\left(-\frac{d^2}{\mu\tau_b V}\right)\right]\left(\frac{1}{1 + d\,S/V\mu}\right), \tag{52}$$

where S is the electron surface recombination velocity (equal to d/τ_s) which is a measure of the rate of recombination between electrons and holes at the surface. For $S=0$, Eq. (52) reduces to the Hecht equation. Fitting Eq. (52) to the data shown in Fig. 7.21 results in much closer values of $\mu\tau_b$ for irradiation on the two surfaces but different values for S/μ. The results are listed in Table 7.4.

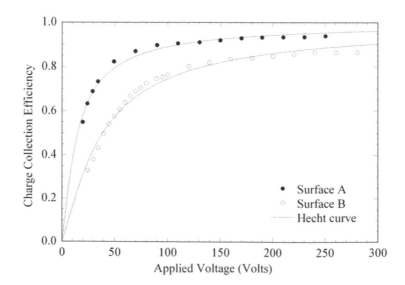

FIGURE 7.21 Charge collection efficiencies measured for two sides, labelled A and B, of a CdZnTe sample as a function of bias for the 59.54 keV photopeak of ^{241}Am (from [33]). The fitted parameters of a Hecht curve are given in Table 7.4.

TABLE 7.4

Mobility-lifetime ($\mu\tau$) products derived from the Hecht equation (Eq. 47) and taking surface recombination (S/μ) into account for the two 10×10 mm^2 surfaces (labelled A and B) of a $10\times10\times2$ mm^3 CdZnTe crystal (Eq. 52, from [33]). Without surface recombination, the measured $\mu\tau$ products are widely different for A and B. Taking surface recombination into account the $\mu\tau$ products are now consistent with each other.

Surface	Charge collection efficiency		
	Eq. (47)	Eq. (52)	
	$\mu\tau$ (cm^2/V)	$\mu\tau$ (cm^2/V)	S/μ (V/cm)
A	1.7×10^{-3}	2.5×10^{-3}	17
B	6.4×10^{-4}	2.1×10^{-3}	110

7.6.5 Measuring the Charge Collection Efficiency

Ionizing radiation absorbed in the sensitive volume excites electron–hole pairs in direct proportion to the energy deposited. The amount of charge created is given by

$$Q_o = q\,(E_o/\varepsilon),\tag{53}$$

where q is the electronic charge, E_o is the energy deposited and ε is the energy consumed to create an electron–hole pair. This charge induces a mirror charge on an electrode as it moves through the device. Because of trapping, some charge is lost, and the collected charge is now dependent on the path length, x, through which the carriers travel and thus the interaction location. The collected charge $Q(x)$ is given by

$$Q(x) = Q_o CCE(x),\tag{54}$$

where $CCE(x)$ is defined as the charge collection efficiency. The signal from the electrode is fed to a charge sensitive preamplifier, which converts it to a voltage signal proportional to its charge gain G_c (equal to the inverse of its feedback capacitance). The preamplifier output voltage is then amplified and shaped by a voltage amplifier voltage of gain G before being fed to a multi-channel analyser (MCA). The input voltage at the MCA is given by

$$V(x) = Q_x\,G_c G = (Q_x G)/C_f.\tag{55}$$

The MCA then converts this signal to a channel number, $Ch(x)$, given by

$$Ch(x) = m\,V(x) + c,\tag{56}$$

where m is the conversion gain of the MCA, in number of channels per volt, and c is an intercept term that corrects for any voltage offsets in the amplifier. Combining Eqs. (53) – (56) and solving for $CCE(x)$, we find

$$CCE(x) = \left(\frac{\varepsilon}{q\,Eo}\right)\frac{(Ch(x) - c)}{m\,G}.\tag{57}$$

Since all terms on the right side of Eq. (57) are known or measurable, the CCE can be determined directly from a measured spectrum.

7.7 Defect Characterization

Crystal defects are invariably introduced throughout the growth and detector fabrication process. These defects may arise from impurities, grain boundaries or interfaces and result in the creation of traps and recombination centers that capture free electrons and holes. Even at very low concentrations, these trapping centers can dramatically alter device performance.

For example, defects in the electronically active part of an integrated circuit (roughly within the first 5 μm – 10 μm) can easily destroy its performance. Therefore, an understanding of which trapping centers are present in a semiconductor is necessary in order to devise mitigation techniques to improve detector performance.

7.7.1 Thermally Stimulated Current (TSC) Spectroscopy

Before the introduction of deep-level transient spectroscopy (DLTS), thermally stimulated current (TSC) spectroscopy [35] was a popular technique for studying active defects or traps in semiconductors and insulators. The energy levels associated with traps are first filled by optical or electrical injection usually at a low temperature. The levels are then emptied by heating the sample to a higher temperature resulting in the emission of electrons or holes. The sample is then scanned in temperature at a given rate and the emitted current recorded. The resultant curve consists of a series of peaks that give information on the trap energy levels. For each peak, the trap depth, E_T (also known as the activation energy), can be determined from the approximate relationship [36],

$$E_T = kT_m ln \frac{T_m^4}{\beta}. \tag{58}$$

Here, k is Boltzmann's constant, T_m is the TSC peak temperature and β is the heating rate for the thermal scan.

While DLTS has replaced TSC for investigating traps in Schottky or p-n junctions, conventional capacitance-mode DLTS cannot be used for wide bandgap materials or insulators due to the difficulty of filling the levels by a change in electrical bias. For these materials, TSC is still used. Recently, however, the sensitivity of the technique has been markedly improved with the introduction of a new class of TSC spectrometers in which the emitted light is measured as a function of both temperature and wavelength [37].

7.7.2 Deep Level Transient Spectroscopy

DLTS is a pulsed bias capacitance transient technique used to investigate energetically "deep" trapping levels in semiconductor space charge structures [38,39,40]. These may be either *pn* junctions or Schottky barriers. By monitoring capacitance transients produced by pulsing the junction at different temperatures, a spectrum is generated that exhibits a peak for each deep level on a flat baseline. The height of the peak is proportional to trap density, and its sign allows one to distinguish between minority and majority traps. The position of the peak on the temperature axis leads to the determination of fundamental parameters governing thermal emission and capture (activation energy and cross-section).

DLTS relies on the fact that the RF capacitance of a sample (usually measured at 1 MHz under reverse bias) depends on the charge state of deep levels in the space charge region. The RF capacitance of a sample having a homogeneous doping concentration is given by

$$C_o = A\sqrt{\frac{q\varepsilon_s(N_d - N_a)}{2(V_r - V_d)}}, \tag{59}$$

assuming it is fully depleted. Here A is the sample area, N_d -N_a is the total net charge density in the space charge layer, V_r is the reverse bias, ε_s (= $\varepsilon\varepsilon_o$) is the permittivity of the semiconductor material and q is the electronic charge. If the sample is a *pn* junction, N_d – N_a refers to the lower doped side of the junction. V_d is the built-in diffusion voltage of the space charge structure, which is the crossing point of the extrapolated $1/C^2$ plot versus V_r with the V_r -axis (see Fig. 7.10). If A is measured in units of mm^2, N_d -N_a in cm^{-3}, C_o in pF and V_r and V_d in volts, Eq. (59) can be rewritten as:

$$N_d - N_a = 1.41 \times 10^{12} \frac{C_o(V_r - V_d)}{A^2\varepsilon_s}. \tag{60}$$

If charged trapping levels exist in the space charge layer, their space charge has to be added to N_d -N_a. On the assumption of a donor-like trap level of concentration N_t in an n-type sample biased under a reverse bias V_r, the capacitance change by recharging these levels is

$$\Delta C = A\sqrt{\frac{\varepsilon_s q (N_d - N_a)}{2(V_r - V_d)}} - A\sqrt{\frac{\varepsilon_s q (N_d - N_a + N_t)}{2(V_r - V_d)}} \cong C_o \frac{N_t}{2(N_d - N_a)}. \tag{61}$$

The last identity holds approximately if $N_t \ll N_d - N_a$. In this case, the trap concentration can be calculated from the change in capacitance ΔC,

$$N_t = \frac{2\Delta C}{C_o} (N_d - N_a). \tag{62}$$

The practical implementation of DLTS may be found in Lang [38]. The main advantage of DLTS is that it is a very sensitive technique, measuring defect concentrations down to a level of 10^{10}cm^{-3}. It is also non-destructive. The main disadvantages are:

1) it cannot be used for insulating materials,

2) the defects must be electrically active,

3) their concentration must be < 10% of the doping concentration,

4) the sample must have a depletion region (Schottky or *pn* junction) and

5) the identification of levels usually requires comparison with other techniques.

7.7.3 Photo-Induced Current Transient Spectroscopy (PICTS)

For insulating or very high resistivity materials, photo-induced current transient spectroscopy (PICTS) [41] is widely used. It was first proposed by Hurtes *et al.* [42] as a "modification" to DLTS to accommodate high resistivity materials and relies on measuring the current released when filled traps empty following optical excitation. The sample is irradiated from the cathode side and the photo-generated charges drift in the bulk semiconductor and fill the deep levels. After equilibrium between generation, trapping/de-trapping and collection is reached, the optical excitation is abruptly stopped. The photocurrent transient at the end of the light pulse consists of a rapid drop followed by a slow decay. The initial rapid drop is due to electron–hole pair recombination, and the slow decay is due to carriers thermally ejected from the traps [41,43].

Consider a saturated semiconductor, with no re-trapping of the emitted carriers. The current at time t after excitation is given by

$$I(t) \propto \xi \mu \tau e_n \exp(-e_n t), \tag{63}$$

where ξ is the electric field, μ and τ are the mobility and the lifetime of the carriers, and e_n the emission rate of the trapped carriers. In the case of a discrete trap, e_n can be written as

$$e_n(T) \propto \sigma_n T^2 \exp\left(\frac{-E_t}{kT}\right), \tag{64}$$

where T is the sample temperature, σ_n and E_t are the cross-section and the energy of the trap, respectively. To carry out a PICT analysis, semiconductor samples are excited with a pulsed light source (usually a LED) of wavelength above the bandgap energy and the thermal relaxation time calculated from the time dependence of the photo-induced current as a function of sample temperature [43,44]. Around certain temperatures, peaks in the PICTS signal can be seen. The temperatures corresponding to the peak of the PICTS current signal are then plotted as a function of the relaxation times on an Arrhenius plot[9] and the energy and emission cross-section of the trap determined from the intercept of the curve, using Eq. (64). The measurements are repeated at different sample temperatures to probe different trapping levels. The main advantage of PICTS is that measurements are relatively simple to carry out. However, it is difficult to determine trap densities and measurements can be unreliable if the trapped carriers recombine or are re-trapped. In addition, where DLTS measurements can be carried out, it is found to be much more sensitive than PICTS. For example, in a study of deep trapping levels in undoped, semi-insulating and p-type CdTe single crystals, Kremer and Leigh [45] found that DLTS measurements revealed several trapping levels in all samples, whereas the PICTS data was generally only sensitive to one level.

---

[9] An Arrhenius plot displays the logarithm of a parameter in a thermally activated process against inverse temperature.

7.8 Photon Metrology

Synchrotron light sources are widely used in materials science, protein crystallography and biomicroscopy applications. They provide a unique stable source of high intensity photons, extending over a broad energy range from the far infrared to the γ-ray region. However, they have also proven invaluable for carrying out detailed metrology of radiation detectors by making available highly collimated, monochromatized beams of synchrotron radiation [46]. Light sources are only accessible at synchrotron research facilities and a number of specialized laboratories (for example the Physikalisch-Technische Bundesanstalt radiometry laboratories in Berlin, Germany [47]) have been established specifically to carry out photon metrology from the UV to the X-ray range using primary source standards in conjunction with primary detector standards.

7.8.1 Synchrotron Radiation

Classically an electron moving in a magnetic field will execute a spiral trajectory and radiate as a dipole [48]. The emission is isotropic at the Larmor frequency,

$$\nu_L = \frac{eB}{2\pi m_o c} = 2.8\,\text{MHz per Gauss}, \tag{65}$$

where B represents the magnetic field component perpendicular to the particle velocity vector. If the electron is non-relativistic, the radiation is isotropic and is emitted only at the Larmor frequency. This is known as cyclotron radiation. In the relativistic case, synchrotron radiation is emitted in a relativistically narrow cone of angle, $\theta \sim \gamma^{-1}$, where γ is the particle energy in units of its rest energy (typically 10^3–10^4). The frequency distribution is no longer discrete as in the non-relativistic case but is an asymmetric distribution with a maximum of the envelope at

$$\nu_m = 2/3\gamma^2\nu_L, \tag{66}$$

or in terms of energy

$$E_m = 5 \times 10^{-9}\gamma^2 B \quad (\text{keV}). \tag{67}$$

For the magnetic fields used to steer particles in accelerators, E_m will be in the UV to X-ray range, and indeed synchrotron radiation was first observed emanating from early electron accelerators over 60 years ago [49]. At first the phenomena was considered an inconvenient waste product of particle acceleration and was only really exploited in the early 1970s when it was realized that highly collimated, monochromatic photon beams make an excellent tool for probing the electronic structure of matter from the sub-nanometer to the millimetre level. This led to the construction of a number of dedicated synchrotron facilities based on storage rings.

The unique properties of synchrotron radiation can be summarized as follows:

- Unique production mechanism in that it can be precisely described
- High brightness and high intensity, many orders of magnitude more than that of X-rays produced in conventional X-ray tubes
- Wide tuneability in energy/wavelength by monochromatization (from sub-eV up to MeV)
- High collimation, small angular divergence of the beam
- Low emittance, the product of source cross-section and small solid angle of emission
- High level of polarization (linear or elliptical)
- Pulsed light emission with durations of 1 ns or less

Which of these attributes are exploited by the various scientific disciplines and how are described in [50].

7.8.2 Light Sources

At present, there are over 50 synchrotron light sources operating worldwide [51]; they can be broadly grouped into three categories, or generations. The fourth generation is currently under development. In chronological order, these are as follows:

- First generation synchrotron radiation light sources were essentially parasitic on other programs, such as high-energy physics. Examples include the Synchrotron Ultraviolet Radiation Facility (SURF) in Maryland, USA, and the 6 GeV Deutsches Elektronen-Synchrotron (DESY) in Hamburg, Germany.

- Second generation synchrotron radiation light sources are dedicated synchrotron radiation facilities but not designed for low emittance or with straight sections for insertion devices. Examples include the Daresbury SRS in the UK and HASYLAB at DESY in Germany.

- Third generation synchrotron radiation light sources are dedicated synchrotron radiation facilities designed for low emittance and have many straight sections for incorporating insertion devices. Examples include the ESRF and Soleil in France, the ALS in the USA, BESSY II in Germany and the Diamond light source in the UK.

Each generation differs from the previous generation by innovation and is improved by at least an order of magnitude in performance, usually quantified by the flux and the brilliance of the source. The flux is defined as

$$\Phi = \frac{N_{\mathrm{p}}}{0.1\% \, mrad} \left(\text{photons} / \left(\text{second, } 0.1\,\% \text{ energy spread, mrad horizontally} \right) \right), \tag{68}$$

where N_{p} is the number of photons emitted per second for a given stored beam current. The brilliance, B, is the peak flux density in phase space,

$$B = \frac{N_{\mathrm{p}}}{0.1\%, mm^2, mrad^2}. \tag{69}$$

The flux is a function only of the electron current and energy, while the brilliance takes into account the phase space defined by diffraction effects and the electron beam emittance.

Considerable effort is now underway developing fourth-generation light sources, which will most likely combine a hard X-ray (wavelength less than 1Å) free-electron laser (FEL) with a very long undulator in a high-energy electron linear accelerator. Such a device would have a peak brightness many orders of magnitude beyond that of the third-generation sources, pulse lengths of 100 fs or shorter and would be fully coherent.

7.8.3 Synchrotron Radiation Facilities

In general, a synchrotron facility consists of several sub-systems: an electron gun, a linear accelerator, a booster ring, a storage ring and a set of beamlines, which feed specialist experimental stations, as illustrated in Fig. 7.22. Synchrotron light is produced by accelerating electrons to relativistic energies. To achieve this, free electrons are first generated in an electron gun by a heated cathode and accelerated through a hole at the end of the gun by a powerful electric field. They are then fed into a linear accelerator, where they are accelerated from an initial energy of ~50 keV to ~50 MeV and injected into a booster ring in which the electrons are confined to travel in a circle. They are continuously energized by microwaves, before being injected at GeV energies into the main storage ring. The trajectory of the electrons is steered in the storage ring by a series of bending magnets. These define the ring and maintain the electron trajectory within the ring. However, whenever high energy, relativistic electrons are forced to travel in a curved path by a magnetic field, synchrotron radiation is emitted in a fan beam in a plane parallel to the electrons orbit in the ring. The synchrotron light produced at the bending magnets is extracted by a suitable beamport and directed to an experimental station by a vacuum beampipe. An RF cavity, located in a straight section of the ring, is used to replenish synchrotron energy losses at the bending magnets.

For third-generation machines, long straight sections are incorporated into the storage ring for the inclusion of insertion devices. These are periodic arrays of magnets designed to produce a series of deflections of the primary electron beam in the straight-line section of the orbit. They consist of one array of magnets above the electron beam path and one co-aligned array below. The poles alternate so that instead of one magnet deflecting the electron beam and generating a single fan of light, an entire array of alternating magnets now deflect the beam such that the electrons follow a wiggling or undulating path. Each deflection produces a kink in the electron trajectory adding to the intensity of the light from that point. There are two basic types of insertion device: wigglers and undulators. A wiggler can be considered a concatenation of N bending magnets and its brilliance scales as N, emitted over a wide bandwidth, whereas in an undulator, the magnets are arranged such that the emitted radiation adds in phase and its brilliance scales as N^2, emitted over a narrow bandwidth. A comparison of the on-axis brilliance for bending magnets, wigglers and undulators is given in Fig. 7.23.

The beam emerging from the bending magnets is known as "white light" and has a well-defined energy spectrum extending from the microwave through to the UV, VUV, soft and hard X-ray regions of the electromagnetic spectrum. The spectrum is

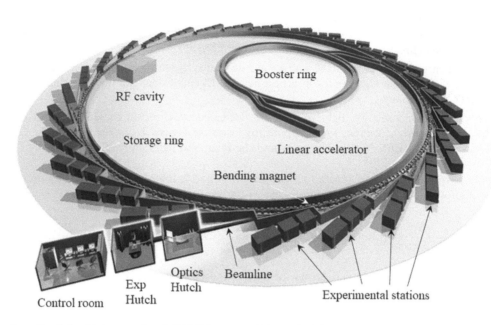

FIGURE 7.22 A schematic of the third-generation 3 GeV light source, Diamond Light Source [52] at Harwell Oxford, UK, showing its major components. Diamond is composed of three main machines in order to speed up and accelerate the electrons. The main ring is not circular, but a 50-sided polygon with beamlines emerging at the vertices. Each beamline is optimized to support an experimental station that specializes in a specific area of science, including the life, physical and environmental sciences. The beamlines themselves are typically comprised of an optics hutch, experimental hutch and a control room as shown (Image courtesy Diamond Light Source 2017).

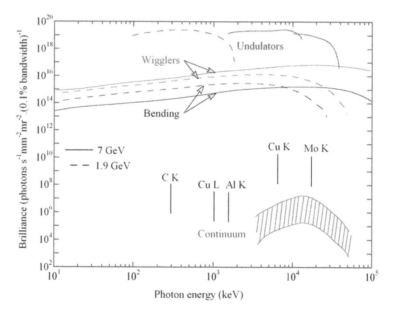

FIGURE 7.23 Spectral brilliance for several synchrotron radiation sources and conventional X-ray sources. The data for conventional X-ray tubes should be taken as rough estimates only, since brightness strongly depends on operating conditions (adapted from [53], courtesy LBNL).

usually characterized by its critical energy, E_c, which is defined as the energy at which half the radiant power is carried by photons above E_c. The critical energy is given by

$$E_c(\text{keV}) = 0.665 E^2/\rho, \tag{70}$$

where E is electron beam energy in GeV and ρ is the bending radius. Critical energies generally range from ~10 keV for second-generation machines to ~25 keV for third-generation machines. A useful rule of thumb is that for general synchrotron work, the practical working energy range is $4E_c$ while for single photon counting detector work it is $8E_c$.

7.8.4 Properties of the Beam

White light direct from an extraction point is extremely intense and has a brilliance (defined as the number of photons per second per mm^2 per mrad^2 in $\Delta\lambda/\lambda = 10^{-3}$) of ~$10^{15}$ photons s^{-1}mm^{-2}mrad^{-2} (second-generation machines) up to 10^{20} photons s^{-1}mm^{-2}mrad^{-2} (third-generation machines). To put things into context, a rotating anode X-ray generator has a brilliance of 10^9 photons s^{-1}mm^{-2}mrad^{-2} in $\Delta\lambda/\lambda = 10^{-3}$. The white light directly from the synchrotron is, in fact, so intense that the raw beam will seriously damage a detector and will even melt most metal components if they are not cooled (*e.g.*, beam shutters). Fortunately, it is possible to lose many orders of magnitude of intensity by (1) monochromatizing the beam using a double crystal monochromator, (2) detuning the RC of the first, or downstream, crystal and (3) greatly reducing the area of the beam on the detector, For the type of photon metrology carried out on detectors' incident flux rates at the sample need to be in the range ~10^2 to 10^4 photons s^{-1}.

7.8.5 Beamline Design

Beamlines are usually tailored for particular experimental disciplines using sub-systems to filter, intensify or otherwise manipulate the light to generate a specific set of characteristics suitable for the needs of the experimental station, which are normally application specific. A typical beamline layout suitable for detector metrology is shown in Fig. 7.24 – the X1 beamline at the Hamburger Synchrotronstrahlungslabor (HASYLAB) radiation facility [54] located at DESY in Hamburg, Germany. This beamline utilizes a double Si crystal monochromator to produce highly monochromatic X-ray beams across the energy range 10–100 keV. Depending on the energy range of interest (and the amount of acceptable harmonic pollution) a choice of Si(111), Si(311) and Si(511) crystal pairs can be selected. Because mechanically, the range of adjustable monochromator angles and crystal separation is finite, the order of the reflection usually sets the useable energy range in a particular system. The range of Si(511) is generally larger than Si(311) which in turn is larger than Si(111). Also the energy resolution tends to be higher for high-order crystals, ranging from 10^{-4} for Si(111) to 10^{-5} for Si(511). To cover the entire energy range 10–100 keV, a [511] reflection is usually used, yielding an intrinsic energy resolution of ~1 eV at 10 keV, rising to 20 eV at 100 keV.

The white beam extracted from the synchrotron passes through a set of entrance slits, which serve to define the beam profile incident on the first, or upstream, monochromator crystal. In normal operation, the slit width is set to ~10 mm in the horizontal plane and ~(0.1–1) mm in the vertical plane. The vertical width is much smaller, since it defines the energy resolution of the system. The upstream crystal can rotate in the vertical plane and translate in the horizontal plane along the beam axis and is used to direct the Bragg reflected beam on to the second or downstream crystal. The second mono-chromator crystal is free to rotate about a single axis and thus in combination with the translation and rotation on the first crystal allows the diffracted beam to pass through a fixed exit point, which is parallel to the incident beam but displaced in the vertical plane. The monochromator serves two purposes. The first is to select the photon energy that will be incident on the sample. The second is to reduce significantly the flux of incident photons (by some six orders of magnitude), since the pass-band is only a few eV. The flux can be further reduced by detuning the parallelism of the two crystals. This is achieved in practice by using a piezoelectric piston to rotate finely the upstream crystal. Detuning this crystal has the added advantage of suppressing higher order harmonics that may otherwise contaminate or compromise the measurements. Note that it is usually not possible to reduce the intensity by simply inserting absorbers. In the case of monochromatic radiation, absorbers "amplify" the higher harmonics relative to the fundamental and generate non-negligible secondary radiation. In the case of white light, in order to reduce the intensity of highest energy component of the white light continuum to an acceptable level for detector operation, the absorber becomes so thick that all spectral information is lost. Following the monochromator, the

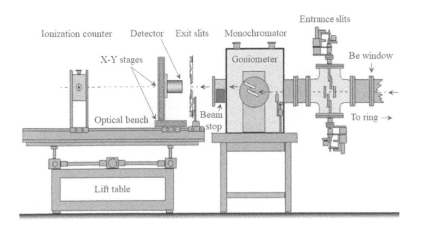

FIGURE 7.24 The X1 hard X-ray beamline at the HASYLAB synchrotron radiation source at DESY [54].

diffracted beam passes through a pair of precision stepper-driven exit slits positioned immediately in front of the detector (see Fig. 7.24) which define the beam size at the detector and reduces stray light contributions. For the majority of detector measurements this is usually set to a size of $50 \times 50 \ \mu m^2$. The monochromator, slits and X-ray stages are all controlled from the stations computer, which is preprogrammed to carry out an extensive set of operations, such as energy scans for X-ray Absorption Fine Structure (XAFS) measurements and 1-D and 2-D spatial scans for detector characterization. The station computer also logs housekeeping data, such as ring current and motor settings, as well as experimental data.

7.8.6 Installing the Detector

Detectors are usually mounted on an X–Y stage capable of positioning a detector to a precision of <1 μm in each axis over a range of ± 10 cm. The mechanical interface between the X-Y stage and the detector is a CNC machined plate with a number of predefined locations for installing different detectors. The detectors are located by means of guide pins and quick release screws. A reference detector can be installed at one of a number of locations on the interface, all of which are precisely located with respect to principal detection axis. This is defined by the position of the detector being tested. After the detector has been mounted (see Fig. 7.25), a laser assembly is attached to the front of the detector. The laser beam axis is precisely aligned with the center of the detector aperture and points towards the center of the X-Y slits. The reference detector can now be used to locate the beam precisely with respect to the principle detection axis. This is achieved by carrying out a series of X, Y scans across the beam, in which the reference detector count rate is recorded as a function of position. The center of the beam is located by centroiding. Next, the X-Y stage is driven to position the beam at the centre of the detector under test and X-Y scans again carried out. Once the center of the detector has been located the X-Y coordinates are now zeroed and measurements can begin.

The reference detector also serves several other functions. Firstly, it is robust, so it is used to adjust the beam structure and flux between experiments before a sensitive test detector is exposed to the beam. The detector is also fully spectroscopic, so it can check on the spectral purity of the beam that may have been contaminated by scattering and fluorescence backgrounds caused by slight misalignments of the beam. For new experimental detectors, a lack of a signal does not necessarily mean the detector has malfunctioned – it could mean there is no beam. The reference detector is also used to check this.

7.8.7 Harmonic Suppression

The monochromator transmits not only the desired fundamental energy but also higher harmonics of that energy. The suppression of these higher harmonics is an important consideration for X-ray experiments because in the presence of any absorbing material, such as the path length in air or detector windows, the fundamental can be easily suppressed, effectively amplifying the higher harmonics. Braggs law of X-ray reflection states that

$$n\lambda = 2d \sin \theta, \tag{71}$$

FIGURE 7.25 Photograph showing the installation of a detector ready for characterization on beamline X1 at the HASYLAB synchrotron research facility at DESY. A laser attached to the front of the detector is used to align its aperture with the centre of the slits. A co-aligned reference detector is then used to establish precisely the position of the beam with respect to the detector principle axis prior to scanning.

where d is the spacing of the atomic planes of the crystal parallel to its surface, θ is the angle of the crystal with respect to the incident white light beam, λ is the wavelength of the diffracted X-ray, and n is an integer. The fundamental X-ray energy corresponds to $n=1$ and X-rays of higher harmonic energies correspond to $n>1$. Harmonic X-rays that are diffracted from the crystal depend on the crystal lattice and the cut of the crystal. It is useful to use a crystal that does not diffract the second harmonic ($n=2$) because the intensity of the second harmonic is usually much greater than the intensities of the higher harmonics. Si crystals with a diamond structure (space group Fd$\underline{3}$m(O_h^7)) will not allow the harmonics that satisfy the eq. $h+k+l=n$, where n is twice an odd number. Hence, for example, Si(111) crystals do not diffract the second or sixth harmonic. When working at high X-ray energies, it is possible that the energy of higher order harmonics exceeds the maximum energy produced by the machine in which case no harmonic rejection is needed. Common methods of reducing harmonic X-ray content include detuning the second crystal or using a harmonic rejection mirror. To detune the monochromator, the angle of the second crystal is slightly offset, or misaligned, with respect to the angle of the first crystal using a piezoelectric transducer. This has the effect of reducing the harmonic content much more than the fundamental. For example, when two Si(111) crystals are detuned by 50% on the rocking curve, the intensity of the third harmonic is reduced by a factor of 10^3. For detector work, harmonics rarely present a problem, since to operate in single photon counting mode, the monochromator has to be detuned by as much as 95% at some energies. Another common method for removing harmonics X-rays is to use a harmonic rejection mirror. This mirror is usually made of Si for the lower XUV energies, Rh for X-ray energies and Pt for the high X-ray energies. The mirror is placed at the grazing angle of the beam such that the X-rays with the fundamental energy are reflected towards the sample, while the harmonics are not. Slits placed downstream of the mirror are also used to block the direct beam containing the harmonics.

7.8.8 Extending the Energy Range

Most conventional synchrotron experiments use the increased brilliance produced by insertion devices purely to increase the flux incident within a very small area, for example, in protein crystallography to compensate for weak diffraction from very small samples. For detector metrology, wigglers are particularly useful, because the increased flux coupled with the increase spectral bandwidth allow measurements to be carried out outside the useful energy range of bending magnets (subject of course to a suitably "sized" monochromator). Even though the photon flux falls off exponentially above the critical energy, for single photon counting applications there may still be significant flux at energies as high as an MeV. For example Owens *et al.* [55] reported measurements over the energy range 6 keV to 800 keV,[10] carried out on the ID15 high-energy scattering beamline [56] at the ESRF. To achieve usable photon fluxes over such an extreme energy range, two insertion devices were used, an asymmetrical multipole wiggler followed by a superconducting wavelength shifter. The spectral resolution was typically around 20 eV at 30 keV rising to ~1 keV at 1 MeV. Higher-order harmonics were not an issue since the exponential fall in intensity above the critical energy ensured that the flux was vanishingly small.

7.8.9 Detector Characterization

The use of highly collimated and tuneable photon beams allows a wide range of detector measurements to be carried out, of which we list a few examples. We will consider primarily monolithic detectors and material characterization. For the specific characterization of area detectors (*e.g.,* measurements of QE, DQE, MTF, LPI, *etc.*), the reader is referred to the review of Ponchut [57].

For X-ray and gamma-ray applications, the energy resolution of a detector, its linearity and uniformity of response can be directly measured at any energy within the energy range of the monochromator. For example, in Fig. 7.26 we show a composite of energy-loss spectra measured with a 250×250 μm^2, 40 μm thick GaAs pixel detector, which perfectly illustrates the high-quality data that can be obtained with a synchrotron radiation source. From the figure, we see that the energy resolution is so good that the escape peaks are also resolved. The enhancement in low energy events is due to Compton scattering at the higher energies. The pixel was part of a 32×32 array and the purpose of the tests was to establish the uniformity of response and isolation of pixels within the array – the data from which would be used by the foundry for process control. Towards this end, four pixels were instrumented with resistive feedback pre-amplifiers.

In Fig. 7.27 (a) we show the linearity curves for all four pixels from which we can see that the detected charge response is linear with photon energy. The regression coefficients for best-fit linear functions to the peak channel number versus incident X-ray energy were in excess of 99.99% across the entire energy range. For all pixels, the rms non-linearity was $\leq 0.14\%$ consistent with the statistical error in the fit. The lower panel of Fig. 7.27(a) shows the individual residuals of the fit, that is (measured energy – energy)/energy × 100% from which we can see there are no systematic trends with energy. The energy resolution function was derived from the pulse height data and is shown in Fig. 7.27(b). The resolutions range from 290 eV FWHM at 10 keV to 780 eV FWHM at 100 keV. As can be seen, the measured resolutions for all four pixels agree within

---

[10] It is possible to generate higher energies. However, above 800keV, photon grazing angles become so shallow that it is exceedingly difficult to separate the monochromatic beam from the white beam without recourse to a custom-built collimation system.

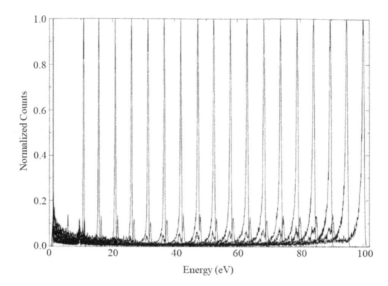

FIGURE 7.26 Composite of measured energy-loss spectra from a 250×250 μm², 40 μm thick GaAs pixel detector measured at HASYLAB [58]. The beam size was 20×20 μm², incident at the center of the pixel.

20% – the differences almost certainly arise from the difference in stray capacitance of the front-end components. Because large energy ranges can be sampled frequently in energy space and to arbitrary precision, it is possibly to de-couple the various noise contributions by best fitting the expected revolution function [59], that is

$$\Delta E = 2.355\sqrt{F\varepsilon E + (\Delta E_e/2.355)^2} + a_1 E^{a_2} \quad \text{keV,} \tag{72}$$

where ε is the energy required to create an electron-hole pair, F, is the Fano factor, ΔE_e is the electronic noise component measured directly using a precision pulser and a_1 and a_2 are treated as semi-empirical constants determined by best fitting. The first term in the square root accounts for noise due to carrier generation, the second term for the electronic noise of the system and the third term for trapping noise. Note that the noise contribution due to the beam divergence at the monochromator is of the order of tens of eV or less, which is many orders of magnitude less than other sources; what is measured is solely due to the detector. The results of best fitting Eq. (72) are shown in Fig. 7.27(b), in which we show the best-fit resolution function and its principal components. Because of the uncertainly in the value of F, it was also treated as an additional semi-empirical constant yielding a best-fit value of 0.15±0.04. From the figure, we see than Fano noise dominates above 10 keV, while electronic noise dominates below.

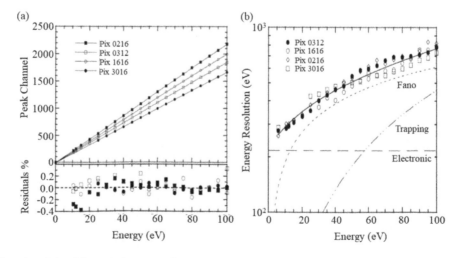

FIGURE 7.27 (a) The linearity of the all four pixels measured over the energy range 10 keV to 100 keV (from [58]. The solid lines show best-fit linear regressions. The lower panels shows the residuals (*i.e.*, (measured energy – energy)/energy × 100%. (b) The energy-dependent FWHM energy resolutions of all four pixels measured at HASYLAB under pencil beam illumination. The solid line represents a best-fit resolution function to the average. For completeness, the individual components of this curve are also shown.

7.8.10 Spatially Resolved Spectroscopy

Cross talk and gain maps can be easily and quickly assessed in pixel detectors across a detector or array. As an example, in Fig. 7.28(b) we show the X-ray spatial response in the form of a gain map of a 4×4 GaAs pixel array (shown in Fig. 7.28(a) [60]. The pixel sizes are 350×350 μm^2 with an inter-pixel gap of 50 μm. The array thickness is 325 μm. The map was produced using a 15 keV pencil beam of 20×20 μm^2, normally incident on the pixels. The beam was raster scanned across this area with 10 μm spatial resolution in both dimensions. Spectra were accumulated at each position for a fixed-time interval and the total count rate above a 3 keV threshold, the peak centroid position and the FWHM energy resolution were determined by best fitting. Typically for this type of characterization, it is found that for 3% counting statistics the centroid (essentially the gain) can be located to a precision of ~0.1% and the fitted FWHM energy resolution to a precision of ~20%. The measurements show that apart from a slight decrease in these parameters directly under the bond wires, their spatial distributions were very uniform over the surface of each pixel and the entire array. In fact, the average non-uniformity is typically no worse than a few percent and is consistent with a flat response. In other words, the variations seen in each distribution are consistent with the expected statistical variations. The fact that both the count rate and centroid responses are zero in the inter-pixel gaps, with no evidence of cross talk, implies that it is possible to replicate isolated and identical pixels and thus in principle mega pixel arrays also.

7.8.11 Probing Depth Dependences

In addition to measuring basic response functions such as energy resolution and linearity, it is relatively easy to directly measure the energy dependent efficiency function by comparison with a calibrated reference detector. The active depletion depth can be determined using the same method [61]. Alternatively, fine spatial scans at various energies can probe the depth dependent structure in the detector. For example, at low energies the spatial morphology and lateral uniformity of the contacts can be mapped, while at high energies the spatial uniformity of the internal electric field can be probed. In fact, for detectors with good charge collection efficiency, these measurements can be carried out simultaneously by exciting the higher harmonics of the Bragg reflection. This can be archived by "tuning-up" the RC and inserting thin pieces of absorber into the beam.

7.8.12 Pump and Probe Techniques

Another powerful technique that can be exploited at synchrotron facilities is the so-called "pump and probe" technique in which an intense finely focused monochromatic X-ray beam is used to "pump" a region of interest on a detector and a much weaker "probe" beam used to map the resulting creation and subsequent dispersal of charge. As an example of the power of such a technique, Kozorezov *et al.* [63] used it to study the onset and subsequent evolution of polarization effects in TlBr radiation detectors.

Polarization in semiconductor X-ray detectors can cause significant degradation in detector performance, particularly at high radiation levels. The term polarization in this context means time-dependent variations in the detector properties, such as count rate, charge collection and resolving power that seem to be correlated with incident radiation fluence and material properties, such as purity, stoichiometry, high resistivity and dielectric constant. To date, the treatment of polarization has

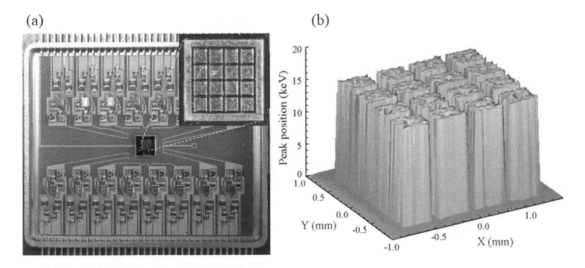

FIGURE 7.28 (a) Photograph of a 4×4, GaAs pixel array and its associated front-end electronics. The inset shows a blow up of the array before wire bonding. The pixel size is 350 μm×350 μm and the inter-pixel gap 50 μm. (b) A surface plot of the spatial variation of the gain (the fitted centroid position) across the array measured at HASYLAB using a 15 keV, 20×20 μm^2 pencil beam (from [60]). The spatial sampling in X and Y, was 10 μm.

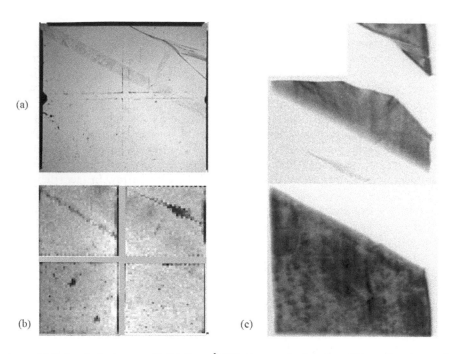

FIGURE 7.29 A comparison of defect metrologies on a 25×25×2 mm³ CdZnTe crystal using three different imaging modalities (taken from Carini *et al.* [62]). These are (a) an IR transmission map, (b) an X-ray response map and (c) a WBXRT topograph. From (a) and (b) we can see a strong correlation between structure in the IR map (corresponding to inclusions and decorated grain boundaries) and detector performance, reflected in the X-ray map. This is not so apparent in the WBXRT topograph, which is primarily responsive to individual domains. Topographical contrast inside each domain should be related to tellurium inclusions (not clear in this case).

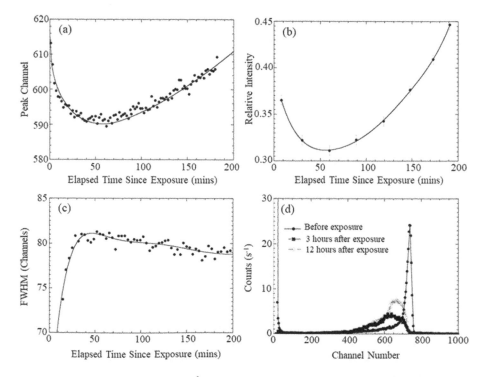

FIGURE 7.30 Defect diagnostics carried out on a 15×15×10 mm³ CdZnTe coplanar grid detector (from [55]). The figure shows spectrally resolved count rate maps obtained by raster scanning a 20×20 µm², 180 keV normally incident beam in 40 µm steps across the detector. Crystal defects are plainly evident in the detector count rate response and account for ~2% of the active area of the detector (light colour corresponds to photopeak events, black to a lack of events). The lower right-hand image shows that the counts in the defects do not originate from the photopeak and in fact mainly emanate from energies 120 to 170 keV (see lower left-hand image). In this energy region there are virtually no events from the rest of the detector area.

been largely anecdotal and qualitative. Recently, Bale and Szeles [64] have presented a quantitative dynamical model of polarization, which is based on the supposition that polarization is a consequence of strong carrier trapping by deep impurity levels. As a result, a space charge region with a high charge density can be created, which causes a significant change in the profile of the internal electric drift field. At some point in the bulk of a detector (the "pinch point"), the polarization field cancels the electric field created by the detector biasing, leading to a catastrophic deterioration of detector performance. In this regard, the measurements of Camarda *et al.* [65] are particularly interesting. They used a finely collimate, high intensity X-ray beam to study the correlation between microscopic defects (inclusions) and variations in the collected charges in CZT detectors. They were able to demonstrate that single Te inclusions trap a significant amount of charge in an electron cloud and that polarization effects could be correlated with these defects and their surrounding areas.

The model of Bale and Szeles [64] is simplified in that it considers only uniform illumination, which means that the profile of the internal electric field is one-dimensional and depends only on the coordinate normal to the plane of the detector. As such, many details of the effect are still not understood, particularly those relating to the mechanism of the polarization process itself – the extent of the polarized region, the energy dependence of the process and details of the trapping processes responsible. As reported by Kozorezov *et al.* [63], this work was recently extended in a series of "pump and probe" scanning measurements with tightly focused monochromatic pencil beams at the HASYLAB synchrotron research facility. By combining these 2-D scans for a range of X-ray energies and intensities, a 3-D picture of polarization phenomena within the bulk material could be constructed, which could then be used to extend the model to three dimensions. These results confirm that local polarization close to the exposed spot results in distortion of electric fields both inside and around the affected volume and that the charging of deep traps by one of the carriers perturbs the internal electric field sufficiently to affect the collection of the other carrier. The relaxation of the polarized region was also studied by regularly repeating the probe-pulse scans over a period of three hours after the initial "pump".

An example of the correlative information that can be gained is shown in Fig. 7.31, in which we show data obtained from the pulse height spectra obtained for scans at 60 keV incident energy. We see that while the peak channel position and charge collection efficiency experience moderate signs of recovery, the spectral resolution and particularly the line shape remain degraded for much longer, leading to the conclusion that full relaxation of the polarized volume at -20°C is a very slow process. By modelling these results, it was found that de-polarization proceeds through thermal recombination and cannot be achieved effectively through modulation of the bias voltage, as previously thought [66]. The model can also be used as a diagnostic tool with which to understand and ultimately mitigate or eliminate the effects of polarization in semiconductor detectors. For example, by simulating pump-and-pulse synchrotron data, trap densities, occupancy rates and ionization

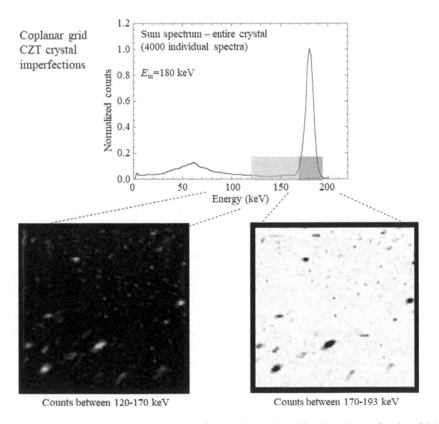

FIGURE 7.31 Pulse height data from repeated 60 keV "probe" scans of a polarized volume following a "pump" pulse of 2.4×10^6 60 keV photons. The figures show the evolution of the 60 keV energy–loss peak: (a) Peak channel position (initial value 740), (b) Relative efficiency (initial value unity), (c) FWHM energy resolution (initial value 27) and (d) Pulse height spectrum as a function of elapsed time (taken from [63]).

energies can be derived together with a host of previously experimentally illusive transport properties, such as hole mu-tau products. Such data may aid in the identification of the particular traps involved. Even if material shortcomings cannot be avoided or compensated for during the growth process, it should now be possible to use the simulation to custom design susceptible detection systems to operate in specific radiation environments.

7.8.13 Defect Metrology

Several high-resolution imaging techniques based on X-ray diffraction [67] are particularly well suited to the study of structural defects in materials.[11] For example, double crystal RC (DCRC) topography is particularly effective in studying local stochiometric variations by probing X-ray reciprocal space at discrete points across the crystal surface. This is achieved by raster scanning a finely collimated monochromatic X-ray beam across a crystal sample and recording individual topographs at discrete points by oscillating the crystal through the Bragg diffraction condition. Generally, the crystal is scanned across a few cm^2 area and the RC peak position and FWHM measured as a function of position on the crystal. These can then be correlated with local stoichiometry variations and other local defects. However, this technique is very time consuming since the quality of the data obtained, both statistically and spatially, depends on the RC acquisition time and the spatial sampling density (which is generally dependent on the size of the beam).

More commonly used is X-ray tomography (XRT), which is a variant of X-ray imaging that makes use of diffraction contrast, rather than absorption contrast. In this technique, a spatially extended X-ray beam incident on a sample gives rise to a diffracted beam that can be viewed in reflection (Bragg geometry) or transmission (Laue geometry) by a large area imaging detector. Reflection XRT is sensitive to the microstructure of the surface region (within ~1–10 μm) and is particularly useful in assessing the effectiveness of chemical and mechanical treatments used during detector fabrication. High-energy transmission XRT, on the other hand, provides a powerful method for studying the bulk properties of thick (up to several mm) crystals; from these detectors are fabricated and can thus directly probe the relationship between growth and process-induced structural defects and detector performance. XRT works as follows: A homogeneous sample with a perfect crystal lattice will give rise to a homogeneous intensity distribution across the topograph. Intensity modulations (topographic contrast) arise from deviations in long-range atomic order, which in turn arise from irregularities in the crystal lattice, caused by various defects – such as grain boundaries, sub-grain boundaries, dislocations, voids, cracks, secondary phases and particularly strain fields. However, it should be noted that for many defects such as dislocations, topography is not directly sensitive to the defects themselves, but to the strain field surrounding them. The method is non-destructive and relatively rapid. Measurements are usually carried out using the white beam (WBXRT) which has a number of advantages: Sample orientation is not necessary, several reflections are obtained with one exposure and all crystal parts are visible simultaneously with good geometrical resolution. Another particular advantage of transmission XRT is that measurements can be carried out after the contacts have been formed, enabling simultaneous *in-situ* X-ray and XRT maps to be acquired. However, there are several drawbacks, namely, a limited sensitivity to weak distortions, higher harmonics images overlap and not the least, the heat load on the sample. In addition, images can be difficult to interpret without comparison with standard analytical techniques such as chemical etching, ultrasound, infrared (IR) transmission, TEM or cathodoluminescence. This is illustrated in Fig. 7.29, in which we show a comparison of defect metrology data from a 25×25×2 mm^3 CdZnTe crystal taken using three different imaging modalities [62]. These are (a) an IR transmission map, (b) an X-ray response map and (c) a WBXRT topograph. The X-ray map was obtained by operating the crystal as a simple planar detector and raster scanning with a collimated beam while recording pulse height data as a function of spatial position. Thus, detector performance can be correlated with known defects revealed in the IR transmission map. From Fig. 7.29(a) and (b) we see a strong correlation between structure in the IR map (corresponding to inclusions and decorated grain boundaries) and detector performance, reflected in the X-ray map. It is not so apparent in the WBXRT topograph, which is primarily responsive to individual domains, which, in this case, exhibit a uniform X-ray response, separated by grain boundaries. Topographical contrast inside each domain should be related to tellurium inclusions (not clear in this case).

For detectors with sufficient spectral acuity, fine spatial scanning provides higher quality defect information *via* their impact on local charge collection efficiency. This is illustrated in Fig. 7.30 in which we show spatial maps in two spectra regions of a large 15×15×10 mm^3 CdZnTe coplanar grid detector [55]. The map was obtained by raster scanning a 20×20 $μm^2$, normally incident 180 keV beam in 40 μm steps across the detector. From the figure, crystal defects (mainly Te inclusions) are plainly evident in the detector count rate response and account for ~1% of the active area of the detector. What is interesting is that spatially resolved spectroscopy shows that spectra acquired within the defects are essentially the same as those acquired outside the defects but shifted to lower energies. In other words, the effect of the defects is not to distort the spectra but to reduce the charge collection efficiency. Coupled with cross edge, differential absorbtometry, the type of inclusion can even be determined. For example, den Hartog *et al.* [68] used such a technique to probe defects in a CdZnTe ring detector and showed that localized enhancements of Te could be recognized down to a level of < 0.1% by composition.

---

[11] Tomographic techniques are not considered here since they tend to be complicated, time consuming, expensive and more importantly, not readily available for routine detector metrology.

Furthermore, by using spatial and depth information, they tentatively identify one extended defect to be a decorated Te grain boundary with a volume of extent 40 μm thick, located 45 μm below the surface.

7.8.14 X-Ray Absorption Fine Structure (XAFS) Metrology

X-ray Absorption Fine Structure measurements (XAFS) are routinely carried out at synchrotron facilities to probe both short- and long-range order in materials. XAFS is a generic term and can be broken down into structure originating far from an absorption edge and structure originating close to the edge, which arise from different processes. Extended X-ray Absorption Fine Structure (EXAFS) is a diagnostic of short-range order by means of which details in the local geometry (atom types, bond lengths and bond angles) around the photo-absorbing atom can be extracted from far-edge spectra. X-ray Absorption Near Edge Structure (XANES), on the other hand, is a diagnostic of long-range order through which details of atom types and how they are structured collectively (the coordination environment) can be extracted from near-edge spectra. XAFS measurements, both EXAFS and XANES, can also be applied to detector metrology. For example, several authors have noted the wide spread in the radiation detection properties of $Cd_{1-x}Zn_xTe$ crystals and have attempted to use the structural information embedded in XAFS to find a link between performance and structural perfection.

7.8.15 Structural Studies

Wu *et al.* [69] carried out XAFS studies around the Zn K-edge for a number of alloys with x ranging from 0.005 to 1.0 to assess the effects of zinc segregation and defect formation on the local atomic structure of CdZnTe. Ideally, the alloy $Cd_{1-x}Zn_xTe$ can be thought of as a CdTe crystal with Zn atoms randomly substituting a fraction x of Cd atoms. By Vegard's law [6], the substitution of Zn atoms should be accompanied by a change in the lattice constant linearly varying between CdTe ($x=0$) and ZnTe ($x=1$). However, a Fourier analysis of EXAFS data in wavevector phase space revealed a bimodal distribution of bond lengths, suggesting distortion of the Te sub-lattice. If true, this suggests that a linear interpolation of lattice constants with zinc fraction is, at best, an approximation.

Duff *et al.* [70] carried out a similar analysis on two CdZnTe crystals of predetermined spectral quality – one of high spectral performance and one of low spectral performance. Unlike previous analyses, the crystals used were intact and the XAFS measurements carried out in transmission. No significant differences were found in the local atomic structure of the three primary elements comprising the crystals. The derived Debye–Waller factors, a measure of disorder in crystalline material, are essentially the same for each element. The authors conclude that spectral performance in CdZnTe is more intimately linked to other factors, such as the presence of secondary phases (precipitates and inclusions), which although known to limit local electron mobility, do not significantly degrade their bulk performance. In a follow-up study of secondary phases, Duff *et al.* [71] identified two dominant secondary phase morphologies. The first consists of numerous pyramidal shaped empty voids of extent 20 μm. The other consists of 20 μm hexagonal shaped bodies, which are composites of metallic Te layers that contain tear-drop shaped cores of polycrystalline CdZnTe.

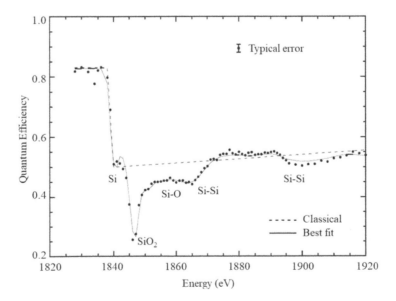

FIGURE 7.32 The measured quantum efficiency across the Si K-edge of an X-ray CCD [72]. Individual edges and bonds are identified. For comparison, we show our calculated values based on new linear absorption coefficients abstracted from photocurrent measurements along with the classical predictions of Cromer and Liberman [73].

7.8.16 Topographical and Surface Studies

An XAFS analysis of a semiconductor surface is a powerful tool with which to examine nanostructures such as interfacial layers and heterostructures. For example, Owens [72] described a method for determining the compositions, abundances and thicknesses of dead layers above a silicon CCD soft X-ray detector using the structural information embedded in XANES around the various K-edges. This was achieved fitting the expected efficiency function, given by

$$QE = \frac{1}{p}\left[1 - \exp(-\mu_d d)\right] \int_0^P \prod_{i=1,n} \exp\left[-a_i(x)(\mu_e)_i t_i\right] dx, \tag{73}$$

to the measured quantum efficiency curve shown in Fig. 7.32. Here, p is the pixel size, $(\mu_e)_i$ is the absorption coefficient of the i_{th} component of the electrode structure and t_i its thickness. The first term in brackets in Eq. (73) accounts for interactions in the active depletion depth of the device while the second product term accounts for absorption in the various overlying materials, such as gates, dielectrics and passivation layers. Because of the complicated nature of the pixel structure (see Fig. 7.33), the variation in material across a pixel is accounted for by a weighting factor, a_i. The absorption cross-sections (including XAFS) used in the calculation were derived from the photocurrent measurements of Owens *et al.* [74] using the following expression:

$$P(E) = \mu(E)[a_1 + a_2 E], \tag{74}$$

where $P(E)$ is the total photo-yield at energy, E and $\mu(E)$ is the corresponding absorption coefficient. The constants a_1 and a_2 are determined by normalizing $\mu(E)$ to the classical values far enough above and below the edges to be free of the effects of XAFS. The applicability of Eq. (74) has been verified by Owens *et al.* [75] who showed that the linear relationship holds over wide energy ranges to a precision of at least the few percent level. The derived linear attenuation coefficients for Si-c, Si-a, SiO_2 and Si_3N_4 are shown graphically in Fig. 7.34.

Eq. (73) was then fitted to the measured quantum efficiency using a non-linear minimization routine allowing the thicknesses of overlying materials to be free parameters. The manufacturer's values were used as the initial inputs, which although only known to an accuracy of $\sim\pm20\%$, can be reproduced to precisions of a percent. The calculated quantum efficiency based on best-fit values is shown graphically in Fig. 3.32, from which we can see there is excellent agreement with measurement. In fact, the residuals display no global systematic trends with energy yielding an average error of $\sim1.4\%$. The errors in the fitted layer thicknesses were typically < 10%, which is about 50% of that derived from process control metrology.

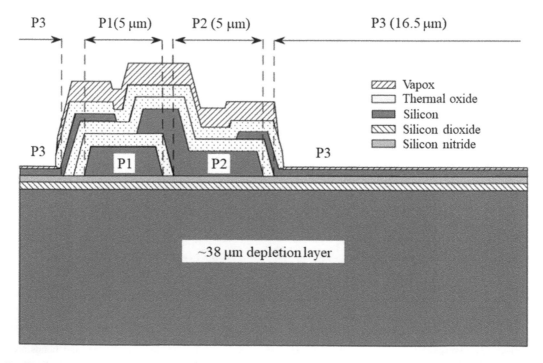

FIGURE 7.33 Detailed cross-section through the overlying dead layers (electrodes, gate dielectrics, polysilicon gates and passivation layers) above the active depletion region (from [72]).

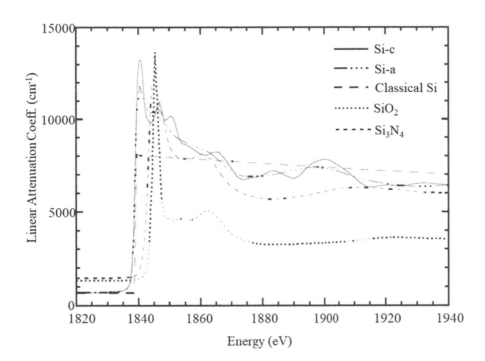

FIGURE 7.34 The derived linear attenuation coefficients across the Si K-edge. For Si, the letters *c* and *a* refer to crystalline and amorphous (taken from [74]). We also show the "classical" Si curve based on the calculation of Cromer and Liberman [73].

The above analysis, illustrates the power of XANES metrology. In addition, in contrast to scanning electron microscopy (SEM), which only yields information on the linear extent of surface features, XANES, can also provide elemental and bonding information – potentially to a precision of 1 atom in 10^{10}. By combining the structural information contained in the quantum efficiency measurement with X-ray Photoelectron Spectroscopy (XPS), it should be possible to isolate and image surface features using XANES tomography by focusing on a specific near edge structure. Such a technique is a powerful diagnostic tool with which to explore macroscopic surface structures, offering distinct advantages over traditional techniques, such as SEM, in that it is non-invasive and non-destructive.

REFERENCES

[1] M.H. Yang, M.L. Lee, J.J. Shen, H.L. Hwang, "Precise compositional and trace-elemental analysis by chemical methods in compound semiconductors", *Mater Sci Eng: B*, Vol. **12**, no. 3 (1992), pp. 253–260.

[2] J.R. Dean, "*Practical Inductively Coupled Plasma Spectroscopy*", John Wiley & Sons, Hoboken, NJ (2005) ISBN-13: 978-0-470-09348-1.

[3] T. Nelis, R. Payling, "*Glow Discharge Optical Emission Spectroscopy: A Practical Guide*", Royal Society of Chemistry, Cambridge UK (2004) ISBN 0-85404-521-X.

[4] W.L. Bragg, "The diffraction of short electromagnetic waves by a crystal", *Proc. Camb. Phil. Soc.*, Vol. **17** (1913), pp. 43–57.

[5] J. Faber, T. Fawcett, "The powder diffraction file: Present and future", *Acta Cryst.*, Vol. **58**, part 3, no. 1 (2002), pp. 325–332.

[6] L. Vegard, "Die Konstitution der Mischkristalle und die Raumfüllung der Atome", *Z. Phys.*, Vol. **5** (1921), pp. 17–26.

[7] M.S. Goorsky, H. Yoon, M. Schieber, R.B. James, D.S. McGregor, M. Natarajan, "X-ray diffuse scattering for evaluation of wide bandgap semiconductor nuclear radiation detectors", *Nucl. Instr. Meth.*, Vol. **A380** (1996), pp. 6–9.

[8] E.H. Rhoderick, "*Metal-Semiconductor Contacts*", Clarendon Press, Oxford, 2nd edition (1978), p. 87.

[9] C.R. Crowell, V.L. Rideout, "Normalised thermionic-field emission in Schottky barriers", *Solid State Electron.*, Vol. **12** (1969), pp. 89–105.

[10] M. Missous, E.H. Rhoderick, "New way of plotting current/voltage characteristics of Schottky diodes", *Electronic Letts.*, Vol. **22**, no. 9 (1986), pp. 477–478.

[11] A.A. Bergh, P.J. Dean, "Light-emitting diodes", *Proc. IEEE*, Vol. **60** (1972), pp. 156–224.

[12] S. Montanari, "Fabrication and characterisation of planar Gunn diodes for monolithic microwave integrated circuits", PhD thesis, University of Aachen RWTH (2005).

[13] F. Wenner, "A method of measuring earth resistivity", *Bur. Stand. (U.S.), Bull.*, U.S. Bur. of Stand., Washington, D.C, Vol. **12** (1915), pp. 469–478.

[14] G.S. Marlow, M.B. Das, "The effects of contact size and non-zero metal resistance on the determination of specific contact resistance", *Solid-State Electron.*, Vol. **25** (1982), pp. 91–94.

[15] G. Dresselhaus, A.F. Kip, C. Kittel, "Cyclotron resonance of electrons and holes in Silicon and Germanium crystals", *Phys. Rev.*, Vol. **98** (1955), pp. 368–384.

[16] E.H. Hall, "On a new action of the magnet on electric currents", *Am. J. Math.*, Vol. **2** (1879), pp. 287–292.

[17] L.J. van der Pauw, "A method of measuring specific resistivity and Hall effect of discs of arbitrary shape", *Technical Report, Philips Res. Reports*, Vol. **13** (1958), pp. 1–9.

[18] H.A. Lorentz, "La Théorie Électromagnétique de Maxwell et son Application aux Corps Mouvants", *Arch. Ne´Erl.*, Vol. **25** (1892), pp. 363–552.

[19] *"Appendix A: Hall Effect Measurements"*, Lake Shore 7500/9500 Series Hall System User's Manual, Lake Shore Cryotronics, Inc., Westerville, OH (1996).

[20] S. Ramo, "Currents induced by electron motion", *Proc. IRE*, Vol. **27** (1939), pp. 584–585.

[21] W. Shockley, "Currents to conductors induced by a moving point charge", *J. Appl. Phys.*, Vol. **9** (1938), pp. 635–636.

[22] P. De Antonis, E.J. Morton, F.J.W. Podd, "Infra-red microscopy of Cd(Zn)Te radiation detectors revealing their internal electric field structure under bias", *IEEE Trans. Nucl. Sci.*, Vol. **43**, no. 3 (1996), pp. 1487–1490.

[23] S. Namba, "Electro-optical effect of zincblende", *J. Opt. Soc. Am.*, Vol. **51** (1961), pp. 76–79.

[24] A. Cola, I. Farella, "Electric field and current transport mechanisms in Schottky CdTe X-ray detectors under perturbing optical radiation", *Sensors*, Vol. **13** (2013), pp. 9414–9434.

[25] A. Cola, I. Farella, A.M. Mancini, A. Donati, "Electric field properties of CdTe nuclear detectors", *IEEE Trans. Nucl. Sci.*, Vol. **54**, no. 4 (2007), pp. 868–872.

[26] D. Blackie, C. Shenton-Taylor, V. Perumal, A. Lohstroh, "Exploration of the pockels effect for radiation detection applications", *2012 IEEE Trans. Nucl. Sci., NSS/MIC Conf. Rec.*, paper **NI-189** (2012), pp. 402–404.

[27] J.R. Haynes, W. Shockley, "The mobility and life of injected holes and electrons in Germanium", *Phys. Rev.*, Vol. **81**, no. 5 (1951), pp. 835–843.

[28] J.E. Baciak, Z. He, "Spectroscopy on thick HgI_2 detectors: A comparison between planar and pixelated electrodes", *IEEE Trans. Nucl. Sci.*, Vol. **50**, no. 4 (2003), pp. 1220–1224.

[29] K. Hecht, "Zum Mechanismus des lichtelektrischen Primärstromes in isolierenden Kristallen", *Z. Physik*, Vol. **77** (1932), pp. 235–245.

[30] H. Kim, L. Cirignamo, A. Churliov, G. Ciampi, W. Higgins, F. Olschner, K. Shah, "Developing larger TlBr detectors – Detector performance", *IEEE Trans. Nucl. Sci.*, Vol. **56**, no. 3 (2009), pp. 819–823.

[31] M.C. Veale, P.J. Sellin, A. Lohstroh, A.W. Davies, J. Parkin, P. Seller, "X-ray spectroscopy and charge transport properties of CdZnTe grown by the vertical Bridgman method", *Nucl. Instr. Meth.*, Vol. **A576** (2007), pp. 90–94.

[32] A. Ruzin, Y. Nemirovsky, "Statistical models for charge collection and variance in semiconductor spectrometers", *J. Appl. Phys.*, Vol. **82**, no. 6 (1997), pp. 2754–2758.

[33] Y. Cui, G.W. Wright, X. Ma, K. Chattopadhyay, R.B. James, A. Burger, "DC photoconductivity study of semi-insulating $Cd_{1-x}Zn_xTe$ crystals", *J. Of Elect. Mat.*, Vol. **30**, no. 6 (2001), pp. 774–778.

[34] A. Many, "High-field effects in photoconducting cadmium sulphide", *J. Phys. Chem. Solids*, Vol. **26**, no. 3 (1965), pp. 575–578.

[35] M.G. Buehler, "Impurity centers in PN junctions determined from shifts in the thermally stimulated current and capacitance response with heating rate", *Solid-State Electron.*, Vol. **15** (1972), pp. 69–79.

[36] K.H. Nicholas, J. Woods, "The evaluation of electron trapping parameters from conductivity glow curves in cadmium sulphide", *British. J. Appl. Phys.*, Vol. **15** (1964), pp. 783–795.

[37] P.D. Townsend, Y. Kirsh, "Spectral measurement during Thermoluminescence - an essential requirement", *Contemporary Physics*, Vol. **30** (1989), pp. 337–354.

[38] D.V. Lang, "Deep level transient spectroscopy: A new method to characterize traps in semiconductors", *J. Appl. Phys.*, Vol. **45** (1974), pp. 3023–3032.

[39] P.M. Mooney, "Defect identification using capacitance spectroscopy", in *Identification of Defects in Semiconductors*, (Semiconductors and Semimetals), ed. M. Stavola, Academic Press, San Diego (1999), pp. 93–152.

[40] P.M. Mooney, "Deep donor levels (DX centers) in III-V semiconductors", *J. Appl. Phys.*, Vol. **67** (1990), pp. R1–R26.

[41] R.H. Bube, *"Photoconductivity of Solids"*, Wiley, New York (1960).

[42] C. Hurtes, M. Boulou, A. Mitonneau, D. Bois, "Deep-level spectroscopy in high-resistivity materials", *Appl. Phys. Lett.*, Vol. **32**, no. 12 (1978), pp. 821–823.

[43] J.C. Balland, J.P. Zielinger, C. Noguet, M. Tapiero, "Investigation of deep levels in high-resistivity bulk materials by photo-induced current transient spectroscopy. I. Review and analysis of some basic problems", *J. Phys. D: Appl. Phys.*, Vol. **19** (1986), pp. 57–70.

[44] M. Tapiero, N. Benjelloun, J.P. Zielinger, S. El Hamd, C. Noguet, "Photoinduced current transient spectroscopy in high resistivity bulk materials: Instrumentation and methodology", *J. Appl. Phys.*, Vol. **64** (1988), pp. 4006–4012.

[45] R.E. Kremer, W.B. Leigh, "Deep levels in CdTe", *J. Crys. Growth*, Vol. **86**, no. 1–4 (1990), pp. 490–496.

[46] A. Owens, "Synchrotron light sources and detector metrology", *Nucl. Instr. Meth.*, Vol. **A695** (2012), pp. 1–12.

[47] R. Klein, G. Ulm, M. Abo-Bakr, P. Budz, K. Bürkmann-Gehrlein, D. Krämer, J. Rahn, G. Wüstefeld, "The metrology light source of the Physikalisch-Technische Bundesanstalt in Berlin-Adlershof", in *Proc. EPAC 2004*, L. Rivkin, Chairman of the EPAC'04 Organizing Committee, Lucerne, Switzerland (2004), pp. 2290–2292.

[48] J. Schwinger, "On the classical radiation of accelerated electrons", *Phys. Rev.*, Vol. **75** (1949), pp. 1912–1925.

[49] F.R. Elder, A.M. Gurewitsch, R.V. Langmuir, H.C. Pollock, "Radiation from Electrons in a Synchrotron", *Phys. Rev. B.*, Vol. **71** (1947), pp. 829–830.

[50] A. Hofmann, "*The Physics of Synchrotron Radiation*", Cambridge University Press, Cambridge (2004) ISBN-10 9780521308267.

[51] H. Winick, D. Attwood, "X-Ray Data Booklet", Section 2.3 Operating and planned facilities, *LBNL/PUB-940 Rev.*, Vol. **3** (2009), pp. 2–29.

[52] www.diamond.ac.uk/

[53] K. Kim, "X-ray data booklet", Section 2.1 Characteristics of Synchrotron Radiation, *LBNL/PUB-940 Rev.*, Vol. **3** (2009), pp. 2–15.

[54] A. Owens, A.J.J. Bos, S. Brandenburg, P. Dorenbos, W. Drozdowski, R.W. Ostendorf, F. Quarati, A. Webb, E. Welter, "The hard X-ray response of Ce-doped lanthanum halide scintillators", *Nucl. Instr. Meth.*, Vol. **A574** (2007), pp. 158–162.

[55] A. Owens, T. Buslaps, V. Gostilo, H. Graafsma, R. Hijmering, A. Kozorezov, A. Loupilov, D. Lumb, E. Welter, "Hard X- and γ-ray measurements with a large volume coplanar grid CdZnTe detector", *Nucl. Instr. Meth.*, Vol. **A563** (2006), pp. 242–248.

[56] www.esrf.eu/UsersAndScience/Experiments/StructMaterials/ID15

[57] C. Ponchut, "Characterization of X-ray area detectors for synchrotron beamlines", *J. Synch. Rad.*, Vol. **13** (2006), pp. 195–203.

[58] D. Martin, A. Owens, C. Erd, S. Andersson, A. Peacock, H. Andersson, V. Lamas, S. Nenonen, "High Resolution X-ray spectroscopy using a large format GaAs array", *Proc. SPIE*, Vol. **4507** (2001), pp. 152–161.

[59] A. Owens, M. Bavdaz, H. Andersson, T. Gagliardi, M. Krumrey, S. Nenonen, A. Peacock, I. Taylor, L. Tröger, "The X-ray response of CdZnTe", *Nucl. Instr. Meth.*, Vol. **A484** (2002), pp. 242–250.

[60] A. Owens, H. Andersson, M. Bavdaz, G. Brammertz, C. Erd, T. Gagliardi, V. Gostilo, N. Haack, I. Lisjutin, S. Nenonen, A. Peacock, H. Sipila, I. Tay, S. Zataloka, "Development of compound semi-conductor arrays for X- and Gamma-ray spectroscopy", *Proc. SPIE*, Vol. **4507** (2001), pp. 42–49.

[61] C. Erd, A. Owens, G. Brammertz, D. Lumb, M. Bavdaz, A. Peacock, S. Nenonen, H. Andersson, "Measurements of the quantum efficiency and depletion depth in gallium-arsenide detectors", *Proc. SPIE*, Vol. **4784** (2002), pp. 386–393.

[62] G. Carini, G. Camarda, Z. Zhong, D. Siddons, A. Bolotnikov, G. Wright, B. Barber, C. Arnone, R. James, "High-energy X-ray diffraction and topography investigation of CdZnTe", *J. Elec. Mat.*, Vol. **34** (2005), pp. 804–810.

[63] A. Kozorezov, V. Gostilo, A. Owens, F. Quarati, M. Shorohov, M.A. Webb, J.K. Wigmore, "Polarization effects in thallium bromide x-ray detectors", *J. Appl. Phys.*, Vol. **108** (2010), pp. 1-1–1-10.

[64] D.S. Bale, C. Szeles, "Nature of polarization in wide-bandgap semiconductor detectors under high-flux irradiation: Application to semi-insulating $Cd_{1-x}Zn_xTe$", *Phys. Rev. B*, Vol. **77** (2008), pp. 035205–035221.

[65] G.S. Camarda, A.E. Bolotnikov, Y. Cui, A. Hossain, R.B. James, "Polarization studies of CdZnTe detectors using synchrotron X-ray radiation", *IEEE Trans. Nucl. Sci.*, Vol. **55**, no. 6 (2008), pp. 3725–3730.

[66] K.G. Mckay, "Electron bombardment conductivity in diamond", *Phys. Rev.*, Vol. **74** (1948), pp. 1606–1621.

[67] S. Weissman, F. Balibar, J.-F. Petroff, "*Applications of X-Ray Topographic Methods to Materials Science*", Plenum Press, New York (1984).

[68] R. Den Hartog, A.G. Kozorezov, A. Owens, J.K. Wigmore, V. Gostilo, A. Loupilov, V. Kondratjev, M.A. Webb, E. Welter, "Synchrotron study of charge transport in a CZT ring-drift detector", *Nucl. Instr. Meth.*, Vol. **A648** (2011), pp. 155–162.

[69] Y.L. Wu, Y.-T. Chen, Z.C. Feng, J.-F. Lee, P. Becla, W. Lu, "Synchrotron radiation x-ray absorption fine-structure and Raman studies on CdZnTe ternary alloys", *Proc. SPIE*, Vol. **7449** (2009), pp. 74490Q-1–74490Q-11.

[70] M.C. Duff, D.B. Hunter, P. Nuessle, D.R. Black, H. Burdette, J. Woicik, A. Burger, M. Groza, "Synchrotron X-ray based characterization of CdZnTe crystals", *J. Elec. Mat.*, Vol. **36**, no. 8 (1010), pp. 1092–1097.

[71] M.C. Duff, D.B. Hunter, A. Burger, M. Groza, V. Buliga, J.P. Bradley, G. Graham, Z. Dai, N.E. Teslich, D.R. Black, H. Burdette, A. Lanzirotti, "Characterization of spatial heterogeneities in detector grade CdZnTe", *J. Mat. Res.*, Vol. **24**, no. 4 (2009), pp. 1361–1367.

[72] A. Owens, "XANES fingerprinting: A technique for investigating CCD surface features and measuring dead layer thicknesses", *Nucl. Instr. Meth.*, Vol. **A526** (2004), pp. 391–398.

[73] D.T. Cromer, D. Liberman, "Relativistic calculation of anomalous scattering factors for x rays", *J. Chem. Phys.*, Vol. **53** (1970), pp. 1891–1898.

[74] A. Owens, G.W. Fraser, S.J. Gurman, "Near K-edge linear attenuation coefficients for Si, SiO_2 and Si_3N_4", *Rad. Phys. Chem.*, Vol. **65** (2002), pp. 109–121.

[75] A. Owens, S. Bayliss, G.W. Fraser, S.J. Gurman, "On the relationship between total electron photoyield and X-ray absorption coefficient", *Nucl. Instr. Meth.*, Vol. **A358** (1997), pp. 556–558.

8

Radiation Detection and Measurement

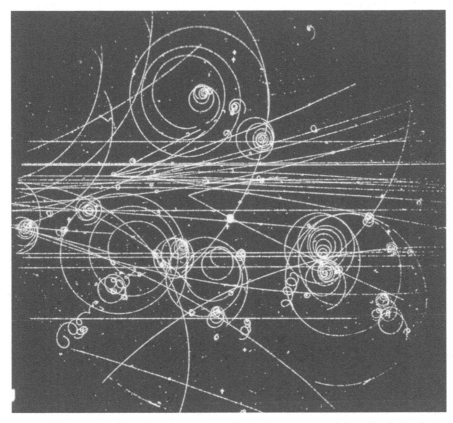

Credit: CERN Photolab: Particle tracks from the decay of 16 GeV pions captured in a liquid hydrogen bubble chamber.

CONTENTS

8.1   Interaction of Radiation with Matter

Detectors detect radiation by recording energy deposition in their active components. For most detectors, this energy deposition is in the form of ionization produced in the detection medium (which may be solid, liquid or gas) by charged particles. The choice of which detection medium to use depends to a large extent on what the detector will be used for. For example, in a tracking detector one wishes to detect the presence of a particle without affecting its trajectory, so the medium will be chosen to minimize energy loss and particle scattering (thus, low density). Conversely, if one wishes to measure the total energy deposition by calorimetry or spectroscopy, the absorber will be chosen to maximize energy loss, for example, by high density or high atomic number. Energy is then converted into an electrical signal, either directly or indirectly. In direct energy conversion, the incident radiation produces charge in the detector which is directly proportional to the energy absorbed and is collected by an electrode system. For example, in a gas counter the radiation ionizes the atoms/molecules of the gas and the resulting charge is collected by electrodes. Similarly, in a semiconductor detector, the ionization produced by the radiation will create electron-hole pairs that are swept towards the electrodes by an electric field. In indirect conversion, incident radiation excites atomic or molecular states that decay by the emission of light, as in the case of scintillation detectors. This light is then converted into an electrical signal using a photosensitive sensor, such as a photomultiplier tube.

The spectroscopic power of these systems or their ability to resolve different energies depends directly on how many "information carriers" are generated. For a scintillator, it takes ~20 to 500 eV of energy deposition to create a single scintillation photon, whereas in a gas counter it takes 30 eV of energy to create an electron-ion pair and in a semiconductor between 3 and 10 eV of energy to create an electron-hole pair.

8.2   Charged Particles

When charged particles pass through matter, they lose energy primarily by the ionization and excitation of the atoms and molecules of the medium and through nuclear interactions with the atoms and nuclei of the material. The former processes dominate at low and intermediate energies whilst the latter dominates at the highest energies. It is the process of ionization we consider here. The basic theory was first developed by Bohr [1] using a classical approach and later using quantum mechanics by Bethe [2] and Bloch [3]. Physically, the electrostatic interactions between the traversing particles and the electrons of the atoms composing the medium result in the electrons being ionized. Eventually, the particles are slowed down and brought to rest by the small, but almost continuous, transfer of kinetic energy to the electrons. The energy loss per unit path length of a charged particle of charge z in any material can be described by the Bethe [2] equation

$$-\left(\frac{dE}{dx}\right) = \frac{4\pi\, e^4 z^2}{m_0 v^2} NZ \left[\ln\frac{2m_0 v^2}{I} - \ln\left(1 - \beta^2\right) - \beta^2\right] \quad \text{ergs\,cm}^{-1}, \tag{1}$$

where $v = \beta c$ is the velocity of a particle whose mass, M, is much larger than m_0, the electron mass, in a medium containing N atoms cm^{-1} of atomic number Z and I is a constant for the material derived from the Thomas-Fermi model [4,5], which is proportional to Z and close to the mean ionization potential. The quantity dE/dx is often referred to as the stopping power of the material.

In Fig. 8.1 we show dE/dx plotted as a function of β for a number of common materials and for three particle types (protons, pions and muons). Note the similarity of the curves. The various terms in the formula arise as follows. For non-relativistic energies ($\beta << 1$), only the first term in the square brackets is significant and the energy loss is proportional to $1/\beta^2$, that is, it decreases with energy, whereas for relativistic particles the rate of energy loss increases with $\ln \beta$. For small values of β, the two relativistic correction terms in the square brackets can be approximated by $\beta^4/2$. As β increases, the loss

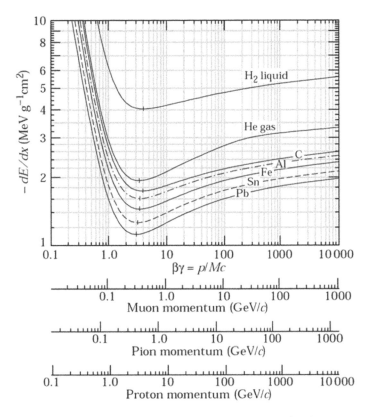

FIGURE 8.1 The mean energy loss of a charged particle (dE/dx) per unit path length (in MeV g^{-1} cm^2) for a range of materials, as derived from the Bethe-Bloch formula (from [6]). To a first approximation, energy loss in a material is simply characterized by Z/A.

dE/dx decreases, mainly due to the $1/v^2$ term outside the brackets. As v approaches c the ionization losses pass through a broad minimum and then increase logarithmically with energy. It is interesting to note that the minimum ionization rate occurs at a Lorentz factor of $\gamma \sim 3$ for all materials that correspond to $E = Mc^2$. A good approximation for $(dE/dx)_{min}$ for many nuclear species is given by $\sim 0.2\ z^2$ MeV (kg m^{-2})$^{-1}$. A particle at this energy is known as a minimum ionizing particle, and in fact most relativistic particles have energy loss rates close to the minimum. For singly charged particles at minimum ionization, the energy loss varies from approximately 2 MeV g^{-1}cm^{-1} for light elements to 1 MeV g^{-1}cm^{-1} for heavy elements. Thus we can see why light materials are more efficient per gm in stopping protons than high mass materials. The Bethe formula begins to break down at both low energies and high energies. At low energies ($\beta\gamma \leq 0.05$) charge exchange between the particle and absorber becomes important. Positively charged particles will tend to pick up electrons from the absorber, which effectively reduces their charge and consequently their linear energy loss. At high energies($\beta\gamma \geq 1000$), radiative effects become important, particularly for pions and muons.

In comparing different materials as absorbers, dE/dx depends primarily on NZ, which lies outside the logarithmic term. High atomic number, high-density materials will result in greater stopping powers. The total energy loss in any material is calculated by solving Eq. (1) iteratively for many differential thicknesses. In practice it is found that the solution converges to < 5% for differential thicknesses less than a few mm. Note: the thicknesses of the active regions of a lot of semiconductor detectors rarely exceed a few mm. However, for X-ray detection, the active regions are usually less than 100 microns thick, and it has been known for some time that the measured distribution of energy losses caused by minimally ionizing particles in thin absorbers is far broader than that predicted by Landau [7] (see reference [8]). The energy spectrum observed in these layers resembles a Gaussian distribution with long upper tail resulting from a small number of delta electrons that have experienced a large energy transfer from the primary particle. For protons, the effect is most pronounced for energies > 1 GeV and absorbers < 200 microns, which is particularly troublesome for X-ray sensitive Si CCDs as the active regions are typically only ~30 μm – 100 μm thick [9]. Based on the classical theory of Landau [7], it had been assumed that charged particles would deposit far more energy in the active volume than the upper X-ray energy threshold and could therefore be easily separated from valid X-ray events. However, since the actual distribution of energy losses is about three times broader than predicted by classical theory, there is a real possibility that a substantial fraction of charged particles can deposit energy within the operating energy range of the CCD and thus masquerade as X-rays. Fortunately, charged particle contamination can be reduced to negligible proportions by taking into account the spatial distribution of the deposited charge, which tends to be different from that produced by X-rays [9].

The discrepancy between the predicted and measured energy loss distributions can be attributed to atomic electron binding effects and, to a lesser extent, δ-ray escape. The question of energy loss by fast particles in thin layers was first solved by Landau [7], who derived the expected energy-loss distribution by solving the integral transport equation

$$\frac{df}{dx}(x, \Delta) = \int_0^\infty w(E)\left[f(x, \Delta - E) - f(x, \Delta)\right] dE, \tag{2}$$

where $f(x, \Delta)$ represents the distribution function (the probability that the incident particle will lose an amount of energy Δ on traversing a layer of thickness x and $w(E)\, dE$ denotes the probability per unit path length of a collision transferring energy E to an electron in the material. Hall [10] pointed out that Eq. (2) may be conveniently solved by the Laplace transform method and the results written as

$$f(x, \Delta) = \frac{1}{2\pi i} \int_{c-i\infty}^{c+i\infty} e^I \, dp, \tag{3}$$

where

$$I = d\Delta - x \int_0^\infty w(E)\left(1 - e^{-pE}\right) dE. \tag{4}$$

Unfortunately, $w(E)$ is a complicated function of E and x and cannot be simply derived. Landau [6] was able to arrive at an approximate solution based on the free electron Rutherford cross-section. However, this was later found to break down for certain regions of parameter space, notably highly relativistic particles and thin absorbers. Close agreement with experiment was eventually achieved by applying kinematic constraints on the energy transfer, corrections for the density effect and for taking into account the fact that the electrons in the material are not free. As expected, the revised formula converges to the Landau form for most of parameter space. By accounting for the above corrections semi-empirically, Hall [10] was able to show that solution of Eq. (3) reduces to a simple convolution of the Landau distribution, f_L, with a normal distribution

$$f(x, \Delta) = \frac{1}{\sqrt{2\pi x \delta_2}} \int_{-\infty}^{+\infty} f_L(\Delta - y)\, e^{-y^2/2x\delta^2} \, dy, \tag{5}$$

whose variance $x\delta_2$ may be derived from the "bound electron" corrections of Shulek *et al.* [11] or Blunck and Leisegang [12] or alternately from the photoabsorption ionization model of Hall [10].

8.2.1 Energy Loss of Secondary Electrons – Collisional and Bremsstrahlung

In practical detection systems we must also consider energy losses by electrons, since these are the secondary electrons produced by photoelectron, Compton and pair processes, which transfer energy from the primary photons to the detection medium. For electrons and positrons, the Bethe-Bloch formula has to be modified to include radiative energy losses in the form of bremsstrahlung as well as collisional losses, due to their small masses. Thus, electrons lose energy through two processes. The energy loss per unit path length due to collisions is given by

$$-\left(\frac{dE}{dx}\right)_c = \frac{2\pi e^4 z^2}{m_0 v^2} NZ \left[\ln \frac{2m_0 v^2}{2I^2(1 - \beta^2)} - (\ln 2)\left(2\sqrt{1 - \beta^2} - 1 + \beta^2\right)\right.$$
$$\left. + (1 - \beta^2) + \frac{1}{8}\left(1 - \sqrt{1 - \beta^2}\right)^2\right]^2. \tag{6}$$

Radiative losses occur because the electron is constantly scattered and therefore deflected along its trajectory. The deflections cause acceleration, and therefore the electron will radiate electromagnetically from any position along the electron trajectory. The energy loss per unit path length in this case is given by

$$-\left(\frac{dE}{dx}\right)_{\rm r} = \frac{NEZ(Z+1)\,e^4}{137m_{\rm o}^2 c^4}\left[4\ln\frac{2E}{m_{\rm o}c^2}-\frac{4}{3}\right].\tag{7}$$

These losses are most important for high energies and heavy absorbers. The total electron energy loss is therefore the sum of the collisional and radiative losses

$$\left(\frac{dE}{dx}\right)_{\rm T} = \left(\frac{dE}{dx}\right)_{\rm c} + \left(\frac{dE}{dx}\right)_{\rm r}.\tag{8}$$

We can assess the relative importance of these two loss mechanisms in a detector by examining the ratio of the two:

$$\frac{(dE/dx)_{\rm r}}{(dE/dx)_{\rm c}} \approx \frac{EZ}{700},\tag{9}$$

where E is in units of MeV. For semiconductor detectors, the operational energy range of the detector will be < 1 MeV, and the average Z of the detection medium will be ~ 30–60. Therefore, radiative losses will be a factor of ~10 to 20 times lower than collisional losses and can generally be ignored.

8.3 Neutron Detection

Since neutrons are uncharged, they cannot be easily detected in conventional detectors and can only be detected indirectly by conversion into charged particles. In fact, they do not even interact directly with the electrons in matter, as gamma-rays do. It is the ionization in the detecting medium caused by these secondary charged particles that is detected. For fast neutrons, this can be achieved by inelastic scattering of detector nuclei, transferring some of their kinetic energy to the nuclei. If enough energy is transferred, the recoiling nucleus can ionize the material surrounding the point of interaction. The maximum energy transferred to a nucleus of atomic weight A by a neutron of kinetic energy E is given by

$$E_{\rm max} = \frac{4AE}{(A+1)^2}.\tag{10}$$

Thus, it can be seen that this mechanism is only efficient for neutrons interacting with light nuclei. For a single scattering event the actual energy transferred to the recoiling nucleus lies between 0 and $E_{\rm max}$ depending on the scattering angle and has equal probability for any value in this range. Because the neutron is uncharged, the reactions are essentially billiard ball collisions with geometric cross-sections. As such, scattering cross-sections do not vary very much from nucleus to nucleus and typically have values of a few barns (10^{-24} cm^2).

While neutrons do not interact with the charge distribution in atoms by means of long-range Coulomb interactions, they do interact with atomic nuclei through short-range nuclear force interactions. Slow or thermal neutrons ($E < 1$ eV) can be captured or absorbed by detector nuclei exciting them to a more energetic state. The capture probability varies inversely to the neutrons' velocity. The de-excitation products from these reactions, such as protons, alpha particles, gamma-rays, and fission fragments, can then initiate the detection process. Because the energy of the captured neutron is small compared to the Q-value of the reaction, the reaction products carry away an energy corresponding to the Q-value, which means that any knowledge of the incident neutron energy is lost. Capture or absorption cross-sections vary widely depending upon the distribution of nuclear energy levels so that nuclear resonance effects may be encountered. For example, the absorption cross-section for thermal neutrons on ^{12}C is about 0.0034 barns, whereas for ^{157}Gd it is 254,000 barns. In Table 8.1, we list thermal neutron capture reactions commonly exploited in conventional neutron detectors. Most of these materials are used either as converter coatings on gas counters or as dopants in scintillation materials. The table lists the Q value, or the total energy available from each reaction to create charged particles, as well as the total cross-section.

For most semiconductor detection media, both scattering and capture mechanisms are fairly inefficient. However, there are a number of boron compounds (BN, BP, B$_x$C), cadmium compounds (CdZnTe, CdTe, CdMnTe) and mercury compounds (HgI$_2$, HgCdTe) that may potentially be used for neutron detection by virtue of the high thermal neutron capture cross-sections of ^{10}B (3837 barns), ^{113}Cd (20,600 barns) and ^{199}Hg (2150 barns). Note, however, these are all isotopes – the effective cross-sections for "normal" isotopic compositions are for boron 749 barns, for cadmium 2450 barns and for mercury 384 barns.

TABLE 8.1

Nuclear reactions commonly exploited for thermal neutron detection.

Reaction products	Reaction Q value (MeV)	X-section [13,14] (barns)
$^3He + {}^1n \Rightarrow {}^3H(0.191\,MeV) + {}^1p(0.573\,MeV)$	0.76	5333
$^{10}B + {}^1n \Rightarrow \begin{cases} {}^7Li^*(0.84\,MeV) + {}^4\alpha(1.470\,MeV) + \gamma(0.480\,MeV) \\ {}^7Li^*(1.015\,MeV) + {}^4\alpha(1.777\,MeV) \end{cases}$	2.310 (1st excited state) 2.792 (to ground state)	3837
$^6Li + {}^1n \Rightarrow {}^3H(2.73\,MeV) + {}^4\alpha(2.05\,MeV)$	4.780	940
$^{113}Cd + {}^1n \Rightarrow {}^{114}Cd + \gamma(0.56MeV) + conv.\ electrons$	9.04	20,600
$^{155}Cd + {}^1n \Rightarrow {}^{156}Cd + \gamma(0.09\ 0.20, 0.30MeV) + conv.\ electrons$	8.54	60,900
$^{157}Cd + {}^1n \Rightarrow {}^{158}Cd + \gamma(0.08,\ 0.18, 0.28MeV) + conv.\ electrons$	7.94	254,000
$^{235}U + {}^1n \Rightarrow$ fission fragments	201	583
$^{239}Pu + {}^1n \Rightarrow$ fission fragments	160	748

8.4 X- and Gamma-Rays

Unlike charged particles, a well-collimated beam of photons shows a truly exponential absorption in matter, described by the well-known Lambert-Beer law [15,16]. This is because photons can only be absorbed or scattered in a single interaction. Both the type of scattering and absorption process depend on the frequency and therefore the quantum energy of the electromagnetic radiation involved. In essence, infrared, visible light and UV scattering depend on the vibrational and/or rotational properties of molecules, whereas X- and gamma-ray scattering is predominantly with atomic electrons. Similarly, the absorption of infrared radiation is restricted to exciting molecular vibrational and rotational states, whereas visible and UV light usually cause electron transitions in materials. XUV and higher frequency radiation (X- and gamma-ray), tend to be absorbed by ionization processes. In this section, we concentrate on X-and gamma-ray radiation in the range 1 keV to 10 MeV.

Fano [17,18] categorized the possible processes by which the electromagnetic field of an X- or gamma-ray photon can interact with matter. With reference to Table 8.2, there are 12 ways of combining columns 1 and 2; thus in theory there are 12 different processes by which X- or gamma-rays can be absorbed or scattered. Many of these are quite infrequent, and indeed some have yet to be observed. It turns out that X- and gamma-rays interact primarily through only four of the twelve processes listed. These are the photoelectric effect 1(a), elastic scattering 1(b), inelastic or Compton scattering 1(c) and pair production 3(a).

8.4.1 Photoelectric Effect

The photoelectric process involves the complete transfer of an incident photon's energy to an atomic electron, which is then ejected from the atom. The atom is left in a highly excited ionized state, which it then corrects for by spontaneously filling the hole by means of an electron transition from a higher energy level. Since the whole sequence of events occurs within 100 fs, the two processes, absorption and ejection, may be considered as a single interaction as illustrated in Fig. 8.2. Note that a free electron cannot absorb a photon and become a photoelectron, since a third body (the nucleus) is required to conserve momentum. Because the entire atom participates, the photoelectric effect may be visualized as an interaction of the primary photon with the atomic electron cloud, so that the entire energy of the incident photon is absorbed and an electron is ejected

TABLE 8.2

Possible photon interaction mechanisms in matter and the net effect.

Interaction	Effect
1. Interaction with electrons	(a) Complete absorption
2. Interaction with nuclei	(b) Elastic scattering (coherent)
3. Interaction with the electric field surrounding nuclei or electrons	(c) Inelastic scattering (incoherent)
4. Interaction with the meson field surrounding nucleons	

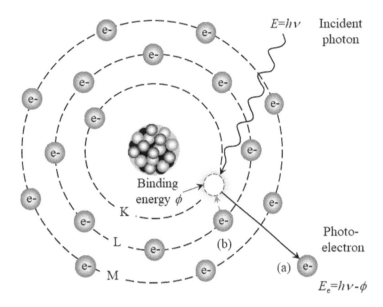

FIGURE 8.2 Illustration of the photoelectric effect. Here, an incident photon ejects electron (a) from the K-shell of an atom leaving a vacancy. The atom is now ionized. The vacancy is subsequently filled by the capture of electron (b) from one of the outer shells.

with a kinetic energy, E_e, equal to the difference between that of the incident photon, hv, and that of the binding energy[1] of the electron, ϕ,

$$E_e = hv - \phi. \tag{11}$$

This electron goes on to produce the ionization that generates the information carriers (*i.e.,* the signal) in a detector. The remainder of the energy, $hv - E_e$, appears as characteristic X-rays or Auger electrons resulting from the filling of the vacancy.

8.4.1.1 The Ejection Process

Photoelectrons may be ejected from any of the K, L, M ... shells of an atom, providing the photon energy exceeds the binding energy, ϕ, which is different for each shell. It decreases as we proceed towards the outer shells according to Moseley's [19] law. For the K, L and M shells, respectively,

$$\phi(\text{K}) \cong R_y(z - 1)^2 \quad \text{eV}, \tag{12}$$

$$\phi(\text{L}) \cong 1/4 R_y(z - 5)^2 \quad \text{eV},$$

and

$$\phi(\text{M}) \cong 1/9 R_y(z - 13)^2 \quad \text{eV},$$

where R_y is the Rydberg constant of value 13.61 eV. Note that as a result of quantum mechanical constraints, the L and higher shells are split into subshells with slightly different binding energies (typically within about 10% for detector materials). In fact, each shell has 2*n*-1 subshells, where *n* is the principle quantum number (equal to 1 for the K-shell, 2 for the L shell, 3 for the M shell and so on). Thus, the L shell consists of three subshells: L1 (inner), L2 (middle) and L3 (outer), and the M shell of five subshells. From the inner to the outer subshell, these are denoted by M1 to M5. Experimentally, for the intermediate and high mass materials typically used for detector applications, it is found that about 80% of photoelectric absorptions take place at the K-shell and ~10% at the L-shell (mainly the L1 subshell), providing, of course, that the incident photon energy, hv, exceeds the binding energy of the participating shell.

The angular distribution of photoelectrons depends on the incident photon energy, hv, and is illustrated in Fig. 8.3. With reference to the figure, the photoelectron emission angle, θ, is defined as the angle between the incident photon direction and

---

[1] The electron binding energy is the energy required to free an electron from its atomic or molecular orbital. It is more commonly known as the ionization energy.

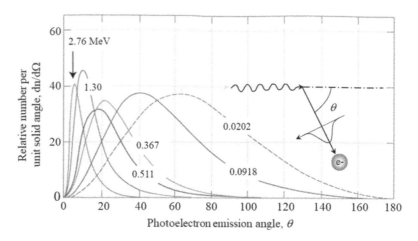

FIGURE 8.3 Angular distribution of photoelectrons for various incident photon energies (adapted from [20]).

the direction of the emitted photoelectron. At low energies (of the order of 10 keV) photoelectrons tend to be emitted at angles close to 90° to the incident photon direction, in the direction of the photon's electric vector. However, as the photon energy increases, the photoelectron emission peak moves progressively to smaller non-zero emission angles ($\theta = 0°$ is forbidden); that is, the emission begins to peak in the forward direction.

8.4.1.2 The De-Excitation Process

Following the ejection of the photoelectron, the atom is left in a highly excited state by virtue of an incomplete shell. It subsequently de-excites by filling the vacancy with an electron captured from one of its outer shells (see Fig. 8.4a). This results in a release of energy, which can be dissipated either by the emission of a photon (radiative emission) or absorption by a bound electron of a higher shell, causing its ejection (non-radiative emission), as illustrated in Fig. 8.4b. For radiative emission, the emitted photon is known as a characteristic X-ray, while for non-radiative emission, the ejected electron is known as an Auger electron.

Radiative transitions lead to the X-ray emission of a single energy. For example, when an electron from the higher L-shell makes a transition to fill a K-shell electron vacancy, as in Fig. 8.4b, a photon is emitted with energy $\phi(K) - \phi(L)_i$, where i is the subshell number. Note that not all transitions from outer shells or subshells are allowed, only those obeying the quantum mechanical selection rules (*e.g.*, see reference [21]). Briefly, this theory states that every electron in an atom moves on an orbital that is characterized by four quantum numbers summarized in Table 8.3.

The matrix of all the possible electronic states is obtained by considering all possible combinations of the quantum numbers under the restriction that no two electrons can have identical values for all four of their quantum numbers. This is

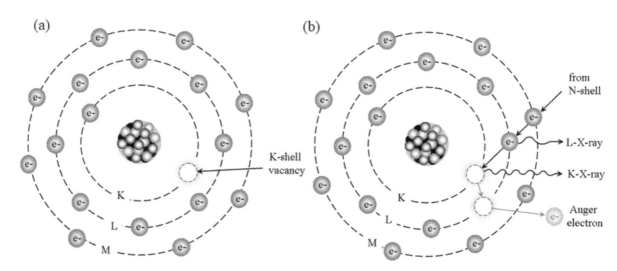

FIGURE 8.4 The photoelectric de-excitation process illustrating both radiative and non-radiative processes. In (a) the vacancy left by the ejected photoelectron leaves the atom in an unstable state that readjusts itself by capturing an electron from a higher energy shell. In (b) the atom de-excites by emitting either a characteristic photon or one or more Auger electrons.

TABLE 8.3

The principal quantum numbers that determine the number of electrons and energy levels assigned to each orbital shell. Each electron in an atom can be described by four quantum numbers. The first three (n, l, m) specify the particular orbital of interest while the forth (s) dictates how many electrons can occupy that orbital.

Quantum number	Physical interpretation
n	The principal quantum number, n, is associated with successive orbitals; n is a positive integer 1, 2, 3, ... that designates the K, L, M, N, shells, respectively. It essentially describes the energy of an electron and the size of the orbital. All orbitals that have the same value of n are said to belong to the same shell.
l	The angular momentum (sometimes known as the secondary azimuthal) quantum number, is a measure of the orbital angular moment and accounts for the existence of circular $(l = 0)$ and elliptical orbitals. In essence, it describes the shape of an orbital with a particular principal quantum number, n. It can be thought of as dividing a shell into smaller groups of suborbitals, known as subshells or sublevels. Since l can take all the integer values between 0 and $(n\text{-}1)$, usually a letter is used to identify a particular l to avoid confusion with n. For example $l = 0, 1, 2, 3, 4, 5$ is denoted as the s, p, d, f, g and h sublevels, respectively. In this nomenclature, the subshell with $n=2$ and $l=1$ is known as the $2p$ level and the subshell with $n=3$ and $l=0$ is known as the 3s level. Furthermore, an upper index after the number indicates the number of electrons that have these two quantum numbers. Thus, the symbol $3d^6$ means that there are six electrons with $n = 3$ and $l = 2$. These electrons differ between each other by the values of the two other quantum numbers m and s.
m	The magnetic quantum number, m, essentially describes the orientation in space of an orbital of a given energy (n) and shape (l). m can take all the integer values between $-l$ and $+l$.
s	The spin quantum number, s, specifies the orientation of the spin axis of an electron. s can only have one of two values – $+1/2$ and $-1/2$ and for electrons in the same orbital the spins must be opposite.

known as the "Pauli exclusion principle". Different electrons within the same shell do not all have exactly the same energy. This is taken into account by a fifth quantum number, j, which is known as the total quantum number and represents the total angular momentum obtained by coupling the orbital and spin angular momenta of the electron. j has the values $l \pm 1/2$ but with the restriction that it cannot be negative.

8.4.1.3 Characteristic Lines and Selection Rules

The quantum mechanical criteria described in the previous section governs the configuration of electrons in the shells and subshells of atoms. However, quantum mechanics impose additional restrictions when it comes to the transfer of electrons from different energy levels. In fact, any transition between levels must also satisfy the following selection rules:

$$\Delta n \geq 1$$

$$\Delta l \pm 1 \tag{13}$$

$$\Delta j = 0 \text{ or } \pm 1$$

Using these rules, we note that K to n_i transitions, where $i=1$ are forbidden. However, it should also be noted that the last two rules are not always obeyed, leading to so-called "forbidden" transitions. The transitions that are "allowed" by the selection rules are depicted in Fig. 8.5, which shows the allowed transitions leading to a series of characteristic lines from the K and L shells. The various emission lines are identified by a capital letter indicating the final level of the transition involved with an α, β or γ subscript denoting the starting orbital and a further subscript to account for the differences in energy due to the subshell involved. For example, lines emanating from L to K transitions are known as K_α emission; lines emanating from K to M transitions are known as K_β emission. Similarly, lines originating from M to L transitions are known as L_α emission; lines emanating from N-L transitions are known as L_β emissions. These X-rays are called "characteristic" because their energies are unique for each element, since every element has its own distinct energy levels.

The probability that a vacancy will result in X-ray emission is called the fluorescence yield and is denoted by ω. The fluorescence of an atomic shell or subshell is defined as the probability that a characteristic X-ray is emitted when one of the outer electrons fills the vacancy in an inner shell. It is defined as

$$\omega = \frac{n_f}{n}, \tag{14}$$

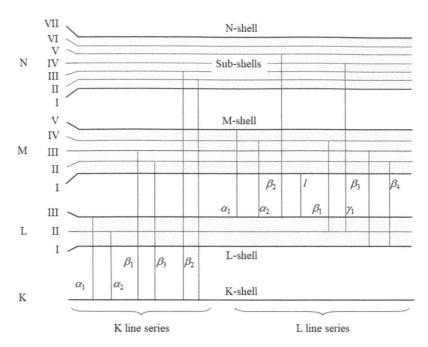

FIGURE 8.5 The energy level diagram of the main emission lines from the K and L shells. Here, K, L, M and N denote the main atomic shells and the Roman numerals, I, II, III, *etc.* denote the subshells.

where n is either the number of primary photons that have induced the ionization in a given level or the number of secondary photons that are subsequently emitted, n_f is the number of these secondary photons that effectively leave the atom. So for example, the yield of characteristic X-ray photons emitted from the K-shell is

$$\omega_K = \frac{\text{Number of K vacancies created}}{\text{Number of KX} - \text{rays emitted}}. \tag{15}$$

ω_K is very small (<0.01) for elements below fluorine ($Z = 9$). For high Z elements (> ~40), ω_K approaches unity. The difference $n - n_f$ is the number of secondary photons absorbed within the atom on their way out (the *Auger effect*).

Radiation-less transitions can be of two types: Auger and Coster-Kronig. The basic mechanisms in terms of energy transitions are illustrated in Figs. 8.6 (b) and (c). For comparison we also show radiative emission in Fig. 8.6(a). The Auger effect is a spontaneous process following photoelectric absorption in which an inner shell electron vacancy is filled by an electron from a higher energy and a further electron from the same atom ejected instead of a photon, with the result that the residual atom now has two electron vacancies. This second ejected electron is called an Auger electron and the process may be repeated as the new vacancies are filled; otherwise, X-rays will be emitted. In fact, an atom can emit a number of Auger electrons more or less simultaneously in a kind of chain reaction. The atom thus exchanges one energetically "*deep*" inner shell vacancy for a number of "*shallow*" outer shell vacancies that are eventually neutralized by conduction band electrons. Upon ejection, the kinetic energy of the Auger electron is equal to the difference between the binding energy of the shell containing the original vacancy and the sum of the binding energies of the two shells having vacancies at the end. Thus, the kinetic energy of the Auger electron emitted in Fig. 8.6(b) is $L_\phi - 2M_\phi$ (ignoring small differences in the M substate energies). Coster-Kronig transitions on the other hand, occur within subshells of the same shell [22]. For example, a vacancy in the L1 subshell can be filled by an electron from a higher subshell, like the L2 subshell. As a result, a new vacancy is created, which in turn can be filled by any of the Coster-Kronig, Auger or fluorescent relaxation modes. This makes it possible for a given subshell to receive contributions from different higher subshells to produce the same group of lines. Note that the separation of these levels depend on the type of atom and the chemical environment in which the atom is located, and these dependencies are exploited in Auger electron spectroscopy for compositional surface analysis [23]. The Coster-Kronig yield, f, is defined as the fraction of events that undergo Coster-Kronig transitions.

The probability of non-radiative transitions with the emission of Auger electrons is larger for low Z material and can be described by the Auger yield, a, which is the fraction of events that end in Auger emission, such that the sum of the fluorescence, Auger and Coster-Kronig yields is unity, that is, $\omega + a + f = 1$. The Auger yield decreases with increasing Z and at $Z \sim 30$ the probabilities of X-ray emission from the innermost shell and the emission of Auger electrons are almost equal. Auger electrons have extremely short ranges and are re-absorbed; however, the characteristic X-rays (and particularly K X-rays) may escape in detectors of small size giving rise to an escape peak. This is especially true for compound semiconductor detectors for two reasons. Firstly, they have limited volumes because of poor transport parameters, and

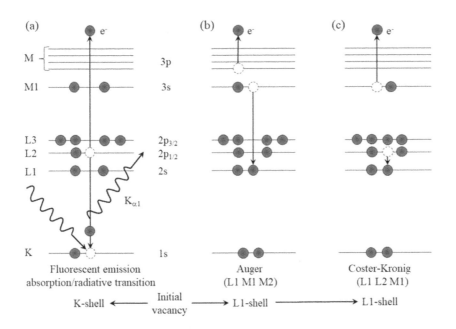

FIGURE 8.6 Transition-level diagram illustrating both radiative and non-radiative emission. In (a) a fluorescent photon is emitted following K-shell photoelectric absorption. The vacancy is filled by an L2 electron with the subsequent emission of a Kα₁ photon. In (b) and (c) we show examples of non-radiative de-excitation processes. Auger emission is shown in (b) in which a L1-shell vacancy is filled by an electron from the M1 shell (3s level), and the transition energy is imparted to a 3p electron, which is ejected. This is represented by the spectroscopic nomenclature (L1 M1 M2). The final atomic state thus has two holes, one in the M1 (3s orbital) and the other in the M2 (3p orbital). In (c) we shown a Coster-Kronig transition in which an L1 vacancy is filled by an L2 electron and an M1 electron emitted. This is denoted by (L1 L2 M1).

secondly detectors are generally fabricated from high Z materials so the fluorescent escape X-rays have high enough energies (*e.g.*, 60 keV and above) that they may not be contained in the detection volume.

If the energy of the photon is sufficiently small to render relativistic effects insignificant but large enough so that the binding energy of the electrons in the K-shell may be neglected (*i.e.*, $E = h\nu$), then the cross-section per atom, $_a\tau_K$, for photoelectric absorption is given by [21],

$$_a\tau_K = \sigma_T Z^5 \alpha^4 2^{5/2} \left(\frac{m_0 c^2}{h\nu}\right)^{7/2} \quad \text{cm}^2/\text{atom}, \qquad (16)$$

where σ_T is the Thomson cross-section, α is the fine structure constant ($\alpha = 1/137$) representing the strength of the interaction between electrons and photons, $m_0 c^2$ is the rest mass energy of the electron and Z is the atomic number of the absorbing material. The most important feature of the cross-section is the strong dependence on the atomic number $\left(\sim Z^5\right)$ and on the energy of the incident photon $\left(\sim E^{-3.5}\right)$, from which it follows that this process is especially efficient in the absorption of low-energy photons by heavy atoms. The cross-sections for the L and higher-order shells are more difficult to calculate based on a pure Coulomb potential model, due to the effects of screening, which are appreciable.

8.4.2 Coherent Scattering – Thomson and Rayleigh Scattering

If the energy of the incident photon is insufficient to produce ionization or excitation of the atom, either Thomson or Rayleigh scattering can occur depending on whether the photon interacts with free or bound electrons. For both mechanisms, the photon is elastically scattered with no change in internal energy of the scattering atom or of the X-ray photon. In fact, the practical difference between the two is in the angular distributions of the scattered photons.

In Thomson scattering [24], the electromagnetic wave vector associated with a passing photon interacts with a free electron, which then responds by oscillating classically. This causes the electron to emit electromagnetic radiation (in the form of photons) at the same frequency as the incident wave. The net effect is a redirection of the incident photon with no transfer of energy to the medium. The total cross-section is given by

$$\sigma_T = \frac{8\pi}{3} \left(\frac{e^2}{4\pi\varepsilon_0 m_0 c^2}\right) = \frac{8\pi}{3} r_0^2 = 6.652 \times 10^{-25} \quad \text{cm}^2, \qquad (17)$$

where ε_o is the permittivity of free space and the quantity r_o is the classical radius of the electron $(=2.818\times10^{-15}$ m). Note that both the differential and the total Thomson scattering cross-sections are completely independent of the frequency (or wavelength) of the incident radiation. The angular distribution for unpolarized incident radiation is given by

$$\frac{d\sigma_T}{d\Omega} = r_o^2\left(\frac{1+\cos^2\theta}{2}\right) \quad \text{cm}^2\text{sr}^{-1}\text{per electron,} \tag{18}$$

where θ is the scattering angle. Thus photons are scattered mainly in the forward and reverse directions. If the beam is unpolarized, there is a cylindrical symmetry about the beam axis. For polarized incident radiation, the cross-section vanishes at 90° in the plane of polarization.

In contrast to Thompson scattering, Raleigh scattering results from electron interactions with the atom as a whole. In this case, the atom recoils to conserve momentum; however, given the relative masses of the electron and atom there is no appreciable transfer of energy from the photon to the atom – the electron is merely deflected. For Rayleigh scattering, the cross-section per atom is a perturbation of the classical Thomson cross-section

$$_a\sigma_{\text{coh}} = \frac{3}{8}\sigma_T \int\limits_{-1}^{+1} \left(1+\cos^2\theta\right) [F(x,Z)]^2 d\cos\theta \quad \text{cm}^2/\text{atom}, \tag{19}$$

where $F(x,Z)$ is the atomic form factor and $x = \sin(\theta)/\lambda$. The form factor tends towards Z as $\theta \to 0$ and 0 as $\theta \to \pi$. Thus, Rayleigh scattered photons have an angular distribution much more peaked in the forward direction than Thomson scattering, especially at high energies. Rayleigh scattering also tends to be much more efficient in the forward direction than Thomson scattering, since the interaction probability varies as $\sim Z^2$; however, the interaction cross-section decreases significantly with increasing energy (in fact as $\sim E^{-2}$). For higher energy X-rays Eq. (19) can be approximated by

$$_a\sigma_{\text{coh}} = k\frac{Z^2}{(h\nu)^2} \quad \text{cm}^2/\text{atom}, \tag{20}$$

where k is a constant. For practical detector design, elastic scattering is never more than a minor contribution to the absorption coefficient and has little effect in photon detection other than to scatter photons into and out of the detection volume. It can, however, cause acute problems in imaging systems, leading to a blurring of images.

8.4.3 Incoherent Scattering – Compton Scattering

At energies much greater than the binding energy of the electrons, photons can be scattered as if the electrons were free and at rest. The energy of the incident photon is shared between the scattered photon, which may escape, and the kinetic energy of the recoil electron, which is most likely absorbed. This is the Compton effect [25] which at 1 MeV is the dominant mode of interaction in most detector materials.

Fig. 8.7 shows the scattering geometry for a Compton interaction. The energy of the scattered photon is given by

$$E'_\gamma = h\nu' = \frac{E_\gamma}{1+(E_\gamma/m_oc^2)(1-\cos\theta)}, \tag{21}$$

where E_γ is the energy of the incident photon and θ is the angle through which it is scattered. For small scattering angles, θ, very little energy is transferred.

The kinetic energy of the scattered electron is,

$$E_e = E_\gamma - E'_\gamma = \frac{E_\gamma(1-\cos\theta)}{m_oc^2(1+(E_\gamma/m_oc^2)(1-\cos\theta))}. \tag{22}$$

The kinetic energy of the electron has its maximum value when $\cos\theta = -1$ or $\theta = 180°$. This electron will produce the ionization that generates the information carriers in a detector. Note the scattered photon may then continue to Compton scatter again in the detector or be completely absorbed *via* the photoelectric effect or escape the detection volume totally. The latter is particularly true for small detectors.

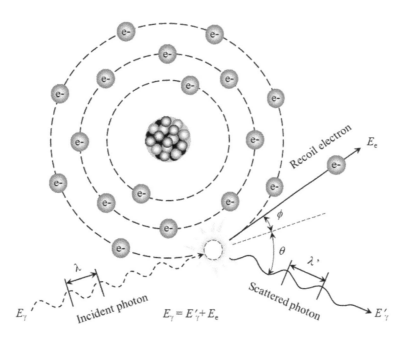

FIGURE 8.7 Compton scattering geometry. The electron is assumed to be initially at rest and is ejected by the incident photon at angle ϕ with kinetic energy E_e. The initial photon in turn is scattered through angle θ but with reduced energy, E'_γ, which by conservation of energy is equal to $E_\gamma - E_e$.

The differential scattering cross-section (that is, the probability of scattering a photon through angle θ within a solid angle Ω) is given by the Klein-Nishina [26] formula, which was derived by applying Dirac's relativistic treatment of the electron [27] to [28] the Compton effect. The energy scattered from unpolarized radiation by stationary, unbound electrons is given by

$$\frac{d\sigma_{\text{incoh}}}{d\Omega} = r_e^2 \left(\frac{(1 + \cos^2\theta)}{2} \right) \frac{1}{(1 + \gamma(1 - \cos\theta)^2)} \times \left[\frac{\gamma^2(1 - \cos\theta)^2}{(1 + \cos^2\theta)(1 + \gamma(1 - \cos\theta))} \right], \tag{23}$$

where γ is the ratio of the incident photon energy to the rest mass energy of the electron (E_γ/m_0c^2). Equation 23 can be written more simply in terms of the energies of the primary and scattered photons,

$$\frac{d\sigma_{\text{incoh}}}{d\Omega} = \frac{r_e^2}{2} \left(\frac{E'_\gamma}{E_\gamma} \right)^2 \left(\frac{E'_\gamma}{E_\gamma} + \frac{E_\gamma}{E'_\gamma} - \sin^2\theta \right) \quad \text{cm}^2\text{sr}^{-1}\text{per electron.} \tag{24}$$

This is the most useful form for detector design unless polarization is important. The Klein-Nishina formula can be thought of as the classical Thomson cross-section multiplied by a form factor. Whereas the Thomson cross-section gives a symmetric angular distribution of scattered photons, which is symmetric about 90°, the Klein-Nishina formula predicts a strongly forward peaked cross-section as γ increases. This is illustrated in Fig. 8.8(a) in which we plot the differential angular distribution of Compton scattered photons as a function of the angle of scattering, θ, for various incident photon energies. The actual angular distribution of scattered events is most easily visualized by the polar diagram shown in Fig. 8.8(b). From the figure we see that it becomes increasingly peaked in the forward scatter direction ($\theta \to 0$) at high energies. At low energies, $E_\gamma \approx E'_\gamma$ and the angular distribution described by Eq. (24) reduces to the Thomson form of $(1 + cos^2 \theta)$, which is symmetrical about the angle $\theta = 90°$

For most detection systems where the energy of the incident particle is determined by detecting the scattered photon and electron, a more useful quantity is the total Compton cross-section obtained by summing the Klein-Nishina differential cross-section over all angles and polarizations of the scattered photon. The solution leads to a lengthy formula which is given approximately by

$$\sigma_{\text{incoh}} = 8\pi r_e^2 \frac{(1 + 2\gamma + 1.2\gamma^2)}{3(1 + 2\gamma)^2} = 3\sigma_T \frac{(1 + 2\gamma + 1.2\gamma^2)}{3(1 + 2\gamma)^2} \quad \text{cm}^2/\text{atom}. \tag{25}$$

For very low energies ($\gamma \to 0$) the formula reduces to the classical Thomson cross-section; this represents the probability of removal of the photon from a collimated beam while passing through an absorber containing one electron cm^{-2}. As the

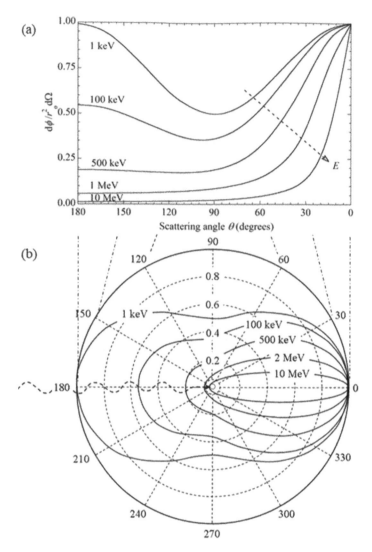

FIGURE 8.8 (a) Differential scattering cross-sections for Compton scattered photons as a function of the angle of scattering, θ, for various incident photon energies, γ, expressed in units of the electron rest mass energy (*i.e.*, $\gamma = E/m_0c^2$). (b) Polar diagram illustrating the angular distribution of scattered photons as a function of energy. The cross-section is represented by the radius vector, which is expressed in units of r_0^2. In this diagram, photons are incident from the left, so scattering in the forward direction is to the right.

photon energy increases and eventually becomes comparable with the rest mass energy of the electron ($\gamma \sim 1$), the Klein-Nishina formula predicts that forward scattering of photons becomes increasingly favoured relative to backward scattering. The probability of interaction is found to be proportional to the density, ρ, of the material and inversely proportional to the incident gamma-ray energy, E_γ.

$$\sigma_{incoh} = C\left(\frac{Z}{A}\right)\rho\frac{1}{E_\gamma}\quad \text{cm}^2/\text{atom}\,,\qquad (26)$$

where C is a constant.

8.4.4 Pair Production

Pair production is a phenomenon where energy in the form of a photon is converted directly into mass in a Coulomb forcefield, producing an electron and by charge conservation, its antiparticle a positron. It generally takes place in the field near an atomic nucleus. However, it can also take place (albeit with a much lower probability) in the field of an atomic electron in a process known as triplet production. The pair production process in the field of the nucleus is illustrated in Fig. 8.9. It is assumed that the electrons surrounding the nucleus play no part in the kinetics of the reaction, their only effect being to screen the field of the nucleus if the reaction takes place at distances comparable with the orbits of the atomic

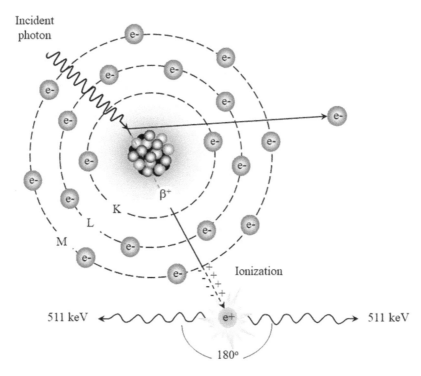

FIGURE 8.9 Pair production in the field of the nucleus when incident gamma-ray energies exceed 1.022 MeV. The presence of the nucleus is required to conserve momentum.

electrons. Due to the need of simultaneous energy and momentum conservation, pair production needs a recoiling particle, in this case a nucleus. The minimum energy required for pair production is given by the relation

$$E_\gamma \geq 2m_e c^2 (1 + m_e/m_n), \tag{27}$$

where m_n is the mass of the recoiling nucleus, and $m_e c^2$ is the electron rest mass. For heavy atoms $m_n >> m_e$ and thus the threshold energy[2] for the reaction is simply $2m_e c^2 = 2 \times 0.511 \, \text{MeV} = 1.022 \, \text{MeV}$. Above the threshold, the surplus energy appears as kinetic energy of the two particles, such that

$$E_\gamma = (E_- + m_0 c^2) + (E_+ + m_0 c^2), \tag{28}$$

where E_- and E_+ denote the energies of the electron and positron, respectively. The excess energy above the threshold is randomly shared between the electron and positron. The information carriers in a detector are then generated by the ionization trail of the electron and positron. The electron is usually completely absorbed. However, if the positron slows sufficiently, it will annihilate with an electron in the detector giving rise to two characteristic 511 keV gamma-rays. These may then go onto interact in the detector medium *via* a Compton interaction or photoelectric absorption. In order to conserve momentum, the two photons are emitted 180° to each.

Pair production is related to bremsstrahlung, and theoretical calculations generally follow the same approach for both processes [29]. The cross-section is difficult to derive and begins with the Bethe and Heitler [30] Born-approximation unscreened pair-production cross-section as an initial approximation, to which Coulomb, screening and radiative corrections are applied. Marmier and Sheldon [31] summarize the useful forms of the total cross-section as,

$$_a\kappa_p = \frac{r_0^2 Z^2}{137} \left[\frac{28 \ln}{9} \left(\frac{2h\upsilon}{m_0 c^2} \right) - \frac{218}{27} \right] \quad \text{cm}^2/\text{atom}, \tag{29}$$

for $1 << E_\gamma << 1/\gamma Z^{1/3}$ and for complete screening,

---

[2] The threshold for triplet production is $4m_e c^2$, Even though the energy being converted in still $2m_e c^2$, the higher threshold is required by conservation of momentum arguments [28].

$$_a\kappa_p = \frac{r_o^2 Z^2}{137} \left[\frac{28 \ln}{9} \left(\frac{183}{Z^{0.3}} \right) - \frac{2}{27} \right] \quad cm^2/atom, \tag{30}$$

for $E_\gamma/m_oc^2 >> 1/\gamma Z^{1/3}$. Note that the latter expression is independent of energy. For both cases, the probability of interaction scales as Z^2 and unlike σ_p and σ_c, increases with energy. Since the useful operating energy range of a detector is dictated by its active volume, or more specifically by its ability to contain the incident radiation, pair production effects can be ignored in compound semiconductor detectors since the small active volumes generally limit operating energy ranges to < 500 keV.

8.4.5 Attenuation and Absorption of Electromagnetic Radiation

When a monoenergetic beam of photons of energy E_o traverses a homogenous medium, a number of photons will be removed from the beam through the interactions described previously. Unlike heavy charged particles, which slow down and continuously lose energy without losing intensity, a well-collimated beam of photons maintains its energy but continuously loses intensity. This is a direct consequence of the stochastic, one-shot nature of photon interaction mechanisms. Basically, if a photon interacts, it is removed from the beam – either by complete absorption or by scattering out of the beam. The number of photons removed, dI, when passing through an absorber is proportional to the thickness of absorber traversed, dx, and is given by

$$dI = -\mu_o I dx, \tag{31}$$

where the constant of proportionality, μ, is called the linear attenuation coefficient, which is, in fact, the probability per unit path length that a photon will be removed from the beam. The linear attenuation coefficient, in turn, is simply related to the photoelectric, Rayleigh, Compton and pair production cross-sections derived in previous sections,

$$\mu_o = \tau + \sigma_{incoh} + \sigma_{coh} + \kappa \tag{32}$$

and to the mean free path of photons, λ_o, in the absorber at that energy by,

$$\lambda_o = \frac{1}{\mu_o}. \tag{33}$$

In semiconductors, typical values of λ_o range from microns at X-ray energies to tens of cm at gamma-ray energies. Integration of Eq. (31) yields

$$I = I_o \exp(-\mu_o x), \tag{34}$$

which gives the probability that a collimated photon beam of energy E_o of initial intensity, I_o, will have a residual intensity, I, of unaffected primary photons after traversing a thickness x of absorber. It is in fact the product of the survival probabilities for each individual type of interaction as reflected in Eq. (32). Note that it is not necessary to include a Thomson scattering term in σ_{coh}, since it is the limiting case of Compton scattering and therefore already implicitly included in σ_{incoh}.

Absorption coefficients are usually expressed as mass attenuation coefficients, which are the linear coefficients divided by the density, ρ g/cm^3. As Evans [32] points out, mass attenuation coefficients[3] are of more fundamental value for detector work than the linear coefficients because all mass attenuation coefficients are independent of the actual density or physical state of the absorber, whether in gaseous, liquid or solid form. This is because the fundamental interactions are expressible as cross-sections per atom, $_a\tau$, $_a\sigma$ and $_a\kappa$, and when these are multiplied by the number of atoms per gram, the mass attenuation coefficient is obtained directly. Thus, for mixtures or compounds the mass attenuation coefficient is simply the sum of μ_o/ρ of the different elements multiplied with the corresponding weight fractions w_i, thus

---

[3] As pointed out by Hubble [33], the quantity μ/ρ has often been referred to in the literature as the "mass absorption coefficient". However, the term "mass absorption coefficient" has also been used to refer to the mass energy-transfer coefficient (*e.g.*, Evans [32]) and mass energy-absorption coefficient both having to do with photon energy deposition in the target material. We follow the International Commission on Radiation Units and Measurements' "mass attenuation coefficient" as used to refer to the total probability of the photon interaction processes.

$$\left(\frac{\mu}{\rho}\right) = \sum_i w_i \left(\frac{\mu_o}{\rho}\right) \quad \text{cm}^{-1}. \tag{35}$$

The total mass-attenuation coefficients for two common compound semiconductors, GaAs and CdTe, are shown in Figs. 8.10(a) and (b). In (a) we show the global mass absorption coefficients from 1 keV to 10 MeV, while (b) shows an expansion around the absorption edges. For completeness, in (a) we also show the individual photoelectric, Compton and pair contributions to the total mass attenuation coefficient for CdTe. Because of its strong Z^5 absorption dependence, we see that the photoelectric effect dominates all other absorption process at low energies. Since photoelectric absorption can occur at any of the excitable levels of the atom, the total photoelectric cross-section τ is actually the sum of all the (sub) shell-specific contributions and can be expressed as such

$$\tau = \tau_K + \tau_L + \tau_M = \tau_K + (\tau_{L1} + \tau_{L2} + \tau_{L3}) + (\tau_{M1} + \ldots + \tau_{M5}) + \ldots \tag{36}$$

The discontinuities in the photoelectric cross-sections are called the absorption edges, and the ratio of the cross-section just above and below an edge is called the jump ratio. Taking the K-edge as an example, below the edge the two K-shell electrons cannot participate in the photoelectric effect because their binding energy is too great. Only the L, M and higher-shell electrons can do so. Just above the K-edge the K-electrons can also participate, as evidenced by the abrupt increase in the absorption coefficient. In fact, for CdTe this is almost a factor of 5. Except at the absorption edges, μ is more or less proportional to $Z^4 E^{-3}$. Above a few hundred keV, Compton becomes the dominant interaction mechanism. Pair production only becomes significant above a few MeV. Rayleigh scattering plays a minor role and only at low energies. Interestingly, for the heavy materials usually used for gamma-ray detection media, the ratio of the Rayleigh scattering cross-section to the total interaction cross-section is approximately constant at 5% for energies < 100 keV.

As illustrated in Fig. 8.10, the absorption of radiation in matter is the cumulative effect of several types of photon matter interaction processes that take place in parallel, and in Fig. 8.11 we show the regions in energy space where each of the three primary energy deposition processes dominates. The solid lines show the boundaries where the photoelectric (left) and pair production (right) become equal to the Compton cross- section. We can see that Compton dominates the entire energy range for $Z<10$, but becomes increasingly "squeezed" by photoelectric and pair effects as Z increases. For the intermediate and high-mass materials used to fabricate detectors, photoelectric interactions dominate for energies < ~500 keV, Compton scattering dominates above ~500 keV but less than ~5 MeV. Pair production dominates above ~5 MeV. To illustrate the benefit of using materials with a high effective Z, the horizontal dotted lines show the bounds for four semiconductor detector materials: a light material, Si; a relatively light material, GaAs; an intermediate mass material, CdTe and a relatively

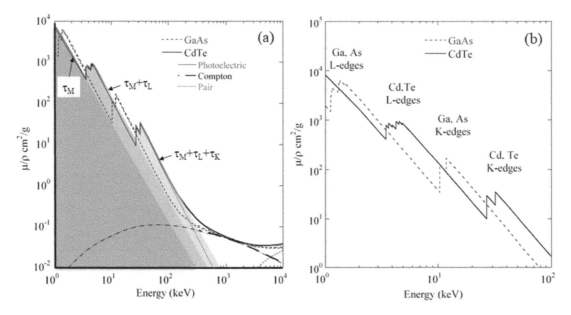

FIGURE 8.10 The mass attenuation coefficients for two common compound semiconductors – GaAs and CdTe. (a) shows the global absorption coefficient from 1 keV to 10 MeV, while (b) shows an expansion around the absorption edges. For completeness, in (a) we also show the individual photoelectric, Compton and pair contributions to the total mass attenuation coefficient for CdTe from which we see that the photoelectric contribution dominates the attenuation coefficient for energies up to a few hundred keV, whereupon Compton becomes the dominant interaction mechanism. The shaded regions are used to guide the eye between the different photoelectric contributions. Pair production only becomes significant above a few MeV. In (b) we see that between the absorption edges, the mass attenuation coefficients have an almost constant slope with μ being proportional to $Z^4 E^{-3}$.

heavy material, HgI_2. We see that when comparing, for example, HgI_2 with GaAs, the photoelectric effect is effective over three times the energy range for a factor of two increase in Z_{eff}. Comparing with Si is more dramatic – the range of photoelectric dominance is seven times larger. In this case the increase in Z_{eff} is a factor of five.

8.5 Radiation Detection Using Semiconductors

In terms of radiation detection, semiconductor detectors are usually operated in one of two modes, either by exploiting their photo conduction properties or by operating as a solid state ionization chamber. The former method was widely used in early work but has now been largely superseded by the latter, which offers many advantages, not least being the combined ability to resolve precisely the spectral and temporal signatures of a wide range of radiation types. However, photodetectors are still the detectors of choice when considering "light buckets" [34] – used, for example, in infrared, visible and UV astronomy.

8.5.1 Photodetectors

The simplest type of semiconductor radiation detector is the photodetector, which can be subdivided into photoconductive and photovoltaic devices (most commonly used as solar cells). Photoconductive detectors utilize the increase the conductivity caused by the photo generation of charge carriers (electrons, holes or electron-hole pairs) in a photoconductive media. When suitably biased, the light intensity is then proportional to the current flow through the device. Photovoltaic devices, on the other hand, require an internal potential barrier with a built-in electric field in order to separate photo-generated electron-hole pairs. Such potential barriers can be created by the use of p-n junctions or Schottky barriers. Because it is possible to form a deep depletion region, photovoltaic devices are generally used to detect short wavelength radiation, such as UV and EUV, whereas photoconductors are the detectors of choice for wavelengths ranging from the near UV to the far IR.

8.5.2 Photoconductors

Photoconductors are the more commonly used device. They differ from crystal counters[4] in that they usually have a limited bandwidth and are primarily integrating devices. This is because their large intrinsic conductivity makes them insensitive to individual particles or quanta. A schematic of such a device is shown in Fig. 8.12 for both transverse electrode (a) and front electrode (b) configurations. When a bias voltage and a load resistor are connected in series with the semiconductor to form

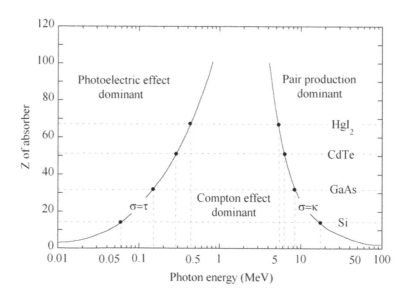

FIGURE 8.11 Figure illustrating the regions in energy space where each of the three primary energy deposition processes dominates. The lines delineate where Compton absorption is equal to photoelectric absorption (left-hand curve) and absorption due to pair production (right-hand curve). The regions between the lines are regions in which the photoelectric effect, Compton effect and pair production dominate for a given Z (from Evans [32]). The bounds for Si, GaAs, CdTe and HgI_2 are shown explicitly, from which we see that the photoelectric effect in HgI_2 is effective over three times the energy range than in GaAs and seven times the range in Si.

---

[4] An early name for a semiconductor radiation detector operated as a "solid state ionization chamber".

a potential divider, an increase in light intensity causes a corresponding increase in electrical conductivity and a current to flow through the circuit. The signal is registered as a change in voltage drop across the resistor. In essence, the device functions as a resistor whose resistance depends on the light intensity.

Photoconductors can either be intrinsic or extrinsic. An intrinsic semiconductor can only source its own charge carriers and is therefore not efficient, since the only electrons available for conduction are in the valence band. The spectral response is thus limited to photons that have energies equal to, or exceeding, the bandgap of the detector material (*i.e.*, of energy $E \geq E_g$). Intrinsic photoconductors have traditionally been used for optical applications, mainly at visible and infrared wavelenghts, since optical photon absorption depths are compatible with typical device thicknesses and photon energies are comparable with the bandgap energies of common semiconductors. For the traditional semiconductors, such as silicon and germanium, these energies correspond to maximum wavelengths of 1.1 μm and 1.8 μm, respectively (*i.e.*, in the near infrared). It is possible to extend the response to the far infrared using semiconductors with smaller bandgaps. However, in compound materials it is difficult to achieve the extremely high impedances needed to reduce Johnson noise (see Section 8.5.2.2) to levels necessary for low light level operation. In fact in narrow gap materials, intrinsic resistivities are rarely greater than the ohm-cm level. Therefore, small bandgap photoconductors are generally restricted to signal dominated applications, where a rapid time response is important. Even so, the performance of intrinsic photoconductors rapidly degrades as the wavelength extends beyond about 15 μm due to poor stability of the materials, difficulties in achieving high uniformity in material properties and problems in making good electrical contacts. Detector operation deep into the infrared must therefore be achieved by altering the conduction properties of the bandgap, either by stressing the semiconductor to the extent of narrowing the bandgap (not practical for most materials) or more commonly by doping to introduce additional active energy levels into the bandgap.

Extrinsic devices have impurities added whose ground state energy is closer to the conduction band, allowing conductivity to be induced by freeing the impurity-based charge carriers. Since the electrons do not have as far to jump, lower energy photons are sufficient to trigger the device, extending the spectral response to long infrared wavelengths. Both intrinsic and extrinsic photoconductors operate on similar physical principles and share the same sources of noise. However, there are some important

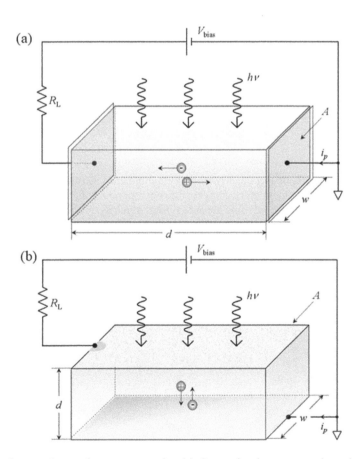

FIGURE 8.12 Schematic of two photoconductor detector geometries. (a) Conventional transverse electrode geometry, consisting of a highly conductive semiconductor bar of cross-sectional area, A. Radiation incident on the device creates additional electron hole pairs leading to an increase in conductivity. This in turn leads to a change in voltage drop across the load resistor R_L. (b) An example of top-illuminated geometry. For this electrode configuration, the top electrode is transparent to allow low-energy photons to enter the device. For optical wavelengths, a reflective back contact is sometimes used to increase the photon absorption depth by a factor of two.

differences. For example, in an intrinsic device a photon creates both a mobile electron and a mobile hole, whereas in an extrinsic device either the electron or hole is in a localized impurity state so there is only one carrier. Also, because lower excitation energies also allow for large thermally excited dark currents, extrinsic devices must be operated at low temperatures.

Depending on the wavelength region of interest, photoconductive materials commonly used include Si, Ge, InSb, HgCdTe, PbS and PbSe. Detectors in the infrared are usually operated at cryogenic temperatures to minimize noise (down to 77 K for liquid nitrogen and 4 K for liquid helium). This cooling requirement can be a major disadvantage in some applications. However, in others (*e.g.*, thermal imaging) this limitation is far outweighed by superior electronic performance. The achievable wavelength limits for a number of commonly used photoconduction materials are listed in Table 8.4.

8.5.2.1 Current Generation in Photoconductors

A comprehensive treatment of the photoconductive process is given in Rose [35]. With reference to Fig. 8.12, light incident on the crystal generates electron-hole pairs, which increase the carrier concentration and in turn, its conductivity. If a small bias is now applied across the crystal via two contacts, a measurable current can be induced to flow through it.

The conductivity σ in the absence of illumination may be determined from the current density J, velocity ν, charge density ρ, electric field, ξ and mobility μ. The current density is given by

$$J = \sigma\xi = \rho\nu, \tag{37}$$

since, $\nu = \mu\xi$, the conductivity can be expressed in terms of the mobility

$$\sigma = \mu\rho. \tag{38}$$

Now, the current density, ρ, is equal to eN, where N is the number of charge carriers. If q is the electrical charge of the charge carriers, μ_e is the electron mobility, μ_h the hole mobility, N_e the density of electrons in conduction band and N_h the density of holes in the valence band, the conductivity is given by the sum of electron and hole charge components

$$\sigma = q\left(\mu_e N_e + \mu_h N_h\right). \tag{39}$$

Consider an incoming photon flux $\phi\gamma\,s^{-1}$. The number of carriers in equilibrium is

TABLE 8.4

The achievable limits of spectral sensitivity for a number of intrinsic and extrinsic semiconductor photoconductors. For extrinsic photoconductors, the nomenclature is as follows; *primary semiconductor: dopant*. Typical impurity concentrations are 10^{15} to 10^{16} cm^{-3} for Si and $\sim 10^{14}$ cm^{-3} for Ge.

Wavelength region	Material	Bandgap energy (eV)	Cutoff wavelength (μm)
Intrinsic photoconductors			
UV	C (diamond)	5.4	0.23
UV, Vis	CdS	2.38	0.48
NIR	GaAs	1.52	0.82
UV, V, NIR	Si	1.17	1.06
NIR	Ge	0.74	1.68
MIR	InSb	0.24	5.17
NIR, MIR	$Hg_{1-x}Cd_xTe$	0.06–1.6	0.77–25
Extrinsic photoconductors			
IR	Si:Be	0.150	8.3
MIR	Si:As	0.054	23
MIR	Ge:Cu	0.040	31
MIR	Ge:Be	0.024	52
Thermal IR	Ge:Ga	0.011	115
Thermal IR	(stressed)	0.006	200

$$N_e = N_h = \phi\eta\tau, \tag{40}$$

where η is the quantum efficiency and τ is the mean lifetime before recombination. Typically, $\tau \sim 1/$ impurity concentration. The number of carriers per unit volume, n (electrons), p (holes), is

$$n = p = \frac{\phi\eta\tau}{Ad}, \tag{41}$$

where d is the total length the carriers drift through. The resistance is related to the conductivity by

$$R = \frac{d}{\sigma A}. \tag{42}$$

Substituting Eqs. (39) and (41) for σ and A we obtain

$$R = \frac{d^2}{q\left(\mu_e + \mu_h\right)\varphi\eta\tau}, \tag{43}$$

which is useful for optical applications, as it relates the resistivity of the detector to the photon arrival rate. The time for an electron to drift from one electrode to the other, τ_t, is given by

$$\tau_t = \frac{d}{\langle v_d \rangle} = \frac{d}{\mu_e \xi}, \tag{44}$$

where v_d is the average drift velocity between the electrodes. The photoconductive gain is given by

$$G = \frac{\tau\mu_e\xi}{d} = \frac{\tau}{\tau_d} \tag{45}$$

and is the number of carriers passing through the contacts per one generated electron-hole pair. In effect, it is the ratio of the carrier lifetime to carrier transit time, implying that it is possible to have a gain greater than unity so that the absorption of a single photon produces τ/τ_d carrier pairs. This is understood by realizing that τ_d represents the time for the charge carriers to traverse the conductor and enter the external circuit. Current continuity and charge conservation require that new carriers enter the semiconductor from the external circuit and eventually recombine after time τ. Thus, the charge carriers effectively traverse the semiconductor τ/τ_d times resulting in gain. The quantity G is, in fact, the probability that a generated charge carrier will traverse the extent of the detector and reach an electrode. If $G \ll 1$ the majority of charge carriers recombine before reaching an electrode, and if $G \gg 1$ most reach the electrode. To optimize G, the detector should be as thin as possible to reduce the transit time but thick enough to maximize η. The efficiency can also be improved by increasing the bias voltage or improving material quality by reducing defects and impurities.

8.5.2.2 Noise in Photoconductors

For a detailed treatment of noise in photoconductors, the reader is referred to Rieke [36]. There are four main components of noise in photoconductors:

1. Generation-recombination noise. The fundamental limitation of any detector is the Poisson noise implicit in the incoming photon stream. In a photoconductor, these photons are converted into free electrons and holes, so that the incident photon noise is now modified by the random fluctuations in the generation and recombination of charge carriers. If a photoconductor absorbs φ photons per sec in time Δt, it will create

$$N = \eta\varphi\Delta t \tag{46}$$

electrons-hole pairs, where η the quantum efficiency. Since $\mu_e \gg \mu_h$, only electrons will contribute to the signal, so the *rms* variation in the generated events will be $\sqrt{N}$ in the Poisson limit. The recombination of electrons with photons is also a random process, and under steady-state conditions there is an average of N recombination events in time Δt so that the total *rms* variation in the number of carriers (generated and annihilated) is thus $\sqrt{2N}$. The associated noise current is

$$\left\langle I_{\text{G-R}}^2 \right\rangle^{1/2} = \frac{q \sqrt{2N} G}{\Delta t}. \tag{47}$$

The mean detector current due to the incident photons is

$$I_{\text{ph}} = \eta q \phi G. \tag{48}$$

Squaring Eq. (47) and combining with Eq. (48) we get,

$$\left\langle I_{\text{G-R}}^2 \right\rangle = \frac{q^2 2N G^2}{(\Delta t)^2} = \left(\frac{2q}{\Delta t} \right) \left(\frac{qNG}{\Delta t} \right) G = \frac{2q}{\Delta t} \left\langle I_{\text{ph}} \right\rangle G. \tag{49}$$

This noise can be minimized by keeping any DC component to the current small, especially the dark, current and by keeping the bandwidth of the amplification system small.

   2. Dark current noise, I_{DC}, arises from the thermal generation of charge carriers and can be thought of as the signal produced in the absence of photon illumination. It is usually only significant if the detector is operated at room temperature, where the thermal generation of charge can be comparable to that generated by photon absorption. The dark current is therefore proportional to the concentration of carriers generated by thermal excitation across the bandgap, *i.e.*,

$$I_{\text{DC}} \propto \sqrt{N_c N_v} \exp \left(-E_g/2kT \right), \tag{50}$$

where N_c and N_v are the conduction and valence band density of states and E_g is the bandgap energy. Although, N_v, N_c and E_g are slightly temperature dependent, the dominant factor is the exponential term, which suggests that even moderate cooling can be effective in reducing dark current in intrinsic photoconductors. For extrinsic devices, the effective bandgap is much smaller, and there are many more impurity-based charge carriers available for conduction, depending on the doping density. In this case, the dark current can only be effectively suppressed by more severe cooling, usually to liquid nitrogen temperatures or below. Generally, in the optical and near IR regimes, devices are operated at such a temperature that I_{DC} is negligible.

   3. Johnson (Nyquist) noise is due to the fluctuations in the voltage across a dissipative circuit element (the photo-conductor and its load resistor, R_L) caused by the thermal/random motion of the charge carriers. It is given by

$$\left\langle I_{\text{DC}}^2 \right\rangle = \frac{2 \left\langle P \right\rangle}{R_L} = \frac{2 \left(\frac{1}{2} kT \right)}{R_L t} = \frac{4kT}{R_L} \Delta f. \tag{51}$$

The magnitude of Johnson noise can be reduced by cooling the system (especially the load resistor) or by limiting the bandwidth of the amplification system.

   4. 1/*f* noise. Most electronic devices exhibit a component of noise, known as 1/*f* noise, which varies inversely with frequency. It is also known as excess noise, since it can greatly exceed the shot noise at frequencies below a few hundred Hz, dominating system performance. It is usually attributed to the front end-components of the amplification system. However, bad electrical contacts, temperature fluctuations, surface effects (damage) and crystal defects can contribute significantly. Since, there is no general understanding of 1/*f* noise, it is usually described by an empirical formula,

$$I_{\text{DC}} \propto \frac{2 \left\langle P \right\rangle}{R_L} = c \frac{I^a}{f^b} \Delta f, \tag{52}$$

where I is the current flowing through the device, Δf is the bandwidth over which the measurement is made, $a \sim 2$ and $b \sim 1$ and c is a constant. In the limiting case, when $b = 0$, this component is independent of frequency and is known as white noise. As with generation-recombination noise, 1/*f* noise can be minimized by keeping any DC current components small (*e.g.*, dark current).

   Since all four noise components vary randomly in phase they can be added in quadrature and thus the total system noise, I_N, is,

$$\langle I_N^2 \rangle = \langle I_{G-R}^2 \rangle + \langle I_{DC}^2 \rangle + \langle I_j^2 \rangle + \langle I_{1/f}^2 \rangle. \tag{53}$$

8.5.2.3 Photoconductor Performance Metrics

Several performance metrics or figures of merit can be used to characterize photoconductors. These are (1) the responsivity, S, which is a measure of how much signal is generated for a given incident photon power, (2) the noise equivalent power, *NEP*, which is a measure of the limiting sensitivity of a specific device and (3) the detectivity, D^*, which is a device-independent parameter of a material's effectiveness in detecting a weak signal in the presence of background.

8.5.2.3.1 Responsivity

A key performance parameter for a photoconductor is its responsivity, S, which is the detector output per unit input power, P. It is usually expressed as

$$S = \frac{I}{P}, \tag{54}$$

where I is the photo-generated current (not including the dark current). If we assume that $h\nu$ is low enough that a single photon can produce at most one conduction electron and every absorbed photon generates a conduction electron, then the quantum efficiency can be written as

$$\eta = 1 - \exp\left(-\mu_0 d\right), \tag{55}$$

where μ_0 is the absorption coefficient of the material and d is the device thickness. The number of photons incident per second is

$$N_p = \frac{P}{h\nu}, \tag{56}$$

of which a fraction produce electron-hole pairs. If the electron has an average lifetime of τ before recombining with a hole, then the average density of electrons n and holes p is given by

$$n = p = \frac{P\eta\tau}{h}, \tag{57}$$

where A is the active area of the device. Once an electron is created, it is accelerated by the electric field ζ with a mean velocity

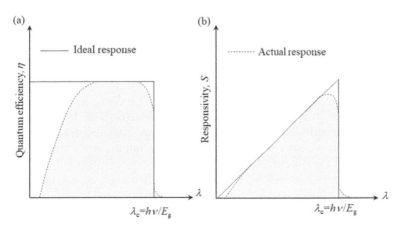

FIGURE 8.13 Schematic showing (a) the wavelength dependence of the quantum efficiency and (b) the responsivity of an intrinsic photoconductor. From the figures, we see that the efficiency is constant up to the cutoff wavelength dictated by the bandgap, whereas the responsivity increases proportionally with wavelength up to the cutoff.

$$\langle v_\text{e} \rangle = \mu_\text{e} \xi, \tag{58}$$

where μ_e is its mobility. Assuming that the electric field and current are parallel to the incident beam, the photocurrent is given by

$$I = enA \langle v_e \rangle + enA \langle v_h \rangle = P \frac{e\eta}{h\nu} \frac{\tau(\mu_e + \mu_h)}{d} \xi. \tag{59}$$

Therefore,

$$S = \frac{e\eta}{h\nu} G = \frac{e\eta \, G}{hc} \lambda, \tag{60}$$

where G is the photoconductive gain given by Eq. (45), *i.e.*,

$$G = \frac{\tau \mu_e \xi}{d}. \tag{61}$$

Note that we have dropped the hole mobility term, which is generally much smaller in comparison. To maximize the responsivity we see that we need to maximize G and τ. Fig. 8.13 shows the behaviour of the quantum efficiency and the responsivity with wavelength. From Eq. (55) we see that the quantum efficiency depends only on the absorption coefficient and the active depth of the device; typically, it is found to be very nearly independent of wavelength up to the cutoff wavelength – which in turn, is dictated by the bandgap. The responsivity, on the other hand, increases proportionally with wavelength up to the cutoff (see Eq. 60).

8.5.2.3.2 *Noise Equivalent Power* (NEP)

A second figure of merit is the noise equivalent power (NEP), which depends on noise characteristics and is thus device specific. It is a useful metric because it sets the detection limit of the detector. It is defined as the optical power that produces a signal voltage (or current) equal to the noise voltage (or current) of the detector. The noise is dependent on the bandwidth of the measurement, so that bandwidth must be specified. Frequently it is taken as 1 Hz. The equation defining NEP is

$$NEP = \frac{PAV_n}{V_s \sqrt{\Delta f}}, \tag{62}$$

where P is the irradiance incident on the detector of area A, V_n is the root mean square noise voltage within the measurement bandwidth Δf and V_s is the root mean square signal voltage. From the definition, it is apparent that the lower the value of the *NEP*, the better the characteristics of the detector for detecting a small signal in the presence of noise.

8.5.2.3.3 *Detectivity*

The NEP of a detector is a useful measure of the sensitivity of a specific device. To provide a general figure of merit that is dependent only on the intrinsic properties of the detector and not its size, we define the detectivity, D^*,

$$D^* = \frac{\sqrt{A}}{NEP}, \tag{63}$$

which is equal to the square root of the detector area per unit value of NEP. Since many detectors have *NEPs* proportional to the square root of their areas, D^* is independent of the area of the detector. The detectivity thus gives a measure of the intrinsic quality of the detector material itself. A high value of D^* means that the detector is suitable for detecting weak signals in the presence of noise. Fig. 8.14 shows some typical detectivity values for a number of available photoconductors, such as CdS, PbS, InAs, InSb, HgCdTe and doped Ge.

8.5.3 The Solid-State Ionization Chamber

For energies much greater than the bandgap energy, semiconductor radiation detectors are usually operated as solid-state ionization chambers. Until recently, this was most commonly realized by making a pn-junction or a metal-semiconductor junction (surface barrier diode) on very low-doped material. The junction is reverse biased so that a thick depletion layer is

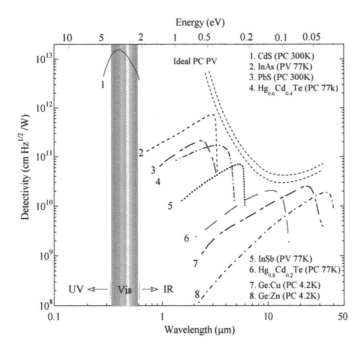

FIGURE 8.14 Detectivity as a function of wavelength for a number of different types of photodetectors operating in the infrared spectrum. The temperature of operation is indicated. Photovoltaic detectors are denoted PV; photoconductive detectors are denoted PC. The curve for an ideal photoconductive detector assumes a 2π steradian field of view and a 295 K background temperature.

formed where the field is high and at the same time the leakage current is low. In Fig. 8.15 we illustrate a simple planar detector geometry. The device can be thought of as a parallel plate capacitor of separation L with ohmic contacts. We assume that space charge within the detector is negligible, implying that the electric field, E, is constant in the material. Ionizing radiation absorbed in the sensitive volume excites electron-hole pairs in direct proportion to the energy deposited (that is, $n = E_0/\varepsilon$, where n is the number of electron-hole pairs generated, E_o is the energy deposited and ε is the average energy consumed to create an electron-hole pair). Note that ε will be greater than ε_g the bandgap energy; the difference being due to the energy lost in producing phonons, which is required to conserve both energy and momentum. Remarkably, the ratio, $\varepsilon/\varepsilon_g$, is constant (=2.8) for most materials and is independent of the type of radiation [38]. Applying an electric field across the detector causes the liberated carriers to separate. The electrons drift towards the anode and the holes towards the cathode with drift velocities v_e and v_h.

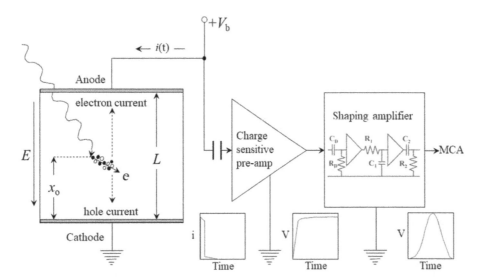

FIGURE 8.15 Schematic of a simple planar detection system (from [37], reproduced with permission of the International Union of Crystallography). For completeness, we also show the signal chain and the evolution of currents and voltages. Ionizing radiation absorbed in the sensitive volume generates electron-hole pairs in direct proportion to the energy deposited. These are subsequently swept towards the appropriate electrode by the electric field established by the bias.

The motion of the carriers creates an induced charge on the electrodes and by current continuity into the electronics. The induced charge Q_{ind} is given by the Shockley-Ramo theorem [39,40], which makes use of the concept of a weighting potential. The weighting potential is defined as the potential that would exist in the detector with the collecting electrode held at unit potential, while holding all other electrodes at zero potential. The theorem states that regardless of the presence of space charge, the change in the induced charge ΔQ_{ind} and the current i at an electrode caused by a charge q moving from x_i to x_f are given by

$$\Delta Q_{ind} = \int_{xi}^{xf} q\vec{E}_w d\vec{x} = -q[\Phi_w(x_f) - \Phi_w(x_i)] \tag{64}$$

and

$$i = \frac{dQ}{dt} = q\vec{v}\vec{E}_w, \tag{65}$$

$\vec{v}$ is the carrier velocity and Φ_w and $\vec{E}_w$ are the weighting potential and electric field at position x. Note that the charge considered here need not be physically collected by an electrode to produce a signal. The motion of the charges through the bulk produces the signal by induction, so the signal begins to form immediately when energy deposition starts. It is only when the last carrier arrives at the collecting electrode that charge induction ceases and the signal is now fully formed. The total induced charge will be the sum of the induced charges due to the electron and holes. The integration of these charges along their respective pathlengths gives the total charge, Q. In the absence of trapping, $Q=Q_o$, the original charge is created, which in turn is proportional to the energy of the incident photon. However, in any semiconductor some density of electron and hole traps are always present, and these result in a loss of carriers and therefore of charge at the electrodes. For compound semiconductors, crystal growth techniques lead to a much higher density of traps than in the elemental semiconductors and hence to shorter lifetimes. For example, in CdZnTe, typical lifetimes are $\tau_e = 3 \times 10^{-6}$ sec and $\tau_h = 5 \times 10^{-8}$ sec. Because the hole lifetime is much shorter than the hole transit time the induced charge is significantly reduced and is now dependent upon the depth of interaction, x_o. For a uniform electric field and negligible de-trapping, the fraction of charge that is induced at the electrodes is best described by the charge collection efficiency (CCE), given by the Hecht equation [41]

$$CCE = \frac{Q}{Q_o} = \frac{\lambda_e}{L}\left[1 - \exp\left(-\frac{(L - x_o)}{\lambda_e}\right)\right] + \frac{\lambda_h}{L}\left[1 - \exp\left(-\frac{x_o}{\lambda_h}\right)\right], \tag{66}$$

where L is the detector thickness, x_o is the distance from the cathode to the point of charge creation and λ_e and λ_h are the carrier drift lengths[5] in the applied electric field, E, given by, $\lambda_e = \mu_e \tau_e E$ and $\lambda_h = \mu_h \tau_h E$. Here, μ_e and μ_h are the electron and hole mobilities and t_e and t_h are the corresponding lifetimes. It follows from Eq. (66) that the CCE depends not only on λ_e and λ_h, but also on the location where the charge was created. Since the interaction points of incident photons are essentially random at intermediate and high energies, being weighted by the classical exponential absorption law, the width of the peak in the energy spectrum broadens to an extent governed by the ratios λ_e/L and λ_h/L.

In characterizing materials, the most useful figure of merit is the mobility-lifetime product ($\mu\tau$), which is directly related to the drift length. Low $\mu\tau$ products result in shorter drift lengths and therefore smaller λ/L, which in turn limits the maximum size and energy range of detectors. For typical electric field strengths of ~ 1 kV/cm, electron drift lengths range from ~ 1 m for Si and Ge, ~ 1 cm for $\sim$CdZnTe and CdTe, 1 mm for GaAs and HgI$_2$ and ~ 1 μm for GaP, InP and PbI$_2$. The hole drift lengths are usually 1 to 2 orders of magnitude smaller. For the elemental semiconductors $\mu\tau$ is of the order of 1 cm^{2}V^{-1} for both electrons and holes, whereas for compound semiconductors it is typically a few times 10^{-4} cm^{2}V^{-1} for electrons and 10^{-5} cm^{2}V^{-1} for holes becoming smaller with increasing Z (see Table 8.5). The degradation can usually be traced to trapping centers caused by impurities, vacancies, structural irregularities, such as dislocations, or for the softer materials, plastic deformation caused by mechanical damage during fabrication.

8.5.3.1 Spectral Broadening in Radiation Detection Systems

In addition to broadening caused by poor charge collection, the width of the full energy peak, ΔE, is broadened by the statistics of carrier generation (Fano noise) and by electronic noise. For most compound semiconductors, however, these components are generally far less important than the noise due to incomplete charge collection, except for thin detectors where $l_e/L \gg 1$. The energy

---

[5] λ is also known as the trapping length, "schubweg" or charge-collection distance and is the mean distance travelled by a charge carrier before trapping or collection occurs.

TABLE 8.5

A condensed list of the room temperature properties of wide band gap compound semiconductor materials suitable for hard X- and g-ray detectors (taken from [37]). The abbreviations are: FZ Float Zone, Cz Czochralski, CVD Chemical Vapor Deposition, LEC Liquid Encapsulated Czochralski, THM Traveller Heater Method, B-S Bridgman-Stockbarger method, HPB High Pressure Bridgman and VAM Vertical Ampoule Method, cp cubic close packed, hcp hexagonal close packed.

Material	Crystal Structure	Growth method	Atomic Number	Density (g/cm^3)	Band Gap (eV)	Epair (eV)	Resistivity (W-cm)	μt(e) Product (cm^2/V)	μt (h) Product (cm^2/V)
Si	cp	FZ	14	2.33	1.12	3.66	$<10^4$	>1	~ 1
Ge	cp	Cz	32	5.33	0.66	2.96	50	1	1
4H-SiC	hcp	CVD	14,6	3.29	3.27	7.28	$>10^3$	4×10^{-4}	8×10^{-5}
GaAs	cp	CVD	31,33	5.32	1.43	4.18	10^{10}	1×10^{-4}	4×10^{-6}
InP	cp	LEC	15,49	4.79	1.34	4.2	10^8	2×10^{-5}	1×10^{-5}
CdTe	cp	THM	48,52	5.85	1.44	4.43	10^9	3×10^{-3}	2×10^{-4}
Cd$_{0.9}$Zn$_{0.1}$Te	hcp	HPB	48,30,52	5.78	1.57	4.64	10^{11}	1×10^{-2}	2×10^{-4}
PbI$_2$	layered	B-S	82,53	6.2	2.32	4.9	10^{13}	1×10^{-5}	1×10^{-6}
HgI$_2$	layered	VAM	80,53	6.4	2.13	4.15	10^{13}	1×10^{-4}	4×10^{-5}
TlBr	cp	B-S	81,35	7.56	2.68	6.5	10^{12}	3×10^{-3}	6×10^{-5}

resolution, ΔE, of the system is defined as the full width at half-maximum of the energy-loss distribution resulting from exposure to monoenergetic radiation. The width of the energy-loss spectrum, in turn, results from the convolution of the probability distributions of the various noise components. For most detection systems, three components tend to dominate,

$$\Delta E = f(\sigma_F^2, \sigma_e^2, \sigma_c^2) \quad (keV), \qquad (67)$$

where, σ_F^2 is the variance of the noise due to carrier generation or Fano noise, σ_e^2 is the variance of the noise due to the leakage current and amplifier noise and σ_c^2 is the variance of the noise due to incomplete charge collection due to carrier trapping.

8.5.3.1.1 Fano Noise

In semiconductors, only about one-tenth of the initial energy deposition results in the creation of electron-hole pairs, the remainder being lost to lattice vibrations, phonons and plasmon production. Consequently, unlike scintillation or gas counter detection systems, the limiting spectroscopic resolution achievable is not a simple function of total energy deposition. Until recently, the only realistic descriptions were semi-empirical, based on the so-called Fano factor [42]. This was originally introduced to quantify the departure of the observed statistical fluctuations in the number of charge carriers in a gas from that expected from pure Poisson statistics. It is generally expressed as

$$F = \frac{\sigma_{exp}^2}{var_{poisson}}, \qquad (68)$$

where F is the Fano factor, σ_{exp}^2 is the experimentally observed variance and $var_{poisson}$ is the Poissonian variance (equal to the mean number of events, N). Obviously, for purely Poisson statistics σ_{exp}^2 is equal to the variance and $F = 1$. Thus, the Fano factor essentially describes the energy partition, and in semiconductors, the fraction of the total energy that goes into the production of electron-hole pairs. Its contribution to the resolution function, ΔE_F, can be calculated from

$$\Delta E_F = 2.355\sqrt{\sigma_F^2} = 2.355\sqrt{FE\varepsilon} \quad keV, \qquad (69)$$

where ε is the energy to create an electron-hole pair and E is the incident energy. The experimental determination of the Fano factor is extremely difficult for reasons outlined in Owens *et al.* [43] and has only been determined to reasonable accuracy for a few semiconductors (see Table 8.6). The first reason is that the determination is only accurate when trapping noise and electronic noise are so low that the resolution is already close to the Fano limit. This is because Fano noise is not normally distributed, so a simple Gaussian decomposition of Eq. (67) is incorrect (see discussion in [43]). The second reason is that Fano noise is a multiplicative function of the Fano factor and the electron-hole pair energy – the latter being usually

TABLE 8.6

Measured Fano factors in various materials. Except for the measurement of Owens, Fraser and McCarthy [43] for the Fano factor in Si, the other experimental values were derived from a Gaussian decomposition of the resolution function and as pointed out in [43] such an approach will lead to systematically low values of the Fano factor, because Fano noise is not normally distributed.

Material	Bandgap (eV)	Pair energy @300K (eV)	Fano factor	Temp. (K)	Excitation	Ref.
Si	1.12	3.66 ± 0.03	0.155 ± 0.002	170	2–4 keV X-rays	[44,45]
Ge	0.66	2.96	0.112 ± 0.001	77	14–6129 X, γ-rays	[46]
4H-SiC	3.27	7.28	0.128 ± 0.0? ‡	300	5.49 MeV α-rays	[47]
GaAs	1.43	4.184 ± 0.025	0.140 ± 0.05†	233	10–100 keV X-rays	[48,49]
GaN	3.39	8.33 ± 0.013	0.07 ± ? ‡	300	5.49 MeV α-rays	[50]
HgI_2	2.13	4.15	0.19 ± 0.03†	253	5.9 keV X-rays	[51]
CdTe	1.47	4.43	0.09 ± 0.03†	253	18–122 keV X-rays	[52]
$Cd_{0.9}Zn_{0.1}Te$	1.57	4.64	0.099 ± 0.02	253	10–100 keV X-rays	[53]
$In_{0.5}Ga_{0.5}P$	1.9	4.95 eV ± 0.07	0.13 ± ?	303	4.95, 21.1 keV X-rays	[56]

† Derived from a Gaussian decomposition of the resolution function.
‡ Derived from an empirical model.

determined by assuming the Fano factor *a priori*. A further complication was revealed by the theoretical work of Fraser *et al.* [54] supported by the experimental work of Owens *et al.* [44] on Si, which showed that both quantities are energy and temperature dependent and that the dependences are different. The same has been shown for Ge [55].

8.5.3.1.2 *Electronic Noise*

The electronic noise term in Eq. (67) arises largely from shot noise generated by the leakage current of the detector and noise generated in the front-end electronics of the amplification system. Both types of noise are random in amplitude and time over a wide frequency range, and their contribution to measured energy fluctuations depends on how the signal is processed. A typical detector front-end schematic is given in Fig. 8.16(a). The detector can be represented by a capacitor, since in essence it consists of two parallel conductors with a dielectric in between. The detector is coupled to the bias supply through a bias resistor R_b. The bias supply is decoupled by capacitor, C_b, whose primary function is to act as a filter for the supply. Under AC conditions this capacitor allows R_b essentially to shunt the detector. The signal from the detector is then fed to the preamplifier *via* the blocking capacitor, C_c, whose function is to pass the signal to the preamp unimpeded while isolating the preamp from the bias supply. The series resistance R_s represents the sum of all resistances present in the input signal path (such as contact and any parasitic resistances). The preamplifier buffers the signal and sends it to a pulse shaper, which filters the overall frequency response to optimize signal to noise. At the same time, it limits the duration of the signal pulse to accommodate the expected signal rate. Generally, Gaussian or pseudo Gaussian shaping is used, designed to avoid overshoot to a step function input while also minimizing the rise and fall time of the signal.

The electronic noise expected from the front-end can be derived from its equivalent electrical circuit shown in Fig. 8.16(b), in which various contributors to the noise are represented by current or voltage sources depending on

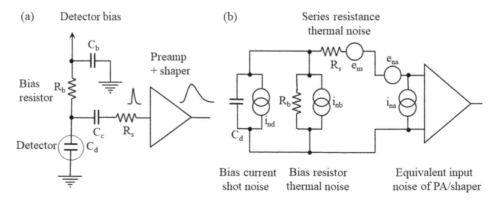

FIGURE 8.16 (a) A typical detector front-end detector-preamplifier-shaper circuit and (b) the equivalent circuit for noise calculations (from Spieler [57]). Here e_n and i_n represent the various series (voltage) and parallel (current) noise sources, respectively.

whether they are in parallel (current) or series (voltage) with the detector capacitance. By convention, those in parallel are known as parallel noise sources and those in series as series noise sources. The various noise sources add in quadrature. Since the energy deposited in the detector translates directly into charge, it makes sense to express the electronic noise in terms of an Equivalent Noise Charge (ENC), defined as the number of electron charges which if applied to the input would give rise to the same RMS output voltage. The ENC and FWHM energy resolution are simply related by

$$\Delta E_e = \frac{2.355\,\varepsilon\,\mathrm{ENC}}{e}. \tag{70}$$

The parallel component, Q_p, is comprised of the shot noise due to the detector leakage current, I, and Johnson noise from the bias resistor, R_b, which under signal conditions effectively shunts the detector. It is given by the expression

$$Q_\mathrm{p}^2 = \left(2eI + \frac{4kT}{R_\mathrm{b}} + i_\mathrm{na}^2\right)\tau. \tag{71}$$

Here e is the electronic charge, k the Boltzmann constant, T the temperature and i_na^2 represents the current noise due to the preamplifier. The leakage current is the sum of two main components, the bulk leakage current and the surface leakage current, which depends on detector design. Bulk leakage currents are generated internally within the detector volume and arise from a number of sources. Firstly, because of the finite resistivity of the bulk material, a standing current will always be present. In junction-type detectors, bulk leakage also arises as a natural consequence of the junction, since while majority carriers are repelled from the boundaries of the depletion region under bias, minority carriers are attracted and free to diffuse across it. Minority carrier current is usually small and can generally be neglected compared to surface leakage currents. Bulk leakage current also arises due to the thermal generation of carriers within the depletion region, the rate of which depends on its volume. Clearly this component is largely dependent on the bandgap energy and can be significantly reduced by cooling. Surface leakage takes place at the surface of the detector and is particularly acute at the edges where potential gradients are the greatest. Its magnitude is very dependent on material processing, surface preparation and passivation techniques as well as material type.

Elemental semiconductor detectors are usually junction devices operated in a reverse bias mode to reduce bulk leakage currents and maximize active depths. However, for most semiconducting materials the bulk resistivity is still too low to allow low noise operation. Therefore, a Schottky barrier or blocking contact is usually employed. Assuming surface leakage components can be neglected, the leakage current in this case is

$$I = A^* T^2 e^{-e\Phi/kT}, \tag{72}$$

where A^* is a constant related to Richardson's constant and Φ is the barrier height of the Schottky contact. We see that the leakage current is a strong function of temperature. For this reason, it is usually necessary to cool the detector to achieve low noise performance. At the extremes, Fano limited operation requires leakage currents in the pA range, whereas MeV alpha particles can barely be detected above the noise for leakage currents in the mA range.

The series noise contribution, Q_s, is largely due to the voltage noise due to the series resistance, R_s and noise due to the amplifier. It is given by the expression

$$Q_\mathrm{s}^2 = \left(4kTR_\mathrm{s} + v_\mathrm{na}^2\right)C_\mathrm{d}^2\frac{1}{\tau} + 4A_\mathrm{f}C_\mathrm{d}^2, \tag{73}$$

where v_na^2 is voltage noise due to the preamplifier and A_f is a noise coefficient that depends on the particular amplifier design. Normally the input stage of the preamp is a FET. In this case, the square of the shot noise of the first stage of the preamplifier is proportional to

$$\frac{T}{g_\mathrm{m}\tau}C_\mathrm{d}^2, \tag{74}$$

where g_m is the transconductance of the FET. The last term in Eq. (73) represents $1/f$ or "pink" noise due to the amplifier. Note that the amplifier noise sources are not present at the amplifier input but originate within the amplifier and appear at the output. In order to take account of this component, the noise can be referred back to the input by dividing by the gain so that it looks like a voltage noise generator. For simple RC-RC shaping, Spieler [57] gives the following formulation for the total ENC.

$$Q_n^2 = \left(\frac{\varepsilon^2}{8}\right)\left[\left(2eI + \frac{4kT}{R_b}\right)\tau + \left(4kTR_s + v_{na}^2\right)C_d^2\frac{1}{\tau} + 4A_fC_d^2\right].$$

(75)

$$\underset{\text{current noise}}{\uparrow} \qquad \underset{\text{voltage noise}}{\uparrow} \qquad \underset{1/f \text{ noise}}{\uparrow}$$

We see from Eq. (75) that current or parallel noise increases with shaping time and is independent of detector capacitance. It can be reduced by (a) measuring at shorter shaping time, (b) reducing the current through the detector or (c) choosing a feedback resistor with a high resistance or by avoiding it altogether and using a different reset mechanism. Voltage or series noise can be reduced by (a) measuring at a longer shaping time and (b) minimizing detector and stray capacitance. For an FET input stage, voltage noise can be reduced by selecting a low noise FET with a larger transconductance. The $1/f$ noise is independent of shaping time and can only be lowered by reducing the detector capacitance or the noise coefficient of the amplifier. As the shaping time τ is changed, the total noise goes through a minimum, where the current and voltage contributions are equal. This is illustrated in Fig. 8.17 from which we see that at short shaping times voltage noise dominates, whereas at long shaping times current noise takes over. Consequently, the total noise has a minimum value at the shaping time constant where the series noise is equal to the parallel noise. This time constant is called the noise corner time constant. Note that the effect of $1/f$ noise is merely to smooth out this minimum. The time constant for minimum noise will depend on the characteristics of the detector, the preamplifier and the amplifier pulse shaping network. A spectrometer will normally be operated with $\tau = \tau_{opt}$, if event rate considerations do not require shorter shaping times, as in high-energy physics applications. For most compound semiconductor detectors, τ_{opt} is usually in the range 0.5 to 6 μs, while for Ge detectors, it can be as long as 20 μs. Such long time constants impose a severe restriction on the counting rate capability – for Ge detectors this usually means a count rate limitation of around 10 kHz for to preserve good energy resolution. In practice, the resolution is often compromised by selecting shorter shaping time constants in order to handle higher counting rates.

Generally, we know little of the preamplifier and shaper. For noise evaluation purposes, Eq. (75) can be simplified if either the Johnson noise contribution is negligible for the bias resistor (therefore high resistance and low temperature), the series resistance is negligible (which is usually true) or the remaining voltage noise terms can be treated as a single empirical constant. In this case, the FWHM energy resolution due to the electronic noise component, ΔE_e, reduces to

$$\Delta E_e = 2.355\sqrt{I\tau\varepsilon^2 A/e + \sigma_a^2} \quad \text{keV},$$

(76)

where A is a constant depending on the type of signal shaping and σ_a^2 is the variance of the amplifier shot noise, which depends on the actual design and specifically the capacitive loading of the front-end electronics. We can treat this term as an empirical constant. For the Gaussian shaping typically used in spectroscopy, $A = 0.875$.

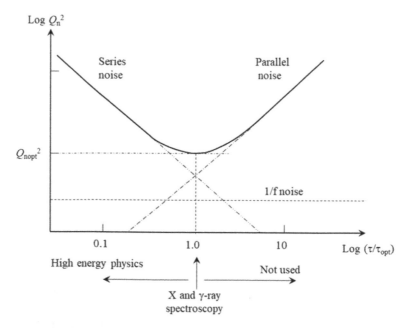

FIGURE 8.17 Equivalent noise charge versus shaping time. At low shaping times, series (voltage) noise dominates whereas at high shaping times parallel (current) noise dominates.

From equations (75) and (76) we note that for given temperature and operating conditions, electronic noise is independent of the energy of the incident radiation, which makes it relatively easy to measure directly. This is achieved by injecting a precise amount of charge into the input of the preamplifier and measuring the degree by which it is broadened. The empirical term in Eq. (76) can then be determined by best fitting.

8.5.3.1.3 Trapping Noise

The functional form of noise due to incomplete charge collection, ΔE_c, is much more difficult to predict, since it is intimately dependent on the trap density distribution as well as the charge diffusion and collection properties of the detector. For the case of $\lambda_e \neq \lambda_h$, which is invariably true for compound semiconductors, Iwanczyk *et al.* [58] have arrived at the following analytic form for the relative broadening on the assumption of a uniform electric field:

$$\left(\frac{\sigma_c^2}{E}\right)^2 = \frac{2\lambda_e^2\lambda_h^2}{L^3(\lambda_e-\lambda_h)}\left(e^{-L/\lambda_e} - e^{-L/\lambda_h}\right) - \frac{1}{L^4}\left[\lambda_e^2\left(e^{-L/\lambda_e} - 1\right) + \lambda_h^2\left(e^{-L/\lambda_h} - 1\right)\right]^2.$$
$$- \frac{\lambda_e^3}{2L^3}\left(e^{-2L/\lambda_e} - 1\right) + \frac{\lambda_h^3}{2L^3}\left(e^{-2L/\lambda_h} - 1\right) \qquad (77)$$

However, while Eq. (77) can be used to calculate the width ΔE_c, the shape of the pulse height distribution can only be realistically evaluated by a more detailed approach such as that of Trammell and Walter [59], in which the individual pulse heights in infinitesimal slices through the detector are summed over the detector thickness. For low trapping, the summed pulse height distribution will be nearly symmetric at low and intermediate energies, and we can therefore assume ΔE_c is normally distributed. For this case, several semi-empirical formulas have been proposed for the summed response. For example, Henck *et al.* [60] derived an expression for planar detectors in which $\Delta E_c \propto E^{\frac{1}{2}}$ whereas Owens [61] predicted that $\Delta E_c \propto E$ for coaxial detectors. Because of the uncertainties in the functional form of ΔE_c and our imprecise knowledge of F, a semi-empirical approach is generally used to describe the resolution function in the form

$$\Delta E = 2.355\sqrt{F\varepsilon E + (\Delta E_e/2.355)^2 + a_1 E^{a_2}} \quad \text{keV}, \qquad (78)$$

where the electronic noise component ΔE_e is measured directly using a precision pulser and F, a_1 and a_2 are treated as semi-empirical constants determined by best-fitting. Eq. (78) is found to fit the resolution functions of compound semiconductors reasonably well with the charge trapping exponent, a_2, varying between 2 and 3. Recently, the validity of the functional form of the trapping term has been rigorously tested by Kozorezov *et al.* [62], who derived a general analytical expression that includes the effects of geometry. Interestingly, it can be expressed in the form $G(E)E^2$, in which the function $G(E)$ reduces to a constant when the *1/e* absorption lengths become comparable with the detector thickness – which is invariably true for spectroscopic detectors.

The relative magnitudes of the various noise components are illustrated in Fig. 8.18. Here we show the measured ΔE for a 3.1 mm^2, 2.5 mm thick CdZnTe detector as a function of energy deposition [53]. The total leakage current at the nominal

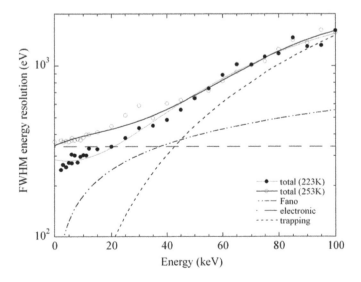

FIGURE 8.18 The energy resolution ΔE of a 3.1 mm^2, 2.5 mm thick CdZnTe detector measured at two temperatures (from [53]). The solid line shows the best-fit resolution function to −20°C data. The individual components to the FWHM are also shown for this curve. These are the noise due to carrier generation or Fano noise ΔF, electronic noise due to leakage current and amplifier shot noise, Δe, and incomplete charge collection or trapping noise Δc.

bias was < 1nA. In the low energy region, electronic noise dominates the response, while at intermediate energies Fano noise takes over. At high energies, trapping noise becomes the dominant component.

REFERENCES

[1] N. Bohr, "On the theory of decrease of velocity of moving electrified particles on passing through matter", *Philos. Mag.*, Vol. **25**, no. 145 (1913), pp. 10–31.

[2] H.A. Bethe, "Zur Theorie des Durchgangs schneller Korpuskularstrahlen durch Materie", *Ann. Physik.*, Vol. **397** (1930), pp. 325–400.

[3] F. Bloch, "Bremsvermögen von Atomen mit mehreren Elektronen", *Z. Phys.*, Vol. **81** (1933), pp. 363–376.

[4] L.H. Thomas, "The calculation of atomic fields", *Proc. Camb. Phil. Soc.*, Vol. **23**, no. 5 (1927), pp. 542–548.

[5] E. Fermi, "Un Metodo Statistico per la Determinazione di alcune Priorieta dell 'Atome", *Rend. Accad. Naz. Lincei*, Vol. **6** (1927), pp. 602–607.

[6] C. Patrignani, et al., "(Particle data group)", *Chin. Phys. C*, Vol. **40**, no. 100001 (2016), pp. 1–1808.

[7] L. Landau, "On the energy loss of fast particles by ionisation", *J. Phys. (USSR)*, Vol. **8** (1944), pp. 201–205.

[8] R. Bailey, C.J. Damerell, R.L. English, A.R. Gillman, A.L. Lintern, S.J. Watts, F.J. Wickens, "First measurement of efficiency and precision of CCD detectors for high-energy physics", *Nucl. Instrum. Methods*, Vol. **213** (1983), pp. 201–215.

[9] A. Owens, K.J. McCarthy, "Energy deposition in X-ray CCDs and charged particle discrimination", *Nucl. Instrum. Methods*, Vol. **A366** (1995), pp. 148–154.

[10] G. Hall, "Ionization energy losses of highly relativistic charged particles in thin silicon layers", *Nucl. Instrum. Methods*, Vol. **220** (1984), pp. 356–362.

[11] P. Shulek, B.M. Golovin, L.A. Kulyukina, S.V. Medved, P. Pavlovich, "Fluctuations of ionization loss", *Sov. J. Phys.*, Vol. **4** (1967), pp. 400–401.

[12] O. Blunck, S. Leisegang, "Zum Energieverlust schneller Elektronen in dünnen Schichten", *Z. Physik*, Vol. **128** (1950), pp. 500–505.

[13] S.F. Mughabghab, M. Divadeenam, N.E. Holden, *"Neutron Cross Sections, Volume 1, Neutron Resonance Parameters and Thermal Cross Sections, Part A, Z=1-60"*, Academic Press, New York (1981).

[14] S.F. Mughabghab, *"Neutron Cross Sections, Volume 1, Neutron Resonance Parameters and Thermal Cross Sections, Part B, Z=61-100"*, Academic Press, New York (1984).

[15] J.H. Lambert, "Photometria sive de mensura et gradibus luminis, colorum et umbrae", *Augustae Vindelicorum, Sumptibus Vidvae Eberhardi Klett, typis Christophori Petri Detleffsen* (1760).

[16] A. Beer, "Bestimmung der Absorption des rothen Lichts in farbigen Flüssigkeiten", *Annalen Der Physik Und Chemie*, Vol. **86** (1852), pp. 78–88.

[17] U. Fano, "Gamma-ray attenuation, part 1", *Nucleonics*, Vol. **11**, no. 8 (1953a), pp. 8–12.

[18] U. Fano, "Gamma-ray attenuation, part 2", *Nucleonics*, Vol. **11**, no. 9 (1953b), pp. 55–60.

[19] H.G.J. Moseley, "The high frequency spectra of the elements", *Philos. Mag.*, Vol. **26** (1913), pp. 1024–1034.

[20] C.M. Davisson, R.D. Evans, "Gamma-ray absorption coefficients", *Rev. Mod. Phys.*, Vol. **24**, no. 2 (1952), pp. 79–107.

[21] W. Heitler, *"The Quantum Theory of Radiation"*, Oxford University Press, Oxford, 3rd ed. (1954).

[22] D. Coster, R.D.L. Kronig, "New type of Auger effect and its influence on the X-ray spectrum", *Physica*, Vol. **2**, no. 1–12 (1935), pp. 13–24.

[23] M. Thompson, M.D. Baker, A. Christie, J.F. Tyson, *"Auger Electron Spectroscopy"*, John Wiley & Sons, Chichester (1985) ISBN 0-471-04377-X.

[24] J.J. Thomson, "On the scattering of rapidly moving electrified particles", *Proc. Camb. Philos. Soc.*, Vol. **75** (1910), pp. 465–471.

[25] A.H. Compton, "A quantum theory of the scattering of X-rays by light elements", *Phys. Rev.*, Vol. **21**, no. 5 (1923), pp. 483–502.

[26] O. Klien, Y. Nishina, "Über die Streuung von Strahlung durch freie Elektronen nach der neuen relativistischen Quantendynamik von Dirac", *Z. Physik*, Vol. **52** (1929), pp. 853–868.

[27] P.A.M. Dirac, "The quantum theory of the electron", *Proceedings of the Royal Society A: Mathematical, Physical and Engineering Sciences*, Vol. **117**, no. 778 (1928), pp. 610–624.

[28] F. Perrin, "Possibilité de matérialisation par interaction d'un photon et d'un électron", *Comptes Rendus Hebdomadaires Des Séances De l'Académie Des Sciences*, Vol. **197** (1933), pp. 1100–1102.

[29] H. Davies, H.A. Bethe, L.C. Maximon, "Theory of Bremsstrahlung and pair production. II. Integral cross section for pair production", *Phys. Rev.*, Vol. **93** (1954), pp. 788–795.

[30] H.A. Bethe, W. Heitler, "On the stopping of fast particles and on the creation of positive electrons", *Proc. Roy. Soc. A*, Vol. **146** (1934), pp. 83–112.

[31] P. Marmier, E. Sheldon, *"Physics of Nuclei and Particles, Volume I"*, Academic Press, New York (1969).

[32] R.D. Evans, *"The Atomic Nucleus"*, McGraw-Hill, New York (1955), p. 712.

[33] J.H. Hubbell, "Review and history of photon cross section calculations", *Phys. Med. Biol.*, Vol. **51** (2006), pp. R245–R262.

[34] R.M. Genet, A.A. Henden, B.D. Holenstein, "Light bucket astronomy", *The Society for Astronomical Sciences 29th Annual Symposium on Telescope Science*, Pub. by the Society for Astronomical Sciences (2010), pp.117–122.

[35] A. Rose, *"Concepts in Photoconductivity and Allied Problems"*, Interscience Publishers, New York (1963).

[36] G. Rieke, "*Detection of Light: From the Ultraviolet to the Submillimeter*", Cambridge University Press, Cambridge, 2nd ed. (2002) ISBN:9780521017107.

[37] A. Owens, "Semiconductor materials and radiation detection", *J. Synchrotron Radiation*, Vol. **13**, part 2 (2006), pp. 143–150.

[38] C.A. Klein, "Bandgap dependence and related features of radiation ionization energies in semiconductors", *J. Appl. Phys.*, Vol. **4** (1968), pp. 2029–2033.

[39] W. Shockley, "Currents to conductors induced by a moving point charge", *J. Appl. Phys.*, Vol. **9** (1938), pp. 635–636.

[40] S. Ramo, "Currents induced by electron motion", *Proc. IRE*, Vol. **27** (1939), pp. 584–585.

[41] K. Hecht, "Zum Mechanismus des lichtelektrischen Primärstromes in isolierenden Kristallen", *Z. Physik*, Vol. **77** (1932), pp. 235–245.

[42] U. Fano, "Ionization yield of radiations. II. The fluctuations of the number of ions", *Phys. Rev.*, Vol. **72** (1947), pp. 26–29.

[43] A. Owens, G.W. Fraser, K.J. McCarthy, "On the experimental determination of the Fano factor in Si at soft X-ray wavelengths", *Nucl. Instrum Methods*, Vol. **A491** (2002), pp. 437–443.

[44] A. Owens, G.W. Fraser, A.F. Abbey, A. Holland, K. McCarthy, A. Keay, A. Wells, "The X-ray energy response of silicon (B): Measurements", *Nucl. Instrum. Methods*, Vol. **A382** (1996), pp. 503–510.

[45] F. Scholze, H. Henneken, P. Kuschnerus, H. Rabus, M. Richter, G. Ulm, "Determination of the electron–Hole pair creation energy for semiconductors from the spectral responsivity of photodiodes", *Nucl. Instrum Methods*, Vol. **A439** (2000), pp. 208–215.

[46] S. Croft, D.S. Bond, "A determination of the Fano factor for germanium at 77.4 K from measurements of the energy resolution of a 113 cm³ HPGe gamma-ray spectrometer taken over the energy range from 14 to 6129 keV", *Int. J. Radiat. Appl. Instrum. A.*, Vol. **42**, no. 11 (1991), pp. 1009–1014.

[47] S.K. Chaudhuri, K.J. Zavalla, K.C. Mandal, "Experimental determination of electron-hole pair creation energy in 4H-SiC epitaxial layer: An absolute calibration approach", *Appl. Phys. Letts.*, Vol. **102** (2013), pp. 031109-1–031109-4.

[48] G. Bertuccio, D. Maiocchi, "Electron-hole pair generation energy in gallium arsenide by x and γ photons", *J. Appl. Phys.*, Vol. **92** (2002), pp. 1248–1255.

[49] A. Owens, M. Bavdaz, A. Peacock, A. Poelaert, H. Andersson, S. Nenonen, H. Sipila, L. Tröger, G. Bertuccio, "High resolution X-ray spectroscopy using GaAs arrays", *J. Appl. Phys.*, Vol. **90** (2001), pp. 5376–5381.

[50] P.L. Mulligan, "*Fabrication and Characterization of Gallium Nitride Schottky Diode Devices for Determination of Electron-Hole Pair Creation Energy and Intrinsic Neutron Sensitivity*", PhD thesis, Ohio State University (2015).

[51] G.R. Ricker, J.V. Vallerga, A.J. Dabrowski, J.S. Iwanczyk, G. Entine, "New measurement of the Fano factor of mercuric iodide", *Rev. Sci. Instrum.*, Vol. **53** (1982), pp. 700–701.

[52] L. Abbene, G. Gerardi, "High resolution X-ray spectroscopy with compound semiconductor detectors and digital pulse processing systems", *X-Ray Spectrosc.*, Vol. **2012-02** (2012), pp. 39–64.

[53] A. Owens, M. Bavdaz, H. Andersson, T. Gagliardi, M. Krumrey, S. Nenonen, A. Peacock, I. Taylor, L. Troger, "The X-ray response of CdZnTe", *Nucl. Instrum. Methods*, Vol. **A484** (2002), pp. 242–250.

[54] G.W. Fraser, A.F. Abbey, A. Holland, K. McCarthy, A. Owens, A. Wells, "The X-ray energy response of silicon Part A. Theory", *Nucl. Instrum. Methods*, Vol. **A350** (1994), pp. 368–378.

[55] B.G. Lowe, "Measurements of Fano factors in silicon and germanium in the low-energy X-ray region", *Nucl. Instrum. Methods*, Vol. **A399** (1997), pp. 354–364.

[56] G. Lioliou1, A.B. Krysa, A.M. Barnett, "Energy response characterization of InGaP X-ray detectors", *J. Appl. Phys.*, Vol. **124** (2018) pp. 195704-1–195704-8.

[57] H. Spieler, "*Semiconductor Detector Systems*", Oxford University Press, Oxford (2005) ISBN: 9780198527848.

[58] J.S. Iwanczyk, W.F. Schnepple, M.J. Masterson, "The effect of charge trapping on the spectrometric performance of HgI₂, gamma-ray detector", *Nucl. Instrum. Methods.*, Vol. **A322** (1992), pp. 421–426.

[59] R. Trammell, F.J. Walter, "The effects of carrier trapping in semiconductor gamma-ray detectors", *Nucl. Instrum. Methods*, Vol. **76** (1969), pp. 317–321.

[60] R. Henck, D. Gutknecht, P. Siffert, L. De Laet, W. Shoenmaekers, "Trapping effects in Ge(Li) detectors and search for correlation with characteristics measured on the p-type crystals", *IEEE Trans. Nucl. Sci.*, Vol. **NS-17** (1970), pp. 149–159.

[61] A. Owens, "Spectral degradation effects in an 86cm³ Ge(HP) detector", *Nucl. Instrum. Methods*, Vol. **A235** (1986), pp. 473–478.

[62] A. Kozorezov, K. Wigmore, A. Owens, R. Den Hartog, A. Peacock, H.A. Al-Jawhari, "Resolution degradation of semiconductor detectors due to carrier trapping", *Nucl. Instrum. Methods*, Vol. **A546** (2005), pp. 209–212.

9

Materials Used for General Radiation Detection

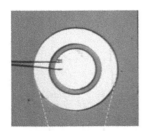

Evolution of GaAs X-ray detector technology at the European Space Agency, starting with test diodes, small format arrays and large arrays towards an ultimate goal of spatially resolved spectroscopy for X-ray astrophysics and planetary space missions (photographs courtesy of Oxford Instruments Analytical Oy and Thales Alenia Space).

CONTENTS

9.1 Semiconductors and Radiation Detection

Two and a half decades ago, Armantrout *et al.* [1] produced a rank-ordered listing of the most promising materials for further development as radiation detectors. However, out of a list of nine compounds, only CdSe, HgI$_2$ and CdTe were investigated, and of these only HgI$_2$ and CdTe (and its alloy CdZnTe) are still under active development. Unfortunately, it soon became clear that the development of semiconductors based on compounds would be far more difficult than that of those based on group IV elements, for three fundamental reasons, as follows:

1) Resistivities are not as high as expected from the bandgap; indeed, it is only possible to get resistivities above 1 MΩ for a few materials.
2) Carrier trapping greatly reduces charge collection, giving rise to poor spectroscopic capability or none at all. The problem is further exacerbated by the large miss-match between the electron and hole transport properties. Generally, the $\mu\tau$ products for holes are at least an order of magnitude worse than for electrons.
3) Polarization effects degrade poor performance even further.

All these effects can be directly attributed to imperfect material or surface problems, which arise from the difficulty in growing monocrystalline material with exact stoichiometry and in its processing. In fact, of the 50 or so compounds available, less than half have been investigated as possible detection media. The situation is summarized in Table 9.1. The materials superscripted with * have shown some response to radiation, usually MeV alpha particles. Those superscripted with ++ have shown a spectroscopic response to X-rays (thus $E/\Delta E > 1$); of these, only CdTe, CdZnTe and HgI$_2$ have matured sufficiently to produce commercially viable detection systems, and another four are tantalizingly close. Nevertheless, CdTe is now used in less and less applications as CdZnTe has largely superseded it in view of its higher resistivity, lower dislocation density and lower susceptibility to polarization effects. We describe these materials in this chapter; their physical and electrical properties are listed in Appendix F. We normally quote results for simple monolithic detectors with planar electrodes, because these tend to be dominated by basic material properties and not by electrostatic effects such as the small pixel effect. Techniques to mitigate against poor transport will be presented in a later chapter.

9.2 Group IV and IV–IV Materials

This grouping contains the classical elemental semiconductors, Si, Ge, C (diamond) and gray tin (α-Sn), which crystallize in the diamond structure and are unique in the periodic table in that their outer shells are exactly half filled. Consequently, they only bond covalently. An examination of the properties of group IV elements shows that bandgap, hardness and melting points all decrease with increasing Z, while charge carrier mobilities, densities and lattice constants generally increase. These trends may be attributed to the progressive metalization of the elements with increasing Z within the group. One can also combine two different group IV semiconductors to obtain compounds such as SiC and SiGe whose physical and electronic properties are intermediate.

9.2.1 Silicon

Silicon (Si) is the eighth most common element in the universe by mass and the second most abundant element in the Earth's crust, manifesting itself in various forms of silicon dioxide (silica) and silicates. It is a tetravalent metalloid, less reactive than its chemical analog carbon, the non-metal directly above it in the periodic table, but more reactive than germanium, the metalloid directly below it. Several compounds of silicon exist; the most commonly known are silicon dioxide and silicon carbide. Apart from its use in glass production, silicon dioxide (silica) is widely used in electronics manufacturing for passivation and isolation in planar processes or as an intermetallic dielectric. Historically, silicon carbide (SiC) has been used as an abrasive. As a wide gap semiconductor, it has been, and is being, actively explored for high-temperature, high-power, high-frequency and radiation-hardened applications [2].

Silicon is a group IVA material that crystallizes in a diamond cubic crystal lattice structure at a temperature of 1414 °C. It has a low density of 2.39 gcm^{-3} and a relatively small indirect bandgap of 1.12 eV. Whilst it has modest electron and hole mobilities (*i.e.*, 1500 cm^{2}V^{-1}s^{-1} and 480 cm^{2}V^{-1}s^{-1}, respectively), its mu-tau products are extremely high ($\mu_e\tau_e \sim \mu_h\tau_h \geq 1$ cm^{2}V^{-1}). Two allotropes of silicon exist at room temperature: amorphous and crystalline. Amorphous Si is widely used in the production of thin-film solar cells, while crystalline Si is used extensively in the semiconductor industry for the production of electronic components and radiation detectors. More than 75% of all single crystal silicon wafers are grown by the Czochralski (CZ) method. Higher quality material can be produced using the float-zone technique, but at greater expense. For radiation detection applications, the best spectral performances have been achieved using epitaxial materials produced by liquid phase epitaxy (LPE) or vapor phase epitaxy (VPE).

TABLE 9.1

Compound semiconductor materials listed by group and in order of increasing bandgap ranging from the near IR to XUV wavelengths. We include the elemental semiconductors for completeness. Materials subscripted with a ++ are those for which spectroscopic measurements (that is, $E/\Delta E > 1$) have been made at X-ray wavelengths. Here, ΔE is the FWHM energy resolution at energy E. The compounds subscripted with a * have shown some response to radiation.

Bandgap energy (eV)	Elemental group IVB	Binary IV-IV compounds	Binary III-V compounds	Binary III-VI compounds	Binary II-VI compounds	Binary I-VII compounds	Binary IV-VI compounds	Binary n-VIIA, n-VIIB compounds	Ternary Compounds
0.00–0.25	Sn		*InSb		HgTe			Bi_2Te_3	$Hg_{0.8}Cd_{0.2}Te$
0.25–0.50			*InAs		HgSe		PbSe, PbS, PbTe		
0.50–0.75	++Ge		++GaSb						InGaAs
0.75–1.00		SiGe							$Hg_{0.2}Cd_{0.8}Te$
1.00–1.25	++Si						SnS		
1.25–1.50			++GaAs, +InP		++CdTe				AlInAs
1.50–1.75			AlSb	*GaTe	++CdSe				$^{++}Cd_{0.9}Zn_{0.1}Te$, $^{++}Cs_2Hg_6S_7$, $^{++}Cd_{0.95}Mg_{0.05}Te$, $^{++}Cd_{0.9}Mn_{0.1}Te$
1.75–2.00			$*B_4C$, BP, InN	*GaSe				$^{++}BiI_3$	$^{++}Pb_2P_2Se_6$, $^{++}Cd_{0.7}Zn_{0.3}Se$, $^{++}Tl_6I_4Se$, $^{++}Cd_{0.3}Zn_{0.7}Te$, InAlP, $^{++}TlGaSe_2$ InGaP^{++}
2.00–2.25		3H-SiC	AlAs		HgS			$*SbI_3$, *HgS, *HgO *InI, $^{++}HgI_2$, $^{++}Hg_2I_2$	$Pb_2P_2S_6$
2.25–2.50			++GaP, AlP		ZnTe, CdS			$^{++}PbI_2$, $^{++}Hg_2Br_2$, $^{++}TlBr_{0.35}I_{0.65}$	$*TlPbI_3$, $^{++}PbGa_2Se_4$
2.50–2.75					++ZnSe	AgBr		$^{++}TlBr$, $^{++}TlBr_{0.99}Cl_{0.01}$	$^{++}Hg(Br_{0.2}I_{0.8})_2$
2.75–3.00					MnSe	AgI, CuBr, CuI			$^{++}LiInSe_2$
3.00–3.25		*6H-SiC			MnTe	++AgCl, CuCl			
3.25–3.50		++4H-SiC	*GaN		MgTe, MnS				
3.50–3.75					MgSe, ZnS				$^{++}Cs_2Hg_6S_7$
3.75–4.00								Hg_2Cl_2	
4.00–4.25									
4.25–4.50					MgS				
4.50–4.75									
4.75–5.00									
5.00–5.25									
5.25–5.50	*C								
5.50–5.75									
5.75–6.00			*BN						
6.00–6.25			AlN						
6.25–6.50									
6.50–6.75									
6.75–7.00									

Apart from electronic component applications, silicon has been widely used in radiation detection applications since the early 1960s. Initially, silicon diodes were used almost exclusively for charged particle spectroscopy because depletion depths were limited to ~ 1 mm. Detector sizes ranged up to a few cm^2 – the maximum size being dictated by noise due to detector capacitance. In 1960, the lithium ion-drifting process was introduced [3] to compensate for impurities, allowing semi-insulating material to be produced. This allowed higher biases to be applied and thicker detectors to be produced. In addition, because of the increase in detector thickness, capacitance could be reduced, which meant lower noise levels could be achieved. By 1970, Si(Li) detectors became a laboratory standard for X-ray spectroscopy applications. However, in order to achieve low noise levels, detectors have to operate at liquid nitrogen temperature. At present, large area devices (up to 200 mm^2) with depletion depths of up to 5 mm are commercially available with a useful energy range for energy dispersive spectroscopy from a few hundred eV to 50 keV and FWHM energy resolutions of < 250 eV at 5.9 keV.

Planar technology was introduced in 1982, which led to rapid progress in the development of silicon detectors, particularly for particle accelerator applications. The well-defined properties of silicon–silicon dioxide interfaces facilitated the fabrication of detector architectures far more advanced than simple classical diodes structures, allowing "custom" detectors to be produced on a commercial basis. The main advantages of planar technology are easier production, smaller capacitance, lower noise (between a tenth and a thousandth of that of surface barrier or diffused junction detectors), larger area coverage (up to 10 cm^2) and room temperature operation or reduced cooling requirements for very low noise operation. Planar technology is widely used in the construction of strip and hybrid pixel detectors used in particle physics experiments for tracking particles. On the spectroscopy side, in Fig. 9.1 we show the response of a 0.8 mm^2, 500 μm thick Si planar detector to ^{241}Am and ^{55}Fe radioactive sources [4]. The bias was +100V and the operating temperature -15°C. The FWHM energy resolutions recorded were 250 eV at 5.9 keV and 524 eV at 59.54 keV. From the figure, we note that all the Np K X-rays from ^{241}Am are clearly resolved, as well as the ^{55}Fe Kβ line. Although spectra could not be measured at room temperature (20°C) because of excessive leakage currents, at +15°C, measured energy resolutions of 750 eV and 800 eV were recorded at 5.9 keV and 59.54 keV, respectively.

Later advances in CCDs, drift diodes, Depleted P-channel Field Effect Transistor (DEPFET) and 3D technology have led to X-ray detectors with near Fano limited operation. For example, Bertuccio *et al.* [5] measured FWHM energy resolutions of 124 eV at -20°C and 136 keV at room temperature with a 13 mm^2, 450 μm thick silicon drift detector (the Fano limited resolution is 118 eV).

Recently, a 256 × 256 pixel matrix array of sensitive area 1.92 × 1.92 cm^2 has been produced with a DEPFET readout for X-ray astrophysics applications [6,7]. The pixel size was 75 μm × 75 μm and the thickness 450 μm. An energy resolution of 130 eV FWHM at 5.9 keV was achieved for single events and 150 eV FWHM for all events (*i.e.*, including those events in which charge was shared amongst pixels [7]. The array had no dead or noisy pixels, and the uniformity of the gain was at 2% level, consistent with statistics. Of particular note, the device was operated with only moderate cooling of -5 °C.

9.2.2 Germanium

Germanium (Ge) is a group IVA material that crystallizes in a diamond cubic crystal lattice structure at a temperature of 937 °C. It has a density of 5.35 gcm^{-3} and a narrow indirect bandgap of 0.66 eV. Its electron and hole mobilities are two to three times greater than those of Si (*i.e.*, 3800 cm^{2}V^{-1}s^{-1} and 1850 cm^{2}V^{-1}s^{-1}, respectively), and its mu-tau products ($\mu_e\tau_e \sim \mu_h\tau_h > 1$ cm^{2}V^{-1}) are the largest of any semiconductor. Ge is a metalloid, having characteristics of both metals and non-metals. It has five naturally occurring isotopes, ^{70}Ge, ^{72}Ge, ^{73}Ge, ^{74}Ge and ^{76}Ge with abundances of 21%, 28%, 8%, 36% and 7%, respectively. Of these, ^{76}Ge is radioactive with a half-life of 1.78×10^{21} years. In Ge gamma-ray spectrometers, these isotopes generate numerous background

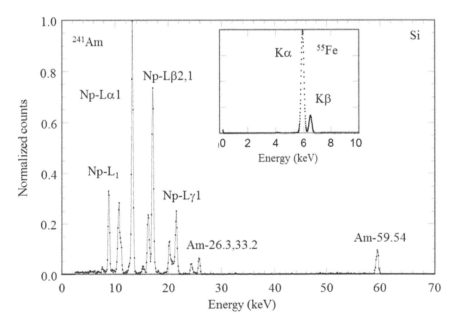

FIGURE 9.1 The response of a 1 mm dia, 500 micron thick Si planar detector to ^{241}Am and ^{55}Fe radioactive sources (from [4]). The detector operating temperature was -15°C. The FWHM energy resolutions are 245 eV at 5.9 keV and 524 eV at 59.54 keV.

gamma-ray lines when exposed to radiation, which can be used for calibration purposes. Until the invention of the transistor in the 1950s, there were no important uses for germanium. Germanium was extensively used to produce transistors until the 1970s. Apart from producing high-resolution gamma-ray detectors, today Ge is used mainly for fiber-optic systems and infrared optics.

Germanium was one of the first materials to be exploited for radiation detection and now represents the gold standard for high-resolution gamma-ray spectroscopy. In 1949, McKay [8] successfully measured the polonium alpha-ray spectrum using a reverse biased, point contact Ge diode. Low resistivity and charge trapping issues were circumvented by constructing a very thin detector with an evaporated metal surface layer, which acts as a rectifying contact whilst simultaneously creating a very high electric field across this layer using a point contact. McKay and McAfee [9] subsequently extended this work producing Ge p-n junction detectors. Schottky barrier detectors were introduced in 1955, followed by Au/Ge surface barrier detectors in 1959. At this time, depletion depths were limited to the order of microns.

In 1960, the lithium drifting technique [3] was introduced to compensate for naturally occurring p-type excess impurities in Ge. This allowed higher biases to be applied, which in turn allowed much larger detectors to be fabricated. However, since the bandgap is so small, detectors had to be cooled to cryogenic temperatures to reduce thermally generated leakage current sufficiently to allow for high-resolution operation. Note also that because of the high diffusivity of Li ions in Ge, Li-drifted detectors also had to be stored at low temperature; otherwise, the detector lost its compensation in a matter of hours. For planar detectors, Li drifting increased the maximum depletion depth from microns to 1–2 cm, making them particularly useful in the hard X-ray band. In fact, present planar detectors are typically sensitive from a few keV to a few hundred keV with a FWHM energy resolution of ~0.5–0.7 keV at 122 keV (see Fig. 9.2). However, for the efficient detection of MeV gamma-rays, active depths of ~5 cm are required; this was achieved by employing the coaxial electrode configuration in which the detector is basically a cylinder of germanium with one contact on the outer surface and the other contact on the surface of an axial well, drilled through the center of the Ge. In this way, depletion depths were effectively doubled.

In the late 1970s, High-Purity Ge (HPGe) detectors became available commercially with a net impurity concentration of a few times 10^{10}cm^{-3}. At this impurity concentration, compensation is no longer required and maximum biases of a few thousand volts can be applied, sufficient to deplete the full volume of a large (~100 cm^3) coaxial diode. Whilst the detector still has to be operated at temperatures < 110 K, it could be now stored at room temperature. In Fig. 9.3 we show ^{137}Cs and ^{60}Co spectra measured with a 110 cm^3 intrinsically pure Ge crystal in a closed-end coaxial configuration, which dramatically demonstrates the superlative energy resolution expected from Ge detectors at gamma-ray energies. The crystal diameter is 49 mm and its length 67 mm. The bias voltage was 3000V, which gave an optimum FWHM spectral resolution of 1.9 keV at 662 keV and 2.5 keV at 1332 keV. At 1333 keV the total interaction efficiency was 17.2% (relative to a standard 3 inch NaI(Tl) crystal). Now, HPGe detectors with volumes in excess of 500 cm^3 volume are readily available with FWHM energy resolutions of ~0.2% at 662 keV and 1333 keV.

9.2.3 Carbon (Diamond)

Diamond has been proposed for use in hostile, corrosive and very high radiation environments [11]. It was originally investigated as a UV photoconductor in the 1920s [12] and later as a crystal counter in the 1940s [13]. Early work was carried out using natural diamonds, which led to highly variable results, mainly due to the compositional variations in geologically derived diamonds. In addition, the small size and in particular the cost of natural diamonds further limited

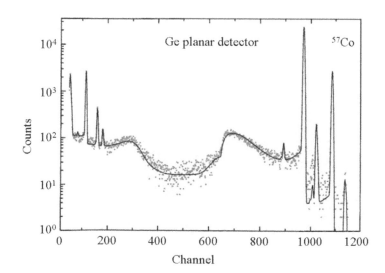

FIGURE 9.2 A ^{57}Co spectrum measured with a 32 mm diameter, 10 mm thick planar Ge detector (from [10]). The active volume is 8 cm^3. The solid line is a fitted spectrum using a 13 parameter response function. The FWHM energy resolution at 122 keV is 740 eV.

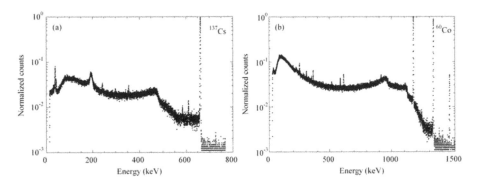

FIGURE 9.3 (a) ^{137}Cs and (b) ^{60}Co spectra measured with a 110 cm^3 intrinsically pure Ge crystal in a closed-end coaxial configuration. The bias is 3000V and the amplifier shaping time 10 μs. The FWHM energy resolutions are 1.9 keV at 662 keV and 2.5 keV at 1332 keV, respectively.

work. Consequently, progress pretty much stagnated until the first reproducible synthetic diamonds became available in the mid-1950s [14]. However, real advances only occurred after the 1980s with the introduction of CVD techniques [15,16], which allowed the production of large crystals of a quality (at least in terms of the electrical and thermal properties) of natural diamond. Recent developments have been largely driven by the high-energy physics community [17] and more recently by human-tissue dosimetry applications [18] (for a general review see [19]).

Diamond is an allotrope of carbon. Carbon is the lightest Group VI element with a half-filled valence shell and an s^2p^2 electronic configuration.[1] Depending on temperature and pressure, it can form one of a number of allotropes, the most well-known being graphite and diamond. In graphite, the *s* orbital mixes with two *p* orbitals, forming an sp^2 hybridization. Physically, each carbon atom shares one electron with two of its neighbours and two electrons with the third neighbour, structurally forming two-dimensional hexagonal lattice planes stacked on top of each other. Bonds within planes are strong (covalent) and weak between planes (Van der Waals). Graphite is therefore a layered, soft material. At normal temperatures and pressures, this is the thermodynamically favoured configuration. However, at high temperatures and pressures (*i.e.,* $T > 1000°C$, $P > \sim50$ kBar), the *s*- and *p*- states can hybridize to form extremely strong sp^3 bonds. Structurally, the atoms bond tetrahedrally into the aptly named *diamond* lattice structure (two interpenetrating face-centered cubic lattices with a displacement of one-quarter body diagonal). This bond structure, in conjunction with the low atomic number of carbon, gives diamond the highest atom density of any material, a feature that is responsible for many of its superlative properties. It is chemically inert, strong, an excellent thermal conductor and extremely hard. In fact, the word diamond is derived from the Greek "αδάμας" meaning invincible.

For radiation detection purposes, diamond has a number of advantages over other materials. With reference to Table 9.2, these can be summarized as follows. Its wide bandgap (5.47 eV) results in extremely low thermally generated leakage currents at room temperature as well as a maximum operating temperature of ~ 1000°C, compared to ~ 200°C for silicon. In terms of wavelength response, the wide bandgap allows for the construction of detectors that are solar blind. Diamond's electric field breakdown strength is more than 60 times higher than the corresponding values for silicon, allowing much higher biases to be applied across a detector. Together with its high saturation drift velocity, high carrier mobilities, relatively long carrier lifetime and high charge collection efficiency, this leads to the possibility of rapid and complete extraction of the generated

TABLE 9.2

A comparison of the mechanical, electronic and thermal properties of diamond with other common semiconductor materials at room temperature. In almost every category, diamond outperforms other semiconductors.

Parameter	C	Ge	Si	GaAs	CdTe	HgI$_2$	4H-SiC	GaN	Units
Hardness	**10,400**	780	1150	750	50	<10	3980	1830	kg mm^{-3}
Melting Point	**4,100**	1210	1687	1513	1366	532	3103	3246	K
Thermal conductivity	**24**	0.58	1.5	0.55	0.06	0.004	3.7	1.3	Wcm^{-1}K^{-1}
Resistivity	**10^{13}**	50	<10^4	10^{10}	10^9	10^{13}	>10^3	10^6	Ω cm
Bandgap	**5.5**	0.7	1.1	1.4	1.5	2.1	3.2	3.4	eV
Breakdown field	**21.5**	0.1	0.3	0.6	0.8	2	5	5	MV cm^{-1}
Electron mobility	**4500**	3900	1450	8500	1050	100	900	2000	cm^{2}s^{-1}V^{-1}
Hole mobility	**3800**	1900	480	400	100	4	120	200	cm^{2}s^{-1}V^{-1}

---

[1] Strictly 1s22s22p2.

charge (*i.e.*, a radiation sensor with ultra-fast response). Its large bandgap, high thermal conductivity and maximal operating temperature allow diamond to be used in hot and strong particle flux environments [20]. Its displacement energy (that is the threshold energy to displace atoms from their lattice sites) is ~ 45 eV, which is approximately 2–3 times that of Si. In addition, the generated vacancies are immobile at room temperature, unlike in silicon where the vacancies are free to migrate to form complex defects particularly with dopants. Therefore, the effects of radiation damage in diamond are much less severe than the equivalent in silicon. A last advantage of diamond follows from its atomic number, which is nearly tissue equivalent ($Z = 6$ as opposed to 7.4 for biological tissue). This leads directly to applications in medical dosimetry (*e.g.*, hadron therapy), when in-vivo measurements and a signal directly proportional to the absorbed dose rate are required, alleviating the need for dose corrections [21]. However, at present, for most applications, the use of diamond is still limited by the inability to grow material of sufficiently high grade and for some applications, the difficulty in doping this extremely stable material.

In terms of radiation detection, initial studies show that both natural and CVD material respond to all forms of radiation [19] and are extremely radiation hard [22]. While they are not spectroscopic to photons at this time, Kaneko *et al.* [23] achieved an energy resolution of 0.4% FWHM for 5.486 MeV alpha particles with a single CVD crystal diamond detector, having a size of $2.0 \times 2.0 \times 0.7$ mm^3. In fact, the resolution was so good that the peaks of 5.443 MeV and 5.389 MeV resulting from two of the excited states of ^{247}Np from ^{241}Am could be clearly resolved. The detector was fabricated from a single diamond crystal grown by plasma-assisted CVD on to a [100] surface of a type Ib diamond substrate. An Al Schottky contact and a Ti/Pt ohmic contact were applied to the crystal by evaporation and sintering.

9.2.4 Silicon Carbide

Silicon carbide (SiC) is currently being widely explored as a high-temperature Si alternative that is also chemical and radiation tolerant. It has several distinct advantages over Si: it has twice the thermal conductivity and eight times the maximum breakdown electric field. The former property is important for producing thermally stable or high-power semiconductor devices, while the latter means that much higher biases can be applied, resulting in higher drift velocities and more efficient charge collection. It also has a high saturated electron drift velocity (almost twice that of Si), which ensures a low trapping probability as well as a high-displacement threshold energy of 21.8 eV. This ensures a high radiation tolerance. The wide bandgap (three times that of Si) means that dark currents are extremely low, which in principle should allow high temperature operation up to +700 °C. Indeed, SiC n-MOS devices have been successfully operated and thermally cycled up to 630°C [24].

SiC belongs to a family of materials that display a one-dimensional polymorphism called polytypism [25]. Polytypes differ by the stacking sequence of each tetrahedrally bonded Si–C bi–layer, crystallizing into cubic, hexagonal or rhombohedral structures. In this respect, SiC can be thought of as an ordered alloy. Although more than 200 polytypes have been discovered, the main building block for each is a tetrahedron consisting of a carbon atom bonded to four silicon atoms and *vice versa*. This allows polytypes to be easily categorized using the notation proposed by Ramsdell [26]. Essentially all SiC polytypes may be divided into three groups: cubic (β-SiC, B3 type), denoted 3C-SiC with crystal lattice similar to one of sphalerite (space group T2d-F43m); hexagonal polytype (denoted 2H-SiC) with the wurtzite type crystal lattice, space group C46v-P63mc; and the third group that includes all other known polytypes, the most studied being polymorphs with the rhombohedral unit cell (15R and 21R). Here the number refers to the number of silicon carbide double layers in a unit cell, while the letters H, C and R refer to the type of lattice (hexagonal, cubic or rhombohedral). SiC is a polar crystal in that the outer most atoms in the [0001] direction are Si-atoms and in the [000$\bar{1}$] direction, C-atoms. As a result, the Si and C-terminated faces have different chemical and growth behaviours. A summary of the properties of the three main polytypes used in the semiconductor industry is given in Table 9.3. The significantly higher electron mobility observed in 4H–SiC makes this the preferred polytype for radiation detector applications.

The growth of high resistivity SiC for detector applications can be achieved using two different techniques, either as bulk material grown as a single crystal or as an epitaxial layer. However, it should be noted that controlling the crystal structure of SiC presents a major growth problem, since many different polymorphs can grow under apparently identical conditions. At present, the standard technique for bulk growth is the seeded sublimation method, since unlike most semiconductor materials, growth from the melt is not possible in view of the extremely high pressures and temperatures (100,000 atm and 3200°C) required to produce stoichiometric material. Bulk SiC is currently the only route to produce thick wafers (100–500 μm thickness), although the quality is relatively poor. Conversely, epitaxial SiC grown on to a substrate wafer can produce high purity material, but thicknesses are currently limited to ~ 150 μm. In any case, high residual doping concentrations ($\geq 5 \times 10^{13}$ cm^{-3}) currently limit depletion depths to ~100 μm. At the present time, only epitaxial material is suited for detector use and even then only for radiation having a mean penetration depth of less than 100 μm. Fortunately, both growth methods are undergoing rapid evolution and the quality of SiC material continues to improve. However, it is likely that thick epitaxial SiC will provide the best route to detector grade material, provided that a fast and cost effective growth route can be commercially developed.

SiC radiation detectors were originally developed four decades ago for applications in the nuclear power industry, specifically for extreme environment instrumentation and control systems for direct monitoring of reactor cores and waste

TABLE 9.3

Material properties of selected silicon carbide polytypes. The number refers to the number of silicon carbide double layers in a unit cell; the letters H, C and R refer to the type of lattice (hexagonal, cubic or rhombohedral).

Parameter	SiC Polytype		
	3C	4H	6H
Lattice structure	face-centered cubic	hexagonal	hexagonal
Lattice constants a, c (Å)	4.359	3.079, 10.115	3.0817, 15.117
Eg (eV) @ T< 5K	2.20	3.29	3.02
Eg (eV) @ T = 300K	2.39	3.26	3.05
Ecrit (MV/cm)	2.1	2.2	2.5
Θ_κ (Wcm^{-1}K^{-1})@ 300K	3.6	3.7	4.9
μ_e (cm^2 V^{-1}s^{-1})*	800	1000	400
μ_h (cm^2 V^{-1}s^{-1})	40	120	100
Sat. e–drift vel. (cm/s)	2.5×10^7	2.0×10^7	2.0×10^7
$\varepsilon_r(0)$	9.7	10.0	10.0
Knoop Hardness (kgmm^{-2})	2900	3980	2460

* parallel to c-axis

management. Babcock *et al.* [27,28] studied the response of small SiC p–n junction diodes to alpha particles up to 700°C [29]. Ferber and Hamilton [29] were later able to demonstrate neutron detection using ^{235}U converter layers. They found good agreement between flux profile measurements made with the diode and those made with conventional gold foil activation techniques in a low-power reactor. In addition, the diode response as a function of neutron flux was linear over four decades of reactor power output. X-ray detection with SiC diodes was first demonstrated by Bertuccio *et al.* [30] who fabricated Schottky junctions on epitaxial SiC. The devices had a junction area of 3 mm^2 and were grown on a 30 µm thick *n*-type 4H–SiC layer with a dopant concentration of 1.8×10^{15} cm^{-3}. At 300 K, the reverse current density of the best device varied between 2 pA/cm and 18 pA/cm for mean electric fields of 40 kV cm^{-1} and 170 kV cm^{-1}. The devices showed

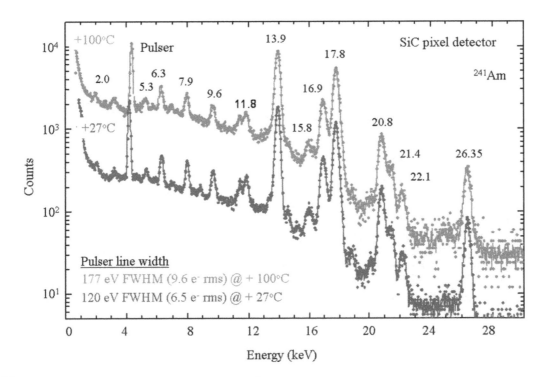

FIGURE 9.4 ^{241}Am X-ray and γ-ray spectra recorded by an 0.03 mm^2, 70 µm thick epitaxial 4H–SiC detector at +27°C and at +100°C (taken from [31]). The spectrum at +100°C is displaced on the y-axis for clarity. At 26.3 keV, the FWHM energy resolution is 670 eV at 27°C and is essentially unchanged at 100°C operating temperature.

a spectroscopic response to X- and gamma-rays. At 60 keV measured energy resolutions of 2.7 keV FWHM were recorded at room temperature, limited mainly by detector capacitance.

Since this time, X-ray performances have steadily improved [32]. For a 0.03 mm^2, 70 μm thick device, Bertuccio *et al.* [33] recorded a FWHM resolution at 60 keV of 693 eV at 27°C increasing to 1.1 keV at 100°C operating temperature. Phlips *et al.* [34] tested the radiation response of a 1 mm diameter SiC PIN^2 diode originally developed for high-power applications. The device consisted of a 100 μm thick 4H-SiC layer grown epitaxially on a SiC substrate. The measured room temperature resolution at 60 keV was ~550 eV FWHM, limited by an electronic noise of 28 electrons rms. Later measurements by Bertuccio *et al.* [31], on gold contacted 4H SiC, 200 μm diameter Schottky diodes, show current densities of 17 pA/cm^2 at 340 K – more than two orders of magnitude lower than commercial silicon devices. Good spectroscopic performance was demonstrated at low energies, with energy resolutions of 196 eV FWHM being recorded at 5.9 keV at +30°C and 233 eV FWHM at +100°C. The former value is reasonably close to the expected Fano limited value of 160 eV, assuming a Fano factor of 0.1. In Fig. 9.4, we show the measured spectral response to a ^{241}Am source, at room temperature and at 100°C. The electronic noise corresponds to 120 eV FWHM (6.5 electrons rms) at +27°C and 177 eV FWHM (9.6 electrons rms) at +100°C. At 26.3 keV, the resolution is ~400 eV and is essentially the same at operating temperatures of both 27°C and 100°C.

9.3 Group VI Materials

9.3.1 Selenium

Selenium is a member of the chalcogen^3 family and along with tellurium is the only other member that possesses semiconductor properties. It exists in six stable isotopes (^{74}Se, ^{76}Se, ^{77}Se, ^{78}Se, ^{80}Se and ^{82}Se) that exhibit strong photovoltaic and photoconductive properties. Selenium is a metalloid that has several allotropic forms: an amorphous red powder and several crystalline forms that are distinct only in the stacking order of Se "molecules". The main crystalline forms are the monoclinic alpha and beta modifications, that exist as red crystals, and a gray trigonal form, which has metallic properties. Only gray Se is thermodynamically stable below the melting point at atmospheric pressure, and in fact the monoclinic forms will convert into the gray modification if moderately heated. Most selenium is recovered from the electrolytic copper refining process – usually in the form of the red allotrope.

Gray Se has a melting point of 490 K, a density of 4.8 g cm^{-3} and a bandgap of 1.8 eV. Its most outstanding physical property is its photoconductivity. As such, it was widely used in solar and photocell applications. After the 1980s, its use in photoconductor applications began to decline as Se was gradually replaced by other more sensitive and/or less expensive materials, such as organic photoconductors. It was also extensively used for the production of electrical rectifiers, although silicon-based devices have largely replaced these. At present, the main commercial uses for Se are in glassmaking, pigments and xerographic applications.

Amorphous Se (a-Se) is a semiconductor with a resistivity at room temperature of 10^{12} Ω-cm, a density of 4.3 g cm^{-3} and an energy gap of 1.7 eV. It has an acceptable X-ray absorption coefficient for low energy X-rays, good charge transport properties and a low dark current. In view of these qualities, it is the most highly developed photoconductor for large area X-ray imaging [35] for medical applications and is used in the only commercially available large area direct conversion flat panel X-ray detector (for a review of flat panel development, see [36]).

9.4 Group III–V Materials and Alloys

These are compounds which combine an anion from Group V (from nitrogen or below) and a cation from Group III (usually, Al, Ga or In). Each Group III atom is bound to four Group V atoms and *vice versa* so that each atom has a filled (eight-electron) valence band. Although bonding would appear to be entirely covalent, the shift of valence charge from the Group V atoms to the Group III atoms induces a component of ionic bonding to the crystal. This ionicity causes significant changes in semiconducting properties. For example, it increases both the Coulomb attraction between the ions and the forbidden bandgap. When grown epitaxially (MBE, MOCVD and variants), III-V materials usually take up a zincblende (ZB) structure so that in their basic electronic and crystal structures, they are completely analogous to the Group IV elements. The stable bulk allotrope often has a wurtzite structure.

---

^{2} PIN≡P-type–Intrinsic–N-type diode. Essentially a refinement of the PN junction in which a layer of intrinsic material is inserted between the P and N layers. As a result, PIN diodes have a high breakdown voltage, a low level of junction capacitance and a larger depletion region than a simple junction – ideal for detector applications.

^{3} The chalcogens are elements in Group 16 (VIA) of the periodic table. The name is derived from the Greek word chatkos, meaning "ore" since the most common chalcogens (oxygen and sulfur), are found in most ores.

9.4.1 Gallium Antimonide

Gallium antimonide (GaSb) is a relatively dense ($\rho = 5.81$ g cm^{-3}), narrow gap (0.72 eV at 300K), Group III-V material with properties similar to those of Ge but with the added advantage of a direct bandgap. GaSb crystallizes in zincblende structure that is identical to that of the diamond lattice except that each Ga atom has four tetrahedrally arranged Sb neighbours and vice versa. Since its melting point is as low as 712°C, bulk GaSb is usually grown using the Czochralski method, whilst epitaxial growth is usually achieved using LPE. Its fairly large lattice constant (0.61 Å), low electron mass, high mobilities and lattice match with various ternary and quaternary III-V compounds make it a promising material with significant electro-optical potential in the near IR range ($\lambda > 1.5$ μm^{-1}). In fact, InAsSb heterojunctions and GaAlSb/AlSb and InAs/InGaSb superlattices are seen as potential competitors to present-day state-of-art IR detectors based on HgCdTe. The physical and electronic properties of GaSb and its various application areas have been reviewed by Milnes and Polyakov [37] and Dutta and Bhat [38].

Juang *et al.* [39] have proposed using GaSb as a high-resolution X and gamma-ray detection medium based on its reasonably high density and small bandgap. They fabricated a prototype detector by growing a 2 μm intrinsic GaSb layer on an *n*-type substrate using MBE. Bulk leakage currents in the intrinsic region were reduced by compensating with Te and surface leakage reduced using ammonium sulfide ((NH$_4$)$_2$S) passivation. The anode signal was extracted using a Ti/Pt/Au electrode, which defined an active detection area of 200 μm diameter. The cathode signal was extracted using a Ni/Ge/Au contact. To reduce leakage current, the device was cooled to 140K, which at a bias of 2V, was ~6 μAcm^{-2}. The detector was found to be spectrally responsive to X- and gamma-rays. This is illustrated in Fig. 9.5 in which we show the measured energy loss spectra after exposure to ^{55}Fe (a) and ^{241}Am (b) gamma-ray sources. The measured FWHM energy resolutions were 1.24 keV at 5.9 keV and 1.8 keV at 59.54 keV. The corresponding noise floor and FWHM electronic noise of the system were measured to be 3 keV and 1.23 eV FWHM, respectively. By comparing spectrum (b) with an ^{241}Am spectrum obtained with a GaAs detector of similar p-i-n structure, the pair creation energy was estimated to be 2.92 eV at 140K.

9.4.2 Gallium Arsenide

Gallium arsenide (GaAs) is a direct bandgap III–V material with a simple cubic lattice structure. It is widely used in the manufacture of red LEDs, infrared windows and laser diodes. Compared to many semiconductors it has a number of attractive attributes. For example, its density (5.32 g cm^{-3}) is more than twice that of Si and thus it has better stopping power, especially in the hard X-ray energy range. Its bandgap (1.43 eV) is wide enough to permit room temperature operation but small enough that its Fano limited spectroscopic resolution is close to that of Si. Its electron mobility (8500 cm^{2}V^{-1}s^{-1}) is about a factor of 6 higher than Si, allowing faster operation, while its bulk resistivity (>10^3 Ω-cm) is much larger than Si allowing higher fields to be used for charge collection which in turn allow thicker devices to be fabricated. In fact, high frequency MESFETs with usable gain up to 40 GHz are now commercially available. From a manufacturing point of view, devices fabricated on semi–insulating material can be self-isolating, and therefore GaAs is ideally suited to integrated circuit fabrication and replication techniques. Unlike most of the II–VI compounds, which are grown specifically for radiation detectors, bulk grown semi-insulating GaAs is produced mainly for substrates for the electronics industry. The preferred growth method is the liquid encapsulated Czochralski (LEC) technique. For the best spectral performances, epitaxial techniques are employed, both LPE and VPE.

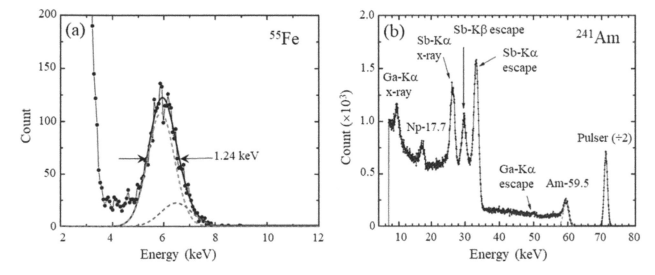

FIGURE 9.5 The response of a 0.03 mm^2, 2 μm thick GaSb planar detector to ^{55}Fe (a) and ^{241}Am (b) radioactive sources (from Juang *et al.* [39]). The detector was operated at a bias of 2V and a temperature of 140K. The FWHM energy resolutions were 1.24 keV at 5.9 keV and 1.8 keV at 59.54 keV.

Harding *et al.* [41] reported particle detection with semi-insulating GaAs bulk photoconductivity devices. The first radiation detectors to show reasonable energy resolution at X-ray wavelengths were fabricated in the 1970s using liquid phase epitaxy. Eberhardt *et al.* [42,43] reported measuring an energy resolution of 1 keV FWHM at 59.5 keV at 130 K with a 60 μm thick, 1.5 mm^2 surface barrier device. Since this time, performances have steadily improved, primarily due to improvements in both bulk material properties and detector fabrication techniques. Schottky barrier detectors fabricated using semi-insulating GaAs have consistently demonstrated reasonable room temperature performances. For example, McGregor and Herman [44] report measuring a room-temperature energy resolution of 8 keV FWHM at 59.5 keV with a 130 μm thick, 0.5 mm^2 device. Owens *et al.* [4,45] fabricated a series of detectors from ultra-high purity GaAs grown by chemical vapor phase deposition (CVPD). From I/V characteristics, the typical current densities at –100V were < 0.04nA/mm^2 at room temperature. This is a factor of 3 to 100 lower than normally obtained with CVPD detectors and a factor of 10^3–10^4 lower than bulk detectors. The measured energy resolutions of a 3.142 mm^2, 40 μm thick device operated at -40°C were 435 eV FWHM at 5.9 keV and 670 eV FWHM at 59.54 keV (see Fig. 9.6). At room temperature, they were 572 eV and 780 eV, respectively. In Fig. 9.7, we show the response of a pixel detector (250 × 250 × 40 μm^3), fabricated from the same base material to an ^{55}Fe radioactive source [40]. At 5.9 keV, a room temperature (23°C) resolution of 266 eV FWHM was achieved and 219 eV FWHM with only modest cooling (–31°C). The corresponding pulser resolutions were 163 eV (13 electrons rms equivalent) and 242 eV (19 electrons rms), respectively. The expected Fano noise at this energy is ~ 130 eV. The energy resolution of the 59.5 keV ^{241}Am nuclear line was measured to be 487 eV FWHM at room temperature.

Owens *et al.* [46] fabricated a 64 × 64 pixel array from the same base material. The pixel size was 170 μm and the pitch 200 μm. The array was back-thinned, contacted and flip-chip bump-bonded onto a MEDIPIX–1 Application Specific Integrated Circuit readout chip [47]. Its analogue front end comprises a charge sensitive pre-amplifier and a shaper. Incoming charge from a semiconductor sensor is amplified and compared with a threshold in a comparator. If the signal exceeds this threshold, the event is counted. In X-ray tests, the bump yield was determined to be 99.9%. In Fig. 9.8 we show an image of a "Swatch" illustrating the imaging quality of the GaAs chip. The data were taken with a conventional X-ray tube using a tungsten target. The objects were mounted on a sample holder and the detector on a precision *x–y* stage. The detector was then scanned past the object in steps of 0.8 of the detector width using the so-called move and tile method. The resulting composite images have been flat field corrected and a 3 × 3 median filter applied locally around defective pixels. After flat field corrections, the spatial uniformity of the array was commensurate with Poisson noise.

9.4.3 Gallium Phosphide

Gallium phosphide (GaP) is an indirect, wide bandgap (2.26 eV) semiconductor with a cubic (zincblende) crystal structure. Its density is 4.1 g cm^{-3}, intermediate between Si and Ge. Very little information is available on its transport properties. It was originally investigated as an optical material and for high temperature component applications. However, its main use since the 1960s has been in the manufacture of low and standard brightness red, orange and green light-emitting diodes, the various colours being achieved by doping. Industrially, crystalline material is generally

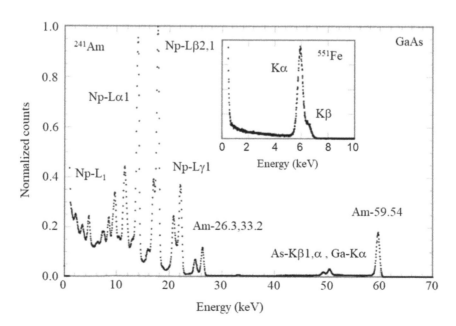

FIGURE 9.6 The response of a 1 mm dia, 40 micron thick GaAs detector to ^{241}Am and ^{55}Fe radioactive sources (from [4]). The detector operating temperature was – 40°C. The FWHM energy resolutions are 435 eV at 5.9 keV and 670 eV at 59.54 keV.

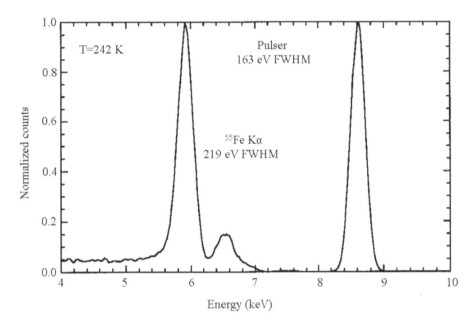

FIGURE 9.7 ^{55}Fe spectrum acquired with a 200 × 200 µm^2 GaAs pixel detector at –31°C. At room temperature the same pixel shows 242 eV FWHM on the pulser line and 266 eV FWHM at 5.9 keV (from [40]).

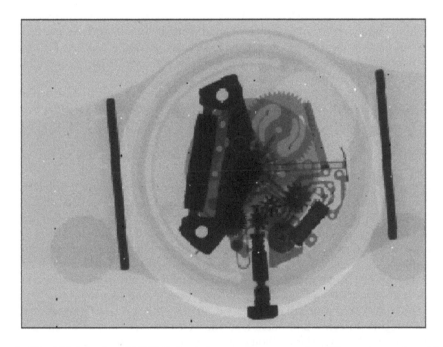

FIGURE 9.8 X-ray image of a "Swatch", taken by a 64 × 64 GaAs pixel array using a conventional X-ray set operating at 55 kV with a tungsten target and a 2.5 mm Al filter (from [46]).

grown by the LEC method. At an early stage, it was identified as a possible candidate for room temperature gamma-ray detection [48]. However, this application is unlikely to be realized, in view of its low electron and hole mobilities (<200 cm^{-2}V^{-1}s^{-1}) coupled with the fact that carrier lifetimes in III–V materials rarely exceed 100 ns. Most recently, Litovchenko *et al.* [49] noted that the IV characteristics of GaP LEDs are permanently altered in strong neutron and electron radiation fields, making them candidates for some dosimetry applications. Recent experiments using a commercial GaP Schottky diode [50] have shown a spectroscopic response to alpha particles. The device consists of a 10 nm thick Au Schottky layer (anode) deposited on a 30 µm n–type GaP layer ($n_d < 10^{16}$cm^{-3}) grown on an n–type GaP (100) substrate ($n_d \sim 5 \times 10^{17}$cm^{-3}). A further metallization forms the rear contact (cathode). The measured energy resolution to 5.5 MeV alpha particles from an ^{214}Am source was 3.5% at room temperature. The device was also found to be responsive to X-rays in the range 11–100 keV. Although individual energies are not spectrally resolved, there was a proportionality of response to increasing X-ray energy.

9.4.4 Gallium Nitride

Gallium nitride (GaN) is a refractory material that has two structural modifications: a metastable cubic form (space group F$\underline{4}$3m) and a stable wurtzite form (space group P63ma). The more common stable form has a melting temperature of 2500°C. Its physical properties, such as its wide bandgap (3.39 eV), large density (6.15 g cm^{-3}), large displacement energy (~ 20 eV) and thermal stability, make it an ideal candidate material for extreme environment applications. It also forms continuous solid solutions with AlN and InN allowing the bandgap energy to be engineered to between 2 and 6 eV. Although advances in GaN growth technology have led to the realization of epitaxial growth of GaN layers, hetero-structures and nanocrystallites, its future development is currently limited by the lack of a suitably latticed matched substrate. Consequently, only thin layers can presently be grown. Until recently, much of the work focused on optoelectronic applications, since only GaN, by virtue of its direct bandgap, can emit efficiently in the blue region of the spectrum making possible the solid-state generation of white light. Besides optoelectronics, GaN is attracting considerable attention for high-temperature/high-power electronic device applications, since its material properties offer an order of magnitude improvement in power amplifier performance over, for example GaAs and Si, particularly at microwave frequencies as well as several system-level benefits. These include compact size, high power per mass ratio, lower combiner losses, high bandwidths, facilitated thermal management and robustness.

GaN has also been explored as a high temperature, radiation hard particle and X-ray detection media (for a review see [51]). Both PIN and Schottky devices have been fabricated by depositing epitaxial films on sapphire or silicon carbide substrates, despite the mismatch in lattice constants. For example, Vaitkus *et al.* [52] grew epitaxial GaN layers by MOCVD on Al$_2$O$_3$ substrates. The structure consisted of a 2.5 μm layer of semi-insulating GaN on a 2 μm thick n*-GaN layer with Au Schottky contacts. At room temperature, the devices showed a spectroscopic response to 5.5 MeV alpha particles (25–40% FWHM) but no response to 60 keV photons. Owens *et al.* [53] produced a number of GaN PIN diodes by MOCVD. The devices consisted of a 2 μm GaN layer epitaxially grown on an *n*–type Al$_x$Ga$_{1-x}$N nucleation layer which in turn was deposited on a *p*-type 4H–SiC substrate. Au ohmic contacts were applied to both the top of the GaN layer and the bottom of the SiC substrate that act as the anode and cathode electrodes, respectively. A number of devices with contact radii varying from 0.4 mm to 0.7 mm were tested. All showed good diode behaviour with reverse leakage currents in the tens to hundreds of μA range. C–V measurements showed that the GaN layer was fully depleted for nominal biases > 20V. The devices also showed spectroscopic responses to 5.5 MeV alpha particles with typical energy resolutions of ~25% FWHM at room temperature. No response to 60 keV photons was observed.

Although spectroscopic X-ray responses have not been observed, a number of groups have reported X-ray photocurrent measurements using GaN detectors based on Schottky metal-semiconductor-metal (MSM) [54] Schottky diode [55,56] and p-i-n structures [57].

9.4.5 Indium Phosphide

Indium phosphide (InP) is widely used in optoelectronic [58] and high-speed microelectronic applications ([59] and references therein) and has even been proposed as a neutrino detector [60] because of the large neutrino capture cross-section on indium (ν + ^{115}In → ^{115}Sn* + e$^-$). It is a Group III–V direct bandgap material whose electronic properties are similar to Si and GaAs. It has a zincblende crystal structure, with a single non-destructive phase transition below the melting point making it amenable to standard crystal growth techniques. It has one of the highest electron mobilities of any semiconductor material (~3 times that of silicon, *i.e.,* 4600 cm^{2}V^{-1}s^{-1}), making it particularly suitable for applications where high-count rate operation is desirable. Its bandgap of 1.35 eV implies that detectors should operate at room temperature with a Fano limited spectroscopic resolution close to that of Si, and its relatively large density (approximately twice that of Si) ensures a high X-ray detection efficiency above 10 keV. From a materials point of view, InP is also of great interest since it is structurally suitable for the creation of integrated devices and micro-machines. However, in spite of all these desirable attributes, very little work has been carried out on this material even though it was first proposed as a radiation detector over two and a half decades ago [1]. In fact, it has only recently been possible to resolve photon peaks clearly at hard X- and gamma-ray wavelengths. The main difficulty in fabricating radiation detectors is related to the formation of an ohmic electrode system, due to a high concentration of surface states and the high chemical reactivity of the free InP surface. The net result is that the Fermi energy level lies close to the middle of the bandgap and is essentially pinned. It has not been possible to fabricate true blocking contacts.

Experimentally, InP was initially studied as a high-speed photoconductor for time-resolved measurements of synchrotron X-ray pulses [61,62]. The first single photon-counting experiments were reported by Lund *et al.* [60] who fabricated a number of gamma-ray detectors from semi-insulating, iron-doped, InP. Doping with Fe increases resistivity by compensating for the background conductivity in bulk material. The detectors consisted of a ~2 mm thick single crystal of InP with two identical gold contacts on opposite sides. The detectors were found to be sensitive but not spectroscopic to 662 keV photons. Suzuki *et al.* [63] fabricated a p–n junction diode by diffusing Zn into an *n*-type InP layer grown by liquid phase epitaxy. The epitaxial layer was 25 μm thick and had a sensitive area of 3.4 mm^2 with Au contacts deposited on both sides. The detector was operated at room temperature and found to be sensitive to both α-particles and γ-rays. For 2.2 MeV α-particles, a measured resolution of 10% (FWHM) was obtained, whilst a resolution of 55% was obtained for 60 keV photons. Olschner *et al.* [64] fabricated a number of gamma-ray detectors from single crystals of iron-doped, zinc-doped and copper-doped InP. All detectors were tested using [207]Bi, [57]Co and ^{137}Cs radioactive sources. The detectors had active volumes ranging from 2.5×10^{-3} cm^3 to 1 cm^3 and thicknesses from 0.26 mm to 3 mm. They were all responsive to radiation, but in terms of performance, the iron-doped detectors gave the best results, with peaks clearly discernible but not resolved for the [57]Co, ^{137}Cs and [207]Bi sources. Jayavel *et al.* [65] grew Fe-doped InP single crystals using the LEC technique. The crystals were cut into 0.5 mm thick wafers which were then lapped and polished to a thickness of 350 μm. An Au/Ge/Ni-alloyed ohmic contact was deposited on the backside of the sample and an Au Schottky contact applied to the top, resulting in a response to the [57]Co and ^{137}Cs sources similar to that measured by Olschner *et al.* [64].

Dubecky *et al.* [66] fabricated a number of radiation detectors from semi-insulating InP wafers doped with Fe. After lapping and polishing, the detectors had thicknesses of ~200 μm. An Au electrode was evaporated onto the top of the detector samples and a non-alloyed AuGeNi eutectic onto the bottom. The Au forms a pseudo blocking contact and the eutectic an ohmic contact. The devices were tested with [241]Am, [133]Ba, [57]Co and ^{137}Cs radioactive sources. They were found to be responsive to X-rays at room temperature but not spectroscopic. However, at an operating temperature of 216K, FWHM energy resolutions of 7 keV at 59.54 keV, 11 keV at 122 keV and 93 keV at 662 keV were recorded. Owens *et al.* [67] reported FWHM spectral resolutions at –60°C of ~ 2.5 keV and 9.2 keV at 5.9 keV and 59.54 keV respectively, with a small Fe-doped device of area 3.1 mm^2 and thickness 180 μm (Fig. 9.9). At -170°C, these figures improved considerably to 911 eV at 5.9 keV and 2.5 keV at 59.54 keV [4].

Gorodynskyy *et al.* [68] and Yatskiv *et al.* [69] have argued that while Fe doping increases material resistivity, it also introduces deep electron and hole traps which limit charge collection efficiency (CCE) and therefore energy resolution. Attempts to increase the CCE by lowering the doping concentration have the same result, since lowering the Fe doping concentration also lowers the resistivity, which in turn degrades the energy resolution. In fact, the best room temperature CCEs obtained with bulk material are ~75% when measured with alpha particles emitted by an [241]Am source [70]. The authors proposed to increase the CCE by co-doping with Ti and Mn [68] or Ti and Zn [69] instead of Fe. The presence of Ti produces high resistivity material with much reduced hole trap capture cross-sections compared to Fe doping [71]. Co-doping with Mn has the effect of suppressing the electron traps. While a significant improvement in CCE was found using Mn-Ti co-doped SI material (91% for 5.5. MeV alpha particles at 230 K), an X- or gamma-ray response could not be measured due to a high level of system noise. Subsequent work by Yatskiv *et al.* [69] using a Ti-Zn co-doped InP system has achieved a CCE of 99.9% and a spectral resolution of 0.9% FWHM for 5.5 MeV alpha particles at 230 K. No results were given for photons.

9.4.6 Indium Iodide

Indium iodide (InI) is a wide bandgap, base-centered orthorhombic crystal with a layered structure. It has relatively low melting point (351°C) and exhibits no solid-solid phase transition between its melting point and room temperature. Therefore, high quality crystals may be obtained by using simple melt-based processes. Due to the high atomic numbers

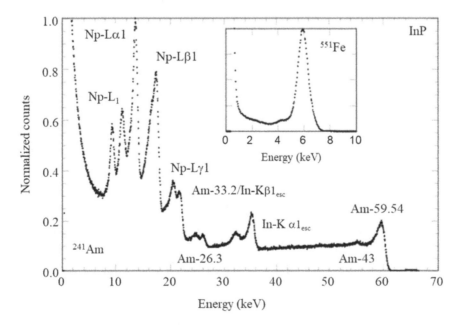

FIGURE 9.9 The measured response of a 3.1 mm^2, 200 μm thick InP detector to [241]Am under full area illumination at –170°C. The inset shows its response to [55]Fe. The measured FWHM energy resolutions were 911 eV at 5.9 keV and 2.5 keV at 59.54 keV (from [4]).

of its constituent elements ($Z_{In} = 49$ and $Z_I = 53$) and high density (5.31 g cm^{-3}), InI exhibits a photon stopping power similar to that of CdTe. Its bandgap is 2.0 eV, which offers the potential for low-noise operation at and above room temperature.

Squillante *et al.* [72] fabricated a radiation detector on a single crystal wafer grown by the Bridgman process. Carbon electrodes were formed by painting a lacquer–graphite suspension onto InI slices. The contact area was 5×5 mm^2 and the thickness was 0.5 mm. The resistivity was found to be $> 10^{11}$ Ω cm. The device showed a clear, but not totally spectroscopic, response to 22 keV X-rays from a ^{109}Cd radioactive source at room temperature. At 120°C, the spectroscopic performance degraded but the device was still able to function as a counter with a counting efficiency almost as high as at lower temperatures.

Onodera *et al.* [73] fabricated radiation detectors from InI crystals grown by the travelling molten zone (TMZ) method. Au electrodes of area 1 mm^2 were deposited on to the top and bottom of InI wafers of thickness $\sim$ 0.4 mm. The response of the devices to 22 keV X-rays from a ^{109}Cd radioactive source at room temperature were very similar to that observed by Squillante *et al.* [72], despite the resistivity of the device being two orders of magnitude lower ($\sim 3 \times 10^9$ Ω cm). All devices showed clear polarization effects, which manifested themselves as a time-dependent deterioration of energy resolution and peak position at room temperature. The effects could be suppressed by cooling the detectors to −20 °C.

Bhattacharya *et al.* [74] grew InI ingots using both a vertical Bridgman and a vertical gradient freeze technique using zone-refined commercially available material and material synthesized from vapor, respectively. The ingots were sliced into 2 mm thick wafers and Pd metal electrodes deposited by RF sputtering. The current-voltage characteristics were found to be linear in either polarity of applied voltages, with resistivities of 2×10^9 Ω cm for the zone refined and 1×10^8 Ω cm for the vapor synthesized starting materials, respectively. Both detectors showed a spectroscopic response to alpha particles from a ^{241}Am source with a best-recorded energy resolution of $\sim$50% at 4°C operating temperature and a non-spectroscopic response to 662 keV gamma-rays at room temperature.

9.4.7 Narrow Gap Materials

Narrow-gap refers to a semiconductor with a forbidden energy gap of less than $\sim$0.5 eV. Examples include InAs, InSb, PbS, PbSe, PbTe, Bi$_2$Te$_3$ and HgCdTe. They are used primarily in terahertz source, thermoelectric and infrared applications. At X-ray and gamma-ray wavelengths, several narrow bandgap materials have attracted attention as possible replacements for Si and Ge. These materials, InAs and InSb in particular, offer the possibility of spectral resolution beyond that of the elemental semiconductors, closer to the high-temperature superconductors, despite the disadvantage of requiring substantially lower operating temperatures. Potentially, InAs can achieve twice the energy resolution of silicon and InSb three times. In addition, both have very high electron mobilities permitting low bias operation and both are already extensively used in the semiconductor industry. From a radiation detection point of view, both InSb and HgCdTe have been successfully used for infrared focal planes [75,76].

9.4.7.1 Indium Arsenide

Indium arsenide, or indium monoarsenide (InAs) is a direct bandgap material with a cubic crystal structure, a bandgap of 0.35 eV, an electron mobility of $\sim$33,000 cm^{2}V^{-1}s^{-1} and a density of 4.68 gcm^{-3}. It has conventionally been used in infrared detection in the wavelength range of (1–3.8) μm and to produce infrared diode lasers for telecom applications. Theoretically, the energy resolution of InAs based X-ray detectors can be expected to exceed that of Si by a factor of two by virtue of its low bandgap. Recently, Säynätjoki *et al.* [77] fabricated a 3×3 pixel array onto a commercially available substrate. The active pixel volumes were defined by diffusing Zn through a mask using metal–organic vapor phase epitaxy (MOVPE). The pixels themselves were then defined by chemically etching mesas and contacts applied by metallization. The pixel sizes were 250 μm square and the thickness of the active layer $\sim$1 μm. Typically, reverse leakage currents were of the order of a few mA at liquid nitrogen temperature. The device was found to be responsive, but not spectroscopic, to 5.5 MeV alpha particles.

9.4.7.2 Indium Antimonide

Like InAs, indium antimonide (InSb) is commonly used to fabricate infrared detectors that are sensitive in the wavelength range 1–5 μm. It has a zincblende crystal structure and a bandgap of 0.165 eV, almost 10 times less than silicon. Its density of 5.78 g cm^{-3} is twice that of Si. Its electron mobility is exceedingly high (78,000 cm^{2}V^{-1}s^{-1} at 77K), and although its hole mobilities are considerably lower (750 cm^{2}V^{-1}s^{-1}), they are still higher than those measured in HgI$_2$ and CdTe. These are very attractive attributes for a potential high-resolution radiation detector. Although initially suggested as an X-ray detection medium by Harris [78] in 1986, very little work has been carried out until recently. Kanno *et al.* [79] fabricated both Schottky and *p–n* junction diodes on *p*-type InSb grown by the vertical Bridgman method. A 3 mm diameter, 10-micron thick mesa was etched into the substrate and doped with Sn to form an *n*-type layer. An evaporation of Au–Pd formed a Schottky contact for the Schottky diode and an Al–Sn evaporation formed the ohmic contact for the p-n device. The I–V characteristics show typical diode behaviour with reverse leakage currents in the tens of microampere range. The estimated resistances of the Schottky and pn junctions at 4.2 K were 50 kΩ and 250 kΩ respectively. Both detectors were found to be responsive to 5.5 MeV alpha rays with FWHM energy resolutions of ~25% at 20 K. The Schottky diode detector operated up to 77 K, the p-n junction device up to 115 K. Although not spectroscopic, undoped Schottky devices were found to be responsive to 60 keV and 81 keV gamma-rays [80].

Following on from this work, Sato *et al.* [81] fabricated a detector by depositing a ~115 μm thick *p*-type InSb layer on a 0.4 mm thick commercially available InSb substrate by LPE. A 1 mm diameter Al electrode was deposited on the epitaxial side of the device forming a rectifying contact. The substrate was indium soldered to a Cu plate, which served as an ohmic contact. At 4.4 K, the resistance of the device was measured to be 680 kΩ, substantially higher than achieved with Schottky and pn devices fabricated from material grown by the vertical Bridgman method. From the I-V characteristics, the typical reverse leakage currents were lower than in previous work, being of the order of ~4 μA. In addition, the breakdown voltage of ~–2 V represents a factor of 2 improvement on previous work. The detector was spectroscopic to alpha particles with a measured FWHM energy resolution of 3.1% at 5.5 MeV. The measured energy loss spectrum is shown in Fig. 9.10. The device also showed a response to gamma-rays from a ^{133}Ba radioactive source, shown by the inset in Fig. 9.10. While it is tempting to ascribe the apparent feature near channel 250 to 81 keV gamma-rays from the source, transport simulations have indicated that it is more likely due to a "shoulder" of 356 keV decay gamma-rays from the source [82].

9.4.8 Aluminum Gallium Arsenide

Aluminum Gallium Arsenide (AlGaAs) has been explored as a potential X-ray sensor by a number of groups. The material was initially investigated by Leuter *et al.* [83] as an avalanche region in an AlGaAs/GaAs separate absorption and multiplication region (SAM) APD in which a graded $Al_xGa_{1-x}As$ graded layer (*x* varying linearly from 1% to 45%) acts as a multiplication layer. Absorption and generation of the carriers takes place in a thicker (4.5 μm) overlying GaAs layer. Measurements with X-rays from a ^{241}Am source showed that the device was spectroscopic. Without avalanche gain, a FWHM energy resolution of 1.95 keV (14%) at 13.96 keV was achieved at room temperature. With an internal gain of 4, the FWHM energy resolution improved to 0.9 keV (6.5%).

Lees *et. al.* [84] constructed a set of $Al_{0.8}Ga_{0.2}As$ p^+–p^-n^+ circular mesa diodes. MBE was used to produce a layered structure consisting of an n+ 350 μm GaAs substrate by MBE on which was grown a 1 μm thick and n-doped GaAs contact layer, an n-$Al_{0.8}Ga_{0.2}As$ buffer layer, followed by a 0.58 μm thick $Al_{0.8}Ga_{0.2}As$ intrinsic layer, a p-$Al_{0.8}Ga_{0.2}As$ buffer layer and a 0.01 thick p-GaAs contact layer. Circular mesa devices were then fabricated using standard optical lithography and wet chemical etching. Each device had an Au/Zn/Au ohmic contact on the top p-GaAs layer, and an In/Ge/Au alloy was deposited on the back n+ substrate forming an n-metal ohmic contact. 100 μm diameter and 200 μm diameters devices were

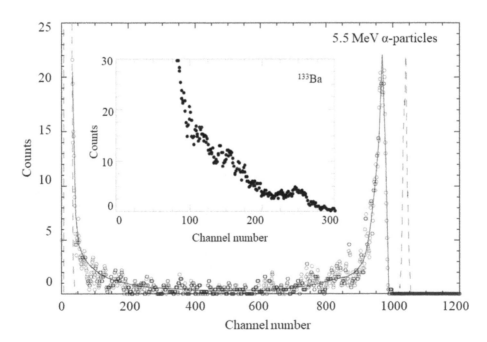

FIGURE 9.10 Energy spectrum of 5.5 MeV alpha particles recorded at 5.5 K by a 1 mm diameter InSb detector grown by LPE (data courtesy of I. Kanno). The dotted line shows a pulser spectrum. The inset shows the response of the detector to gamma-rays from a ^{133}Ba radioactive source. The amplifier gain is 1.75 times that of the gain used to acquire the alpha particle spectrum.

produced. For nominal biases < 26V the measured leakage currents were found to be < 10 nA at room temperature. The devices were found to be spectroscopic to a ^{55}Fe source with a measured FWHM energy resolution of 1.5 keV at 5.9 keV for the 200 μm diameter device for an optimal reverse bias of 20V. At higher biases, avalanche multiplication was observed, which while it improved the signal to noise ratio, it also degraded the spectral performance. Barnett *et al.* [85] explored the temperature characteristics of similar diodes from the same batch. Operating in non-avalanche photon counting mode, the temperature-dependent FWHM energy resolutions at 5.9 keV were measured to be 0.9 keV at –30°C, 1.1 keV at room temperature and 2.5 keV at +90°C.

Whitaker *et al.* [86] fabricated a set of AlGaAs p-i-n mesa photodiodes with an Al fraction of 20%. The mesa diameter was 200 μm and the thickness of the intrinsic region was 3 μm. The measured room temperature FWHM energy resolution at 5.9 keV was 1.2 keV, comparable with the resolution obtained with p-i-n diodes with an Al fraction of 0.8. Subsequent measurements [87] showed that the resolution could be improved by cooling the detector. A FWHM energy resolution of 830 eV at 5.9 keV was obtained at -20°C.

9.4.9 Aluminum Indium Phosphide

Butera *et al.* [88] report soft X-ray measurements with AlInP p-i-n mesa photodiodes. The devices were grown by MOCPE and consisted (from the bottom up) of a 0.1 μm n+ $Al_{0.52}In_{0.48}P$ layer grown on a n+GaAs substrate, followed by a 2 μm thick $Al_{0.52}In_{0.48}P$ epilayer and then a 0.2 μm p+ $Al_{0.52}In_{0.48}P$ layer. Ohmic metallic contacts were deposited by evaporation. An annular Ti/Au (20 nm/200 nm) contact was applied to the top of the mesa and an InGe/Au (20 nm/200 nm) contact deposited on the rear of the substrate. For a 200 μm diameter device, a FWHM energy resolution of 930 eV at 5.9 keV was achieved at room temperature. In subsequent work [89], the temperature characteristic of the device was investigated. At 100°C, the FWHM energy resolution at 5.9 keV was measured to be 1.57 keV, falling to 770 eV at –20°C.

Auckloo *et al.* [90] report the performance of 200μm diameter $Al_{0.52}In_{0.48}P$ avalanche photodiodes for soft X-ray spectroscopy. The devices consisted of a of a p-i-n structure where the intrinsic layer was subdivided into a 0.5 μm thick detection layer followed by a 1.0 μm avalanche region. By varying the bias across the device, the internal gain of the avalanche region can be changed. This gain is the end result of successive impact ionization events whereby an energetic electron (or a hole) can create another electron and a hole, which can be used to increase the amount of charge created for each ionization event. This in turn can improve the signal-to-noise ratio and hence energy resolution. For example, increasing the reverse bias from 30V to 65V (corresponding to a gain change of a factor of 4) improves the FWHM energy resolution at 5.9 keV from 2 keV to 682 eV.

9.4.10 Indium Gallium Phosphide

Butera *et al.* [91] fabricated InGaP (GaInP) mesa p-i-n photodiodes. The devices were fabricated by successively depositing an n+ InGaP (0.1 μm), an intrinsic $In_{0.5}Ga_{0.5}P$ (5 μm) and a p+ InGaP (0.2 μm) layer onto a GaAs substrate by MOVPE to produce a p+-i-n+ structure. The top layers were etched to fabricate circular mesa photodiodes of 200 μm and 400 μm diameter. Circular Ti/Au (20 nm/200 nm) layers were then deposited on top of the mesa and InGe/Au (20 nm/200 nm) layers on the bottom of the GaAs substrates to form ohmic contacts. At 5.9 keV, room temperature FWHM energy resolutions of 1.2 keV and 0.9 keV were achieved for the 400 μm and 200 μm devices, respectively.

9.5 Group II–VI Materials and Alloys

Group II-VI materials have attracted a lot of attention because of the wide range of compounds available and the possibility to engineer the bandgap over an almost continuous range. These are compounds which combine a Group IIb metal (such as Zn, Cd, and Hg, in periods 3, 4 and 5, respectively) with a Group VIa cation. The latter is usually S, Se or Te. Structurally, it forms when atomic elements from one type bonds to the four neighbours of the other type, as shown in Fig. 3.7c. A major motivation for developing II-VI semiconductors is their broad range of bandgaps (from 0.15 eV for HgTe to 4.4 eV for MgS), high effective Z and the capability for making MBE and MOCVD grown heterostructures, in the same way as for III-V systems. Additionally, all II-VI binaries have direct bandgaps that make them particularly attractive for optoelectronic applications. Compounds generally crystallize naturally in a hexagonal or NaCl-structure. Representative compounds are CdTe and HgTe. Pseudo-binary alloys with Zn, Se, Mn or Cd are also common – particularly for radiation detector and optoelectronic applications (*e.g.*, $Cd_{(1-x)}Zn_xTe$, $Cd_{(1-x)}Mn_xTe$ and $Hg_{(1-x)}Cd_xTe$). Group II-VI compounds typically exhibit a larger degree of ionic bonding than III-V materials, since their constituent elements differ more in electron affinity due to their location in the periodic table. A major limitation of these compounds is the difficulty in forming *n*-type and *p*-type material of the same compound, preventing the formation of a *p-n* junction. In fact, it is only recently that *p*-type doping of ZnSe has been achieved [92]. In addition, it has been difficult to control the defect state density within the bandgap due to self-compensation. However, despite these limitations, Group II-VI compounds are the most widely used for radiation detection. Group II-VI semiconductors can also be created in ternary and quaternary forms, although less common than III-V varieties. As with III-V materials, a major problem is controlling the multiplication of stoichiometric errors during the growth process, which leads to poor transport properties.

9.5.1 Cadmium Telluride

Cadmium telluride (CdTe) was one of the first compound semiconductors to be synthesized in the late 1800s. However, until the 1940s its only use was as a pigment, whereupon it was used in the production of photocells. It is a Group II–VI material that crystallizes in a cubic zincblende structure. In its bulk crystalline form it is a direct bandgap semiconductor with a bandgap of 1.56 eV at 300 K which is high enough to allow room temperature operation. It has a density of 5.85 gcm^{-3}. The large atomic numbers of its constituent atoms ($Z_{Cd} = 48$, $Z_{Te} = 52$) ensure a high stopping power and therefore high quantum efficiency, in comparison with Si and Ge. For example, for gamma-ray detection, 2 mm of CdTe is equivalent to 10 mm of Ge.

CdTe is generally grown by the travelling heater method (THM). To achieve high resistivity (10^8–10^{10} Ω cm) THM grown crystals are sometimes doped with Cl to compensate for impurities. However, Cl doping can also introduce a number of other problems, namely polarization effects and long-term reliability issues. As a rule, CdTe is easier to produce in large quantities than CnZnTe, since the latter is invariably grown by the High Pressure Bridgman method. Large wafers of CdTe can be produced up to 50 mm in diameter without grain boundaries with good reproducibility and homogeneity. CdTe is currently used primarily in the manufacture of photovoltaics, particularly solar cells, nanoinks and nanorods. Other applications include electro-optical modulators, IR windows and photorefractive materials.

Since the late 1960s, CdTe has also been regarded as a promising semiconductor material for hard X-ray and γ-ray detection in view of its relatively high carrier mobilities ($\mu_e \sim 10^3$ cm^{2}V^{-1}s^{-1} and $\mu_h \sim 100$ cm^{2}V^{-1}s^{-1}) and high stopping power. Early experiments by Akutagawa *et al.* [93] showed that it was possible to resolve spectrally both alpha particles and photons using ~1 mm thick detectors, and a FWHM spectral resolution of ~40% for 160 keV photons was achieved at room temperature. Over the next three decades detector performances improved incrementally. For small ~25 mm^3 planar detectors, room temperature energy resolutions are in the range 2–6 keV.

Recently, however, high-resolution CdTe detectors with FWHM energy resolutions better than 1 keV at 60 keV have become available. This has been achieved by significantly lowering leakage currents by fabricating diode structures, either using a blocking electrode [94] or a P–I–N type structure [95,96]. Leakage currents of the order of several nA mm^{-2} have been obtained at room temperature, which compares well with commercially available CdZnTe detectors. Takahashi *et al.* [97] report a room temperature FWHM energy resolution of 1.8 keV for the ^{241}Am nuclear line at 59.54 keV using a Schottky contact detector[4] of dimensions $2 \times 2 \times 1$ mm^3. For comparison, the same sized detector with conventional ohmic contacts yielded a resolution of 3.3 keV. However, the main disadvantage of CdTe Schottky contact diodes is their susceptibility to polarization effects. While the conventional detector was stable, the Schottky diode was unstable on a timescale of minutes and the spectra disappeared after 30 mins. Polarization effects can be reduced or even eliminated by operating the detector at a much higher bias or reduced temperature. At a reduced temperature of –25°C, the performance of the Schottky diode detector improved considerably to 810 eV and was stable over 24 hours. A similar diode of thickness 0.5 mm gave FWHM energy resolutions of 830 eV at 59.54 keV and 2.1 keV at 662 keV [94] when operated at a bias of 1400V and a temperature of –40°C. This bias is roughly 3 times that of an equivalent CdTe detector with ohmic electrodes. These resolutions are close to those expected from liquid nitrogen-cooled high purity Ge detectors.

Niemela *et al.* [96] and Khusainov *et al.* [98] produced CdTe diodes with a P-I-N structure. For a 16 mm^2, 1 mm thick device, they achieved FWHM energy resolutions of 1.1 keV at 59.54 keV and 2.5 keV at 662 keV at an operating temperature of –30°C. Khusainov *et al.* [97] obtained similar results with a 7 mm^2, 1 mm thick detector. At 662 keV and an operating temperature of –30°C, they measured a FWHM energy resolution of 2.5 keV. In Fig. 9.11 we show ^{57}Co and ^{137}Cs spectra taken with a larger $11.3 \times 9.1 \times 2.13$ mm^3 P-I-N detector cooled to –35°C [97].

In contrast to HgI$_2$, CdTe does not undergo an irreversible phase change below its melting point, which could prevent high-temperature operation. The maximum temperature of CdTe is actually limited to ~120°C by the onset of migration of the chlorine dopant used to compensate impurities. Mahdavi *et al.* [99] repeatedly operated eight sample CdTe detectors for extended time periods at temperatures up to 100°C. The detectors had simple planar geometries of size $5 \times 5 \times 2$ mm^3 with ohmic contacts. The most obvious change in performance was an exponential increase in leakage current with temperature. At 100°C the current was ~10 μA – about 1,000 times the value at room temperature. This resulted in a progressive degradation in energy resolution with temperature. For example, for 122 keV photons, the FWHM energy resolution at room temperature was ~5 keV, rising to 10 keV at 60°C and 35 keV at 100°C.

---

[4] Also known as M–π–n detectors.

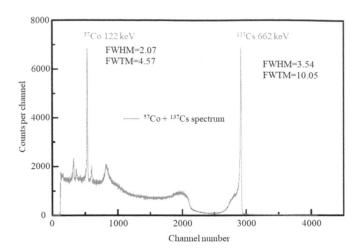

FIGURE 9.11 ^{57}Co and ^{137}Cs spectra taken with a 1 cm^2, 2.1 mm thick P-I-N CdTe detector cooled to -35°C (from [97], ©2001 IEEE). The operating bias was -3000V. The FWHM energy resolution at 662 keV was 3.5 keV.

9.5.2 Cadmium Selenide

Cadmium selenide (CdSe) crystallizes into a hexagonal (wurtzite) close packed structure at 1239°C, which contributes to its relatively high photon attenuation coefficients at hard X-ray wavelengths. These are comparable to CdTe despite a lower effective Z and density (5.8 g cm^{-3}). Its bandgap is 1.73 eV, which should be high enough to allow room temperature operation, and its electron and hole mobilities (720 cm^{2}V^{-1}s^{-1} and 75 cm^{2}V^{-1}s^{-1}, respectively) are amongst the highest of II–VI compounds. CdSe was first explored as a room-temperature radiation detector by Burger *et al.* [100] who grew high purity single crystals of CdSe by the vertical unseeded vapor growth method. The crystals were sliced in 250 μm disks. Although the as-grown crystals had a resistivity of only ~10 Ω cm, after annealing the resistivity of the disks increased considerably to 10^{12} Ω cm. The disks were polished and etched to a thickness of 200 μm then washed and processed into detectors by evaporating 4 mm^2 Au contacts on to both sides. The finished detectors were found to be responsive to X-ray photons over the energy range 10 keV to 660 keV, although the energy resolutions were poor. For example, at 60 keV it was 45 keV (75%) FWHM. Roth [101] produced high resistivity (~10^9 Ω-cm) CdSe crystals by the temperature gradient solution zoning technique. The crystals were sliced and fabricated into detectors of thicknesses ranging from 0.2 mm to 1.0 mm. The contacts were formed by painting Aquadag or evaporating Pd onto the crystal surfaces. A room-temperature energy resolution of 1.4 keV FWHM was obtained for the ^{55}Fe line at 5.9 keV and 8.5 keV FWHM for the ^{241}Am nuclear line at 59.54 keV. Chen *et al.* [102] report measurements on undoped CdSe single crystals grown by an unspecified method. The measured resistivities were in the range 10^{10} Ω cm – comparable to that of CdZnTe grown by the High Pressure Bridgman method. One crystal was cleaved into smaller samples having dimensions 4×4 × 1.2 mm^3. The samples were polished, etched and rinsed leaving final detector thicknesses of 0.15, 0.3 and 1.2 mm. Au contacts were deposited by thermal evaporation immediately after chemical treatment to minimize surface oxidation. The devices were found to be spectroscopic to both alpha particles and photons. For example, at a photon energy of 60 keV, measured FWHM energy resolutions of 30% and 42 % were recorded for the 0.15 mm and 0.3 mm thick devices, respectively.

9.5.3 Zinc Selenide

Zinc selenide (ZnSe) belongs to the family of zinc chalcogenides[5] (ZnS, ZnSe and ZnTe) whose wide bandgaps make them promising materials for many optoelectronic applications. Of the three, only the selenide has shown a spectroscopic response to X-rays, although ZnTe has been widely used in scintillator applications. Zinc selenide crystallizes in the zincblende configuration. It is a direct bandgap semiconductor with an energy gap of 2.7 eV at room temperature and a density of 5.3 g cm^{-3}. As-grown ZnSe is insulating due to self-compensation but can be made semiconducting by annealing the crystal in molten zinc. It is an interesting material for several reasons. Firstly, its lattice constant is 5.667 Å, which is almost lattice matched to GaAs. This makes epitaxial growth possible. It is also one of the few II-VI compounds for which both *n*- and *p*-type materials are available, which has opened up new avenues in optoelectronic applications, such as the production of blue LEDs and laser diodes. For radiation detector applications, ZnSe has been explored as a high-temperature alternative to CdTe and CdZnTe [103] since practical operation of CdTe and CdZnTe detectors is limited to ~70°C before irreversible

---

[5] Chalcogenides are materials containing one or more elements from group 16 of the periodic table (*e.g.*, S, Se or Te) as a substantial constituent.

damage sets in. Similarly, the maximum storage temperatures are around 100°C. ZnSe, by virtue of its large bandgap (2.7 eV as opposed to 1.47 eV for CdTe and 2.0 eV for CdZnTe), should operate to much higher temperatures, up to ~200°C.

Eissler and Lynn [103] grew ZnSe crystals using the High Pressure Bridgman method. The crystals were sliced and detectors fabricated. Au or Pt electrodes were applied by sputtering to form a MSM configuration. The electrode dimensions were 10 mm × 10 mm and the detector thickness 2 mm. The detectors were found to be spectroscopic to photons with room temperature FWHM energy resolutions of 25% at 22.1 keV. The device was found to function across the temperature range –70°C to +170°C. Up to ~100°C, very little change was seen in the spectral performance. Above this temperature, a steady degradation was observed. Although the peak width was more or less constant, the system gain systematically decreased while simultaneously the noise floor increased. At 130°C, the FWHM energy resolution at 22.1 keV was 35%. Above 175°C, the system failed, although it was not clear exactly which component failed. However, these experiments demonstrate that ZnSe can function as a high-temperature spectroscopy medium.

9.5.4 Cadmium Zinc Telluride

Cadmium zinc telluride (variously denoted by CZT, CdZnTe, (Cd, Zn)Te or $Cd_{(1-x)}Zn_xTe$, where x is the mole fraction of zinc) is a pseudo ternary compound[6] that has been extensively studied at X- and γ-ray wavelengths (see [104] for a review) and is probably the most widely used compound semiconductor. It was originally produced as a lattice-matched substrate for $Hg_{(1-x)}Cd_xTe$ epilayers for IR rather than X- or gamma-ray applications. It has a cubic, zincblende type lattice with atomic numbers close to that of CdTe and a density ~3 times that of Si (5.8 gcm^{-3}). It is generally produced by the high-pressure Bridgman method, although low pressure Bridgman material is becoming increasingly available and appears to offer good uniformity and comparable electronic properties to the high-pressure Bridgman CdZnTe. CdTe was originally the focus of experimental study in the 1960s, until it was discovered that the addition of a few percent of zinc to the melt results in an increased bandgap as well as the energy of defect formation [105] and the virtual elimination of polarization effects normally associated with CdTe [106]. This in turn increases bulk resistivities and reduces the dislocation density, resulting in lower leakage currents. Specifically, resistivities of CdZnTe are between one and two orders of magnitude greater than that for CdTe, and thus leakage currents are correspondingly lower. Alloying of CdTe with Zn or ZnTe has additional benefits in that it strengthens the lattice mechanically with a resulting lowering of defect densities. Depending on the zinc fraction, its bandgap is typically 1.45–1.65 making it suitable for room- or even elevated temperature operation. In fact, Egarievwe *et al.*, [107] have operated CdZnTe detectors up to 70°C with little degradation in performance over that at room temperature. The variation of bandgap with zinc fraction, x, can be described empirically [108] by,

$$E_g(x) = 1.510 + 0.606x + 0.139x^2 \quad (eV). \tag{1}$$

Whilst an increase in E_g increases Fano noise due to carrier generation statistics, it simultaneously reduces shot noise due to thermal leakage currents. A trade-off can lead to a noise minimum at a given operating temperature and therefore an optimization of the spectral performance [109]. For example, a zinc fraction of 10% is optimum for operation at –30°C, whereas 70% is optimum at room temperature. For 10% zinc fractions, typical energy resolutions at 59.54 keV are in the keV range at room temperature and decrease with decreasing temperature to a minimum resolution at ~–30°C. This is demonstrated in Fig. 9.12 in which we show FWHM energy resolutions for a 3.1 mm^2, 2.5 mm thick CdZnTe detector at 5.9 and 59.54 keV as a function of detector temperature [110]. Although the leakage currents are low enough to allow room-temperature operation, it can be seen that there is a marked improvement in both resolution functions with only a modest reduction in temperature. In fact, the energy resolution improves by a factor of 3 with a temperature reduction of only 20°C, compared with that measured at room temperature. At temperatures below –20°C, there is relatively little improvement if any, in ΔE at both incident energies. We also note that while the resolution at 5.9 keV steadily improves with decreasing detector temperature, the resolution at 59.54 keV shows a minimum near –30°C. This may be due to the holes freezing out, which would explain why it is only observable at the higher energies. At 60 keV, the drift lengths of holes are typically a few hundred microns, whereas at 6 keV, they are only a few microns. The optimum overall performance was found at a detector temperature of –37°C at which the measured FWHM energy resolutions were 311 eV at 5.9 keV and 824 keV at 59.54 keV. The measured energy loss spectra are shown in Fig. 9.13.

9.5.5 Cadmium Manganese Telluride

The pseudo ternary compound cadmium manganese telluride (usually denoted by CMT, CdMnTe, (Cd, Mn)Te or $Cd_{(1-x)}Mn_xTe$) is a magneto-optical material with spintronic properties [111] that has found application in the fabrication of Faraday rotators, optical isolators and magnetic field sensors. It has traditionally been used in the production of LEDs, solar

---

[6] Strictly speaking, it is an alloy of cadmium telluride and zinc telluride.

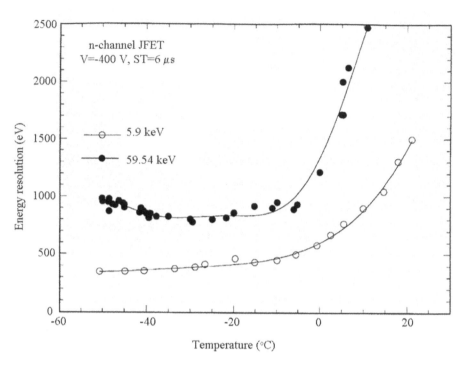

FIGURE 9.12 The temperature dependence of the FWHM energy resolutions measured with a 3.1 mm², 2.5 mm thick CdZnTe at 5.9 keV and 59.54 keV under full-area illumination (from [110]). The solid lines are best-fit polynomials.

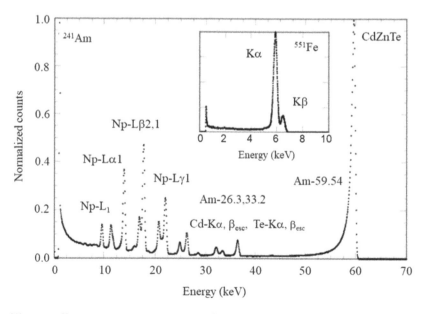

FIGURE 9.13 Composite of ²⁴¹Am and ⁵⁵Fe spectra taken with a 3.1 mm², 2.5 mm thick CdZnTe detector (from [4]). The detector temperature was –37°C and the applied bias +320V. The measured FWHM energy resolutions were 311 eV at 5.9 keV and 824 eV at 59.54 keV. The corresponding pulser widths were (a) 260 keV and 370 eV, respectively.

cells and visible to mid-infrared tuneable lasers [112] and has been proposed as a radiation detection medium [113]. Recently it has attracted attention as an inexpensive alternative for CdZnTe, since it is easier to produce. CdZnTe can only be reliably grown using high pressure Bridgman (HPB), which is complex and expensive, whereas CdMnTe can be grown by the modified Bridgman techniques, which are considerably less expensive than HPB. In addition, the segregation coefficient of Zn along the growth axis in CdZnTe is large ($k = 1.35$) and leads to substantial compositional variations along the growth direction which in turn can lead to substantial variations in the performance in fabricated detectors. In contrast, the segregation coefficient of CdMnTe is almost unity, which means that large homogeneous crystals of the same composition as the starting liquid phase can be grown from melt. Finally, it is easier to engineer the bandgap in CdMnTe due to the large compositional influence of manganese. While the energy gap of CdZnTe increases only by 9.7 meV per atomic percent of Zn, the corresponding value for the Mn concentration in CdMnTe is 13 meV [114]. This means that less Mn is required than Zn to produce a specific bandgap, so, for example, for $x = 0.05$, the bandgap of $Cd_{0.95}Mn_{0.05}Te$ is about 1.60 eV, which is the same as that of $Cd_{0.9}Zn_{0.1}Te$ (the standard composition for spectrometer-grade CdZnTe material). The room-temperature energy gap of $Cd_{1-x}Mn_xTe$ has been found to be linear with Mn fraction x [115],

$$E_g(x) = 1.526 + 1.316x \quad (eV). \tag{2}$$

In principle, it is possible to adjust the bandgap from 1.5 eV to 3.4 eV (in CdZnTe the range is 1.5 eV to 2.2 eV). Thus, CdMnTe could potentially operate over a wider temperature range than CdZnTe. As mentioned previously, the practical operation of CdZnTe detectors is limited to temperatures below ~70°C before irreversible damage sets in.

At present, CdMnTe suffers from several major material problems. Firstly, it is difficult to synthesize due to the high reactivity of Mn, which tends to bond with the residual oxides on the surface of the container. Secondly, compared to CdZnTe, the bond-ionicity of CdMnTe is higher, resulting in a tendency to crystallize in the hexagonal form, or to produce a high degree of twinning in the zincblende modification. Thirdly, the resistivity of "as grown" CdMnTe crystals can be quite low ($<10^3$ Ω-cm) due to a high concentration of cadmium vacancies which invariably result from Bridgman growth. These act as acceptors, and thus the material is p-type. However, the resistivity can be substantially increased ($>10^9$Ω-cm) by annealing the crystals in a Cd ambient or more commonly doping with a donor impurity, usually indium, chlorine or vanadium. Mycielski *et al.* [116], in a critical study of the growth and preparation of CdMnTe for radiation detector applications, concluded that 6% atomic weight Mn content minimizes the density of twins and maximizes the grain size in "as-grown" crystals and that annealing in Te vapor at the temperature gradient followed by annealing in Cd vapor minimizes impurity concentrations, Te precipitates and cadmium vacancies. In an assessment of three contacting schemes they further concluded that amorphous layers of heavily doped semiconductors provide the best electrical contacts to semi-insulating (Cd,Mn)Te crystals.

Photoconduction measurements were carried out by Burger *et al.* [113] on a number of $Cd_{0.85}Mn_{0.15}Te$ and $Cd_{0.55}Mn_{0.45}Te$ crystals that were highly doped (~10^{19}cm^{-3}) with vanadium. The resistivity of the crystals exceeded 1×10^{10} Ωcm, but the mobility–lifetime product ($\mu\tau$) was only 1×10^{-6} cm^{2}V^{-1}. A number of detectors were fabricated and one device was found to be spectroscopic to photons. The device had an area of 1.8 mm^2 and a thickness 0.5 mm. A FWHM energy resolution of 40% at 59.54 keV was achieved for a detector bias of 500V and a 1 μs peaking time. Parkin *et al.* [117] have produced detectors grown by the modified Bridgman technique. The detectors were found to be spectroscopic (~50% FWHM) to alpha particles but not to photons. Cui *et al.* [118] fabricated a number of $Cd_{0.94}Mn_{0.06}Te$ detectors from Bridgman grown material. The "as-grown" crystals were annealed in a Cd ambient and doped with 5×10^{16} cm^{-3} of vanadium. One detector, of active area 28.3 mm^2 and thickness 1.8 mm had a measured electron mu-tau product of 2.1×10^{-4} cm^{2}V^{-1}, over two orders of magnitude larger than the value achieved by Burger *et al.* [113]. From I-V measurements, the resistivity was 3.2×10^{12} Ω cm. The detector was found to be spectroscopic to photons. When it was biased to collect electrons, the measured FWHM energy resolution was 27% at 60 keV. When biased to collect holes, a peak in the spectrum was also obtained, although poorly resolved. The hole mu-tau product could not be directly determined due to system noise. Using vertical Bridgman growth, Kim *et al.* [119] grew $Cd_{0.9}Mn_{0.1}Te$ crystals compensated with In (10^{17}cm^{-3}). The "as-grown" ingots were cut into 2 mm thick slices and mechanically and chemically polished down to a thickness of 1 mm. Detector blanks of area 2×2 mm^2 were diced from the wafers and contacted with Au electrodes deposited using thermal evaporation. Three detectors, of active areas of 0.25, 1 and 4 mm^2, were tested and resistivities, determined from I-V measurements, were all found to be in the range $1–3 \times 10^{10}$ Ω cm. The electron mobility–lifetime products were determined from a Hecht plot to be 1×10^{-3} cm^{2}V^{-1}, which is an order of magnitude higher than previous measurements. The detectors showed a spectroscopic response to both alpha particles and photons with room temperature FWHM energy resolutions of 3% for 60 keV photons.

9.5.6 Cadmium Magnesium Telluride

Cadmium magnesium telluride $Cd_{1-x}Mg_xTe$ (also labelled as CdMgTe or CMgT) is currently being investigated as an alternative room temperature X- and gamma-ray detection medium to $Cd_{1-x}Zn_xTe$ and $Cd_{1-x}Mn_xTe$, since it has a number of mechanical and physical advantages. Its density is 5.83 g cm^{-3} and depending on the Mg fraction, the bandgap can be varied between 1.5 and 3.2 eV. For Mg fractions < 0.7, the bandgap is a direct gap semiconductor – above, it is indirect. In terms of structure, CMgT possesses a high degree of crystallinity due to the near-similar constants of CdTe (6.48 Å) and MgTe (6.42 Å), reducing stress and as such has a low energy of defect formation. It displays good homogeneity, since the Mg segregation coefficient in CdTe is very nearly unity. Furthermore, crystal in-homogeneities due to alloying effects can be minimized, since a given energy bandgap can be achieved using a smaller fraction of Mg in CMgT as compared to the fractions of Zn and Mn needed in $Cd_{1-x}Zn_xTe$ and $Cd_{1-x}Mn_xTe$. In fact, the bandgap as a function of magnesium fraction, x, can be approximated by [120].

$$E_g(x) = 1.52 + 1.7\,x \tag{3}$$

For comparison, the bandgap variation in $Cd_{1-x}Mn_xTe$ is 1.3 eV/x, and in $Cd_{1-x}Zn_xTe$ it is 0.8 eV/x.

Hossain et al. [121] grew both an undoped and a doped ingot of $Cd_{1-x}Mg_xTe$, characterized its material properties and tested its detection performance. For x = 0.05, the bandgap was measured to be 1.61 eV at room temperature. The yield was predominantly single crystals with about two orders of magnitude fewer Te inclusions and other growth defects compared to CdZnTe and CdMnTe crystals. Planar detectors were fabricated, although no details are given. The resistivity of the undoped annealed crystal was measured to be $\sim 10^7$ Ω-cm, which after doping, increased by 2–3 orders of magnitude to 10^9–10^{10} Ω-cm. While the undoped detectors showed no response to an ^{241}Am radioactive source, the doped as-grown crystal were found to be barely spectroscopic with fractional energy resolutions of $\sim 75\%$ at 60 keV. The estimated electron mu-tau value was 8×10^{-4} cm^2V^{-1}.

Travedi et al. [122] grew $Cd_{1-x}Mg_xTe$ crystals using the vertical Bridgman technique with a nominal 8% magnesium composition. Various sized crystals were produced and processed and a number of detector configurations fabricated, including planar, pseudo Frisch-grid, and pseudo-hemispherical. From I/V measurements on a $4 \times 4 \times 2$ mm^3 In-doped planar device, a resistivity of 3×10^{10} Ω-cm was obtained. The detector was found to spectroscopic to a ^{241}Am gamma-ray source with a measured room temperature FWHM energy resolution of 12.8% at 59.54 keV. From a Hecht analysis of bias versus peak channel data, the electron mu-tau product was determined to be 5.3×10^{-3} cm^2V^{-1}, which is comparable to CdZnTe. High-energy measurements were carried out with a $4 \times 4 \times 10$ mm^3 pseudo Frisch-grid detector using a ^{137}Cs gamma-ray source. An energy resolution of 3.4% FWHM at 662 keV was achieved at room temperature, without any additional depth correction.

9.5.7 Cadmium Zinc Selenide

In addition to CdSe, Burger et al. [107] investigated the ternary system $Cd_{0.7}Zn_{0.3}Se$. It has a wurtzite crystal structure with a bandgap of 1.73–2.67 eV and a density of 5.4–5.8 g cm^{-3}, depending on the zinc fraction. As with adding Zn to CdTe, the addition of Zn to the CdSe system results in a higher bandgap enabling the fabrication of detectors with lower leakage currents and sufficiently low noise to detect X-ray photons. In fact, leakage currents in a small $Cd_{0.7}Zn_{0.3}Se$ detector were measured to be approximately one order of magnitude lower than measured in a similar sized CdSe detector [24]. Burger et al. [123] grew single crystals by the Temperature Gradient Solution Zoning technique. The as-grown crystals had resistivities of the order of 10^6 Ω-cm. The crystals were sliced into platelets with thicknesses varying from 0.1 mm to 1 mm. After post-growth treatment, the measured resistivities increased to $\sim 10^{10}$ Ω-cm. Gold or carbon-based Aquadaq contacts were then applied. Measured FWHM energy resolutions of 1.8 keV at 5.9 keV and 4 keV at 27 keV were obtained for a 0.45 mm thick device. The 59.54 keV nuclear line from ^{241}Am could not be resolved because of poor resolution and the overlap of the Cd K_α and K_β escape peaks at 23 keV and 26 keV, respectively.

9.5.8 Cadmium Telluride Selenide

$CdSe_xTe_{(1-x)}$ (or CdTeSe, Cd(Te, Se)) is also seen as another attractive alternative to CdZnTe, since the binding energy of the CdSe system is 1.3 times higher than the corresponding CdTe system and the lattice constant 0.9 times shorter. Thus, it is expected that CdSeTe crystals should have fewer Cd vacancies and increased hardness. In addition, the segregation coefficient (k) of Se in CdTe is ≤ 1 while that of Zn in CdTe is ≥ 1, which means that crystals of CdSeTe should be more uniform in terms of stoichiometry than CdZnTe.

Fiederle et al. [124] investigated the ternary systems (Cd,Zn)Te and Cd(Te,Se). Single crystals of CdTe, $Cd_{0.9}Zn_{0.1}Te$ and $CdTe_{0.9}Se_{0.1}$ were grown by the vertical HPB method. Simple monolithic detectors were then fabricated and tested. While spectroscopic results could be expected for CdTe and CdZnTe, $CdTe_{0.9}Se_{0.1}$ was also found to be spectroscopic to both alpha particles and photons. For a 1.34 mm thick device, a FWHM energy resolution of $\sim 10\%$ was achieved for 60 keV photons [124]. Kim et al. [125] grew semi-insulating CdTeSe:Cl crystals using the vertical Bridgman method. Chlorine doping was used to obtain semi-insulating material by compensating for Cd vacancies. The ingot was sliced into 2 mm thick wafers that were then chemically and mechanically polished down to a thickness of 1 mm. Detector blanks were diced out of the wafer. Based on I-V and Hall measurements, the resistivity of the material was found to be 5×10^9 Ω-cm and the electron and hole mobilities to be 59 and 33 $cm^{-2}V^{-1}s^{-1}$, respectively. The corresponding electron and hole mu-tau products were surprisingly high at 6.6×10^{-2} and 8.1×10^{-2} cm^2V^{-1}, respectively. Au electrodes of area 1×1 mm^2 were deposited on to the wafers by evaporation and a bias of 100V applied. Although the detector material had high resistivity and mobility–lifetime products, the response of the detector to a ^{241}Am source yielded a FWHM energy resolution of only $\sim 30\%$ at 59.54 keV. The authors attribute this poor result to high leakage currents, which limited the applied bias to 100V, which in turn led to poor charge collection efficiencies.

9.5.9 Mercury Cadmium Telluride

Mercury cadmium telluride ($Hg_{(1-x)}Cd_xTe$), variously known as cadmium mercury telluride, MCT, MerCadTel or CMT, is a narrow, direct bandgap, zincblende II-VI ternary alloy of CdTe and HgTe. It is a relatively soft material (HV ~20–80 kgmm^{-2} depending on composition) with the lower values occurring for higher Hg fractions (Hg forms weaker bonds with Te than Cd). Initial difficulties in growing $Hg_{(1-x)}Cd_xTe$ due to the high-vapor pressure of Hg, were largely overcome with the adoption of epitaxial techniques following the development of suitably lattice-matched substrates (mainly $Cd_{(1-y)}Zn_yTe$ with $y \sim 0.04$). The bandgap of CdTe is ~1.5 eV at room temperature while the bandgap of HgTe is actually negative, since it is a semimetal. By alloying these two compounds it is possible to create a new compound, $Hg_{(1-x)}Cd_xTe$, with a bandgap between –0.14 and 1.49 eV depending on the Cd fraction, x. Hansen, Schmit and Casselman [126] give the following temperature-dependent empirical form for the bandgap energy,

$$E_g(x) = -0.302 + 1.93x - 0.81x^2 + 0.832x^3 + 5.35 \times 10^{-4}(1-2x)T \quad (eV), \tag{4}$$

which is valid over the full composition range and for temperatures, T, ranging from 4.2K to 300K. At 77K, the crossover between semimetal and semiconductor (*i.e.*, $E_g > 0$) occurs when $x > 0.15$ while at 300K it occurs when $x = 0.09$. The electron mobility of HgCdTe is governed largely by the Hg content and ranges from ~2000 cm^{2}V^{-1}s^{-1} for $Hg_{0.2}Cd_{0.8}Te$, to ~10,000 cm^{2}V^{-1}s^{-1} for $Hg_{0.8}Cd_{0.2}Te$ at 300 K. At 77K, the mobilities are an order of magnitude higher. For the nominal Cd fractions used 0.2–0.4, only InSb and InAs have greater electron mobilities at room temperature. Hole mobilities at room temperature range from about 40 to ~80 cm^{2}V^{-1}s^{-1} and are also one order of magnitude higher at 77K.

HgCdTe is unique amongst semiconductors in that its bandgap is continuously tuneable across the entire IR waveband – from 0.7 micron to 25 microns. As such, it is widely used for IR applications and in particular for ground-based astronomy since it can detect infrared radiation in both of the accessible atmospheric windows (the mid-wave infrared window, from 3–5 microns and the long-wave infrared window, from 8 to 12 μm). HgCdTe is also widely used in the shortwave infrared band from 1–3 microns, and near infra-red band from 0.7 to 1 micron, for remote sensing applications, for example atmospheric moisture measurements.

At present, the highest quality material produced is used for IR astronomy and is generally grown by MBE with the composition selected to produce the cut-off wavelength[7] appropriate for the desired IR operational band. Active depths range in thickness from ~1 to 8 microns. Other than IR, HgCdTe has not been used at other wavelengths, primarily because of the difficulty in growing thick, defect-free material, coupled with its low intrinsic resistivity. In addition, until recently the 800 μm CdZnTe substrates used to produce the high-quality IR arrays were not removed, which, while transparent at IR

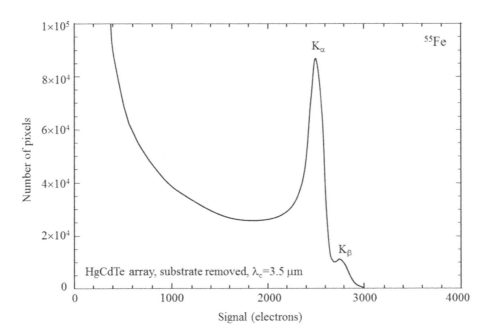

FIGURE 9.14 A spectrum of ^{55}Fe taken with a HAWAII-2RG array designed for short wavelength infrared imaging applications [127]. From the figure we see that the K_α line at 5.9 keV is clearly resolved with a FWHM energy resolution, of ~10%. The K_β line at 6.4 keV is also apparent in the spectrum.

---

[7] The longest detectable wavelength, λ_c, which is related to the bandgap by $E_g = hc/\lambda_c$.

wavelengths are opaque at other wavelengths. However, the present generation of 2D large format arrays are now back-thinned, making them suitable for soft X-ray imaging applications, although at present, exposure to soft X-rays is used only for calibration purposes [127,128]. In Fig. 9.14 we show the spectrum of a ^{55}Fe radioactive source taken with a HgCdTe HAWAII-2RG mosaic array [127]. These arrays are composed of 2K × 2K pixels of size 18μm. The array was operated at a temperature of 80K and a detector bias of ~0.5 V. From the figure we see that the K_α line at 5.9 keV is clearly resolved with a FWHM energy resolution of ~10%. Note, however, that the Fano limit at these energies is much lower at ~0.3%. The K_β line at 6.4 keV is also apparent in the spectrum but not fully resolved.

9.6 Group I–VII Materials

All members of Group I-VII materials display similar, mechanical and electrical properties. They typically have densities in the range 5–6 g cm^{-3}, melting points of ~500°C and bandgaps of ~3 eV. Members of this group bond ionically and consequently; they are characterized by high ionicities. As such they suffer from polarization effects at room temperature. Specific examples of this group include the silver halogenides, AgCl and AgBr, which were some of the first compound semiconductors demonstrated to be sensitive to ionizing radiation back in the 1930s. However, because of polarization issues related to their high ionicities, very little work has been carried out since, although AgCl and AgBr are still used in the production of nuclear emulsions. Recently, there has been renewed interest in the cuprous halide semiconductors (*e.g.,* CuCl, CuBr) as highly efficient alternatives to Group-III nitrides for producing light-emitting diodes [129].

9.6.1 Silver Chloride

Early investigations of silver chloride (AgCl) [130] centered on its photoconductive properties that showed a marked polarization effect characterized by a decrease in both pulse amplitude and counting rate as a function of the irradiation rate [131,132]. The first practical demonstration of AgCl as a radiation detector can be attributed to van Heerden [133], who in 1945 demonstrated that silver chloride crystals when cooled to low temperatures were capable of detecting individual γ-rays, alpha particles and beta particles. Crystals of silver chloride were grown from the melt and diced into 1 cm diameter cylinders of thickness 1.7 mm. Silver electrodes were deposited on each side of the crystal. At room temperature, the crystal showed ionic conductivity but when cooled to liquid air temperature (–186°C) the crystal became a perfect insulator, allowing a bias of up to 2.5 kV to be applied across the crystal. The detector was found to be spectrally responsive to 400 keV beta particles with a measured FWHM energy resolution of ~20% [134]. Subsequent measurements showed that the device was also responsive to alpha particles and gamma-rays, although it was not clear whether the detector was spectroscopic (it was not easy to record a spectrum at that time; in the case of beta rays it was possible to magnetically select the energies and thus build up a spectrum). Subsequent work on AgCl has focused not on active radiation detection but rather on passive detection by observing the decoration of dislocations produced by high-energy nuclear interactions in large single crystals [135]. Since silver chloride is transparent to visible light, the decorated dislocations are observable with an optical microscope at a magnification of about 150–200.

9.7 Group III–VI Materials

Most of the III-VI compounds are chalcogenides, crystallizing in layer type structures. The bonding is predominantly covalent within layers and much weaker van der Waals between layers. These materials are of interest because the behaviour of electrons within the layers is quasi-two-dimensional (Q2D). Characteristic distances in the plane perpendicular to the layer are typically less than the de Broglie wavelength of an electron; hence, the layer has the quantum properties of a lower dimension structure. While many of the features exhibited by these materials reflect those found in bulk materials, the anisotropy of the electron layer gives rise to additional properties and ordering, such as the quantum Hall effect resulting from planar magnetic order (for a review of the electronic properties of Q2D systems, see Ando, *et al.* [136]). Semiconducting examples of this group are at present limited to the gallium- and indium-based chalcogenides: GaS, GaSe, GaTe, InS, InSe and InTe, of which GaSe and GaTe have shown a response to alpha and gamma radiation.

9.7.1 Gallium Selenide

Gallium selenide (GaSe) is a wide bandgap semiconductor ($\varepsilon_g = 2$ eV) with a density of 4.6 gcm^{-3}. Historically, GaSe was of interest because of its photoconducting and luminescence properties. However, recent research has focused on the generation and detection of broadband tuneable terahertz (THz) radiation [137] by exploiting the highly anisotropic properties of its layered structure. GaSe was first investigated as a nuclear detector material in the early 1970s by Manfredotti *et al.* [138], who fabricated simple planar detectors by evaporating Au electrodes onto single crystals grown by the Bridgman method.

A number of detectors were produced with thicknesses ranging from (50–150) μm and surface areas from (20–50) mm^2. The samples were generally *n*-type room-temperature resistivities in the range (10^8 to 10^9) Ω-cm. The detectors were found to be sensitive to α-particles, with measured spectroscopic energy resolutions as low as 6.8% FWHM at 5.5 MeV. Sakai *et al.* [139] and later Nakatani *et al.* [140] reported on measurements taken from thin (~100 μm) detectors fabricated from platelets cleaved from ingots grown by the HPB method. Alpha particle resolutions of about 5% FWHM at 5.5 MeV were obtained. Yamazaki *et al.* [141] explored the properties of GaSe radiation detectors doped with Si, Ge and Sn. They found that doping substantially decreased leakage currents and they were able to realize FWHM energy resolutions as low as 4% for ^{241}Am 5.5 MeV alpha particles. In contrast, most undoped detectors did not function because of excessive leakage currents.

Castellano [142] explored the possibility of using GaSe for X-ray dosimetry. In this application, GaSe detectors were used as photoconduction devices by measuring the direct current induced by high fluxes of 130 and 170 kV X-rays. Mandal *et al.* [143] achieved X-ray detection with material grown by the modified vertical Bridgman technique. The ingots were cut, lapped, polished and cleaved to produce planar detectors of area ~1 cm^2 and thickness 0.8 mm. After the deposition of 3 mm diameter Au contacts, the resistivity was determined from the I-V characteristics to be in excess of 10^{10} Ω-cm. The mobility–lifetime products were determined from a Hecht analysis to be ~ 1.4×10^{-5} cm^{2}V^{-1} for electrons and ~1.5×10^{-5} cm^{2}V^{-1} for holes. The devices showed a spectroscopic response to photons from a ^{241}Am source with a measured energy resolution of ~4% FWHM at 60 keV.

9.7.2 Gallium Telluride

Gallium telluride (GaTe) has attracted significant interest from the electronics industry, since unlike other compounds in this group, single crystals can be grown with low resistivity making it possible to fabricate heterojunctions with negligible series resistance [144]. Recently, effort has been expended in exploring GaTe as a radiation detection medium, as its physical and mechanical properties are similar to GaSe and near ideal. It is non-hygroscopic, has a bandgap of 1.7 eV, a density of 5.44 gcm^{-3} and low melting point (824°C – low evaporation), and high-purity starting materials are readily available

Mandal *et al.* [145] grew a large diameter single crystal by the vertical Bridgman technique. Simple planar devices up to 1 cm^2 in area and 0.8 mm thick were fabricated and contacted with 3mm diameter Au pads. The devices also incorporated a guard ring to reduce leakage currents due to surface recombination. The resistivity of the devices, as determined from the I-V characteristics, was in excess of 10^9 Ω-cm, and the measured mobility–lifetime products of both electrons and holes were found to be similar to that of GaSe; ~1.5×10^{-5} cm^{2}V^{-1}. The devices showed a spectroscopic response to 60 keV photons from a ^{241}Am source with a measured FWHM energy resolution of ~5% when operated with a guard ring. The device was also found to be sensitive to 662 keV photons, although not spectroscopically. Without the guard ring, the devices showed no response.

9.8 Group n–VII Materials and Alloys

Group *n*-VII (where *n* = II,III,IV) materials generally belong to the family of layered structured, heavy metal iodides and tellurides. The Group VII anions form a hexagonal close-packed arrangement while the group *n* cations fill all of the octahedral sites in alternate layers. The resultant structure is a layered lattice with the layers being held in place by van der Waals forces and is typical for compounds of the form AB$_2$. The bonding within the layers is primarily covalent. The fact that layered compounds are strongly bound in two directions (by covalent bonding) and weakly bound in the third direction, along the **c**-axis, leads to an anisotropy in their structural and electronic properties. Materials in these groups tend to be mechanically soft, have low melting points, large dielectric constants and show strong polarization effects. Large crystals may even deform under their own weight.

9.8.1 Mercuric Iodide

Mercuric iodide (HgI$_2$) has been investigated as a room-temperature X- and gamma-ray detector since the early 1970s [146,147]. Its wide bandgap (2.1 eV) in conjunction with the high atomic number (80,53) of its constituent atoms makes it an attractive material for room temperature X- and particularly gamma-ray spectrometers. In fact, since photoelectric absorption varies as Z^5, the specific sensitivity of HgI$_2$ is about 10 times greater than that of Ge for energies > 100 keV. Additional advantages of this material are that it has been demonstrated to operate at elevated temperatures (55°C) with minimal impact on spectral performance [148] and it is extremely radiation hard, at least to a total proton fluence of ~3×10^{11} particles cm^{-2} [149,150]. However, HgI$_2$ suffers from relatively low transport properties ($\mu_e\tau_e = 3 \times 10^{-4}$ cm^{2}V^{-1} and $\mu_h\tau_h = 1 \times 10^{-5}$ cm^{2}V^{-1}) and severe material non-uniformity issues. For conventional planar contacts, these effectively limit detector thicknesses to ~3 mm thick in order to achieve acceptable spectroscopic results.

HgI$_2$ belongs to the family of layered structured, heavy metal iodides. It is a relatively soft material that forms a tetragonal lattice at temperatures below 130°C. This is known as the alpha phase, and the crystals appear red in colour with a bandgap of 2.13 eV. At temperatures above 130°C, HgI_2 undergoes a phase transformation to an orthorhombic lattice (beta phase) appearing yellow in colour with a bandgap of 2.5 eV. However, when cooled below 130°C, the material undergoes a destructive phase transition to alpha–HgI_2. This precludes melt growth. Spectroscopic grade crystals were initially grown from solution growth [151], but now vapor phase processes are the preferred growth technique, specifically, the vertical and horizontal ampoule methods [152]. Therefore, crystal growth is slow and the quality of large crystals inconsistent. Thus, the cost of production is relatively high.

Detectors are generally prepared from bulk crystals by cleaving samples perpendicular to the crystallographic c-axis (*i.e.,* parallel to the [011] planes). The bias is usually applied along the [001] direction, because hole mobilities are significantly higher in this direction. The fabrication of a pn-junction is usually not necessary, as dark currents are generally very low. HgI_2 is highly reactive with many metals, hence only a few materials can be used as electrical contacts, usually colloidal carbon (Aquadag) or Pd. While reported energy resolutions are typically in the keV region ([154,155] as shown in Fig. 9.15, a series of small planar detectors developed for NASA's CRAF[8] space mission yielded near Fano limited performances. For example, Iwanczyk *et al.* [156] reported an energy resolution of 198 eV FWHM at 5.9 keV, obtained with a 5 mm^2, 200 μm thick detector operated at 0°C. These performances have not been equalled or surpassed in over two decades. In Fig. 9.16 we show the X-ray fluorescence spectrum of a sample of the Murchison meteorite[9] taken with such a detector at room temperature [153], illustrating the excellent energy resolution of the system. An X-ray tube with a rhodium anode was used for the excitation source. From the figure, we see that all major elements are clearly resolved.

9.8.2 Mercuric Bromoiodide

Mixed halides of mercury (HgXY) have been explored as possible alternatives to HgI_2. While HgI_2 has many desirable properties (high density, wide bandgap, *etc.*) the presence of a low temperature destructive solid–solid phase transition, makes it difficult to grow large homogeneous crystals. HgXY compounds, on the other hand, are stable at higher operating temperatures, have no phase transformations and can be conveniently grown by the Bridgman-Stockbarger method. Of the available compounds, mercuric bromoiodide ($Hg(Br_xI_{1-x})_2$) appears the most promising. Its bandgap can be tuned by stoichiometry from 2.1 eV ($x = 0$) to 3.5 eV ($x = 1$), and for $x>0.2$ the material is free from phase transitions. When coupled with its large density (6.2 g cm^{-3}) and high resistivity ($\sim10^{12}$ Ω-cm), it is an ideal candidate for room-temperature hard X- and gamma-ray detector applications. The best-measured electron and hole mobilities of 30 and 0.1 $cm^2V^{-1}s^{-1}$ [157], respectively, are comparable to those obtained with HgI_2.

Early work on this compound was carried out by Shah *et al.* [158] who investigated its use as a photodetector to readout scintillators, making use of the fact that the optical response can be tuned to match specific scintillators. Gospodinov *et al.* [159] fabricated simple planar detectors from plates of material cleaved along the (001) plane. The plates had an area of

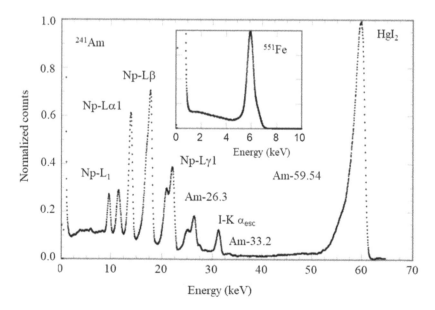

FIGURE 9.15 Composite [241]Am and [55]Fe spectra taken with a 7 mm^2, 0.5 mm thick HgI_2 detector (from [4]). The detector temperature was +24°C and the applied bias +800V. The measured FWHM energy resolutions are 600 eV at 5.9 keV and 2.4 keV at 59.54 keV.

---

[8] Comet Rendezvous Asteroid Flyby (CRAF).
[9] Named after the town of Murchison in Victoria, Australia where the meteorite fell in 1969.

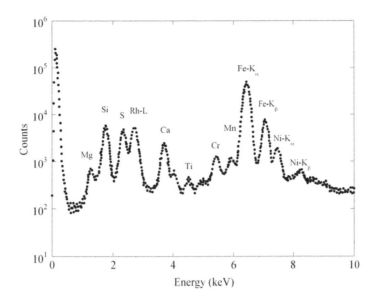

FIGURE 9.16 X-ray fluorescence spectrum of a sample of the Murchison meteorite taken with a 5 mm^2, 200 μm thick HgI$_2$ detector operated at room temperature. The sample was excited using an X-ray tube source (taken from Iwanczyk *et al.* [153], ©1991 IEEE). The energy resolution is ~200 eV FWHM.

16 mm^2 and thickness 1 and 3 mm. Aquadag electrodes were then deposited directly on the upper and lower surface of these plates.

The spectroscopic properties of the detector were assessed using ^{55}Fe and ^{241}Am radioactive sources. It was found that the best spectroscopic results were obtained with a 20% bromine fraction (Hg(Br$_{0.2}$I$_{0.8}$)$_2$ [158]) and that crystals with slightly higher Br concentration gave significantly poorer energy resolutions. In Fig. 9.17 we show the measured room temperature spectra under full area illumination for 1 mm and 3 mm thick devices fabricated from crystals with a 20% Br fraction at a detector bias was 500 V. The FWHM energy resolutions at 5.9 keV and 59.54 keV were 0.9 keV and 6.5 keV, respectively.

9.8.3 Mercuric Sulphide

The hexagonal form of mercuric sulphide (HgS) has received considerable attention as a room-temperature replacement for HgI$_2$, in view of its high density (8.2 g cm^{-3}), favourable bandgap (2.1 eV) and relative chemical stability. HgS can assume

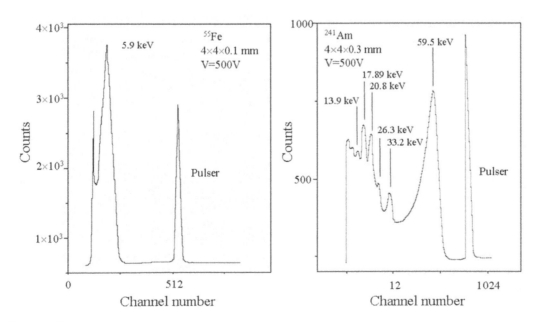

FIGURE 9.17 ^{55}Fe and ^{241}Am spectra, measured at room temperature with Hg(Br$_{0.2}$I$_{0.8}$)$_2$ detectors (taken from Gospodinov *et al.* [159]). The measured FWHM energy resolutions at 5.9 keV and 59.54 keV are 0.9 keV and 6.5 keV, respectively, under full area illumination.

two different crystalline forms, – α-HgS, which crystallizes in a trigonal type, hexagonal structure, and β-HgS, which crystallizes in a zincblende type, cubic structure. Alpha-HgS is more commonly recognized as the mineral cinnabar, which happens to be the main commercial source of the metal mercury. It is also widely used as the high-grade paint pigment, vermillion. Recently, it has been explored for acousto-optical [160], infrared sensing [161] and photoelectronic applications [162-164]. Beta-HgS (also known as metacinnabar) is generally considered a semimetal but has recently been shown to be a strong topological insulator [165]. As a bulk material, it has potential for use in low-power consumption electronic devices [165] and in its nano-particle form, in solid-state solar cell and photo-electrochemical cell applications [164]. Interestingly, metacinnabar is the compound form that mercury evolves into in aged amalgam dental fillings.

Squillante *et al.* [166] followed a "heuristic" approach to material selection. They point out that the deep red colour of α-HgS indicates that it will have a bandgap energy around 2 eV, which, when coupled with the high Z of Hg, a reported resistivity of $\sim 10^{12}$ Ω [167] and a cited electron mobility of 45 cm^{2}V^{-1}s^{-1} [168] makes it a promising room-temperature semiconductor for detecting gamma-radiation (β-HgS on the other hand is black and conducting). Because of the difficulties[10] in growing α-HgS, they used samples of natural cinnabar[11] to produce radiation detectors. Cinnabar has a hexagonal (hP6) crystal structure and grows as deep coloured, transparent rhombohedral crystals. The bandgap is 2.1 eV and the density is 8.2 g cm^{-3}, both promising properties for radiation detection. However, α-HgS also undergoes a destructive phase transition at 344°C and is soft and slightly sectile, although reported hardness values (2.5 Mohs) are comparable to HgI$_2$, CdTe and CdZnTe (Mohs hardness 2.3–2.9).

Devices were fabricated as follows. Well formed, clear, and readily accessible crystals were chosen from the mineral samples. Samples were positioned under a diamond string saw and thin wafers cut, etched and fabricated into detectors. The devices were processed as planar, two terminal devices, of approximate thickness 2.6 mm. Electrodes measuring about 2 mm in diameter were deposited using a carbon paste. Contact to the electrodes was made using palladium wires embedded in the paste. From the IV characteristic, resistivities were found to be $\sim 4 \times 10^{12}$ Ω-cm. The detectors were found to be responsive, but not spectroscopic, to ^{241}Am 60 keV, ^{57}Co 122 keV and ^{137}Cs 662 keV gamma-rays [166]. In all cases, the pulse height response was found to increase with increasing bias.

Kim *et al.* [169] grew α-HgS by physical vapor transport (PVT) and fabricated a detector 0.44 mm thick. The electrodes were formed by an evaporation of Cr/Au through a shadow mask and were 2 mm in diameter. Contact to the electrodes was made using palladium wires attached using a carbon-based suspension. The IV characteristic of the detector was linear with a measured bulk resistivity of 6×10^9 Ω-cm at room temperature. As with the cinnabar detectors, the PVT grown detectors were found to be responsive, but not spectroscopic, to ^{137}Cs 662 keV gamma-rays.

9.8.4 Mercuric Oxide

Mercury oxide has also been proposed as a room-temperature detection medium in view of its high density (11.14 g cm^3) and favourable bandgap (2.2 eV). At normal pressures, HgO has two crystalline forms: an orange orthorhombic form known as montroydite, which is rarely encountered in nature, and a second, red hexagonal, form, which is analogous to the sulfide mineral cinnabar. Both are characterized by Hg-O chains in a zigzag pattern, which in turn are packed to form planar layers. At pressures above 10 GPa both structures convert to a tetragonal form. However, for radiation detection, HgO has several disadvantages – it is soft, sectile, very toxic and decomposes into mercury and oxygen on exposure to light or on heating above 500°C. Because of its toxicity, HgO is used in a limited number of applications – mainly as base in marine and porcelain paints, with graphite as a depolarizer in dry batteries and ironically in some skin ointments.

Thrall [170] produced an HgO detector 0.5 mm thick with 0.6 mm diameter carbon paste electrodes. No details were given in reference [170] on the growth method or crystallographic structure. The I-V characteristic was linear indicating ohmic behaviour with a resistivity of 5×10^9 Ω-cm. The detector showed a very weak response to ^{241}Am alpha particles and to 662 keV gamma-rays.

9.8.5 Mercurous Halides

The mercurous halides Hg$_2$X$_2$ (where X = Cl, Br, I) have been proposed as a detection medium based on a number of distinct advantages. All are relatively easy to grow and crystallize in a tetragonal configuration without problematic phase transitions, as in the case of HgI$_2$. All have wide bandgaps suitable for room temperature operation [171] and densities of $\sim$ 7 gcm^{-3}, which ensure good stopping power. Chen *et al.* [172] have investigated Hg$_2$I$_2$, Hg$_2$Cl$_2$ and Hg$_2$Br$_2$. Detector grade single crystals were grown by the PVT method. The initial precursor materials were commercially sourced (3N) materials, which were then purified to 6N+ by repeated sublimation, prior to crystal growth. For each material, 2 to 6 mm thick platelets were sliced from 2-inch diameter boules and contacted with graphite-loaded epoxy electrodes painted on the top and bottom surfaces.

---

[10] α-HgS undergoes a solid-solid phase transition at 344°C to β-HgS (red to black form), precluding melt growth.
[11] Derived from the Persian for "dragon's blood" – a reference to its deep red colour.

The most promising results were obtained for Hg_2I_2, which has several advantages as a radiation-detection medium. For example, it is easier to grow than the more conventional HgI_2, it is non-hydroscopic, non-toxic and stable for long-term operation. Chen *et al.* [172] fabricated both planar and hemispherical detectors from single crystalline material. The bulk resistivity of the base material was measured to be in the range $\sim10^{12}$ Ω-cm and the electron–hole mu tau product estimated to be 2×10^{-3} cm^{2}V^{-1}, comparable to CdZnTe. The spectroscopic properties of simple planar detectors were investigated using ^{241}Am and ^{137}Cs radioactive sources. Room temperature FWHM energy resolutions of 3% at 60 keV and 1.4% at 662 keV were recorded for 0.5 mm and 2 mm thick detectors, respectively.

Chen *et al.* [172] also reported preliminary measurements from a $20 \times 10 \times 6$ mm^2 Hg_2Br_2 array comprised of 6 mm^2 pixels. At 662 keV, a FWHM energy resolution of 1.8% was measured at room temperature. The applied bias was 1000 V.

9.8.6 Thallium Bromide

For room temperature hard X- and γ-ray applications, thallium bromide (TlBr) has emerged as a particularly interesting material in view of its wide bandgap (2.5 times that of Si) and high atomic numbers (Tl=81, Br=35) of its constituent atoms. In addition, its density (7.5 gm cm^{-3}) is comparable to that of bismuth germanate (BGO), and thus it has excellent stopping power for hard X- and gamma-rays. Hofstadter [13] originally demonstrated it as a radiation detector material in 1949, albeit with limited success due to purity and fabrication problems. Surprisingly, compared to other compound semiconductors, relatively little work has been carried out since this time [173–179]. The material has a CsCl-type simple cubic crystal structure and melts congruently at 480°C, with a single non-destructive phase transition below the melting point. Thus, its physical properties are amenable to easy and rapid purification and growth using standard techniques. Its large bandgap of 2.68 eV suggest that detectors should operate at or above room temperature with low noise performance. The relative softness[12] of TlBr (Knoop hardness 12 kg mm^{-2}) can cause major problems during detector fabrication since any mechanical processing, such as lapping and polishing, can create an inordinate amount of damage, principally in the form of dislocations oriented parallel to the surface being processed. The depth of the damage can be several microns, sufficient to have a significant effect on charge collection efficiencies. The effect is further exacerbated by the high dielectric constant of TlBr, which ensures a low energy of defect formation (1.1 eV [180]). Consequently, post processing of the surface, such as etching with H_2O [181] is usually carried out prior to contacting.

Early TlBr detectors could only be operated at reduced temperatures because of stoichiometric and crystallographic imperfections mainly caused by interstitial impurities [182,183]. Shah *et al.* [174] found that zone refining the base material prior to crystal growth resulted in significantly better performance, in terms of both resistivity and charge transport properties. At room temperature, they measured FWHM energy resolutions of 1.5 keV, for the iron line at 5.9 keV and 8 keV for the americium 59.54 keV nuclear line. Hitomi *et al.* [177,178] used multiple-pass zone refining and measured room temperature FWHM energy resolutions at 5.9 keV and 59.54 keV of 1.8 keV and 3.3 keV, respectively. The detectors they used had areas of 0.8 mm^2 and ~3 mm^2 and thicknesses of < 100 μm and 150 μm. However, detector performances deteriorated with time, which was attributed to polarization effects[13] arising from two main sources, firstly the modification of the internal electric field due to space charge effects caused by deep hole trapping [182], and secondly a gradual increase in leakage currents caused by ionic conduction. TlBr is a mixed electronic-ionic conductor, with the ionic current being significant even at room temperature. Above 250K, the main conduction mechanism is due to Tl$^+$ ions [183]. For biased detectors, this leads to a build-up of Tl+ and Br- ions under the cathode and anode electrodes [184], reducing the internal electric field and therefore signal amplitudes. The effect can be suppressed using sacrificial Tl electrodes coated in Au, in which the Tl electrode is consumed under the anode and formed under the cathode. In fact, Hitomi *et al.* [185] showed that such devices were stable over extended periods (~30 hours) compared to devices with Au contacts. Vaitkus *et al.* [186] propose that ionic conductivity also creates micro-inhomogeneities in the material, which are activated by the electric field and by non-equilibrium carrier generation. The observed threshold-type effects, in the form of transient current spikes, are related to the growth of these structures and can be substantially reduced by lowering the temperature or increasing hydrostatic pressure. For example, Samaru [187] found that conductivities could vary by over a magnitude per decade change in temperature.

Owens *et al.* [163] carried out a series of experiments on prototype monolithic detectors of area ~8 mm^2 and thickness ~800 μm. Room temperature performances of 1.8 keV and 3.3 keV FWHM at 5.9 and 59.54 keV, respectively, were achieved. The measured energy-loss spectra are shown in Fig. 9.18. These detectors were operated at -22°C and showed stable and reproducible results over a time scale of two years. At higher energies, Hitomi *et al.* [189] demonstrated a room temperature energy resolution of 1.3% FWHM at 662 keV using a 2×2 pixel array with depth correction. The pixel size was 1×1 mm^2, and the detector thickness 4.2 mm. Without depth correction, the resolution degraded to 2%. Kim *et al.* [188] fabricated a number of planar and pixel detectors due to repeated and careful purification, electron $\mu\tau$ products of as high as 3×10^{-3} cm^{2}V^{-1} were achieved,

---

[12] About the same as refrigerated butter.

[13] The term polarization effect is commonly used to refer to any change in the performance of a radiation detector with time that is not correlated with changes in operating parameters.

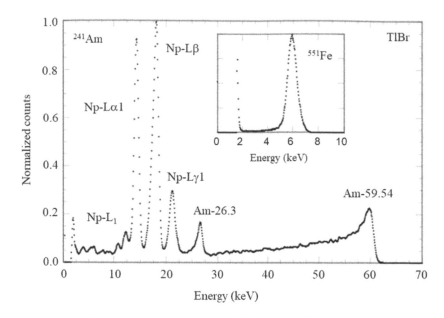

FIGURE 9.18 The response of an ~8 mm^2, 0.8 mm thick TlBr detector ^{241}Am and to ^{55}Fe (insert) using radioactive sources under full area illumination. The detector temperature was –22°C and the pulser noise width 690 eV FWHM (from [4]).

approaching those of CdTe. FWHM spectral resolutions of 5.3% at 122 keV and 1.7% at 662 keV were measured with a 3 mm thick 2 × 2 pixel array (pixel size 1.3 × 1.3 mm^2). To reduce polarization effects, the detectors were operated at -18°C. For a larger 10 × 10 × 10 mm^3, 3 × 3 pixel array, a spectral resolution of 5.5% at 122 keV and 2.5% at 662 keV was achieved. The measured energy loss spectra are shown in Fig. 9.19. The μτ values reported by Kim *et al.* [188] are somewhat surprising, since soft-lattice ionic compounds generally have inferior transport properties compared to covalent compounds, such as CdTe. Du [190] argued that the high μτ values are a consequence of the effective dielectric screening of charged defects and impurities, coupled with the electrically benign nature of the native defects. If true, the effect should be present in similar materials with high dielectric constants and may be used to pre-screen materials.

9.8.7 Thallium Mixed Halides

While TlBr has many desirable properties (such as high density and wide bandgap), its mechanical softness does not lend itself to the growth of large homogeneous crystals or the mechanical processing required in the fabrication of detectors. In addition, the reproducibility of results is still a major problem due to stability and polarization issues. In view of these limitations, there has been recent interest in thallium salts as a potential gamma-ray detection media. Of the available compounds, thallium bromoiodide (Tl(Br,I), TlBr$_x$I$_{1-x}$) and thallium bromochloride (Tl(Br,Cl), TlBr$_x$Cl$_{1-x}$) appear the most promising [191,192]. They were originally developed as optical materials for the transmission, refraction and focusing of

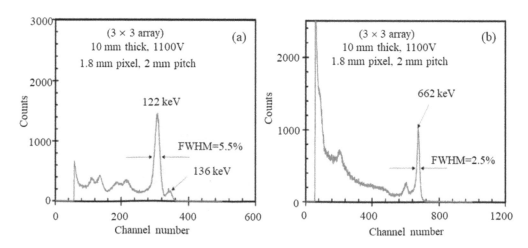

FIGURE 9.19 Room temperature spectra of 122 keV (a) and 662 keV (b) gamma-rays taken with a large 10 × 10 × 10 mm^3 pixel detector (from [188], ©2009 IEEE). The pixel size was 1.8 × 1.8 mm^2. The measured FWHM spectral resolutions at 122 keV and 662 keV were 5.5% and 2.5%, respectively.

infrared radiation in the 0.6 μm – 40 μm range and are commonly known as KRS-5 and KRS-6, respectively [193], where the KRS prefix is an abbreviation of "Kristalle aus dem Schmelz-fluss" (crystals from the melt). The actual bromide fraction is 40% for KRS-5 and 30% for KRS-6. Both halides are about three times mechanically harder than TlBr, making them more tolerant to mechanical processing during device fabrication.

9.8.7.1 Thallium Bromoiodide

Thallium bromoiodide, in the form $TlBr_{0.4}I_{0.6}$, has been used for several decades for infrared (IR) spectroscopy, specifically in the production of attenuated total reflection prisms, IR windows and lenses, where transmission in the 0.6 μm – 40 μm range is required. For non-IR applications, thallium bromoiodide has been investigated as a photoconductor for the readout of scintillators [194], taking advantage of the fact that the bandgap can be tuned by stoichiometry to the peak emission spectrum of almost all common inorganic scintillators, In fact the bandgap can be tuned from 2.15 eV ($x = 0.3$) to 2.8 eV ($x = 1$) and for $x > 0.3$ the material is free from destructive phase transitions. When coupled with its large density (7.4 g cm^{-3}) and high resistivity (~10^{10} Ω-cm), it is also an ideal candidate for room temperature hard X- and gamma-ray detector applications. The best measured electron μτ values approach 10^{-3} cm^{2}V^{-1} [191], which are comparable to those obtained with TlBr.

TlBr$_x$I$_{1-x}$ was investigated for gamma-ray detection by Churilov *et al.* [191] who fabricated simple planar detectors from ingots grown by the TMZ method. The detectors had compositions $x = 0.35$, 0.5 and 0.65. The measured resistivities were found to be > 10^{10} Ω-cm, which ensured that leakage currents were sufficiently low to allow room-temperature operation. One device, a 0.6 mm thick TlBr$_{0.35}$I$_{0.65}$ planar detector, showed a spectroscopic response to a ^{109}Cd source. This device had 2 mm diameter chromium and gold electrodes and was operated at 40V bias and 12μs shaping time. The measured FWHM energy resolution was ~40% at 22 keV. The $(\mu\tau)_e$ product was determined from the Hecht relationship to be ~10^{-3}cm^{2}V^{-1}. However, while this device was stable, other devices fabricated from the same ingot degraded quickly.

9.8.7.2 Thallium Bromochloride

As with thallium bromoiodide, thallium bromochloride is an equally attractive material for X- and gamma-ray detection. Kim *et al.* [192] grew a TlBr$_x$Cl$_{1-x}$ boule out of 5N powder, purified by multiple pass-zone refining in a horizontal furnace. The Br fraction was not specified. They constructed a 4.3 mm thick, 3 × 3 pixel array with 0.9 mm square pixels, mounted on a circuit board. Cr/Au contacts were deposited through a shadow mask by thermal evaporation. The measured resistivity of the material was 2×10^{10} Ω-cm. When exposed to 662 keV gamma-rays from a ^{137}Cs radioactive source, both the 662 keV full energy peak and the Tl escape peak were clearly observed for each pixel at a bias of 400V. One pixel exhibited an energy resolution of 2.3% FWHM at 662 keV.

9.8.8 Lead Iodide

Lead iodide (PbI$_2$) is a layered compound crystallizing in a hexagonal close-packed lattice whose structure displays a large degree of polytypism. Although over 32 polytypes are known, it generally solidifies into the hexagonal 2H form. It has been considered as a suitable material for X-ray and γ-ray detection since the 1970s in view of its high density (6.2 g cm^{-3}) and wide bandgap (2.3–2.5 eV), which should allow detectors to operate at, or even above, room temperature. As a material, it has several advantages over HgI$_2$. For example, it is environmentally very stable (low vapor pressure) and unlike HgI$_2$ does not undergo a phase transformation below the melting point (403°C). This makes it possible to grow lead iodide monocrystals directly from the melt or to use sublimation near the melting point for purification and film deposition. Unfortunately, at the present time carrier mobility–lifetime products are poor, being of the order $\mu_e\tau_e = 1\times10^{-5}$ cm^{2}V^{-1} and $\mu_h\tau_h = 3 \times 10^{-7}$ cm^{2}V^{-1}. This is roughly an order of magnitude lower than HgI$_2$, which effectively precludes the fabrication of thick detectors if spectral performance is to be maintained. However, the high atomic number of its elements ($Z_{Pb} = 82$, $Z_I = 53$) ensures good stopping power well into the hard X-ray region, and so detector thickness can be minimized for a given detection efficiency. For example, at 100 keV, the detector needs only be ~1 mm thick to absorb 90% of the incident radiation.

Detectors produced to date give reasonable results but only for thicknesses < 200 μm (a direct consequence of the poor μτ products). For example, Lund *et al.* [195] fabricated radiation detectors from melt-grown crystals. The detectors exhibited good energy resolution (915 eV FWHM at 5.9 keV at 20°C). Results also indicated they were more stable than HgI$_2$ detectors and capable of operating at temperatures over 100°C. Similar results were found by Deich and Roth [196] who fabricated detectors from boules produced by both the horizontal TMZ and Bridgman methods. The detectors had a planar geometry with painted Aquadag[14] electrodes of area 3 mm^2. For a 107 μm thick detector, a spectroscopic energy resolution

---

[14] A colloidal suspension of ~18% of sub-micron sized graphite powder in a water carrier that dries to form an adherent, conductive film (<300Ω per square) on virtually any surface, including glass and flexible materials.

of 712 eV (12%) FWHM was obtained for 5.9 keV X-rays and 1.8 keV (3%) FWHM for 60 keV gamma-rays. Shah *et al.* [197] used the Bridgman technique to produce 1 mm^2, 150 μm thick detectors. By careful attention to detector fabrication techniques and pre-amplifier electronics design, they were able to achieve room-temperature energy resolutions of 415 eV FWHM and 1.38 keV FWHM at 5.9 keV and 60 keV, respectively.

9.8.9 Antimony Tri-Iodide

Antimony tri-iodide (SbI_3), also known as mono-antimony tri-iodide or triiodostibane, was originally considered by Armantrout *et al.* [1] as a potential room-temperature gamma-ray detection medium in view of its favourable bandgap of 2.2 eV and relatively high density (4.92 g cm^{-3}). Its density coupled with the high atomic numbers of its constituent atoms (Sb = 51, I = 52) ensures it has a high gamma-ray stopping power, in fact, similar to that of CdTe. Depending on temperature and pressure, SbI_3 can solidify into one of four forms, three of which are crystalline (trigonal, rhombic and monoclinic) and one amorphous. At ordinary temperatures and pressures, the trigonal form is the preferred channel.

In view of its relatively low melting point (171°C) and that it does not undergo a phase change below it, trigonal SbI_3 is most easily grown by melt techniques or directly from the gaseous phase. It crystallizes into a reddish-layered structure, composed of strongly bonded I-Sb-I stacks. The I-I interlayers are bonded by van der Waals forces, resulting in relatively weak and clean cleavage planes.

Crystalline SbI_3 has several distinct properties such as a high intrinsic optical anisotropy, high resistivity (> 10^9 Ω-cm) and wide bandgap (2.2 eV), which lend themselves to several application areas. Historically, it was used as a high refractive index immersion medium in mineralogy [198]. Currently, it is commonly used in halogen metallurgy (e.g., the production of halogen lamps for medicine) and as a dopant in the preparation of thermoelectric materials [199]. Thin photosensitive films of SbI_3 are being investigated for non-linear optical applications (frequency-shifting optical modulation, switching, logic and storage) [200].

Ondera *et al.* [201] explored the possibility of using SbI_3 for radiation detection. They synthesized stoichiometric SbI_3 in a sealed quartz ampoule. Following multi-pass zone refining to reduce impurities, single crystals of SbI_3 were grown using the Bridgman method. Simple planar detectors were fabricated by cleaving the as-grown crystal into thin plates of thickness ~ 0.1 to 0.3 mm. Three mm diameter gold electrodes were then deposited on each face by vacuum evaporation and electrical contact made using palladium wires. From I-V measurements, the resistivities were estimated to be 1 × 10^{10} Ω-cm at room temperature. The detectors were found to be responsive, but not spectroscopic, to 5.5 MeV α–particles, from which we deduce that the $\mu\tau$ products in SbI_3 are considerably smaller than 10^{-5} cm^{2}s^{-1}.

9.8.10 Bismuth Tri-Iodide

Bismuth tri-iodide (BiI_3) is a high Z ($Z_{Bi} = 83$ and $Z_I = 53$), relatively high density ($\rho = 5.8$ g cm^{-3}), indirect bandgap semiconductor with a bandgap of 1.67 eV. Although several polymorphs exist, below 408°C BiI_3 crystallizes into a mechanically soft (K_H ~15 kg mm^{-2}), layered, hexagonal form, similar to tetragonal HgI_2. BiI_3 is unusual in that it belongs to a family of iodides (with AsI_3 and SbI_3) that retain their semiconducting properties even in the liquid state [202] and have found applications in photography and holography. Because BiI_3 is materially similar to α-HgI_2, it has been proposed as a gamma-ray detection medium [203]. Nason and Keller [203] grew single crystals by PVT from seed. A detector was fabricated by applying colloidal graphite electrodes and palladium wire leads to an as-grown 1.2 × 1.2 × 0.4 cm^3 crystal. Electrical conductivity measurements indicated a resistivity of 2 × 10^9 Ω-cm, although no α-particle or γ-ray response was detected. Similarly, Dmitriev *et al.* [204] fabricated detectors from polycrystals grown by the vertical Bridgman method. Although resistivities were of the order 1 GΩ-cm and the measured electron mobility–lifetime products were ~10^{-5}cm^{2}V^{-1}, no radiation response was detected. Matsumoto, *et al.* [205] grew BiI_3 crystals by the vertical Bridgman method using commercially available powder. The crystals were fabricated into radiation detectors by cleaving into wafers of thickness approximately 100 μm and depositing Pd electrodes by vacuum deposition on both cleaved surfaces. Electrical signals were extracted using thin Pd wires, bonded to the electrodes using silver epoxy. The measured resistivities were estimated to be 2 × 10^{10} Ω-cm, which is about one order of magnitude higher than previously reported. The detector was found to spectroscopic to α–particles with a measured FWHM energy resolution of 2.2 MeV at 5.5 MeV. Fornaro *et al.* [206] produced BiI_3 monocrystals by the travelling molten zone method, which yielded detector material, but not of spectrometric grade. The measured resistivities were up to 2 × 10^{12} Ω-cm and unlike previous work, small detectors of thickness ranging from 50 μm to 80 μm showed a response to 60 keV gamma-rays.

Saito *et al.* [207] grew single crystals of BiI_3 were grown by the PVT method. The starting material was commercial powder (99.99%) that had been purified to sub-ppm levels by repeated sublimation. Detectors were fabricated from hexagonal single crystal platelets of sides 2–5 mm long and thicknesses of 25–150 μm. Au electrodes 1.5 mm in diameter and 50 nm thick were deposited on both sides of the crystal by vacuum evaporation and contacted using Au wires attached with graphite paste. The resistivity and detection properties were improved substantially by post-annealing in an iodine atmosphere at 150°C for 2hr. For example, the resistivity before annealing was 3 × 10^9 Ω-cm and after 1.6 × 10^{11} Ω-cm. At 10V bias, clear peaks were observed in the energy spectrum from ^{241}Am alpha particles from which the electron mu-tau product was determined to be 8.6×10^{-6} cm^{2}V^{-1}. Saito *et al.* [190] also observed a gamma-ray photopeak at 59.5keV, with a FWHM energy resolution of 40%.

Gokhale *et. al.* [208] grew a number of antimony-doped BiI$_3$ single crystals using a modified vertical Bridgman technique. A number of detectors of size 8 mm × 8 mm × ~0.5 mm were fabricated by cutting the as-grown crystal with a diamond saw along the [001] plane and cleaving along the [001] plane, followed by polishing with an abrasive slurry. Planar Au electrodes were deposited on two opposite sides of the crystal by sputtering on both cleaved surfaces. The work followed on from a previous paper [209] that showed that Sb doping could improve the electrical properties of single BiI$_3$ crystals. Gokhale *et al.*, [208] characterized a number of crystals with different doping concentrations ranging from 0.5% at% to 5% at% and found that the lowest leakage currents and therefore highest resistivities (3.5×10^{10} Ω-cm) were obtained for Sb doping concentrations of 0.5 at%. A number of different detector surface treatments were also evaluated after which the detector cathodes were exposed to an ^{241}Am source and the response measured. The best response was found when surfaces had been mechanically polished followed by chemical etching. These detectors were found to be spectroscopic to γ-rays with a measured FWHM energy resolution of 7.5% at 59.95 keV (see Fig. 9.20). By measuring the peak position of 5.5 MeV alpha peaks from ^{241}Am as a function of bias, the electron mobility–lifetime product was estimated to be 5×10^{-4} cm^{2}V^{-1}. The hole mobility–lifetime product could not be measured because the output signal due to holes was too low. It should be noted that the detector also showed a significant decrease in counting efficiency with time – indicative of polarization effects.

In a study of BiI$_3$, Johns *et al.* [210] found that the major cause of performance degradation in detectors produced from material grown by conventional Bridgman was due to the formation of geometric void inclusions introduced into the bulk during growth. The voids are believed to originate from ionically bound aggregates in the solid phase that float in the melt

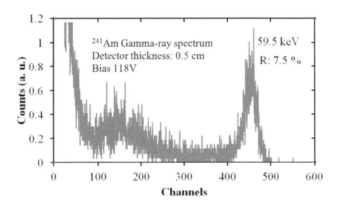

FIGURE 9.20 ^{241}Am spectrum, measured with an antimony doped BiI$_3$ 8 mm×8 mm×0.5 mm detector at room temperature (taken from Gokhale *et al.* [208]). The measured FWHM energy resolution at 59.54 keV is 4.5 keV, under full area illumination.

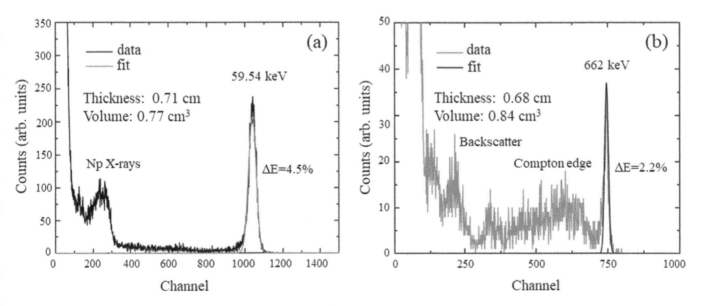

FIGURE 9.21 (a) ^{214}Am and (b) ^{137}Cs spectrum obtained with a 1 mm^2 pixel Sb:BiI$_3$ detector grown by the superheated gradient Bridgman technique (from [213]). The FWHM energy resolutions at 59.54 keV and 662 keV were 4.5% and 2.2%, respectively.

and serve as a nucleation sites for crystallographic defects. Johns *et al.* [211] found that the density and size of these defects could be substantially reduced by adding an additional superheating gradient step to the molten phase, prior to crystal growth. This gradient had been previously shown [212] to induce convection-based mass flow through the melt, which helps dissolve these aggregates improving the quality of the crystal phase. On average, Johns *et al.* [210] found a reduction of void concentration from 4600 voids per cm^3 to 300 voids cm^3. Antimony doped BiI_3 material was synthesized using a "superheating gradient" modified Bridgman technique and processed into ~1 cm^3 single crystal cubes. A number of planar and pixel detectors were then fabricated. The pixel detector consisted of a 3×3 grid of 1×1 mm^2 pixels. Since the pixel size is much less that the pixel depth, the effect of hole trapping will be substantially reduced by the small pixel effect. This was indeed found to be the case. With simple planar electrodes, the detectors exhibited weak evidence of a photopeak when exposed to a 662 keV gamma-ray source, whereas the pixelated detectors showed a clearly resolved photopeak. In Fig. 9.21 we show the responses of two 1 mm^2 pixel $Sb:BiI_3$ detectors to ^{214}Am and ^{137}Cs gamma-ray sources. The FWHM energy resolutions were found to be 4.5% and 2.2% at 59.54 and 662 keV, respectively.

9.8.11 Other Compounds

Numerous other compounds have been proposed and tested (*e.g.,* AlSb [214], Bi_2S_3 [215], $PbBr_2$ [216] and $HgBr_2$ [217]). None has shown a response to alpha particles.

9.9 Ternary Compounds

Other than pseudo ternary alloys of the form AB_xC_{1-x} (*e.g.,* $Cd_{1-x}Zn_xTe$), very little work has been carried out on true ternary materials, like $A^{II}B^{IV}C_2^V$ or $A_2^{II}B^VC^{VII}$, mainly because of the difficulties in maintaining compositional homogeneity during growth. Consequently, these materials are generally grown by MBE or MOCVD, which provide finer control over stoichiometry than other growth techniques. As a rule, ternary compounds based on the heavier metallic elements (such as Hg and Bi) are of lower mechanical strength, poorer phase and chemical stability than their binary derivatives.

9.9.1 Lithium Chalcogenides

Lithium based $A^IB^{III}C^{VI}$ chalcogenides have also been explored as materials for radiation detection. Tupitsyn *et al.* [218] synthesized and tested four Li-based chalcogenide compounds ($LiInSe_2$, $LiGaSe_2$, $LiGaTe_2$ and $LiInTe_2$). These materials are attractive in that they are chemically and physically stable and have a bandgap (~2.8 eV) suitable for room-temperature radiation detection. The disadvantage of these compounds is that they are difficult to synthesize, since lithium is such a highly reactive alkali metal that growth can proceed explosively when it is mixed with non-metallic elements like Se or Te [219]. Fortunately, techniques have been developed for synthesizing single crystals of $LiMX_2$ (where, M = Al, In, Ga; X = S, Se, Te) crystals for non-linear optical applications [220]. Of the four tested compounds, only $LiInSe_2$ held promise as a medium by showing well-pronounced photoconductivity when exposed to blue light, indicating good carrier generation coupled with good carrier mobility.

Single crystals of semiconductor-grade $LiInSe_2$ were grown by the vertical Bridgman method. The room temperature bandgap of the as-grown material was found to be 2.85 eV using optical absorption measurements. From a boule, a 30 mm^2, 0.5 mm thick sample was cut and a simple pad detector fabricated by sputtering 3 mm diameter gold contacts on both faces of the sample. The I/V characteristic was found to be linear, with a room-temperature bulk resistivity of ~6×10^{11} Ω-cm. The device showed a spectroscopic response to alpha particles from an ^{241}Am source with a FWHM of ~80%.

9.9.2 Cesium Thiomercurate

Li *et al.* [221] investigated the cesium thiomercurate $Cs_2Hg_6S_7$ as a potential detection medium for high-energy radiation detection. $Cs_2Hg_6S_7$ crystallizes in the tetragonal crystal system in space group P42nm. It has a high density of 6.94 gcm^{-3} and is composed of elements with high enough atomic numbers (Z_{Cs} = 55, Z_{Hg} = 80, and Z_s = 16) to ensure it has a stopping power significantly greater than that of CdZnTe over a large photon energy range (>10^2 keV). Its bandgap of 1.63 eV makes it a suitable detection medium for room temperature operation.

Li *et al.* [222] synthesized polycrystalline $Cs_2Hg_6S_7$ material using a vapor transport technique, producing single crystals with resistivities ~30 times higher than previously reported for material grown by the Bridgman method [222]. A $Cs_2Hg_6S_7$ ingot was then cut into 7 mm diameter, 1.6 mm thick wafers and then into rectangular blocks, which after polishing had dimensions ~$1 \times 1.4 \times 3.5$ mm^3. Planar gold electrodes were evaporated on the front and back surfaces of the samples in a parallel plate configuration for electrical testing. To improve resistivity, some samples were also doped with In or $HgCl_2$ to compensate for shallow acceptor states, introduced by Cs and Hg vacancies [223]. The resistivities of the samples were found to be low – 6 MΩ-cm for the as-grown samples and 47 MΩ-cm for the $HgCl_2$ doped samples and 3 MΩ-cm for the In doped samples. The electron and hole mobility–lifetime products for the samples were estimated from photocurrent measurements. For the as-grown sample, the mu-tau products were found to be 1.3×10^{-3} for electrons and 9.1×10^{-4} cm^{2}V^{-1} for holes, which is comparable to that of CdZnTe. Doping with HgCl$_2$ had little effect on the electron mu-tau product ($\mu\tau_e = 1.7 \times 10^{-3}$) but improved the hole mu-tau product by over an order of magnitude to 2.4×10^{-3} cm^{2}V^{-1}. The spectroscopic radiation response of two Cs$_2$Hg$_6$S$_7$ crystals (one undoped and one HgCl$_2$ doped) was evaluated using an unfiltered Ag X-ray radiation source (22.4 keV). Both crystals detected X-rays [222], although the spectral response for the undoped sample was not spectroscopic. For the doped sample, a well-resolved peak at 22.4 keV was found with a measured FWHM energy resolution of ~8%.

9.9.3 Cesium-Based Perovskite Halides

Metal halide perovskites[15] of the form CsPbX$_3$, where X is a halide ion (*i.e.*, Cl, Br or I), are currently being explored as tuneable nanomaterials for optoelectronic applications [223]. Recently, Liu *et al.* [224] proposed CsPbBr$_3$ and CsPbCl$_3$ as promising materials for high-energy radiation detection. They point out that both have bandgaps conducive to room temperature operation and are reasonably mechanically robust (Knoop hardness ~30 kg mm^{-2}) when compared to similar heavy-element inorganic compounds like HgI$_2$ and TlBr. In addition, both compounds display strong photoluminescence and measurable photoconductivity at ambient temperature.

CsPbBr$_3$ melts congruently at 567°C and exhibits two non-destructive phase transitions when cooled to lower temperatures. The first transition occurs around 130°C from cubic to tetragonal, which is followed by a second-order transition at around 88°C to the orthorhombic (Pbnm) phase, which is stable at room temperature. CsPbBr$_3$ has a density of 4.86 g cm^{-3} and is a direct gap material with a bandgap of 2.25 eV.

CsPbCl$_3$ has a Pnma orthorhombic structure at room temperature with a mass density of 4.24 g cm^{-3}. It is similar to CsPbBr$_3$ in that in undergoes several non-destructive phase transitions when cooled from the melting point of 610°C to room temperature – a cubic phase above 47°C, a tetragonal structure in the range of 47°C to 42°C and an orthorhombic phase below 42°C. It is also a direct gap material with a bandgap of 2.3 eV.

Liu *et al.* [224] grew samples of both compounds by the modified Bridgman method from which ~7 mm diameter, 2–3 mm thick samples were cut, polished and etched for electrical investigation. The resistivities of both compounds were found to be in the range 10^8 to 10^9 Ω-cm, which is low when compared to TlBr and CdZnTe. The electron and hole mobility–lifetime products for both materials were estimated from photocurrent measurements. For CsPbBr$_3$, the mu-tau products were found to be 1.7×10^{-3} for electrons and 1.3×10^{-3} cm^{2}V^{-1} for holes [225], while those of CsPbCl$_3$ were an order of magnitude lower for both carriers. Of particular note in CsPbBr$_3$ the electron mu-tau product is comparable to those measured in CdZnTe while the hole mu-tau product is an order of magnitude higher. The measured room-temperature spectral response to a Ag X-ray source showed a well-resolved 22.4 keV Kα radiation peak with a FWHM energy resolution of ~25%. For comparison, the measured energy resolution obtained with a commercial CdZnTe reference detector was ~15%. No spectral measurements for CsPbCl$_3$ were reported [224].

Subsequent work by He *et al.* [226] has produced significantly better results for CsPbBr$_3$. This was mainly achieved by improvements in the growth process and by the use of a novel electrode system. This resulted in a marked reduction in leakage current, which was typically of order 10 nA – over 10^4 times smaller than had been previously obtained. Simple single crystal planar detectors of dimensions $3 \times 3 \times 0.9$ mm^3 and $4 \times 2 \times 1.24$ mm^3 were fabricated and the spectral performance determined from exposure to ^{241}Am, ^{57}Co, ^{22}Na and ^{137}Cs radioactive isotopes. The measured FWHM energy resolutions were 9.6% at 59.95 keV, 4.3% at 122 keV and 3.8% at 662 keV. In Fig. 9.22 we show two spectra: (a) the measured response of a $3 \times 3 \times 0.9$ mm^3 single CsPbBr$_3$ crystal to a ^{57}Co γ-ray source and (b) the response of $4 \times 2 \times 1.24$ mm^3 single crystal to a ^{137}Cs γ-ray source. From the figures, we see that the main photopeaks are clearly resolved.

9.9.4 Copper Chalcogenides

Lin *et al.* [227] explored CuI$_2$Se$_6$ as a room temperature radiation detection medium, in view of its chemical stability, reasonably high density (5.3 g cm^{-3}) and wide bandgap (~2 eV). Polycrystalline base material was synthesized by direct combination of elements in an evacuated silica ampoule. Single crystal ingots were then grown from this material using a vertical Bridgman method. The ingot was sliced into wafers and a simple pad detector fabricated from a single 1 mm thick crystalline wafer extracted from the center of the boule. Contacting was achieved using Cu wires attached to carbon paint electrodes of diameter 2 mm. The I-V characteristic displayed ohmic behaviour with a bulk resistivity of ~10^{12} Ω-cm. The device demonstrated high photosensitivity to Ag Kα X-rays (22.4 keV) and a spectroscopic response ($\Delta E/E \sim 50\%$) to ^{241}Am α-particles (5.5 MeV) radiation.

---

[15] A class of compounds that have the same type of crystal structure ($^{XII}A^{2+VI}B^{4+}X^{2-}_3 \equiv ABX_3$) as the mineral ore perovskite (CaTiO$_3$).

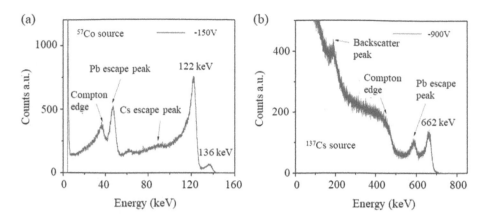

FIGURE 9.22 (a) A room temperature energy-resolved ^{57}Co spectrum measured with a $3 \times 3 \times 0.9$ mm^3 single CsPbBr$_3$ crystal detector at a bias of 150V and shaping time of 2 µs [226]. The measured FWHM energy resolution at 122 keV was ~3.9%. (b) The measured room-temperature energy response of a $4 \times 2 \times 1.24$ mm^3 detector to a ^{137}Cs γ-ray source. The applied bias was 900V and the shaping time 0.5 µs. The measured FWHM energy resolution at 662 keV was 3.9% (reproduced from He *et al.* [226], Creative Commons license BY 4.0).

9.9.5 Mercury Chalcogenides

Li *et al.* [228] explored mercury chalcogenide (Hg$_3$Se$_2$Br$_2$) as a promising candidate for room-temperature X-ray and γ-ray radiation detection in view of its bandgap of 2.22 eV and high density of 7.6 g cm^{-3}. Hg$_3$Se$_2$Br$_2$ melts congruently at a relatively low temperature of 566°C, which allows single crystal growth directly from a stoichiometric melt. Single crystals up to 1 cm long were grown using the Bridgman method. The crystals exhibited a photocurrent response when exposed to Ag X-rays. The resistivity was measured by the two-probe method to be of the order of 10^{11} Ω-cm, and the mobility–lifetime products were estimated from Ag X-ray spectroscopy to be 1.4×10^{-4} cm^{2}V^{-1} for electrons and 9.2×10^{-5} cm^{2}V^{-1} for holes. The single photon response to an unfiltered Ag X-ray source was spectroscopic and similar to that obtained with a CdZnTe reference detector with a FWHM energy resolution of ~40% at 22.1 keV.

9.9.6 Thallium Lead Iodide

The ternary compound, thallium lead iodide (TlPbI$_3$) has been investigated as a room-temperature X- and gamma-ray detection medium, in view of its high Z, high density (6.6 g cm^{-3}) and bandgap of 2.3 eV. Its low melting point (346°C), low vapor pressure and lack of a destructive phase transition between room temperature and its melting point facilitate purification and crystal growth directly from the melt. It crystallizes in a rhombohedral, perovskite-like (ABX_3), structure.

Kocsis [229] synthesized TlPbI$_3$ from the melt and sliced a 0.8 mm thick disk from the ingot. A detector was fabricated by applying colloidal graphite on the top and bottom surfaces of the disk. The measured resistivity was 2.5×10^{12} Ω-cm. While the device gave measureable photoconductivity response when exposed to an X-ray tube, no measurements were reported using α-particle or gamma-rays sources. Hitomi *et al.* [230] grew TlPbI$_3$ crystals using the vertical Bridgman method. Several detectors were fabricated by sawing the as-grown crystals into several 0.24 mm thick wafers and applying 0.1 mm diameter Au electrodes by vacuum evaporation. Pd contact wires were then bonded to the electrodes using silver epoxy. The resistivities, as evaluated from their current-voltage characteristics, were typically greater than 10^{11} Ω-cm. The detectors showed a clear peak when exposed to 5.5 MeV α-particles from a ^{241}Am source. However, when exposed to a ^{137}Cs source, the 662 keV photon peak was not resolved, although an increase in counts above the noise spectrum was observed. It was also reported that the performance of the detectors was unstable with time.

9.9.7 Thallium Chalcohalides

Recently Johnsen *et al.* [231,232] have proposed investigating a number of thallium chalcohalide-based semiconductors, for example, TlGaSe$_2$ and Tl$_6$SeI$_4$. These studies were inspired by the concept of "dimensional reduction" and lattice hybridization [233] in which binary semiconductors with high mass density and narrow bandgaps are combined with those with wide bandgaps to form ternary materials with suitable bandgaps and high densities. For radiation detection, Johnsen *et al.* [231] point out that the heavy metal halide[16] semiconductors tend to have large bandgaps (>2.6 eV) but are mechanically soft, whereas the metal chalcogenides[17] have bandgaps that are too small to allow room temperature operation

---

[16] A binary compound of which one part is a halogen (group 7 elements) and the other part is an element or radical that is less electronegative than the halogen.
[17] Compounds containing sulfur, selenium or tellurium.

but are mechanically robust. By combining binary halides (MX_n, where M is a heavy metal X is a halogen) and binary chalcogenides (M_xQ_y, where Q is a chalcogen), it is possible to form hybrid chalcohalide compounds ($M_xQ_yX_z$) that have energy gaps that lie between the corresponding end members of the binary chalcogenides and binary halides (that is 1.6 eV to 2.0 eV), whilst retaining the mechanical properties of the chalcogenides. In fact, the introduction of the halide disrupts the extensive orbital overlap of the heavy chalcogenide atoms in the structure and narrows the band widths, creating bandgaps that are intermediate between those of pure chalcogenides (too small) and pure halides (too large) as illustrated in Fig. 9.23. The main advantages of this approach over, say, just using a heavy metal halides, such as HgI_2, $TlBr$, PbI_2 and BiI_3 is that they tend to have higher densities and larger mu-tau products and are mechanically stronger (typically 3–6 times harder). The latter quality is particularly important for fabricating detectors, since any mechanical processing introduces defects in soft materials. Two compounds in particular have shown promising results, namely $TlGaSe_2$ and Tl_6I_4Se.

9.9.7.1 Thallium Gallium Selenide

Thallium gallium selenide ($TlGaSe_2$) melts congruently at 350°C and crystallizes in a layered type structure. The bonding is strongly covalent within layers and much weaker between layers. It is an indirect bandgap material ($\varepsilon_g = 1.95$ eV) and has a density of 6.4 gcm^{-3}. Johnsen *et al.* [232] synthesized $TlGaSe_2$ from a stohchiometric combination of TlSe, Ga and Se and grew single crystalline material by a modified vertical Bridgman method. The ingot was cut into wafers from which single crystals were cleaved. A detector of size $3 \times 5 \times 0.87$ mm³ was fabricated from one crystal and contacted with an evaporation of Ti/Au. From I-V measurements, the resistivity was estimated to be in $\sim 10^9$ Ω-cm. The mobility–lifetime products were determined from photoconductivity measurements to be, $\mu_e\tau_e = 6 \times 10^{-5}$ cm²V⁻¹ for electrons and $\mu_h\tau_h = 9 \times 10^{-6}$ cm²V⁻¹ for holes. The X-ray response of the device was investigated using a Ag X-ray tube. The measurements were carried out at room temperature and with an operating bias of 190V. Although the device was spectroscopic, just resolving the characteristic Kα line at 22.2 keV, the Kβ line at 24.9 keV was not resolved.

9.9.7.2 Thallium Iodide Selenide

Thallium iodide selenide (Tl_6I_4Se) is a direct-gap material with a bandgap of 1.86 eV. It melts congruently at 432°C and crystallizes into a dense (7.4 g cm⁻³) tetragonal structure. Johnsen *et al.* [231] grew single crystalline material by a modified

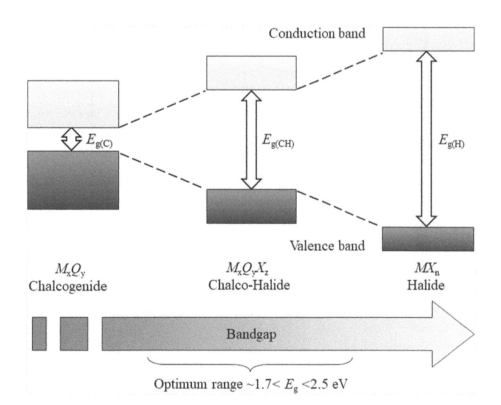

FIGURE 9.23 Illustration of the "lattice hybridization approach", in which a binary halide (MX_n, where M is a heavy metal X is a halogen) and a binary chalcogenide (M_xQ_y, where Q is a chalcogen) are combined to form a hybrid chalcohalide compound ($M_xQ_yX_z$) that has an intermediate energy gap, E_g(CH), between the halide, $E_{g(H)}$, and the Chalcogenide, $E_{g(C)}$, but retains the mechanical properties of the chalcogenide (adapted with permission from Johnsen *et al.* [231], copyright {2011} American Chemical Society).

vertical Bridgman method. The resulting sample shows single-crystalline domains from which wafers were cut perpendicular to the growth direction. A detector was fabricated from a $6 \times 4 \times 2$ mm^3 single crystal diced from a wafer and Ti/Au contacts evaporated on to the top and bottom faces. From I-V measurements, the resistivity was found to be 4×10^{12} Ω-cm along the $\langle 001 \rangle$ crystallographic direction. The mobility–lifetime products were determined from photoconductivity measurements and found comparable to CdZnTe for electrons ($\mu_e \tau_e = 7 \times 10^{-3}$ cm^{2}V^{-1}) and an order of magnitude larger for holes ($\mu_h \tau_h = 6 \times 10^{-4}$ cm^{2}V^{-1}).

Figure 9.24 shows pulse height spectra recorded using a ^{57}Co radioactive source from which we see that the principal line emissions at 14.4, 122.1 and 136.5 keV are clearly resolved. The operating bias was 290V and the measurement carried out at room temperature. The measured FWHM energy resolution was 5.7 keV (4.7%) at 122.1 keV. For comparison, Fig. 9.24 also shows the measured spectrum obtained with a commercial $5 \times 5 \times 5$ mm^3 CdZnTe detector. In this case, the measured FWHM energy resolution is 5.54 keV (4.5%) at 122.1 keV.

9.9.8 Other Thallium Ternary Compounds

Kahler *et al.* [234] have proposed the ternary compounds Tl$_3$AsSe$_3$, (TAS) and Tl$_4$HgI$_6$ (THI) as possible replacements for CdZnTe, based on their high densities and bandgap energies. These materials had previously been investigated for non-linearly optical applications, such as long wave infrared acousto-optical tuneable filters. Suitable ingots of both materials were grown using the vertical Bridgman technique and 3 mm thick planar detectors fabricated with Au, Ag or Al/Ge electrodes deposited on opposite faces. The measured resistivities were 2×10^6 Ω-cm for TAS (mainly limited by leakage currents caused by the low bandgap of 1.3 eV) and 2×10^{12} Ω-cm for the THI detectors. The crystals were then exposed to gamma-radiation from a ^{137}Cs radioactive source. The TAS device was found to be responsive but not spectroscopic to the radiation, whereas the THI device was found responsive and spectroscopic with an estimated energy resolution at 662 keV of 15% FWHM.

Lin *et al.* [235] explored the halide antiperovskite semiconductor TlSn$_2$I$_5$. The compound has an indirect bandgap of 2.14 eV, and a density of 6.1 g cm^{-3}. It melts congruently at 314°C and has no phase transitions between melting and ambient temperature. Centimetre-size TlSn$_2$I$_5$ single crystals were grown from the melt by the Bridgman method. Electrical measurements were carried out on a detector fabricated on a 0.1 cm thick wafer with dimensions of 5 mm × 9 mm. The IV characteristic was found to be linear indicating ohmic behavior. For the slope of the curve, the resistivity was estimated to be 4×10^{10} Ω–cm. The X-ray and gamma-ray responses were evaluated with a simple planar detector, fabricated from a 1 mm

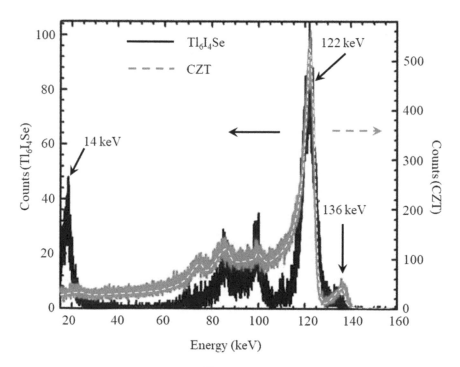

FIG. 9.24 Recorded pulse height spectrum from γ-radiation from a ^{57}Co source using a $6 \times 4 \times 2$ mm^3 $\langle 001 \rangle$ Tl$_6$I$_4$Se wafer (dark solid line) (reprinted with permission from Johnsen *et al.* [231], copyright {2011} American Chemical Society). For comparison, the measured pulse height spectrum from a commercial 5×5 × 5 mm^3 CZT detector[18] is also shown (dashed line). The measurements were carried out at 295 K.

---

[18] SPEAR detector manufactured by EI Detection & Imaging Systems.

thick $TlSn_2I_5$ wafer with a top carbon electrode of ~2 mm in diameter and a full planar carbon paint bottom electrode. Contacting was achieved using Cu wires. The device was found to be sensitive but not spectroscopic to 22 keV Ag Kα X-rays, 122 keV γ-rays from a [57]Co source as well as 5.5 MeV α-particles from a [241]Am source. The electron mobility–lifetime product was estimated from fitting a single carrier Hecht equation to bias dependent pulse height data to be 4.5×10^{-5} cm^2 V^{-1}. The electron mobility was estimated to be 94 cm^{2}V^{-1}s^{-1} by measuring the photoexcited electron drift time using α-particles from a [241]Am source.

TlAu$_4$S$_3$ was investigated by Liu *et al.* [236] in view of its very high density (10.2 g cm^{-3}) and favourable bandgap (1.6 eV). TlAu$_4$S$_3$ crystals were grown by modified Bridgman. From IV measurements, the samples were found to have a resistivity of ~60 GΩ-cm. The mobility–lifetime products of the carriers were determined with photoconductivity measurements using a He-Ne laser. Values of 1.2×10^{-4} cm^{2}V^{-1} and 1×10^{-5} cm^{2}V^{-1} were obtained for electrons and holes, respectively.

Lin *et al.* [237] explored the ternary chalcogenide compound TlSbS$_2$ as a hard radiation detection medium. A simple pad detector made from a $5 \times 4 \times 0.2$ mm^3 cleaved plate with carbon paste electrodes displayed ohmic behaviour with bias. The measured resistivity was $> 10^{10}$ Ω-cm. However, while the device responded to 22 keV Ag X-rays and 5.5 MeV alpha-particles, the response was not spectroscopic.

9.9.9 Lead Chalcogenides

Recently, the Pb-complex chalcogenides have been investigated as gamma-ray detection media in view of their high effective Z and high density and therefore good stopping power. Bandgaps range from 1.6 eV to 2.8 eV, ensuring that detectors fabricated from pure, high-quality crystals will be able to operate at room temperature with low dark current. In addition, these materials can exhibit high electron mobility and no ionic conductivity, important qualities for good charge collection and detector stability.

9.9.9.1 Lead Gallium Selenide

Kargar *et al.* [238] investigated the suitability of a number of compounds (Pb$_2$Ga$_2$S$_5$, Pb$_2$Ga$_2$Se$_5$, Pb$_2$In$_2$Se$_5$, PbGa$_2$S$_4$, PbGa$_2$Se$_4$ and Pb$_2$GeS$_4$) as possible gamma-ray detection media. Of these, lead gallium selenide (PbGa$_2$Se$_4$) was identified as the most promising candidate. It had previously been investigated as a photoconductor in view of its large photosensitivity in the 0.4 to 1.2 μm range [239,240]. It has a bandgap of 2.35 eV, melts congruently at 1053K and crystallizes in an orthorhombic structure. Kargar *et al.* [238] grew single crystalline material by the vertical Bridgman method and fabricated simple planar detectors with gold contacts from cleaved crystals. The resistivity of the samples was measured to be in the 10^{13} Ω-cm range. The electron mobility–lifetime product, as determined from a Hecht analysis, was 10^{-6} cm^{2}V^{-1}. A 250 μm thick planar device was found to be spectrally responsive to photons emanating from both [109]Cd and [241]Am gamma-ray sources. For the [109]Cd source, a full energy peak could be resolved at 22.1 keV. However, the large noise floor prevented the clear determination of its FWHM.

9.9.9.2 Lead Selenophosphate

The heavy metal, chalcophosphate, Pb$_2$P$_2$Se$_6$, has also been investigated as a promising new material for X- and γ-ray detection [241]. Lead selenophosphate has an effective Z of 39.8 and a density of 6.14 g cm^{-3}, which is comparable to the commercial benchmark, CdZnTe. It has an indirect bandgap of 1.88 eV, making it suitable for room-temperature applications, and melts congruently at 812°C. As pointed out by Wang *et al.* [241] stoichiometric Pb$_2$P$_2$Se$_6$ has a simpler phase diagram than CdZnTe which, being an alloy of cadmium telluride and zinc telluride, exhibits a range of solid solutions. This should reduce the technological and engineering challenges in terms of crystal growth and scale-up development, while at the same time offering comparable performance in terms of density, bandgap, *etc.* Also, the precursor growth material is commonly available and far less expensive than Cd or Te and far less toxic than Cd.

Wang *et al.* [241] grew crystals of Pb$_2$P$_2$Se$_6$ by the vertical Bridgman method. The compound was prepared from as-purchased commercial elements without any further purification. The resulting ingots were subsequently cut and polished into detector samples. The average surface area of the wafers ranged from 3×3 up to 5×10 mm, while the thickness varied between 1 and 2 mm. The samples were contacted with copper wires secured to the wafer surfaces by colloidal graphite paint. From the I-V characteristics, the resistivity was measured to be $\sim 1 \times 10^{10}$ Ω-cm. From a Hecht analysis, the electron mobility–lifetime product was estimated to be 3.5×10^{-5} cm^{2}V^{-1}. Single crystal samples displayed a significant photo-conductivity response to optical, X-ray, and γ-ray radiation. When tested with a [57]Co γ-ray source, a 1.8 mm thick detector showed a spectroscopic response sufficient to resolve the 122.1 and 136.5 keV peaks with a bias of 600V. At 122 keV, the room temperature FWHM energy resolution was determined to be 8%, comparable to that obtained with a CdZnTe reference detector.

9.10 Quaternary Compounds

Li *et al.* [242] investigated the heavy metal quaternary compound TlHgInS$_3$ as a gamma-ray detection medium in view of its favourable bandgap of 1.74 eV and high density (ρ = 7.241 g cm^{-3}). The compound melts incongruently, which makes it difficult to grow large single crystals by the Bridgman method so TlHgInS$_3$ crystals were synthesized by a stoichiometric reaction of Tl$_2$S, HgS and In$_2$S$_3$ at 800°C, followed by fast cooling in 3 h. A rod-shaped single crystal (0.8 mm × 0.4 mm × 0.3 mm^3) of TlHgInS$_3$ was selected for resistivity measurements. Silver paste electrodes were applied to the crystals. From I-V measurements, the crystal showed a linear ohmic behaviour. The resistivity of the crystal was determined to be 4.32 GΩ-cm. TlHgInS$_3$ single crystals exhibit photocurrent response when exposed to Ag X-rays. The mobility–lifetime product ($\mu\tau$) of the electrons and holes was estimated from photocurrent measurements to be 3.6×10^{-4} cm^{2}V^{-1} for electrons and 2.0×10^{-4} cm^{2}V^{-1} for holes.

Other quaternary compounds that have been investigated are CsHgInS$_3$, CsCdInSe$_3$, CsCdInTe$_3$ [243,244]. All were found to show a strong photoconductive response to laser stimulation and gave carrier mobility–lifetime products comparable to other detector materials such as HgI$_2$ and PbI$_2$.

9.11 Organic Semiconductors

So far we have concentrated on inorganic semiconductors as radiation detection media. However, there is no reason organic semiconductors may not serve the same function (see [245]). In fact, they should be responsive not only to charged particles but also to fast neutrons, since their basic structures are hydrogen rich. These are the so-called "plastic" semiconductors, although strictly only a subset of organic semiconductors can be considered plastic.

The great advantage of organic semiconductors is their diversity and the relative ease of tailoring their properties to specific applications. They also lend themselves to simple manufacturing techniques, for example, in the case of polymers, spin coating as opposed to epitaxy. In addition, the number of known organic semiconductors is huge and even includes structures like plant- and animal-chelates such as carotene and chlorophyll, the so-called edible semiconductors [246], blood pigments and DNA [247]. At present, organic semiconductors are largely being exploited for use in low-cost flexible displays, low-end data storage media and inexpensive solar cells. They are already widely used commercially in xerography [248]. For display applications, they offer the advantage that they emit light directly,[19] rather than relying on a back lighting system as is used in liquid crystal displays. The main disadvantage of polymeric systems at this time is environmental stability; they cannot sustain high current densities for extended periods and degenerate when exposed to H$_2$ and H$_2$O atmospheres. For solar cell applications, efficiencies of up to 10.5% have been achieved for single junction devices [249] and most recently up to 17% for tandem cell devices [250]. They are also of interest for medical applications because they are "tissue-equivalent" materials, in that the low atomic numbers of their constituent atoms closely match those of biological tissue.

9.11.1 Structure

Organic semiconductors are chemically synthesized; as such, an extremely wide variety of molecular types can be physically realized. These compounds, oligomers,[20] polymers[21] and derivatives, are composed of mainly carbon, nitrogen, hydrogen and oxygen. They can be divided into two classes: polymer and molecular (for an overview see [251]). Most work has been carried out on polymeric systems. These materials are composed of large chains based on the repetition of a basic unit structure, called a monomer. The most basic polymer is shown in Fig. 9.25. Carbon atoms form the backbone of the chain. Other substitutes can be attached to the backbone, which can be used to influence the polymer's chemical and electronic properties, for example, to promote solubility in organic solvents or tune the structural properties of the material. Molecular organic materials, on the other hand, are composed of molecules, generally with non-repeating structure[22] – in other words, they are monomers. They can be divided into two sub-classes – pigments, which are insoluble, and dyes, which are soluble. Molecular organic materials can be amorphous or crystalline. When they do crystallize, it tends to be in complicated classes (such as monoclinic or triclinic) whose semiconductor properties are much more aligned with amorphous inorganic

---

[19] Conjugated molecules may conduct charges by breaking double bonds. Recombination of the broken bonds may result in light emission.

[20] The term is derived from the Greek "oligo" meaning few, and "mer" meaning parts. An oligomer is an intermediate mass molecular complex consisting of a few monomer (single) units. This class includes dimmers (two monomers), trimers (three monomers) and tetramers (four monomers).

[21] Similarly, from the Greek "poly" meaning many and "mer" meaning parts. A polymer is a heavy molecular complex consisting of many monomer units (typically > 20) that form a chain.

[22] Hence, they have a very well established chemical structure, while polymers are long chains with a distribution of lengths (number of repeated monomeric units).

TABLE 9.4

Physical and electronic properties of conductive organic and inorganic-organic hybrid semiconductors. For comparison, we also show values for the conventional semiconductors, Si (crystalline and amorphous), GaAs and InSb, from which it is apparent that mobilities are many orders of magnitude higher. This limits the potential use of organics to thin-film applications. A number of low-molecular weight materials can be grown as single crystals, allowing intrinsic electronic properties to be studied directly (for example, note the difference in mobilities between crystalline and synthetically produced pentacene). Note the transport properties for these mareials can vary widely depending on the synthesis technique, impurities and overall material quality.

Material (abbreviation)	Formula	Structure	Space group	Density g·cm⁻³	Bandgap eV	Melting point K	Molecular mass g·mol⁻¹	Conductivity σ S·m⁻¹	Dielectric constant Conductivityσ	Mobility cm²s⁻¹V⁻¹ electron	Mobility cm²s⁻¹V⁻¹ hole
Inorganics											
InSb	InSb	zincblende	cF8	5.78	0.7	797	236.6	10^{-2}	17.7	80,000	1250
Si	Si	diamond	cF8	2.33	1.12	1687	28.1	1.56×10^{-3}	11.9	1500	480
GaAs	GaAs	zincblende	cF8	5.32	1.4	2577	144.6	10^{-8} to 10^{3}	13.1	800	400
a-Si:H	a-Si	amorphous	-	2.33	1.8	1420	28.1	10^{-8}	11.7	1	0.05
Organics											
Anthracene	$C_{14}H_{10}$	monoclinic	P2₁/b	1.28	4.0	489	178.23	$10\text{-}10^{2}$	2.9	1.7	2.1
Napthalene	$C_{10}H_8$	monoclinic	P2₁/b	1.16	5.0	351	128.17	10^{-3}	3.4	0.6	1.5
Perylene	$C_{20}H_{12}$	monoclinic	P2₁/a	1.30	3.1	550	252.32	10^{-3}	12	5.5@60K	87@60K
Pentacene	$C_{22}H_{14}$	triclinic	P-1	1.30	1.6	645[23]	278.36	0.01-2.5			2.0
Polyacetylene	$[C_2H_2]_n$	polymer		0.4	1.5			$10^{3}\text{-}10^{5}$			0.1-0.5
Polythiophene (spin coated)	$[C_4H_2S]_n$	polymer			2.0	235		10^{-3}			$10^{-4}\text{-}10^{-2}$
Polythiophene (printed)	$[C_4H_2S]_n$	polymer[24]			2.0	235		10^{-3}			0.1
Phenothiazine	$C_{12}H_9NS$	ortho	Pnma	1.35	3.4	458	199.27			2.5	0.02
Oligothiophene (Py-4T)		polymer			1.5			100		1.7	0.1-0.5
Polyphenylene		polymer			3.0			$10^{2}\text{-}10^{3}$			
Poly(p-phenylene vinylene) (PPV)	$[C_8H_6]_n$	polymer			2.5			$3\text{-}5\times10^{3}$		10^{-4}	40
Poly(3,4-ethylenedioxythiophene) (PEDOT)	$[C_6H_6O_2S]_n$	polymer			1.1			300			0.8-9.7
Polypyrrole (PPy)	C_4H_5N	polymer		1.46	3.1	250	67.089	$10^{2}\text{-}10^{4}$			10^{-2}
Polyaniline (PANI)	$[C_6H_7N]_n$	polymer		1.36	3.2	603	214.272	30-200			10^{-3}
4-hydrocyanobenzene (4HCB)	$[C_7H_5NO]_n$	polymer		1.20	4.2	385	119.12	10^{-9}			7×10^{-4}

(Continued)

---

[23] Sublimation temperature

[24] We designate the structure here simply as polymer, since in reality organic structures are neither amorphous nor crystalline. The degree of crystallinity typically ranges between 10 and 80%.

TABLE 9.4 (Cont.)

Material (abbreviation)	Formula	Structure	Space group	Density $g \cdot cm^{-3}$	Bandgap eV	Melting point K	Molecular mass $g \cdot mol^{-1}$	Conductivity σ $S \cdot m^{-1}$	Dielectric constant Conductivityσ	Mobility $cm^2 s^{-1} V^{-1}$ electron	hole
Methylammonium Lead Tribromide MAPbBr$_3$	CH$_3$NH$_3$PbBr$_3$	cubic	Pm-3m	3.58	2.18		478.98	5×10^{-6}	25.5	8	115
Methylammonium Lead Triiodide MAPbI$_3$	CH$_3$NH$_3$PbI$_3$	cubic	Pm-3m	3.95	1.55		619.98	5×10^{-6}	32	20–100	165
Methylammonium Tin Triiodide MASnI$_3$	CH$_3$NH$_3$SnI$_3$	cubic	Pm-3m	3.62	1.21		531.46	5	10	2320	200–300
Formamidinium Lead Tribromide FAPbBr$_3$	H$_2$NCHNH$_2$PbBr$_3$	cubic	Pm-3m		2.15		491.98	2×10^{-8}	43.6	14	62
Formamidinium Lead Triiodide FAPbI$_3$	H$_2$NCHNH$_2$PbI$_3$	cubic	Pm-3m		1.41		632.98	2×10^{-5}	49.4	4	35
Formamidinium Tin Triiodide FASnI$_3$	CH(NH$_2$)$_2$SnI$_3$	cubic	Pm-3m	3.58	1.41		544.46	10^{-5}	49.3	103	20

semiconductors. This can be seen in Table 9.4, which compares conductivities and mobilities for amorphous Si with crystalline pentacene. Some polymers can crystallize, either by cooling from the melt, being mechanically stretched, or by solvent evaporation. However, they crystallize in a semi-crystalline form with crystalline regions dispersed within amorphous material. The crystalline regions are associated with the partial alignment of their molecular chains. These chains fold together and form ordered regions called lamellae, which comprise larger spheroidal structures named spherulites. This is illustrated in Fig. 9.25. Common imperfections, such as disorder in the chains or misalignment, lead to amorphous material since twisting, kinking and coiling prevent the strict ordering required in the crystalline state. The fraction of ordered monomers in a polymer is characterized by the degree of crystallinity, which typically ranges between 10% and 80% with most polymers being around 10–30% crystalline. Higher values are only achieved in materials having small monomeric units, which are usually brittle. Crystalline polymers are denser than amorphous polymers, so the actual degree of crystallinity can be determined from a density measurement. Although there are no 100% crystalline polymers, some polymers may be 100% amorphous under certain conditions.

An important difference between molecular and polymer organic materials lies in how they are processed to form thin films. Because of insolubility issues, molecular organic materials are usually deposited from the gas phase by sublimation or evaporation, both relatively complex processes. However, polymers can usually be processed directly from solution leading to greatly simplified production techniques, for example, spin coating or ink-jet printing. It should be noted that a number of low-molecular materials can be grown as single-crystals (*e.g.,* pentacene) allowing intrinsic electronic properties to be studied directly.

9.11.2 Bonding

Organic materials are characterized by dipole-dipole bonds (van der Waal interactions) between molecular chains resulting in considerably weaker bonding when compared to covalently bonded semiconductors like Si or GaAs. The consequences are reflected in their poorer mechanical and thermodynamic properties such as reduced hardness and lower melting point and more importantly in a much weaker delocalization of electronic wavefunctions amongst neighbouring molecules. This has direct implications for charge transport properties manifesting itself in low drift velocity and poor mobilities (see Table 9.4). In molecular materials, the bonding is almost exclusively van der Waals between molecules, whereas in polymeric materials it is van der Waals between chains and a mixture of covalent and ionic bonding within chains; each carbon atom is covalently bonded to another carbon atom and by partial ionic interactions to atoms of other elements. In addition, the electronic interaction between adjacent chains is usually also weak in this class of materials. Consequently, material properties are improved compared to molecular materials but still poorer when compared to covalently bonded materials.

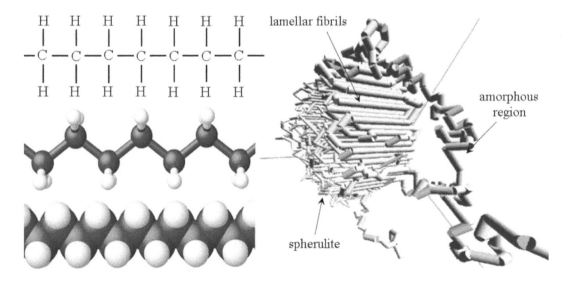

FIGURE 9.25 Left: common depictions of a polymer structure (in this case a polyethylene molecule [C_2H_4]$_n$). (top) Lewis diagram[25] which shows the bonding between atoms, (middle) ball-and-stick model and (bottom) space-filling model. Right: diagram illustrating crystallization in polymers. The crystalline regions are associated with partial alignment of the molecular chains, which when folded together form ordered regions called lamellae. These in turn compose larger spheroidal structures called spherulites that sit in an amorphous soup of disordered linking molecules (Reprinted with permission from Hu *et al.* [252]. Copyright {2003} American Chemical Society).

---

[25] Named after Gilbert Newton Lewis (October 23, 1875 – March 23, 1946) an American physical chemist, who introduced the Lewis diagram in 1916 to show in a simplified way the bonding between atoms of a molecule and the lone pairs of electrons that may exist in the molecule.

9.11.3 Electronic Organic Materials

All electronic organic materials, whether monomer, oligomer or polymer, contain a backbone of "conjugated" bonds in which the p-orbitals overlap, allowing the de-localization of π electrons across adjacent p-orbitals. This results in the creation of alternating single and double bonds between the carbon atoms by *pz–pz* bonding (–C=C–C=C–) and the creation of a bandgap of between 1.5 and ~3 eV. It also facilitates the transport of charge. In Fig. 9.26(a) we show a schematic for the structure of conjugated *trans*[27] polyacetylene. Generally, the conductivity of organic materials is extremely poor,[28] since most organic polymers do not have intrinsic charge carriers. However, it can be greatly improved by doping, although the term "doping" in this context is misleading, since it reflects not an intrinsic ability to control conductivity by adding substitutional impurities to generate additional charge carriers, but rather how easily holes and electrons can be injected from their contact electrodes. Through this process, charged defects (*e.g.*, a polaron, bipolaron or soliton) are introduced, which may then act as charge carriers.

The main type of carrier produced depends on the polymer structure; however, in most of the conjugated conducting polymers, conduction occurs *via* polarons or bipolarons. These can move along the backbone by intersite "hopping" and are responsible for the macroscopically observed conductivity of the polymer – although the actual transport mechanism(s) are not well understood and the reader is referred to reference [253] for a review. Unlike inorganic semiconductors where doping is achieved by introducing specific atoms as substitutional impurities, in organic semiconductors "doping" is achieved chemically – either by partial oxidation of the polymer chain with electron acceptors (*p*-type doping) or partial reduction of the chain with electron donors (*n*-type doping). During the process, the dopant is ionized and is interstitially bound by the Coulombic interaction with the charge it created on the polymer chain. In polyacetylene, for example, *p*-type doping is achieved by exposing the polymer to vapors of an electron-attracting halogen, usually iodine (I_2). In this case, the iodine molecule attracts an electron from the double bond of the polymer chain as shown in Fig. 9.26(b) *via* the reaction

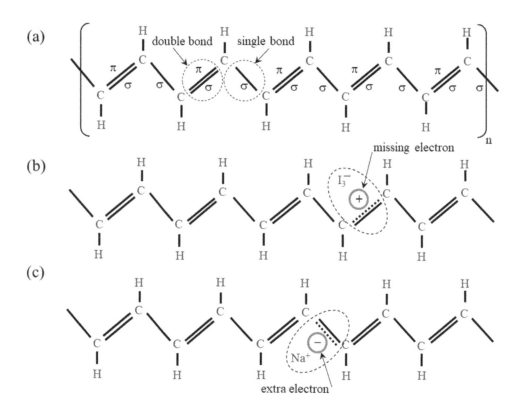

FIG. 9.26 (a) Schematic diagram of the simplest fully conjugated polymer molecule – *trans* polyacetylene ($[C_2H_2]_n$). A single localized σ bond, common to all atoms, forms a strong covalent bond between the carbon atoms and provides strength to the chain. The σ bonds are fixed and immobile. Alternate bonds also contain a less strongly localized π bond, which is weaker. It is through this bond that conduction takes place. Generally, the conductivity of polymers is poor. However, it can be greatly improved by doping. This is illustrated in figures (b) and (c) and is achieved by removing electrons from the π-system by oxidation ("p-doping") as shown in (b) or adding electrons into the π-system by reduction ("n-doping") as shown in (c). A charged unit called a bipolaron[26] is then formed and is responsible for the macroscopically observed conductivity.

---

[26] Essentially a quasi-particle consisting of two polarons. In organic chemistry, it is a molecule or a part of a macromolecular chain containing two positive charges in a conjugated system.

[27] *Trans* and *cis* are two geometric isomers of the polymer, which as a consequence of the slightly different physical arrangement of C and H bonds have different physical properties.

[28] Most organic materials are electrical insulators with room temperature conductivities in the range 10^{-9}–10^{-14} S cm^{-1} (10^9–10^{14} Ω-cm resistivity).

$$[CH]_n + \frac{3x}{2} I_2 \rightarrow [CH]_n^{x+} + x I_3^-, \tag{4}$$

where x is the dopant concentration. The polyacetylene molecule, now positively charged, is termed a radical cation, or polaron. The remaining electron of the double bond can easily move by a process of intersite hopping. Consequently, the double bond successively moves along the molecule. The positive charge, on the other hand, is fixed by electrostatic attraction to the iodide ion and does not move so easily. For n-type doping, a solution method is generally used in which the reductant and the by-products from the doping process are soluble, but the polymer is not. The reduction takes place using an alkali metal, for example, Na, K, Li, *etc.* In the case of Na, the reaction proceeds as follows,

$$[CH]_n + x Na \rightarrow [CH]_n^{x-} + x Na^+. \tag{5}$$

In this case, it is the reducing agent that is oxidized (*i.e.*, the Na loses (or "donates") an electron to the polymer chain). The resulting positive ion is interstitially fixed by the Coulombic interaction with negative charge attached to the polymer chain as shown in Fig. 9.26(c). Doping is usually performed at much higher levels (20–40%) in conducting polymers than in inorganic semiconductors (which is typically much less than 0.1%).

To summarize, the important qualities in an organic semiconductor for functional applications (electronics, optoelectronics, photovoltaic and radiation detection) are:

a) the presence of a conjugated system,

b) π-electron clouds as overlapped as possible,

c) good thin film structural properties,

d) chemical purity,

e) material stability.

9.11.4 Polyacetylene

In terms of radiation detection, Beckerle and Strobele [254] pointed out that semiconducting polymer foils could, in principal, be used as drift detectors for tracking charged particle trajectories. They demonstrated that stretched polyacetylene foils were sensitive to 5.5 MeV alpha particles. The foils had dimensions of (1×1) cm^2 and (1×2) cm^2 with thicknesses of 10, 50 and 100 μm. Electrical contacts were applied by sputtering 200 μm thick gold layers onto the foils. Without stretching, the mobility of the free carriers is ~10^7–10^9 times smaller than in silicon leading to very low drift velocities. However, stretching the film creates a degree of alignment of the polymer chains in the stretched direction in which both charge mobilities increase markedly with the degree of stretching. This is because the alignment of the polymer chains with respect to the electric field should reduce the number of hopping steps per unit length. In an unstretched foil, the efficiency for detecting 5.5 MeV α-particles in a 10 μm film is around 35% for drift lengths < 3 mm. Stretching the foil by a factor of 3 increased the detection efficiency to ~70% and electron mobilities by a factor of ~5. However, drift velocities were still very low, which limited their use to low-event rate applications. In addition, long-term material stability and radiation damage issues may prohibit the use of polymer foils for many other applications.

9.11.5 Poly(3,4-ethylenedioxythiothene) (PEDOT)

Blakesley *et al.* [255] investigated the potential for using polymeric semiconductors in medical X-ray imaging applications. They found that polymer photodiodes coupled to phosphor screens show an efficient response to X-ray radiation. The photodiodes were fabricated on glass substrates with indium–tin oxide (ITO) bottom contacts. A layer of poly(3,4-ethylene-dioxythiothene) poly(styrenesulphonate) (PEDOT–PSS) was spin coated on to the substrate and dried. Finally, aluminum contacts were evaporated on to the top surface. Mobilities were typically of the order of 0.01–0.1 cm^{2}V^{-1}s^{-1}. When coupled to phosphor screens, the diodes are were found to be responsive to radiographic X-rays with a phosphor-to-diode quantum efficiency of 14% and a dark current of ~100 nA cm^{-2}. The measured photocurrents were found to be linear with X-ray doses up to 8 mGys^{-1} and showed no measurable change in performance with total doses of up to 500 Gy.

9.11.6 Polyaniline (PANI)

Suzuki *et al.* [256] investigated the conjugated polymer polyaniline (PANI) as a material for real-time radiation detection. The base material was synthesized from solution and a thick circular polymer plate produced by placing the solution in a cylindrical container and evaporating the solvent. The plate was then flattened with a stainless steel piston before being

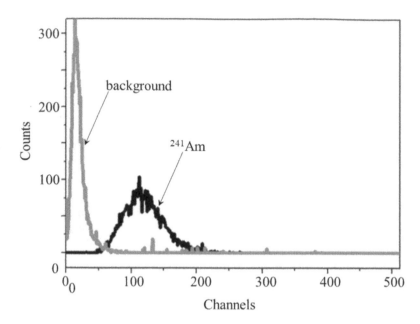

FIGURE 9.27   The response of 2–4 mm^2, 200 μm thick 4-hydrocyanobenzene (4HCB) detector to ~5 MeV alpha particles from an ^{241}Am source at room temperature (from [257]). The applied bias was 300 V and the shaping time 50 μs.

ground flat. The detector had a diameter of 10 mm and a thickness of 1–2 mm. A gold wire attached to the top surface served as the anode and a planar 0.15 μm thick Au cathode applied to the bottom surface by sputtering. The I/V characteristic was found to be ohmic. From the curve, a resistivity of ~ 10^{11} Ω-cm was determined. The detector was found to be responsive, but not spectroscopic, to both alpha particles from a ^{241}Am source and beta particles from a ^{90}Sr source. The corresponding detection efficiencies were determined to be 30% for alpha particles and 0.01% for beta particles.

9.11.7 4-hydrocyanobenzene (4HCB)

Ciavatti *et al.* [257] investigated the suitability of the conductive polymer 4-hydrocyanobenzene (4HCB) for charged-particle spectroscopy. As well as being environmentally robust, 4HCB was chosen because single crystals have already demonstrated a photocurrent response and optimized charge collection efficiency for X-ray illumination [258]. Single crystals of 4HCB were grown from solution. Two detectors of thickness 200 μm and diameter 2–4 mm^2 were fabricated. Sputtered Au electrodes were deposited on the crystals forming a MSM device structure. Two electrode configurations were produced: the first a simple planar sandwich configuration (anode on top, cathode on bottom) and the second a so-called co-planar configuration that essentially consisted of anode and cathode pad electrodes on the top crystal surface. The current-voltage characteristics of both were found to be ohmic with measured resistivities in the range (10^{10}–10^{11}) Ω-cm. The detectors were irradiated on their top surfaces with a ^{241}Am alpha source. Both detectors were found to be responsive, with a well-resolved full energy peak of FWHM ~70% (see Fig. 9.27). The peak channel was found to vary with bias and from a subsequent Hecht analysis, mobility–lifetime products for both detectors were determined to be 1.9 × 10^{-6} cm^{2}s^{-1} for the sandwich detector and 5.5 × 10^{-6} cm^{2}s^{-1} for the coplanar detector. In both detectors, the alpha particles are stopped close to the crystal surface. Thus, both electrons and holes are collected by the two surface electrode system of the co-planar detector but only holes for the sandwich configuration. For the latter configuration, it was possible to measure the hole drift velocity as a function of bias using a time of flight technique, from which a hole mobility of (7.4±0.3) × 10^{-4} cm^{2}V^{-1}s^{-1} was deduced.

9.12 Hybrid Organic-Inorganic Semiconductors

In order to mitigate some of the shortcomings of organic semiconductors, a new class of semiconductor that combines the best attributes of both the organic and inorganic semiconductors is being actively investigated. Conventional radiation detectors are fabricated from inorganic materials because of their robust physical and electronic properties. However, the use of these materials is also characterized by expensive processing and fabrication processes. Organic materials, on the other hand, are less robust but are easily and inexpensively fabricated by means of spin coating and printing. Another advantage of the organic semiconductors, over say Si, is that they can have a direct bandgap, which makes them more suitable for optoelectronic applications. Organic-inorganic hybrid materials combine the robustness of inorganics with the ease of production of organics. For this reason, the organic-inorganic hybrids have been actively studied in the search for high mobility soluble semiconductors.

9.12.1 Hybrid Organic-Inorganic Perovskites

Hybrid organic-inorganic perovskites (HOIPs) are crystals with the structural formula ABX_3, where A, B and X are organic and inorganic ions, respectively. They are most commonly synthesized by combining a metal salt with an organic halide salt in a single process, either by spin-coating from a solution of both salts or by co-evaporation. They have emerged in recent years as an extremely promising semiconducting material for solar energy applications. In particular, power conversion efficiencies of HOIP-based solar cells have improved significantly over the last decade surpassing 20%. The outstanding performance in photovoltaic devices is made possible by a combination of long charge-carrier diffusion lengths and carrier lifetimes and high $\mu\tau$ products. However, long lifetimes imply slow recombination and low trapping probabilities but do not automatically imply high mobilities, since these are limited by scattering.

9.12.1.1 Methylammonium Lead Halides

Methylammonium lead halides (MALHs) are solid compounds with a perovskite structure and the chemical formula $CH_3NH_3PbX_3$ ($MAPbX_3$), where X = I, Br or Cl. At present, they are being explored for solar cell[29] [259], optoelectronic [260] and magneto-optical data storage [261] and other applications [262]. For single crystals, mobilities of ~1 cm^2 $V^{-1}s^{-1}$ and carrier lifetimes in excess of 15 μs have been reported [263,264]. When coupled with the high atomic numbers of the constituent metal Pb and halides I, Br and Cl, MALHs are ideal candidates as X-ray and gamma-ray detection media. Náfrádi *et al.* [265] reported photocurrent generation by X-ray irradiation of methylammonium lead iodide $MAPbI_3$ single crystals. In millimeter-sized samples prepared by precipitation from aqueous solution, they measured a 75% charge collection efficiency when exposed to 20–35 keV X-rays from a Mo X-ray tube. More importantly, the material was stable against X-ray radiation with dose, leading the authors to propose direct electricity production in highly energetic radiation environments.

Yakunin *et al.* [266] demonstrated that 0.3–1 cm solution-grown single crystals (SCs) of semiconducting hybrid lead halide perovskites ($MAPbI_3$ and I-treated $MAPbBr_3$,) can serve as solid-state gamma-detecting media. Simple planar detectors were formed by contacting the crystals on the top and bottom surfaces with silver paste. Using Cu Ka 8 keV X-ray excitation, the $\mu\tau$ product of $MAPbI_3$ was determined from a Hecht analysis to be 1×10^{-2} cm^2V^{-1} comparable with CdZnTe and consistent with previously reported results obtained for $MAPbBr_3$ single crystals [262]. The noise in the detector was low enough for single photon counting, with individual pulses from [241]Am and ^{137}Cs observable on an oscilloscope. When cooled to ~220 K, the iodine-treated $MAPbBr_3$ showed a spectroscopic response to 59.95 keV gamma-rays from a [241]Am radiation source, although the 60 keV peak could not be fully resolved.

Wei *et al.* [267] grew single crystals of methylammonium lead bromide ($MAPbBr_3$) perovskite by the anti-solvent method. The crystals had areas of ~80 mm^2 and thicknesses of 2–3 mm. Simple planar detectors were fabricated by applying contacts to the top and bottom surfaces by thermal evaporation. The anode consisted of a 25 nm Au layer applied to the top surface, and the cathode was a layered structure consisting of 20 nm of C^{60}, 8 nm BCP (2,9-dimethyl-4,7-diphenyl-1,10-phenanthroline) and a 14 nm Ag or 25 nm Au layer. The I-V characteristics showed small dark-current density of 29 nA cm^{-2} at −0.1V bias, which gives a bulk resistivity of 1.7×10^7 Ω-cm. The hole carrier mobilities were measured by a time-of-flight method to be ~217 $cm^2V^{-1}s^{-1}$. The $\mu\tau$ product of the single crystals was measured by the photoconductivity method to be 1.0×10^{-2} cm^2V^{-1} and was determined to be dominated by the contribution from holes. Photocurrent measurements showed that the devices were sensitive to X-rays with a measured efficiency of 16.4% to continuum X-ray energies up to 50 keV. When compared to α-Se X-ray plate detectors (which is a standard for medical X-ray imaging), the sensitivities of the $MAPbBr_3$ single crystal detectors were found to be four times lower.

In later work, Wei *et al.* [268] noted that $MAPbBr_3$ and $MAPbCl_3$ show p- and n- type conduction, respectively, and argued that by alloying Cl with $MAPbBr_3$ to produce $MAPbBr_{3-x}Cl_x$, compensated material can be produced. In fact, they found that near-intrinsic material can be produced for a chlorine fraction of around 6%, increasing the bulk resistivity ten-fold, from 2×10^8 Ω-cm to 4×10^9 Ω-cm. In addition, both the electron and hole the mobilities increased. In the case of electrons, from 140 cm^2 $V^{-1}s^{-1}$ for x = 0 to 340 cm^2 $V^{-1}s^{-1}$ for x = 0.06, and in the case of holes, from 220 cm^2 $V^{-1}s^{-1}$ for x = 0 to 560 cm^2 $V^{-1}s^{-1}$ for x = 0.06. Above x = 0.06, the mobilities decrease with increasing chlorine fraction, approaching that of $MAPbCl_3$ (*i.e.*, μ_e = 62 cm^2 $V^{-1}s^{-1}$, μ_h=145 cm^2 $V^{-1}s^{-1}$) for x~20%. At x = 0.06, the $\mu\tau$ for holes was determined to be 1.8×10^{-2} cm^2V^{-1}. A number of planar detectors were fabricated of area ~1 cm^2 and ~0.5 cm thick and were found to be spectrally responsive to 662 keV gamma-rays from a ^{137}Cs radioactive source. Typical FWHM energy resolutions of 12% were achieved. In Fig. 9.28 we showed the best recorded ^{137}Cs spectrum with a measured FWHM energy resolution of 6.5%.

---

[29] In fact, Sha *et al.* [257] have shown that the theoretical efficiency limit of perovskite cells is about 31%, comparable with that achieved with GaAs.

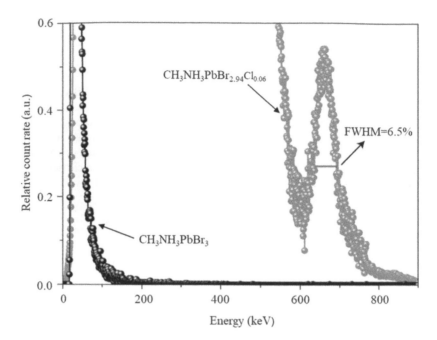

FIGURE 9.28 ^{137}Cs spectra recorded with a MAPbBr$_3$ planar detector and a dopant compensated MAPbBr$_{3-x}$Cl$_x$ planar detector of size 1.44 × 1.37 × 0.58 cm^3 (adapted from [268]). The applied bias is ~10V. A clear peak at 662 keV (FWHM = 6.5%) is apparent in the compensated detector. No peak or Compton edge could be discerned in the uncompensated detector.

For comparison we also show the spectra from an uncompensated MAPbBr$_3$ planar detector for which no peak or Compton edge is apparent.

9.12.1.2 Formamidinium Lead Halides

Formamidinium (FA)-based lead halides (FALHs) are perovskites with the chemical formula H$_2$NCHNH$_2$PbX$_3$ (simplified to FAPbX$_3$), where X = I, Br or Cl. They are similar to MALHs but are more chemically and thermally robust and have a comparatively narrow bandgap, which means broader solar absorption spectrum. This has facilitated the fabrication of photovoltaic devices with power conversion efficiencies in excess of 20% [269]. FALHs, however, have one major short-coming; they are thermodynamically unstable, undergoing a phase change from the desired narrow bandgap 3D cubic form to wide bandgap 1D hexagonal form, on a timescale of a day.

Wei *et al.* [267] grew 3 mm single crystals of FAPbI$_3$ from solution. A simple planar detector was formed by contacting the crystal on the top and bottom surfaces with silver paste. The noise in the detector was low enough for single photon

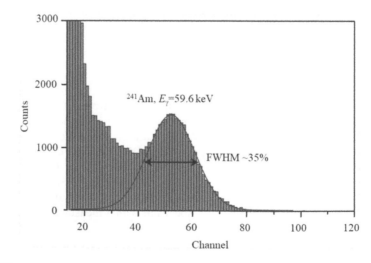

FIGURE 9.29 Energy-resolved ^{241}Am spectrum measured with a FAPbI$_3$ 3 mm single crystal detector at room temperature and a bias of 23V (from [267]). The measured FWHM energy resolution at 60 keV was ~35%.

counting, with individual proportional pulses from ^{241}Am and ^{137}Cs observable on an oscilloscope. The detector showed a spectroscopic response to 59.95 keV gamma-rays from a ^{241}Am radiation source. The measured room-temperature energy spectrum obtained with the detector at a bias of 23V is shown in Fig. 9.29. The measured energy resolution at 60 keV is ~35% FWHM. In an attempt to solve the instability problem of FAPbI$_3$, Nazareko *et al.* [270] grew single crystals doped with Cs and Br from solution using the inverse temperature crystallization technique. Substitutional doping with Cs and Br increases the crystalline phase stability, and in fact, Nazareko *et al.* [270] found that the shelf lives of cm-sized single crystals (the time before hexagonal phase impurities could be detected) improved from days to at least several months. To test the gamma-ray response, a simple planar detector was fabricated from a ~25 mm^2, 2 mm thick single crystal of composition Cs$_{0.1}$FA$_{0.9}$PbI$_{2.8}$Br$_{0.2}$ and contacted on the top and bottom surfaces by silver paste. From photocurrent measurements, a $\mu\tau$ of 4×10^{-2} cm^{2}V^{-1} was obtained. The detector was spectrally responsive to gamma-rays from ^{241}Am and ^{137}Cs radioactive sources, but in both cases the photopeaks could not be fully resolved. The long-term stability of the device was at least several months.

9.13 Discussion

Although compound semiconductor radiation detectors have been under development for almost four decades, progress has been incremental. In all cases, this can be directly traced back to a single material issue, namely, the difficulty of producing crystallographically perfect crystals of high purity and exact stoichiometry. For III–V materials, the problems usually occur during the growth process (*e.g.,* defects introduced due to lattice-mismatched substrates), while for II–VI materials detector handling and fabrication techniques can play a significant role in the soft or layered nature of some materials. In Table 9.5, we summarize some of the best reported spectral resolutions at soft and hard X-ray wavelengths for the discussed materials.

TABLE 9.5

Some of the best energy resolutions achieved with simple planar detectors (updated from [271]). Note the figures are quoted for the collection of both carriers. Spectral enhancement techniques involving single carrier collection have not been employed. The measurements were carried out under uniform illumination using ^{55}Fe and ^{241}Am radioactive sources. For completeness, we also list the resolutions at room temperature (RT) where possible, since there are many applications in which resolving power is not a primary requirement.

Material	Detector size area, thickness	ΔE @5.9 keV (eV)	Energy resolution ΔE @ 59.5keV (eV)	References
Si	0.8 mm^2, 500 μm	245 @ $-15°$C 750 @$+15°$C	524 800	Owens *et al.* [4]
Ge	8.0 cm^2, 1 cm	595 @ $-196°$C	670 eV@-196°C	Zevallos-Chevas *et al.* [10]
GaSb	0.03 mm^2, 2 μm	1240 @ $-133°$C	1800 @ $-133°$C	Juang *et al.* [40]
GaAs	0.8 mm^2, 40 μm	450 @ $-22°$C 572 @ RT	670 780	Owens *et al.* [4]
GaAs pixel	$250 \times 250 \times 40$ μm^3	219 @ $-30°$C 266 @ RT	468 487	Owens *et al.* [4]
SiC–4H pixel	0.03 mm^2, 25 μm	196 @ $+30°$C 233 @ $+100°$C	not measured not measured	Bertuccio *et al.* [32]
SiC–4H	0.79 mm^2, 100 μm	not measured	550 @ RT	Phlips *et al.* [35]
InP	3.142 mm^2, 200 μm	911 @ $-170°$C 2480 @ $-60°$C	3050 9200	Owens *et al.* [4]
CdTe	16 mm^2, 1.2 mm	310 @ $-60°$C	600	Loupilov *et al.* [272]
CdZnTe	4 mm^2, 2mm 3.142 mm^2, 2.5 mm	240 @ $-40°$C 1508@ RT	1200@ $-30°$C 2900	Niemela & Silipa [273] Owens *et al.* [110]
CdMgTe	16 mm^2, 2 mm	not measured	7700 @ RT	Trivedi *et al.* [122]
HgI$_2$	5 mm^2, 200 μm	198 @ 0°C	650	Iwanczyk *et al.* [156]
HgBrI	16 mm^2, 1 mm 16 mm^2, 3 mm	860 eV @ RT not measured	not measured 6500 eV @ RT	Gospodinov *et al.* [159] Gospodinov *et al.* [159]
Hg$_2$I$_2$	6 mm^2, 0.5 mm	Not measured	1800 eV @ RT	Chen *et al.* [172]
PbI$_2$	1 mm^2, 50 μm	415 @ RT	1380	Shah *et al.* [197]
TlBr	3.142 mm^2, 800 μm	800 @ $-30°$C 1800 @ RT	2300 3300	Owens *et al.* [179]
BiI$_3$	$8 \times 8 \times 0.5$ mm^3	not measured	4500@ RT	Gokhale *et al.* [208]

REFERENCES

[1] G. Armantrout, S. Swierkowski, J. Sherohman, J. Yee, "What can be expected from high Z semiconductor detectors", *IEEE Trans. Nucl. Sci.*, Vol. **NS–24** (1977), pp. 121–125.

[2] S.E. Saddow, A. Agarwal, *"Advances in Silicon Carbide Processing and Applications"*, Artech House, Boston (2004) ISBN-13: 978-1580537407.

[3] E.M. Pell, "Ion drift in an n-p junction", *J. Appl. Phys.*, Vol. **31** (1960), pp. 291–302.

[4] A. Owens, A. Peacock, M. Bavdaz, "Progress in compound semiconductors", *Proc. SPIE*, Vol. **4851** (2003), pp. 1059–1070.

[5] G. Bertuccio, M. Ahangarianabhari, C. Graziani, D. Macera, Y. Shi, M. Gandola, A. Rachevski, I. Rashevskaya, A. Vacchi, G. Zampa, N. Zampa, P. Bellutti, G. Giacomini, A. Picciotto, C. Piemonte, N. Zorzi, "X-ray silicon drift detector–CMOS front-end system with high energy resolution at room temperature", *IEEE Trans Nucl. Sci.*, Vol. **63**, no. 1 (2016), pp. 400–406.

[6] A. Meuris, F. Aschauer, G. De Vita, B. Guenther, S. Herrmann, T. Lauf, P. Lechner, G. Lutz, P. Majewski, D. Miessner, M. Porro, J. Reiffers, A. Stefanescu, F. Schopper, H. Soltau, L. Strueder, J. Treis, "Development and characterization of new 256 × 256 Pixel DEPFET detectors for X-Ray astronomy", *IEEE Trans. Nucl. Sci.*, Vol. **58**, no. 3 (2011), pp. 1206–1211.

[7] L. Strüder, F. Aschauer, M. Bautz, L. Bombelli, D. Burrows, C. Fiorini, G. Fraser, S. Herrmann, E. Kendziorra, M. Kuster, T. Lauf, P. Lechner, G. Lutz, P. Majewski, A. Meuris, M. Porro, J. Reiffers, R. Richter, A. Santangelo, H. Soltau, A. Stefanescu, C. Tenzer, J. Treis, H. Tsunemi, G. de Vita, J Wilms., "The wide-field imager for IXO: Status and future activities", *Proc. SPIE*, Vol. **7732** (2010), pp. 77321I-1–77321I-9.

[8] K.G. McKay, "A germanium counter", *Phys. Rev.*, Vol. **76** (1949), p. 1537.

[9] K.G. McKay, K.B. McAfee, "Electron multiplication in silicon and germanium", *Phys. Rev.*, Vol. **91** (1953), pp. 1079–1084.

[10] J.Y. Zevallos-Chávez, M.T.F. Da Cruz, M.N. Martins, V.P. Likhachev, C.B. Zamboni, S.P. Camargo, F.A. Genezini, J.A. G. Medeiros, M.M. Hindi, "Response function of a germanium detector to photon energies between 6 and 120 keV", *Nucl. Instr. and Meth.*, Vol. **A457** (2001), pp. 212–219.

[11] P. Bergonzo, D. Tromson, C. Mer, "Radiation detection devices made from CVD diamond", *Semicond. Sci. Tech.*, Vol. **18** (2003), pp. S105–S112.

[12] B. Gudden, R. Pohl, "Das Quantenäquivalent bei der lichtelektrischen Leitung", *Z. Physik*, Vol. **17** (1923), pp. 331–346.

[13] R. Hofstadter, "Crystal counters–I", *Nucleonics*, Vol. **4**, no. 2 (1949), pp. 2–27.

[14] F.P. Bundy, H.T. Hall, H.M. Strong, R.H. Wentorf, "Man-made diamonds", *Nature*, Vol. **176** (1955), pp. 51–55.

[15] P.W. May, "CVD diamond – A new technology for the future?", *Endeavour Magazine*, Vol. **19**, no. 3 (1995), pp. 101–106.

[16] M.H. Nazaré, A.J. Neves (Editors), *"Diamond"*, INSPEC, London, (2001).

[17] W. Adam, *et al.* (the RD42 Collaboration), "Review of the development of diamond radiation sensors", *Nucl Nucl. Instr. and Meth.*, Vol. **434** (1999), pp. 131–145.

[18] G.T. Betzel, S.P. Lansley, F. Baluti, L. Reinisch, J. Meyer, "Clinical investigations of a CVD diamond detector for radiotherapy dosimetry", *Physica Medica*, Vol. **28**, no. 2 (2012), pp. 144–152.

[19] R.J. Tapper, "Diamond detectors in particle physics", *Rep. Prog. Phys.*, Vol. **63** (2000), pp. 1273–1316.

[20] A. Brambilla, D. Tromson, N. Aboud, C. Mer, P. Bergonzo, F. Foulon, "CVD diamond gamma-dose rate monitor for harsh environment", *Nucl. Instr. and Meth.*, Vol. **A458** (2001), pp. 220–226.

[21] B. Górka, B. Nilsson, R. Svensson, A. Brahme, P. Ascarelli, D.M. Trucchi, G. Conte, R. Kalish, "Design and characterization of a tissue-equivalent CVD–Diamond detector for clinical dosimetry in high-energy photon beams", *Physica Medica*, Vol. **24**, no. 3 (2008), pp. 159–168.

[22] A. Mainwood, "CVD diamond particle detectors", *Diamond and Related Materials*, Vol. **7** (1998), pp. 504–509.

[23] J.H. Kaneko, T. Tanaka, T. Imai, Y. Tanimura, M. Katagiri, T. Nishitani, H. Takeuchi, T. Sawamura, T. Iida, "Radiation detector made of a diamond single crystal grown by a chemical vapor deposition method", *Nucl. Instr. and Meth.*, Vol. **A505** (2003), pp. 187–190.

[24] A. Burger, I. Shilo, M. Schieber, "Cadmium selenide: A promising novel room temperature radiation detector", *IEEE Trans. Nuc. Sci.*, Vol. **NS–30** (1983), pp. 368–370.

[25] R.N. Ghosh, R. Loloee, T. Isaacs-Smith, J.R. Williams, "High temperature reliability of SiC n-MOS devices up to 630°C", *Mat. Sci. Forum, Silicon Carbide and Related Materials*, Vols. **527–529** (2006), pp. 1039–1042.

[26] A.R. Verma, P. Krishna, *"Polymorphism and Polytypism in Crystals"*, John Wiley & Sons Inc. New York (1966) ISBN-10: 0471906433.

[27] R.S. Ramsdell, "Studies on silicon carbide", *Am. Mineral.*, Vol. **32** (1947), pp. 64–82.

[28] R.V. Babcock, S.L. Ruby, F.D. Schupp, K.H. Sun, "Miniature neutron detectors", *Westinghouse Elec. Corp., Materials Engineering Report No. 5711–6600–A*, November (1957).

[29] R.V. Babcock, H.C. Chang, "SiC neutron detectors for high temperature operation, neutron dosimetry", *Proceedings of the Symposium on Neutron Detection, Dosimetry and Standardization*, Vol. 1, International Atomic Energy Agency (IAEA), Vienna, December (1962), pp. 613–622.

[30] R.R. Ferber, G.N. Hamilton, "Silicon carbide high temperature neutron detectors for reactor instrumentation", *Westinghouse Research and Development Center*, Document No. 65–1 C2–RDFCT–P3, June (1965).

[31] G. Bertuccio, R. Casiraghi, F. Nava, "Epitaxial silicon carbide for X-ray detection", *IEEE Trans. Nucl. Sci.*, Vol. **NS–48** (2001), pp. 232–233.

[32] G. Bertuccio, R. Casiraghi, A. Cetronio, C. Lanzieri, F. Nava, "Advances in silicon carbide X-Ray detectors", *Nucl. Instr. and Meth.*, **A652** (2010), pp. 193–196.

[33] G. Bertuccio, R. Casiraghi, E. Gatti, D. Maiocchi, F. Nava, C. Canali, A. Cetronio, C. Lanzieri, "SiC X-ray detectors for spectroscopy and imaging in a wide temperature range", *Materials Science Forum*, Vols. **433–436** (2003), pp. 941–944.

[34] G. Bertuccio, R. Casiraghi, A. Cetronio, C. Lanzieri, F. Nava, "Silicon carbide for high resolution X-ray detectors operating up to 100°C", *Nucl. Instr. and Meth.*, Vol. **A522** (2004), pp. 413–419.

[35] B.F. Phlips, K.D. Hobart, F.J. Kub, R.E. Stahlbush, M.K. Das, B.A. Hull, G. De Geronimo, P. O'Connor, "Silicon carbide PiN diodes as radiation detectors", *Materials Science Forum*, Vols. **527–529** (2006), pp. 1465–1468.

[36] S.O. Kasap, J.A. Rowlands, "Direct-conversion flat-panel X-ray image", Sensors for Digital Radiography", *Proc. IEEE*, Vol. **149**, no. 2 (2002), pp. 591–604

[37] H.K. Kim, I.A. Cunningham, Z. Yin, G. Cho, "On the development of digital radiography detectors: A review", *Int. J. Of Prec. Eng. and Man.*, Vol. **9**, no. 4 (2008), pp. 86–100.

[38] A.G. Milnes, A.Y. Polyakov, "Gallium antimonide device related properties", *Solid State Electron.*, Vol. **36** (1993), pp. 803–938.

[39] P.S. Dutta, H.L. Bhat, "The physics and technology of gallium antimonide: An emerging optoelectronic material", *J. Appl. Phys.*, Vol. **81** (1997), pp. 5821–5870.

[40] B.-C. Juang, D.L. Prout, B. Liang, A.F. Chatziioannou, D.L. Huffaker, "Characterization of GaSb photodiode for gamma-ray detection", *Appl. Phys. Express*, Vol. **9** (2016), pp. 086401-1–086401-4.

[41] A. Owens, M. Bavdaz, A. Peacock, A. Poelaert, H. Andersson, S. Nenonen, L. Tröger, G. Bertuccio, "High resolution X-ray spectroscopy using GaAs arrays", *J. Appl. Phys.*, Vol. **90** (2001), pp. 5367–5381.

[42] W.R. Harding, C. Hilsum, M.E. Moncaster, D.C. Northrop, O. Simpson, "Gallium arsenide for γ–ray spectroscopy", *Nature*, Vol. **187** (1960), pp. 405.

[43] J.E. Eberhardt, R.D. Ryan, A.J. Tavendale, "High resolution radiation detectors from epitaxial n–GaAs", *Appl. Phys. Lett.*, Vol. **17** (1970), pp. 427–429.

[44] J.E. Eberhardt, R.D. Ryan, A.J. Tavendale, "Evaluation of epitaxial n–GaAs for nuclear radiation detection", *Nucl. Instr. and Meth.*, Vol. **94** (1971), pp. 463–476.

[45] D.S. McGregor, H. Hermon, "Room–Temperature compound semiconductor radiation detectors", *Nucl. Instr. and Meth.*, Vol. **395** (1997), pp. 101–124.

[46] A. Owens, M. Bavdaz, S. Kraft, A. Peacock, R. Strade, S. Nenonen, H. Andersson, M. A. Gagliardi, T. Gagliardi, H. Graafsma, "Synchrotron characterization of deep depletion epitaxial GaAs detectors", *J. Appl. Phys.*, Vol. **86** (1999), pp. 4341–4347.

[47] A. Owens, H. Andersson, M. Campbell, D. Lumb, S. Nenonen, L. Tlustos, "GaAs arrays for X-ray spectroscopy", *Proc. SPIE*, Vol. **5501** (2004), pp. 241–248.

[48] M. Campbell, H.M. Heijne, G. Meddeler, E. Pernigotti, W. Snoeys, "Readout for a 64×64 pixel matrix with 15–Bit single photon counting", *IEEE Trans. Nucl. Sci.* Vol. **45** (1998), pp. 751–753.

[49] L.R. Weisberg, B. Goldstein, "GaAs and GaP for room temperature gamma–Ray counters", *Nucleonics in Aerospace*, P. Polishuk (ed.), New York, Plenum Press (1968), pp. 182–186.

[50] P. Litovchenko, D. Bisello, A. Litovchenko, S. Kanevskyj, V. Opilat, M. Pinkovska, V. Tartachnyk, R. Rando, P. Giubilato, V. Khomenkov, "Some features of current–Voltage characteristics of irradiated GaP light diodes", *Nucl. Instr. and Meth.*, Vol. **A552** (2005), pp. 93–97.

[51] A. Owens, S. Andersson, R. Den Hartog, F. Quarati, A. Webb, E. Welter, "Hard X-ray detection with a GaP Schottky diode", *Nucl. Instr. and Meth.*, Vol. **A581** (2007), pp. 709–712.

[52] J. Wang, P. Mulligan, L. Brillson, L.R. Cao, "Review of using gallium nitride for ionizing radiation detection", *Appl. Phys. Rev.*, Vol. **2** (2015), pp. 031102-1–031102-12.

[53] J. Vaitkus, W. Cunningham, E. Gaubas, M. Rahman, S. Sakai, K.M. Smith, T. Wang, "Semi–Insulating GaN and its evaluation for a particle detection", *Nucl. Instr. and Meth.*, Vol. **509** (2003), pp. 60–64.

[54] A. Owens, A. Barnes, R.A. Farley, M. Germain, P.J. Sellin, "GaN detector development for particle and X-ray detection", *Nucl. Instr. and Meth.*, Vol. **A695** (2012), pp. 303–305.

[55] J.Y. Duboz, M. Lauegt, D. Schenk, B. Beaumont, J.L. Reverchon, A.D. Wieck, T. Zimmerling, "GaN for x-ray detection", *Appl. Phys. Lett.*, Vol. **92**, no. 26 (2008), pp. 263501-1–263501-3.

[56] Y. Duboz, E. Frayssinet, S. Chenot, J.L. Reverchon, M. Idir, "X-ray detectors based on GaN Schottky diodes", *Appl.Phys. Lett.*, Vol. **97**, no. 16 (2010), pp. 163504-1–163504-3.

[57] T. Gohil, J. Whale, G. Lioliou, S.V. Novikov, C.T. Foxon, A.J. Kent, A.M. Barnett, "X-ray detection with zinc-blende (cubic) GaN Schottky diodes", *Scientific Reports*, Vol. **6** (2016), pp. 29535-1–29535-5.

[58] C.S. Yao, K. Fu, G. Wang, G.H. Yu, M. Lu, "GaN-based p–I–N X-ray detection", *Phys. Status Solidi A*, Vol. **209**, no. 1 (2012), pp. 204–206.

[59] F.J. Leonberger, P.F. Moulton, "High–Speed InP optoelectronic switch", *Appl. Phys. Lett.*, Vol. **35** (1979), pp. 712–714.

[60] A.G. Foyt, F.J. Leonberger, R.C. Wiamson, "Picosecond InP optoelectronic switches", *Appl. Phys. Lett.*, Vol. **40** (1982), pp. 447–449.

[61] J. Lund, F. Olscher, F. Sinclair, M.R. Squillante, "Indium phosphide particle detectors for low energy solar neutrino spectroscopy", *Nucl. Instr. and Meth.*, Vol. **A272** (1988), pp. 885–888.

[62] T.F. Deutsch, F.J. Leonberger, A.G. Foyt, D. Mills, "High–Speed ultraviolet and X-ray–Sensitive InP photoconductive detectors", *Appl. Phys. Lett.*, Vol. **41** (1982), pp. 403–405.

[63] D. Kania, R. Bartlett, R. Wagner, R. Hammond, "Pulsed soft X-ray response of InP:Fe photoconductors", *Appl. Phys. Lett.*, Vol. **44** (1984), pp. 1059–1061.

[64] Y. Suzuki, Y. Fukuda, Y. Nagashima, "An indium phosphide solid state detector", *Nucl. Instr. and Meth.*, Vol. **A275** (1989), pp. 142–148.

[65] F. Olschner, J.C. Lund, M.R. Squillante, D.L. Kelly, "Indium phosphide particle detectors", *IEEE Trans. Nucl. Sci.*, Vol., **NS–36** (1989), pp. 210–212.

[66] P. Jayavel, S. Ghosh, A. Jhingan, D.K. Avasthi, K. Asokan, J. Kumar, "Study on the performance of SI–GaAs and SI–InP surface barrier detectors for alpha and gamma detection", *Nucl. Instr. and Meth.*, Vol. **A454** (2000), pp. 252–256.

[67] F. Dubecký, B. Zaťko, V. Nečas, M. Sekáčová, R. Fornari, E. Gombia, P. Boháček, M. Krempaský, P.G. Pelfer, "Recent improvements in detection performances of radiation detectors based on bulk semi–Insulating InP", *Nucl. Instr. and Meth.*, Vol. **A487** (2002), pp. 27–32.

[68] A. Owens, M. Bavdaz, V. Gostilo, D. Gryaznov, A. Loupilov, A. Peacock, H. Sipila, "The X-ray response of InP", *Nucl. Instr. and Meth.*, Vol. **A487** (2002), pp. 435–440.

[69] V. Gorodynsky, K. Zdansky, L. Pekarek, V. Malina, S. Vackova, "Ti and Mn co-doped semi-insulating InP particle detectors operating at room temperature", *Nucl. Instr. and Meth.*, Vol. **555** (2005), pp. 288–293.

[70] R. Yatskiv, K. Zdansky, L. Pekarek, "Room-temperature particle detectors with guard rings based on semi-insulating InP co-doped with Ti and Zn", *Nucl. Instr. and Meth.*, Vol. **598** (2009), pp. 759–763.

[71] P.G. Pelfer, F. Dubecky, R. Fornari, M. Pikna, E. Gombia, J. Darmo, M. Krempaský, M. Sekácová, "Present status and perspectives of the radiation detectors based on InP materials", *Nucl. Instr. and Meth.*, Vol. **A458** (2001), pp. 400–405.

[72] K. Zdansky, L. Pekarek, P. Kacerovsky, "Evaluation of semi-insulating Ti-doped and Mn-doped InP for radiation detection", *Semi. Sci. Tech.*, Vol. **16**, 12 (2001), pp. 1002–1007.

[73] M.R. Squillante, C. Zhou, J. Zhang, L.P. Moy, K.S. Shah, "InI nuclear radiation detectors", *IEEE Trans. Nucl. Sci.*, Vol. **40** (1993), pp. 364–366.

[74] T. Onodera, K. Hitomi, T. Shoji, "Fabrication of indium iodide X– and Gamma–ray detectors", *IEEE Trans. Nucl. Sci.*, Vol. **52** (2006), pp. 2056–2059.

[75] P. Bhattacharya, M. Groza, Y. Cui, D. Caudel, T. Wrenn, A. Nwankwo, A. Burger, G. Slack, A.G. Ostrogorsky, "Growth of InI single crystals for nuclear detection applications", *J. Cryst. Growth*, Vol. **312**, 8 (2010), pp. 1228–1232.

[76] A.W. Hoffman, E. Corrales, P. Love, "2K × 2K InSb for astronomy", *Proc. SPIE*, Vol. **5499** (2004), pp. 59–67.

[77] G. Finger, J.W. Beletic, "Review of the state of infrared detectors for astronomy in retrospect of the June 2002 workshop on scientific detectors for astronomy", *Proc. SPIE*, Vol. **4841** (2003), pp. 839–852.

[78] A. Säynätjoki, P. Kostamo, J. Sormunen, J. Riikonen, A. Lankinen, H. Lipsanen, H. Andersson, K. Banzuzi, S. Nenonen, H. Sipilä, S. Vaijärvi, D. Lumb, "InAs pixel matrix detectors fabricated by diffusion of Zn utilising metal–Organic vapour phase epitaxy", *Nucl. Instr. and Meth.*, Vol. **A563** (2006), pp. 24–26.

[79] W.C. Harris, "InSb as a γ–Ray detector", *Nucl. Instr. and Meth.*, Vol. **242** (1986), pp. 373–375.

[80] I. Kanno, F. Yoshihara, R. Nouchi, O. Sugiura, T. Nakamura, M. Katagiri, "Cryogenic InSb detector for radiation measurement", *Rev. Sci. Instr.*, Vol. **73** (2002), pp. 2533–2536.

[81] I. Kanno, S. Hishiki, O. Sugiura, R. Xiang, T. Nakamura, M. Katagiri, "InSb cryogenic radiation detectors", *Nucl. Instr. and Meth.*, Vol. **A568** (2006), pp. 416–420.

[82] Y. Sato, Y. Morita, T. Harai, I. Kanno, "Photopeak detection by an InSb radiation detector made of liquid phase epitaxially grown crystals", *Nucl. Instr. and Meth.*, Vol. **A621** (2010), pp. 383–386.

[83] Y. Sato, K. Watanabe, A. Yamazaki, I. Kanno, "Charge collection process of a liquid-phase epitaxially grown InSb detector", *Jpn. J. Appl. Phys.*, Vol. **50** (2011), pp. 096401–096405.

[84] J. Lauter, D. Protić, A. Förster, H. Lüth, "AlGaAs/GaAs SAM-avalanche photodiode: An X-ray detector for low energy photons", *Nucl. Instr. and Meth.*, Vol. **A356** (1995), pp. 324–329.

[85] J.E. Lees, D.J. Bassford, J.S. Ng, C.H. Tan, J.P.R. David, "AlGaAs diodes for X-ray spectroscopy", *Nucl. Instr. Meth.*, Vol. **A594** (2008), pp. 202–205.

[86] A.M. Barnett, D.J. Bassford, J.E. Lees, J.S. Ng, C.H. Tan, J.P.R. David, "Temperature dependence of AlGaAs soft X-ray detectors", *Nucl. Instr. and Meth.*, Vol. **A621** (2010), pp. 453–455.

[87] M.D.C. Whitaker, G. Lioliou, S. Butera, A.M. Barnett, "$Al_{0.2}Ga_{0.8}As$ X-ray photodiodes for X-ray spectroscopy", *Nucl. Instr. and Meth.*, Vol. **A840** (2016), pp. 168–173.

[88] M.D.C. Whitaker, S. Butera, G. Lioliou, A.M. Barnett, "Temperature dependence of $Al_{0.2}Ga_{0.8}As$ X-ray photodiodes for X-ray spectroscopy", *J. Appl. Phys.*, Vol. **122**, (2017), pp. 034501-1–034501-11.

[89] S. Butera, G. Lioliou, A.B. Krysa, A.M. Barnett, "Characterisation of $Al_{0.52}In_{0.48}P$ mesa p-i-n photodiodes for X-ray photon counting spectroscopy", *J. Appl. Phys.*, Vol. **120** (2016), pp. 024502-1–024502-6.

[90] S. Butera, T. Gohil, G. Lioliou, A.B. Krysa, A.M. Barnett, "Temperature study of $Al_{0.52}In_{0.48}P$ detector photon counting X-ray spectrometer", *J. Appl. Phys.*, Vol. **120** (2016), pp. 174503-1–174503-6.

[91] A. Auckloo, J.S. Cheong, X. Meng, C.H. Tan, J.S. Ng, A. Krysa, R.C. Tozer, J.P.R. David, "$Al_{0.52}In_{0.48}P$ avalanche photodiodes for soft X-ray Spectroscopy", *Jinst*, Vol. **11** (2016), pp. P03021-1– P03021-5.

[92] S. Butera, G. Lioliou, A.B. Krysa, A.M. Barnett, "InGaP (GaInP) mesa p-i-n photodiodes for X-ray photon counting spectroscopy", *Scientific Reports*, Vol. 1 (2017), pp. 10206-1–10206-8.

[93] R.M. Park, M.B. Trofer, C.M. Rouleau, J.M. Depuydt, M.A. Haase, "P-Type ZnSe by nitrogen atom beam doping during molecular beam epitaxial growth", *Appl. Phys. Lett.*, Vol. **57** (1990), pp. 2127–2129.

[94] W. Akutagawa, K. Zanio, J. Mayer, "CdTe as a gamma–Detector", *Nucl. Lnstr. and Meth.*, Vol. **55** (1967), pp. 383–385.

[95] T. Takahashi, S. Watanabe, "Recent progress in CdTe and CdZnTe detectors", *IEEE Trans Nucl. Sci.*, Vol. **48** (2000), pp. 950–959.

[96] T. Takahashi, T. Mitani, Y. Kobayashi, M. Kouda, G. Sato, S. Watanabe, K. Nakazawa, Y. Okada, M. Funaki, R. Ohno, K. Mori, "High–resolution Schottky CdTe diode detector", *IEEE Trans. Nucl. Sci.*, Vol. **49** (2002), pp. 1297–1303.

[97] A. Niemela, H. Sipila, V.I. Ivanov, "High–resolution p–I–N CdTe and CdZnTe X-ray detectors with cooling and rise–Time discrimination", *IEEE Trans. Nucl. Sci.*, Vol. **43** (1996), pp. 1476–1480.

[98] A. Khusainov, J.S. Iwanczyk, B.E. Patt, A.M. Pirogov, D.T. Voa, P.A. Russo, "Approaching cryogenic Ge performance with Peltier cooled CdTe", *Proc. SPIE*, Vol. **4507** (2001), pp. 50–56.

[99] A. Khusainov, R. Arlt, P. Siffert, "Performance of a high resolution CdTe and CdZnTe P–I–N detectors", *Nucl. Instr. and Meth.*, Vol. **A380** (1996), pp. 245–251.

[100] M. Mahdavi, K.L. Giboni, S. Vajda, J.S. Schweitzer, J.A. Truax, "First Year PIDDP Report on gamma-ray and X-ray spectroscopy X-ray remote sensing and in situ spectroscopy for planetary exploration missions and gamma-ray remote sensing and in situ spectroscopy for planetary exploration missions", *NASA document ID 19950009501* (1994).

[101] M. Roth, "Advantages and limitations of cadmium selenide room temperature gamma–ray ray detectors", *Nucl. Instr. and Meth*, Vol. **A283** (1989), pp. 291–298.

[102] H. Chen, M. Hayes, X. Ma, Y.–F. Chen, S.U. Egarievwe, J.O. Ndap, K. Chattopadhyay, A. Burger, J. Leist, "Physical properties and evaluation of spectrometer grade CdSe single crystal", *Proc. SPIE*, Vol. **3446** (1998), pp. 17–28.

[103] E.E. Eissler, K.G. Lynn, "Properties of melt–Grown ZnSe solid–State radiation detectors", *IEEE Trans. Nucl. Sci.*, Vol. **42** (1995), pp. 663–667.

[104] R.B. James, T.E. Schlesinger, J. Lund, M. Schieber, "Cd$_{1-x}$Zn$_x$Te spectrometers for gamma and X-ray applications", in *Semiconductors for Room Temperature Nuclear Detection Applications*, eds. T.E. Schlesinger, R.B. James, Academic press, New York (1995), pp. 335–384.

[105] A.W. Webb, S.B. Quadri, E.R. Carpenter, E.F. Skelton, "Effects of pressure on Cd$_{1-x}$Zn$_x$Te alloys (0≤ x<0.5)", *J. Appl. Phys.*, **61** (1987), pp. 2492–2494.

[106] J.F. Butler, C.L. Lingren, F.P. Doty, "Cd$_{1-x}$Zn$_x$Te gamma ray detectors", *IEEE Trans. Nucl. Sci.*, Vol. **39** (1992), pp. 605–609.

[107] U. Egarievwe, L. Salary, K.T. Chen, A. Burger, R.B. James, "Performances of CdTe and Cd$_{1-x}$Zn$_x$Te gamma–Ray detectors at elevated temperatures", *Proc. SPIE*, Vol. **2305** (1994), pp. 167–173.

[108] D. Olega, J. Faurie, S. Sivananthan, P. Raccah, "Optoelectronic properties of Cd$_{1-x}$Zn$_x$Te films grown by molecular beam epitaxy on GaAs substrates", *Appl. Phys. Lett.*, Vol. **47** (1985), pp.1172–1174.

[109] J.E. Toney, T.E. Schlesinger, R.B. James, "Optimal bandgap variants of Cd$_{1-x}$Zn$_x$Te for high–resolution X-ray and gamma-ray spectroscopy", *Nucl. Inst. and Meth.*, Vol. **A428** (1999), pp. 14–24.

[110] A. Owens, M. Bavdaz, H. Andersson, T., Gagliardi, M. Krumrey, S. Nenonen, A. Peacock, I. Taylor, "The X-ray response of CdZnTe", *Nucl. Instr. and Meth.*, Vol. **A484** (2002), pp. 242–250.

[111] J. Frey, R. Frey, C. Flytzanis, R. Triboulet, "Theoretical and experimental investigation of nonlinear Faraday processes in diluted magnetic semiconductors", *J. Opt. Soc. Am. B*, Vol. **9**, no. 1 (1992), pp. 132–142.

[112] V.V. Fedorov, W. Mallory, S.B. Mirov, U. Hőmmerich, S.B. Trivedi, W. Palosz, "Iron-doped Cd$_x$Mn$_{1-x}$Te crystals for mid-IR room-temperature lasers", *J. Of Cryst. Gowth*, Vol. **310** (2008), pp. 4438–4442.

[113] A. Burger, K. Chattopadhyay, H. Chen, J.O. Ndap, X. Ma, S. Trivedi, S.W. Kutcher, R. Chen, R.D. Rosemeier, "Crystal growth, fabrication and evaluation of cadmium manganese telluride gamma ray detectors", *J. Of Cryst. Growth*, Vol. **198/199** (1999), pp. 872–876.

[114] A. Mycielski, A. Burger, M. Sowinska, M. Groza, A. Szadkowski, P. Wojnar, B. Witowska, W. Kaliszek, P. Siffert, "Is the (Cd,Mn)Te crystal a prospective material for X-ray and γ-ray detectors?", *Phys. Stat. Sol. (C)*, Vol. **2**, no. 5, (2005), pp. 1578–1585.

[115] R. Triboulet, A. Heurtel, J. Rioux, "Twin-free (Cd,Mn)Te substrates", *J. Cryst. Growth*, Vol. **101** (1990) pp. 131–134.

[116] A. Mycielski, D. Kochanowska, M. Witkowska, R.J. Baran, A. Szadkowski, B. Witkowska, W. Kaliszek, B. Kowalski, A. Reszka, P. Łach, K. Izdebska, A. Suchocki, R. Jakieła, V. Domukhovski, T. Wojtowicz, M. Wiater, M. Węgrzycki, Ł. Kilański, "Studies of (Cd,Mn)Te crystals as a material for X- and gamma-ray detectors: Where we are?", Invited paper presented at the *IEEE NSS/MIC and 17th RTSD workshop*, October 30 - November 6, Knoxville (2010).

[117] J. Parkin, P.J. Sellin, A.W. Davies, A. Lohstroh, M.E. Őzsan, P. Seller, "α Particle response of undoped CdMnTe", *Nucl. Instr. and Meth.*, Vol. **A573** (2007), pp. 220–223.

[118] Y. Cui, A. Bolotnikov, A. Hossain, G. Camarda, A. Mycielski, G. Yang, D. Kochanowska, M. Witkowska-Baran, R. James, "CdMnTe in X-ray and gamma-ray detection: Potential applications", *Proc. SPIE*, Vol. **7079**, SPIE (2008), pp. 70790N-1–70790N-9.

[119] K. Kim, S. Cho, J. Suh, J. Hong, S. Kim, "Gamma-ray response of semi-insulating CdMnTe crystals", *IEEE Trans. Nucl. Sci.*, Vol. **56**, no. 3, no. 2 (2009), pp. 858–862.

[120] "Optical Properties. Part 2", *Landolt-Börnstein - Group III Condensed Matter*, ed. C. Klingshirn, Springer, Berlin, Heidelberg, Vol. **34C2** (2004), pp. 13–48.

[121] A. Hossain, V. Yakimovich, A.E. Bolotnikov, K. Bolton, G.S. Camarda, Y. Cui, J. Franc, R. Gul, K-H Kim, H. Pittman, G. Yang, R. Herpst, R.B. James, "Development of Cadmium Magnesium Telluride ($Cd_{1-x}Mg_xTe$) for room temperature X- and gamma-ray detectors", *J. Cryst. Growth*, Vol. **379** (2013), pp. 34–40.

[122] S.B. Trivedi, S.W. Kutcher, W. Palsoz, M. Berding, A. Burger, *"Next Generation Semiconductor-Based Radiation Detectors Using Cadmium Magnesium Telluride"*, U.S. Department of Energy Final Report, no. DOE/11172015-Final (2014).

[123] A. Burger, M. Roth, M. Schieber, "The ternary $Cd_{0.7}Zn_{0.3}Se$ compound, a novel room temperature X-ray detector", *IEEE Trans. Nuc. Sci.*, Vol. **NS–32** (1985), pp. 556–558.

[124] M. Fiederle, D. Ebling, C. Eiche, D.M. Hofmann, M. Salk, W. Stadler, K. Benz, B.K. Meyer, "Comparison of CdTe, $Cd_{0.9}Zn_{0.1}Te$ and $CdTe_{0.9}Se_{0.1}$ crystals: Application for γ- and X-ray detectors", *J. Cryst. Growth*, Vol. **138** (1994), pp. 529–533.

[125] K. Kim, J. Hong, S.U. Kim, "Electrical properties of semi-insulating $CdTe_{0.9}Se_{0.1}$: Clcrystal and its surface preparation", *J. Cryst. Growth*, Vol. **310**, no. 1 (2008), pp. 91–95.

[126] G.L. Hansen, J.L. Schmit, T.N. Casselman, "Energy gap versus alloy composition and temperature in $Hg_{1-x}Cd_xTe$", *J. Appl. Phys.*, Vol. **53** (1982), pp. 7099–7101.

[127] G. Finger, R.J. Dorn, S. Eschbaumer, D.N.B. Hall, L. Mehrgan, M. Meyer, J. Stegmeier, "Performance evaluation, readout modes and calibration techniques of HgCdTe HAWAII-2RG mosaic arrays", *Proc. SPIE*, Vol. **7021** (2008), pp. 70210P-1–70210P-13.

[128] O. Fox, A. Waczynski, Y. Wen, R.D. Foltz, R.J. Hill, R.A. Kimble, E. Malumuth, B.J. Rauscher, "The ^{55}Fe X-ray energy response of mercury cadmium telluride near-infrared detector arrays", *Pasp*, Vol. **121** (2009), pp. 743–754.

[129] D. Ahn, S.-H. Park, "Cuprous halides semiconductors as a new means for highly efficient light-emitting diodes", *Scientific Reports*, Vol. **6**, no. 20718 (2016), pp.1–23.

[130] W. Lehfeldt, "Ober die elektrische. Leitfahigkeit von Einkristallen", *Z. Physik*, Vol. **85** (1933), pp. 717–726.

[131] L.F. Wouters, R.S. Christian, "Effects of space charge on the detection of high energy particles by means of silver chloride crystal counters", *Phys. Rev.*, Vol. **72** (1947), pp. 1127–1128.

[132] H.O. Curtis, "Fluctuations in the pulse of a silver-chloride crystal counter", PhD dissertation, Harvard University (1948).

[133] P.J. Van Heerden, "The crystal counter: A new instrument in nuclear physics", PhD Dissertation, Rijksuniversiteit Utrecht, July (1945).

[134] P.J. Van Heerden, J.M.W. Milatz, "The crystal counter: A new apparatus in nuclear physics for the investigation of β and γ-rays. Part II", *Physica*, Vol. **16**, 6 (1950), pp. 517–527.

[135] C.B. Childs, L. Slifken, "A new technique for recording heavy primary cosmic radiation and nuclear processes in silver chloride single crystals", *IRE Trans. Nucl. Sci.*, Vol. **9**, 3 (1962), pp. 413–414.

[136] T. Ando, A.B. Fowler, F. Stern, "Electronic properties of two-dimensional systems", *Rev. Mod. Phys.*, Vol. **54** (1982), pp. 437–672.

[137] W. Shi, Y.J. Ding, N. Fernelius, K. Vodopyanov, "Efficient, tunable and coherent 0.18–5.27-THz source based on GaSe crystal," *Opt. Lett.*, Vol. **27** (2002), pp. 1454–1456.

[138] C. Manfredotti, R. Murri, L. Vasanelli, "GaSe as nuclear particle detector", *Nucl. Instr. and Meth.*, Vol. **115** (1974), pp. 349–353.

[139] E. Sakai, H. Nakatani, C. Tatsuyama, F. Takeda, "Average energy needed to produce an electron-hole pair in GaSe nuclear particle detectors", *IEEE Trans. Nucl. Sci.*, Vol. **35**, no. 1 (1988), pp. 85–88.

[140] H. Nakatani, E. Sakai, C. Tatsuyama, F. Takeda, "GaSe nuclear particle detectors", *Nucl. Instr. and Meth.*, Vol. **A283** (1989), pp. 303–309.

[141] T. Yamazaki, H. Nakatani, N. Ikeda, "Characteristics of impurity-doped GaSe radiation detectors", *Jap. J. Appl. Phys.*, Vol. **32**, 4R (1993), pp. 1857–1858.

[142] A. Castellano, "GaSe detectors for x-ray beams", *Appl. Phys. Lett.*, Vol. **48** (1996), pp. 298–299.

[143] K.C. Mandal, M. Choi, S.H. Kang, R.D. Rauh, J. Wei, H. Zhang, L. Zheng, Y. Cui, M. Groza, A. Burger, "GaSe and GaTe anisotropic layered semiconductors for radiation detectors", *Proc. SPIE*, Vol. **6706** (2007), pp. 67060E-1–67060E-10.

[144] V.N. Katerinchuk, M.Z. Kovalyuk, "Gallium telluride heterojunctions", *Tech. Phys. Letts.*, Vol. **25**, no.1 (2007), pp. 54–55.

[145] K.C. Mandal, private communication.

[146] W.R. Willig, "Mercury iodide as a gamma-ray spectrometer", *Nucl. Instr. Meth.*, Vol. **96** (1971), pp. 615–616.

[147] H.L. Malm, "A mercuric iodide gamma–Ray spectrometer", *IEEE Trans. Nucl. Sci.*, Vol. **19** (1972), pp. 263–265.

[148] L. van Den Berg, A.E. Proctor, K.R. Pohl, "Spectral performance of mercuric iodide gamma-ray detectors at elevated temperatures", *Proc. SPIE*, Vol. **5198** (2004), pp. 144–149.

[149] A. Owens, L. Alha, H. Andersson, M. Bavdaz, G. Brammertz, K. Helariutta, A. Peacock, V. Lämsä, S. Nenonen, "The effects of proton–induced radiation damage on compound–semiconductor X-ray detectors", *Proc. SPIE*, Vol. **5501** (2004), pp. 403–411.

[150] B.E. Patt, R.C. Dolin, T.M. Devore, J.M. Markakis, J.S. Iwanczyk, N. Dorri, "Radiation damage resistance in mercuric iodide X-ray detectors", *Nucl. Instr. and Meth.*, Vol. **A299** (1990), pp. 176–181.

[151] I.F. Nicolau, J.P. Joly, "Solution growth of sparingly soluble single crystals from soluble complexes-III. Growth of α–HgI_2 single crystals from dimethylsulfoxide complexes", *J. Cryst. Growth*, Vol. **48** (1980), pp. 61–73.

[152] M. Schieber, W.F. Schnepple, L. van Den Berg, "Vapor growth of HgI$_2$ by periodic source or crystal temperature oscillation", *J. Cryst. Growth*, Vol. **33** (1976), pp. 125–135.

[153] J.S. Iwanczyk, Y.J. Wang, N. Dorri, A.-J. Dabrowski, T.E. Economou, A.L. Turkevich, "Use of mercuric iodide X-ray detectors with alpha backscattering spectrometers for space applications", *IEEE Trans. Nucl. Sci.*, Vol. **NS–38** (1991), pp. 574–579.

[154] A.M. Gerrish, L. van Den Berg, "Perspectives on mercuric iodide as a radiation detector material for space measurements", *Conference on the High Energy Radiation Background in Space, Cherbs, The IEEE Nuclear and Plasma Sciences Society and The Institute of Electrical and Electronic Engineers, Inc.* (1998), pp. 94–98.

[155] A. Owens, M. Bavdaz, G. Brammertz, M. Krumrey, D. Martin, A. Peacock, L. Tröger, "The hard X-ray response of HgI$_2$", *Nucl. Instr. and Meth.*, Vol. **A479** (2002), pp. 535–547.

[156] J.S. Iwanczyk, Y.J. Yang, J.G. Bradley, J.M. Conley, A.L. Albee, T.E. Economou, "Performance and durability of HgI$_2$ X-ray detectors for space missions", *IEEE Trans. Nucl. Sci.*, Vol. **NS–36** (1989), pp. 841–845.

[157] V. Marinova, I. Yanchev, M. Daviti, K. Kyritsi, A.N. Anagnostopoulos, "Electron and hole-mobility of Hg(Br$_x$I$_{1-x}$)$_2$ crystals (x = 0.25, 0.50, 0.75)", *Mat. Res. Bull.*, Vol. **37** (2002), pp. 1991–1995.

[158] K.S. Shah, L. Moy, J. Zhang, F. Olschner, J.C. Lund, M.R. Squillante, "HgBr$_x$I$_{2-x}$ photodetectors for use in scintillation spectroscopy", *Nucl. Instr. and Meth.*, Vol. **A322** (1992), pp. 509–513.

[159] M.M. Gospodinov, D. Petrova, I.Y. Yanchev, M. Daviti, M. Manolopoulou, K.M. Paraskevopoulos, A.N. Anagnostopoulos, E.K. Polychroniadis, "Growth of single crystals of Hg(Br$_x$I$_{1-x}$)$_2$ and their detection capability", *Journal of Alloys and Compounds*, Vol. **400** (2005), pp. 249–251.

[160] J. Sapriel, "Cinnabar (α HgS), a promising acousto-optical material", *Appl. Phys. Lett.*, Vol. **19** (1971), pp. 533–535.

[161] K.A. Higginson, M. Kuno, J. Bonevich, S.B. Qadri, M. Yousuf, H. Mattoussi, "Synthesis and characterization of colloidal β–HgS quantum dots", *J. Phys. Chem.*, Vol. **B106** (2002), pp. 9982–9985.

[162] I. Chakraborty, D. Mitra, S.P. Moulik, "Spectroscopic studies on nano-dispersions of CdS, HgS, their core–shells and composites prepared in micellar medium", *J. Nanopart. Res.*, Vol. **7** (2005), pp. 227–236.

[163] S.V. Kershaw, M. Harrison, A.L. Rogach, A. Kornowski, "Development of IR-emitting colloidal II–VI quantum-dot materials", *Proc. IEEE, J. Sel. Top. Quantum Electron.*, Vol. **6** (2000), pp. 534–543.

[164] G.G. Roberts, E.L. Lind, E.A. Davis, "Photoelectronic properties of synthetic mercury sulphide crystals", *J. Phys. Chem. Solids*, Vol. **30** (1969), pp. 833–844.

[165] F. Virot, R. Hayn, M. Richter, J. van Den Brink, "Metacinnabar (β-HgS): A strong 3D topological insulator with highly anisotropic surface states", *J. Phys. Review Lett.*, Vol. **106** (2011), pp. 236806-1–236806-4.

[166] M.R. Squillante, W.M. Higgins, H. Kim, L. Cirignano, G. Ciampi, A. Churilov, K. Shah, "HgS: A rugged, stable semiconductor radiation detector material", *Proc. SPIE*, Vol. **7449** (2009), pp. 74491U-1–74491U-6.

[167] M.D. Tabak, G.G. Roberts, "Electron-drift mobility in single crystal HgS", *J. Appl. Phys.*, Vol. **39** (1968), pp. 4873–4874.

[168] L.I. Berger, "*Semiconductor Materials*", Physical Sciences References, CRC Press, Boca Raton, FL (1997).

[169] H. Kim, L. Cirignano, A. Churilov, G. Ciampi, A. Kargar, W. Higgins, P. O'Dougherty, S. Kim, M.R. Squillante, K. Shah, "Continued development of room temperature semiconductor nuclear detectors", *Proc. SPIE*, Vol. **7806** (2010), pp. 780604-1–780604-13.

[170] C.L. Thrall, "Alternative wide-band-gap materials for gamma-ray spectroscopy", PhD thesis, The University of Michigan (2013).

[171] M.J. Weber, "*The Handbook of Optical Materials*", CRC Press, Laser and Optical Science and Technology Series, Boca Raton (2002) ISBN-13 9780849335129.

[172] H. Chen, J.-S. Kim, P. Amarasinghe, W. Palosz, F. Jin, S. Trivedi, A. Burger, J.C. Marsh, M.S. Litz, P.S. Wijewarnasuriya, N. Gupta, J. Jensen, J. Jensen, "Novel semiconductor radiation detector based on mercurous halides", *Proc. SPIE*, Vol. **9593** (2015), pp. 95930G-1–95930G-11.

[173] I.U. Rahman, W.A. Fisher, R. Hofstadter, J. Shen, "Behavior of thallium bromide conduction counters", *Nucl. Instr. and Meth.*, Vol. **A261**, (1987), pp. 427–439.

[174] K.S. Shah, F. Olschner, L.P. Moy, J.C. Lund, W.R. Squillante, "Characterization of thallium bromide nuclear detectors", *Nucl. Inst. and Meth.*, Vol. **A299** (1990), pp. 57–59.

[175] F. Olscher, K. Shah, J. Lund, J. Zhang, K. Daley, S. Medrick, W.R. Squillante, "Thallium bromide semiconductor X–ray and γ–ray detectors", *Nucl. Instr. and Meth.*, Vol. **A322** (1992) pp. 504–508.

[176] K. Shah, J. Lund, F. Olschner, L. Moy, M. Squillante, "Thallium bromide radiation detectors", *IEEE Trans. Nucl. Sci.*, Vol. **36** (1989) pp. 199–202.

[177] K. Hitomi, T. Murayama, T. Shoji, T. Suehiro, Y. Hiratate, "Improved spectrometric characteristics of thallium bromide nuclear radiation detectors", *Nucl. Instr. and Meth.*, Vol. **A428** (1999) pp. 372–378.

[178] K. Hitomi, O. Muroi, T. Shoji, T. Suehiro, Y. Hiratate, "Room temperature X– and gamma–ray detectors using thallium bromide crystals", *Nucl. Instr. and Meth.*, Vol. **A436** (1999) pp. 160–164.

[179] A. Owens, M. Bavdaz, G. Brammertz, V. Gostilo, H. Graafsma, A. Kozorezov, M. Krumrey, I. Lisjutin, A. Peacock, A. Puig, H. Sipila, S. Zatoloka, "The X–ray response of TlBr", *Nucl. Instr. and Meth.*, Vol. **A497** (2003) pp. 370–380.

[180] A.K. Shukla, S. Radmas, C.N.R. Rao, "Formation energies of Schottky and Frenkel defects in thallium halides", *J. Phys. Chem. Solids*, Vol. **34**, no. 4 (1973) pp. 761–764.

[181] L.F. Voss, A.M. Conway, R.T. Graff, P.R. Beck, R.J. Nikolic, A.J. Nelson, S.A. Payne, H. Kim, L. Cirignano, K. Shah, "Surface processing of TlBr for improved gamma spectroscopy", *IEEE Nucl. Sci. Symp. Conference Record*, NSS/MIC (2010) pp. 3746–3748.

[182] A. Kozorezov, V. Gostilo, A. Owens, F. Quarati, M. Shorohov, M.A. Webb, J.K. Wigmore, "Polarization effects in thallium bromide x-ray detectors", *J. Appl. Phys.*, Vol. **108** (2010) pp. 064507-1–064507-10.

[183] J. Vaitkus, J. Banys, V. Gostilo, S. Zatoloka, A. Mekys, J. Storasta, A. Žindulis, "Influence of electronic and ionic processes on electrical properties of TlBr crystals", *Nucl. Instr. and Meth.*, Vol. **546** (2005) pp. 188–191.

[184] V. Kozlov, M. Kemell, M. Vehkamaki, M. Leskela, "Degradation effects in TlBr single crystals under prolonged bias voltage", *Nucl. Instr. and Meth.*, Vol. **A576** (2007) pp. 10–14.

[185] K. Hitomi, T. Shoji, Y. Niizeki, "A method for suppressing polarization phenomena in TlBr detectors", *Nucl. Instr. and Meth.*, Vol. **A585** (2008) pp. 102–104.

[186] J. Vaitkus, V. Gostilo, R. Jasinskaite, A. Mekys, A. Owens, S. Zataloka, A. Zindulis, "Investigation of degradation of electrical and photoelectrical properties in TlBr crystals", *Nucl. Instr. and Meth.* **A531** (2004) pp. 192–196.

[187] G.A. Samara, "Pressure and temperature dependences of the ionic conductivities of the thallous halides TlCl, TlBr, and TlI", *Phys. Rev. B*, Vol. **23**, no. 2 (1981) pp. 575–586.

[188] H. Kim, L. Cirignamo, A. Churliov, G. Ciampi, W. Higgins, F. Olschner, K. Shah, "Developing larger TlBr detectors – detector performance", *IEEE Trans. Nucl. Sci.*, Vol. **56**, no. 3, no. 2 (2009) pp. 819–823.

[189] K. Hitomi, T. Onodera, T. Shoji, Z. He, "Investigation of pixellated TlBr gamma–ray spectrometers with the depth sensing technique", *Nucl. Instr. and Meth.*, Vol. **A591** (2008) pp 276–278.

[190] M.-H. Du, "First-principles study of native defects in TlBr: Carrier trapping, compensation, and polarization phenomenon", *J. Appl. Phys.*, Vol. **108** (2010) pp. 053506-1–053506-4.

[191] A.V. Churilov, G. Ciampi, H. Kim, W.M. Higgins, L.J. Cirignano, F. Olschner, V. Biteman, M. Minchello, K.S. Shah, "TlBr and TlBr$_x$I$_{1-x}$ crystals for γ-ray detectors", *J. Cryst. Growth*, Vol. **312** (2010) pp. 1221–1227.

[192] H. Kim, A. Churilov, G. Ciampi, L. Cirignano, W. Higgins, S. Kim, P. O'Dougherty, F. Olschner, K. Shah, "Continued development of thallium bromide and related compounds for gamma-ray spectrometers", *Nucl. Instr. and Meth.*, Vol. **A629** (2011) pp. 192–196.

[193] W.J. Tropf, "*Cubic Thallium(I) Halides*", in Handbook of optical constants of solids, Vol. III, ed. E.D. Palik, Academic Press, Cambridge, MA (1997) ISBN10 0125444230.

[194] K.S. Shah, J.C. Lund, F. Olschner, J. Zhang, L.P. Moy, M.R. Squillante, W.W. Moses, S.E. Derenzo, "TlBr$_x$I$_{1-x}$ photo detectors for scintillation spectroscopy", *IEEE Trans. Nucl. Sci.*, Vol. **NS-41** (1994) pp. 2715–2718.

[195] J.C. Lund, K.S. Shah, M.R. Squillante, L.P. Moy, F. Sinclair, G. Entine, "Properties of lead iodide semiconductor radiation detectors", *Nucl. Instr. and Meth.*, Vol. **A283** (1989) pp. 299–302.

[196] V. Deich, M. Roth, "Improved performance lead iodide nuclear radiation detectors", *Nucl. Instr. and Meth.*, Vol. **A380** (1996) pp. 169–172.

[197] K.S. Shah, F. Olschner, L.P. Moy, P. Bennett, M. Misra, J. Zhang, M.R. Squillante, J.C. Lund, "Lead iodide X–ray detection systems", *Nucl. Instr. and Meth.*, Vol. **A380** (1996) pp. 266–270.

[198] H.G. Fisk, "Preparation and purification of the tri-iodides of antimony and arsenic for use in immersion media of high refractive index", *American Mineralogist*, Vol. **15**, no. 7 (1930) pp. 263–266.

[199] D.-Y. Chung, T. Hogan, P. Brazis, M. Rocci-Lane, C. Kannewurf, M. Bastea, C. Uher, M.G. Kanatzidis, "CsBi$_4$Te$_6$: A High-Performance Thermoelectric Material for Low-Temperature Applications", *Science*, Vol. **287**, no. 5455 (2000) pp. 1024–1027.

[200] A. Samoc, M. Samoc, P.N. Prasad, A. Krajewska-Cizio, "Second-harmonic generation in the crystalline complex antimony triiodide – sulfur", *J. Opt. Soc. of Am. B*, Vol. **9**, no. 10 (1992) pp. 1819–1824.

[201] T. Onodera, K. Mochizuki, N. Nakamura, K. Hitomi, T. Shoji, "Evaluation of Antimony Tri-Iodide Crystals for Radiation Detectors", *Science and Technology of Nuclear Installations*, Vol. **2018**, Article ID 1532742 (2018) pp. 1–7.

[202] G. Fischer, "The Electrical Resistivity of Solid and Liquid Tri-iodides of Antimony and Bismuth", *Helv. Phys. Acta*, Vol. **34** (1961) pp. 827–833.

[203] D. Nason, L. Keller, "The growth and crystallography of bismuth tri-iodide crystals grown by vapor transport", *J. Cryst. Growth*, Vol. **156**, no. 3 (1995) pp. 221–226.

[204] Y.N. Dmitriev, P.R. Beimett, L.J. Cirignano, M. Kiugerman, K.S. Shah, "Bismuth Iodide Crystals as a Detector Material: Some Optical and Electrical Properties", *Proc. of the SPIE*, Vol. **3768** (1999) pp. 520–529.

[205] M. Matsumoto, K. Hitomi, T. Shoji, Y. Hiratate, "Bismuth Tri-Iodide Crystal for Nuclear Radiation Detectors", *IEEE Trans. Nucl. Sci.*, Vol. **49** (2002) pp. 2517–2520.

[206] L. Fornaro, A. Cuña, A. Noguera, M. Pérez, L. Mussio, "Growth of Bismuth Tri–Iodide Platelets for Room Temperature X–ray Detection", *IEEE Trans. Nucl. Sci.*, Vol. **51** (2004) pp. 2461–2465.

[207] T. Saito, T. Iwasaki, S. Kurosawa, A. Yoshikawa, T. Den, "BiI$_3$ single crystal for room-temperature gamma ray detectors", *Nucl. Instr. and Meth.*, Vol. **A806** (2016) pp. 395–400.

[208] S.S. Gokhale, H. Han, O. Pelaez, J.E. Baciak, J.C. Nino, K.A. Jordan, "Fabrication and Testing of Antimony Doped Bismuth Tri-Iodide Semiconductor Gamma-Ray Detectors", *Rad. Meas.*, Vol. **91** (2016) pp. 1–8.

[209] H. Han, M. Hong, S.S. Gokhale, S.B. Sinnott, K.A. Jordan, J.E. Baciak, J.C. Nino, "Defect engineering of BiI$_3$ single crystals: enhanced electrical and radiation performance for room temperature gamma-ray detection", *J. Phys. Chem. C*, Vol. **118** (2014) pp. 3244–3250.

[210] P.M. Johns, *"Materials Development for Nuclear Security: Bismuth Triiodide Room Temperature Gamma Ray Sensors"*, PhD dissertation, University of Florida, Gainsville (2017).

[211] P.M. Johns, S. Sulekar, S. Yeo, J.E. Baciak, M. Bliss, J.C. Nino, "Superheating Suppresses Structural Disorder in Layered BiI_3 Semiconductors Grown by the Bridgman Method", *J. Cryst. Growth*, Vol. **433** (2016) pp. 153–159.

[212] P. Rudolph, H.J. Koh, N. Schäfer, T. Fukuda, "The crystal perfection depends on the superheating of the mother phase too — experimental facts and speculations on the "melt structure" of semiconductor compounds", *J. Cryst. Growth*, Vol. **166** (1996) pp. 578–582.

[213] P.M. Johns, J.E. Baciak, J.C. Nino, "Enhanced Gamma Ray Sensitivity in Bismuth Triiodide Sensors through Volumetric Defect Control", *Appl. Phys. Lett.*, Vol. **109** (2016) pp. 092105-1–092105-4.

[214] V.E. Kutny, A.V. Rybka, A.S. Abyzov, L.N. Davydov, V.K. Komar, M.S. Rowland, C.F. Smith, "AlSb single–crystal grown by HPBM", *Nucl. Instr. and Meth.*, Vol. **A458** (2001) pp. 448–454.

[215] F.V. Wald, J. Bullitt, R.O. Bell, "Bi_2S_3 as a high–z material for γ–ray detectors", *IEEE Trans. Nucl. Sci.*, Vol. **NS–22** (1975) pp. 246–250.

[216] M. Giles, A. Cuna, N. Sasen, M. Llorente, L. Fornaro, "Growth of lead bromide polycrystalline films", *Cryst. Res. Technol.*, Vol. **39** (2004) pp. 906–911.

[217] L. Fornaro, N. Sasen, M. Pérez, A. Noguera, I. Aguiar, "Comparison Of Mercuric Bromide And Lead Bromide Layers as Photoconductors For Direct X-ray Imaging Applications", *IEEE Trans. Nucl. Sci.*, Conference Record of the 15th Workshop on Room Temperature Semiconductor Detectors, San Diego, Vol. **6** (2006) pp. 3750–3754.

[218] E. Tupitsyn, P. Bhattacharya, E. Rowe, L. Matei, M. Groza, B. Wiggins, A. Burger, A. Stowe, "Single crystal of $LiInSe_2$ semiconductor for neutron detector", *Appl. Phys. Lett.*, Vol. **101** (2012) pp. 202101-1–202101-3.

[219] A.C. Stowe, J.S. Morrell, P. Bhattacharya, E. Tupitsyn, A. Burger, "Synthesis of a potential semiconductor neutron detector crystal $LiGa(Se/Te)_2$: Materials purity and compatibility effects", *Proc. of the SPIE*, Vol. **8142** (2011) pp. 81421H-1–81421H-8.

[220] L. Isaenko, A. Yelisseyev, S. Lobanov, A. Titov, V. Petrov, J.J. Zondy, P. Krinitsin, A. Merkulov, V. Vedenyapin, J. Smirnova, "Growth and properties of $LiGaX_2$ (X = S, Se, Te) single crystals for nonlinear optical applications in the mid-IR", *J. Cryst. Res. Technol.*, Vol. **38** (2003) pp. 379–387.

[221] H. Li, J.A. Peters, Z. Liu, M. Sebastian, C.D. Malliakas, J. Androulakis, L. Zhao, I. Chung, S.L. Nguyen, S. Johnsen, B. W. Wessels, M.G. Kanatzidis, "Crystal Growth and Characterization of the X-ray and γ-ray Detector Material $Cs_2Hg_6S_7$", *Cryst. Growth Des.*, Vol. **12**, no. 6 (2012) pp. 3250–3256.

[222] H. Li, C.D. Malliakas, Z. Liu, J.A. Peters, M. Sebastian, L. Zhao, D.Y. Chung, B.W. Wessels, M.G. Kanatzidis, "Investigation of Semi-Insulating $Cs_2Hg_6S_7$ and $Cs_2Hg_{6-x}Cd_xS_7$ Alloy for Hard Radiation Detection", *Cryst. Growth Des.*, Vol. **14**, no. 11 (2014) pp. 5949–5956.

[223] L. Protesescu, S. Yakunin, M.I. Bodnarchuk, F. Krieg, R. Caputo, C.H. Hendon, R.X. Yang, A. Walsh, M.V. Kovalenko "Nanocrystals of cesium lead halide perovskites ($CsPbX_3$, X = Cl, Br, and I): Novel optoelectronic materials showing bright emission with wide color gamut", *Nano Lett.*, Vol. **15** (2015) pp. 3692–3696.

[224] Z. Liu, J.A. Peters, H. Li, M.G. Kanatzidis, B.W. Wessels, "Heavy metal ternary halides for room-temperature x-ray and gamma-ray detection", *Proc. of the SPIE*, Vol. **8852** (2013) pp. 88520A-1–88520A-7.

[225] C.C. Stoumpos, C.D. Malliakas, J.A. Peters, Z. Liu, M. Sebastian, J. Im, T.C. Chasapis, A.C. Wibowo, D.Y. Chung, A. J. Freeman, B.W. Wessels, M.G. Kanatzidis, "Crystal Growth of the Perovskite Semiconductor $CsPbBr_3$: A New Material for High-Energy Radiation Detection", *Cryst. Growth Des.*, Vol. **13**, no. 7 (2013) pp. 2722–2727.

[226] Y. He, L. Matei, H.J. Jung, K.M. McCall, M. Chen, C.C. Stoumpos, Z. Liu, J.A. Peters, D.Y. Chung, B.W. Wessels, M. R. Wasielewski, V.P. Dravid, A. Burger, M.G. Kanatzidis, "High spectral resolution of gamma-rays at room temperature by perovskite $CsPbBr_3$ single crystals", *Nature Communications*, Vol. **9**, Article number 1609 (2018) pp. 1−8.

[227] W. Lin, C.C. Stoumpos, O.Y. Kontsevoi, Z. Liu, Y. He, S. Das, Y. Xu, K.M. McCall, B.W. Wessels, M.G. Kanatzidis, "$Cu_2I_2Se_6$: A Metal–Inorganic Framework Wide-Bandgap Semiconductor for Photon Detection at Room Temperature", *J. Am. Chem. Soc.*, Vol. **140**, no. 5 (2018) pp. 1894–1899.

[228] H. Li, F. Meng, C.D. Malliakas, Z. Liu, D.Y. Chung, B.W. Wessels, M.G. Kanatzidis, "Mercury Chalcohalide Semiconductor $Hg_3Se_2Br_2$ for Hard Radiation Detection", *Cryst. Growth Des.*, Vol. **16** (2016) pp.6446−6453.

[229] M. Kocsis, "Proposal for a new room temperature X–ray detector–thallium lead iodide", *IEEE Trans. Nucl. Sci.*, Vol. **47** (2000) pp. 1945–1947.

[230] K. Hitomi, T. Onodero, T. Shoji, Y. Hiratate, "Thallium lead iodide radiation detectors", *IEEE Trans. Nucl. Sci.*, Vol. **50** (2003) pp. 1039–1042.

[231] S. Johnsen, Z. Liu, J.A. Peters, J.-H. Song, S. Nguyen, C.D. Malliakas, H. Jin, A.J. Freeman, B.W. Wessels, M. G. Kanatzidis, "Thallium Chalcohalides for X-ray and γ-ray Detection", *J. Am. Chem. Soc.*, Vol. **133**, no. 26 (2011) pp. 10030–10033.

[232] S. Johnsen, Z. Liu, J.A. Peters, J.-H. Song, S.C. Peter, C.D. Malliakas, N.K. Cho, H. Jin, A.J. Freeman, B.W. Wessels, M. G. Kanatzidis, "Thallium Chalcogenide-Based Wide-Band-Gap Semiconductors: $TlGaSe_2$ for Radiation Detectors", *Chem. Mater.*, Vol. **23**, no. 12 (2011) pp. 3120–3128.

[233] J. Androulakis, S.C. Peter, H. Li, C.D. Malliakas, J.A. Peters, Z. Liu, B.W. Wessels, J.-H. Song, H. Jin, A.J. Freeman, M. G. Kanatzidis, "Dimensional Reduction: A Design Tool for New Radiation Detection Materials", *Adv. Mater.*, Vol. **23** (2011) pp. 4163–4167.

[234] D. Kahler, N.B. Singh, D.J. Knuteson, B. Wagner, A. Berghmans, S. McLaughlin, M. King, K. Schwartz, D. Suhre, M. Gotlieb, "Performance of novel materials for radiation detection: Tl_3AsSe_3, $TlGaSe_2$, and Tl_4HgI_6", *Nucl. Instr. and Meth.*, Vol. **A652** (2011) pp. 183–185.

[235] W. Lin, C.C. Stoumpos, Z. Liu, S. Das, O.Y. Kontsevoi, Y. He, C.D. Malliakas, H. Chen, B.W. Wessels, M.G. Kanatzidis, "$TlSn_2I_5$, a Robust Halide Anti-perovskite Semiconductor for γ-Ray Detection at Room Temperature", *ACS Photonics*, Vol. **4**, no. 7 (2017) pp. 1805–1813.

[236] Z. Liu, J.A. Peters, C. Zang, N.K. Cho, B.W. Wessels, S. Johnsen, S. Peter, J. Androulakis, M.G. Kanatzidis, J.-H. Song, H. Jin, A.J. Freeman, "Tl-based wide gap semiconductor materials for x-ray and gamma ray detection", *Proc. of SPIE*, Vol. **8018** (2011) 80180H-1–180180H-9.

[237] W. Lin, H. Chen, J. He, C.C. Stoumpos, Z. Liu, S. Das, J.-I.L. Kim, K.M. McCall, B.W. Wessels, M.G. Kanatzidis, "$TlSbS_2$: a Semiconductor for Hard Radiation Detection", *ACS Photonics*, Vol. **4**, no. 11 (2017) pp. 2891–2898.

[238] A. Kargar, J. Tower, H. Hong, L. Cirignano, H. Kim, K. Shah, P.R. Beck, A.M. Conway, O.B. Drury, L.F. Voss, R.T. Graff, A.J. Nelson, R.J. Nikolic, S.A. Payne, V. Badikov, "$PbGa_2Se_4$ Semiconductor for Gamma-Ray Detection", *Conference Record of the IEEE/MIC/RTSD 19th International Workshop on Room Temperature Semiconductor Detectors*, Anaheim, CA, USA, 27 October – 3 November (2012) pp. 4258–4261.

[239] N.N. Musayeva, O.B. Tagiyev, R.B. Jabbarov, "The studying of photo-conduction mechanism of the photosensitive crystals by the type of $PbGa_2S_4(Se_4)$", *Fizika*, Vol. X, no. 3 (2004) pp. 61–65.

[240] D.I. Bletskan, V.M. Kabatsiia, M. Kranichetsb, V.V. Frolova, E.G. Gulec, Photoconductivity and photoluminescence of $PbGa_2Se_4$ crystals", *Chalcogenide Lett.*, Vol. **3**, no. 12 (2006) pp. 125–132.

[241] P.L. Wang, Z. Liu, P. Chen, J.A. Peters, G. Tan, J. Im, W. Lin, A.J. Freeman, B.W. Wessels, M.G. Kanatzidis, "Hard Radiation Detection from the Selenophosphate $Pb_2P_2Se_6$", *Adv. Funct. Mater.*, Vol. **25** (2015) pp. 4874–4881.

[242] H. Li, C.D. Malliakas, F. Han, D.Y. Chung, M.G. Kanatzidis, "$TlHgInS_3$: An Indirect-Band-Gap Semiconductor with X-ray Photoconductivity Response", Chem. Mater., Vol. 27 (2015) pp. 5417−5424.

[243] H. Li, C.D. Malliakas, Z. Liu, J.A. Peters, H. Jin, C.D. Morris, L. Zhao, B.W. Wessels, A.J. Freeman, M.G. Kanatzidis, "$CsHgInS_3$: a New Quaternary Semiconductor for γ-ray Detection", *Chem. Mater.*, Vol. **24**, no. 22 (2012) pp. 4434–4441.

[244] H. Li, C.D. Malliakas, J.A. Peters, Z.F. Liu, J. Im, H. Jin, C.D. Morris, L.D. Zhao, B.W. Wessels, A.J. Freeman, M. G. Kanatzidis, "$CsCdInQ_3$ (Q = Se, Te): New Photoconductive Compounds as Potential Materials for Hard Radiation Detection", *Chem. Mater.*, Vol. **25** (2013) pp. 2089–2099.

[245] D. Natali, M. Sampietro, "Detectors based on organic materials: status and perspectives", *Nucl. Instr. and Meth.*, Vol. **A512** (2003) pp. 419–426.

[246] P. Frenger, "Edible Organic Semiconductors", *IEEE Green Techn. Conf.*, Tulsa, OK (2012) p. 197.

[247] H.-W. Fink, C. Schönenberger, "Electrical conduction through DNA molecules", *Nature*, Vol. **398** (1999) pp. 407–410.

[248] G. Malliarias, R. Friend, "An Organic Electronics Primer", *Physics Today*, Vol. **58**, no. 5 (2005) pp. 53–58.

[249] M.C. Scharber, N.S. Sariciftci, "Efficiency of bulk-heterojunction organic solar cells", *Prog. in Polym. Sci.*, Vol. **38**, no. 12 (2013) pp. 1929–1940.

[250] L. Meng, Y. Zhang, X. Wan, C. Li, X. Zhang, Y. Wang, X. Ke, Z. Xiao, L. Ding, R. Xia, H.-L. Yip, Y. Cao, Y. Chen, "Organic and solution-processed tandem solar cells with 17.3% efficiency", *Science*, Vol. 361, no. 6407 (2018) pp. 1094–1098.

[251] *Organic Electronic Materials*, eds. R. Farchioni, G. Grosso, Springer, New York (2001) ISBN: 978-3-642-63085-9.

[252] W. Hu, D. Frenke, V.B.F. Mathot, "Sectorization of a Lamellar Polymer Crystal Studied by Dynamic Monte Carlo Simulations", *Macromolecules*, Vol. **36**, no. 3 (2003) pp 549–552.

[253] V. Coropceanu, J. Cornil, D.A. Da Silva Filho, Y. Olivier, R. Silbey, J.-L. Brédas, "Charge Transport in Organic Semiconductors", *Chem. Rev.*, Vol. **107**, no. 4 (2007) pp.926−952.

[254] P. Beckerle, H. Strobele, "Charged particle detection in organic semi-conductors", *Nucl. Instr. and Meth.*, Vol. **A449** (2000) pp. 302–310.

[255] J.C. Blakesley, P.E. Keivanidis, M. Campoy–Quiles, C.R. Newman, Y. Jin, R. Speller, H. Sirringhaus, N.C. Greenham, J. Nelson, P. Stavrinou, "Organic semiconductor devices for X-ray imaging", *Nucl. Instr. and Meth.*, Vol. **A580** (2007) pp. 774–777.

[256] T. Suzuki, H. Miyata, M. Katsumata, S. Nakano, K. Matsuda, M. Tamura, "Organic semiconductors as real-time radiation detectors", *Nucl. Instr. and Meth.*, Vol. **A763** (2014) pp. 304–307.

[257] A. Ciavatti, P.J. Sellin, L. Basiricò, A. Fraleoni-Morgera, B. Fraboni, "Charged-particle spectroscopy in organic semiconducting single crystals", *Appl. Phys. Lett.*, Vol. **108** (2016) pp. 153301-1–153301-5.

[258] A. Ciavatti, E. Capria, A. Fraleoni-Morgera, G. Tromba, D. Dreossi, P.J. Sellin, P. Cosseddu, A. Bonfiglio, B. Fraboni, "Toward Low-Voltage and Bendable X-Ray Direct Detectors Based on Organic Semiconducting Single Crystals", *Adv. Mater.*, Vol. **27**, no. 44 (2015) pp. 7213–7220.

[259] C. Momblona, O. Malinkiewicz, C. Roldan-Carmona, A. Soriano, L. Gil-Escrig, E. Bandiello, M. Scheepers, E. Edri, H. J. Bolink, "Efficient methylammonium lead iodide perovskite solar cells with active layers from 300 to 900 nm", *APL Mat.*, Vol. **2** (2014) pp. 081504-1–081504-7.

[260] W.E.I. Sha, X. Ren, L. Chen, W.C.H. Choy, "The Efficiency Limit of $CH_3NH_3PbI_3$ Perovskite Solar Cells", *Appl. Phys. Lett.*, Vol. **106**, no. 22 (2015) pp. 221104-1–221104-5.

[261] B. Náfrádi, P. Szirmai, M. Spina, H. Lee, O.V. Yazyev, A. Arakcheeva, D. Chernyshov, M. Gibert, L. Forró, E. Horváth, "Optically switched magnetism in photovoltaic perovskite $CH_3NH_3(Mn: Pb)I_3$", *Nature Comm.*, Vol. 7, article no. 13406 (2016).

[262] B. Saparov, D.B. Mitzi, "Organic–Inorganic Perovskites: Structural Versatility for Functional Materials Design", *Chem. Rev.*, Vol. 116, no. 7 (2016) pp. 4558–4596.

[263] B. Maynard, Q. Long, E.A. Schiff, M. Yang, K. Zhu, R. Kottokkaran, H. Abbas, V.L. Dalal, "Electron and hole drift mobility measurements on methylammonium lead iodide perovskite solar cells", *Appl. Phys. Lett.*, Vol. 108, article no. 173505 (2016)

[264] Y. Bi, E.M. Hutter, Y. Fang, Q. Dong, J. Huang, T.J. Savenije, "Charge Carrier Lifetimes Exceeding 15 μs in Methylammonium Lead Iodide Single Crystals", *J. Phys. Chem. Lett.*, Vol. 7, no. 5 (2016) pp. 923–928.

[265] B. Náfrádi, G. Náfrádi, L. Forró, E. Horváth, "Methylammonium Lead Iodide for Efficient X-ray Energy Conversion", *J. Phys. Chem. C*, Vol. 119 (2015) pp.25204−25208.

[266] S. Yakunin, D. Dirin, Y. Shynkarenko, V. Morad, I. Cherniukh, O. Nazarenko, D. Kreil, T. Nauser, M. Kovalenko, "Detection of gamma photons using solution-grown single crystals of hybrid lead halide perovskites", *Nature Photonics*, Vol. 10 (2016) pp. 585–589.

[267] H. Wei, Y. Fang, P. Mulligan, W. Chuirazzi, H.H. Fang, C. Wang, B.R. Ecker, Y. Gao, M.A. Loi, L. Cao, J. Huang, "Sensitive X-ray detectors made of methylammonium lead tribromide perovskite single crystals", *Nature Photonics*, Vol. 10 (2016) pp. 333–339.

[268] H. Wei, D. DeSantis, W. Wei, Y. Deng, D. Guo, T.J. Savenije, L. Cao, J. Huang, "Dopant compensation in alloyed $CH_3NH_3PbBr_{3-x}Cl_x$ perovskite single crystals for gamma-ray spectroscopy", *Nature Materials*, Vol. 16 (2017) pp. 826–833.

[269] W.S. Yang, J.H. Noh, N.J. Jeon, Y.C. Kim, S. Ryu, J. Seo, S.I. Seok, "High-performance photovoltaic perovskite layers fabricated through intramolecular exchange", *Science*, Vol. 348, no. 6240 (2015) pp. 1234–1237.

[270] O. Nazarenko, S. Yakunin, V. Morad, I. Cherniukh, M.V. Kovalenko, "Single crystals of caesium formamidinium lead halide perovskites: solution growth and gamma dosimetry", *NPG Asia Materials*, Vol. 9, paper e373 (2017) pp. 1–8.

[271] A. Owens, A. Peacock, "Compound semiconductor radiation detectors", *Nucl. Instr. and Meth.*, Vol. A531 (2004) pp. 18–37.

[272] L. Loupilov, A. Sokolov, V. Gostilo, "X–ray peltier cooled detectors for X–ray fluorescence analysis", *Rad. Phys. and Chem.*, Vol. 61 (2001) pp. 463–464.

[273] A. Niemela, H. Sipila, "Evaluation of CdZnTe detectors for soft X–Ray applications", *IEEE Trans. Nucl. Sci.*, Vol. 41 (1994) pp. 1054–1057.

10

Current Materials Used for Neutron Detection

Frontespiece. The Core (20 ^{235}U MTR fuel assemblies, 4 control assemblies) of the 2 MW research reactor at the Technical University of Delft used by the European Space Agency to characterize solid state neutron detectors.

CONTENTS

10.1 Neutron Detection

Since neutrons are uncharged, the only practical detection method is through the observation of their reaction products following neutron capture, scattering or nuclear interaction processes. We will mostly consider thermal neutron detection in this chapter, since fast neutrons are usually detected by first moderating them to thermal energies. The most common detector type utilizing this technique is the Bonner sphere spectrometer [1], which consists of a standard thermal neutron sensor (usually an ^{3}He proportional counter) that can be located at the center of a number of different diameter, moderating spheres (usually made of polyethylene). The neutron sensitivity of each sphere will peak at a particular energy depending on the sphere's diameter. By comparing the response of each sphere, the incident neutron energy spectrum can be unfolded [2]. If energy *and* direction information are required, then other detection technologies are usually applied (*e.g.*, time of flight techniques/recording proton recoils in counter telescopes using multiple plastic scintillator planes, for a review see refs [3,4]). For thermal neutron detection, neutron capture is perhaps the simplest and most efficient method. However, it is noteworthy that very few elements have high enough capture cross-sections to be useful in practical detection systems. These are listed in Table 8.1 along with their principal capture processes (*e.g.*, (n,γ), (n,p), (n,α), (n,β) or $(n,$ fission$)$ reactions). We see from the table that for these few elements, cross-sections are typically $\sim$1,000 barns or more. In comparison, the majority of elements in the periodic table have cross-sections less than a barn. At present, neutrons are detected either indirectly by detecting the reaction products leaving a material with high neutron capture characteristics in a separate semiconductor detector or directly in a detector fabricated from materials that have a neutron sensitivity (*i.e.*, elemental components with high neutron capture cross-sections [5]). In the latter case, since the stopping medium is a solid, the volumetric efficiency of such a detector is a thousand times greater than conventional ^{3}He or BF$_3$ gas detectors.

10.2 Indirect Neutron Detection

Until recently, work has concentrated on so-called indirect techniques, which normally involve depositing a thin layer of a highly neutron absorbing material (containing ^{6}Li, ^{10}B or ^{157}Gd) on top of an active detector – usually a planar device (see for example [6–8]). The neutrons are converted into charged particles in this layer following neutron capture, and these secondary products are detected in the detector (see McGregor *et al.* [9] for a review). Generally, the most commonly used converter materials are based on films of lithium fluoride (lithium is too reactive to be used on its own) or boron. In the case of LiF, the ^{6}Li component of natural lithium (which occurs with an abundance of 7.5%) is the active component. The other component of naturally occurring Li is ^{7}Li, which is essentially inert. In the case of ^{6}Li, the reaction proceeds as follows:

$$^1n + {}^6Li \Rightarrow {}^7Li^* \Rightarrow {}^3H \ (2.73 \ \mathrm{MeV}) + {}^4He(2.05 \ \mathrm{MeV}) \qquad Q = 4.78 \ \mathrm{MeV}. \tag{1}$$

Like Li, naturally occurring boron is composed of two main isotopes, ^{10}B and ^{11}B, which occur with abundances of 20% and 80%, respectively. Here, the ^{10}B isotope is the active component. There are actually two decay modes, one to the first excited state and the other directly to the ground state,

$$^1n + {}^{10}B \ \Rightarrow \ {}^{11}B^* \ \Rightarrow \ \begin{cases} {}^7Li \ (0.84 \ \mathrm{MeV}) + {}^4He \ (1.47 \ \mathrm{MeV}) + \gamma(0.48 \ \mathrm{MeV}) & Q = 2.31 \mathrm{MeV} \\ {}^7Li \ (1.02 \ \mathrm{MeV}) + {}^4He \ (1.78 \ \mathrm{MeV}) & Q = 2.79 \mathrm{MeV} \end{cases} \tag{2}$$

with branching ratios of 94% and 6%, respectively. In the reactions described in Eqs. (1) and (2), the charged reaction products are emitted at 180 degrees relative to one another, which means that for an indirect detection system employing a converter layer, only one product has the ability to reach the semiconductor and create electron-hole pairs. Gadolinium is a particularly interesting converter material that has been proposed by a number of authors. It is a naturally occurring rare earth metal composed of six stable isotopes: ^{154}Gd, ^{155}Gd, ^{156}Gd, ^{157}Gd, ^{158}Gd and ^{160}Gd. Of these, ^{157}Gd and ^{155}Gd have the largest thermal neutron capture cross-sections of any stable isotope, namely, 255,000 barns and 65,000 barns, respectively. It is not economically viable to use isotopically separated ^{157}Gd as a converter material for most practical applications. However, it should be noted that together ^{157}Gd and ^{155}Gd constitute 15.7% and 14.8% of natural Gd, presenting an average overall cross-section of 49,000 barns – which is still more than 10 times larger than ^{10}B and 50 times large than ^{6}Li. In addition, the cost of natural Gd is not overly prohibitive. Thermal neutron absorption in these isotopes proceeds as follows:[1]

$$^1n + {}^{155}Gd \Rightarrow {}^{156}Gd \ + \gamma(89, \ 199 \ \mathrm{keV}, \dots) + \mathrm{c.e.}(39, \ 81, \ 88 \ \mathrm{keV}, \dots)^1 Q \ = \ 8.54 \ \mathrm{MeV} \tag{3}$$

$$^1n + {}^{157}Gd \Rightarrow {}^{158}Gd \ + \gamma(80, \ 182, .. \ 944, \ 960, \ 975, ..) + \mathrm{c.e.}(29, 72, 132, 174, \ .. \ \mathrm{keV})^1 Q = 7.94 \ \mathrm{MeV}$$

---

[1] Only those emissions with an intensity in natural Gd of > 1% are listed.

and results in the emission of an assortment of γ-rays over a wide keV energy range and a cascade of conversion electrons (denoted by c.e.). The primary energies of the conversion electrons occur at approximately 30 keV, 80 keV and 130 keV, with a quantum yield near unity.

A typical detection system is illustrated in Fig. 10.1, in which we have separated the converter layer and detector for clarity so that the trajectories and interactions of the reaction products are more apparent. The advantage of such a technique is its relative simplicity; the disadvantage is that its detection efficiency is limited. To be efficient in the transconductance process (*i.e.*, converting neutrons into charged particles), the conversion layer has to be as thick as possible. However, if it is too thick, the reaction products may be absorbed before reaching the semiconductor or have insufficient energy for adequate electron-hole pair generation. For example, in the case of ^{10}B, the average ranges for the 1.47 MeV alpha particle and 840 keV ^{7}Li ion in the boron layer are 3.6 μm and 1.6 μm, respectively, and as already noted only one can reach the detector for geometric reasons. In practice, a maximum efficiency of 4–5% can be achieved. Note that although the reactions products of ^{6}Li have higher energies and therefore larger ranges (34 μm for the triton and 7.8 μm for the alpha particle), the reduced thermal neutron capture cross-section of ^{6}Li (940 b as opposed to 3840 b for B) means that in reality the sensitivities of ^{6}Li and ^{10}B based detection systems are comparable. In fact, the simulations of McGregor *et al.* [9] of representative systems show that the maximum detection efficiencies that can be achieved for ^{10}B and ^{6}LiF converter layers are 4% and 4.4%, respectively. The corresponding layer thicknesses to achieve these efficiencies are 2.8 μm (^{10}B) and 27 μm (^{6}LiF).

For Gd-based detection systems, the 72 keV conversion electrons are the most significant reaction products, as they are emitted in 39% of capture reactions. The range of these particles is typically ~30 μm, which is much greater than the range of the reaction products generated in Li or B converters. This means that converters can be thicker and therefore more efficient. For an adjacent detector, the neutron conversion efficiency can be as high as 30%, much larger than the ~3–4% more typical of ^{6}Li- and ^{10}B-coated devices. However, in most converter/detector systems this advantage is largely negated by the much lower energies of the conversion electrons, compared to the reaction products generated in Li and B converters. This means that in a practical detection system, the low energy threshold set on the electronics to discriminate against noise and general background limits detection efficiencies to ~12%. In fact, for planar devices, noise arising from the detector capacitance will ultimately limit the size of the detector that can be used.[2] For single Si-Gd planar detection systems, reported detection efficiencies are generally around 5%. For example, Aoyama *et al.* [10] reported an efficiency of 5.6% using a system consisting

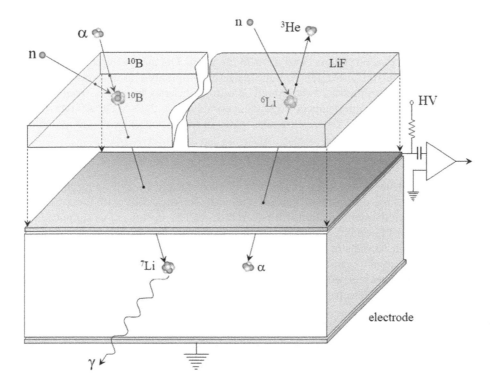

FIGURE 10.1 Schematic depiction of an indirect neutron detection detector. The detector consists of a thin layer of a material containing elements with high thermal neutron capture cross-sections, mounted directly on top of a simple planar detector (we show an expanded view in the diagram for clarity). Generally, boron or lithium fluoride layers are used for the conversion layer. Both reactions are shown. Neutrons are converted into charged particles in this layer following neutron capture; it is one of the secondary products detected in the underlying detector.

---

[2] This also leads to an interesting consequence that the efficiency-area product for a given detector technology should be constant, as demonstrated by Schulte *et al.* [8]

of a 1 cm^2 Si PIN diode and a 25 μm thick natural gadolinium converter. The system was novel in that it included a second neutron-insensitive detection channel to correct for gamma-ray contamination.

Miyake *et al.* [11] circumvented the threshold problem by proposing to detect the prompt gamma-rays emitted following neutron capture in the Gd conversion layer, rather than the conversion electrons. To achieve this, they used a 0.5 mm thick CdTe detector as the active detection element and showed that the energy resolution of the detector at ~90 keV (~4% FWHM) was sufficient to fully resolve the capture gamma-rays emanating from the converter layer from the general background. They also argued that by careful design, the CdTe can be made thick enough to efficiently detect the prompt capture gamma-rays but thin enough to be insensitive to general gamma-ray contamination associated with neutron sources. However, no efficiency values were reported. It should be noted that in applications where the associated gamma-ray background is low, the efficiency of the system can be increased by using a thicker CdTe detector, to detect not only higher energy Gd capture gamma-rays near 1 MeV, but also the 558 keV and 651 keV capture gamma-rays from Cd, *via* the reaction, ^{113}Cd(n, γ)^{114}Cd.

10.2.1 Increasing Efficiency

In order to improve detection efficiency, several solutions have been proposed. Perhaps, the simplest is to stack planar detectors. However to achieve a good neutron detection efficiency, many layers are required.[3] One way to reduce this number significantly is to sandwich the absorbing layer between two active detectors or apply a converter layer to the top and bottom of an active detector, potentially doubling the thickness of the conversion layer. For example, Pappalardo *et al.* [12] achieved an efficiency of 5.2% with a 3 cm×3 cm, 300 μm thick Si planar detector with 16 μm thick ^{6}LiF layers placed on the top and bottom surfaces. A stack of two such detectors achieved an efficiency of 10%. Schulte *et al.* [8] recorded a thermal neutron detection efficiency of 18% with a 5.2 cm^2 Si-Gd-Si sandwich detector.

10.2.2 Shaped Converters

An improvement in efficiency can also be achieved by creating dips or pores (3-D structures) in the active detector body and backfilling them with a suitable neutron converter, in essence increasing the contact area between the converter and the detector. The dips or pores can be quite deep and when added to the overlaying converter layer can provide a meaningful increase in the amount of absorbing material and therefore detection efficiency. For example, Uher *et al.* [13] created a 3-D microstructure of inverted pyramidal dips in a 5×5 mm^2, 300 μm thick planar Si detector by anisotropic etching with KOH. The pyramid bases had a size of 60×60 μm^2 and were 28 μm deep. The gap between pyramids was 23 μm. The dips double the surface between the neutron converter and the detector. Compared to a simple planar 7.5 μm thick ^{6}LiF converter, the addition of pyramidal dips increased the overall detection efficiency by ~30%, from 4.9% to 6.3%. In addition, contrary to the planar detector case, the spectrum now contains events above 2.73 MeV, as both reaction particles (alpha and triton) can be detected simultaneously if the capture takes place in the region close to the pyramid tip.

10.2.3 Three-Dimensional Structures

A much larger improvement in efficiency has been made over the last decade by exploiting advances in micromachining technologies to produce custom three-dimensional cavities that can be backfilled with neutron reactive material. The geometry of these cavities and the active semiconductor between them can be optimized to significantly increase the surface between the neutron converter and the detector sensitive volume, increasing the probability that the reaction products are detected. The simplest configuration consists of a common p-n junction diode that has been microstructured, by etching or 3-D reactive ion beam milling, to create cavities that are then filled with converter material. The geometry of the active detection medium and converter can then be optimized to ensure a high probability that neutron absorption will take place and that both reaction products will be detected in the active semiconductor. Generally, three geometries are considered:

a) perforated structures, where deep holes are created in the semiconductor and backfilled with converter material such as LiF,

b) the semiconductor is etched to produce pillar-like structures, which are then surrounded by converter material and

c) trench geometries, in which deep channels are etched into an active detector and then backfilled with neutron converter material.

These geometries are illustrated in Fig. 10.2.

---

[3] Although it should be pointed out that the efficiency does not increase linearly with the number of layers, since each layer absorbs part of the neutron flux, so successive layers receive an attenuated neutron flux.

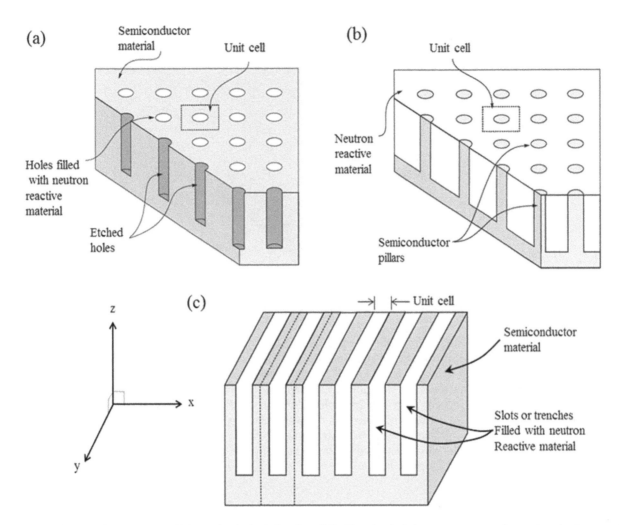

FIGURE 10.2 Schematic of 3-D micromachining technologies (taken from [14]). Here the shaded areas represent active semiconducting material and the white regions neutron reactive material. (a) Perforated structure in which deep holes are etched into a planar detector and then backfilled with neutron converter material. (b) A pillar-type detector in which a silicon wafer is micromachined to create high aspect ratio p-i-n pillars, which are then surrounded by converter material. (c) A channel or trench geometry in which deep channels are etched into an active detector and backfilled with neutron converter material.

10.2.3.1 Perforated Structures

Huang *et al.* [15] report on the fabrication and characterization of a scalable solid-state thermal neutron detector. The detector is illustrated in Fig. 10.3 and consists of a microstructured Si diode. A honeycomb of 2.8 μm wide, 45 μm deep, hexagonal holes was created in the active Si by deep reactive ion etching. The holes were then filled with ^{10}B by low pressure CVD. A continuous p-n junction formed over the entire surface of the microstructure helped to achieve a low leakage current. An intrinsic thermal neutron detection efficiency of up to 26% was measured for a 2.5×2.5 mm^2 detector module and up to 24% for a 1 cm^2 detector module. These measurements were obtained under zero bias voltage using a moderated ^{252}Cf source. The relative efficiency remains almost the same when scaling the detector area up to 8 cm^2 by connecting 1 cm^2 detector modules in series. However, it decreases to 22% and 20%, respectively, when scaled up to areas of 12 cm^2 and 16 cm^2.

Bellinger *et al.* [16–18] constructed a series of perforated detectors with large aspect ratio trenched microstructures, backfilled with ^{6}LiF. The detectors were fabricated from diffused p-n junction diodes, each of 1 cm^2 area. The detector was then etched to produce straight 60 μm deep trenches, 25 μm wide with a pitch of 50 μm. For a 6×6 array of such detectors, an overall efficiency of 6.8% was achieved. Later work with two stacked devices with 250 μm deep trenches achieved an intrinsic thermal neutron detection efficiency of 42%.

10.2.3.2 Pillar Structures

Pillar structured detectors are fabricated by etching a Si diode to leave a periodic array of high aspect ratio p-i-n pillar structures that are then surrounded by neutron converter material. Since a large volume of converter material can be

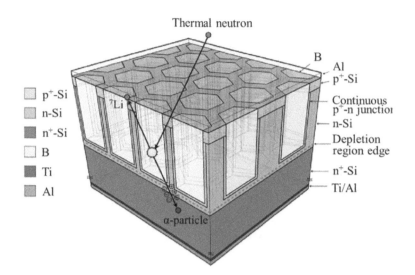

FIGURE 10.3 Schematic of the 3-D honeycomb structured neutron detector of Huang *et al.* [15]. Note that, the surface boron layer and boron etch stop layer are not shown for clarity.

accommodated in a pillar array, surprising high detection efficiencies can be achieved. For example, Shao *et al.* [19] fabricated a high-aspect-ratio Si p-i-n diode pillar array filled with ^{10}B. A schematic of the detector is shown in Fig. 10.4. The detector was produced by epitaxially growing p+ and i layers on an n+ silicon substrate by chemical vapor deposition. The pillar diameters and spacing were defined lithographically, followed by plasma etching to create the high-aspect-ratio structures. Two geometries were produced – one with 26 μm high Si pillars and another with 50 μm high Si pillars. Both had pillar diameters of a few microns. A conformal coating of ^{10}B was then deposited on the pillar array by chemical vapor deposition filling the voids. The top surface was then back etched to expose the pillar tops and Al/Cr/Au contacts applied to the top and bottom surfaces by sputtering. As reported by Shao *et al.* [19] the efficiency ranges from 22% for the 26 μm high pillar device to 48.5% for the 50 μm high pillar device.

10.2.3.3 Trenched Structures

Trench structures are essentially channels etched into an active detector (usually by deep reactive etching) and backfilled with a neutron reactive material. In a study of cylindrically perforated, pillar and trench geometries, Shultis and McGregor [14], concluded that the trench structure gives the best efficiencies, potentially exceeding 20% using simple linear geometries backfilled with ^{6}LiF and over 35% for more complicated structures, such as "sinusoidal" or "chevron" shaped trenches. For a single-sided 4 cm^2 Si device with straight trenches, 400 μm deep, 20 μm wide and spaced 40 μm apart, backfilled filled with ^{6}LiF, Ochs *et al.* [20] measured a thermal neutron detection efficiency of 30%.

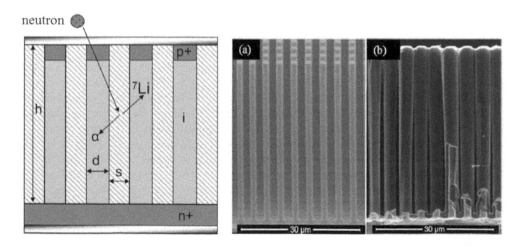

FIGURE 10.4 Schematic of a pillar structured thermal neutron detector (from Shao *et al.* [19]). Here, h is the pillar height (26/50 μm), d is the pillar size (2 μm) and s is the pillar separation (2 μm). Scanning electron microscopy images of the 50 μm silicon pillar structures: (a) as fabricated by etching, (b) after ^{10}B deposition by CVD.

10.3 Direct Neutron Detection

The efficiency of solid state neutron detectors can be greatly improved by combining neutron capture and charge collection layers in a single material. A schematic of such a detector is shown in Fig. 10.5. Direct neutron detection in a highly absorbent semiconducting material offers several major advantages over indirect detection techniques. For example:

1) high detection efficiency, since the device acts as both absorber and detector,
2) highly compact, since the range of the reaction products is of the order of microns, the thickness of the detector can be a fraction of a mm,
3) detection systems can operate without sophisticated amplification and noise reduction circuitry, since such a large amount of charge[4] is generated locally per absorbed event ($\geq 10^5$ e/h per neutron) and
4) the existence of a built-in potential in Schottky and p-n devices means that they could even operate without bias.

10.3.1 Choice of Semiconductor

The ideal candidate for direct solid-state thermal neutron detection should satisfy the following criteria.

1. It should have a high probability of interaction with a thermal neutron, and the reaction should give an easily identifiable and unambiguous signal.
2. There should be a low probability of interference due to other radiation. It is worth noting that most nuclear materials emit 10 or more times as many gamma-rays as neutrons.
3. The characteristics of the neutron-sensitive element or one of its compounds should be semiconductor-like.
4. It should be sufficiently abundant to make it affordable for the application.
5. The detector should have chemical and physical compatibility with a suitable substrate.
6. It should also have chemical and electrical compatibility with a contacting system.

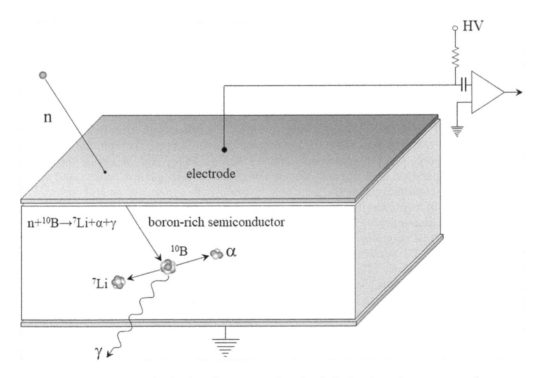

FIGURE 10.5 Schematic of the interaction mechanism in a direct neutron detection device based on a boron compound.

---

[4] This attribute could also be exploited for power generation, potentially leading to an alternate to radioisotope thermoelectric generators for remote locations or extreme environments (*e.g.*, outer planetary missions).

10.3.2 Cadmium Based Devices

Attempts at direct neutron detection using cadmium compounds date back to the 1960s. Johnson [21] reported the detection of induced radioactivity produced by fast neutrons in a CdS crystal operating as both neutron absorber and detector. Following irradiation, changes in the post-irradiation resistivity were observed. This was attributed to self-ionization of the sample by radioactive decay of transmuted atoms produced during the irradiation, specifically from the decays of ^{115}Cd and ^{32}P [22]. Since the measured spectra and its temporal properties are unique for a given material, Johnson [21] points out that neutron spectroscopy could be performed by using semiconducting materials with different neutron-reaction cross-sections.

Fasasi *et al.* [23] exposed two cadmium telluride detectors to a known thermal neutron beam in an attempt to detect decayed gamma-rays from ^{113}Cd *via* the reaction ^{113}Cd(n,γ)^{114}Cd. ^{113}Cd has a large thermal neutron cross-section of 20,000 barns. Since the isotopic abundance of ^{113}Cd is 12.6%, this leads to an effective absorption cross-section of 2450 barns for natural cadmium. Almost 500 gamma-ray transitions are possible, of which the most dominant produce prompt emission lines at 559 keV (100%), 651 keV (19%), 806 keV (7%), 1209 keV (6%) and 1364 keV (6%) [24]. Satisfactory agreement was observed between calculations and experimental data, and spectral line features were detected at 96 keV and 560 keV. Fasasi *et al.* [23] point out that since the atomic density of CdTe is much higher than in gas detectors, its efficiency for small detection volumes can be comparable to that of a ^{3}He tube, when the entire pulse-height spectrum is used. For a 2 mm thick detector, the authors report an efficiency of about 5%, when using the counts in the entire spectrum, 0.5% when only the 96 keV gamma-ray line is used and 0.38% when the 558 keV line is measured. However, it is worth noting that most nuclear materials emit many more gamma-rays than neutrons. Thus, given the relatively high density of CdTe, in a high radiation field the gamma-ray response may overwhelm its neutron response. Miyake *et al.* [11] approached the latter problem by proposing the use of a CdTe detector in conjunction with a Gd converter layer. They argue that in this case the CdTe can be much thinner since the gamma-rays generated in the Gd layer by neutron capture are emitted at a much lower energy than the gamma-rays produced in ^{113}Cd(n,γ) reactions (*i.e.*, approximately 90 keV as opposed to ~600 keV). In addition, the energy resolution of a CdTe detector (~4%) is sufficient to discriminate neutron capture gamma-rays in the converter layer from background gamma emission.

McGregor *et al.* [25] proposed using an active CdZnTe detector to detect thermal neutrons. In an exposure of a 10×10×3 mm^3 CdZnTe detector to a thermal neutron source, they found clear signatures of the 586 keV and 651 keV gamma-ray lines. The detection sensitivity is, however, dependent on the gamma-ray absorption efficiency, which in turn depends on the detector active volume. In this case, the detection sensitivity using the 586 keV line is ~4%. Also, as the authors point out, since CdZnTe is an efficient X- and gamma-ray detection medium, a weak thermal neutron source may easily be masked by a high gamma-ray background.

10.3.3 Mercury Based Devices

Beyerle and Hull [26] and Melamud *et al.* [27] have pointed out that mercuric iodide can act as an efficient neutron detector *via* the reaction ^{199}Hg(n, γ)^{200}Hg, which has a capture cross-section of 2150 barns. Taking into account the isotopic abundance of ^{199}Hg, the effective cross-section of natural Hg is 374 barns. The main gamma-ray lines emitted are prompt of energies, 368 keV (81%) and 1694 keV (14%). However, neutron absorption can also proceed *via* the production of radioactive ^{128}I, which produces a strong beta continuum with a half-life of 25 mins. As such, the build-up of this induced background can swamp the prompt neutron response of Hg. As with CdZnTe, HgI$_2$ is also a very efficient gamma-ray detection medium, and so its operation as a thermal neutron detector in a high gamma background environment will be compromised. Bell *et al.* [28] proposed improving the sensitivity by coating the surfaces of an HgI$_2$ detector with boron. The boron film acts as a converter by capturing neutrons entering the HgI$_2$ crystal and generating 478 keV gamma-rays *via* the transition to the first excited state of the ^{10}B(n, α)^{7}Li* reaction. This occurs in 94% of thermal captures on boron and is the only gamma-ray emitted in this reaction. An HgI$_2$ detector of a few mm thickness would be expected to have good photopeak efficiency at this energy. The presence of both the 368 keV and 478 keV gamma-rays in the HgI$_2$ detector with the correct ratio results in an improvement of the neutron capture signature. Bell *et al.* [29] extended this analysis and demonstrated that the information in both lines could be used to estimate the average energy of incident neutrons and also to distinguish between unmoderated radioactive and fission sources. It should be noted that since the energies of these gamma-rays are lower than the main lines from Cd observed in a CdZnTe detector, the probability of gamma-ray detection is higher in an HgI$_2$-based device. For a 25 mm×25 mm× 2.6 mm HgI$_2$ crystal, Bell *et al.* [29] estimated the thermal neutron capture efficiency of a boron clad detector to be 4%.

10.3.4 Lithium Based Devices

Lithium containing semiconductors have also been explored as materials for neutron detection. For example, Tupsitsyn *et al.* [30] synthesized and tested four Li based A^{I}B^{III}C^{VI} chalcogenide compounds, namely, LiInSe$_2$, LiGaSe$_2$, LiGaTe$_2$ and LiInTe$_2$. These materials are attractive, in that they are chemically and physically stable, have a bandgap suitable for room-temperature radiation detection and offer the possibility of a thermal neutron response by detecting the reaction products following neutron capture on Li (see Table 8.1). The disadvantage of using these compounds is that they only contain about 3% by weight of Li and as such the overall neutron detection efficiency is relatively low – even if isotropically enriched ^{6}Li is used. Of the four tested compounds only LiInSe$_2$ showed a response to alpha particles from a ^{241}Am source with a FWHM energy resolution of ~80%. More importantly, a simple pad detector fabricated out of LiInSe$_2$ also showed a clear response to a moderated ^{252}Cf neutron source but no response to an unmoderated source, indicating that the device is indeed sensitive to thermal neutrons [30]. Wiggins *et al.* [31] pointed out that indium also has a reasonably large neutron capture cross section (~200 barns) *via* the reaction

$$^{115}\text{In} + \text{n} \Rightarrow {}^{116\text{m}}\text{In}^* \Rightarrow {}^{116}\text{Sn} + \beta + \gamma \tag{3}$$

and in fact, for a fully absorbing sample of LiInSe$_2$, 18% of captures take place *via* this channel. However, these events add little to the neutron response, since the reactions products consist of betas and gammas, rather than alpha particles which are more easily converted into a detection signal. The authors suggest that by replacing the In with Ga to form another compound in the chalcopyrite family, for example LiIn$_{1-x}$Ga$_x$Se$_2$, the relative upper bound for neutron detection efficiency can be increased from 80% to 97% simply by removing the non-productive indium capture channel and allowing capture to proceed through the ^{6}Li(n,α) channel instead. An additional benefit of the quaternary system is that it has a lower melting point than the ternary system, which promotes easier growth.

Montag *et al.* [32] explored the Nowotny-Juza compounds [33], LiZnP and LiZnAs, as possible solid state neutron detecting materials. The Nowotny-Juza compounds, A^{I}B^{II}C^V (*e.g.*, LiZnP, LiZnAs, LiZnN, LiCdP, LiCdAs, LiMgN, LiMgP and LiMgAs) comprise a special class of filled tetrahedral semiconductors that were originally considered photonic materials (in view of their direct bandgap) and continue to be studied for solar cell [34] and Li ion battery applications [35]. Montag *et al.* [32] synthesized LiZnP and LiZnAs by compounding equimolar portions of Li, Zn and P or As in quartz ampoules. The base material was then purified by a static sublimation process and bulk crystals grown by high temperature Bridgman. LiZnP and LiZnAs wafers were cut from bulk ingots and diced into rectangular detector pieces of various sizes up to $2.1 \times 4.1 \times 4.2$ mm^3. After polishing, Ti/Au metallic contacts were applied to the top and bottom surfaces by evaporation. Both materials were found to have high resistivities in the range $10^6 - 10^{11}$Ω-cm. However, while both materials showed a response to a ^{241}Am alpha source and a weak response to neutrons, the overall performance was unstable. This was attributed to poor material quality.

10.3.5 Uranium Based Devices

Meek [36] pointed out that a number of ^{235}U and ^{238}U compounds are semiconducting materials. In particular, the oxides, such as uranium oxide,[5] (UO$_2$) and triuranium octoxide (U$_3$O$_8$), have intrinsic electrical and electronic properties similar to those of Si, Ge and GaAs with room-temperature resistivities and bandgaps in the range ~10^3 Ω-cm and 1–2 eV, respectively. In addition, the refractory nature of the oxides makes them particularly attractive for use in high-temperature and/or high-radiation applications. Uranium-based films are relatively easy to grow and can be used in indirect-conversion devices or for active semiconductor layers.

Kruschwitz *et al.* [37] proposed direct neutron detection using depleted uranium dioxide (UO$_2$). In such a device, a neutron interacts with a uranium nucleus, inducing fission during which a compound nucleus is formed and then splits into fast-moving lighter elements (fission fragments). In the case of ^{235}U the reaction proceeds as follows,

$$^1\text{n}+{}^{235}\text{U} \rightarrow {}^{236}\text{U}^* \ \rightarrow \ \text{fission fragments, e.g.,} \begin{cases} {}^{87}\text{Br}+{}^{146}\text{La} + 3\,{}^1\text{n} \\ {}^{90}\text{Sr}+{}^{114}\text{Xe} + 2\,{}^1\text{n} \\ {}^{96}\text{Rb}+{}^{137}\text{Cs} + 3\,{}^1\text{n} \\ {}^{137}\text{Te}+{}^{97}\text{Zr} + 2\,{}^1\text{n} \\ {}^{139}\text{Ba}+{}^{94}\text{Kr} + 3\,{}^1\text{n} \end{cases} \quad Q = 201 \text{ MeV} \tag{4}$$

Note that ~50 sets of reactions are possible leading to a range of fission products, typically one isotope with a mass number around 85–105, and another with a mass number about 50% larger or about 130–150. However, unlike capture reactions in which the total energy released by the reaction products is of the order of a few MeV, the average total energy of fission fragments is enormous (*i.e.*, >165 MeV). In some reactions, prompt and delayed gamma-rays (of energies ~8 MeV) are also emitted. However, even though uranium compounds have large stopping powers (by virtue of their high effective z), such a high-energy release during fission means that gamma-rays and other particles can be effectively discriminated against,

---

[5] also known as urania or uranous oxide

merely by energy thresholding. Thus, noise discrimination in such a device should be excellent, particularly when operated in mixed high-radiation fields. Another unique feature of uranium compounds is that they are efficient devices for detecting both thermal and fast neutrons. ^{235}U is sensitive to thermal neutrons, whereas ^{238}U is sensitive to fast neutrons. Whilst there is very little ^{235}U in depleted uranium, the cross-section for thermal neutron fission in ^{235}U is so much larger (580 barns) than the cross-section for MeV neutron fission in ^{238}U(~2 barns) that as a consequence, efficiencies for thermal and 10 MeV neutrons are expected to be comparable.

Kruschwitz *et al.* [37] fabricated direct neutron detection devices from single-crystal depleted UO_2 samples grown using an arc-fusion technique. The samples were approximately 1.3 cm in diameter and ~1.3 mm thick. Thin layers (50 nm) of Au and Ag were sputtered onto each surface of the samples to create Schottky and ohmic contacts, respectively. Current-voltage characteristics showed that, in general, the samples exhibit Schottky diode behaviour with reverse-bias breakdown voltages in the range 5 to 10 Volts. The devices were found to be sensitive to neutrons from a thermalized ^{252}Cf source showing clear enhancements in high-energy events when exposed to the source. However, expected spectral signatures were not resolved. This was attributed to electronics problems and the poor quality of the Schottky diodes.

10.3.6 Boron Based Devices

Of the available compounds, those based on boron are arguably the most promising in view of boron's high-thermal neutron absorption cross-section and the large number of boron compounds available. The electron deficiency of boron together with its small atomic size allow boron atoms to coalesce in three-dimensional atomic networks, consisting of boron icosahedra linked together to form the various polymorphs. Boron bonds with the p-block elements of the periodic table to form a variety of binary and ternary covalently bonded compounds like BN (both hexagonal and cubic), BP (cubic and rhombohedral), BxC, BAs, BCxOy and similar compounds. Since the densities of boron compounds are low, the background from gamma-rays, which invariably accompany neutron production, will also be low. In Table 10.1 we list the properties of the some of the more promising common boron compounds for direct detection applications. For efficient neutron detection, BP, BN, B_4C and B are particularly interesting since they are refractory, chemically inert, low density and already being actively developed for high-temperature/high-power electronic applications (*e.g.*, [38]). The phosphide has a cubic zincblende structure, while the nitride and carbide are available in three phases – cubic, hexagonal and amorphous. Until recently, impurities, high-growth temperatures and the lack of suitable substrates prevented the production of detector quality material. However with improvements in growth techniques and particularly in CVD, great strides forward in the production of monocrystalline material for all three boron compounds have been made. Zhang *et al.* [39] have reported the production of cubic BN films, Kumashiro *et al.* [40] reported the production of single crystalline BP wafers and Lee *et al.* [41] reported the production of electronic grade boron carbide films.

10.3.6.1 Boron Doping and Alloying

Perhaps the simplest implementation is to dope a conventional semiconductor detector material with boron, and in fact, thermal neutron detection has been claimed using boron doped diamond [42,43]. We also note, that such a device will have a sensitivity to fast neutrons *via* the reactions, $^{12}C(n,\alpha)^9$Be and $^{12}C(n,n')^{12}C^*$. However, since the active component is a dopant, it will constitute a very small fraction of the overall detector volume, with a corresponding impact on detection efficiency. In fact, even though boron doping concentrations can be very high in diamond (up to 10^{23} atoms cm^{-3}), for detector applications there is a limit imposed by the increasing metallization of the host with doping concentration. The transition from insulator to metal occurs for boron concentrations of ~2.2% and the conductivity becomes quasi-metallic at room temperature.

Along similar lines, Atsumi *et al.* [44] proposed creating neutron sensitivity in a conventional semiconductor, simply by alloying with a neutron-absorbing compound such as ^{10}B. As an example, they chose GaN, which is also a refractory material, capable of operating in extreme environments with a low sensitivity to gamma-rays. They proposed alloying GaN with BN to form $B_xGa_{1-x}N$, although alloying AlN with BN to form $B_xAl_{1-x}N$ is another possibility.

BGaN growth techniques had already been studied for producing solar blind UV MSM detectors, since adding boron to GaN increases the bandgap from the blue to the UV spectral region [45]. Atsumi *et al.* [44] proposed introducing the boron component for neutron sensitivity rather than bandgap tuning. However, while an inherently simple technique for creating neutron sensitivity, the maximum attainable neutron efficiency will be low primarily because a phase separation occurs for boron contents in excess of 2% as a consequence of the large lattice mismatch between BN and GaN. In fact, the phase diagram displays large miscibility gaps in the temperature range usually adopted to grow the alloy. A similar problem exists for $B_xAl_{1-x}N$. In a study of the electrical and structural characterizations of MOCVD grown $B_xGa_{1-x}N$, Baghdadli *et al.* [46] found that both the resistivity and mobility increased strongly with boron composition, x. Specifically, the resistivity increased from 3×10^{-3} Ω-cm for $x=0\%$ to ~10^5 Ω-cm for $x=1.75\%$, whilst electron mobility increased from 60 cm^2 V^{-1} s^{-1} ($x=0\%$) to 290 cm^2 V^{-1} s^{-1} ($x=1.75\%$).

TABLE 10.1

General properties of boron compounds suitable for direct detection applications.

Material	β-B	B$_4$C	BP	B$_{12}$P$_2$	BAs	B$_{12}$As$_2$	B$_x$Ga$_{1-x}$N	c-BN	h-BN
Lattice structure	Rhomb	Rhomb	ZB	Rhomb	ZB	Rhomb	ZB	ZB	H
Lattice constant (nm)	1.019, 2.38	0.59,1.185	0.454	0.599,1.184	0.477	0.616,1.193	0.308[1]	0.362	0.250, 0.666
Molecular weight (g mol^{-1})	10.8	55.26	41.8	191.1	85.7	279.6	82.6[1]	24.8	24.8
Density (g cm^{-3})	2.35	2.52	2.94	2.60	5.22	3.58	6.07[1]	3.45	2.18
Melting point (°C)	2092	2763	1373	2393	2075	2027	-	2973	2600
Hardness (kg mm^{-2})	2110–2580	2900–3580	3263	2644	1937	2493	-	5400	3400
Thermal conductivity (Wcm^{-1} K^{-1})	0.27	0.35	4.6	0.38	11.4	1.2	-	7.4	0.20
Resistivity (Ω-cm)	10^6	10	20	10^5	0.01	10^4	~10^5 (x = 2%)	~10^{15}	10^{13}
Breakdown field (MV cm^{-1})	0.01–0.06	0.05–0.1	0.4–1.0	-	-	-	3.8	2–6	30($\perp$), 12(ıı)
Dielectric constant (ε_o)	10.6	10.0	11.0	6.6	9.9	7.8($\perp$), 9.0(ıı)	-	7.1	6.9($\perp$), 5.1(ıı)
Bandgap @ RT (eV)	1.5 (I)	2.1 (D)	2.1 (I)	3.4 (I)	1.46 (I)	3.2–3.47 (I)	$2.097+2.672x$ (D)[2]	6.4 (I)	5.2 (D)
Electron mobility (cm^2 s^{-1} V^{-1})	< 300	1	500	50	400	50–100	300[1]	500	50
Hole mobility (cm^2 s^{-1} V^{-1})	2	2	70	30	50–100	25	-	<500	20

[1] x=2%
[2] for x<0.75

Atsumi et al. [44] grew B$_x$Ga$_{1-x}$N crystals by MOCVD using natural boron in the triethyleneboron source gas. A 1 μm thick GaN buffer layer was first grown on a sapphire substrate followed by a 200 nm BGaN layer. The boron content of the BGaN layer was determined by SIMS to be x=1.2%. A Schottky diode was then fabricated by depositing a 10 nm thick GaN contact layer onto the BGaN followed by 1.5 mm diameter Au electrodes deposited by vacuum deposition. The device's response to radiation was determined by measuring the change in conduction of the device at a nominal bias of 3V. Exposure to a 5.4 MeV alpha particle source increased the dark current through the device by ~40%. No response was found to gamma-rays. The device also showed a response to individual neutron events – the neutrons being generated by a moderated ^{252}Cf source.

10.3.6.2 Boron

It should be noted that boron is itself a semiconductor, so it should be possible to fabricate a detector from epitaxially grown boron films, thus greatly simplifying the growth and fabrication process. In fact, boron is the only element in Group IIIb that possesses semiconducting properties with a bandgap of 1.50 or 1.56 eV. In principle, a depleted thickness of only ~200 μm is needed for 100% detection efficiency. Elemental boron can exist in several allotropes, the most common being brown amorphous boron and crystalline boron, which exists in three major polymorphic forms. These are red crystalline α-rhombohedral boron, black crystalline β-rhombohedral boron (the most thermodynamically stable allotrope) and black crystalline β-tetragonal boron. All are based on various modes of condensation of the B$_{12}$ icosahedron. For example, α-rhombohedral boron has a unit cell of 12 boron atoms. The structure consists of icosahedra in which each boron atom has five nearest neighbours within the icosahedron. β-rhombohedral boron has a unit cell containing 105–108 atoms. Most atoms form B12 discrete icosahedra.

Crystalline boron is grey/black and very hard (comparable to cubic boron nitride) with a melting point of 2,080°C and is a weak conductor at room temperature. Growth difficulties have so far prevented the use of boron as a semiconductor, so it is hardly surprising that little work has been reported in the literature. The electron and hole mobilities are reported to be 0.7 and 2 cm^2 V^{-1} s^{-1}, which are very low. The carrier lifetimes are unknown and if poor would severely limit the thickness of an active boron layer – thus capping the achievable detection efficiency. Wald and Bullitt [47] proposed using red alpha-rhombohedral boron for use as a semiconductor neutron detector. They prepared microscopic crystals by a number of methods including the vapor-liquid-solid and travelling solvent method variations of CVD as well as precipitation from copper-gold solutions. They concluded that trap densities were too high and free carrier mobilities were too low, precluding its use for direct radiation detection.

Tomov et al. [48] fabricated detectors from crystalline beta-boron films grown on Si substrates by PECVD. The film thicknesses ranged from 500 nm to 3.8 μm. Simple planar detectors were formed by evaporating 3 mm^2 diameter Au/Cr contacts onto the top and bottom surfaces of the films. All devices showed strongly rectifying behaviour and a spectroscopic response to alpha particles ($\Delta E/E \sim 5\%$). They also showed a response to neutrons. This is illustrated in Fig. 10.6. For the neutron response, the individual components of the spectrum can be identified. Since the diodes were fabricated from natural

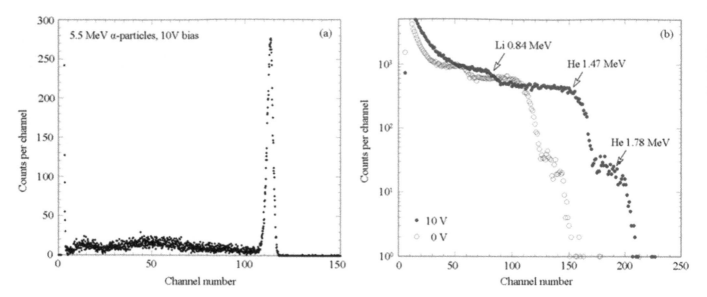

FIGURE 10.6 (a) The response of a 12.6 mm², 1.7 μm thick β-boron diode to a 5.5 MeV alpha particles. The bias was 10V. (b) The response to thermal neutrons from the Delft 2MW nuclear research reactor. Even under zero bias the diode shows a clear response.

boron, efficiencies are expected to be low because of the low isotropic concentration of ¹⁰B. In fact, the measured efficiency of a 2.5 μm thick diode was 0.5%. Coupled with the relatively thin boron layers, it is not clear if the response is due to an active boron layer, a boronated Si p-n junction or both. The latest generation of devices now have such low enough leakage currents (~100 nA) that they also show a response to 60 keV gamma-rays. This is illustrated in Fig. 10.7 in which we show the measured energy loss spectrum of a 15 mm², 2.8 μm thick device when exposed to a ²⁴¹Am radioactive source. A 60 keV peak is clearly resolved (shown in the inset) with an energy resolution of ~30% FWHM. The FWHM energy resolution of the alpha peak at 5.5 MeV is 12%. Note that the source used is a "covered" source, in that the active material is coated with a thin protective film. The width of the 5.5 MeV line is largely due to straggling in this film. Subsequent thermal neutron measurements [49] at the Rez nuclear reactor, near Prague, indicated a neutron detection efficiency of 1%.

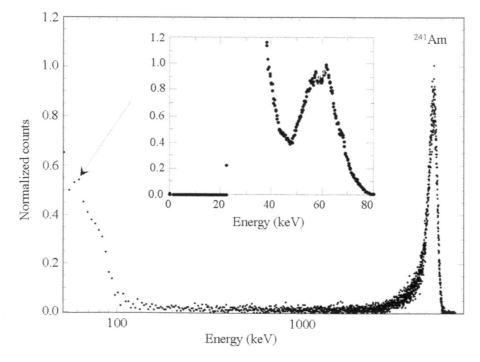

FIGURE 10.7 Alpha and photon response of a 16 mm², 2.8 μm thick boron diode to a ²⁴¹Am radioactive source. The inset contains an expansion of the spectrum in the vicinity of 60 keV showing that the 60 keV gamma-ray line is clearly resolved. The detector bias was 6 V and shaping time 2 μs.

10.3.6.3 Boron Carbide

Robertson *et al.* [50] reported direct neutron detection using icosahedral B_5C as the detection medium. A detector was fabricated from a 276 nm B_5C film deposited by PECVD on an n-type Si(111) substrate. One mm^2 Cr/Au contacts were then applied to the top and bottom surfaces by sputtering through a mask. Whilst the detector showed a clear response to neutrons, its spectral signature was unusual, leading to the suggestion that the neutron signal actually originated in a boronated Si layer at the B_5C/Si substrate interface [51] rather than the B_5C bulk. In fact, two known compounds of boron and silicon exist, B_6Si and B_4Si, both of which are semiconducting with a bandgap close to 0.5 eV. Later work with boron-only-based heterostructures (*i.e.*, homojunctions) demonstrated that B_5C is indeed an active semiconductor [52], and weak neutron capture sum peaks from the reaction products were eventually observed [53]. However, measured detection efficiencies were very low. This was attributed to a high concentration of traps and/or recombination centered resulting in poor charge collection.

10.3.6.4 Boron Phosphide

In the early 1970s, Ananthanarayanan *et al.* [54] demonstrated that small 6.4 mm diameter, 1 mm thick polycrystalline BP and BN detectors were responsive to thermal neutron fluxes of 10^8 $cm^{-2}s^{-1}$. Lund *et al.* [55] fabricated radiation detectors from BP films grown by chemical vapor deposition on (100) orientated n-type Si substrates. The BP films were typically 1–10 mm thick. The detectors showed a spectroscopic response to 5.5 MeV alpha-particles but were surprisingly unresponsive to thermal neutrons. Kumashiro *et al.* [40] grew single crystal (10 mm×20 mm×300 μm thick) wafers of ^{10}BP on both (100) and (111) orientated Si substrates using CVD. In exposure to thermal neutrons of fluence ~10^4 n cm^{-2} s^{-1}, no change in electrical properties was found for the wafers grown on Si(100) substrates and only a slight change for those grown on Si (111) substrates.

10.3.6.5 Boron Nitride

Boron nitride (BN) is a refractory compound of boron and nitrogen, which crystallizes in three allotropic modifications. Like carbon, the hexagonal form (h-BN) is the most stable and consists of layers of tiny platelets similar to graphite. The cubic variety (c-BN) is analogous to diamond, and while it is softer (it is the second hardest material known), its thermal and chemical stability properties are superior. Cubic-BN has an indirect bandgap of 6.4 eV at room temperature and has in fact the largest indirect bandgap of the diamond-like semiconductors. However it is metastable at standard conditions and converts to h-BN when heated to 1700 K in vacuum. A third wurtzite modification also exists (w-BN) although it is metastable at all conditions in the solid state.

Hexagonal BN (h-BN) is the most widely used polymorph. Unlike the cubic allotrope, it is a direct bandgap semiconductor with a bandgap of 6.0 eV. Owing to its low dielectric constant, high thermal conductivity and chemical inertness, h-BN has found use across a wide range of applications. For example, it is extensively used by the cosmetics industry [56] as a base for foundations, in electronics as a substrate, in tribology as an additive for high temperature lubrication and in microwave applications as a transparent window. The cubic form has potential applications in high temperature/power electronic devices while the hexagonal form is currently being explored as a deep ultraviolet photonic material [57] and as a building block for van der Waals heterostructures [58]. The potential for using BN for the detection of thermal neutrons has been proposed by a number of authors. BN formed from natural boron has an effective thermal neutron absorption coefficient of 42 cm^{-1}. This increases significantly to 211 cm^{-1} if enriched ^{10}B is used and when coupled with its weak sensitivity to gamma-rays by virtue of its relatively low density (2.2 g cm^{-3}) makes it an attractive material for thermal neutron detection.

Kaneko *et al.* [59] fabricated single and polycrystalline cubic boron nitride (cBN) crystals synthesized using a high-pressure and high-temperature method. Single crystal samples of area ~1–2 mm^2 and polycrystalline samples of area 20 mm^2 were mechanically polished to a thickness of 0.3 mm and Al electrodes deposited on the top and bottom surfaces by evaporation. Surprisingly, they found that, while the single crystal device did not detect alpha particles due to high leakage currents, the polycrystalline device did (although not spectroscopic). It also showed a direct sensitivity to neutrons.

McGregor *et al.* [60] reported a series of experiments on commercially available pyrolitic[6] boron nitride (p-BN) material. The devices were constructed from 5 mm×45 mm×1 mm thick samples of p-BN. Circular contacts of 2.5 mm diameter were applied to both sides by evaporating 600 Å of Ti followed by a further evaporation of 1,000 Å Au through a shadow mask. The devices were mounted on sapphire substrates and operated as traditional planar detectors. The I-V characteristics show non-rectifying behaviour with a bulk resistivity of ~10^{12} Ω-cm. At a nominal bias of 400V the devices showed a clear response to thermal neutrons from a reactor (see Fig. 10.8). The thermal neutron counting efficiency was found to vary from 1.2% to 7.2% between samples – the variation was attributed to the fact that the bulk of the response is confined to a few channels of the spectrum above the noise and is therefore extremely sensitive to threshold and polarization effects. Based on

---

[6] Pyrolytic in this case, meaning an ordered solid of individual crystallites showing strong anisotropic properties.

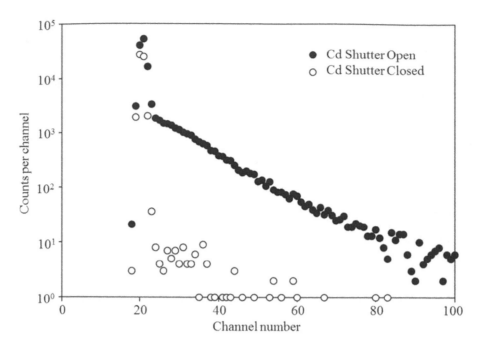

FIGURE 10.8 Pulse height spectra from a p-BN detector exposed to a thermal neutron beam from a reactor demonstrating a clear response (taken from McGregor *et al.* [60]).

the ^{10}B content of the samples, the expected efficiencies should be ~98%, indicating poor charge transport, probably caused by non-uniform electric fields.

Li *et al.* [61] have proposed direct detection using h-BN. Epitaxial layers of thickness 1 μm were synthesized on a sapphire substrate by MOCVD using triethylboron and ammonia precursors. To alleviate the effects of strain caused by lattice mismatch, a 20 nm BN/AlN buffer layer was first grown on the substrate prior to depositing the BN epi-layer. In order to optimize charge collection, a microstrip contact system was used to collect charge. The strips were formed on the BN epi-layer by an evaporation of Ni/Au – the strip width being 5 μm and the spacing between strips also 5 μm. The overall detector size was 1 mm×1.2 mm. The detector was found to be responsive to neutrons from a reactor. Although the efficiency for detecting individual neutrons that interact in the epi-layers was expected to be 77% (which indicates direct neutron detection), the overall system efficiency was extremely low partly because of the thinness of the epi-layer (1 μm), partly that natural boron was used and partly poor charge collection. The authors estimate that an active layer of 200 μm of boron-enriched h-BN would be sufficient to capture 98.5% of thermal neutrons. However, it should be borne in mind that the growth of such a thick layer by MOCVD is commercially impractical in view of cost driven by low growth rates that are typically 3 μm per hour or less.

In later work by Doan *et al.* [62], two new detectors were fabricated using the same general format. The first had a 2.7 μm thick h-BN epi-layer using ^{10}B enriched material. The detector size was 5 × 5 mm^2. The contacts were formed by an evaporation of Ti/Ni microstrips onto the active area. The leakage current was measured to be 10^{-11}A cm^{-2}. A mu-tau product was determined by Many's equation [63] using photo-stimulation to be 5 × 10^{-8} cm^2 V^{-1}. Because the measurements are taking place in the lateral direction between strips and both holes and electrons have similar effective masses in single sheet h-BN, it is assumed that the mu-tau products are also similar. The CCE was determined to be 83%. The device was found to be sensitive to thermal neutrons from a ^{152}Cf source, with a measured detection efficiency of ~4%. Doan *et al.* [62] also fabricated a 4.5 μm thick h-BN epi-layer of natural boron. The contact system was a microstrip consisting of 10×10 strips each 10 μm wide in an inter-digitated finger pattern. The detector size was 6×6 mm^2. In this case, the detector was found to be spectrally responsive to thermal neutrons.

The measured spectrum is shown in Fig. 10.9 (a) and exhibits well resolved peaks corresponding to the energies of ^{10}B and thermal neutron reaction described in Eq. (2) – namely the Li*, Li, α* and α peaks as well as the sum peaks of Li* + α* (2.31 MeV) and the Li + α peak at 2.785 MeV. By plotting the peak energies against channel number, the linearity curve was found be linear (see Fig. 10.9b). The measured efficiency was ~0.64%. The expected efficiency for this detector is 0.95% – the difference being attributed to poor collection efficiency (67%) and the fact that natural boron had been used for the growth. Both the 2.7 μm thick and 4.5 μm thick detectors were found to be insensitive to gamma-rays. Maity *et al.* [64] grew a 43 μm thick freestanding h-BN epi-layer using ^{10}B-enriched material from which they fabricated a 1×1 mm^2 planar detector. Instead of using a microstrip contact system as used previously [61,62], they applied conventional Ni/Au planar ohmic contacts to the top and bottom surfaces by e-beam evaporation. The measured resistivity as determined from the I/V

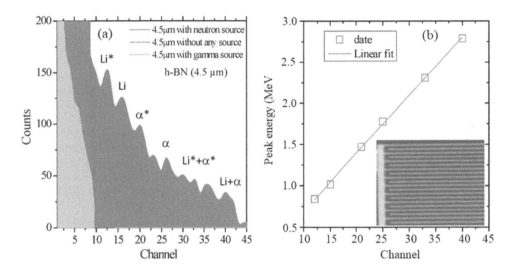

FIGURE 10.9 (a) Pulse height spectrum obtained with a 6 mm×6 mm, 4.5 µm thick h-BN detector after exposure to a [152]Cf neutron source (from [62]). The contact system is formed by a 10×10, interdigital finger metal contact array, shown in the inset of (b). The dark yellow and red column bars in (a) show the total counts measured in the absence and in the presence of the [252]Cf source, respectively. The blue-green column bars represent the measured counts while the detector was irradiated by a [137]Cs γ-ray source. (b) The linearity curve in which the expected energies of the nuclear reaction products given in Eq. (2) are plotted against peak channel. The open squares in (b) are the data points and the solid line is a linear fit.

characteristic was $> 10^{13} \Omega$-cm. The mu-tau products were again derived using Many's equation. These were 8.3×10^{-7} cm^2 V^{-1} for electrons and 2.2×10^{-5} cm^2 V^{-1} for holes. Since $m_e^* \sim m_h^*$, this means that $\tau_h \gg \tau_e$. At a bias of 400V, the charge collection efficiency was determined to be 86%. The detector was found to be responsive to thermal neutrons with a measured detection efficiency of 51%. However, the spectral quality was poor, reflecting the poor transport properties coupled with the relatively large thickness of the detector.

10.3.6.6 Boron Arsenide

Icosahedral boron arsenide, B$_{12}$As$_2$, has attracted attention as a material for both neutron detection and energy generation in radioisotope batteries [65]. For neutron detection it has several advantages over other boron-rich compounds. For example, apart from the high boron content, hole mobilities are reported to be relatively high (of the order of 50–100 cm^2 s^{-1} V^{-1}), which when coupled with its wide bandgap of 3.2 eV (and potentially low leakage currents) should result in efficient neutron detection, particularly if synthesized from enhanced [10]B boron. The interest in using B$_{12}$As$_2$ for radioisotope battery applications is driven by the fact that it is a refractory material, mechanically hard (with a microhardness similar to boron nitride) with a very high melting point. More significantly, it is intrinsically radiation hard with the unique ability to "self-heal". Specifically, in a study of irradiated boron-rich solids, Carrard *et al.* [66] and Emin [67] found no evidence of defect clusters or amorphization in B$_{12}$As$_2$ following heavy doses of ionizing radiation. They concluded that radiation-induced atomic vacancies and interstitials spontaneously recombine as a direct consequence of the unusual structural and electronic stability of boron icosahedra – even when in a degraded state.

The very high melting temperature of B$_{12}$As$_2$ (2400°C) makes melt crystal growth difficult. Similarly, its large coefficients of thermal expansion makes CVD growth problematic, since only a limited number of substrate materials can be used at such high temperatures, and these have quite different thermal properties. Alternative growth techniques such as the flux growth method [68] have proved particularly successful for the icosahedral borides [69,70]. Like melt growth, flux growth is based on crystallization from a liquid. However, in flux growth, a molten metal solution saturated with elemental boron and arsenic is used, and crystals are formed by precipitation rather than by freezing. The method has a distinct advantage for refractory materials in that the temperature at which the crystals form can be much lower than the melting point of the source compounds. In the case of B$_{12}$As$_2$, this is ~1150°C as opposed to 2,400°C. Whiteley *et al.* [65] used this technique to grow single B$_{12}$As$_2$ crystals of length 10–15 mm. These were cleaved into flat platelets, polished and diced into crystals of size $5 \times 4 \times 1$ mm^3. Metallic electrodes of size 0.5–1.0 mm diameter were then applied to the planar sides by evaporation. I/V measurements displayed rectifying behavior with leakage currents of ~1 pA and Hall effect measurements showed that hole mobilities were around 25 cm^2 s^{-1} V^{-1}. However, despite the low leakage currents and favorable mobilities, the devices showed no response to alpha particles from a [241]Am source. Since mu-tau products could not be measured, Whiteley *et al.* [65] concluded that the crystals suffered from very low carrier lifetimes caused by high impurity concentrations, subsequently verified by SIMS and XPS.

In Table 10.2 we summarize the relative advantages and disadvantages of the described compounds. The advantages mostly relate to boron content and the refractory nature of the materials, while the disadvantages mostly relate to growth.

TABLE 10.2

A comparison of the pros and cons of potential materials for direct neutron detection.

Material	Advantages	Disadvantages	Reported eff. (%)	Ref.
LiInSe$_2$	Chemically and physically stable, has a bandgap suitable for room temperature radiation detection. Weak response to thermal neutrons reported [30].	Overall neutron detection efficiency is relatively low since material only contains about 3% by weight of Li.	Not reported	[30]
LiZnP	A^{I}B^{II}C^V compound with zincblende structure. High concentration of Li. Neutron detection claimed [32]	Difficult to synthesize single crystals. Material quality issues leading to unstable behaviour.	Not reported	[32]
LiZnAs	A^{I}B^{II}C^V compound with zincblende structure. High concentration of Li. Neutron detection claimed [32]	Difficult to synthesize single crystals. Material quality issues leading to unstable behaviour.	Not reported	[32]
Depleted uranium dioxide (UO$_2$)	Large signal, efficient noise rejection. Fast neutron response. Not a SNM. Neutron detection claimed [37]	Procurement and handling issues. Expensive, radioactive.	Not reported	[37]
Boron (α-B, β-B)	Simple, 100% ^{10}B composition possible. Neutron detection claimed [48]. Spectral structure observed.	Difficult to grow, slow growth, no obvious substrate. α-B only thermodynamically stable at temperatures below 1,100°C. Low charge carrier mobility, high crystal defect densities.	1	[49]
Boron arsenide (BAs)	Ultra-high thermal conductivity.	Difficult to synthesize single crystals. Stable only at low temperatures (<920°C) due to high As vapor pressure. At higher temperatures decomposes to B$_{12}$As$_2$.	-	[71]
Boron subarsenide (B$_{12}$As$_2$)	High boron content, radiation hard – self-heals. Relatively high hole mobilities. Potential to self-heal radiation damage [66].	High melting point (~2,030°C) and large thermal expansion makes melt and CVD crystal growth difficult.	-	[65]
Boron nitride (h-BN)	Benign process. Low thermal expansion. Non-toxic. Neutron detection claimed [61]. Spectral structure observed.	Stacking fault energy is low, making crystal growth difficult, slow growth, diamond substrate required. Susceptible to radiation damage.	56	[61]
(c-BN)	Very high hardness and thermal conductivity. Low thermal expansion. Non-toxic. Response to neutrons claimed [59].	Extremely high pressures required (>5 GPa), limiting crystal sizes to ~mm. Less stable than the hexagonal form.	Not reported	[59]
Boron phosphide (BP)	Simple CVD, reasonable growth rates, cubic and rhombohedral forms look promising.	Phosphine gas precursor, no obvious substrate, high temperature growth. High vapor pressure of phosphorus for temperatures > 1,000°C, tends to decompose to B$_{12}$P$_2$.	-	[40]
Icosahedral boron subphosphide (B$_{12}$P$_2$)	High boron content. Self-healing properties [66].	High melting point temperature (2,120°C) complicates crystal growth from the melt.	-	[70]
Boron carbide (B$_4$C, B$_5$C)	Low temperature deposition, single, solid precursor, no lattice-matching required, neutron detection claimed [50]	Expensive precursor, many polytypes, heterostructure may be required. Very high melting temperature (2,450°C). Propensity for crystal twinning. Highly variable stoichiometry.	1	[50]
Boron gallium nitride B$_x$Ga$_{1-x}$N	Relatively simple growth. Neutron detection claimed [44].	Neutron detection efficiency limited as boron content has to be less than 2% for phase stability.	Not reported	[44]
Doped diamond C:B	Neutron detection claimed [43].	Expensive, single uninterested vendor, low detection efficiency.	Not reported	[43]

REFERENCES

[1] R.L. Bramblett, R.I. Ewing, T.W. Bonner, "A new type of neutron spectrometer", *Nucl. Instr. and Meth.*, Vol. **9** (1960), pp. 1–12.

[2] D.J. Thomas, A.V. Alevra, "Bonner sphere spectrometers—A critical review", *Nucl. Instr. and Meth.*, Vol. **476**, no. 1–2 (2002), pp. 12–20.

[3] G.F. Knoll, "*Radiation Detection and Measurement*", John Wiley & Sons Inc., New York, NY, 3rd ed. (2000), pp. 537–576. ISBN 0-471-07338-5.

[4] A.J. Peurrun, "Recent developments in neutron detection", *Nucl. Instr. and Meth.*, Vol. **443**, no. 2–3 (2000), pp. 400–415.

[5] A.N. Caruso, "The physics of solid-state neutron detector materials and geometries", *J. Phys. Condens. Mat.*, Vol. **22**, no. 44 (2010), p. 443201 (32 pp.).

[6] S. Pospíšil, B. Sopko, E. Havránková, Z. Janout, J. Koniček, I. Mácha, J. Pavlů, "Si diode as a small detector of slow neutrons", *Rad. Prot. Dosim.*, Vol. **46**, no. 2 (1993), pp. 115–118.

[7] P. Chaudhari, A. Singh, A. Topkar, R. Dusane, "Fabrication and characterization of silicon based thermal neutron detector with hot wire chemical vapor deposited boron carbide converter", *Nucl. Instr. and Meth.*, Vol. **A779** (2015), pp. 33–38.

[8] R.L. Schulte, F. Swanson, M. Kesselman, "The use of large area silicon sensors for thermal neutron detection", *Nucl. Instr. and Meth.*, Vol. **353**, no. 1–3 (1994), pp. 123–127.

[9] D.S. McGregor, M.D. Hammig, Y.-H. Yang, H.K. Gersch, R.T. Klann, "Design considerations for thin film coated semiconductor thermal neutron detectors—I: Basics regarding alpha particle emitting neutron reactive films", *Nucl. Instr. and Meth.*, Vol. **A500** (2003), pp. 272–308.

[10] T. Aoyama, Y. Oka, K. Honda, C. Mori, "A neutron detector using silicon PIN photodiodes for personal neutron dosimetry", *Nucl. Instr. and Meth.*, Vol. **A314**, no. 3 (1992), pp. 590–594.

[11] A. Miyake, T. Nishioka, S. Singh, H. Morii, H. Mimura, T. Aoki, "A CdTe detector with a Gd converter for thermal neutron detection", *Nucl. Instr. and Meth.*, Vol. **654**, no. 1 (2011), pp. 390–393.

[12] A. Pappalardo, M. Barbagallo, L. Cosentino, C. Marchetta, A. Musumarra, C. Scirè, S. Scirè, G. Vecchio, P. Finocchiaro, "Characterization of the silicon+^{6}LiF thermal neutron detection technique", *Nucl. Instr. and Meth.*, Vol. **A810** (2016), pp. 6–13.

[13] J. Uher, C. Fröjdh, J. Jakůbek, C. Kenney, Z. Kohout, V. Linhart, S. Parker, S. Petersson, S. Pospíšil, G. Thungström, "Characterization of 3D thermal neutron semiconductor detectors", *Nucl. Instr. and Meth.*, Vol. **A576** (2007), pp. 32–37.

[14] K. Shultis, D.S. McGregor, "Design and performance considerations for perforated semiconductor thermal-neutron detectors", *Nucl. Instr. and Meth.*, Vol. **A606** (2009) pp. 608–636.

[15] K.-.C. Huang, R. Dahal, J.J.Q. Lu, A. Weltz, Y. Danon, I.B. Bhat, "Scalable large-area solid-state neutron detector with continuous p-n junction and extremely low leakage current", *Nucl. Instr. and Meth.*, Vol. **A763** (2014), pp. 260–265.

[16] S.L. Bellinger, R.G. Fronk, W.J. McNeil, T.J. Sobering, D.S. McGregor, "Enhanced variant designs and characteristics of the microstructured solid-state neutron detector", *Nucl. Instrum. Meth.*, Vol. **A652** (2011), pp. 387–391.

[17] S.L. Bellinger, R.G. Fronk, T.J. Sobering, D.S. McGregor, "High-efficiency microstructured semiconductor neutron detectors that are arrayed, dual-integrated, and stacked", *Appl. Radiat. Isot.*, Vol. **70** (2012), pp. 1121–1124.

[18] S.L. Bellinger, R.G. Fronk, W.J. McNeil, T.J. Sobering, D.S. McGregor, "Improved high efficiency stacked microstructured neutron detectors back filled with nanoparticle Lif", *IEEE Trans. Nucl. Sci.*, Vol. **59** (2012), pp. 167–173.

[19] Q. Shao, L.F. Voss, A.M. Conway, R.J. Nikolic, M.A. Dar, C.L. Cheung, "High aspect ratio composite structures with 48.5% thermal neutron detection efficiency", *Appl. Phys. Lett.*, Vol. **102** (2013), pp. 063505-1–063505-4.

[20] T.R. Ochs, S.L. Bellinger, R.G. Fronk, L.C. Henson, C.J. Rietcheck, T.J. Sobering, R.D. Taylor, D.S. McGregor, "Fabrication of present-generation microstructured semiconductor neutron detectors", *Proc. IEEE, Nuclear Science Symposium and Medical Imaging Conference (NSS/MIC)*, Seattle, WA (2014), pp. 1–3.

[21] R.T. Johnson, "Semiconductor-neutron detectors utilizing radioactive decay", *Jr., Nucl. Instr and Meth.*, Vol. **77** (1970), pp. 189–196.

[22] R.T. Johnson Jr., "Fast–neutron irradiation effects in CdS", *J. Appl. Phys.*, Vol. **39** (1968), pp. 3517–3526.

[23] M. Fasasi, M. Jung, P. Siffert, C. Teissier, "Thermal neutron dosimetry with cadmium telluride detectors", *Radiat. Prot. Dosim.*, Vol. **23** (1988), pp. 429–431.

[24] J.K. Tuli, "Thermal neutron capture gamma rays", *BNLNCS–5 1647 Report*, UC–34–C, Brookhaven National Laboratory (1983).

[25] D.S. McGregor, J.T. Lindsay, R.W. Olsen, "Thermal neutron detection with cadmium$_{1-x}$ zinc$_x$ telluride, semiconductor detectors", *Nucl. Instr. and Meth.*, Vol. **A381** (1996), pp. 498–501.

[26] A.G. Beyerle, K.L. Hull, "*Neutron* detection with mercuric iodide detectors", *Nucl. Instr. and Meth.*, Vol. **A256** (1980), pp. 377–380.

[27] M. Melamud, Z. Burshtein, A. Levi, M.M. Schieber, "New thermal–Neutron solid state electronic based on HgI$_2$ single crystals", *Appl. Phys. Lett.*, Vol. **43** (1983), pp. 275–277.

[28] Z.W. Bell, K.R. Pohl, L. van Den Berg, "Neutron detection with mercuric iodide", *IEEE Trans. Nucl. Sci.*, Vol. **51** (2004), pp. 1163–1165.

[29] Z.W. Bell, W.G. West, K.R. Pohl, L. van Den Berg, "Monte Carlo analysis of a mercuric iodide neutron/gamma detector", *IEEE Trans. Nucl. Sci.*, Vol. **52** (2005), pp. 2030–2034.

[30] E. Tupitsyn, P. Bhattacharya, E. Rowe, L. Matei, Y. Cui, V. Buliga, M. Groza, B. Wiggins, A. Burger, A. Stowe, "Lithium containing chalcogenide single crystals for neutron detection", *J. Cryst. Growth*, Vol. **393** (2014), pp. 23–27.

[31] B. Wiggins, J. Bell, A. Burger, K. Stassun, A.C. Stowe, "Investigations of ^{6}LiIn$_{1-x}$Ga$_x$Se$_2$ semi-insulating crystals for neutron detection", *Proc. SPIE*, Vol. **9593** (2015), pp. 95930B-1–95930B-9.

[32] B.W. Montag, P.B. Ugorowski, K.A. Nelson, N.S. Edwards, D.S. McGregor, "Device fabrication, characterization, and thermal neutron detection response of LiZnP and LiZnAs semiconductor devices", *Nucl. Instr. and Meth.*, Vol. **A836** (2016), pp. 30–36.

[33] H. Nowotny, K. Bachmayer, "Die Verbindungen LiMgP, LiZnP und LiZnAs", *Monatsh. Chem. und Verwandte Teile Anderer Wissenschaften*, Vol. **81**, no. 4 (1950), pp. 488–496.

[34] D. Kieven, A. Grimm, A. Beleanu, C.G.F. Blum, J. Schmidt, T. Rissom, I. Lauermann, T. Gruhn, C. Felser, R. Klenk, "Preparation and properties of radio-frequency-sputtered half-heusler films for use in solar cells", *Thin Solid Films*, Vol. **519**, no. 6 (2011), pp. 1866–1871.

[35] M.P. Bichat, L. Monconduit, J.L. Pascal, F. Favier, "Anode materials for lithium ion batteries in the Li-Zn-P system", *Ionics*, Vol. **11**, no. 1 & 2 (2005), pp. 66–75.

[36] T. Meek, "Semiconductive properties of uranium oxides", *Proc. Waste Management 2001 Symposium*, Tucson, Arizona February 25–March 1 (2001).

[37] C.A. Kruschwitz, S. Mukhopadhyay, D. Schwellenbach, T. Meek, B. Shaver, T. Cunningham, J.P. Auxier, "Semiconductor neutron detectors using depleted uranium oxide", *Proc. SPIE*, Vol. **9213** (2014), pp. 92130C-1–92130C-9.

[38] R. Kirschman, ed., *"High–Temperature Electronics"*, John Wiley & Sons, Ltd., Surrey, UK (1998) ISBN: 978-0-7803-3477-9.

[39] W. Zhang, H.-G. Boyen, N. Deyneka, P. Ziemann, F. Banhart, M. Schreck, "Epitaxy of cubic boron nitride on (001)–oriented diamond", *Nat. Mat.*, Vol. **2** (2003), pp. 312–315.

[40] Y. Kumashiro, K. Kudo, K. Matsumoto, Y. Okado, T. Koshiro, "Thermal neutron irradiation experiments on ^{10}bp single-crystal wafers", *J. Less-Common Metals*, Vol. **143** (1988), pp. 71–75.

[41] S. Lee, J. Mazurowski, G. Ramseyer, P.A. Dowben, "Characterization of boron carbide thin films fabricated by PECVD from boranes", *J. Appl. Phys.*, Vol. **72** (1992), pp. 4925–4933.

[42] A.J. Whitehead, *"Diamond radiation detector"*, United States Patent application 6952016 (2005).

[43] C. Mer, M. Pomorski, P. Bergonzo, D. Tromson, M. Rebisz, T. Domenech, J.-C. Vuillemin, F. Foulon, M. Nesladek, O. A. Williams, R.B. Jackman, "An insight into neutron detection from polycrystalline CVD diamond films", *Diamond and Related Materials*, Vol. **13**, no. 4–8 (2004), pp. 791–795.

[44] K. Atsumi, Y. Inoue, H. Mimura, T. Aoki, T. Nakano, "Neutron detection using boron gallium nitride semiconductor material", *APL Mat.*, Vol. **2** (2014), pp. 032106–1–032106–6.

[45] H. Srour, J.P. Salvestrini, A. Ahaitouf, S. Gautier, T. Moudakir, B. Assouar, M. Abarkan, S. Hamady, A. Ougazzaden, "Solar blind metal-semiconductor-metal ultraviolet photo detectors using quasi alloy of BGaN/GaN superlattices", *Appl. Phys. Letts.*, Vol. **99** (2011), pp. 221101-1–221101-3.

[46] T. Baghdadli, S. Ould Saad Hamady, S. Gautier, T. Moudakir, B. Benyoucef, A. Ougazzaden, "Electrical and structural characterizations of BGaN thin films grown by metal-organic vapor-phase epitaxy", *Phys. Status Solidi C*, Vol. **6**, no. S2 (2009), pp. S1029–S1032.

[47] F. Wald, J. Bullitt, "Semiconductor neutron detectors," U.S. Nat. Tech. Inform. Serv., AD-771526/1 (1973), pp. 1–36.

[48] R. Tomov, R. Venn, A. Owens, A. Peacock, "The development of a portable thermal neutron detector based on a boron rich heterodiode", *Proc. SPIE*, Vol. **7119** (2008), pp. 71190H-1–71190H-9.

[49] Z. Kohout, C. Granja, M. Kralik, A. Owens, R. Venn, L. Jankowski, S. Pospisil, B. Sopko, J. Vacik, "Characterization and calibration of novel semiconductor detectors of thermal neutrons for ESA space applications", *Proc. IEEE, Conf. Record NPl. M-1932011* (2011), pp. 400–404.

[50] B.W. Robertson, S. Adenwalla, A. Harken, P. Welsch, J.I. Brand, P.A. Dowben, J.P. Claassen, "A class of boron based solid state neutron detectors", *Appl. Phys. Letts.*, Vol. **80** (2002), pp. 3644–3646.

[51] D.S. McGregor, J.K. Shultis, "Spectral identification of thin-film-coated and solid-form semiconductor neutron detectors", *Nucl. Instr. And Meth*, Vol. **A517** (2004), pp. 180–188.

[52] S.-D. Hwang, K. Yang, P.A. Dowben, A.A. Ahmad, N.J. Ianno, J.Z. Li, J.Y. Lin, H.X. Jiang, D.N. McIlroy, "Fabrication of n-type nickel doped $B_5C_{1+\delta}$ homojunction and heterojunction diodes", *Appl. Phys. Lett.*, Vol. **70** (1997), pp. 1028–1030.

[53] A.N. Caruso, P.A. Dowben, S. Balkir, N. Schemm, K. Osberg, R.W. Fairchild, O. Barrios Flores, S. Balaz, A.D. Harken, B. W. Robertson, J.I. Brand, "The all boron carbide diode neutron detector: Comparison with theory", *Mater. Sci. Eng. B*, Vol. **135** (2006), pp. 129–133.

[54] K.P. Ananthanarayanan, P.J. Gielisse, A. Choudry, "Boron compounds for thermal neutron detection", *Nucl. Instr. and Meth.*, Vol. **118** (1974), pp. 45–48.

[55] J.C. Lund, F. Olschner, F. Ahmed, K.S. Shah, "Boron phosphide on silicon for radiation detectors", *Proc. Mat. Res. Soc. Symp.*, Vol. **162** (1990), pp. 601–604.

[56] M. Engler, C. Lesniak, R. Damasch, B. Ruisinger, L. Eichler, "Hexagonal boron nitride (hBN) – Applications from metallurgy to cosmetics", *Ceramic Forum International, DKG*, Vol. **84**, no. 12 (2007), pp. E49–E53.

[57] S. Majety, J. Li, X.K. Cao, R. Dahal, B.N. Pantha, J.Y. Lin, H.X. Jiang, "Epitaxial growth and demonstration of hexagonal BN/AlGaN p-n junctions for deep ultraviolet photonics", *Appl. Phys. Lett.*, Vol. **100** (2012), pp. 06112-1–06112-4.

[58] A.K. Geim, I.V. Grigorieva, "Van der Waals heterostructures", *Nature*, Vol. **499** (2013), pp. 419–425.

[59] J.H. Kaneko, T. Taniguchi, S. Kawamura, K. Satou, F. Fujita, A. Homma, M. Furusaka, "Development of a radiation detector made of a cubic boron nitride polycrystal", *Nucl. Instr. and Meth.*, Vol. **A576** (2007), pp. 417–421.

[60] D.S. McGregor, T.C. Unruh, W.J. McNeil, "Thermal neutron detection with pyrolytic boron nitride", *Nucl. Instr. and Meth.*, Vol. **A591** (2008), pp. 530–533.

[61] J. Li, R. Dahal, S. Majety, J.Y. Lin, H.X. Jiang, "Hexagonal boron nitride epitaxial layers as neutron detector materials", *Nucl. Instr. and Meth.*, Vol. **A654** (2011), pp. 417–420.

[62] T.C. Doan, J. Li, J.Y. Lin, H.X. Jianga, "Growth and device processing of hexagonal boron nitride epilayers for thermal neutron and deep ultraviolet detectors", *AIP Adv.*, Vol. **6** (2016), pp. 075213–1–075213–11.

[63] A. Many, "High-field effects in photoconducting cadmium sulphide", *J. Phys. Chem. Solids*, Vol. **26** (1965), pp. 575–578.

[64] A. Maity, T.C. Doan, J. Li, J.Y. Lin, H.X. Jianga, "Realization of highly efficient hexagonal boron nitride neutron detectors", *Appl. Phys. Lett.*, Vol. **109** (2016), pp. 072101–1–072101–4.

[65] C.E. Whiteley, Y. Zhang, Y. Gong, S. Bakalova, A. Mayo, J.H. Edgar, M. Kuball, "Semiconducting icosahedral boron arsenide crystal growth for neutron detection", *J. Cryst. Growth*, Vol. **318**, no. 1 (2011), pp. 553–557.

[66] M. Carrard, D. Emin, L. Zuppiroli, "Defect clustering and self-healing of electron-irradiated boron-rich solids", *Phys. Rev.*, Vol. **B51** (1995), pp. 11270–11274.

[67] D. Emin, "Unusual properties of icosahedral boron-rich solids", *J. Solid State Chem.*, Vol. **179** (2006), pp. 2791–2798.

[68] Z. Fisk, J.P. Remeika, "Growth of single crystals from molten metal fluxes", in *Handbook on the Physics and Chemistry of Rare Earths*, eds. K.A. Gschneidner Jr., L. Eyring, North Holland, Amsterdam, Vol. **12** (1989), pp. 53–70.

[69] G.A. Slack, T.F. McNelly, E.A. Taft, "Melt growth and properties of B_6P crystals", *J. Phys. Chem. Solids*, Vol. **44**, no. 10 (1983), pp. 1009–1013.

[70] T.L. Aselage, "Preparation of boron-rich refractory semiconductors", *Mater. Res. Soc. Symp. Proc.*, Vol. **97** (1987), pp. 101–113.

[71] T.L. Chu, A.E. Hyslop, "Preparation and properties of boron arsenide films", *J. Electrochem. Soc.*, Vol. **121** (1974), pp. 412–415.

11

Performance Limiting Factors

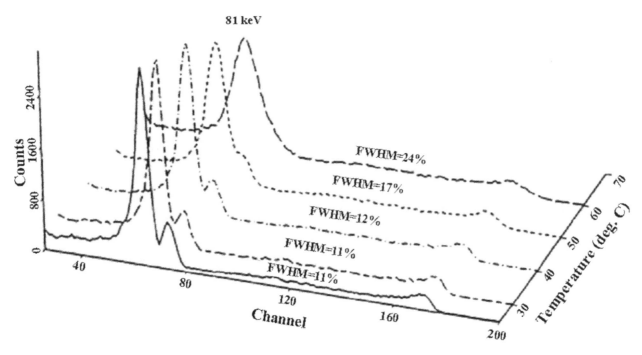

Frontispiece. Figure illustrating the progressive degradation of spectroscopic performance of a CdZnTe detector with increasing temperature (from Egarievwe, et al., *Proc. of the SPIE*, Vol. **2305** (1994), pp. 167–173). In this particular case, the resolution degrades by about 0.5 % per degree C.

CONTENTS

11.1 Introduction

A number of factors can adversely affect the performance of semiconductor detectors, usually manifesting themselves as a degradation in energy resolution and/or short- and long-term stability effects. The most common of these are (a) temperature effects that are directly related to the generation of thermal carriers, (b) polarization effects that are a consequence of space charge effects resulting from ionic conductivity and/or deep trapping and (c) radiation effects that ultimately stem from damage to the lattice structure and the introduction of trapping and recombination centers. The interrelationship among these effects is illustrated in Fig. 11.1. While thermal effects can be reduced by cooling the detector and front end, components (b) and (c) ultimately stem from material imperfection and specifically from the presence of trapping centers. For component (b), these centers may be introduced by material impurities and stoichiometric imbalances introduced during growth and material processing detector fabrication. The component is more or less permanent and can only really be offset by judicious settings of bias and shaping time. In the case of component (c), deep traps are invariably created by lattice damage caused by bombarding radiation – so-called displacement damage. In this case atoms are simply knocked out of their equilibrium positions in the lattice. This effect can be offset to some extend by annealing a detector – a process in which the detector temperature is raised high enough so that the displaced atoms can "rattle" themselves back into their initial positions but low enough that the detector, contacting system and front-end electronics are not damaged. The effects of temperature, polarization and radiation are discussed in detail in the following sections, along with mitigation.

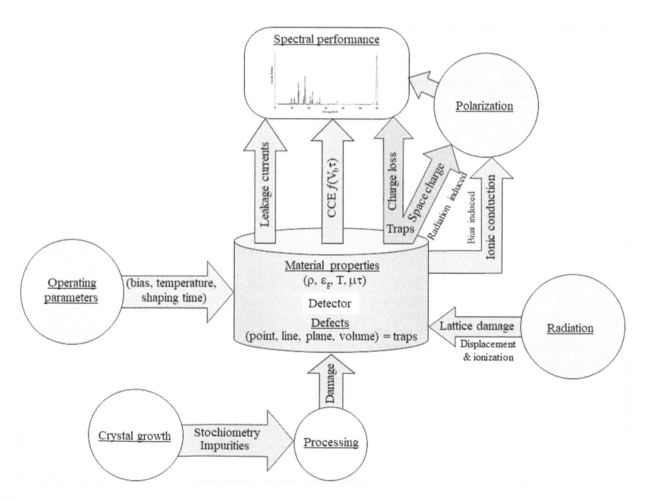

FIGURE 11.1 Diagram showing the interrelationship among material, environmental and operational parameters that lead to performance limitations.

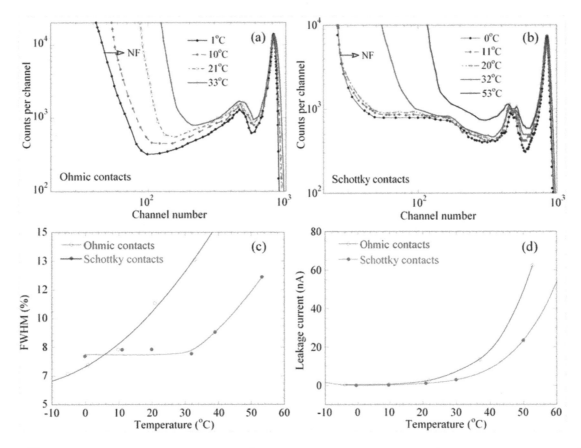

FIGURE 11.2 ^{241}Am energy spectra measured with two 5×5 mm^2 CdZnTe detectors at various temperatures. (a) Ohmic contact detector, (b) Schottky contact detector. Panel (c) shows the corresponding energy resolutions as a function of temperature and (d) the leakage current of both detectors as a function of temperature (redrawn from Park *et al.* [1]). Here NF is the noise floor, which is essentially the lower limit of the useful energy range of the detector.

11.2 Temperature Effects

Temperature can adversely affect the performance of a detection system through several internal and external mechanisms. All seem to be related, to some extent, and in general have inverse temperature dependences in that the higher the temperature, the worse the performance. By internal mechanisms, we mean the temperature dependencies of intrinsic semiconductor properties, such as structure, bandgap and carrier mobilities. By external mechanisms, we mean those temperature dependencies that arise from the system as a whole and are reflected in the system series and parallel noise components, the magnitude of which depends largely on how the detector was fabricated and its front-end design. At system level, both internal and external components lead to degraded performance, most notably in an increase in noise, resulting in spectral broadening or indeed a complete loss of spectroscopy.

In Fig. 11.2, we show spectra obtained with two 5×5 mm^2 CdZnTe detectors – one (2 mm thick) fitted with resistive (ohmic) contacts and the other (5 mm thick) fitted with rectifying (Schottky) contacts (Park *et al.* [1]). The effect of temperature on the measured spectral responses to a 60 keV ^{241}Am gamma-ray source can be clearly seen in both detectors in Fig. 11.2(a) and (b), both in terms of extent of the noise floor and spectral broadening (quantified in Fig. 11.2(c)). From the figure, we see that for ohmic detectors the resolution degrades steadily with increasing temperature by approximately 0.2% per degree C, whereas for Schottky detectors the resolution is initially insensitive to increasing temperature up to ~30°C whereupon it degrades in a fashion similar to ohmic detectors. In Fig. 11.2(d), we show the corresponding leakage current in both detectors as a function of temperature. Although the two detectors are not exactly identical, the overall magnitudes and trends are representative when comparing the two contacting systems – Schottky detectors tend to have lower noise over a wider temperature range.

11.2.1 Intrinsic Components

Temperature affects several fundamental semiconductor properties with significant impacts on carrier generation, transport and collection. For example, as the temperature of a semiconductor increases, the amplitude of atomic vibrations increase, which in turn increases the interatomic spacing expanding the lattice. This decreases the potential seen by the electrons in the material, leading to a reduction in the size of the energy gap [2]. As a consequence, there is an increase in the number of thermal carriers that can cross the bandgap and contribute to the system noise. Fortunately for semiconductors like Si, GaAs and CdTe the bandgap changes by only ~0.3–0.5 meV per degree K and so for the temperature excursions normally encountered by working detection systems (say < 50°C), the effect is negligible. In reality, the dominant effect of temperature on semiconductor properties is through the thermal generation of charge carriers in the semiconductor bulk, which vary exponentially with temperature (see Chapter 2, section 2.8.1.1) and have a direct and much larger impact on system noise through the leakage current. This occurs when the temperature is high enough for electrons to be thermally excited over the bandgap producing electron-hole pairs. These carriers are now free to respond to the internal electric field established by the bias, producing electrical noise. The probability of an electron-hole pair being thermally created is given by the equation,

$$P(T) \sim cT^{\frac{3}{2}} \exp\left(-\frac{E_g}{2kT}\right), \tag{1}$$

where T is the absolute temperature, $E_g(T)$ is the bandgap energy that is in itself temperature dependent [2], k is Boltzmann's constant and c is a material constant. In Fig. 11.3, we plot the thermal generation probability as a function of temperature for a number of elemental and compound semiconductors and as can be seen, the thermal generation probability is strongly temperature dependent. For example, for wide bandgap materials $(E_g \sim 1.4\,\text{eV})$ the flux of thermally generated carriers increases by four orders of magnitude between 0°C and 100°C, and this will be reflected in the leakage current. From the figure, we note that all the curves have similar shapes and the main effect of the bandgap is to determine when temperature effects become important. In essence, the lower the bandgap, the lower the temperature at which this occurs. For example, in Fig. 11.3, we see that to operate an InAs detector with the same thermal carrier generation rate as a GaAs detector operating at room temperature $(300K)$ $(E_g = 1.44\,\text{eV})$, the InAs detector should be cooled to less than 100K $(E_g = 0.41\,\text{eV})$.

Carrier mobilities are also strongly affected by temperature (see Chapter 2, section 2.5). In compound semiconductors, mobilities fall at low temperatures due to impurity scattering and at high temperatures due to deformation potential scattering. Maximum mobilities occur around 100K. Since the operating temperatures of most detectors are higher than this,

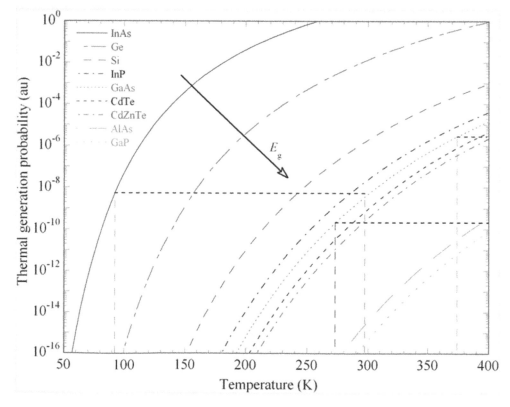

FIGURE 11.3 The thermal generation probability of electron hole pairs at different temperatures for InAs $(E_g = 0.36\,\text{eV}, \text{T} = 300\text{K})$, Ge$(E_g = 0.66\,\text{eV}, \text{T} = 300\text{K})$, Si $(E_g = 1.12\,\text{eV}, \text{T} = 300\text{K})$, InP $(E_g = 1.34\,\text{eV}, \text{T} = 300\text{K})$, GaAs $(E_g = 1.43\,\text{eV}, \text{T} = 300\text{K})$, CdTe $(E_g = 1.51\,\text{eV}, \text{T} = 300\text{K})$, Cd$_{0.9}Zn_{0.1}$Te $(E_g = 1.57\,\text{eV}, \text{T} = 300\text{K})$, AlAs $(E_g = 2.16\,\text{eV}, \text{T} = 300\text{K})$ and GaP $(E_g = 2.27\,\text{eV}, \text{T} = 300\text{K})$. The arrow depicts the direction of increasing bandgap.

mobilities therefore decrease with increasing temperature. The fall-off is gradual near the maximum, becoming more rapid near room temperature. For covalent semiconductors where acoustic phonon scattering dominates, electron mobilities fall off as $T^{-3/2}$. However, for some compound semiconductors the fall-off tends to be steeper due to the effects of polar optical scattering. In GaAs, Blakemore [3] gives a useful approximation of the temperature dependent electron mobility near room temperature,

$$\mu_e \sim 8000 \left(\frac{300}{T}\right)^{2.3} \quad \mathrm{cm^2V^{-1}s^{-1}}. \tag{2}$$

The functional form is similar for holes and for other semiconductors [4,5]. In GaAs, the electron mobility decreases by a factor of two as the temperature increases from 0°C to 100°C. The net effect (and in fact in all compound semiconductors) is an overall reduction in speed, charge collection and resistivity. Secondary effects include an increase in power dissipation and a decrease in reliability.

11.2.2 External Components

For external components, the effects of temperature manifest themselves primarily though the series and parallel noise components of the system (see section 8.5.2.1.2). Series noise is largely composed of contributions arising from the series resistance of the detector preamplifier and from the input FET. Both are positively proportional to temperature – meaning an increase in T causes an increase in noise. The parallel noise component, Q_p, is composed of Johnson noise from the bias resistor, R_b, and shot noise due to the detector leakage current. Again, both are positively proportional to temperature. The temperature dependence of Johnson noise is given by,

$$Q_p^2 \sim 4kT/R_b, \tag{3}$$

from which we see that R_b needs to be as large as possible, preferably in the GΩ range to minimize this component. However, this can be difficult to achieve in practice and is only practical if the resistivity of the detection medium is itself 10^9 Ω-cm or larger. The other component of parallel noise is due to the shot noise of the leakage current. While Johnson noise is proportional to T, the leakage current is found to be much more strongly dependent on temperature and, as expected, follows the form for thermally generated carriers,

$$I(T) \sim T^2 \exp\left(-\frac{E_g(T)}{2kT}\right), \tag{4}$$

where E_g is the bandgap. For uncooled detection systems, this component can quickly dominate the overall noise of the system, even for small temperature increases. For example, from Eq. (4) we note that for a bandgap of 1.43 eV (*e.g.*, GaAs) the leakage current increases by a factor of ~10 between 0°C and room temperature (+20°C) and a further factor of ~40 between +20°C and +60°C. For CdZnTe, Butler *et al.* [6], report that the leakage current in an 8×8×1.6 mm^3 detector increased from 1 pA at −40°C to 1 μA at 100°C – a factor of 10^6 increase. Park *et al.* [1] carried out both leakage current and spectral resolution measurements as a function of temperature on the two CdZnTe detectors described in Section 11.2. At 0°C, the leakage current and energy resolution in the Schottky detector were 0.5 nA and 8% FWHM, respectively, at 60 keV rising to 25 nA and 12% FWHM at 50°C. The corresponding values for the ohmic contact detector were 3 nA and 7% FWHM at 0°C rising to 55 nA and ~17% FWHM at 50°C.

Leakage current can be broken down into two main components: (a) noise due to bulk leakage currents arising from crystal imperfections and noise due to the thermal generation of carriers and (b) noise due to surface leakage currents.

11.2.3 Bulk Leakage Currents

For high resistivity materials, such as CdZnTe, HgI$_2$ and TlBr, ohmic or resistive contacts are generally used. In these cases, the thermal generation of carriers described in the previous section ultimately determines the magnitude of the leakage current. For ohmic detectors, the bulk leakage current should depend only on the bulk resistivity and the applied bias voltage (*i.e.*, basically Ohms law). The temperature dependence should then be driven by the thermal generation of carriers in the bulk. However, for most semiconducting materials the bulk resistivity is too low to allow low-noise operation. To overcome this, Schottky barrier or blocking contacts are generally employed, in which case the leakage current is given by

$$I(T) = A^* T^2 \exp\left(-\frac{e\phi_b}{kT}\right) \tag{5}$$

where A^* is a constant related to Richardson's constant[1] and ϕ_b is the barrier height of the Schottky contact. From the strong temperature dependences implicit in Eqs. (4) and (5), we see why detector cooling can be very effective in reducing bulk leakage currents.

11.2.4 Surface Leakage Currents

Surface leakage currents are usually attributed to electron states that develop on the surface of air-exposed semiconductor surfaces. Such states do not exist in the bulk and provide an additional parallel conduction path – thus increasing noise. In addition, for compounds, non-stoichiometric material is typically formed on the surface following dicing and chemical polishing. For example, in the case of CdZnTe, a Te-rich, low-resistivity layer can be left on the surface following bromine etching. As well as increasing leakage, low surface resistivity can affect the electric field line distribution near the contacts, leading to poor charge collection efficiency, further degrading performance. Fortunately, in many cases, this non-stoichiometric layer can be removed by selective chemical etching.

Data on the temperature dependence of surface leakage is limited. Recently Bolotnikov *et al.* [7]. measured both bulk and surface conductivities in a high resistivity CdZnTe pixel detector. They found that the surface leakage follows a similar temperature dependence as the bulk leakage. Both could be fit by the following function:

$$I(T) = c\, T^2 \exp\left(-\frac{\phi_b}{kT}\right) \tag{6}$$

which has the same functional form as that expected for a diffusion-limited saturation current crossing a Schottky barrier. Here c is a constant and ϕ_b is the potential barrier, for which values of 0.57 and 0.84 eV were obtained for the surface and bulk leakage currents, respectively.

In general, the surface leakage component is found to vary considerably between detectors and is very much dependent on the crystal processing techniques employed during fabrication. Attempts to control the surface component generally follow two approaches, either by passivation of the surfaces and/or the use of guard rings as described in Chapter 6, Section, 6.5.1.3. Both techniques can be quite effective.

11.3 Polarization Effects

Polarization effects in semiconductor X-ray detectors have long been known to be the cause of significant degradation in detector performance, particularly for high resistivity materials at high radiation levels. The term polarization in this context means a time dependent variation in the detector's properties, such as count rate, charge collection and resolving power, that seems to be correlated with detector operating conditions (such as bias and incident radiation fluence), as well as material properties, such as purity, stoichiometry, high resistivity and dielectric constant. Many materials, for example CdTe, HgI$_2$ and TlBr, exhibit such effects to some extent and until recently, the treatment of polarization has been largely anecdotal and qualitative. In fact, the phenomena was so little appreciated or understood that almost all anomalous behaviour was attributed to it.

The polarization phenomenon was first reported in the 1930s, when it was noted that an *"anomalous behaviour in conductivity"* often appears in insulating media when electric currents pass through them [8,9]. Later, it was anticipated that similar effects would also occur in the newly introduced "crystal counters", which was indeed found to be the case. For example, early investigations of AgCl [10,11] and diamond [12], showed a decrease in both pulse amplitude and counting rate as a function of the irradiation rate. Even at this time, several techniques were proposed to suppress this effect. For example, in diamond detectors, Wouters and Christian [10] and McKay [13] made use of an alternating bias voltage, whereas, Chynoweth [14] applying the results of Gudden and Pohl [15], found that polarization could be substantially reduced by illumination with red light. The study of polarization effects then stagnated for almost two decades, mainly because of the decline in interest in crystal counters. It re-emerged in the 1970s [16] with the widespread availability of CdTe as a detection medium [17,18], whose early development was hampered by polarization effects.

---

[1] Modified form of the Richardson constant for thermionic emission from a metal where the free electron mass m, is substituted by the semiconductors effective electron mass m^*.

It was realized early on, that polarization effects arise as a consequence of the build-up of space charge inside a crystal during irradiation (*e.g.* [10],). This space charge results from the trapping of signal charge carriers generated by the interaction of incident radiation. The time-dependent build-up of these trapped carriers creates the space charge, which in turn generates an electric field that opposes the applied field. The weakened field causes the height of the charge pulses produced by the incident radiation to decrease, leading to inferior charge collection and deteriorated spectral resolution in the detector. It also increases the charge collection time, which means that the electronic circuitry requires longer shaping times, resulting in increased electronic noise (mainly through 1/f noise). Irradiation of the material with sub-bandgap light (usually in the IR region) can be very effective in reducing polarization effects by stimulating the deep traps to release trapped carriers, which in turn supresses the build-up of space charge, especially near the electrodes. Experimentally, the mechanics of polarization (*e.g.*, the internal electric field distribution, trap concentrations and associated energies) have been studied using a number of techniques, including ion beam induced current [19] the Pockels effect [20] and Time-of-Flight drift mobility measurements [21].

11.3.1 Polarization Taxonomy

In general, two forms of polarization effects can be observed – bias-induced polarization and radiation/ionization-induced polarization. In bias-induced radiation, the detector instability occurs as soon as the bias is applied, becoming progressively worse with time with characteristic timescales of the order of minutes to hours. This type of behaviour is commonly observed in materials that display ionic conductivity, such as AgI and TlBr. Radiation-induced polarization, as the name suggests, occurs when detectors are exposed to incident radiation – the higher the incident count rate, the higher the induced ionization and the more pronounced the effects of polarization are. In both the bias-induced and radiation-induced cases, the effects also manifest themselves as a reduction in count rate and charge collection efficiency. In spectroscopic detectors, this translates to a reduction of pulse height and spectral broadening.

11.3.2 Polarization in CdTe Detectors

Most work has been carried out for CdTe, for which a basic understanding exists. Polarization effects were first observed at low temperatures in 1970 [18] and later at room temperature in 1973 [16]. For electron collection (which is generally the case), the effect is attributed to deep acceptor levels introduced by vacancies and impurities, which when filled with electrons supplied from the cathode create a negative space charge at or near the positive contact. Since the ionization rates associated with these levels are much slower than the deionization rates (hours as opposed to seconds), a slow build-up of space charge occurs. When charge densities are high enough, they will effectively screen the bias, causing a significant change in the internal electric field profile. In other words, a polarized layer is dissipating a significant fraction of the externally applied bias. The origin of the acceptor levels has been traced to mid-gap region traps associated with residual impurities (such as Ag, Au or Cu) with energies around 0.8 eV [22] and in particular, a deep cadmium vacancy acceptor level at 0.64 eV [23,24].

11.3.2.1 Bias-Induced Polarization

An early classical description of polarization effects in CdTe was given by Bell *et al.* [25] who found a time-dependent exponential decrease in the detector count rate and charge collection efficiency, starting from the time of the first application of the electric field. The effect is illustrated in Fig. 11.4. In spectrum (a) we show a pulse height spectrum acquired from

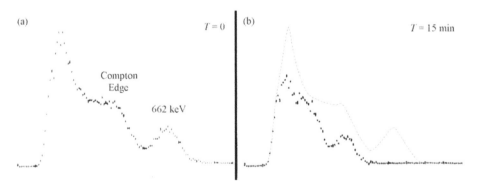

FIGURE 11.4 Demonstration of bias-induced polarization effects in a 1 cm³ CdTe detector in which we show pulse height spectra measured in response to a ¹³⁷Cs source (from Bell *et al.* [25]). In (a) we show a spectrum measured immediately after the application of bias. The accumulation time was 10 s. The position of the photopeak and Compton edge are indicted. In (b) we show a 10 sec accumulation taken after waiting 15 min. For comparative purposes, the outline of spectrum (a) is illustrated by the dotted line.

a CdTe planar detector after exposure to a ^{137}Cs source immediately after the application of bias. The accumulation time was 10 s. The active detector volume is approximately 1 cm^3. Spectrum (b) shows a 10 sec accumulation taken after waiting 15 mins. The outline of spectrum (a) is indicated by the dotted line for comparative purposes. The decrease in count rate continues with a time constant typically of the order of 1 to 2 hr until it asymptotically approaches a limit. After 18 hours, the charge collection efficiency had reduced by 90%. This behaviour depends only on the length of time the bias is applied and not on the presence or absence of nuclear radiation. Removing the bias for a short period (~10 s), at any time, returns the detector to its initial state. Bell *et al.* [25] interpret this as being due to significant de-trapping of holes, which occurs on times scales of seconds, leading to a decrease in the thickness of the space charge region and a corresponding increase in the internal electric field, thus improving charge collection.

11.3.2.2 Radiation-Induced Polarization

Radiation-induced polarization in CdTe detectors was initially studied by Siffert *et al.* [26] who found that, unlike other insulating crystal counter materials (*e.g.* [27],), the effect was only significant for count rates in excess of 10^5 s^{-1}. Hodgkinson [28] obtained similar results, but noted that radiation-induced polarization only occurred in detectors that exhibited a bias-induced polarization, which was contrary to the observations of Vartsky *et al.* [29] who found significant polarization in detectors that were free from bias-induced polarization. It was observed that although the detector did not suffer from the classical polarization effect under bias (*i.e.* decrease in efficiency at low rates above a certain radiation exposure level, corresponding to a detector count rate of approximately 200 kcps), the rate decreased with time. The magnitude of the effect was dependent on the exposure level. For example, increasing the irradiation level by a factor of four reduced the count rate by approximately a factor of two. The characteristic time scales involved are in the seconds and minutes range and when the radiation intensity was decreased back to a low level, the time necessary for the recovery of the detector was about 15 min. The decrease in detector count rate with time was also dependent on the applied bias – the higher the bias the less the effect, indicating that the phenomenology and probable mechanism underlying radiation-induced polarization are substantially different from bias-induced polarization. The effect can be understood based on an increase in the mean free path of the electron and holes with the increase of the electric field. Thus, the probability of charge carriers being trapped in their path towards the electrodes decreases with increasing field. In addition, it was observed that turning off the bias for less than one minute and then reapplying it results in only partial recovery of detector efficiency; however, the decrease in count rate rapidly resumes when the bias is switched back on. In contrast, bias-induced polarization is almost instantaneously and entirely reset when the bias is switched off and then reapplied.

11.3.2.3 Eliminating Polarization in CdTe

Initially, it was difficult to untangle the effects of polarization from a number of interrelated dependencies (such as detector bias, operating temperature, fabrication technique) and so its mitigation was largely multimodal and therefore prescriptive. For example, Malm and Martini [16] suggested a number of operational changes to help reduce the effects but concluded that *"the operating conditions required for satisfactory stability in these detectors may be inconvenient to the user"*. Of significance, at this time, Bell *et al.* [25] found that electrode systems could have a profound effect on polarization effects. For example, polarization in CdTe detectors with ohmic contacts (*e.g.*, Au or Pt) was substantially less than in detectors with Schottky contacts (*e.g.* In or Al). Okada *et al.* [30] compared ohmic (Pt) and Schottky (In) contact systems fabricated on p-type material. In detectors with ohmic contacts, the devices were temporally stable, but the resolution was poor for both electron and hole collection. By contrast, Schottky devices were found to have high spectral resolution but were temporally unstable for electron collection with the signal decreasing by a factor 4 over 1,000s. Furthermore, the effect was dependent only on bias and not irradiation and significantly was not observed for hole collection suggesting that electrons are trapped in the vicinity of the Schottky contact whereas holes are trapped and de-trapped throughout bulk. The results are interpreted as follows. In a Schottky detector, band-bending might be so severe near the metal-semiconductor interface that the deep acceptor levels now lie below the Fermi level, thus allowing the filling of the traps from the contact. This in turn decreases the width of the depletion region, modulating the space charge region and increasing polarization. In contrast, in ohmic detectors the Fermi level remains fixed and thus the degree of polarization is only dependent on the relative trapping and de-trapping rates. Further work by Serreze *et al.* [31] has shown that by starting with high quality material and combining improved surface preparation techniques with higher work function contacting systems, polarization effects can be effectively eliminated.

Polarization in high quality CdTe detectors can also be significantly reduced if not completely eliminated by cooling [32,33]. For example, Niraula *et al.* [33] found that by cooling a high quality, Al-contacted detector down to ~0°C polarization effects were completely eliminated. They attribute this phenomenon to a change in the ionization time of the deep acceptor levels, which increases with decreasing temperature – in essence "freezing out" the traps.

11.3.3 Polarization in CdZnTe Detectors

Polarization in CdZnTe occurs, but the effects are substantially less pronounced than in CdTe and unlike CdTe are generally not bias induced but only occur when irradiated at high photon flux levels. For example, Strassburg *et al.* [34] assessed the effects of polarization on CdZnTe imaging detectors used for CT scanning. In the hard X-ray region, they found that polarization effects only began to become apparent for CT fluxes in excess of 10^6 photons $s^{-1}cm^{-2}$. For input fluxes of 10^9 photons $s^{-1}cm^{-2}$, the mean pulse heights of detected events decreased by ~6% over a period of 10 s and doubled for every decade increase in flux. Sellin *et al.* [35] performed real-time Pockels imaging on semi-insulating CdZnTe to measure the electric field profile in the material bulk. In steady-state room temperature conditions, the measured electric field profile was uniform, consistent with a low space charge concentration. However, at temperatures < 270K a significant non-uniform electric field profile is observed, which is explained in terms of temperature-induced band bending at the metal-semiconductor interface, causing the formation of positive space charge in the bulk. A similar distortion of the internal field profile was also observed at room temperature when the samples were irradiated by high fluxes of X-rays. The resulting high rate of photo-induced charge ensures that charge trapping dominates over thermal emission and recombination processes, leading to a significant build-up of space charge in the bulk. Elhadidy *et al.* [36] also studied polarization in Schottky, n-type CdZnTe detectors using Pockels imaging. An analysis of these data showed that a deep electronic trap with energy E_c−0.92 eV, density of 2×10^{11} cm^{-3} and an electronic capture cross-section of 3.5×10^{-13} cm^2 was responsible for the observed polarization phenomena. The authors suggest that the origin of this level may be the same one that is active in p-type material, in which case the level emits holes when the bands are bent sufficiently downwards. The physical location of the trapping centers has been attributed by Strassburg *et al.* [34] to defects that decorate dislocations and the crystal interfaces between Te inclusions and precipitates, which are invariably introduced during the growth process. Recently, Abbene *et al.* [37] fabricated CdZnTe planar detectors of active area 2×2 mm^2, surrounded by a guard ring to suppress surface leakage currents. The detectors show good spectral room temperature performance up to an event rate of 10^6 cps. However, while no bias-induced polarization was observed, at high rates a collapse of the pulse height spectra was observed when the detector was cooled to less than 10°C. This indicates that at low temperatures the characteristic time scale of the de-trapping process increases, leading to a build-up of space charge and a subsequent increase in radiation induced polarization.

11.3.4 Polarization in HgI$_2$ Detectors

Bias-induced polarization is a persistent feature of HgI$_2$ detectors and is generally attributed to trapping in the bulk of the crystal and non-optimal fabrication techniques [38,39]. However, in contrast to many other polarizable materials, the effects tend to be most severe when bias is first applied and steadily decrease over time scales of hours. In a study of a large number of detectors, Holzer and Schieber [40] noted that polarization effects were only marked in those detectors that had a FWHM energy resolution greater than approximately 5% at 59.5 keV. The authors suggest that this

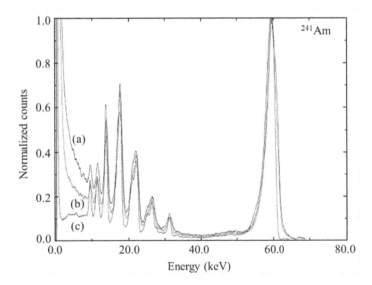

FIGURE 11.5 Figure showing the temporal evolution of ^{241}Am spectra following the application of bias on a 7 mm^2, 0.5 mm thick planar HgI$_2$ detector with Pd contacts (taken from [42]). Curves (a) and (b) show 2,000s spectral accumulations started after 1 and 2 hr after the application of bias at room temperature. Curve (c) shows the case when the bias has been applied for more than 12 hr before measurement. In this case, polarization effects are not apparent and results are reproducible.

occurs because the main bulk defect appears to be a deep hole trap, which for detectors that display no polarization effects is either absent or compensated by other native defect centers. Polarization effects also tend to be absent from detectors with Pd contacts [40] or detectors thinner than ~400 microns [41]. Figure 11.5 shows the temporal evolution of ^{241}Am spectra following the application of bias on a 7 mm^2, 0.5 mm thick planar HgI$_2$ detector with Pd contacts [42]. The spectral resolution is ~2% FWHM at 59.5 keV, so the effects of polarization are expected to be small. Measurements were taken at room temperature. Spectra (a) and (b) were taken 1 and 2 hr after bias was applied, from which it can be seen that the spectra gradually improve. Spectrum (c) shows the case when the bias has been applied for several days before measurements began and is clearly markedly better than cases (a) and (b). The spectral limit is in fact (c) and it was subsequently found, that reproducible results could be obtained by applying bias a minimum of 12 hr before measurement. Further measurements have shown that the effects will only return if the bias is completely removed or is decreased significantly ($\leq$ 200V as opposed to 800 V). From the figure, it can be seen that the effects of polarization are more pronounced at the lower energies. For example, the variation in ΔE between (a) and (c) was factor of 1.7 at 5.9 keV falling to 1.2 at 59.5 keV.

11.3.5 Polarization in TlBr Detectors

In TlBr, the effects of polarization are further exacerbated by the fact that TlBr is a mixed electronic-ionic conductor, with an ionic current being significant even at room temperature. Therefore, in addition to a polarization component arising from space charge effects caused by deep trapping [43], there is an additional component arising from the gradual build-up of Tl ions and Br ions, which accumulate under the cathode and anode electrodes [44], respectively. This leads to an internal electric field that also opposes the applied field, which reduces signal amplitudes as well as increasing the leakage current. TlBr has relatively low defect formation energies and thus large equilibrium concentrations of vacancies are expected at room temperature. It has been suggested that these vacancies are primarily responsible for mediating the ionic conductivity [45,46]. Vaitkus *et al.* [47] has shown that above 250K, the main conduction mechanism is due to Tl$^+$ ions. The effect can be suppressed using sacrificial Tl electrodes coated in Au, in which the Tl electrode is consumed under the anode and formed under the cathode [48]. Vaitkus *et al.* [49] propose that ionic conductivity also creates micro-inhomogeneities in the material, which are activated by the electric field and by non-equilibrium carrier generation. The observed threshold-type effects [49], in the form of transient current spikes, are related to the growth of these structures and can be substantially reduced by lowering the temperature or increasing hydrostatic pressure. For example, Samara [45] found that conductivities could vary by over a magnitude per decade change in temperature Kozlov *et al.* [50] studied the aging of TlBr crystals for a range of applied voltages (100–200 V) and temperatures (277–353 K). At 277K the signal and associated current decreased slowly to a minimum value after ~1.5 hr, whereupon the current grew steadily over the next 15 hr to a value that exceeded the minimum value by a factor of ~15. This was interpreted as a transfer of polarization effects caused by electrons and holes only, to a large ionic component due to ion collection at the electrodes. At higher temperatures, Tl$^+$ cations discharged and deposited on the cathode electrode. The process was followed by the formation of compounds on the Ti negative electrode and further corrosion. These are irreversible chemical reactions. To summarize, the ionic current causes the steady degradation of both the sample and electrodes and is a critical problem for TlBr detectors working at ambient temperatures. However, in general, it is found that effects of the ionic conduction can be significantly reduced by cooling the detectors to below 0°C. For example, Onodera *et al.* [51] measured the long-term spectral stability of TlBr detectors by acquiring successive ^{137}Cs energy spectra for an extended period of time at room temperature and at −20°C. Although the detectors exhibited the polarization phenomena at room temperature, no significant degradation in spectral response was measured at −20°C over a period of 100 hr.

In Fig. 11.6, in which we show the observed polarization effects in a 5×5x2 mm^3 TlBr planar detector when exposed to a 50 μm diameter 60 keV, 1 kHz, synchrotron beam [43]. The detector was operated at −20°C and was quite stable at low count rates (10s of Hz)– thus the ionic component can be neglected.The count rate was then increased to 1 kHz. Fig. 11.6(a) shows the evolution of the pulse height spectra as a function of exposure time, and in Fig. 11.6(b) we show the corresponding degradation of the FWHM (open squares) and normalized peak channel position (full circles) as a function of exposure time. Over a period of 1 hour the peak channel position (and hence charge collection efficiency) declined by 10%, whilst the FWHM resolution deteriorated from 2.0 to 4.5 keV and overall the response curve became badly asymmetric.

Several techniques to reduce polarization effects in TlBr have been proposed, including ultra-purification [52], operation at low temperatures [53], using Tl contacts [48], employing surface treatments [54], engineered device geometry [55] and making larger crystals [56]. None of these techniques, however, solved the polarization problem indefinitely. Recently, Leão and Lordi [57] proposed a new approach for reducing ionic current through the introduction of supervalent dopants that form neutral pairs with the electrically charged vacancies, since the mechanism for ionic migration is mediated by vacancies (or at least facilitated by them in materials with Frenkel defects). Theoretical calculations [57] indicate that co-doping TlBr with Pb plus Se or S can achieve the desired result, and additionally the dopant complexes that form do not significantly affect carrier scattering and hence mobility.

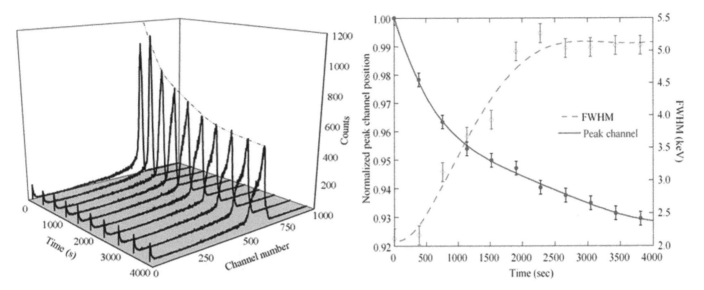

FIGURE 11.6 Polarization effects in a 5×5×2 mm³ TlBr planar detector observed when exposed to a 50 μm diameter 60 keV, 1 kHz, synchrotron beam [43]. In (a) we show the evolution of the pulse height spectra as a function of exposure time. In (b) we show the corresponding degradation of the FWHM (open squares) and normalized peak channel position (full circles).

11.4 Radiation Effects

While radiation detectors are designed to detect radiation, the same radiation can also alter or destroy their detection properties as well as the operation and characteristics of associated electronics. Radiation affects the performance of semiconducting devices through two main mechanisms. These are

a. <u>ionization</u>, in which charged particles ionize the atoms in a device causing an instantaneous release of charge. In bulk materials this charge dissipates by either diffusion or drift in an electric field. However, in structured devices the build-up of localized charge can damage devices – particularly in the vicinity of an interface.

b. <u>displacement</u>, in which device atoms are literally displaced and removed from their equilibrium positions introducing disorder into the crystal lattice structure. The consequences are the introduction of additional energy levels into the bandgap, which act as trapping and generation/recombination centers, altering the semiconductor's electronic properties.

Both mechanisms are important and affect all active elements of the sensor chain, although individual elements (such as the detector, front-end FET, *etc.*) are usually more sensitive to one type of mechanism than the other. For example, the gate and insulating structures of CCDs and CMOS components are more susceptible to ionization effects, whereas active detection media are more sensitive to displacement damage. In general, ionization tends to be responsible for surface damage while displacement is responsible for degradation of the bulk. Both mechanisms are described in detail below.

11.4.1 Ionization Damage and Its Effects

Ionization effects can be divided into two categories – transient and long term. Transient effects are already familiar, since in semiconductors they generate the electron-hole pairs upon which device operation and particle detection are based. After the initial excitation and charge generation, neutrality is eventually restored and the semiconductor returns to its original state. Long-term effects, on the other hand, usually involve the build-up of a space charge, which affect local electric fields and subsequently charge transport and collection. This mechanism is usually only important in insulators – for example, the metal oxide interfaces used in the CMOS construction or the SiO_2 layers used in Si X-ray detector fabrication. As such, the ionization effects are only pronounced in Si devices where oxide interfaces are routinely used and are not relevant for compound semiconductor devices since few compounds have native oxides.

In Si devices, the ionization energy absorbed in oxide layers liberates charge carriers that migrate through the layer by drift and diffusion. The electrons are far more mobile than the holes and quickly leave the layer. The holes, by contrast, are transported through the layer by a slow "hopping" motion. As a consequence, they are likely to be captured by traps in the oxide volume, which in the process, builds up a fixed positive charge. The holes that do make it to the oxide-semiconductor interface may then be captured by the numerous traps (*e.g.*, MIGS[2] induced at the contacts) that congregate at the interfaces. This, in turn, also builds up a localized charge, generating parasitic electric fields, which affect charge collection. As well as a static component, trapped oxide charges can also be mobile, so that the charge distribution is time dependent and therefore so is the electric field across the oxide. In addition, the charge state of a trap depends on the local quasi-Fermi level, so the concentration of trapped charge will vary with changes in the applied voltage and state-specific relaxation times. As charge states also anneal, ionization effects depend not only on the dose, but also on the dose rate. In Si components, the net effect is an increase in dark current and subsequently noise for moderate doses and in fact, doses up to 100 krad can be tolerated to some extent, with only a change in operating conditions. However, for large doses, catastrophic failure can occur in CMOS devices due to field oxide inversion. In Si PIN detectors, the accumulation of space charge at the oxide interface can also cause the adjacent p-channel to type invert, resulting in a significant tunneling current that swamps the pre-amplifier. Normal operation can only be restored by removing the bias for an extended period [58].

Ionization effects depend primarily on the absorbed energy and are independent of the type of radiation. For charged particle energies that are less than the minimum ionizing energy (say < 100 MeV), ionization damage can be expressed in terms of energy absorption per unit volume and is usually expressed in rad or gray $\left(1 \text{ rad} = 100 \text{ erg g}^{-1}, 1 \text{ Gy} = 1 \text{ J kg}^{-1} = 100 \text{ rad}\right)$. Since the charge liberated by a given dose depends on the absorber material, the ionizing dose must be referred to a specific absorber, for example, 1 rad in Si, 1 rad in GaAs or in SI units 1 Gy in Si, *etc.* These are different. For example, for 10 MeV protons, 1 rad absorbed dose in Si is approximately equivalent to 0.5 rad absorbed dose in GaAs. While ionization damage is nearly always associated with charged particles, it should be pointed out that even non-ionizing particles (e.g., neutrons) can contribute to the ionization dose *via* the recoils of collision products. However, in terms of dose, this contribution tends to be very small, since the damage results from a secondary process. For example, in Si neutrons contribute $\sim 10^{-13}$ rad cm^{-2} MeV^{-1} per neutron, whereas protons contribute $\sim 10^{-6}$ rad cm^{-2} MeV^{-1} per proton.

11.4.2 Displacement Damage

Displacement damage results from non-ionizing energy loss (NIEL) interactions of a primary particle with the atoms of the bulk material. Unlike ionization, it results in permanent damage to the lattice structure by displacing atoms from their equilibrium position. If the displaced atom escapes, a vacancy is created. If however it assumes an interstitial position in the lattice, a vacancy-interstitial complex is created, known as a Frenkel defect (see Chapter 3, section 3.5.2). As well as altering the structural properties of the lattice these defects change the electronic characteristics of the crystal by introducing additional energy levels within the bandgap. These levels facilitate the transition of electrons from the valence to the conduction band, which in the active depletion layers of a detector leads to a generation current and in turn electronic shot noise. The effect can be particularly acute in indirect bandgap semiconductors, since these levels can act as a conduit to facilitate electronic transitions between the conduction and valence bands – again leading to a generation current. In addition, those states created close to the band edges act as efficient trapping centers, facilitating charge loss if de-trapping times are longer than amplifier time constants. A measure of the effectiveness of creating a Frenkel defect is given by the threshold displacement energy E_d, which is the minimum kinetic energy an atom in a solid needs, to be permanently displaced from its lattice site to a defect position.

The threshold energy for causing a displacement is closely related to the crystal binding energy. In Table 11.1 we list the measured displacement threshold energies [59,60], E_d, for a number of semiconductors. Empirically, E_d, has been found to be to be inversely proportional to the volume of the unit cell and a roughly quadratic dependence with bandgap. Assuming an inverse cubic dependence, a best fit to reciprocal lattice data gives,

$$E_d(\text{eV}) = 1.7926 \times (1/a_o(\text{nm}))^3, \quad r = 0.98. \tag{7}$$

In general, semiconductors with high threshold energies are more resistant to displacement damage.

Consider the collision of an incident particle of mass m_p and energy E_p with a semiconductor lattice atom of mass M_l. The energy, ΔE, transferred in the interaction in the non-relativistic case, is given by,

$$\Delta E = 4E_p \left[\frac{m_p M_l}{(m_p + M_l)^2} \right] \sin^2\left(\frac{\theta}{2}\right), \tag{8}$$

where θ is the scattering angle (*i.e.*, the angle the incident particle is scattered off the lattice atom). The energy transferred to the lattice atom is constant up to a maximum, E_{max}, which occurs for head-on collisions ($\theta = 180$) and is given by,

---

[2] Metal-Induced Gap States (see Chapter 5, section 5.3.6)

TABLE 11.1

Displacement threshold energies, (E_d) for several semiconductors ordered by inverse lattice spacing (from refs. [59,60]). In compounds, different threshold energies exist for the two constituents of the lattice because of the disparity in masses. An average for the two is given here. Similarly, a separate threshold displacement energy exists for each crystallographic direction.

Material	Inverse lattice spacing (nm^{-1})	Bandgap (eV)	Displacement threshold energy (eV)
InSb	1.543	0.17	6.3
CdTe	1.543	1.48	6.7
ZnTe	1.639	2.28	6.7
GaSb	1.639	0.73	7.0
CdSe	1.647	1.74	8.5
InAs	1.650	0.35	7.4
InP	1.704	1.34	7.8
CdS	1.718	1.74	8.0
ZnSe	1.764	2.63	10.9
Ge	1.767	0.66	14.5
GaAs	1.773	1.43	9.5
Si	1.842	1.12	12.9
ZnS	1.848	3.58	12.5
GaN	2.217	3.42	19.5
6H-SiC	2.260	3.05	21.8
4H-SiC	2.270	3.26	21.3
C	2.801	5.48	37.5

$$\Delta E_{\max} = 4E_p \left[\frac{m_p M_1}{(m_p + M_1)^2} \right]. \tag{9}$$

While Frenkel pair formation energies in crystals are typically around 5–10 eV, the average threshold displacement energies are much higher, 20–50 eV since in reality defect formation is a complex multi-body collision process. The minimum incident particle energy, E_p, required to transfer enough energy to a lattice atom to create a defect is about 80–100 eV for particles of mass 1 amu (*e.g.*, neutrons or protons) and about 200–300 eV for electrons. However, incident particle energies encountered in radiation detection and measurement are usually orders of magnitude higher and consequently so is $E_{\max}$. The recoil energy of the displaced atom[3] can be up to 130 keV of which approximately 50% is deposited *via* ionization. Displacements dominate when the recoil atom loses its final 5 to 10 keV. When such a large amount of energy is transferred (>1,000 eV), the displaced atom can initiate a "cascade", displacing many thousands of other atoms from their equilibrium lattice positions over many lattice sites. The result is the creation of extended disordered regions (as shown in Fig. 11.7) which can extend up to 100 microns or more [61]. These regions can then accumulate a charge of 10−100 times the electronic charge that can electrostatically attract and trap carriers and are generally unstable except at low temperature. In detectors, the net effect is a change of the internal electric field, an increase in leakage current, changes in capacitance and resistivity and (if trapping times are longer than amplifier time constants) charge loss. Low levels of damage can be recognized by an increase in leakage current. For moderate damage, the effects can be most easily identified and characterized by measuring either the leakage current or the broadening of the detector's energy resolution function. Note that it is possible to make use of the deleterious effects of displacement damage. For example, the rise in leakage current can be used for dosimetry and commercial diodes are available specifically for this purpose (for example, RADFETs).

Displacement damage depends only on the non-ionizing energy loss, that is, the energy and momentum transfer to the lattice atoms. As a consequence, it depends on both the incident particle type and its energy but not on the total energy absorbed in the material. While X-rays and gamma-rays cannot cause displacements directly, they can indirectly through the collisions of their secondary interaction products. for example, Compton electrons, photoelectrons, pair-products and delta-rays. However, since these are produced in secondary processes, they are relatively inefficient and as a consequence, photons are about three orders of magnitude less damaging per photon than, say, a 1 MeV neutron. This is reflected in Table 11.2, which gives a comparison of relative displacement damage for different types of radiation at four energies. The values were derived from the NIEL curves shown in Fig. 11.8, normalized to the Si NIEL value for 10 MeV protons. The units of displacement dose are

---

[3] known as the Primary Knock-on Atom (PKA).

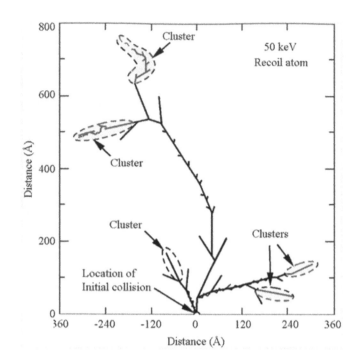

FIGURE 11.7 Displacement damage in silicon for the case where the energy transferred to the lattice site is more than 1,000 times greater than the threshold energy for displacement (from [61]). When such a large amount of energy is transferred, the displaced atom can initiate a "cascade", displacing many thousands of other atoms from their equilibrium lattice positions over many lattice sites. The result is the creation of extended disordered regions. Neutron-induced damage is mainly in the form of defect clusters. For gamma and electron radiations, it is mainly in the form of deep level single defects. For charge particles (protons, pions, *etc.*), it is the mixture of the two, depending on energy.

TABLE 11.2

Relative displacement damage factors for various particles and energies normalized to proton NIEL in Si at 10 MeV.

Particle	Proton				Neutron				Electron			
Energy (MeV)	1	10	100	1000	1	10	100	1000	1	10	100	1000
Si	7	1	0.3	0.4	0.10	0.20	0.33	0.23	0.003	0.01	0.016	0.020
Ge	5	0.7	0.4	0.4	0.04	0.15	0.4	0.40	0.001	0.09	0.015	0.022
GaAs	6	0.7	0.32	0.5	0.06	0.10	0.30	0.40	0.003	0.01	0.020	0.030
CdTe	4	0.6	0.17	0.2					0.0008	0.007	0.001	0.001

MeV per g (normalized to the material density). From the Table we note that in general the most damaging particles are protons. Below 10 MeV proton damage follows a $1/E$ dependence, because of the increasing importance of electromagnetic interactions at low energies. At higher energies, proton NIEL is roughly independent of energy, primarily due to the increasing importance of nuclear reactions, particularly for compounds for which more numerous and complicated reactions are possible. Low energy neutrons are at least 10 times less effective than protons of the same energy. However, above, say, 30 MeV this is no longer the case since at these energies the Coulomb interactions between the protons and lattice nuclei become less important and both protons and neutrons behave like classical billiard balls in collisions. Electrons are the least damaging particles, being at least an order of magnitude less than protons or neutrons.

11.4.2.1 Quantifying Displacement Damage – The NIEL Hypothesis

It has been observed experimentally that effects due to displacement damage in a given device are primarily a function of the amount of non-ionizing energy loss in that device and not of the particular type or energy of the incident particles. In other words, even though the non-ionizing energy loss deposited in a device is dependent on the specific particle and its energy, the overall effects on device performance only depend on the total amount of non-ionizing energy loss deposited in the device. In many cases this results in a surprisingly good and simple relationship between NIEL and device degradation (*e.g.*, dark current, efficiency, energy resolution), although modifications must be made for compound semiconductors because of the different sizes and therefore threshold energies of their constitute atoms. Most of the discrepancies arise at higher energies, where nuclear collisions are involved, which are more complex in compounds than in the elemental semiconductors. For a particle passing through a given material, the NIEL is given by

$$\frac{dE}{dx} = \frac{N_a}{A} \sum_{Z,A} \int E_r \frac{d\sigma}{dE_r} L(E_r) \, dE_r, \tag{10}$$

where $d\sigma/dE_r$ is the differential cross-section for producing a recoil atom or fragment with energy E_r, atomic weight A and atomic number Z. $L(E_r)$ is the Lindhard partition function [64], which describes the fraction of the recoil energy E_r that contributes to displacement. N_a and A are Avogadro's number and the atomic weight of the material, respectively. The sum extends over all types of recoil atoms and fragments.

Whereas NIEL values exist for most materials, data for compound semiconductors is sparse, although the computational tools are now freely available [65]. In Fig. 11.8, we show proton, neutron and electron NIEL curves for Si, Ge, GaAs and CdTe. From the curves, we see that for each particle type, the curves and values of NIEL are similar with the higher values of NIEL being correlated with smaller bandgaps. The curves are so useful because they allow the inter-comparison of damage produced by different radiation types as a function of incident particle energy. If measurements of detector performance (gain, energy resolution, *etc.*) are now made as a function of particle fluence at a particular energy, the NIEL curve can then be used as a predictive tool of radiation damage and subsequent detector performance for any radiation environment evaluated by integration of the curves. Measurements can even be carried out using one radiation type and predictions made for another by suitable scaling between the NIEL curves. For example, 100 MeV neutrons cause approximately the same degradation in a detector's performance as 100 MeV protons whereas 100 MeV electrons cause roughly 20 times less damage for the same fluence.

11.4.3 Radiation Damage – Effects on Performance

Compared to the elemental semiconductors Ge and Si, wide-gap compounds are particularly interesting from a radiation tolerance point of view since the larger bandgap energies increase the energy of defect formation making compound semiconductors intrinsically radiation hard. Additionally, because of their higher effective Zs and therefore X-ray stopping powers, detectors can be made thinner to maintain a given efficiency, with resulting reductions in radiation damage, which is a bulk effect. Lastly, as mentioned previously, because oxide interfaces are not used, they do not suffer from ionization radiation damage or from the space charge effects observed in Si detectors during large transient events (*e.g.*, solar flares

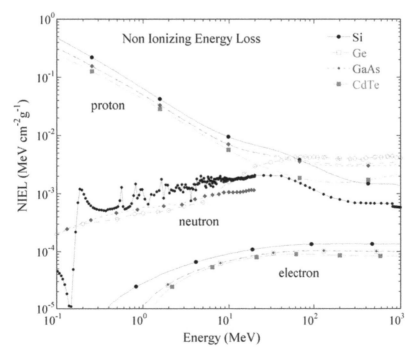

FIGURE 11.8 Non-ionizing energy loss in Si, Ge, GaAs and CdTe as a function of energy for protons, neutrons and electrons. Note resonances in the neutron data in the energy range 0.1 to 1 MeV are not shown for clarity. The Si and GaAs data were taken from the review of Poivey and Hopkinson [62] and references therein. For Ge, the data were taken from Owens *et al.* [63] and references therein.

[58]). Although it has been claimed that some compound semiconductor materials are extremely radiation hard, withstanding proton fluences as high as 10^{12}cm^{-2}, the tolerance of a large number of compounds to radiation has yet to be effectively demonstrated. For example, Franks *et al.* [66] surveyed over 300 abstracts on radiation damage studies on CdTe, CdZnTe and HgI$_2$. Because the initial measurements and results were derived from a mishmash of detectors, detector technologies, radiation sources and energies, direct inter-comparisons were not possible. However, some general conclusions could be drawn, such as, HgI$_2$ appears to relatively immune to proton- and neutron-induced radiation damage. No resolution degradation was found from intermediate energy protons (say 10-100 MeV) for fluences up to 10^{12} p cm^{-2}. Similarly, the material is apparently not susceptible to intermediate energy neutrons. The situation for cadmium telluride was less clear. No data were available for the effects of high-energy protons although the results for intermediate energy protons (33 MeV) suggested vulnerability beginning in the region of 10^8 p cm^{-2}. Given the general flattening in proton NIEL curves in this energy region, we can expect similar results for high-energy protons. Neutron data are also incomplete although at lower energies (8 MeV) the damage threshold for resolution degradation is relatively high (10^{10} n cm^{-2}). However, with reference to the general shape of neutron NIEL curves (see Fig. 11.8), we can expect a marked decrease in tolerance at high-neutron energies. Data on CdZnTe also suggests it is radiation susceptible with evidence of resolution degradation from 200 MeV protons beginning in the region of 10^9 p cm^{-2}.

Owens *et al.* [67] carried out a series of experiments designed to assess the relative radiation hardness of a range of semiconductor X-ray detectors to medium energy protons by measuring the energy resolution at a function of radiation dose in a set of representative planar devices. The specific compounds tested were GaAs, InP, CdZnTe, HgI$_2$ and TlBr, along with an elemental Si device. To allow meaningful comparisons, all devices were of a similar size and, with the exception of the InP detector, had sub-keV energy resolution at 5.9 keV. The irradiations were carried out using a Cyclone 10/5 10 MeV proton cyclotron. Each detector was subjected to six, logarithmically spaced, consecutive exposures – the integral fluences being 2.66×10^9 p cm^{-2}, 7.98×10^9 p cm^{-2}, 2.65×10^{10} p cm^{-2}, 7.97×10^{10} p cm^{-2}, 1.59×10^{11} p cm^{-2} and 2.65×10^{11} p cm^{-2}, respectively. In Si, these correspond to absorbed radiation doses of 2, 6, 20, 60, 120 and 200 krads. The radiation history of the devices is tabulated in Table 11.3. During the exposures, the detectors were kept unbiased and at room temperature. After each irradiation, the effects of the exposure were assessed, both at room temperature and at a reduced temperature (typically - 20°C) using ^{55}Fe, ^{109}Cd and ^{241}Am radioactive sources. These measurements were carried out between 3 and 24 hours later, depending on the level of the leakage current and detector stability. Indeed, after some of the larger exposures (> $\sim$10^{11} p cm^{-2}), measurements could not be carried out for weeks on the Si and CdZnTe detectors. Other than cycling between room and operating temperature, the detectors were not annealed. It was found that with the exception of the HgI$_2$ and TlBr detectors, all materials showed varying degrees of damage effects.

In Fig. 11.9, we plot the measured FWHM energy resolutions at 22.1 keV as a function of proton fluence. The absorbed doses (krads) in Si and GaAs are also indicated. The immediate observable effect of damage is an increase in leakage current, resulting in shorter amplifier shaping times in order to optimize energy resolution. These values are plotted in Fig. 11.9. Note: the detector biases were kept at the same values throughout this study. Simply stated, Si began degrading immediately at the first irradiation of 2.7×10^9 cm^{-2}, followed by CdZnTe at a fluence of a few times 10^{10} cm^{-2}, followed by GaAs at a fluence of 1.6×10^{11} cm^{-2}. The other detectors maintained their resolutions within statistics. The data are tabulated in Table 11.4 in which the fractional energy resolutions at 22 keV are listed (*i.e.*, those measured at a particular fluence

TABLE 11.3

Irradiation history of the devices. For each irradiation, we list the number of incident protons as well as the absorbed dose in krads. Device GaAs2 was given the same Si dose equivalent, which corresponds to about twice the Si fluence.

Compound	Irradiation history		Accumulated dose p cm^{-2}			
	2.66×11^9	7.98×11^9	2.65×10^{10}	7.97×10^{10}	1.59×10^{11}	2.65×10^{11}
	Total dose krads					
Si	2.00	6.00	20.0	60.0	120.0	not irradiated
InP	1.31	3.97	13.1	39.3	78.6	130.7
GaAs1	1.15	3.44	11.44	34.3	68.7	114.2
HgI$_2$	1.05	3.16	10.5	31.6	63.1	105.0
TlBr	0.60	1.80	6.0	18.0	36.0	60.2
CdZnTe	0.31	0.85	3.2	9.30	18.6	31.0
Total dose p cm^{-2}	4.64×10^9	1.39×10^{10}	4.63×10^{10}	1.39×10^{11}	2.78×10^{11}	4.64×10^{11}
GaAs2	2.0	6.0	20.0	60.0	120.0	200.0

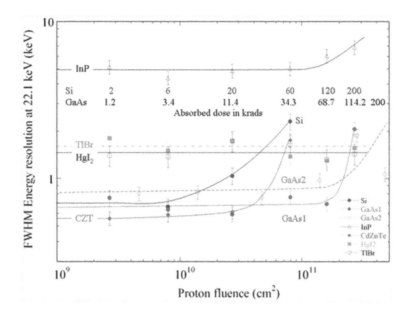

FIGURE 11.9 The measured FWHM energy resolutions at 22 keV as a function of proton fluence (from [67]). The absorbed doses (krads) in Si (blue) and GaAs (black) are also indicated. Note the detectors were not annealed after each irradiation. The resolutions were taken at the optimum shaping time.

divided by their pre-irradiation values) for each irradiation – thus damaged detectors should have a ratio > 1. The last column gives the average value by which the data are ranked. Note, for simplicity the values of GaAs2 have been offset by one column, since this detector was given approximately twice the fluence per irradiation than the others. We summarize the individual responses for each detector material.

Si: Silicon exhibited a factor of ~two degradation in energy resolution after the first irradiation of 2×10^9 p cm^{-2}. After each subsequent exposure, it took the detector an increasingly longer time to recover – much longer than for the other detectors. Furthermore, while the charge pulses looked nominal on an oscilloscope, the baseline signal was found to vary erratically, resembling telegraph noise.

CdZnTe: Cadmium zinc telluride started showing effects after 2×10^{10} p cm^{-2} becoming virtually unusable after 8×10^{10} p cm^{-2}. After 2×10^{10} p cm^{-2} the recorded spectra showed double peaked structure – the peaks becoming increasing separated with dose.

GaAs: Both GaAs detectors showed little variation up to a fluence of 2×10^{10} p cm^{-2}, other than requiring a progressive decrease in shaping time to maintain energy resolution. The devices then degraded by a factor of ~two after a total dose of 3×10^{11} p cm^{-2}.

TlBr: Thallium bromide was found to withstand radiation proton fluences up to 3×10^{11} cm^{-2}, but after 8×10^{10} protons cm^{-2} it was observed that polarization effects increased significantly. These manifest themselves as gain shifts and spectral broadening that is proportional to the total energy deposition per unit time. In pre-irradiation measurements, it was found that these effects only became evident for energies above 50 keV and count rates above 200 s^{-1}. However, after an exposure of 8×10^{10} protons cm^{-2}, polarization effects were evident at energies as low as 15 keV at 200 counts s^{-1}.

TABLE 11.4

Summary of radiation effects in compound semiconductors. Listed are the ratios of the measured energy resolutions at 22.1 keV after each exposure to their initial pre-irradiation values. The last column gives the average value. The data are ranked by this value. The dashed line delineates the boundary of measurable radiation damage.

Dose p cm^{-2}	2.7×10^9	8.0×10^9	2.7×10^{10}	8.0×10^{10}	1.6×10^{11}	2.7×10^{11}	4.6×10^{11}	Average factor
InP	0.8	0.7	0.8	0.8	1.0	1.1		0.9
TlBr	1.0	1.0	1.2	1.1	0.9	1.0		1.0
HgI$_2$	1.3	1.3	1.1	1.0	1.0	1.0		1.1
GaAs2		1.1	1.1	1.0	1.3	2.5	1.4	1.4
GaAs1	1.2	1.0	0.9	1.2	1.1	3.3		1.5
CdZnTe	0.9	0.9	0.9	3.6	2.9	Unmeasurable		1.8
Si	1.9	1.7	2.6	Unmeasurable	2.6	Unmeasurable		2.1

HgI_2: Mercuric Iodide showed no significant variation or increase in polarization effects (which are commonly observed in undamaged HgI_2 detectors) due to the irradiations.

InP: Likewise, indium phosphide showed no change across the entire dose range. However, it could be argued that because the resolution was so poor to begin with (6.1 keV FWHM at 22 keV), one might not expect to see a significant change. Since its properties are very similar to GaAs, we might reasonably expect it to behave in a similar fashion if its energy resolution was comparable.

After the sixth and final irradiation, the detectors were left for one and a half years before re-testing whereupon some recovery of the performance properties was observed, presumably due to self-annealing at room temperature. The Si detector in particular showed better stability and an improved energy resolution of 1 keV FWHM at 22 keV. The second GaAs (GaAs2) detector also showed a factor two improvement in energy resolution after the same period (from 1.9 keV FWHM to 1.1 keV FWHM), despite the fact that it received the highest radiation dose of 5×10^{11} p cm^{-2}. The other detector materials did not show a room temperature annealing effect and still behaved in the same way they had one and a half years earlier. For those detectors that displayed damage effects, the sensors were further annealed by raising their temperatures to 80°C for several weeks. All detectors were found to recover to some extent. Unfortunately, the Si detector failed at this point so it is not clear what affect further annealing would have had. The second GaAs detector (GaAs2) showed the largest recovery. After spending two weeks at a temperature of 80°C, the energy resolution decreased to 900 eV at 22 keV, which is within ~10% of its pre-irradiation resolution. The residual broadening remaining in the spectra arises from a low energy tail, probably caused by hole trapping. Annealing the detector for another two weeks at 80°C did not improve the resolution further. Continued annealing of the CdZnTe detector did not significantly improve its performance either.

In conclusion, for doses up 2.65×10^{11} protons cm^{-2}, HgI_2 is the most radiation hard material, followed by TlBr. However, the latter begins to suffer from severe polarization effects at 1/50 of this dose. The least radiation hard material is Si followed by CdZnTe. Only GaAs has responded significantly to annealing, returning to within 10% of its pre-irradiation value.

11.4.4 Correlation between Dose, Absorbed Dose and Performance

Coulomb processes dominate the absorbed radiation dose in solid-state matter, which means that the stopping power for charged particles is dependent on (a) the average atomic mass, (b) the nuclear charge of the target nuclei and to a lesser extent (c) the atomic density of the target material. In fact, the absorbed dose decreases with increasing density of the nuclear charge and decreases with the average atomic density of the medium. This can be seen in Table 11.5 where, for example, we note that the absorbed dose for CdZnTe is low because it has a high average nuclear charge density combined with a reasonably high atomic density. This can be compared to other compounds that have higher absorbed doses for the same particle fluences, for example, HgI_2 due to its relatively a quite low atomic density and InP due to its low average nuclear charge density. However, while these parameters can be used as a guide, they do not uniquely identify which compounds are the most susceptible. For example, while experimentally, HgI_2 is the most radiation-hard material tested, CdZnTe is the worst (apart from Si) yet the latter has a much lower absorbed dose. Nor does hardness correlate well with bandgap energy. Finally, the last column of Table 11.5 is labelled "Tolerance" and is ranked from 1 to 6 with 1 being the most tolerant. It is based on the last column of Table 11.5, with two subjective modifications. The first is that TlBr was moved from second position to third position because of the much increased polarization effects and secondly, that InP should be moved from 1st to 4th position because of its extremely poor initial energy resolution and its electronic and structural similarity to GaAs.

TABLE 11.5

The factors affecting the calculated absorbed dose. For inter-comparison we list the absorbed doses for a proton irradiation of 2.66×10^9 p cm s^{-1}. The data are ranked by the product of the average nuclear charge and atomic density. The last column gives the "Tolerance" index for each of the materials, which does not track the absorbed dose.

Compound	Bandgap eV	Atomic density cm^{-3}	Av. Nucl. Charge	Product cm^{-3}	Absorbed dose krads	Tolerance
CdZnTe	1.57	1.57×10^{22}	109.5	17.15×10^{23}	0.23	5
TlBr	2.68	1.60×10^{22}	53.2	8.52×10^{23}	0.60	3[‡]
HgI_2	2.13	0.84×10^{22}	92.1	7.81×10^{23}	1.05	1
GaAs	1.43	2.21×10^{22}	32.0	7.08×10^{23}	2.00	2
InP	1.35	1.98×10^{22}	27.1	5.38×10^{23}	1.31	4[†]
Si	1.12	4.97×10^{22}	5.29	2.63×10^{23}	2.00	6

‡ moved from 2nd position to 3rd because of much increased polarization effects.
† moved to 4th because of its very poor initial energy resolution and its electronic and structural similarity to GaAs.

11.4.5 Mitigation Techniques

Although little can be done to reduce radiation damage in a given detector and its front-end electronics during irradiation, a number of techniques can be applied to reduce its effects, both at the sensor and system level. Clearly, the best mitigation is to avoid the problem in the first place. However, if this is not possible, several simple measures can be implemented and result in a significant improvement in radiation tolerance or spectral performance, for example, by shielding sensitive areas or in the case of the front-end electronics, by reducing the number of components in the radiation environment to the minimum necessary to fulfil its function. In this regard, it is useful to note that digital circuitry is far more robust to radiation damage since it only considers two states. Therefore, another approach is to digitize the signal as close as possible to the detector – the approach employed in active pixel sensors.

For a given radiation environment, radiation damage can be also minimized by careful selection of the detector material. Choosing a material that has a larger bandgap will result in increased radiation tolerance, since non-ionizing energy losses, and therefore damage, are usually less in these materials. Also, choosing a direct bandgap material will improve radiation tolerance, since in an indirect bandgap material, the introduction of mid-gap states caused by radiation damage results in a new leakage current channel *via* generation–recombination sites, which can swamp spectral performance. Alternately, choosing a detection medium with a high stopping power means that thinner detectors can be fabricated to achieve a given detection efficiency. For example, Si detectors of thickness 500 µm are commonly used in soft-ray spectroscopy. A GaAs diode of only 40 µm has a similar detection efficiency in this energy range but is 12.5 times thinner. Since radiation damage is a bulk effect, there will be a resulting improvement in radiation tolerance, irrespective of the increased bandgap of GaAs.

11.4.5.1 Cooling

Since most the degrading effects of detector radiation damage stem from an increase in leakage current, a simple compensation approach is to cool the detector, since leakage current decreases exponentially with temperature. In undamaged Si detectors, reducing the detector temperature from room temperature to 0°C, reduces the leakage current by a factor of ~6. However, for damaged detectors the reduction can be considerably more. For example, after an irradiation of 50 krads, the leakage current in a Si PIN diode can increase by more than 500 times. Cooling the detector by an additional 10°C can reduce leakage currents by as much as a factor of 10, which depending on the application, may return performance to acceptable limits. In Si PIN diodes, Spieler [68] has shown that the main cause of detector performance degradation from radiation damage is primarily due to the introduction of inter-bandgap states, which act as efficient generation-recombination centers, generating a generation-recombination leakage current. For radiation-damaged devices, this becomes the dominant component of the leakage current. From the basic functional form for leakage current in a reverse biased diode, that is,

$$I_R(T) \sim T^2 \exp\left(-\frac{E_g(T)}{2kT}\right), \tag{11}$$

where the symbols have their usual meaning, Spieler [68] gives the following useful functional form

$$\frac{I_R(T_2)}{I_R(T_1)} = \left(\frac{T_2}{T_1}\right)^2 \exp\left[-\frac{E_a}{2k}\left(\frac{T_1 - T_2}{T_1 T_2}\right)\right], \tag{12}$$

which can be used to predict the leakage current at a new operating temperature T_1 based on the measured leakage current at T_2. Here, E_a is the activation energy (equal to 1.15 eV for undamaged devices and 1.2 for damaged devices). Tabulated values for the ratio of currents given by Eq. (12) are presented in Table 11.6. The table is read vertically. For example, taking the leakage current measured at an initial operating temperature of 0°C, then the expected leakage current operating at a new reduced temperature of −20°C (looking down the column) will be 12% of that at 0°C (*i.e.*, an eight-fold reduction in leakage current). Conversely, operating the detector at room temperature will result in an eight-fold increase in leakage current. Lämsä [69] measured leakage currents at a variety of temperatures in a damaged Si PIN diode and calculated ratios of leakage currents assuming an activation energy of 1.2 eV, Lämsä [69] showed that the measured ratios were in very good agreement with those calculated using Eq. (11), indicating that generation-recombination noise is indeed the dominant component of leakage current for radiation damaged Si PIN detectors.

The efficacy of operating at a reduced detector temperature on spectral performance after radiation damage has occurred is illustrated in Fig. 11.10. Here we show the measured energy resolution in a Si diode as a function of detector temperature after an irradiation of 10 MeV protons with a total absorbed non-ionizing dose of 13 rads. The measured FWHM energy resolution at 5.9 keV at −20°C degraded from its pre-irradiated value of ~280 eV to ~600 eV. Reducing the operating temperature by 10°C improves the energy resolution by a factor of 2, restoring it to its pre-irradiated value. The effect is

TABLE 11.6

Ratio of leakage currents in a radiation-damaged, reverse-biased Si PIN diode which can be used to improve spectral performance once radiation damage has occurred. From the table, a new operating temperature can be selected that will reduce the leakage current sufficiently to allow spectral performance to return to acceptable limits. The table is read vertically, for example, taking the leakage current measured at an initial operating temperature of 0°C, then the expected leakage current operating at a new reduced temperature of −20°C (looking down the column) will be 12% of that at 0°C (*i.e.*, an eight-fold reduction).

		Leakage current at initial operating temperature T_2 (°C)					
		−20	−10	0	+10	+20	+30
Leakage current at new	+20	57.11	18.59	6.55	2.48	1	0.43
operating temperature T_1 (°C)	+10	23.04	7.50	2.64	1	0.40	0.17
	0	8.72	2.84	1	0.39	0.15	0.065
	−5	5.22	1.70	0.60	0.23	0.092	0.039
	−10	3.07	1	0.35	0.13	0.054	0.023
	−15	1.77	0.58	0.20	0.077	0.031	0.013
	−20	1	0.33	0.12	0.043	0.018	0.0075
	−25	0.55	0.18	0.063	0.024	0.0097	0.0041
	−30	0.30	0.097	0.034	0.013	0.0052	0.0022
	−35	0.16	0.051	0.018	0.0068	0.0027	0.0012
	−40	0.080	0.026	0.0092	0.0035	0.0014	0.0006
	−45	0.040	0.013	0.0046	0.0017	0.0007	0.0003
	−50	0.019	0.0063	0.0022	0.0008	0.0003	0.0001

similar in other materials. However, we have specifically chosen Si as an example because (a) the effect is more noticeable because of its small bandgap[4] and (b) much more experimental data is available.

11.4.5.2 Annealing

Thermal annealing of radiation-induced defects is probably the most effective method for restoring detector performance after displacement damage has occurred. Annealing may occur spontaneously, depending on temperature, or it may be accelerated by either current or photons (known as injection dependent annealing). Procedurally two types of thermal annealing are usually carried out – isothermal and isochronal. In isochronal annealing, annealing is carried out in discrete time intervals. The time intervals are of constant duration and annealing is carried out by altering the temperature within time intervals. In isothermal annealing the anneal temperature is kept constant and applied for variable lengths of time. Isochronal annealing is usually used as a tool to isolate and identify the various temperature-dependent damage processes that have occurred in a material during irradiation, whereas isothermal annealing is generally carried out over long periods purely to remove the effects of radiation damage.

Silicon anneals at room temperature, over time scales of weeks, although annealing at a temperature of 100°C for 2 to 3 hours can result in a ~60% reduction in leakage current after proton exposures up to 30 krads [70]. The majority of GaAs defects are effectively stable for temperatures below 200°C for unbiased devices, although they are strongly affected by injection-enhanced annealing [71,72] for active devices. However, even for unbiased devices, room temperature annealing occurs, but over a timescale of years, for example, the energy resolution of a 4 mm^2 GaAs diode irradiated to a level of 114 krads using 10 MeV protons, degraded from 600 eV FWHM at 22.1 keV to 2 keV [67]. After 2 years in storage, it was re-measured. Room temperature annealing had improved detector performance to about 1 keV at 22.1 keV. However, subsequent isothermal annealing at a temperature of 80°C for 12 weeks was completely ineffective and resulted in no additional improvement.

Fraboni *et al.* [73] carried out a study of annealing in CdTe and CdZnTe using a combination of photo-inducted current transient spectroscopy (PICTS) and gamma-ray spectroscopy. As pointed out by Fraboni *et al.* [73], PICTS and gamma-ray spectroscopy are complementary techniques, since the first identifies electrically active defects and the latter describes the macroscopic behaviour of the charge collection process. By combining results from these two analyses in the study of the evolution with time and temperature of the detector recovery processes, they were able to identify the deep traps that affected their transport properties and correlate the effect on the overall spectroscopic capabilities. In a controlled set of irradiations

---

[4] Except for Si and Ge, low bandgap (*i.e.*, < 1 eV) semiconductors are barely spectroscopic in the X-ray regime at this time.

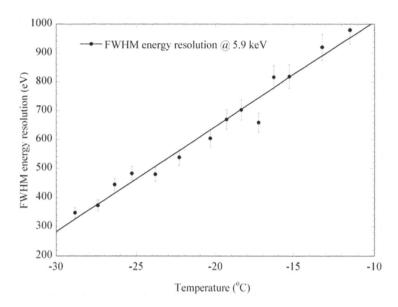

FIGURE 11.10 The effect of cooling on the spectroscopic performance of a radiation damaged 3 mm^2 Si PIN diode. The total absorbed non-ionizing dose was 13 rads. The initial undamaged energy resolution measured at a temperature of −20°C was 280 eV FWHM. The shaping time was 3μs.

using both protons and neutrons, up to fluences of typically 10^{13} cm^{-2}, it was found that for both high energy and low energy neutrons the effects were similar, although the induced damage was greater in CdTe. In both cases, room temperature annealing took place. However the degree to which they recovered depended only on how severely damaged they were, as determined from the PICTS analysis. For example, in CdZnTe, for an irradiation of 2×10^{11} 2 MeV protons almost complete spectroscopic recovery took place after 2 years in storage at room temperature, and for an irradiation of 10^{13} cm^{-2} 1 MeV neutrons, the detector recovered 95% of its CCE after 4 years. In CdTe, the effects were similar but more pronounced. For an irradiation of 2×10^{11} cm^{-2} 2 MeV protons, the detector recovered to 90% of its pre-irradiation CCE value after 2 years in storage at room temperature, whereas for an irradiation of 10^{13} cm^{-2} 1 MeV neutrons, the detector only recovered 35% of its CCE after 4 years in storage. Annealing at a raised temperature gave much more promising results. Fraboni *et al.* [73] found that annealing CdZnTe detectors at temperatures less than 380K had no effect. Above this temperature, annealing the detector for only 3 hours was sufficient to completely restore its performance for particle fluences up to 10^{13} cm^{-2}. However, in CdTe detectors, annealing for 6 hours at 380K did not fully restore spectral performance. The latter results are supported by Ahoranta *et al.* [74] who found that 6×6×1 mm^3 CdTe detectors can effectively be annealed at temperatures of 100°C for 60 hours, but only for proton doses less than 10 rads absorbed dose (*i.e.*, 10^{11} cm^{-2}, 22 MeV protons). Higher doses result in a permanent loss of CCE. The detectors were irradiated with 22 MeV protons in logarithmically increasing fluences, at 10^9, 10^{10}, 10^{11}, and 10^{12} cm^{-2}, corresponding to absorbed doses of 0.1, 1, 10 and 100 rads). The irradiations were at room temperature, and measurements were carried out at a detector temperature of −20°C. The initial FWHM energy resolution, ΔE, was 320 eV at 5.9 keV. No significant change was found for absorbed doses of 0.1 rad. A small effect was found for absorbed doses of 1 rad ($\Delta E = 386$ eV), which could be annealed out after 17 hours of annealing at 100°C. Doses of 10 rads ($\Delta E = 750$ eV) required a total anneal time of 60 hours to restore performance. At 100 rads absorbed dose ($\Delta E = 2200$ eV), no measurable recovery was evident after 16 hours of annealing.

The restorative powers of annealing are dramatically demonstrated in Fig. 11.11 in which we show the effects of prolonged annealing at elevated temperatures on a severely radiation-damaged 55 cm^3 closed-end coaxial high purity germanium detector in the reverse electrode configuration. The study [75] was carried out in support of the BepiColombo mission to Mercury. Owens *et al.* [75] undertook a comprehensive series of tests to assess radiation effects on a wide range of sensors. Towards this end, a series of controlled irradiations were carried out using simulated solar proton spectra in the energy range ~55 MeV to 185 MeV with increasing fluence, namely, 8×10^8 protons cm^{-2} (~800 rad equivalent in Si), 6×10^9 protons cm^{-2} (~6 krads equivalent in Si) and 6×10^{10} protons cm^{-2} (~60 krads equivalent in Si). After each irradiation the detector's performance was assessed in terms of energy resolution, efficiency and activation. The detector was then annealed, as described in [76], and the measurements repeated before the next irradiation. The detector was multiply annealed at 100°C in block periods of 7 days. After each anneal cycle it was cooled to 77K and the relative efficiency, peak channel location and FWHM energy resolution measured at 6 gamma-ray energies. In Fig. 11.11 we show the evolution of photopeak at 1,332 keV with annealing cycle after the largest irradiation of 6×10^{10} protons cm^{-2}. The pre-irradiation peak is offset in channels for clarity. The measurements were taken using a ^{60}Co source placed 25 cm above the detector's forward detection face. After the irradiation, the photopeak efficiency dropped to about 3% of its initial value and FWHM energy resolution increased to ~40 keV. The

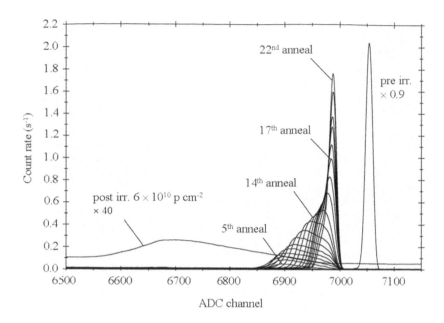

FIGURE 11.11 Illustration of radiation damage and subsequent annealing in a 55 cm³ HPGe detector (from Owens *et al.* [75]). Composite of spectral peak shapes measured at 1,332 keV after an irradiation of 6×10¹⁰ protons cm⁻². The pre-irradiation peak is offset in channels for clarity. The measurements were taken using a ⁶⁰Co source placed 25 cm above the detector's forward detection face. Note, the initial pre-irradiation peak is Gaussian shaped whereas the peaks measured after the irradiation are initially saw-tooth shaped eventually becoming more Gaussian. The effect of reduced efficiency post-irradiation is clearly apparent. The restoration of the detector's performance with 10% of its initial values took 4.5 months of annealing.

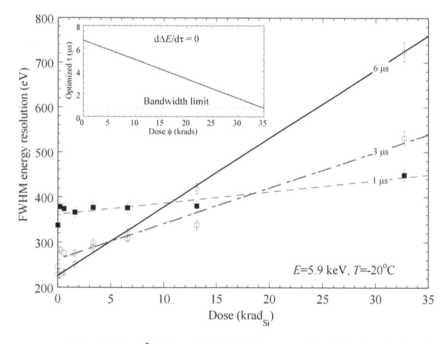

FIGURE 11.12 The X-ray energy resolution of a 3 mm², 500 μm thick, Si PIN-detector evaluated for three shaping time constants as a function of proton dose (krads equivalent into Si) [70]. From the graph, it is clear that long shaping times are optimal for undamaged detectors, while short shaping times are optimal for damaged detectors. The insert shows the optimized shaping as a function of dose for this particular detector/amplifier system. The dotted line in the inset shows the limiting shaping time that can be applied primarily to the bandwidth limit of the amplifier.

effect of reduced efficiency is clearly apparent. The key conclusion of the study was that even after exposure to modest proton fluences the detector required extensive annealing to restore spectroscopic performance to pre-irradiated values. Specifically, after exposure to an event of integral fluence 8×10⁸ protons cm⁻² this amounted to ~1 week duration at 100°C, whereas for a fluence of 6×10¹⁰ protons cm⁻², the detector required 22 annealing cycles (corresponding to a total annealing time of 4.5 months) to return the energy resolution to <3 keV FWHM at 1,332 keV.

11.4.5.3 Electronic Measures

Radiation effects also result in an increase in series noise (see Chapter 8, section 8.5.3.1.2). Therefore reducing signal integration time can reduce both baseline changes due to the integrated effects of detector leakage current and shot noise. However, any improvement in performance is limited because of the finite duration of the signal. If the signal sampling time is too short then charge is lost during the integration – so-called ballistic deficiency effects. As an example, Laukkanen *et al.* [70] radiated a 3 mm^2 Si PIN diode with 10 MeV protons up to an accumulated dose level of 32.5 krads. The pre-irradiated FWHM energy resolutions were 230 eV and 700 eV with shaping times of 1 μs and 6 μs, respectively. After irradiation, the FWHM energy resolutions were now 720 eV and 800 eV, respectively, at 1 μs and 6 μs shaping times. The data are plotted in Fig. 11.12 in which we can see that long shaping times are optimal for undamaged detectors, while short shaping times are optimal for damaged detectors. Thus, the measured energy resolution can be improved by adopting a shorter shaping time. In the case of Fig. 11.12, at the largest dose, the energy resolution can be improved by a factor of 1.6 by switching the shaping time constant from 6 μs to 1 μs. The inset of Fig. 11.12 shows the optimized shaping time as a function of dose for this particular detector/amplifier system. This was calculated by minimizing the function $d\Delta E/d\tau$ at a particular dose, ϕ. The dotted line in the inset shows the limiting shaping time that can be applied primarily to the bandwidth limit of the amplifier.

Alternately, circuitry can be designed to accommodate large baseline shifts, for example, using AC coupling, or correlated double sampling techniques if DC coupling is required. However, such techniques usually result in an increase in complexity and power consumption, which may present problems for some applications, for example, in space.

REFERENCES

[1] S.H. Park, J.H. Ha, J.H. Lee, H.S. Kim, Y.H. Cho, S.D. Cheon, D.G. Hong, "Effect of temperature on the performance of a CZT radiation detector", *J. Korean Phys. Soc.*, Vol. **56**, no. 4 (2010), pp. 1079–1082.

[2] K.P. O'Donnell, X. Chen, "Temperature dependence of semiconductor band gaps", *Appl. Phys. Letts.*, Vol. **58**, no. 25 (1991), pp. 2924–2926.

[3] J.S. Blakemore, "Semiconducting and other major properties of gallium arsenide", *J. Appl. Phys.*, Vol. **53** (1982), pp. R123–R181.

[4] ece-www.colorado.edu/~bart/book/book/chapter2/ch2_7.htm

[5] Z. Burshtein, H.N. Jayatirtha, A. Burger, J.F. Butler, B. Apotovsky, F.P. Doty, "Charge-carrier mobilities in Cd$_{0.8}$Zn$_{0.2}$Te single crystals used as nuclear radiation detectors", *Appl. Phys. Lett.*, Vol. **63**, no. 1 (1993), pp. 102–104.

[6] J.F. Butler, F.P. Doty, C. Lingren, "Recent developments in CdZnTe gamma ray detector technology", *Proc. SPIE*, Vol. **1734** (1992), pp. 140–145.

[7] A.E. Bolotnikov, C.M.H. Chen, W.R. Cook, F.A. Harrison, I. Kuvvetli, S.M. Schindler, "Effects of bulk and surface conductivity on the performance of CdZnTe pixel detectors", *IEEE Trans. Nucl. Sci.*, Vol. **49**, no. 4 (2002), pp. 1941–1949.

[8] G. Jaffé, "Theorie der Leitfähigkeit polarisierbarer Medien I", *Ann. Physik*, Vol. **408**, no. 2 (1933), pp. 217–248.

[9] G. Jaffé, "Theorie der Leitfähigkeit polarisierbarer Medien II", *Ann. Physik*, Vol. **408**, no. 3 (1933), pp. 249–284.

[10] L.F. Wouters, R.S. Christian, "Effects of space charge on the detection of high energy particles by means of silver chloride crystal counters", *Phys. Rev.*, Vol. **72** (1947), pp. 1127–1128.

[11] H.O. Curtis, *"Fluctuations in the Pulse of a Silver-Chloride Crystal Counter"*, PhD dissertation, Harvard University (1948).

[12] A. Chynoweth, "Behavior of space charge in diamond crystal counters under illumination I", *Phys. Rev.*, Vol. **83** (1951), pp. 254–263.

[13] K.G. McKay, "Electron bombardment conductivity in diamond", *Phys. Rev.*, Vol. **74** (1948), pp. 1606–1621.

[14] A. Chynoweth, "Removal of space-charge in diamond crystal counters", *Phys. Rev.*, Vol. **76** (1949), p. 310.

[15] B. Gudden, R.W. Pohl, "Über lichtelektrische Wirkung und Leitung in Kristallen", *Z. Phys.*, Vol. **16**, no. 1 (1923), pp. 170–182.

[16] H.L. Malm, M. Martini, "Polarization phenomena in CdTe: Preliminary results", *Can. J. Phys.*, Vol. **51**, no. 22 (1973), pp. 2336–2340.

[17] R.O. Bell, N. Hemmat, F. Wald, "Cadmium telluride, grown from tellurium solution, as a material for nuclear radiation detectors", *Phys. Stat. Sol. (A)*, Vol. **1**, no. 3 (1970), pp. 375–387.

[18] K. Zanio, J. Neeland, H. Montano, "Performance of CdTe as a gamma spectrometer and detector", *IEEE Trans. Nucl. Sci.*, Vol. **17**, no. 3 (1970), pp. 287–295.

[19] C. Manfredotti, F. Fizzotti, P. Polesello, P.P. Trapani, E. Vittone, M.S. Jaksic, S. Fazinic, I. Bogdanovic, "Investigation on the electric field profile in CdTe by ion beam induced current", *Nucl. Instr. Meth.*, Vol. **A380** (1996), pp. 136–140.

[20] M.A. Hossain, E.J. Morton, M.E. Özsan, "Photo-electronic investigation of CdZnTe spectral detectors", *IEEE Trans. Nucl. Sci.*, Vol. **49** (2002), pp. 1960–1965.

[21] K. Suzuki, T. Sawada, K. Imai, "Effect of DC bias field on the time-of-flight current waveforms of CdTe and CdZnTe detectors", *IEEE Trans. Nucl. Sci.*, Vol. **58** (2011), pp. 1958–1963.

[22] B.M. Vul, V.S. Vavilov, V.S. Ivanov, V.B. Stopachinski, V.A. Chapnin, "Investigation of doubly charged acceptors in cadmium telluride", *Sov. Phys. Semicond.*, Vol. **6** (1973), p. 1255.

[23] M. Ayoub, M. Hage-Ali, J.M. Koebel, A. Zumbiehl, F. Klotz, C. Rit, R. Regal, P. Fougères, P. Siffert, "Annealing effects on defect levels of CdTe:Cl materials and the uniformity of the electrical properties", *IEEE Trans. Nucl. Sci.*, Vol. **50** (2003) pp. 229–237.

[24] H. Toyama, A. Higa, M. Yamazato, T. Maehama, R. Ohno, M. Toguchi, "Quantitative analysis of polarization phenomena in CdTe radiation detectors", *Jpn. J. Appl. Phys.*, Vol. **45** (2006), pp. 8842–8847.

[25] R.O. Bell, G. Entine, H.B. Serreze, "Time-dependent polarization of CdTe gamma-ray detectors", *Nucl. Instr. Meth.*, Vol. **117** (1974), pp. 267–271.

[26] P. Siffert, J. Berger, C. Scharager, A. Cornet, R. Stuck, R.O. Bell, H.B. Serreze, F.V. Wald, "Polarization in cadmium telluride nuclear radiation detectors", *IEEE Trans. Nucl. Sci.*, Vol. **NS-23** (1976), pp. 159–170.

[27] R. Hofstadter, "Crystal counters-I and II", *Nucleonics*, Vol. **4** (1949), pp. 2–27; 29–43.

[28] J.A. Hodgkinson, "Non-linearity effects in the response of cadmium telluride nuclear radiation detectors", *Nucl. Instr. Meth.*, Vol. **A164**, no. 3 (1979), pp. 469–475.

[29] D. Vartsky, M. Goldberg, Y. Eisen, Y. Shamai, R. Dukhan, P.M. Siffert, J.M. Koebel, R. Regal, J. Gerber, "Radiation induced polarization in CdTe detectors", *Nucl. Instr. Meth.*, Vol. **A263** (1988), pp. 457–462.

[30] K. Okada, Y. Sakurai, H. Suematsu, "Characteristics of both carriers with polarization in diode-type CdTe x-ray detectors", *Appl. Phys. Lett.*, Vol. **90** (2007), pp. 063504–063504.

[31] H.B. Serreze, G. Entine, R.O. Bell, F.V. Wald, "Advances in CdTe gamma-ray detectors", *IEEE Trans. Nucl. Sci.*, Vol. **NS-21** (1974), pp. 404–407.

[32] A. Kh. Khusainov, A.L. Dudin, A.G. Ilves, V.F. Morozov, A.K. Pustovoit, R.D. Arlt, "High performance p–i–n CdTe and CdZnTe detectors", *Nucl. Instr. Meth.*, Vol. **A428** (1999), pp. 58–65.

[33] M. Niraula, A. Nakamura, T. Aoki, Y. Tomita, Y. Hatanaka, "Stability issues of high-energy resolution diode type CdTe nuclear radiation detectors in a long-term operation", *Nucl. Instr. Meth.*, Vol. **A491** (2002), pp. 168–175.

[34] M. Strassburg, C. Schroeter, P. Hackenschmied, "CdTe/CZT under high flux irradiation", *JINST*, Vol. **6** (2011), pp. C01055-1–C01055-11.

[35] P.J. Sellin, G. Prekas, J. Franc, R. Grill, "Electric field distributions in CdZnTe due to reduced temperature and x-ray irradiation", *Appl. Phys. Lett.*, Vol. **96** (2010), pp. 133509-1–1133509-3.

[36] H. Elhadidy, V. Dedic, J. Franc, R. Grill, "Study of polarization phenomena in n-type CdZnTe", *J. Phys. D: Appl. Phys.*, Vol. **47** (2014), pp. 055104-1–055104-5.

[37] L. Abbene, G. Gerardi, A.A. Turturici, G. Raso, G. Benassi, M. Bettelli, N. Zambelli, A. Zappettini, F. Principato, "X-ray response of CdZnTe detectors grown by the vertical Bridgman technique: Energy, temperature and high flux effects", *Nucl. Instr. Meth.*, Vol. **A835** (2016), pp. 1–12.

[38] V. Gerrish, "Electronic characterization of mercuric iodide gamma ray spectrometers", in *Semiconductors for Room-Temperature Radiation Detector Applications*, eds. L. Franks, R.B. James, T.E. Schlesinger, P. Siffert, *Mat. Res. Soc. Symp. Proc.*, San Francisco CA, Vol. **302** (1993), pp. 129–138.

[39] V. Gerrish, "Polarization and gain in mercuric iodide gamma-ray spectrometers", *Nucl. Instr. Meth.*, Vol. **A322** (1992), pp. 402–413.

[40] A. Holzer, M. Schieber, "Reduction of polarization in mercuric iodide nuclear radiation detectors", *IEEE Trans. Nucl. Sci.*, Vol. **NS-27** (1980), pp. 266–271.

[41] M. Slapa, G. Huth, W. Seibt, M. Schieber, P. Randtke, "Capabilities of mercuric iodide as a room temperature x-ray detector", *IEEE Trans. Nucl. Sci.*, Vol. **NS-23** (1976), pp. 102–111.

[42] A. Owens, M. Bavdaz, G. Brammertz, M. Krumrey, D. Martin, A. Peacock, L. Tröger, "The hard X-ray response of HgI_2", *Nucl. Instr. Meth.*, Vol. **A479** (2002), pp. 535–547.

[43] A. Kozorezov, V. Gostilo, A. Owens, F. Quarati, M. Shorohov, M.A. Webb, J.K. Wigmore, "Polarization effects in thallium bromide x-ray detectors", *J. Appl. Phys.*, Vol. **108** (2010), pp. 064507-1–064507-10.

[44] V. Kozlov, M. Kemell, M. Vehkamaki, M. Leskela, "Degradation effects in TlBr single crystals under prolonged bias voltage", *Nucl. Instr. Meth.*, Vol. **A576** (2007), pp. 10–14.

[45] G.A. Samara, "Pressure and temperature dependences of the ionic conductivities of the thallous halides TlCl, TlBr, and TlI", *Phys. Rev. B*, Vol. **23**, no. 2 (1981), pp. 575–586.

[46] S.R. Bishop, W. Higgins, G. Ciampi, A. Churilov, K.S. Shah, H.L. Tuller, "The defect and transport properties of donor doped single crystal TlBr", *J. Electrochem. Soc.*, Vol. **158**, no. 2 (2011), pp. J47–J51.

[47] J. Vaitkus, J. Banys, V. Gostilo, S. Zatoloka, A. Mekys, J. Storasta, A. Žindulis, "Influence of electronic and ionic processes on electrical properties of TlBr crystals", *Nucl. Instr. Meth.*, Vol. **546** (2005), pp. 188–191.

[48] K. Hitomi, T. Shoji, Y. Niizeki, "A method for suppressing polarization phenomena in TlBr detectors", *Nucl. Instr. Meth.*, Vol. **A585** (2008), pp. 102–104.

[49] J. Vaitkus, V. Gostilo, R. Jasinskaite, A. Mekys, A. Owens, S. Zataloka, A. Zindulis, "Investigation of degradation of electrical and photoelectrical properties in TlBr crystals", *Nucl. Instr. Meth.*, Vol. **A531** (2004), pp. 192–196.

[50] V. Kozlov, M. Kemell, M. Vehkamäki, M. Leskelä, "Degradation effects in TlBr single crystals under prolonged bias voltage", *Nucl. Instr. Meth.*, Vol. **A576** (2007), pp. 10–14.

[51] T. Onodera, K. Hitomi, T. Shoji, "Spectroscopic performance and long-term stability of thallium bromide radiation detectors", *Nucl. Instr. Meth.*, Vol. **A568** (2006), pp. 433–436.

[52] A.V. Churilov, G. Ciampi, H. Kim, L.J. Cirignano, W.M. Higgins, F. Olschner, K.S. Shah, "Thallium bromide nuclear radiation detector development", *IEEE Trans. Nucl. Sci.*, Vol. **56** (2009), pp. 1875–1881.

[53] T. Onodera, K. Hitomi, T. Shoji, "Temperature dependence of spectroscopic performance of thallium bromide x- and gamma-ray detectors", *IEEE Trans. Nucl. Sci.*, Vol. **54** (2007), pp. 860–863.

[54] I.B. Oliveira, F.E. Costa, P.K. Kiyohara, M.M. Hamada, "Influence of crystalline surface quality on TlBr radiation detector performance", *IEEE Trans. Nucl. Sci.*, Vol. **52** (2005), pp. 2058–2062.

[55] V. Gostilo, A. Owens, M. Bavdaz, I. Lisjutin, A. Peacock, H. Sipila, S. Zatoloka, "Single detectors and pixel arrays based on TlBr". *IEEE Trans. Nucl. Sci.*, Vol. **49** (2002), pp. 2513–2516.

[56] H. Kim, L. Cirignano, A. Churilov, G. Ciampi, W. Higgins, F. Olschner, K. Shah, "Developing larger TlBr detectors–detector performance", *IEEE Trans. Nucl. Sci.*, Vol. **56** (2009), pp. 819–823.

[57] C.R. Leão, V. Lordi, "Simultaneous control of ionic and electronic conductivity in materials: Thallium bromide case study", *Phys. Rev. Letts*, Vol. **108** (2012), pp. 246604-1–246604-5.

[58] R. Starr, P.E. Clark, L.G. Evans, S.R. Floyd, T.P. McClanahan, J.I. Trombka, J.O. Goldsten, R.H. Maurer, R.L. McNutt Jr., D. R. Roth, "Radiation effects in the Si-PIN detector on the Near Earth Asteroid Rendezvous mission", *Nucl. Instr. Meth.*, Vol. **A428**, no. 1 (1999), pp. 209–215.

[59] A. Ionascut-Nedelcescu, C. Carlone, A. Houdayer, H.J. von Bardeleben, J.-L. Cantin, S. Raymond, "Radiation hardness of gallium nitride", *IEEE Trans. Nucl. Sci.*, Vol. **49**, no. 6 (2002), pp. 2733–2738.

[60] A. Johnston, "*Reliability and Radiation Effects in Compound Semiconductors*", World Scientific Press, Hackensack, NJ (2010), p. 241, ISBN13: 978-9814277105.

[61] V.A.J. Van Lint, "The physics of radiation damage in particle detectors", *Nucl. Instr. Meth.*, Vol. **A253** (1987), pp. 453–459.

[62] C. Poivey, G. Hopkinson, "Displacement damage mechanism and effects", in *Space Radiation and Its Effects on EEE Components*, ESA-CERN-SSC Workshop, Chairperson. G. Bourban, EPFL, Lausanne, Switzerland (2009).

[63] A. Owens, S. Brandenburg, H. Kiewiet, F. Quarati, R.W. Ostendorf, "*An Assessment of Radiation Damage in Germanium Gamma-Ray Detectors Due to Solar Proton Events. A Study in Support of the BepiColombo and Solar Orbiter Missions*", ESA Internal Document, Noordwijk, The Netherlands (2008)

[64] J. Lindhard, M. Scharff, H.E. Schiott, "Range concepts and heavy ion ranges", *Kgl. Dan. Vidensk. Selsk. Mat.-Fys. Medd.*, Vol. **33** (1963), pp. 3–42.

[65] www.spenvis.oma.be/

[66] L.A. Franks, B.A. Brunett, R.W. Olsen, D.S. Walsh, G. Vizkelethy, J.I. Trombka, B.L. Doyle, R.B. James, "Radiation damage measurements in room-temperature semiconductor radiation detectors", *Nucl Instr. Meth.*, Vol. **A428**, no. 1 (1999), pp. 95–101.

[67] A. Owens, L. Alha, H. Andersson, M. Bavdaz, G. Brammertz, K. Helariutta, A. Peacock, V. Lämsä, S. Nenonen, "The effects of proton induced radiation damage on compound semiconductor X-ray detectors", *Proc. SPIE*, Vol. **5501** (2004), pp. 403–411.

[68] H. Spieler, "*Semiconductor Detector Systems*", Oxford University Press, Oxford (2005) ISBN-13: 978-0198527848.

[69] V. Lämsä, "*Soft X-Ray Spectrometer for the SMART-1 Satellite*", M.Sc. Thesis, University of Helsinki (2000).

[70] J. Laukkanen, V. Lämsä, A. Salminen, J. Huovelin, H. Andersson, L. Alha, K. Hamalainen, S. Nenonen, H. Sipila, M. Tillander, "Radiation hardness studies for the X-ray Solar Monitor (XSM) onboard the ESA SMART-1 mission", *Nucl. Instr. Meth.*, Vol. **A538** (2005), pp. 496–515.

[71] R. Loo, R.C. Knechtli, G.S. Kamath, "Enhanced annealing of GaAs solar cell radiation damage", *Proc. of the 15th Photovoltaic Specialists Conference* (1981), pp. 33–37.

[72] D. Stievenard, J.C. Bourgoin, "Defect-enhanced annealing by carrier recombination in GaAs", *Phys. Rev. B.*, Vol. **33**, no. 12 (1986), pp. 8410–8415.

[73] B. Fraboni, A. Cavallini, N. Auricchio, W. Dusi, M. Zanarini, P. Siffert, "Recovery of radiation damage in CdTe and CdZnTe Detectors", *IEEE Trans. Nucl. Sci.*, Vol. **52**, no. 6 (2005), pp. 3085–3090.

[74] J. Ahoranta, M. Uunila, J. Huovelin, H. Andersson, R. Vainio, A. Virtanen, H. Kettunen, "Radiation hardness studies of CdTe and HgI$_2$ for the SIXS particle detector on-board the BepiColombo spacecraft", *Nucl. Instr. Meth.*, Vol. **A605** (2009), pp. 344–349.

[75] A. Owens, S. Brandenburg, E.-J. Buis, H. Kiewiet, S. Kraft, R.W. Ostendorf, A. Peacock, F. Quarati, "An assessment of radiation damage in space-based germanium detectors due to Solar proton events", *Nucl. Instr. Meth.*, Vol. **A583** (2007), pp. 285–301.

[76] A. Owens, S. Brandenburg, E.-J. Buis, A.G. Kozorezov, S. Kraft, R.W. Ostendorf, F. Quarati, "Effect of prolonged annealing on the performance of coaxial Ge gamma-ray detectors", *J. Instr.*, Vol. **2** (2007), pp. P01001-1–P01001-7.

12

Improving Performance

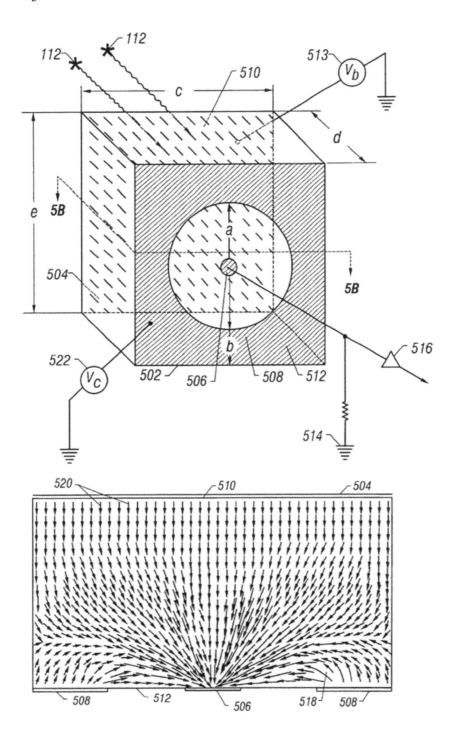

Frontispiece. This planar device uses a novel arrangement of three electrodes (502, 506 and 508) to shape the internal electric field, such that liberated charge in the detector bulk is focused towards a small collecting electrode (506). By suitable biasing, charge tailing can be virtually eliminated whilst maintaining high collection efficiency (Lingren *et al.*, *"Semiconductor radiation detector with enhanced charge collection"*, United States Patent No. US2002/0079456A1, 2002).

CONTENTS

12.1   Introduction

For conventional radiation detectors fabricated from compound semiconductors, the wide disparity between the transport properties of the electrons and holes ensures that detector performances are limited by the carrier with the poorer mobility-lifetime product ($\mu\tau$). In CdZnTe, for example, there is an order of magnitude difference between its electron and hole mobilities, specifically 1,350 cm^{2}V^{-1}s^{-1} for electrons and 120 cm^{2}V^{-1} s^{-1} for holes. Coupled with the fact that the mean free drift times are 5 times smaller for holes, the mu-tau product of holes is thus 50 times worse than it is of electrons. The drift length of the carriers is simply related to the mu-tau product, $\mu\tau$, and the applied electric field, E, by the relationship,

$$\lambda = \mu\tau E. \tag{1}$$

Under typical electric fields of 1 kV/cm, the mean drift lengths of electrons are about 1 cm, while that of holes ~ 0.1 cm. The significance of finite drift lengths is that they introduce an energy-dependent depth term into the charge collection process, which effectively limits the maximum thickness of a detector to a few mm for spectroscopy applications and ~ 1 cm for counter applications. Assuming the electric field is fixed, the only practical way of improving drift length of a carrier is to improve its $\mu\tau$ product. However, mobility is a fundamental material property and so the only practical way of improving $\mu\tau$ is to increase the carrier lifetime, which in turn depends greatly on detector material quality and stoichiometry. Until the specific traps/defects can be identified and corrected, stack geometries, single carrier pulse processing (such as rise time compensation) or sensing techniques (for example coplanar grids) offer the only practical means of obviating the problem, allowing detectors with relatively large active volumes to be constructed. In this chapter, we will review carrier collection in compound semiconductor radiation detectors and examine various approaches to single carrier correction and collection techniques.

12.2   Single Carrier Collection and Correction Techniques

Single polarity sensing and correction techniques have been widely applied to overcome poor hole transport, particularly to CdZnTe detectors. For conventional detection systems, the electrons are almost fully collected for interactions occurring throughout the detector volume. The same is not true for the holes. In fact, spectral tailing can be attributed almost exclusively to holes. Part of the problem can be attributed to their much lower drift mobilities, resulting in inordinately long transit times, and partly to preferential hole trapping exacerbated by the long transit times. The net result is the removal of holes from the signal before they can be collected. In Fig 12.1, we show the temporal distribution of rise times obtained from a 5×5 mm^2, 0.5 mm thick TlBr detector in response to a 662 keV gamma-ray source [1]. We see clearly how the wide disparity in carrier mobilities leads to their separation in the time domain. The figure also illustrates that there will be constraints on amplifier shaping times when collecting the signal. If shaping times are too long, the signal will be degraded

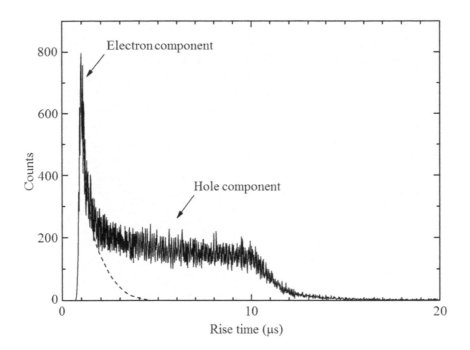

FIGURE 12.1 Distribution of preamplifier output pulse rise times from a 5×5 mm^2, 0.5 mm thick TlBr planar detector in response to stimulation by a 662 keV gamma-ray source, illustrating how the wide disparity in carrier mobilities leads to their clear separation in the time domain (from [1]). For the most statistically efficient charge collection, a trade-off is required between an optimal shaping time to minimize electronic noise and a longer shaping time needed to collect the entire signal.

by electronic noise,[1] whose contribution increases the longer we sample the signal. Conversely, if shaping times are too short, ballistic deficiency effects will occur[2] and not all the liberated charge will be collected, again leading to a degradation of the signal. The magnitude of this effect increases as the inverse square of the shaping time. Obviously an optimal shaping time will exist in which the total degradation due to electronic noise and ballistic deficiency effects can be minimized. However, a significant improvement in spectral acuity can be also be achieved by either discarding the carrier with the poorer transport properties, which is the most common approach, or correcting for it. These two approaches are described.

12.2.1 Directional Illumination

Perhaps the simplest way to improve spectral performance is to ensure that the trapped carrier has the shortest drift distance to travel to its collecting electrode. In its simplest form, for a planar detector with preferential hole trapping, this means illuminating the detector through the cathode side. In this way, the holes have a shorter distance to travel to the cathode than for anode illumination. The electrons have a longer distance to travel to the anode but are more likely to reach it since trapping is less. This is illustrated in Fig. 12.2, in which we show two ^{241}Am spectra obtained with $4 \times 4 \times 2$ mm^3 CdZnTe detector (taken from [2]: ©IEEE, 2001). One spectrum shows the detector response when illuminated from the cathode side (dashed line), and the other shows the response when illuminated from the anode side (solid line). Only the cathode irradiated spectra shows clearly resolved photopeaks, illustrating the effectiveness of directional illumination. Note that in the cathode illuminated spectrum, the effects of trapping are still apparent as witnessed by the tailing on the 60 keV photopeak. This is because not all incident photons interact at the same depth in the detector; they interact at a range of depths as described by the well-known exponential Lambert-Beer Law [3]. Thus, some photons interact at large depths. The generated holes will then be subject to trapping.

12.2.2 Rise Time Discrimination

Rise time discrimination (RTD) is a relatively simple method [4,5] to improve spectral shape and relies on the fact that for widely different transport properties, the rise times of the electron and hole current pulses are quite different. With reference to Fig. 12.3, the times taken for an electron and a hole to traverse the detector width L are

---

[1] Specifically parallel noise due to the leakage current which increases as the shaping time increases.

[2] That is, not all the signal will be collected, since in theory the shaping time should be infinitely long to collect all of the signal.

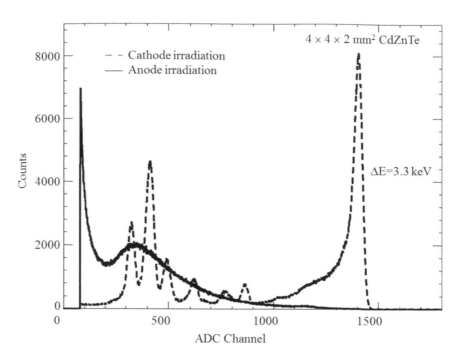

FIGURE 12.2 Typical ^{241}Am spectra obtained with 4×4×2 mm^3 CdZnTe detector irradiated from the cathode side (dashed line) and anode side (solid line) (from Takahashi and Watanabe [2]: ©IEEE, 2001). Only the cathode irradiated spectra shows clearly resolved photopeaks, since the holes only have a small distance to travel and are not lost from the signal induction process. However the effect of trapping is still evident as witnessed by the tailing on the 60 keV photopeak which arises from events that occur deep in the crystal. For anode illumination, no photopeaks are resolved since most of the holes are trapped during the long drift distance to the cathode and therefore do not induce a signal on the cathode.

$$t_e = \frac{L}{\mu_e E} \quad \text{and} \quad t_h = \frac{L}{\mu_h E} , \tag{2}$$

where μ_e and μ_h are the electron and hole mobilities and E is the electric field established by the bias V_b. We assume the radiation is incident on the anode contact. The time profiles for interactions at positions 1 and 3 in Fig. 12.3 (left) are shown in Fig. 12.3 (right). Position 1 illustrates the case where charge is produced very close to the cathode and so the entire *induced* signal is due to the motion of the electrons, while in position 3 the charge is produced very close to the anode and the *induced* signal is now due entirely to the motion of the holes. In the absence of trapping, the induced signal will build up linearly to the value of the initial charge Q_o, at a rate dependent on the hole drift velocity. The current pulse begins when the carriers induce charge on the electrodes as prescribed by the Shockley-Ramo theorem [7,8]. Given the difference in carrier mobilities, there will be two distinct current pulses, one from holes and one from electrons

$$I_{ho} = Q_o \left(\frac{\mu_h E}{L} \right), \qquad I_{eo} = Q_o \left(\frac{\mu_e E}{L} \right). \tag{3}$$

We note from Eq. (3) that for the case when $\mu_e >> \mu_h$ (which is generally true), the current pulse induced by the electrons will have a much larger amplitude and from Eq. (2) a shorter duration than that induced by the holes, as evident in Fig. 12.1. For the general case, illustrated by the intermediate position 2 in Fig. 12.3 (left), the induced signal (shown by time profile 2 in Fig. 12.3 right) will be a composite of electron and hole *components whose relative strengths will depend on the depth of the interaction*. The rapidly rising part of this signal will be due to the electron component (by virtue of its much larger mobility), while the slowly rising part will be due to the holes.

In RTD methods, all pulses whose rise time exceeds a pre-set threshold are rejected, specifically all those events that would normally lie in the tail (see Fig. 12.1). The net effect of RTD is demonstrated in Fig. 12.4, in which we show the measured response of a 3×3×2 mm^3 CdZnTe detector [9]. While the resolution is improved from 1 keV FWHM at 59.54 keV to 700 eV using RTD, the efficiency is lower than would be expected from the physical dimensions of the detector, since many counts are rejected. This is also demonstrated in Fig. 12.4 in which we can see that the amplitude of the lowest peak at ~9 keV is essentially the same in both spectra, since in this detector the sensitive volume is illuminated through the cathode and thus both electrons and holes will be efficiently collected. However, as the energy increases, there is an increasing loss of counts in subsequent peaks of the RTD-on spectrum, caused by hole trapping since the holes now have to transverse an increasing

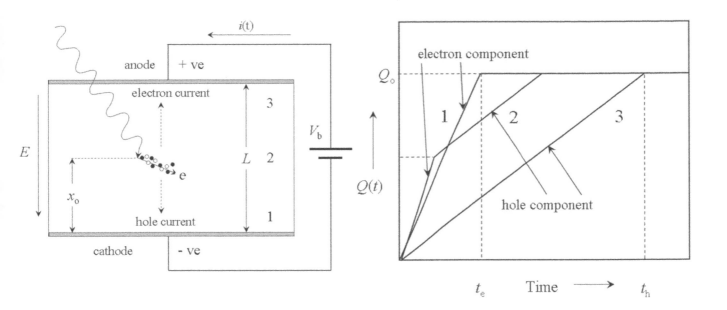

FIGURE 12.3 From [6]. Left: schematic of a simple planar detector showing a photoelectric interaction. Ionizing radiation absorbed in the sensitive volume generates electron-hole pairs in direct proportion to the energy deposited. These are subsequently swept towards the appropriate electrode by the electric field, E induced by the bias voltage, V_b. Right: the time dependence of the induced charge for three different interaction sites in the detector in the absence of trapping: 1) close to the cathode (signal only due to electrons), 3) close to the anode (signal only due to holes) and 2) an intermediate position (signal due to electrons and holes). In the last case, the rapidly rising part of this signal is due to the electron component, while the slowly rising part is caused by the holes.

thickness of detector to get to the cathode. In fact, the ratio of counts in the 60 keV photopeak (*i.e.,* RTD-on/RTD-off) is ~ 0.35 and the ratio of peak heights is ~ 0.65.

12.2.3 Bi-Parametric Techniques

Since both rise time and signal amplitude depend on the depth of interaction, one can use correlative measurements to correct the measured signal for the charge lost on the way to the collecting electrode, thus improving spectroscopic performance. Such methods are termed bi-parametric. In practice, this is achieved by plotting rise time as a function of pulse amplitude and energy from which one can generate a set of correction factors. The measured charge is then multiplied by the appropriate correction factor, yielding a corrected pulse amplitude [10]. The technique permits high resolution with little loss of sensitive volume. For example, using the technique, Verger *et al.* [11], achieved a room temperature FWHM energy resolution of 4.4% at 122 keV using a $4 \times 4 \times 6$ mm^3 CdZnTe detector with standard planar electrodes. The corrected detection efficiency was 82%. Without the correction, the measured energy resolution was 12.9%, and the detection efficiency was 38%.

Tada *et al.* [1] applied a digital pulse processing technique to a simple planar 5×5 mm^2, 0.5 mm thick TlBr detector to improve spectrometric performance. By applying two shaping times – one optimized to minimize electronic noise and the other to minimize ballistic deficiency effects, it was possible to derive an interaction depth correction (basically the ratio of the photopeak pulse heights from the fast and slow shapers) and use it to renormalize the charge collected with optimized system noise shaping. When operated as a single planar detector with optimal shaping, the measured FWHM energy resolution at 662 keV was 5.8%. This improved to 4.2% FWHM with the correction.

Bi-parametric techniques, while intuitively simple, suffer from several disadvantages. Firstly, the electronics is relatively complicated; secondly, since the pulse height deficit varies non-linearly with rise time, the technique is only easily applied over a limited range of interaction depths.

12.2.4 Stack Geometries

One simple way to increase active volumes and improve carrier collection is the use of stack geometries. The basic principle is illustrated in Fig. 12.5. An array of single planar detectors is simply stacked – one on top of another. Each individual element is thin enough to allow good charge transport but with the outputs summed so that the full volume of the stack is used. Charge collection efficiency is thus much higher than in an equivalently sized, planar detector since the travel distance to the collecting electrode can be made much less than the trapping lengths of the carriers. To date, this technique has only

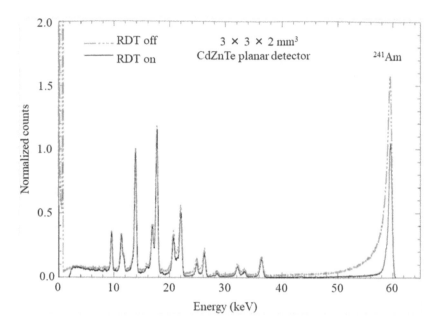

FIGURE 12.4 Two ^{241}Am spectra taken with a 3×3×2 mm^3 CdZnTe detector, illustrating the effectiveness of rise time discrimination (RTD). From the figure, we see that hole tailing is substantially reduced when RTD is employed, and the FWHM energy resolution improves from 1 keV to 700 eV at 59.54 keV. However, an energy dependent decrease in efficiency is also apparent (from [9]).

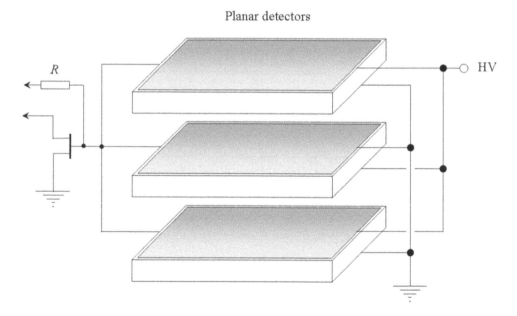

FIGURE 12.5 Stack detector concept. The signal path of each element is essentially or-ed. The bias is common between adjacent planes and alternate between cathodes and anodes.

been implemented for CdTe because hole $\mu\tau$ products are an order of magnitude higher than in other materials and this particular technique relies on the efficient collection of both carriers.

Watanabe *et al.* [12] constructed a CdTe stacked detector with 10 large, thin, CdTe diodes, each with an area of 21.5×21.5 mm^2 and a thickness of 0.5 mm. A measured FWHM energy resolution of 1.2% (7.9 keV) at 662 keV was achieved at a detector operating temperature of $-20°C$. Redus *et al.* [13] describe two stack implementations: one based on three CdTe detector elements and the other on five elements. Each element is 1 mm thick and 25 mm^2 in area. In Fig. 12.6, we show ^{137}Cs spectra taken with both detectors along with a spectrum from a single element planar device, from which we can see a significant improvement in efficiency. For example, the three and five element stacks had 3.8 and 8 times as many counts in the photopeak as the single element detector. The measured energy resolutions of the three and five element devices were 2.5% FWHM at 662 keV.

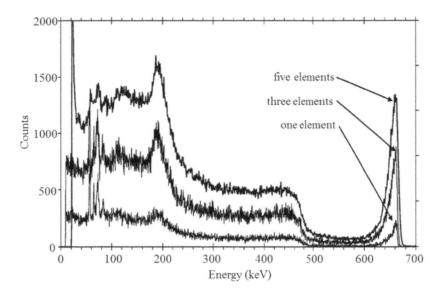

FIGURE 12.6 ^{137}Cs spectra obtained using the first generation stack detectors, with one, three, and five elements. The improvement in both the resolution and efficiency with the number of additional elements is clearly evident (from Redus *et al.* [13], ©2004 IEEE).

12.2.5 Sub-Bandgap Illumination

As we have seen, most spectral performance problems can usually be traced to poor carrier transport due to carrier trapping. An obvious solution is to improve material perfection. However, this is an ongoing technical challenge that has proven particularly difficult for melt grown materials and should be considered a long-term goal. An alternate approach is to try and manipulate the trapping process, either by impairing occupancy or increasing the de-trapping rates. By illuminating an active detector crystal with radiation tuned to wavelengths that correspond to trap level energies within the bandgap, it is possible to excite the traps and free the trapped carriers [14,15]. This will result in an increase in charge collection efficiency, not only from the increase in carriers available for collection but also by modifying and strengthening the internal electric field by reducing localized trapped space charge, especially near the electrodes [16]. In fact, sub bandgap illumination has been shown to be effective in reducing polarization effects in CdZnTe [17]. The optical de-trapping of carriers reduces carrier recombination, improves carrier transport and increases mobility lifetime in the crystal. The technique seems to work particularly well for CdZnTe, for which trapping occurs in mid-gap levels. Illumination is usually carried out using infrared radiation of wavelength around 900–1,000 nm, which corresponds to energies of ~ 1.2–1.3 eV. Generally, in CdZnTe, the optimum Zn fraction used for spectroscopic detectors is ~10%, which from Eq. 1, in Chapter 9, means the bandgap is ~1.6 eV. Thus, the crystal will be essentially transparent to incident IR light in the 1,000 nm wavelength range. Furthermore, the incident light has an energy that is close to the mid-bandgap energy and should stimulate donor levels located ~0.4 below the conduction band. CdZnTe has a number of levels in this region, for example, levels associated with the Te antisite (TeCd) defect and Cd vacancies related to impurities, including excess Te.

Xu *et al.* [14] fabricated a set of six, $5 \times 5 \times 1$ mm³ planar detectors from In-doped CdZnTe single crystals, grown by the modified vertical Bridgman method. The as-grown crystals had resistivities in the range 10^{10}–10^{11} Ω cm. The energy distribution of defect levels in the bandgap was investigated by using the thermally stimulated current method. The main trap level was located 0.58 eV below the conduction band minimum, with a cross section of 1.2×10^{-16} cm² and a trap density of ~7.2×10^{13} cm⁻³. The authors attribute this energy level to the second ionized Te antisite (TeCd), which has the second ionization energy of about 0.59 eV according to theoretical calculations. Electrons and holes can be excited from this level to the conduction and valence bands both optically and thermally. At the same time, free charge carriers can also be trapped. The relative balance of these processes can be obtained by the modified Shockley Read Hall recombination theory [16]. The crystals were illuminated through the contact with IR radiation of wavelength 940 nm through the anode. At this wavelength the optical penetration depth is ~ 2 mm, which is greater than the detector thickness. Under illumination, there was a general ~ 30% increase in the bulk leakage current indicating a net increase in charge carriers with illumination. The corresponding bulk resistivity decreased from ~ 3×10^{10} Ω cm to ~5×10^9 Ω cm, which can be expected since resistivity is inversely proportional to the free carrier concentration.

The effect or sub-bandgap illumination on spectral performance was investigated using an [241]Am radiation source. Figure 12.7 shows a comparison of room temperature energy spectra taken with the same CdZnTe planar detector, with and without sub-bandgap illumination (taken from Xu *et al.* [14]). The operating conditions were kept constant. The effects of the illumination are quite apparent. The FWHM energy resolution at 59.95 keV reduced from 9.7% to 5.7%, and the hole tailing is much less severe. Under illumination, the photopeak position shifted towards higher channel numbers (by ~ 0.02%),

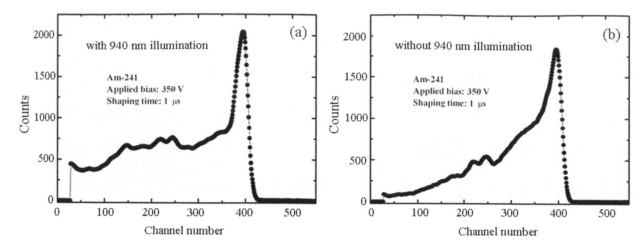

FIGURE 12.7 Comparison of room temperature [241]Am energy spectra taken with the same CdZnTe planar detector under the same operating conditions, without (right) and with (left) sub-bandgap illumination at 940 nm (from Xu *et al.* [14]). The effects of the illumination are quite apparent. With illumination, the FWHM energy resolution at 59.95 keV reduced from 9.7% to 5.7%, and the hole tailing is much less severe.

indicating better charge collection efficiency. In addition, the electron mobility-lifetime product (as determined by fitting the Hecht equation), increased from 6.7×10^{-4} cm^{2}V^{-1} to 1.03×10^{-3} cm^{2}V^{-1} under illumination, illustrating improved transport properties.

Duff *et al.* [15] fabricated two CdZnTe detectors of dimensions approximately $10 \times 10 \times 5$ mm^3. When patterned with simple planar electrodes, both detectors had large leakage currents and gave a relatively poor spectral response with FWHM energy resolutions at 662 keV in the tens of percent range. Consequently, both detectors were re-patterned with hemispherical electrodes to improve the response using single carrier sensing (see Section 12.3.2). When the cathodes were exposed to 662 keV gamma-rays from a ^{137}Cs source, one detector gave a FWHM energy resolution of 1.6% and the other 2%. Both detectors were then illuminated with IR radiation at 950 and 1,000 nm through the anode side. Although the illumination of the detectors was not optimized in terms of wavelength, Duff *et al.* [15], found a clear ~30% improvement in the energy resolution of both detectors. In fact, the best energy resolution recorded was 1% FWHM. It can be concluded that sub-bandgap IR illumination can provide a low-cost effective technique for improving the spectroscopic characteristics of detectors fabricated from relatively low-grade material.

12.3 Electrode Design and the Near Field Effect

Improvements in spectral acuity can also be effected by exploiting the "near field" effect by careful detector and/or electrode design. One way this can be achieved is by designing a detector's geometry such that the part of the detector volume from which the electrons are collected is much larger than that of the holes and structuring the electrodes in such a way that the electrons see a steadily increasing field on their path towards the anode, while the holes see a correspondingly decreasing field on their way to the cathode (assuming of course that the holes are the dominant trapped carrier, which is generally the case).

12.3.1 The Shockley-Ramo Theorem

When radiation interacts in the active region of a detector, it creates a cloud of electron-hole pairs, which, when exposed to an electric field, move towards the electrodes. Naively, one might assume that the signal on electrodes is solely due the collection of charge as it arrives at the electrodes. This is not the case. In fact, the signal is due to the movement of charge through detector bulk. This induces a time dependent charge on the electrodes, which begins as soon as charge starts moving and ends when the last of the charge reaches the electrodes. It is important to note that charge is induced on <u>all</u> electrodes throughout its movement and not just on the collecting electrode. The Shockley-Ramo theorem [7,8] provides a simple relationship between the magnitude of the generated charge cloud and the charge induced on the electrodes. It states that regardless of bias or the presence of space charge, the change in the induced charge Q at an electrode caused by a charge q moving from x_i to x_f is given by

$$\Delta Q = \int_{x_i}^{x_f} q E \vec{E}_w \cdot d\vec{x} = -q[\Phi_w(x_f) - \Phi_w(x_i)] = -q\Delta\Phi, \qquad (4)$$

where Φ_w and $\vec{E}_w$ are known as the weighting potential and weighting field, respectively. These are not the actual potential or permanent electric field, E, established by the bias[3] but representations in which the dependences on factors other than charge induction have been removed. To calculate the weighting potential, one must solve Laplace's equation, $\nabla^2\Phi = 0$, for the geometry of the detector but with artificial boundary conditions. These are, *a)* the voltage on the electrode on which the induced charge is to be calculated is set to unity, *b)* the voltages on all other electrodes are set to zero and *c)* the trapped charge within the detector volume is ignored (otherwise, a formal solution of the Poisson equation would be required). The weighting field is then given by the gradient of the weighting potential, -grad Φ_w. For simple symmetric geometries, such as planar, hemispherical or coaxial analytical solutions may be easily deduced. For more complicated geometries, numerical solutions are usually obtained using 3-D electrostatic codes (see, for example [18]).

We see from Eq. (4) that the induced charge on the electrode is dependent only on weighting potentials. The greater the change in weighting potential the carrier undergoes on its journey to the collecting electrode, the larger the charge induced on that electrode. The instantaneous current, *i*, induced on the electrode is given by

$$i(t) = \frac{dQ}{dt} = q\vec{v}\vec{E}_w = -q\mu E\vec{E}_w. \tag{5}$$

The utility of Eq. (5) is that the field terms have now been separated into two, a static field term, E, which determines the path and the drift velocity of the carrier, and a weighting field term, $\vec{E}_w$, which determines how the motion of the charge is coupled to a particular electrode. The important point to note is that these terms can be altered to improve charge induction. However, while one can shape the internal electric field, it usually affects adversely the leakage current and charge collection efficiency. The same is not true for the weighting field, which can be manipulated by careful electrode design to improve charge induction without affecting the applied field and carrier dynamics. By simulating weighting potentials for a given electrode design and detector geometry, it may be possible to find a self-consistent set of solutions in which the carrier with the poorest transport properties must traverse a region with weak signal induction and be subject to enhanced trapping, while the carrier with the best transport properties must travel through a high field region with high signal induction in which it is less likely to be trapped. The net result is improved spectral performance and a number of detector concepts, such as hemispherical [19], coaxial [20], Frisch grid/ring [21], coplanar grids [22], strip [23] and small-pixel geometries [24] have evolved into practical systems by specifically exploiting this "near-field" effect.

12.3.2 Hemispherical Detectors

Perhaps the simplest implementation is the so-called hemispherical configuration in which the cathode is an extended electrode, of radius r_c, which surrounds a much smaller anode electrode [19]. Since the electric field within the detector varies as r^{-2}, a high field exists close to the anode and a low field near the cathode. This is illustrated in Fig. 12.8 (left). The drift length of an electron will now be a function of its radial distance from the anode. Carriers generated near the cathode, which encompasses the bulk of the active volume for geometric reasons, have to traverse a predominantly low field, with the holes migrating towards the cathode and the electrons towards the anode. As the lifetime for the electrons is typically far larger than that of the holes, the electrons have a high probability of transiting this region and arriving in the high field region. In addition, as they pass through the low field region, they gain energy from the increasing field and are thus less likely to be trapped. Once in the high field region, the electrons will induce a charge dQ on the anode given by

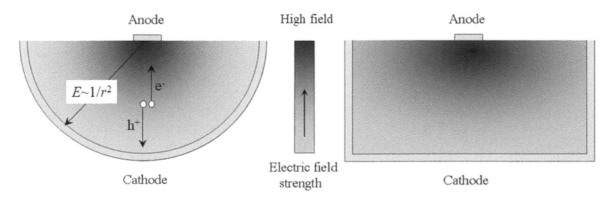

FIGURE 12.8 Distribution of the equipotential field lines in hemispherical (left) and pseudo-hemisphere (right) detector geometries.

---

[3] For two-electrode configurations, the electric field and the weighting field have the same form.

$$dQ = (q/V) E(r) dr,$$ (6)

in the absence of trapped charge. Here $E(r)$ is the electric field at r, and V is the applied bias voltage. In contrast, as the holes head towards the cathode, they encounter an increasingly weaker field region, and due to their low mobility, shorter travel distance and short lifetime will induce a negligible charge on the cathode regardless of trapping. For simplicity, detailed calculations usually proceed by assuming a spherical geometry [19].

Although conceptually straightforward, hemispherically shaped detectors are difficult to fabricate. However, the shape can be approximated by a cubic detector in which the cathode covers five sides, and the anode is a small pad located in the center of the sixth side (see Fig. 12.8 right). In Fig. 12.9 we show a comparison of two simulated ^{137}Cs spectra – one recorded in a 0.5 cm^3 CdZnTe crystal equipped with planar electrodes and an identical 0.5 cm^3 crystal, equipped with quasi-hemispherical electrodes (taken from [25]), which illustrates the effectiveness of the technique. Typically, quasi-hemispherical detector volumes range between 0.01 cm^3 and 1 cm^3. FWHM energy resolutions of $\leq$ 3% at 662 keV have been achieved with 0.5 cm^3 devices [25,26].

12.3.3 Coaxial Geometries

In a variation of the high-field/low-field theme, Lund *et al.* [20] constructed a CdZnTe detector in an open-ended coaxial configuration, similar to that commonly used for HPGe detectors. The detector is illustrated in Fig. 12.10 and has an outer diameter of 0.625 cm, a length of 0.7 cm and an active volume of 0.66 cm^3. It can be thought of as a two-dimensional implementation of a hemispherical detector and, as such, should have similar advantages. While planar detectors offer good resolution for incident energies up to a few hundred keV, they lack sensitivity at higher energies primarily due to the drop in electric field strength with detector thickness. In addition, large area devices suffer from poorer energy resolution, due to the increased capacitance of the collecting electrode. Coaxial configurations, on the other hand, offer a marked increase in field strength near the central collecting electrode, whilst maintaining a relatively low capacitance, resulting in better high-energy performance. However, near the outer surface, the electric field strength will be much weaker and depending on the biasing can be used to suppress the hole current as in hemispherical detectors. However, when the device of Lund *et al.* [20] was exposed to a ^{137}Cs radiation source, the response was found to be no better than that obtained with an equivalently sized planar detector. The response of the device could be predicted with good precision by the formulation of Sakai [27], originally developed for open-ended coaxial Ge detectors,

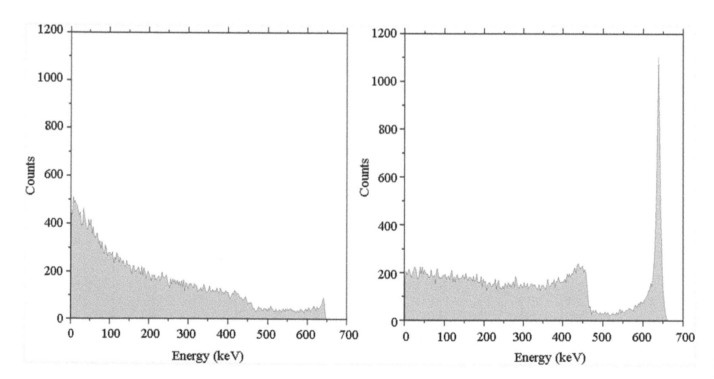

FIGURE 12.9 Comparison between the spectroscopic capabilities of a simple planar detector and a quasi-hemispherical CdZnTe detector of the same size: (a) a simulated ^{137}Cs spectrum for a 10×10×5 mm^3 planar detector; (b) simulated ^{137}Cs spectrum for a 10×10×5 mm^3 quasi-hemispherical detector (from Bale and Szeles [25]).

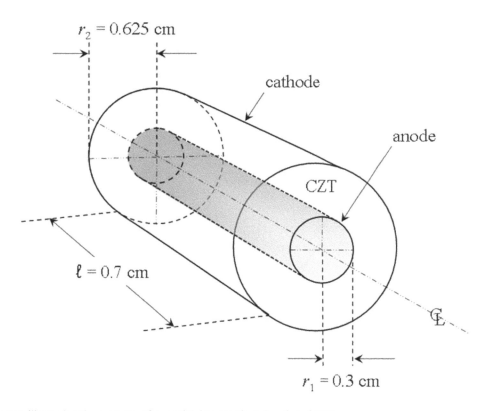

FIGURE 12.10 Diagram illustrating the geometry of a coaxial detector (from Lund *et al.* [20]).

$$Q(V,r_o) = \frac{Q_o}{\ln(r_2/r_1)} \times \left\{ \int_{r_1}^{r_o} \frac{1}{r} \exp\left[-\ln\left(\frac{r_2}{r_1}\right) \frac{(r_o^2 - r^2)}{2\mu_e\tau_e V} \right] dr + \int_{r_0}^{r_2} \frac{1}{r} \exp\left[-\ln\left(\frac{r_2}{r_1}\right) \frac{(r^2 - r_o^2)}{2\mu_h\tau_h V} \right] dr \right\}. \quad (7)$$

Here, $Q(V,r_o)$ is the charge collected as a function of radial position of the interaction, r_o, in the detector, Q_o is the total charge created by the interaction, $\mu_e\tau_e$ and $\mu_h\tau_h$ are the electron and hole mobility-lifetime products, respectively, and V is the potential difference between the contacts at r_1 and r_2. The first term in the parentheses of Eq. (7) represents the charge induced at the output due to the electrons and the second to that of the holes. Using the measured transport parameters in Eq. (7), Lund *el al* [20]. were able to show theoretically that a substantial improvement in performance could be obtained either by improving the quality of the CdZnTe used to manufacture the device, or alternately, if using the existing material by altering its geometry, increasing the hole trapping lifetime by an order of magnitude.

12.3.4 Frisch Grid/Ring Detectors

Both hemispherical and coaxial geometries achieve improved charge collection by shaping the electric field so that the induced charge is generated in a small high field region centered on the collecting electrode within which the signal is nearly independent of where the carrier is. Frisch grid/ring detectors achieve single carrier sensing in a manner emulating the Frisch grid scheme used in gas detectors to reduce the positional dependence of charge collection. The idea has been pursued by McGregor and co-workers in a series of papers (*e.g.*, see [21,28,29]). In a gas detector, the electron mobility is much higher than the positive ion mobility and hence the extraction times of the electrons are considerably less than those for the ions. For typical microsecond integration times, the measured pulse amplitude becomes dependent on the location of the initial interaction in the chamber, and wide variations in pulse amplitude are possible. The effect was significantly reduced by Frisch [30] who incorporated a conductive grid in the chamber near the anode. The grid effectively separates the chamber into two regions – a large region between the grid and cathode where the majority of gamma-ray interactions occur and a much smaller region between the grid and anode in which charge induction takes place. Gamma-rays interacting in the large volume release electron-ion pairs which drift in opposite directions when an external electric field is applied. The electrons drift through the grid and into the measurement region of the device, while the slower moving ions drift away from the grid. By appropriate design of the Frisch grid system, the weighting field, $\vec{E}_w$, can be optimized and as a consequence, an induced charge on the anode will result only from charge carriers moving between the conductive grid and the anode and not from carrier motion between the cathode and the grid, thus greatly reducing the pulse shape dependence on the gamma-

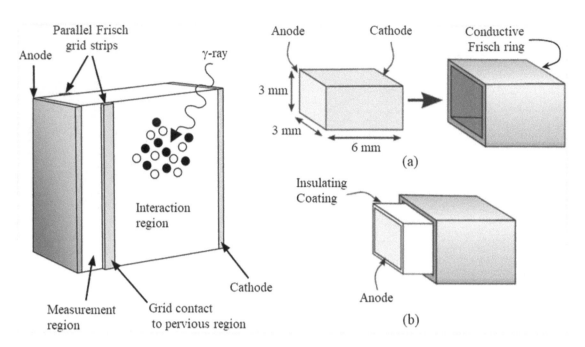

FIGURE 12.11 Two implementations of the Frisch-grid concept. Left: a prototype semiconductor design (from [21]). Two parallel contact strips are fabricated on the sides of the detector between the anode and cathode planes and act as a pseudo-Frisch-grid. Charge carriers are excited in the interaction region, and the electrons are drifted through the parallel grids by an applied electric field. Right: schematic of a capacitively coupled Frisch grid detector, consisting of (a) a bar-shaped detector placed inside an isolated but conductive ring (b) (taken from [29]). This implementation effectively eliminates leakage current between the grid and the anode, since the ring is not actually connected to the detector.

ray interaction position. Simply stated, for semiconductors this is achieved by maximizing $\Delta\Phi_w$ for electrons whilst minimizing it for holes.

While it is impractical to construct a grid inside a semiconductor detector, the concept has been successfully demonstrated using an external pair of "pseudo" grids [21]. In essence, this is an additional electrode system which is patterned on to the detector bulk just above the anode as illustrated in Fig. 12.11 (left). The effectiveness of the technique is demonstrated in Fig. 12.12 (a), in which is shown the response of a $5\times2\times5$ mm^3 CdZnTe device to a ^{137}Cs radioactive source [29]. With the grid disconnected, no full-energy peak is observed. However, with the grid active, a full energy peak at 662 keV is clearly observed with a measured energy resolution of 6% FWHM. The major disadvantages of this design are, firstly, it adds complexity to the fabrication and, secondly, the grid system is difficult to optimize in terms of noise and uniformity of response. At low energies, the system noise is dominated by the grid system itself which introduces additional noise from leakage currents flowing between the anode and the grid. At higher energies, border effects are induced by the grids which lead to non-uniformities in the response. These effects can be largely eliminated whilst still maintaining single-carrier collection using so-called "non-contacting" Frisch grid schemes. For example, Montemont *et al.* [31] and McNeil *et al.* [29] proposed a capacitively coupled grid which used a metal collar as a conduction screen with which to modify charge induction. The screen is grounded but isolated from the sides of the detector, so that it modifies only the transient electric behaviour of the detector given by the weighting field, without perturbing the static applied field. A schematic of such a device is shown in Fig. 12.11 (right). The conductive ring, when correctly biased, confines the largest change in the weighting potential to the vicinity of the anode while effectively screening induction from charge motion elsewhere. As well as being easier to fabricate than a patterned grid system, the design results in much reduced leakage currents. In Fig. 12.12 (b) we show the response of a $3\times3\times6$ mm^3 CdZnTe device with a 5 mm insulated Frisch ring to a ^{137}Cs radioactive source [32]. The thicker lined spectrum shows the results without a Frisch ring, and the thin lined spectrum shows the results with the ring active. With the grid active, the measured spectral resolution was 2.3% FWHM at 662 keV. While this design eliminates ambient grid anode leakage currents at low energies, its major limitation is noise introduced by inter-electrode capacity.

12.3.5 Coplanar Grid Detectors

The coplanar grid technique suggested by Luke [22] is essentially another variant of the classical Frisch grid and is again easily understood using the Shockley-Ramo theorem [7,8]. In the coplanar grid scheme [22,33], the anode electrodes take the form of two inter-digitated grids that are connected to separate charge-sensitive preamplifiers (see Fig. 12.13). An electric field is established in the detector bulk by applying bias, $V_{cathode}$, to the cathode, which is a full-area planar contact located on the side opposite the grid electrodes. For CdZnTe, this is typically of the order of ~ 1–2 kV. The two grid preamplifiers are

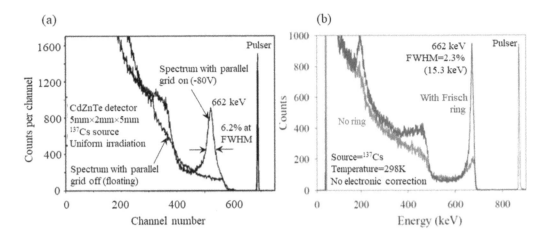

FIGURE 12.12 (a) response of a 5×2×5 mm³ CdZnTe patterned Frisch grid detector to a ¹³⁷Cs radioactive source under full area illumination (from [21]). Shown are spectra taken with the grid turned off and on. No full energy peak is apparent when the parallel grid is off; however, a full energy peak of 6.2% FWHM at 662 keV becomes obvious when the grid is activated. (b) ¹³⁷Cs spectra from a 3×3×6 mm³ CdZnTe device with a 5 mm insulated capacitively coupled Frisch ring (from [32]). The thin lined spectrum shows the response of the device with the ring connected and the thick lined spectrum with it disconnected. With the ring connected, the FWHM energy resolution is 2.3% at 662 keV.

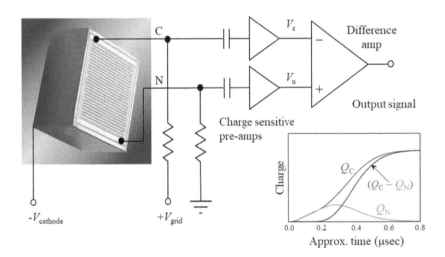

FIGURE 12.13 Schematic of the essential components of a coplanar grid detector (adapted from Luke *et al.* [34]). The anode electrodes take the form of two inter-digitated grids that are connected to separate charge-sensitive preamplifiers which in turn are connected to a subtraction circuit to produce a difference signal. A relatively small additional bias, $(V_{grid} \sim 10 - 100V)$, is applied between the grid electrodes so that one of the grids preferentially collects charge – this is known as the collecting grid (C). The other is known as the non-collecting grid (N). The lower inset shows the signals from the two grid systems and their difference.

connected to a subtraction circuit to produce a difference signal. A relatively small additional bias, $(V_{grid} \sim 10 - 100 \, V)$, is applied between the grid electrodes so that one of the grids preferentially collects charge; this is known as the collecting grid. The other is known as the non-collecting grid. Since the potentials on the two anode grids are similar, charge motion within the bulk of the detector is initially sensed equally by both grid electrodes, so that the difference signal is essentially zero. However, when charge approaches the anode, the grid signals begin to differ due to the difference in grid potentials, and a signal is registered at the output of the difference circuit. The key to effective grid design is to ensure that the weighting potentials of the two anode grids are almost equal inside the bulk of the detector volume where trapping will occur and only differ in the vicinity of the grids. The signal creation process is illustrated in the lower inset diagram of Fig. 12.13. Importantly, the difference signal measures the number of electrons reaching the collecting grid. The holes are essentially ignored, and as a result spectral tailing is largely eliminated. In addition, in a well-designed detector and in the absence of electron trapping, the magnitude of the difference signal is the same no matter where the charge is generated in the device. This results in a large improvement in performance, particularly for gamma-ray spectroscopy when compared to conventional planar device technology. In fact, operating the grids as a simple planar electrode, (*i.e.*, at the same bias voltage) results in no spectroscopic signal, but when operated in the coplanar geometry (*i.e.,* with a small potential difference between the two grids) the results are quite dramatic as is illustrated in Fig. 12.14.

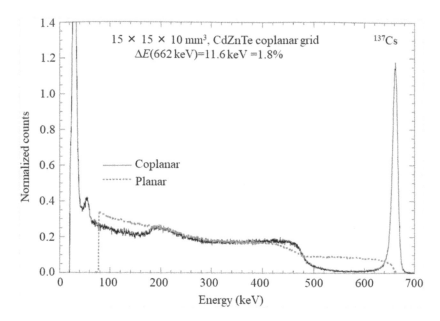

FIGURE 12.14 Comparison of pulse height spectra measured with a 2.25 cm³ device (15×15×10 mm³) when operated as a planar detector and as a coplanar grid detector [35].

The principal strengths of a coplanar grid system are that it offers the potential of large detection volumes with good energy resolution – which cannot be easily achieved by other techniques. For example, a 2.25 cm³ coplanar grid detector fabricated out of CdZnTe had a measured FWHM energy resolution of 1.8% at 662 keV [36]. This value is close to the theoretical limit predicted by Kozorezov *et al.* [37] who showed that the attainable energy resolution is actually limited by broadening due to an intrinsic lateral inhomogeneity inherent in the design. For common designs, this means a limiting energy resolution of about 2% FWHM at 662 keV. Recently, Gostilo *et al.* [38] reported fabricating a 10 cm³ detector. Despite the performance being considered disappointing due to crystal imperfections, spectral resolutions of 8% FWHM at 662 keV were still achieved.

While coplanar grid structures offer superlative performances at γ-ray energies, they do suffer from several limitations. Whereas the technique is essentially immune to hole trapping, it is not immune to electron trapping, which manifests itself as a reduction in the grid signals for electron events near the cathode. Because of the large difference in weighting potentials near the anodes, the effect is seen almost exclusively in the collecting grid. This results in an effective change in gain in the difference circuit, (*i.e.,* between electrons collected near the anode, 100% collected and gain = 1, and electrons generated near the cathode, <100% collected and gain < 1). We can show with the following simple example that electron trapping is not insignificant in the typical thicknesses of materials used to produce coplanar grid detectors. Assuming a uniform distribution of electron traps, the fraction of electrons remaining after travelling a distance x from the point where they were produced to the anode is given by

$$N(x)/N_o = \exp\left(-x/\lambda_e\right), \tag{8}$$

where N_o is the initial number of electrons and λ_e is the mean free drift distance of the electron given by Eq. (1). The fraction of electrons that are removed by traps when travelling a distance x is then given by

$$1 - N(x)/N_o. \tag{9}$$

For the detector described in Owens *et al.* [36], the detector thickness was 1 cm and the applied bias 1,700 V, giving an electric field of 1,700 V cm⁻¹. The electron mobility-lifetime product $\mu\tau_e$ was determined by a modified time of flight method to be 7×10^{-3} cm²V⁻¹ [39]. Therefore, the percentage of electrons trapped (and as a consequence, lost to the signal induction process) when travelling the full detector thickness is ~8%. At the output of the anode amplifiers this will translate to a full-energy peak shift of 8%. For $x/\lambda_e < 1$, the variation in the position of the peak channel with x will be linear. In fact, the peak will shift by an amount, δ, given by $\delta = 0.08x$(cm). Eq. (9) can thus provide the basis of a very simple correction technique first proposed by Luke [33] for compensating for the effects of electron trapping. By measuring the collecting grid signal as a function of input energy, it is possible to determine a linear correction factor with which to renormalize the collecting grid signal to compensate for the effects of electron trapping [40]. However, it should be noted that this method is only effective if the trap density is uniform throughout the detector volume and the variation of gain with thickness is approximately linear with depth (*i.e.,* $x/\lambda_e < 1$). Sturm *et al.* [41] applied the method to a $1.5 \times 1.5 \times 0.95$ cm^3 CdZnTe coplanar grid system. Without correction for electron trapping, the measured energy resolution at 662 keV was 4.3% FWHM. This was reduced to 1.7% with the correction.

He *et al.* [42,43] proposed an alternative method based on depth sensing to correct for electron trapping. This method uses signals from the cathode and coplanar anodes to determine the depth of the gamma-ray interaction, utilizing the fact that the cathode signal behaves identically to a planar electrode signal whose weighting potential is a linear function of depth. In this way, spectra can be mapped as function of interaction depth, which can then be used to compensate for electron trapping for any depth-dependent function. For a $1.5 \times 1.5 \times 0.95$ cm^3 CdZnTe coplanar grid system, Sturm *et al.* [41] measured a FWHM energy resolution of 4.3% without the depth correction and a FWHM energy resolution of 1.8% with this correction.

Additional drawbacks of the technique are that it requires complicated lithography, large single crystals and two sets of electronics for the readout – one set for the collecting grids and another for the non-collecting grids. This not only adds complexity over simpler detection techniques, but it also introduces addition low-level noise, since the two output chains are combined in the difference circuit – thus summing the noise of both circuits in quadrature.

12.3.6 Drift-Strip Detectors

A variation on drift detector electrode design is the drift-strip detector originally based on an idea of Gatti *et al.* [44] and developed by Kammer *et al.* [45] using silicon as the detection medium. The detector geometry is illustrated in Fig. 12.15 and consists of a number of drift detectors with a single anode readout strip. The drift-strip electrodes are biased in such a way that the electrons move towards the anode strip. The electrodes also provide an electrostatic shield so that the movement of the positive charge carriers will induce only a small signal at the anode strip, thus reducing the sensitivity to holes. Patt *et al.* [46] fabricated a prototype 17 electrode device on a $6 \times 6 \times 2$ mm thick HgI$_2$ crystal. Each electrode was approximately 250 μm wide with a 150 μm gap between the electrodes. At 122 keV, the measured FWHM energy resolution was ~3%.

Van Pamelen and Budtz-Jørgensen [47] fabricated a single-sided drift strip detector on a 6 mm^2, 1.5 mm thick, CdZnTe crystal. On one surface they patterned 15 drift strips, each 3 mm long of width 40 μm and pitch 100 μm. On the other surface, they deposited a planar cathode electrode. When operated as a planar detector (by connecting all strips together to form a common anode), no photopeak could be resolved when the device was exposed to 662 keV gamma-rays from a ^{137}Cs source. However, when operated as a drift-strip detector, a FWHM energy resolution of 2.7% was obtained[4] at 662 keV. The authors went on to demonstrate an additional substantial improvement in energy resolution by utilizing depth of interaction information to apply a depth correction for hole trapping as follows. Since the signal from the planar cathode electrode is strongly influenced by the holes, it is possible to use the ratio $r_{pa} = q_{planar}/q_{anode}$ (that is the charge induced on the planar cathode divided by the charge induced on the anode strip) to correct for the residual contribution of the holes at the anode.

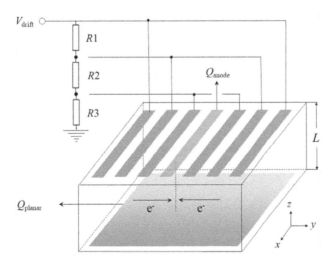

FIGURE 12.15 Drift-strip detector geometry. The drift-strip electrodes are biased in such a way that the electrons in the bulk initially move towards the anode strips whilst at the same time drifting towards the central anode collecting strip.

---

[4] Note the energy resolution in linear drift-strip devices is generally worse than for a circular drift structures due to the presence of uncapped electric fields at the end of the strips.

The corrections are obtained from the bi-parametric distribution of q_{planar} and r_{pa} as described by van Pamelen and Budtz-Jørgensen [47]. Energy resolutions of 1.0% FWHM were obtained at 662 keV. A recent device fabricated on a $10 \times 10 \times 3$ mm^3 discriminator grade[5] CdZnTe crystal had a FWHM energy resolution at 662 keV of 0.6% after depth correction [48]. While drift-strip detectors are capable of giving high energy resolutions they have two additional advantages:

1) Compared to coplanar grid detectors (see Section 12.2.5), they require simpler electronics, since no summing and subtracting circuits are needed, and

2) 2-D positional information is potentially available with one coordinate determined by signals from the strips, the second from the drift time (although at the expense of more complicated electronics).

The main disadvantage is that detector thicknesses are limited to a few mm where depth corrections can still preserve energy resolution.

12.3.7 Ring-Drift Detectors

Most recently a new type of geometry has been introduced – the so-called ring-drift detector [49]. In silicon technology, a considerable reduction in noise can be made by shaping the electric field and channelling charge to a central, small, read-out anode with a capacitance which is much smaller than that of the active surface area. In a drift device [50], this is achieved using a number of concentric ring electrodes designed and biased such that the potential gradient induces a transverse electric field, pushing electrons towards the central anode. A so-called ring-drift detector follows a similar approach for compound semiconductors. However, unlike Si, the motivation is centered not on achieving low readout noise, but on channelling electrons directly to the readout node, whilst incurring minimum charge loss. At the same time, holes drift to their collecting electrodes further away from the near field region of the central anode. This results in their contribution to the detector response being greatly diminished, making the detector essentially a single carrier sensing device. The device is illustrated schematically in Fig. 12.16. An interesting feature of this design is that by the proper adjustment of the ring potentials V_1 and V_2, the detector may be operated in two modes of charge collection – pseudo hemispherical and drift modes. Measurements have been carried out on a small prototype detector fabricated on a CdZnTe crystal of size $5 \times 5 \times 1$ mm^3. On the top face, the crystal is patterned by evaporating gold electrodes consisting of a small circular anode, a double ring electrode structure and finally a guard ring. A planar cathode is deposited on the backside. The electrode geometry is illustrated in Fig. 12.16.

The device was tested in both modes using highly monochromatic X-ray pencil beams across the energy range 10.5 keV to 100 keV. The results showed that it gave simultaneously excellent energy resolution and a wide dynamic range, which makes it particularly attractive in XRF, electron microprobe analysis systems and nuclear medicine applications. For, example, the measured FWHM energy resolutions under pencil beam illumination were 6.3% (drift mode) and 8.1% (hemispherical mode)

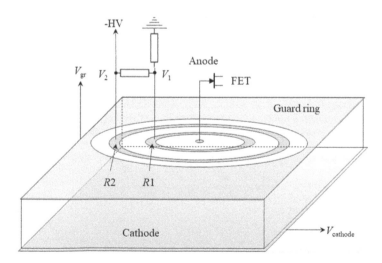

FIGURE 12.16 Schematic image of the prototype ring-drift detector. The crystal has dimensions $5 \times 5 \times 1$ mm^3. The inner anode has a diameter of 80 μm, and the centers of the other two ring electrodes, R1 and R2, are 0.19 and 0.39 mm, away from the center of the anode. The guard ring extends beyond a radius of 0.59 mm.

---

[5] Commercial CdZnTe planar crystals are usually classified as counter, discriminator or spectrometer grade, based on their FWHM energy resolution at 60 keV. Nominally, it should be better than 25% for counter grade, 15% for discriminator grade and 10% for spectrometer grade.

at 10.5 keV falling to 1.4% (drift mode) and 2.0% (hemispherical mode) at 60 keV. Using full area illumination, the measured FWHMs are slightly worse. At 60 keV they are 2.0% (drift mode) and 2.5% (hemispherical mode). The detector was also found to maintain good spectral resolution up to gamma-ray energies. Specifically, at 662 keV, the FWHM energy resolutions were 0.73% (drift mode) and 0.79% (hemispherical mode) [49]. The use of a small readout node ensures that electronic noise due to anode capacitance is low and independent of the active detector area, which implies that bigger detection areas can be fabricated with little loss of performance.

12.3.8 Small Pixel Effect Detectors

For pixel detectors, Barrett, Eskin and Barber [24] have shown that the deleterious effects of hole trapping can be greatly reduced if the pixel dimension, w, is made small relative to the detector thickness, L, as illustrated in Fig. 12.17. This is generally referred to as the small-pixel effect and makes use of highly non-uniform weighting potentials that can be generated within the pixel by the electrode geometry. Eskin, Barrett and Barber [51] describe three methods for calculating the induced signal on a pixel. These comprise a formal solution of the three-dimensional Laplace equation, a solution based on Green's functions and a solution based on the Shockley-Ramo theorem and weighting potentials. It is found that the size of the near-field region depends on the ratio w/L and is roughly hemispherical in shape with the characteristic size of the lateral pixel dimension. Since charge induction takes place within this region, the ratio w/L also determines the relative contribution of electrons and holes. For large values of w/L, the pixel behaves as a planar device, and charge is collected from the entire detector volume, resulting in poor spectroscopy because of a large hole component. For very small values of w/L, the near-field region is so small that spectral acuity is lost due to charge sharing with other pixels. Clearly an optimum ratio exists in which significant charge is registered only when the electrons move into the near field close to the anode while the holes move into an almost exclusively weak field region and contribute little to the induced signal. The near field region can be visualized in Fig. 12.18 for two pixel geometries, one with $w/L = 2$ and the other with $w/L = 0.25$. In the first case (planar), the field is uniform over most of the detector and so charge induction takes place over a large area, in fact 90% of the charge originates from 80% of the detector thickness. In the second case, the near field is much more tightly confined to the anode, and 90% of charge induction now occurs within 40% of the detector thickness from the anode.

The effect of the weighting potential on charge collection can be seen in Fig. 12.18(c), where the accumulation of induced charge is plotted as a function of height above the electrode for different pixel geometries. The curves can be understood as follows. Compared to the planar case ($w/L = 1$) in which the induced charge is directly proportional to the distance of the carriers from the anode, the other curves become increasingly non-linear near the anode as $w/L \to 1$. For interactions close to the cathode ($z = 1$), both carriers move through a region in which the weighting potential changes little and so contributes little to the induced signal. The holes continue to move in decreasing potential and eventually terminate on the cathode. The electrons, on the other hand, approach the anode and eventually enter the near-field region. At this point, the weighting potentials increase rapidly as does the induced signal on the anode which ceases when the electrons reach the anode. The measured energy-loss spectrum consists of a narrow photopeak, which becomes narrower for smaller pixels. However, the relative contributions of the electron and hole signals to the measured energy-loss spectrum depend not only on the ratio w/L but also on the trapping length of the electrons, and this ensures that there is a minimum pixel size. Less than this, the peak will again broaden, since an increasing proportion of electrons is lost to trapping in the low-field region of the detector before they reach the near-field region in which the signal is generated.

To demonstrate the spectral improvement that can be made by exploiting the small pixel effect, Fig. 12.19(a) shows a comparison between two ^{241}Am spectra. The first spectrum was obtained with a prototype 3×3 TlBr pixel array fabricated on mono-crystalline material of size 2.7×2.7×0.8 mm^3. The pixel size was 350 × 350 microns^2, pitch 450 μm and $w/L = 0.35$ (broken line). The second was obtained from a planar detector fabricated from the same wafer material and using the same contacting technology [52]. In this case, $w/L = 3.4$ (solid line). The energy scale and energy-loss spectrum obtained with the

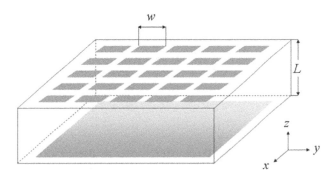

FIGURE 12.17 Figure illustrating the small-pixel geometry. The detector consists of a pixelated array with pixels of size, w, located above a planar cathode. The detector thickness is L.

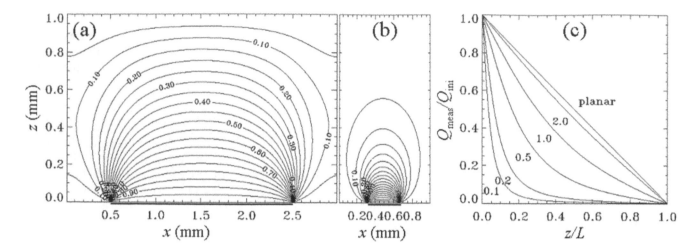

FIGURE 12.18 (a) and (b). Illustration of the weighting potentials $\Phi_w(x)$ for two ratios of pixel size to detector thickness, namely, $w/L = 2.0$ and 0.38 (from [52]). The contour lines in (a) and (b) are equivalent to the fraction of the total charge generated at $z = 1$ that are detected when the electrons are drawn towards the pixel at $z = 0$ mm and all the holes are lost. Thick black lines indicate the location of the pixel electrodes. In (c) the fraction of the generated charge measured as a function of photon absorption depth is shown for different pixel geometries, characterized by w/L values from 0.1 to 1.0.

planar detector have been renormalized to those of the pixel detector for the neptunium $L\alpha$ line at 13.64 keV. At this energy, it is assumed that X-rays interact so close to the cathode (the illuminated electrode) that both devices are essentially 100% efficient. From the figure, it is clear that the gain is different for the planar and pixel detectors and therefore their charge collection efficiencies (CCEs) are different, since the apparent measured energy losses become increasingly divergent at higher energies. At 59.54 keV, the CCE of the pixel detector is 22% larger than that of the planar device. Clearly, the tailing due to hole trapping is greatly reduced and the amplitude of the nuclear line is increased. Note that the bias applied to the planar detector was larger than that applied to the pixel detectors (250V as opposed to 170V). This accounts for its superior spectral resolution at low energies.

In Fig. 12.19(b) we show the calculated spectra for the planar and pixel detectors for the same operating conditions as the experimental values. The photo-absorption sites were distributed according to the expected absorption characteristics of TlBr. The values for the mobility and lifetimes of electrons and holes used in these simulations were $\mu_e = 20$ cm^{2}V^{-1}s^{-1}; $\tau_e = 30\,\mu$s; $\mu_h = 1$ cm^{2}V^{-1}s^{-1}; $\tau_h = 1\,\mu$s. These values reflect the measured $\mu\tau$ products for electrons and holes. Both spectra were convolved with a Gaussian of FWHM 690 eV to account for the additional line broadening due to electronic noise. Although we did not attempt to model the Np lines or the difference in gain between the two detectors, it is clear that the simulated spectra bear credible similarity to experimental measurement.

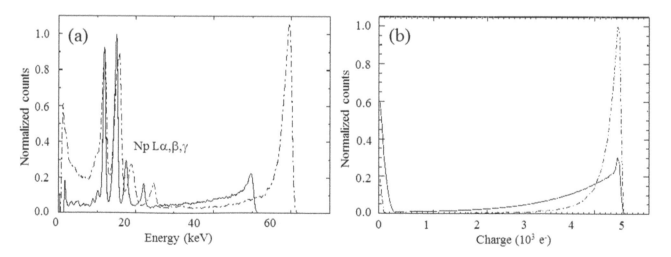

FIGURE 12.19 (*a*) Demonstration of the small pixel effect in TlBr, in which its spectral properties are greatly influenced by pixel geometry and specifically the ratio of the pixel dimension, *w*, to its thickness *L*. We show two ^{241}Am spectra (from [52]), one recorded by a pixel detector with a ratio $w/L = 0.35$ (broken line) and the other by a 2.7×2.7×0.8 mm^3 planar detector with a ratio $w/L = 3.4$ (solid line). (*b*) Modelled 60 keV spectra for the two detectors for the same operating conditions. Both spectra have been convolved with a Gaussian of FWHM 690 eV to simulate the electronic noise of the system.

12.3.9 Other Implementations

If energy resolution is not critical, simpler designs, such as two-terminal[6] [53] and three-electrode geometries [54] may be used to advantage. In essence, these designs generally use a combination of the small pixel effect and steering electrodes to enhance single carrier collection.

TABLE 12.1

Summary of the salient properties of the various single charge sensing/correction schemes (adapted from Owens and Kozorezov [6]). For the purposes of comparison, we list only room temperature results for CdZnTe detectors. Better energy resolution can generally be achieved with cooling. Improvements can also be achieved by employing a combination of techniques, for which examples are given.

Technique	FWHM resolution @ 662 keV, size	Advantages	Disadvantages
Planar [57]	8%, $18 \times 18 \times 2$ mm^3	Simple	Thickness limited to a few mm Resolution heavily dependent on carrier transport properties – particularly holes.
RTD [57]	1.4%, $18 \times 18 \times 2$ mm^3	Good resolution	Loss of counts in photopeaks (~60%) Thickness limited to a few mm Improvement limited to particular energy window Additional electronics
Bi-parametric [58]	1%, $4 \times 4 \times 5$ mm^3	High resolution Little loss of photopeak events	Complicated electronics Complicated calibration Depth correction limited to ~1 cm Expensive
Stack [13]	1%, $5 \times 5 \times 2.25$ mm^3	Relatively simple Good resolution	Volume limited by complexity of stack fabrication
Hemispherical [25] with sub bandgap illumination [15]	1.6%, $10 \times 10 \times 5$ mm^3 1%, $10 \times 10 \times 5$ mm^3	Simple, good resolution Similar performance to planar + RTD Simple to implement	Fabrication difficulties Volume limited to ~ 1 cm^3 See above
Coaxial geometry [20]	~4%, 6.25 mm dia $\times 7$ mm	Conceptually simple, potentially good performance	Difficult to fabricate
Drift strip [48]	0.6%, $10 \times 10 \times 3$ mm^3	Very high resolution 2-D information possible	Thickness limited to ~ few mm
Small Pixel [59] + steering grids and 3D sensing [55]	1.2%, $20 \times 20 \times 10.5$ mm^3 2.46 mm pixel size 0.8%, $15 \times 15 \times 10$mm^3, 11×11 pixels	Simple High resolution 2-D positional information available Very high resolution	Large number of pixels to achieve large area leading to complex electronics. Charge sharing between pixels. More complicated electronics than above
Virtual Frisch-grid [60] +"mixed" electrode and bi-parametric [55]	1.4%, $5 \times 5 \times 14$ mm^3 0.7%, $10 \times 10 \times 10$ mm^3	Simple construction High resolution Large thickness possible Very high resolution	Essentially pixel device Small area without applying other techniques More complicated electronics than above
Coplanar grid [35]	1.8%, $15 \times 15 \times 10$ mm^3	Good performance. Insensitive to hole trapping. Large volumes ~few cm^3	Complicated photolithography Max volume limited by crystal availability
Ring-drift [49]	0.7%, $5 \times 5 \times 1$ mm^3	Simple Very high resolution	Complicated spatial response Only small volumes demonstrated

---

[6] Essentially a hemispherical detector with two asymmetric electrodes.

12.3.10 Combinations of Techniques

As mentioned previously, significant improvements in spectral acuity can be achieved by applying a combination of techniques. For example, Verger *et al* [55]. demonstrated that by using a mixed electrode scheme which combined a non-contacting Frisch grid and the pixel field effect with bi-parametric techniques, an energy resolution of 0.7% at 662 keV could be achieved with a large 1 cm^3 CdZnTe crystal. Similarly, Zhang *et al.* [56] achieved an energy resolution of 0.8% at 662 keV with a large 2.25 cm^3 pixelated detector by merging steering, or focusing, grids and 3D depth correction.

12.4 Discussion and Conclusions

At the present time, problems of hole trapping and material uniformity limit the useful thickness of simple detectors to about 0.2 to a few mm; hence their high-energy performance. Until the particular defects can be identified, single carrier pulse processing and sensing techniques offer the only practical means of obviating the problem. A summary and compilation of the various single carrier sensing and correction schemes is given in Table 12.1. The performance of a particular technique is largely driven by the costs of production and ancillary support equipment, which in turn have impacts on the practicality of the device and its ease of use. We see from Table 12.1 that most techniques work reasonably well for volumes < 0.5 cm^{-3}. However, above this value only coplanar grid techniques can offer a substantial increase in volume whilst still maintaining high spectral acuity. In an analysis of a number of techniques, Luke [40] has shown that, with the exception of coplanar grid designs, electrode structures need to be specifically designed to match material characteristics and detector operating conditions if optimum spectral performances are to be achieved. Coplanar grid detectors, on the other hand, can be optimized by a simple gain adjustment between the collecting and non-collecting grids. However, while coplanar grid detectors would seem to be the obvious candidate for all gamma-ray applications, Kozorezov *et al.* [36] have shown that the energy resolution of coplanar grid detectors is limited by broadening due to an intrinsic lateral inhomogeneity inherent in the design. For common designs, this means a limiting energy resolution of about 2% FWHM at 662 keV.

To achieve even higher resolutions, a combination of techniques can be employed [55,56], but at the expense of increased complexity, which again emphasizes the fact that while novel designs and processing techniques can improve energy resolution, carrier transport still limits the maximum detection volume. Thus, ultimately, the quality of the material has to be improved if a substantial increase in detector volume is to be achieved sufficient to produce an effective MeV gamma-ray detector. However, as Luke [40] has pointed out, with the exception of simple planar geometries, improving hole collection efficiency may actually degrade spectral performance as single carrier collection techniques become less effective.

REFERENCES

[1] T. Tada, K. Hitomi, T. Tanaka, Y. Wu, S.Y. Kim, H. Yamazaki, K. Ishii, "Digital pulse processing and electronic noise analysis for improving energy resolution in planar TlBr detectors", *Nucl. Instr. and Meth.*, Vol. **A638** (2011) pp. 92–95.

[2] T. Takahashi, S. Watanabe, "Recent progress in CdTe and CdZnTe detectors", *IEEE Trans. Nucl. Sci.*, Vol. **48** (2001) pp. 950–959.

[3] A. Beer, "Bestimmung der Absorption des rothen Lichts in farbigen Flüssigkeiten", *Annalen der Physik und Chemie*, Leipzig, Vol. **86** (1852) pp. 78–88.

[4] V.T. Jordanov, J.A. Pantazis, A. Huber, "Compact circuit for pulse rise-time discrimination", *Nucl. Instr. and Meth.*, Vol. **A380** (1996) pp. 353–357.

[5] R. Redus, M.R. Squillante, J. Lund, "Electronics for high resolution spectroscopy with compound semiconductors", *Nucl. Instr. Meth*, Vol. **A380** (1996) pp. 312–317.

[6] A. Owens, A.G. Kozorezov, "Single carrier sensing in compound semiconductor detectors", *Nucl. Instr. and Meth.*, Vol. **A563** (2006) pp. 31–36.

[7] W. Shockley, "Currents to conductors induced by a moving point charge", *J. Appl. Phys.*, Vol. **9** (1938) pp. 635–636.

[8] S. Ramo, "Currents induced by electron motion", *Proc. IRE*, Vol. **27** (1939) pp. 584–585.

[9] A. Owens, T. Buslaps, C. Erd, H. Graafsma, R. Hijmering, D. Lumb, E. Welter, "Hard X- and gamma-ray measurements with a 3 × 3 × 3 mm^3 CdZnTe detector", *Nucl. Instr. and Meth.*, Vol. **A563** (2005) pp. 268–273.

[10] M. Richter, P. Siffert, "High resolution gamma ray spectroscopy", *Nucl. Instr. and Meth.*, Vol. **A322** (1992) pp. 529–537.

[11] L. Verger, M. Boitel, M.C. Gentet, R. Hamelin, C. Mestais, F. Mongellaz, J. Rustique, G. Sanchez, "Characterization of CdTe and CdZnTe detectors for gamma-ray imaging applications", *Nucl. Instr. and Meth.*, Vol. **A458** (2001) pp. 297–309.

[12] S. Watanabe, T. Takahashi, Y. Okada, G. Sato, M. Kouda, T. Mitani, Y. Kobayashi, K. Nakazawa, Y. Kuroda, M. Onishi, "CdTe stacked detectors for gamma-ray detection", *IEEE Trans. Nucl. Sci.*, Vol. **49** (2001) pp. 1292–1296.

[13] R. Redus, A. Huber, J. Pantazis, T. Pantazis, T. Takahashi, S. Woolf, "Multielement CdTe stack detectors for gamma-ray spectroscopy", *IEEE Trans. Nucl. Sci.*, Vol. **51**, no. 5 (2004) pp. 2386–2394.

[14] L. Xu, W. Jie, G. Zha, T. Feng, N. Wang, S. Xi, X. Fu, W. Zhang, Y. Xu, T. Wang, "Effects of sub-bandgap illumination on electrical properties and detector performances of CdZnTe", *Appl. Phys. Lett.*, Vol. **104** (2014) pp. 232109-1–232109-5.

[15] M.C. Duff, A.L. Washington, L.C. Teague, J.S. Wright, A. Burger, M. Groza, V. Buliga, "Use of sub-bandgap illumination to improve radiation detector resolution of CdZnTe", *J. Elec. Mat.*, Vol. **44**, no. 9 (2015) pp. 3207–3213.
[16] R. Guo, W. Jie, Y. Xu, G. Zha, T. Wang, Y. Lin, M. Zhang, Z. Du, "Space-charge manipulation under sub-bandgap illumination in detector-grade CdZnTe", *J. Elec. Mat.*, Vol. **44**, no. 10 (2015) pp. 3229–3323.
[17] J. Franc, V. Dedicˇ, M. Rejhon, J. Zazvorka, P. Praus, J. Tous, P.J. Sellin, "Control of electric field in CdZnTe radiation detectors by above-bandgap light", *J. Appl. Phys.*, Vol. **117** (2015) pp. 165702-1–165702-7.
[18] Z. He, G.F. Knoll, D.K. Wehe, Y.F. Du, "Coplanar grid patterns and their effect on energy resolution of CdZnTe detectors", *Nucl. Instr. and Meth.*, Vol. **A411** (1998) pp. 107–113.
[19] H.L. Malm, C. Canali, J.M. Mayer, M.-A. Nicolet, K.R. Zanio, W. Akutagawa, "Gamma–ray spectroscopy with single-carrier collection in high-resistivity semiconductors", *App. Phys. Letts*, Vol. **26** (1975) pp. 344–346.
[20] J.C. Lund, R.W. Olsen, R.B. James, J.M. Van Scyoc, E.E. Eissler, M.M. Blakeley, J.B. Glick, C.J. Johnson, "Performance of a coaxial $Cd_{1-x}Zn_xTe$ detector", *Nucl. Instr. and Meth.*, Vol. **A377** (1996) pp. 479–483.
[21] D.S. McGregor, Z. He, H.A. Seifert, D.K. Wehe, R.A. Rojeski, "Single charge carrier type sensing with a parallel strip pseudo-Frisch-grid CdZnTe semiconductor radiation detector", *Appl. Phys. Letts.*, Vol. **72**, no. 7 (1998) pp. 792–795.
[22] P.N. Luke, "Single-polarity charge sensing in ionization detectors using coplanar electrodes", *Appl. Phys. Lett.*, Vol. **65** (1994) pp. 2884–2886.
[23] M. van Pamelen, C. Budtz-Jørgensen, "Novel electrode geometry to improve performance of CdZnTe detectors", *Nucl. Instr. and Meth.*, Vol. **A403** (1998) pp. 390–398.
[24] H. Barrett, J. Eskin, H. Barber, "Charge transport in arrays of semiconductor gamma-ray detectors", *Phys. Rev. Letts*, Vol. **75** (1995) pp. 156–159.
[25] D.S. Bale, C. Szeles, "Design of high performance CdZnTe quasi-hemispherical gamma-ray CAPture™ plus detectors", *Proc. SPIE*, Vol. **6319** (2006) pp. 1–11.
[26] www.eurorad.com/PDF/BR_hemisph.pdf
[27] E. Sakai, "Charge collection in coaxial Ge(Li) detectors", *IEEE Trans. Nucl. Sci.*, Vol. **NS-15** (1968) pp. 310–320.
[28] D.S. McGregor, R.A. Rojeski, Z. He, D.K. Wehe, M. Driver, M. Blakely, "Geometrically weighted semiconductor Frisch grid radiation spectrometers", *Nucl. Instr. and Meth.*, Vol. **A422** (1999) pp. 164–168.
[29] W.J. McNeil, D.S. McGregor, A.E. Bolotnikov, G.W. Wright, R.B. James, "Single-charge-carrier-type sensing with an insulated Frisch ring CdZnTe semiconductor radiation detector", *Appl. Phys. Lett.*, Vol. **84** (2004) pp. 1988–1991.
[30] O. Frisch, "Isotope analysis of uranium samples by means of their α-ray groups", *Br. Atom. Ener. Rep.*, Vol. **BR-49**, Atomic Energy Research Establishment: Harwell, UK (1944).
[31] G. Montemont, M. Arques, L. Verger, J. Rustique, "A capacitive frisch grid structure for CdZnTe detectors", *IEEE Trans. Nucl. Sci.*, Vol. **48**, no. 3 (2001) pp. 278–281.
[32] D.S. McGregor, D. Schinstock, M. Harrison, A. Kargar, W. McNeil, A.E. Bolotnikov, G.W. Wright, R.B. James, "*Semiconductor Radiation Detectors with Frisch Collars and Collimators for Gamma Ray Spectroscopy and Imaging*", DOE NEER Grant Number 031D14498, Progress Report for year 1 covering the period June, 2003 – May, 2004, Kansas State University, Manhattan, KS (2006) pp. 1–26.
[33] P.N. Luke, "Unipolar charge sensing with coplanar electrodes – Application to semiconductor detectors", *IEEE Trans. Nuc. Sci.*, Vol. **NS-42** (1995) pp. 207–213.
[34] P.N. Luke, M. Amman, C. Tindall, J.S. Lee, "Recent developments in semiconductor gamma-ray detectors", *J. Radioanal. Nucl. Chem.*, Vol. **264** (2005) pp. 145–153.
[35] A. Owens, T. Buslaps, V. Gostilo, H. Graafsma, R. Hijmering, A. Kozorezov, A. Loupilov, D. Lumb, E. Welter, "Hard X- and gamma-ray measurements with a large volume coplanar grid CdZnTe detector", *Nucl. Instr. and Meth.*, Vol. **A563** (2006) pp. 242–248.
[36] A. Owens, "Semiconductor materials and radiation detection", *J. Synchrotron Radiation*, Vol. **13**, no. part 2 (2006) pp. 143–150.
[37] A.G. Kozorezov, A. Owens, A. Peacock, J.K. Wigmore, "Carrier dynamics and resolution of co-planar grid radiation detectors", *Nucl. Instr. and Meth.*, Vol. **A563** (2006) pp. 37–40.
[38] V. Gostilo, Z. He, V. Ivanov, L. Li, A. Loupilov, I. Tsirkova, "Preliminary results of large volume multi-pair coplanar grid CdZnTe detector fabrication", *IEEE Trans. Nucl. Sci.*, Vol. **NS-35** (2005) pp. 1402–1407.
[39] V. Gostilo, I. Lisjutin, A. Loupilov, V. Ivanov, "Performance improvement of large volume CdZnTe detectors", *IEEE Trans Nucl. Sci. Conf. Rec.*, Vol. **7** (2004) pp. 4590–4595.
[40] P.N. Luke, "Electrode configuration and energy resolution in gamma-ray detectors", *Nucl. Instr. and Meth.*, Vol. **A380** (1996) pp. 232–237.
[41] B.W. Sturm, Z. He, E.A. Rhodes, T.H. Zurbuchen, P.L. Koehn, "Coplanar grid CdZnTe detectors for space science applications", *Proc. SPIE*, Vol. **5540** (2004) pp. 14–21.
[42] Z. He, G.F. Knoll, D.K. Wehe, R. Rojeski, C.H. Mastrangelo, M. Hammig, C. Barrett, A. Uritani, "1-D position sensitive single carrier semiconductor detectors", *Nucl. Instr. and Meth.*, Vol. **A380**, no. 1–2 (1996) pp. 228–231.
[43] Z. He, G.F. Knoll, D.K. Wehe, J. Miyamoto, "Position-sensitive single carrier CdZnTe detectors", *Nucl. Inst. Meth.*, Vol. **A388** (1997) pp. 180–185.
[44] E. Gatti, P. Rehak, A. Longoni, J. Kemmer, P. Holl, R. Klanner, G. Lutz, A. Wylie, F. Goulding, P.N. Luke, N.W. Madden, J. Walton, "Semiconductor drift chambers", *IEEE Trans. Nucl. Sci.*, Vol. **NS-32**, no. 2 (1985) pp. 1204–1208.

[45] J. Kemmer, G. Lutz, E. Belau, U. Prechtel, W. Welser, "Low capacity drift diode", *Nucl. Instr. and Meth.*, Vol. **A253**, no. 3 (1987) pp. 378–381.

[46] B.E. Patt, J.S. Iwanczyk, G. Vilkelis, Y.J. Wang, "New gamma-ray detector structures for electron only charge carrier collection utilizing high-Z compound semiconductors", *Nucl. Instr. and Meth.*, Vol. **380**, no. 1–2 (1996) pp. 276–281.

[47] M. van Pamelen, C. Budtz-Jørgensen, "CdZnTe drift detector with correction for hole trapping", *Nucl. Instr. and Meth.*, Vol. **A411** (1998) pp. 197–200.

[48] I. Kuvvetli, C. Budtz-Jørgensen, L. Gerward, C.M. Stahle, "Response of CZT drift-strip detector to X- and gamma rays", *Rad. Phys. Chem.*, Vol. **61** (2001) pp. 457–460.

[49] A. Owens, R. den Hartog, F. Quarati, V. Gostilo, V. Kondratjev, A. Loupilov, A.G. Kozorezov, J.K. Wigmore, A. Webb, E. Welter, "The hard X-ray response of a CdZnTe ring-drift detector", *J. Appl. Phys. Letts.*, Vol. **102** (2007) pp. 054505-1–054505-9.

[50] P. Lechner, S. Eckbauer, R. Hartmann, S. Krisch, D. Hauff, R. Richter, H. Soltau, L. Strüder, C. Fiorini, E. Gatti, A. Longoni, M. Sampietro, "Silicon drift detectors for high resolution room temperature X-ray spectroscopy", *Nucl. Instr. and Meth.*, Vol. **A377** (1996) pp. 346–351.

[51] J. Eskin, H. Barrett, H. Barber, "Signals induced in semiconductor gamma-ray imaging detectors", *J. Appl. Phys.*, Vol. **85** (1999) pp. 647–659.

[52] R. den Hartog, A. Owens, A.G. Kozorezov, M. Bavdaz, A. Peacock, V. Gostilo, I. Lisjutin, S. Zatoloka, "Optimization of array design for TlBr imaging detectors", *Proc. SPIE*, Vol. **4851** (2003) pp. 922–932.

[53] K. Parnham, C. Szeles, K. Lynn, R. Tjossem, "Performance improvement of CdZnTe detectors using modified two-terminal electrode geometry", *Proc. SPIE*, Vol. **3768** (1999) pp. 49–54.

[54] H. Kim, L. Cirignano, K. Shah, M. Squillante, P. Wong, "Investigation of the energy resolution and charge collection efficiency of Cd(Zn)Te detectors with three electrodes", *IEEE Trans. Nucl. Sci.*, Vol. **51**, no. 2 (2004) pp. 1229–1234.

[55] L. Verger, P. Ouvrier-Buffet, F. Mathy, G. Montemont, M. Picone, J. Rustique, C. Riffard, "Performance of a new CdZnTe portable spectrometric system for high energy applications", *IEEE Trans. Nucl. Sci.*, Vol. **52** (2005) pp. 1733–1738.

[56] F. Zhang, Z. He, G.F. Knoll, D.K. Wehe, J.E. Berry, "3-D position sensitive CdZnTe spectrometer performance using third generation VAS/TAT readout electronics", *IEEE Trans. Nucl. Sci.*, Vol. **NS-52** (2005) pp. 2009–2016.

[57] J.C. Lund, J.M. Van Scyoc, R.B. James, D.S. McGregor, R.W. Olsen, "Large volume room temperature gamma-ray spectrometers from CZT", *Nucl. Instr. and Meth.*, Vol. **A380** (1996) pp. 256–261.

[58] M. Mangun Panitra, A. Uritani, J. Kawarabayshi, T. Iguchi, H. Sakai, "Pulse shape analysis on mixed beta particle and gamma-ray source measured by CdZnTe semiconductor detector by means of digital-analog hybrid signal processing method", *J. Nucl. Sci. Tech.*, Vol. **38** (2001) pp. 306–311.

[59] H. Chen, S.A. Awadalla, K. Iniewski, P.H. Lu, F. Harris, J. Mackenzie, T. Hasanen, W. Chen, R. Redden, G. Bindley, I. Kuvvetli, C. Budtz-Jørgensen, P. Luke, M. Amman, J.S. Lee, A.E. Bolotnikov, G.S. Camarda, Y. Cui, A. Hossain, R.B. James, "Characterization of large cadmium zinc telluride crystals grown by traveling heater method", *J. Appl. Phys.*, Vol. **103** (2008) pp. 014903-1–014903-5.

[60] H. Chen, S.A. Awadalla, J. Mackenzie, R. Redden, G. Bindley, A.E. Bolotnikov, G.S. Camarda, G. Carini, R.B. James, "Characterization of traveling heater method (THM) grown $Cd_{0.9}Zn_{0.1}Te$ crystals", *IEEE Trans. Nucl. Sci.*, Vol. **54**, no. 4 (2007) pp. 811–816.

13

Future Directions in Radiation Detection

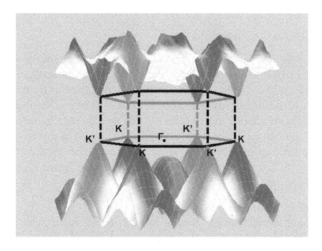

Principle behind valleytronics. Valence band maximum and conduction band minimum for a WS_2 monolayer in the hexagonal Brillouin zone. The corners of the hexagon correspond to K or K' points, and the center of the hexagon corresponds to Γ point. Darker (K) and lighter (K') valley shades indicate the degree of up and down spin polarizations, respectively (from Bussolotti *et al.*, *Nano Futures*, Vol. **2** (2018) 032001). By controlling the number of electrons that occupy a particular valley, it is possible to induce a valley "polarization", which can then be used to transmit/process information without the physical movement of electrons.

CONTENTS

13.1 The Immediate Future

For the immediate future, the requirements of gamma-ray spectroscopy should drive the development of semiconductors, because material properties are critical for thick detectors but not for thin detectors. For example, for X-ray detection, lifetime-mobility products need only be 10^{-4} cm^{2}V^{-1} or less to ensure the efficient collection of carriers and energy resolutions near the Fano limit, provided detector thicknesses are kept below 200 μm. In addition, it should be borne in mind that for most semiconductors, detectors up to 200 microns thick still have nearly unity quantum efficiency for energies up to 20 keV. As such, X-ray applications are beneficial in driving the development of low-noise front-end electronics precisely because the signal levels are so small when compared to those encountered in gamma-ray applications. For gamma-ray applications, the objective should be to produce a detector that will operate at or above room temperature, with a FWHM energy resolution of 1% or less at 500 keV and a usable active volume of 10 cubic cm or more.

13.1.1 General Requirements on Detector Material

In terms of general requirements, the average atomic number Z should be greater than 40 to yield a high stopping power. Structurally, the lattice should have a close packed geometry (such as a face-centered cubic structure) to optimize density. The material should have a low dielectric constant, to ensure low capacitance and therefore system noise. For practical systems, preference should be given to binary or pseudo-alloyed binary systems as opposed to ternary and higher-order compounds, to reduce the multiplication of stoichiometry errors. Additionally, such a restriction would clean up the response function by reducing the number of unwanted absorption and emission features in the measured energy-loss spectra. The selection criteria can be further extended to exclude most II-VI compounds in view of their propensity for toxicity, deep trapping and polarization effects, coupled with their low melting points. The latter makes it difficult to perform thermal annealing or to activate implanted dopants.

a) Bandgap and Pair Creation Energy
From an electronic point of view, the material should have an indirect bandgap to limit radiative recombination processes. The bandgap energy should be greater than 0.14 eV so that there is no thermal generation of carriers at room temperature but less than ~2.2 eV, based on the fact that carrier mobilities tend to drop rapidly with increasing bandgap due to polar lattice scattering[1] [1]. Impurity carrier scattering can also decrease mobilities and effective masses appreciably; therefore, impurity concentrations should be kept to a level of 10^{15}cm^{-3} or less (see [1,2]). Material resistivity should be greater than 10^8 Ω-cm to allow large biases to be applied, resulting in lower leakage currents, faster drift velocities and deeper depletion depths.

In Fig. 13.1 we plot the bandgap energy as a function of electron-hole pair creation energy for a number of compounds. As can be seen, semiconductors align along two curves: the classical curve first reported by Klein (solid line) and a second branch (dashed-dotted line), which is clearly displaced from the Klein curve. The first branch is well described by the empirical formula of Klein [4],

$$\varepsilon_p = 2.8\,\varepsilon_g + r\,(\hbar\omega_R),\tag{1}$$

where r is the average number of optical phonons (or Raman quanta $\eta\omega_R$) emitted during impact ionizations and should lie in the range $0.5 < r < 1.0$. Eq. (1) was fit to the measured data shown in Fig. 13.1 allowing r to be an adjustable parameter. A best fit was obtained for $r = 0.674$. The second branch, however, was found to be better fit by the empirical formula of Que and Rolands [5] derived for amorphous semiconductors, *i.e.*,

$$\varepsilon_p = 2.2\,\varepsilon_g + k\,(\hbar\omega_R),\tag{2}$$

in this case allowing k to be a best-fit adjustable parameter. This occurred, for $k = r/2$. The implication is that more energy is wasted in the pair creation process in the first branch than in the second branch and that optical phonon losses are half in the second branch compared to the first. In fact, choosing semiconductors from the second branch gives a ~30% reduction in the mean pair creation energy for a given bandgap compared to the main branch.

The difference between Eqs. (1) and (2) is that Eq. (1) assumes that momentum is conserved within a reciprocal lattice vector, which is valid for a crystalline solid. However, as pointed out by Que and Rolands [5], this is not the case for an amorphous solid with a non-periodic lattice. From Fig. 13.1 we see that the curve passes through both amorphous solids (denoted by the prefix a- in the figure). However, it should be pointed out that the bulk of semiconductors that lie on the second branch are crystalline, so Eq. (2) should not be valid. In fact, the semiconductors that populate the second branch are derived from all semiconductor groups so it is not clear why some groups populate the main branch and some the second branch, although we can make some general statements. All semiconductors with a bandgap < 1.8 eV belong to the primary branch while all compounds with a bandgap > 2.6 eV belong to the secondary branch. About 40% of semiconductors in each branch lie in the bandgap range 1.8 to 2.6 eV. We note that large fraction of materials on

---
[1] which in turn is due to increasing iconicity, which tends to increase with bandgap (*i.e.*, materials become more polar).

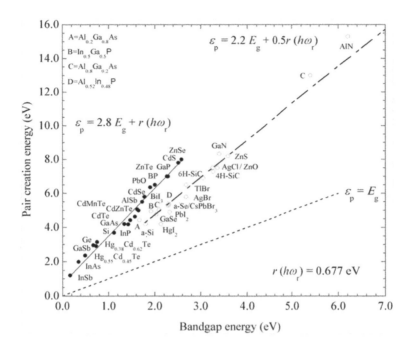

FIGURE 13.1 The average energy to create an electron-hole pair as a function of bandgap energy for a selection of semiconductors (updated from [3]). Note: amorphous semiconductors are denoted by the prefix a- in the figure. Two main bands are evident: the main branch found by Klein [4] (solid line) and a second branch shown by a dashed-dotted line). The dotted line denotes the limiting case when $\varepsilon_p = E_g$. The difference between this curve and the measured curves is due to optical phonon losses and the residual kinetic energy left over from impact ionization thresholding effects. The main branch was found to be best fit by a Klein function of the form $2.8\,E_g + a_1$ in which a_1 accounts for optical phonon losses and is treated as a free parameter. The second branch was found to be well fit by the function of Que and Rolands [5] derived for amorphous semiconductors, *i.e.*, $2.2\,E_g + a_2$, where a_2 is now treated as a free parameter).

the second branch can be characterized by low mobilities and high defect densities – a consequence of poor material quality (in fact, rms residuals of the fit to second branch are twice that of the first branch fit). Perhaps it is not surprising that this branch also contains a number of heavy, mechanically soft, layered compounds, which are not only difficult to handle, but do not lend themselves to standard processing techniques.

b) Transport Properties

In terms of carrier transport properties, the majority of the carrier effective mass should be low ($<0.1\,m_o$) to ensure high-speed operation. Note that effective mass and bandgap are related and that smaller bandgap materials tend to have lower effective masses. So for high-speed and room-temperature operation, bandgap energies should be close to 1.3 eV. The electron and hole mobility-lifetime products should be greater than 10^{-3} and 10^{-4} cm^{2}V^{-1}, respectively, to ensure good carrier transport and therefore spectral performance. This in turn places a limit on the density of typical trapping centers of $< 5 \times 10^{12}$ cm^{-3}. In a study of high performance detector operation, Luke and Amman [6] have suggested that carrier mobilities should have values of at least 500 cm^{2}V^{-1}s^{-1} in order to ensure that the amount of charge induced on the electrodes and collected is reasonably flat with energy. Any non-uniformities will result in signal variations, which in turn will degrade the detector's spectral resolution. The requirement on mu-tau products places a lower limit on carrier lifetimes of ~20 μs. Unfortunately, carrier lifetimes are not an intrinsic property of the material but are determined by the concentration of carrier-trapping defects and their capture cross-sections. Therefore, while we may select compounds with the appropriate mobilities, the resultant mu-tau products are largely unpredictable. An alternative approach is to pursue a single carrier sensing and "kill-off" the carrier with poor transport properties, since it will dominate the spectral response. However, the mu-tau product for this carrier cannot be made too small, as it will lead to polarization effects at high count rates and/or large energy depositions.

In Table 13.1, we summarize desirable material properties in the form of a "wish list". Because of the uncertainty in end applications and compatibility between requirements, some of the recommendations are by necessity subjective. Applying these criteria reduces the number of potential compounds to about three (HgI$_2$, CdTe and CdZnTe).

13.2 Near Term Developments

In the preceding chapters we have concentrated purely on material developments since material issues currently limit the effective exploitation of most semiconductor materials. However, until these are resolved, the short-term future lies in the

TABLE 13.1

Summary of desirable material parameters for the development of the next generation of compound semiconductor detectors, in the form of a list of the minimum material and/or preferred values. Because of the uncertainty in end applications and compatibility between requirements, some of the recommendations are by necessity subjective.

Parameter	Minimum or recommended value	Reason/comments
Composition	Binary (or pseudo binary)	To minimize stoichiometry errors and absorption edges
Structure	Cubic (close packed)	To optimize density and mechanical strength
Density	$> 5\,\text{g}\,\text{cm}^{-3}$	To ensure good stopping power
Growth technique	Epitaxial, FZ	High quality material; allows the possibility of integrated electronic structures
Contact barrier height	$< 0.4\,\text{eV}$	To ensure the contact's "look" ohmic (*i.e.*, with a resistivity $< 10^{-3}\Omega - \text{m}^2$)
Effective atomic number, Z_{eff}	> 40	For high stopping power
Resistivity	$> 10^8\,\Omega\,\text{cm}$	To allow high biases to be applied
Hardness	$> 500\,\text{kgf}\,\text{mm}^{-2}$ (Knoop scale)	Chosen to be high enough to allow the use of a range of mechanical processing and bonding technologies
Bandgap	Indirect	To limit radiative recombination
Bandgap energy	$1.4 < \varepsilon_g < 2.2\,\text{eV}$	Lower limit for room-temperature operation; upper limit imposed by mobility losses due to polar lattice scattering
Static dielectric constant, $\varepsilon\,(0)$	< 5	To ensure low capacitance
Ionicity	< 0.3 (Phillips scale)	Ionicity should be low to prevent problems with ionic conductivity and polarization
Majority carrier trapping center density	$< 5 \times 10^{12}\,\text{cm}^{-3}$	To ensure good charge collection
Impurity concentration	$10^{15}\,\text{cm}^{-3}$	High concentrations adversely affect resistivity, mobilities and effective mass.
Mobility electrons	$> 500\,\text{cm}^2\text{V}^{-1}\text{s}^{-1}$	CCE, Charge induction considerations
Mobility holes	10 or $500\,\text{cm}^2\text{V}^{-1}\text{s}^{-1}$	Lower value if single carrier sensing is used, upper value if not
Majority carrier lifetime	$> 20\,\mu\text{s}$	To ensure reasonably high mu-tau products
Electron mobility-lifetime ($\mu\tau_e$)	$10^{-3}\,\text{cm}^2\text{V}^{-1}$	Minimum value really depends on application, X-rays or γ-rays.
Hole mobility-lifetime ($\mu\tau_h$)	$< 10^{-4}$ or $10^{-3}\,\text{cm}^2\text{V}^{-1}$	($\mu\tau_h$) should be $< 0.1(\mu\tau_e)$ if single carrier sensing is employed.
Majority carrier effective mass	$0.1\,m_o$	To ensure high speed operation
Cost	$1\$\,\text{cm}^{-3}$	To be competitive with other technologies

controlled and directed manipulation of charge to reduce trapping and facilitate low noise readout. Material improvements must be paralleled by corresponding developments in heterostructure and quantum heterostructure technology, provided of course that strain is not a limiting factor. Doping a semiconductor like Si or GaAs provides control over the sign and density of the charge carriers, but by combining different semiconductors in heterostructures, one gains control over many more parameters, including the bandgap, refractive index, carrier mass and mobility, and other fundamental quantities. For example, work on heterostructures may resolve the problem of contacting by the build-up of a series of semiconductor layers until it is possible to satisfy the relationship that the work function of the metal contact is less than that of the semiconductor (*i.e.*, $\phi_m < \phi_s$) for n-type material. Similar arguments can be made for p-type material. An alternative approach is to use a so-called "interface control layer" at the junction between the contact and the semiconductor. It has been shown that such a layer can lower the overall Schottky barrier height (SBH) allowing better contacting to wide bandgap semiconductors. The reduction in barrier height is accomplished largely through the breakdown of one large (Schottky) barrier height into two smaller ones (a SBH and a heterojunction band-offset). HgCdTe contact layers have been used on CdTe detectors to form a heterojunction HgCdTe/CdTe/HgCdTe p-i-n detector [7].

At the present time, the first purpose-built heterostructures for X-ray applications are being reported in the literature. For example, Silenas *et al.* [8] constructed an n-GaAs – p-AlGaAs graded-gap X-ray detector in which the AlGaAs layer functions as the classical absorption and detection layer and the n-GaAs layer as a carrier multiplication zone. Early results show the device is sensitive to α-particles, and gains of up to 100 can be achieved. Lees *et al.* [9] tried a simpler approach and reported the construction of an $Al_{0.8}Ga_{0.2}As$ p^+-p^--n^+ diode of diameter 200 μm and thickness ~3 μm. The detector was found to be spectroscopic to low energy X-rays with a measured FWHM energy resolution of 1.5 keV at 5.9 keV at room temperature. At elevated reverse bias levels (>20 V), avalanche multiplication was observed. In fact, gains of 3 to 4 could be achieved which, while it improved the signal-to-noise ratio, it simultaneously degraded the spectroscopic performance.

In the longer term, developments in nanosemiconductors that further exploit reduced dimensionality and quantum confinement offer a much more controlled method of manipulating charge and low noise operation. Currently, pioneering work is being carried out at the nanoscale, and the first rudimentary efforts have begun to scale up to macroscopic sizes for bulk radiation detection.

13.2.1 Reduced Dimensionality

The properties of the nanoscale and specifically low-dimensional nanoscale structures, such as quantum dots, nanotubes, nanowires, nanobelts and others [10–12], have attracted considerable attention in recent years because at the nanoscale a material's physical and chemical properties behave markedly differently from their bulk values purely because of quantum effects. The intentional manipulation of these effects has led to the potential for exploitation across a large number of areas, particularly in nanoelectronics and optoelectronic nanodevices [13]. The overall impact of nano-science cannot be over-stated – it is truly a platform technology.[2]

The physical and electronic properties of most semiconducting materials can be significantly modified by reducing the physical extent of at least one of the dimensions to the order of the exciton Bohr radius,[3] given by

$$r_{\mathrm{Bohr}} = \frac{\hbar \varepsilon}{e^2} \left[\frac{1}{m_{\mathrm{e}}^*} + \frac{1}{m_{\mathrm{h}}^*} \right], \tag{3}$$

where, ε is the dielectric constant of the material and m_{e}^* and m_{h}^* are the electron and hole effective masses. For typical semiconductors, r_{Bohr} ranges from 1 nm to ~50 nm and is simply related to the bandgap, ε_g, by

$$r_{\mathrm{Bohr}} = A \varepsilon_g^{-\alpha}, \tag{4}$$

where A and α are empirical constants. For III-V and II-VI materials, $A = 13.5 \, \mathrm{nm \, eV}$ and $\alpha = 1.18$. The dimensionality of a material refers to how many dimensions the carriers are free to act in and is illustrated in Fig. 13.2.

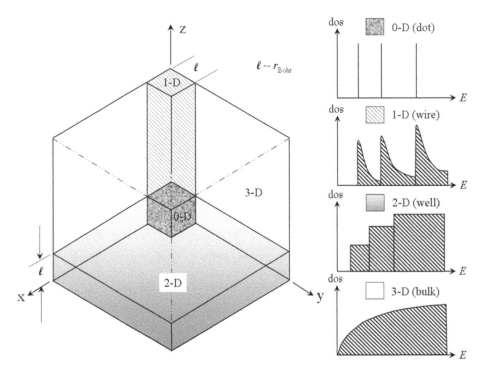

FIGURE 13.2 Schematic illustrating the concept of dimensionality (0-D, 1-D, 2-D and 3-D) in materials at the nanoscale. Here r_{Bohr} is the exciton Bohr radius, which typically ranges from one to tens of nm, depending on the material. Also shown are the density of states (dos) for each system, which become more and more discrete as the dimensionality is reduced. At 0-D this leads to discrete quantized energy levels more like those of an atom than the continuous bands of a bulk semiconductor.

---

[2] A platform technology is a structure, or base, upon which other applications, processes or technologies can be adapted or readily developed with minimum expense.

[3] Essentially the distance between the electron and hole in a bound electron-hole pair.

For example, in bulk materials, the conducting electrons are delocalized and can move freely in all three directions. Accordingly, they are considered 3-D systems precisely because they are not confined in any dimension. A semiconductor in which a particle is free to move in two dimensions but is confined in the third, is described as a two-dimensional (2-D) system and is termed a quantum well (QW). Reducing the dimensionality further results in 1-D quantum wires, which amongst other effects, exhibit quantized conductance. Finally, further reduction of the dimensions leads to a 0-D system known as a quantum dot (QD) in which the motion of electrons and holes (*i.e.*, excitons) is confined quantum mechanically in all three spatial directions. This leads to discrete, quantized energy levels (more like those of an atom than the continuous bands of a bulk semiconductor) and a class of materials whose properties are intermediate between the molecular and bulk forms of matter. The allowed energy levels can be determined from Schrodinger's equation. Assuming an infinitely deep well, we obtain

$$E_n = \left[\frac{\pi^2 \hbar^2}{2m_e^* \ell^2} \right] \left(n_x^2 + n_y^2 + n_z^2 \right) \text{(1-D – Quantum dot)}, \tag{5}$$

where $\hbar = h/2\pi$, h is Planck's constant, m_e^* is the effective mass of the electron, ℓ is the width (confinement) of the well, and n_x, n_y, and n_z are the principal quantum numbers in the three dimensions x, y, and z. The smaller the dimensions of the nanostructure (smaller ℓ), the wider the separation between the energy levels, leading to a spectrum of discreet energies as shown in Fig. 13.2. As a consequence, small changes to the size or composition of a QD allow the energy levels, and therefore the bandgap, to be fine-tuned to specific energies. The sensitivity of bandgap on size can be quite large. For example, by reducing the confinement dimension in a silicon QD to 2 nm, the bandgap increases by a factor of 4 compared to its bulk value. In terms of wavelength, this corresponds to a shift from the infrared, to the UV. Similar equations exist for 1-D and 2-D nanomaterials,

$$E_n = \left[\frac{\pi^2 \hbar^2}{2m_e^* \ell^2} \right] \left(n_x^2 + n_y^2 \right) \text{(1-D – Quantum wire)}, \tag{6}$$

and

$$E_n = \left[\frac{\pi^2 \hbar^2}{2m_e^* \ell^2} \right] \left(n_x^2 \right) \text{(2-D – Quantum well)}. \tag{7}$$

In bulk materials the conducting electrons are delocalized and can now move freely in all three dimensions – they have zero confinement and hence are considered 3-D systems. The energy of the electron is now given by the familiar dispersion relation (see Ch. 2 section 2.3.5.1)

$$E_n = \left[\frac{k^2 \hbar^2}{2m_e^*} \right] \text{(3-D – bulk)}, \tag{8}$$

where k is the wave vector related to the electron's momentum. Note there is no dependence on dimension; the electron's energy is now continuous depending only on wave vector.

13.3 Radiation Detection Using Nano-Technology

The term "nanotechnology" was originally coined by the Japanese scientist, Norio Taniguchi, in 1974 to describe future semiconductor control processes at the atomic level [14], although it was not until the 1980s that the term passed into common usage. This was largely due to (a) the invention of the scanning tunneling microscope, which allowed the direct observation of quantum confinement effects and in particular the strange way nanomaterials behave compared to their bulk counterparts, and (b) the discovery of fullerenes – nanostructured forms of carbon with exceptional chemical, electronic and material properties. However, until about 20 years ago the term was very loosely applied to anything at that was considered to be very small.[4] Since then, the definition of nanotechnology has been refined to mean

---

[4] Indeed, the word "nano" originates from the Greek word "*nanos*", which means dwarf or extremely small.

"research and development at the atomic, molecular, or macromolecular levels using a length scale of approximately one to 100 nanometers in any dimension; the creation and use of structures, devices and systems that have novel properties and functions because of their small size and the ability to control or manipulate matter on an atomic scale".

Although nanotechnology is generally considered a recent development, the use of nanomaterials is not new. In fact, the ancient Egyptians developed chemically synthesised nanocrystal-based cosmetics and hair dyes over 4,000 years ago. At the present time, nanomaterials are widely used in a number of applications, but mainly limited to bulk use as passive materials, for example, nano-lubricants, nano-clays and textile coatings. The application of nanomaterials to radiation detection is still in its infancy and is far less developed than are other areas. However, there have been a number of reports of radiation detection using nanoparticles [16], nano-scintillators [17] and nanostructures [18]. These are described in the following sections.

13.3.1 0-D Materials

Because zero dimensional materials (dots or mostly spherical nanoparticles) are relatively simple to produce by lithography, colloidal[5] synthesis or epitaxy, they have been extensively investigated across a number of application areas such as photovoltaics, solar cells, LEDs, biomedical applications and quantum computing. However, their use in radiation detection is less obvious and is mainly a passive rather than an active material. For example, a novel approach to neutron detection using quasi-0-D particles has been reported by Amaro *et al.* [16] who motivated by the increasing difficulty in obtaining ^{3}He, propose to emulate an ^{3}He proportional gas counter by replacing ^{3}He gas with B_4C nanoparticles dispersed in a standard Ar-CH$_4$ (90–10%) gas mix, forming an aerosol with neutron sensitive properties. Thermal neutrons entering the counter interact with the ^{10}B atoms *via* the ^{10}B(n,α)^{7}Li reaction. If the nanoparticles have dimensions much smaller than the ranges of the α-particle and ^{7}Li reaction products (3.6 µm and 1.8 µm, respectively) then both will escape, depositing a large fraction of their energy in the counter gas. In addition, since boron carbide is essentially inert, it will not interact with gas, wire or the walls of the counters. Consequently, its presence in the gas should not affect the electrical properties of the counter, and the detector response will be similar to that obtained with an ^{3}He proportional counter with the exception that the full energy peak will now be located near 2.3 MeV instead of 0.7 MeV (because of the higher Q value of ^{10}B compared to ^{3}He). In Fig. 13.3, we show the measured response of a prototype detector to a thermal neutron source of flux 7×10^3 n/sec, in which we see two plateaus corresponding to energy losses from the ^{7}Li ion and α reaction products and a single sum peak corresponding to full energy collection in the proportional gas. During detector operation, the gas circulated from the bottom to the top of the proportional counter, supporting the dispersion of the nanoparticles. Although the detector was not performance optimized, an overall thermal neutron detection efficiency of ~4% was achieved. The authors point out that the technique is not restricted to gasses but could be applied to fluids, such as liquid xenon currently used in a number of radiation detection applications.

While colloidal nanocrystals (NCs) have attracted considerable interest over the last few years for a wide range of biomedical, biochemical sensing and optoelectronic applications, their use in radiation detection is limited. Kim *et al.* [19] investigated the use of semiconductor nanoassemblies as a detection medium for ionizing radiation. Nanocrystalline CdTe films were deposited on glass and metallic substrates by the layer-by-layer (LBL) method or by drop casting. The LBL method is well known for efficiently depositing NC colloidal dispersions into high quality and stable thin film layers on a substrate, whilst preserving their distinctive optoelectrical and magnetic properties. The nanocrystals were adsorbed onto a PDDA[6] layer. By an alternately adsorbing procedure, a bilayer consisting of a polymer/nanocrystal composite was developed and the cycle repeated to obtain a multilayer film of the desired thickness. In contrast, the drop casting method is implemented by dropping and drying 0.1 mL of the CdTe NC solution 30 times, producing a film approximately 0.5 mm thick. Assemblies of 6 mm in diameter were synthesized on the metallized substrate. The junction properties of the films were then studied by sandwiching the films between evaporated metallic electrodes. The assemblies were found to be responsive to 5.5 MeV alpha particles from a ^{241}Am radioactive source.

Work was also carried out on PbSe NC films. Initial experiments failed to give a response to alpha particles, primarily due to the absence of a depletion layer. However, later experiments by the same group were more successful. Hammig [20] reports the synthesis of PbSe NC/conductive polymer composite assemblies, deposited on plastic substrates. After contacting, the resulting detectors were 1×1 cm^2 in area and several tens of µm thick. When a detector was exposed to a mixed ^{133}Ba and ^{241}Am radioactive source, it gave a spectroscopic response to both alpha particles and gamma-rays, as shown in Fig. 13.4. The alpha response is shown in Fig. 13.4(a) and an expansion of the ^{133}Ba gamma-ray spectrum, in which both the Pb and Se X-ray escape peaks are also apparent, is shown in Fig. 13.4(b). The measured FWHM energy resolution at 356 keV was found to be 0.42%, comparable with that obtained by a HPGe detector (0.39%). For comparison, the energy resolution of a single crystal CZT detector was 0.96%. Further measurements were carried out with a ^{137}Cs source [20]. At 662 keV, the

---

[5] A colloid is a suspension of microscopically dispersed insoluble particles in another medium.
[6] poly(diallyl-dimethylammonium chloride) [15].

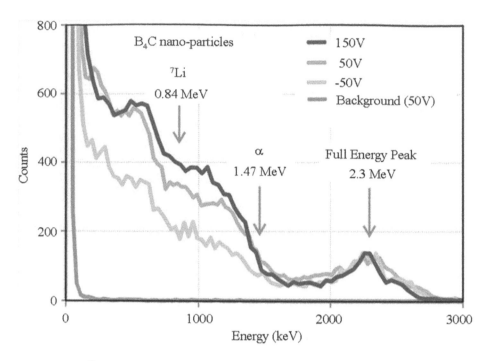

FIGURE 13.3 The response of a using ^{10}B nanoparticle gas detector developed by Amaro *et al.* [16] to a thermal neutron source of flux 7×10^3 n/sec. The acquisition time was 400 sec. From the figure we see a two-step plateau, limited by the ^{7}Li and α-particle edges, and a single sum peak, representing full energy collection in the proportional gas. Several curves are shown for biases applied to an additional set of "gate" electrodes in the vicinity of the counter wall and surrounding the anode. These were used to suppress partial energy loss events from nanoparticles that stick to the walls of the counter, in which case one of the reaction products will be absorbed by the wall and therefore lost from the energy deposition process.

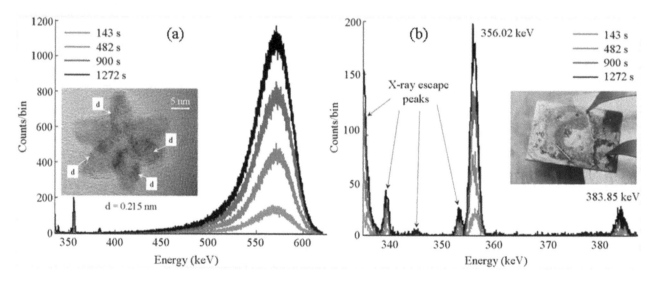

FIGURE 13.4 Energy-loss spectra measured from a ^{133}Ba and ^{241}Am radioactive source (from [21]). The composite alpha particle and gamma-ray response is shown in (a) attenuated through 3.7 cm of air, impinging upon a 1×1 cm thin (10's of μm) composite assembly of a conductive polymer and star-shaped PbSe nanoparticles (upper inset), accumulated for various durations shown in the legend. The lower inset shows a TEM micrograph of a typical PbSe NC in the assembly. (b) An expansion of the ^{133}Ba gamma-ray spectrum, in which the Pb and Se X-ray escape peaks are also apparent.

energy resolution of the PbSe nanocomposite was again comparable (0.32%) with that obtained with a HPGe detector (0.27%).

13.3.1.1 Nano-Scintillators

Nanoparticles in the form of nanocomposites of known scintillators are expected to have improved properties compared to those of the bulk scintillators from which they are derived [17]. Scintillators can be considered to fall into two classes:

organic (plastic and liquid) and inorganic (crystal). Inorganic single crystals are efficient and have a high light output and better energy resolution than organic scintillators. However, they are difficult to grow in large sizes and are expensive. Organic scintillators, on the other hand, have the advantages that they are easy to produce in large sizes; they have a fast decay time but poor energy resolution and stopping power. Nanocomposite scintillators have been proposed in which high-Z inorganic QDs are embedded in a plastic scintillator or other transparent medium or binder in an attempt to combine the best properties of both inorganic and organic scintillators. One of the great advantages of this technique is scalability to produce large detectors. Another advantage is that the size of the nano-particles can be chosen so as to tune the emitted scintillation light to a particular readout device (*e.g.*, APDs, PMTs, *etc.*). A further benefit is that the luminescent yield generally increases as the particle size decreases [22] while the scintillation rise time decreases [23]. At present the main drawback is an increase in quenching[7] of the scintillation light due to a relatively large number of surface states – a direct consequence of the large surface area-to-volume ratios of nano-particles. One mitigating strategy is the core-shell system, in which the nano-particles (the cores) are coated with a passivating layer (the shells) [24]. In actually, the shell has a lot of other practical uses, for example, improving dispensability, protecting the core from hostile host environments, fine tuning optical emission from the core and acting as a container for the controlled release of the core in drug delivery systems (for a review see [25]). The potential performance enhancements gained by nano-scintillators is illustrated by the following examples. Letant and Wang [26] constructed a 1/16 inch thick by 1.5 cm diameter nanocomposite scintillator consisting of CdSe/ZnS core shell QDs in a porous glass matrix. The size of the QDs were tailored such that the bandgap and subsequent peak optical emission were shifted from the red to the green end of the spectrum matching the spectral response of the readout PMT and hence maximizing signal output. The array was found to be responsive to 60 keV gamma-rays from a ^{241}Am source with a measured FWHM energy resolution of 15%. For comparison, the measured energy resolution of a $1'' \times 1''$NaI(Tl) single crystal measured under the same conditions was 30% FWHM. Plumley *et al.* [27] reported the detection of thermal neutrons using gadolinium-doped lanthanide halide nanocrystals (namely, $LaF_3:Gd^{3+}$, GdF_3 and $GdF_3:Ce^{3+}$) produced by colloidal synthesis. However, while all samples were sensitive to neutrons from a ^{252}Cf neutron source, the gamma spectra associated with neutron capture by Gd were not resolved due to a combination of significant self-absorption in the NC material and low NC concentration in the solution.

13.3.2 1-D Materials

To fully exploit reduced dimensionality, current research is focusing on device fabrication based on nano-building blocks [28] and in particular the integration of nanowire elements into more complex functional architectures. For radiation detection, a number of authors have proposed using CdTe and CdZnTe nanowire arrays or nanocrystalline films as detection media [21,29]. The primary advantage of this approach is that it is potentially easier and more cost effective to fabricate a bulk detector out of nanomaterials than it is to synthesize single large volume crystals by conventional Bridgman growth that are crystallographically and stoichiometrically perfect. As-grown Bridgman material displays numerous dislocations, cracks and voids that, along with residual impurities, lead to decreased resistivity and poor charge collection. As a result, the best spectroscopic material has to be "mined" out of the boule. In contrast, high purity CdTe or CdZnTe nanowires can be grown[8] relatively easily by electrodeposition with significant cost advantages over melt growth. For CdTe, this is estimated to be ~\$10 cm^3 as opposed to ~\$500 per cm^3 for Bridgman grown material. In addition, nanowires are one-dimensional conductors and as such, charge collection efficiencies are high, since the generated carriers propagate entirely through drift. Diffusion processes, which can lead to enhanced trapping, are almost entirely absent. Also, phonon-assisted loss processes can be supressed to a larger degree than in single crystal material, so that more of the incident information is converted into information carriers (specifically, light in the case of nano-scintillators or charge in the case of nano-semiconductors). For X-ray imaging, spatial resolution can be improved by preparing nano-scintillators as vertically aligned columns, without and with light guides [30] to reduce cross-scattering of light. Izaki *et al.* [31] recorded micron spatial resolutions with electrochemically grown vertical ZnO nanowires. For X-ray or gamma-ray detection, the biggest remaining technological challenge is to grow nanosemiconductor devices of sufficient size. At present, thicknesses are limited, due to the deleterious effects of colloidal defects that accumulate as an increasing number of layers are cast onto the sample. However, stopping power can be increased by assembling nanostructures into bigger structures. For example, nanowire arrays may be fabricated and then stacked to produce a "bulk" detector.

Ghandi *et al.* [32] grew arrays of CdZnTe nanowires on a TiO_2 nano-porous template by electrodeposition from aqueous solution. The deposition was a single stage process, followed by annealing in an Ar atmosphere. The typical resistivities of the nanowires were ~$10^{10}\Omega$ – cm and the reverse leakage currents were of the order of pA. A five-array stack of nanowires was then assembled – each stack being individually biased (see Fig. 13.5). In this regard it is interesting to note that because

---

[7] Quenching refers to all radiationless de-excitation processes (the energy is usually dissipated as heat).

[8] Although nanoparticles can be grown by MBE or MOCVD, the most common approach is wet chemistry in which atmospheric or low-vac processes are used to create a colloidal dispersion that can then be deposited onto a solid. These will form nanocrystalline structures through self-assembly during the solvent drying phase.

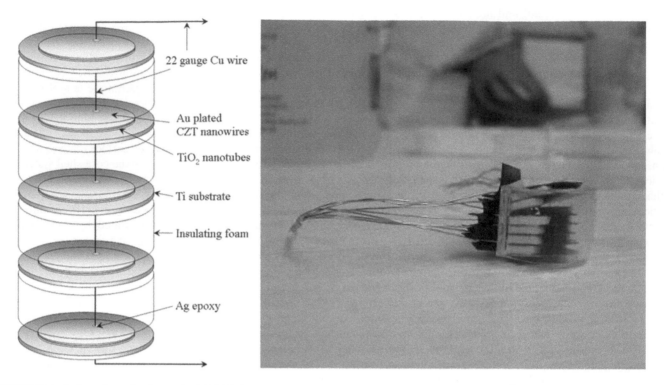

FIGURE 13.5 Left – schematic of a stack of 5 CdZnTe nanowire arrays. Each array is circular, of area ~0.7 cm². Within this area, the individual nanowires are ~80 nm in diameter and 2 microns in length. Right – the completed stack (from [32]).

the nanowires have dimensions of order 80 nm in diameter and 2 μm long, the applied bias across a stack to maintain a given electric field is very low, being typically in the range 0.5 to 2.5V. This is much less than the biases typically applied across bulk detectors produced by conventional means (*i.e.*, ~300 – 500V). Although the stack was not spectroscopic to X-rays, a three-fold increase in conductivity was observed when exposed to a ²⁴¹Am radioactive source.

Most recently, Becchetti *et al.* [33] report the construction of a detector based on CdTe nanowires macroscopically assembled from nanoparticles. The nanoparticles were synthesized in an aqueous solution following the procedures described in Rogach *et al.* [34]. The nanoparticles experienced strong dipole-dipole interactions between themselves and consequently spontaneously aligned themselves into crystalline nanowires. The nanowires were then assembled into films by vacuum filtration. The individual nanowires were determined to be ~300 nm long, and the overall nanowire film thickness was 7 μm. The detector was found to be responsive to 5.5 MeV alpha particles from a ²⁴¹Am radioactive source (showing a transmission peak) as well as a weak response to gamma-rays from a ¹³³Ba source.

It is interesting to note that 1-D freestanding nanowire structures can endure significantly more strain both radially and axially than films. As a consequence, heterostructures based on semiconductor nanowires can be formed without dislocations even in the case of materials that have a large lattice mismatch. This opens up substantial opportunities for various mismatched material combinations, such as InAs/InP and GaAs/GaP [35].

13.3.3 2-D Materials

While much research has been carried out on 0-D and quasi-0-D (*e.g.*, quantum dots, nanoparticles, cage molecules, *etc.* [36–38]), quasi-1-D (*e.g.*, nanowires, nanotubes, nanobelts [11,12]) and, of course, 3-D crystalline objects, relatively little has been done on 2-D systems. By 2-D, we mean planar crystalline materials (thin films) that are essentially one to a few atoms thick. Since all atoms are so close to a surface, vertical quantum confinement plays a significant part in heat and charge transport. What makes these materials particularly attractive for fabricating the next-generation nanoelectronic devices is that it is relatively easy to fabricate 2-D complex structures, by for example MBE, compared to 1-D or 3-D systems [39]. As such, the goal of device research has been to combine 2-D materials, such as a semiconductor, insulator and conductor to create a range of electronic devices, such as field-effect transistors, integrated logic circuits, photodetectors and flexible optoelectronics. Preliminary work has already been carried out on nitrides (*e.g.*, hexagonal boron nitride), dichalcogenides (*e.g.*, molybdenum sulfide) and oxides (*e.g.*, vanadium pentoxide). Indeed, the first operational results have been reported from FETs fabricated on WSe₂ monolayers [40].

Although it was shown decades ago by Frindt *et al.* [41,42] that layered van der Waals materials, such as layered metal dichalcogenides, can be mechanically and chemically exfoliated into a few atomic layers, it was not until the discovery of

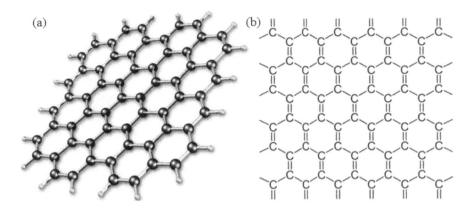

FIGURE 13.6 (a) Idealized structure of a single graphene sheet[9]. (b) Bonding structure of graphene.

graphene, that the first stable single atomic layers could be isolated. As such graphene is the most widely studied two-dimensional material [43]. Recently, its use in radiation detection has been proposed [44,45], albeit not through direct absorption and detection, but rather by using graphene's unique transport properties to "sense" energy absorption in an adjacent medium. The term graphene first appeared in 1987 to describe single sheets of graphite as one of the constituents of graphite intercalation compounds and not as the isolated, two-dimensional carbon structure we understand as graphene today. Graphene is an ambipolar semi-metal. It is an allotrope of carbon whose structure is a single planar sheet of sp^2-bonded carbon atoms that are densely packed in a honeycomb crystal lattice as shown in Fig. 13.6. In essence, it is an isolated atomic plane of graphite and is a truly two-dimensional structure, stable under ambient conditions. Its most important property arises from this structure, which leads to its very unusual electronic properties.

For a 3-dimensional periodic crystal structure, the energy bands appear parabolic in E, k space near the conduction band minimum and valance band maximum and can be approximated by the well-known dispersion relationship

$$E(k) = E_o + \frac{\hbar^2 (k)^2}{2m_e^*},\tag{9}$$

where $E_o = E_c$ is the ground state energy corresponding to a free electron at rest and m_e^* is the effective mass defined by the conduction band curvature, *i.e.*,

$$\frac{1}{m_e^*} = \frac{\partial^2 E(k)}{\hbar^2 \partial k^2}.\tag{10}$$

From Eq. 9 we see that the allowed energies of the carriers are dependent on their effective masses. However, in the two-dimensional structure of graphene, the energy bands near the conduction band minima and valence band maxima actually touch (but do not overlap) at six points in k space as shown in Fig. 13.7. These are known as the K or Dirac points, around which the conduction band minima and valence band maxima are conical.

At the Dirac points, the effective mass tends to zero and the dispersion relationship is linear with the separation

$$E(k) = \pm \hbar v_F \sqrt{k_x^2 + k_y^2},\tag{11}$$

where v_F is the Fermi velocity, corresponding to a kinetic energy equal to the Fermi energy. In graphene v_F is ~10^6 m/s. This means that while in standard semiconductors the charge carriers can be described as electron waves obeying the Schrödinger effective-mass equation, graphene electrons move according to the laws of relativistic quantum physics described by the mass-free Dirac equation; in essence, the electrons and holes behave as massless Dirac fermions.[10] Consequently, in graphene the electron mobilities are remarkably high, with reported values in excess of 15,000 cm^2 V^{-1} s^{-1} at room temperature. Conductance measurements indicate that holes have a similar value. Furthermore, both mobilities are nearly independent of temperature between 10K and 100K, which implies that the dominant scattering mechanism is defect scattering.

---

[9] Download for free at https://cnx.org/contents/eQus82US@4/Graphene.
[10] Sometimes referred to a graphinos.

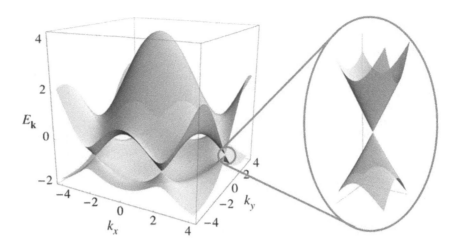

FIGURE 13.7 The band structure of graphene (from [43]). The conduction band and the valence band touch at six discrete points – the so-called K points, which in turn can be divided into two in-equivalent sets of three points each. The points within each set are all equivalent because they can reach each other by reciprocal lattice vectors. The two in-equivalent points are called K and K`, which form the valley pseudo-spin degree of freedom in graphene. The term valley stems from the similarity of the shape of k space in the vicinity of these points with a valley.

A consequence of carriers behaving as massless Dirac fermions is that they respond almost instantaneously to any perturbation in electrostatic potential. The net effect is a marked increase in conductivity. In fact, for small variations in the local electric field, the conductivity shows a tightly peaked response around the charge neutrality point (*i.e.*, no net electric field). This point is also known as the Dirac point; this property can be exploited for radiation detection. In the most commonly used implementation for radiation detection, graphene is used as the readout structure of a FET – a so-called GFET [44,45]. The graphene is deposited as a one atom thick monolayer above a bulk semiconductor substrate separated by an insulator buffer layer that serves as a gate dielectric, as illustrated in Fig. 13.8(a). The gate voltage between the graphene and the back of the absorber results in an electric field across the device. By varying V_g, the field can be adjusted to set the optimum point on the Dirac curve, as illustrated in Fig. 13.8(a). In reality, the device can be biased on either side of the curve in order to collect either holes or electrons depending on whether the absorber is n-type or p-type. This is illustrated in Fig. 13.9 in which we show GFET operation with a thin p-type Si absorber. Here we plot the variation of the drain-source current with gate source voltage.

13.3.3.1 Device Implementation

Under quiescent conditions (*i.e.*, no incident radiation), the absorber acts as an additional insulator; the gate voltage, V_g, is dropped across both the absorber and insulator resulting in a weak electric field $\left(E = V_g/(t_a + t_i)\right)$ where t_a and t_i are the thicknesses of the absorber (semiconductor) and insulator, respectively. Ionizing radiation passing through the bulk

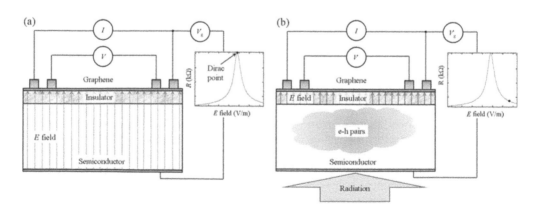

FIGURE 13.8 A graphene field effect transistor (GFET) radiation detector – principle of operation. (a) A GFET consists of an absorber (undoped semiconductor), with an insulator buffer layer, which serves as a gate dielectric. V_g is the gate voltage applied to the sample. Current is supplied across the graphene sample and the resistance of the graphene layer determined. In (b), incoming radiation produces ionization within the semiconductor to create a conducting absorber. The gate voltage now only drops across the insulator. This results in an increased electric field, which is sensed by the graphene and its resistance changes as a result of the change in electric field as shown by the insets in the figure.

semiconductor generates electron-hole pairs and the semiconductor becomes conductive. The resistivity of the absorbing material drops and the gate voltage is now dropped only across the insulator, resulting in an increase in electric field at the interface with the graphene, which by proximity causes a field-induced abrupt change in the electrical conductivity[11] of the graphene layer. This is illustrated in Fig. 13.8(b). This abrupt change forms the basis of the detection mechanism, and since it does not actually depend on the collection of electrons, graphene can be used in conjunction with inexpensive absorber materials of low purity. However, while this scheme can efficiently sense ionization, for single events the correlation between energy deposition and the change in electric field can be poor because of where the charge deposition occurs within the absorber and its morphology. In this case, by suitable biasing, the charge generated can be drifted towards the surface of the absorber, resulting in an electric field response independent of where the ionization occurs spatially in the absorber and dependent only on the number of carriers generated and therefore the energy of the incident radiation. When coupled with narrow-bandgap absorbers, such as InAs and InSb, the potential exists for low-cost, large-volume, high-energy resolution detection systems that operate at room temperature. Initial results from work carried out on Si-based devices look promising. For example, Patil *et al.* [45] observed a 50% change in graphene resistivity when an undoped Si device operated at 4.3K was alternately exposed to high (40kV, 80mA) and low (15kV, 15mA) X-ray fluxes from an X-ray tube. However, no response was observed at room temperature. Foxe *et al.* [47] found that the drain-source current in a doped Si absorber GFET tracked a gate injected HeNe laser beam signal modulated at 100 Hz.

GFETS have also been fabricated with other absorbers such as SiC and GaAs. For example, Patil *et al.* [45] show results from two devices – one constructed using a SiC absorber and the other a GaAs absorber. The response of both devices to radiation sources is shown in Fig. 13.10. In (a) we show the response of a SiC based GFET to gamma-rays and in (b) the response of a GaAs-based GFET to X-rays (from [45]). The GaAs absorber was tested at 4.3 K and shows a faster response to X-rays than Si of SiC-based devices. In this case, a negative gate was applied. The initial rise in the responses corresponds to when radiation sources are abruptly applied. However, when the radiation sources were removed, the graphene resistance could take an inordinate amount of time to return to its pre-irradiation level. The authors attribute this to the fact that the devices lack a mechanism to effectively remove the ionized charges once they are produced. In fact, they accumulate at the interface simply due to the influence of the gate potential. This was confirmed by applying a short voltage pulse across the device with opposite polarity to the gate potential while turning off the radiation source. The resistance returns to its original value upon application of the pulse.

In order to overcome this limitation, the authors have proposed more advanced device architecture in which charge clearing can occur more naturally, namely, a graphene DEPFET (DEpleted P-channel Field Effect Transistor) [48]. A DEPFET [49] is a detector composed of a p-channel FET embedded into a fully depleted substrate simultaneously providing radiation detection and amplification, resulting in very low noise and high resolution. GFETs actually behave in a similar fashion, in terms of combining detection and amplification, except that in a DEPFET the gate channel senses the creation of charge in the bulk following an ionizing event, while in the case of a GFET it is the graphene layer. However, compared to a DEPFET the GFET is missing some essential functional components, such as a p-n junction to deplete the substrate, a potential minimum to confine electrons near the transistor channel (graphene) and a clear contact to drain the

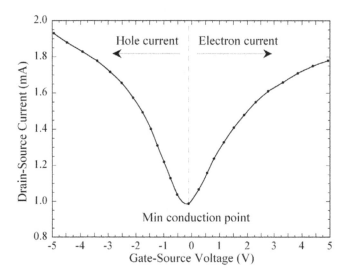

FIGURE 13.9 GFET operation showing the variation of the drain-source current with gate source voltage using a thin p-type Si absorber (from [46]). The curve reflects the Dirac curve, and the device can be biased on either side of the curve to collect either electrons or holes.

---

[11] The effect is similar to the abrupt change in resistance that can be induced in a transition edge sensor (TES) near its critical temperature.

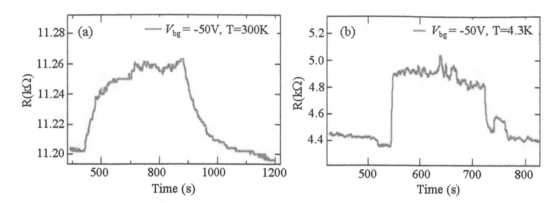

FIGURE 13.10 The response of two GFETs based on different compound semiconductor absorbers. In (a) we show the response of a SiC-based GFET to gamma-rays and in (b) the response of a GaAs-based GFET to X-rays (from [45]). The initial rise in response corresponds to the point in time when a radiation source is abruptly applied, and the fall in response corresponds to when the radiation source is abruptly removed. The slow decay times are attributed to the fact that the devices lack a mechanism to effectively remove the ionized charges once they are produced.

electrons from the potential well after readout. However, these components can be incorporated and simulations [48] have shown that DEPFET-like performances can be achieved. GFET neutron detectors have also been proposed [50]. In this case, a ^{10}B converter layer is inserted between the lower electrical contact and the semiconductor.

13.3.4 Quantum Heterostructures

In 1970, Esaki and Tsu [51] suggested the possibility of growing alternating layers of GaAs and AlGaAs in a periodic array to form a super lattice (SL), which would have remarkably different electronic properties from those of bulk GaAs or AlGaAs. When a layer of GaAs is sandwiched between two "infinite" layers of AlGaAs, the carriers in the GaAs are trapped in the GaAs layer along the growth direction. In this structure, the energy levels in the well are raised in the conduction band for the electrons and lowered in the valence band for the holes. This leads to the confinement of electrons along the growth direction. The composite of many such layers forms a bi-periodic array of rectangular quantum wells whose unique electroabsorption properties can be exploited for radiation detection. For example, Eugster and Hagelstein [52] proposed using a stack of GaAs/AlGaAs QWs to detect soft X-rays. The detection scheme is based on measuring small changes in refractive index of the QWs following photoabsorption using an optical probe. The technique is essentially a variant of a pump and probe method except that carriers are generated by X-ray photoabsorption rather than by optical means. Cooke [53] describes a prototype device in which the QWs are sandwiched in the middle of a p-i-n diode structure. A bias applied across the device is used to sweep out the carriers following a photoionization event. During an event the generated electron-hole pairs partially screen the internal electric field established by the bias, which in turn perturbs the energy position of the first exciton absorption peak. The resulting shift can be detected by probing the QWs with a weak laser beam that has been tuned near the onset of the peak and measuring the laser's change in amplitude. The main advantages of using multiple QWs are that devices are relatively simple to fabricate and the radiation directly modulates the laser probe ensuring high-speed operation. In addition, events can be easily time-gated with tens of picoseconds time resolution, and depending on the probe beam wavelength and imaging optics, micron spatial resolution can be achieved. However, the potential disadvantages are that the scheme is limited to the soft X-ray regime[12] due to the relative inefficiency of X-ray detection at higher energies and the modulation of the probe beam is very weak. For example, Basu [54] simulated a stack of 10 GaAs quantum wells of thickness 9.6 nm, separated by $Al_{0.3}Ga_{0.7}As$ layers of thickness 9.8 nm. Assuming an incident X-ray beam of energy 500 eV, he calculated a 1% change in the optical probe beam intensity in a reflection geometry for an incident flux of 12 X-ray photons per μm^2 (1.2×10^9 photons cm^{-2}). As such, the technique is most suitable for use with strong X-ray sources, such as synchrotron radiation sources and X-ray laser facilities particularly for time-gated X-ray imaging applications; for example, X-ray microscopy, X-ray holography and X-ray laser spectroscopy.

For detector applications, quantum heterostructures could also facilitate ultra-low noise operation. For example, Chang *et al.* [55] noted that whilst leakage currents in a CdZnTe metal-semiconductor-metal (M-S-M) sandwich type detector can be reduced by employing a Schottky barrier between a metal contact and semiconductor, or by fabricating a p-i-n type structure, a marked improvement in noise performance can be achieved by replacing one or both of the contacts with a SL. The SL can be designed to selectively transport one carrier species whilst hindering transport of the other – the carrier with

---

[12] In reality, the optical properties of the quantum wells saturate above ~500 eV and whilst not the upper energy limit of detection, energy resolution is lost.

poorest transport properties. Specifically, one designs a large carrier effective mass for undesired carriers in the electric field direction, which results in low carrier velocities, but with a density of states for undesired carrier that is lower than that of a comparable bulk semiconductor. This results in low carrier concentrations and hence a low current density under an electric field. The opposite carrier species can be designed to have a large velocity and high density of states, hence producing a large current density. In addition to the expected reduction in leakage currents, it has been demonstrated that such SLs prevent propagation of dislocations present in the underlying crystal and smooth out surface roughness, both of which are beneficial for the formation of uniform electrical fields and increased detector resolution.

In the case of a CdTe or CdZnTe detector, a HgTe/HgCdTe SL is grown on a CdTe or CdZnTe bulk detector crystal by MOCVD or MBE. Chang *et al.* [56] modelled the response of an M-S-n device, where M is a metal, S is the compensated CdZnTe bulk material and the n-region is the HgTe/HgCdTe SL. The anode of the detector is the SL with a metal contact. They showed that this structure was superior to HgCdTe/CdZnTe/HgCdTe p–i–n detectors. They found that a SL structure composed of a 2-nm thick HgTe well and a 2-nm thick HgCdTe layer, which acted as a barrier, was about optimum and predicted that such a design would effectively reduce leakage currents by at least two orders of magnitude relative to detectors using the same CdZnTe absorbers but with conventional ohmic metal contacts. To test the model predictions, Chang *et al.* [56] fabricated chess-board-like test structures composed of alternatively-arranged CdZnTe diodes pixels with and without SL layers. The HgTe/HgCdTe SLs was grown by MBE on polished and passivated CdZnTe bulk material. The back metal contact was deposited on the other face of the CdZnTe by electron beam deposition. Current-voltage measurements at room temperature confirmed that diodes with the SL contact layers had significantly reduced leakage currents relative to the diodes with only metal contacts, in fact by over two orders of magnitude. An alternate approach to low noise operation may be to construct a detector composed of a series of quantum valleys, similar to the nano-wire devices described in 13.3.2.1, which have the dimensions of the order of nm's by mm (or even cm) at the contacting surfaces. An applied electric field induces the carriers to be "channeled" towards the contacts. As such, individual readouts could have approximately the areas of a readout node of a Si CCD, which is directly responsible for their ultra-low noise operation.

13.4 New Approaches to Radiation Detection

In this section we describe possible new approaches to radiation detection which by their very nature are speculative. We note that considerable progress has been made in the development of organic semiconductors and nanotechnology, in particular, in the combination of the two when applied to nanofabrication techniques. For example, DNA[13]-origami assembly offers unmatched programmability in producing bespoke 3-D structures on the 10–100 nm scale [57]. More significantly, whereas top-down approaches at the macro-scale rely on external stimuli to create structures, in a bottom-up approach at the nano-scale, the final structure is encoded in the material itself and the precursors "self-assemble" into the desired configuration with features much smaller than currently possible using top-down approaches. However, as pointed out by Isaacoff and Brown [58] in the convergence of these approaches top-down control is used to guide bottom-up processes, where major advances could be made in the fabrication of nano-systems (electronic, photonic and mechanistic). How these technologies can be applied to radiation detection is unclear at this time, but it is assumed that systems at the macro level will be assembled out of nano-building blocks [28,59].

Other areas that may see new approaches to radiation detection include the first exploratory work in exploiting the more obscure internal degrees of freedom of the electron in addition to its charge for non-volatile information processing.

13.4.1 Biological Detection Systems and Intelligent Photonics

Several recent advances in nano-technology have demonstrated the viability of bio-metallic and bio-semiconductor interfacing, leading to the possibility that future detection systems may be directly interfaced to biological systems (the operator), or alternately, biological systems may be nano-engineered to be detection systems. For example, we note reports of porosity in III-V compounds, particularly GaAs, GaP and InP [60]. While one can envision photonic crystal applications, this also leads to the possibility of animal cell:semiconductor interfacing[14] as in the case of nanostructured Si [61]. One idea is to translate the electrical activity of neurons directly into measureable photonic signals or alternatively convert signals from a compound into electrical stimulation of neurons. In this case, the porous semiconductor serves as both the adhesion substrate for the cells and the transducer/receiver. Preliminary work has demonstrated that mammalian cells can be cultured on nanostructured Si that remain viable over time, in terms of both respiration and membrane integrity [62].

In terms of animal cell:semiconductor interfacing, a simpler approach might be to integrate semiconductor and biological content in a more natural way by modifying DNA itself. A number of authors have pointed out that DNA is itself a nano-object

---

[13] DeoxyriboNucleic Acid.
[14] depending on cytotoxicity.

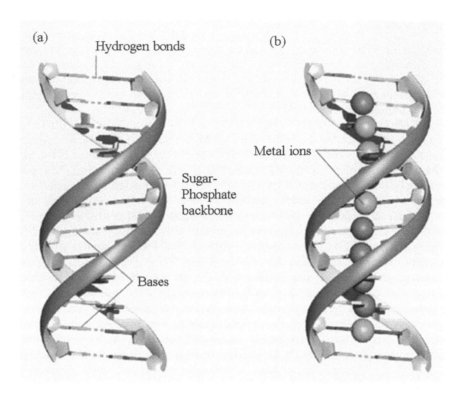

FIGURE 13.11 (a) Single strands of DNA consist of a sugar-phosphate backbone decorated with organic bases. A double helix forms when the bases of one strand form hydrogen bonds to complementary bases on another strand. (b), Tanaka *et al.* [69] have replaced bases in natural DNA with artificial ones that bind specifically to copper ions (Cu2+, darker shaded spheres) or mercury ions (Hg2+, lighter shaded spheres). A duplex forms only when opposing bases bind a metal ion between them. The sequence of metal ions is determined by the sequence of artificial bases (from [67]).

that can provide a scaffold with which to build nanoscale structures from the bottom up [63,64,65]. In this regard, its unique properties have been intensively studied from different points of view, including a drug delivery system, a ladder for ordered arrangements of various nanostructures and a spacer to control distances between nano-objects [65]. Its elongated shape suggests that it can be used as a wire capable of transmitting signals and therefore information along its length, with immediate applications in nanophotonics (*e.g.*, [66]). Although, DNA's electric properties are still under debate (and it is generally considered to be a poor conductor), its structure possesses well-defined binding sites where various molecules, from small dyes to large complex nano-particles (and specifically metal complexes) can be attached with nanometric precision. Depending on the complex, this can produce conductive or semiconductive properties, as outlined by Muller [67] and Clever and Shionoya [68]. The process is illustrated in Fig. 13.11 in which natural bases in DNA have been replaced with artificial ones that bind specifically to metal ions [69]. Charge migration is then considered to occur due to the interaction between π-electrons of stacked base pairs that provide a pathway for channelled one-dimension migration of charge. In fact, Vyawahare *et al.* [66] have demonstrated unidirectional energy transfer over a length of 40 bp[15] (~13 nm) in one DNA chain with an efficiency of 20%. More recently, single-molecule break-junction conductivity measurements of metallo-DNA using single wall carbon nanotubes as point contacts were carried out by Liu *et al.* [70]. They found that the conductive properties could be switched by applying chelating[16] agents, leading to the possibility of building molecular switches or sensors. In addition to wirelike structures, DNA chains can be assembled into origami structures, photonic crystallike assemblies, liquid-crystal phases and thin films [65,71].

In addition to creating single molecular devices, nanotechnology can provide a tool for the organization of photonic components, both serial and parallel, and the possibility of forming branched networks. The modularity of assembly, along with the plethora of DNA functionality available, allows for the construction of entire molecular circuits [72]. As an example, in Fig. 13.12, we show an energy harvesting concept proposed by Pinheiro *et al.* [65] in which different light-harvesting complexes are spatially clustered, leading to optimized channelling efficiency. The spheres and rods represent photonic components that can serve as light-harvesting and energy-transfer materials. They are positioned and aligned using DNA as a molecular pegboard. The uneven spheres represent enzyme or membrane complexes that can be used as final energy or electron acceptors. These act as molecular transducer units, where light is transformed into chemical potential, represented by the transformation of substrate

---

[15] bp = base pairs. The nucleobases (*e.g.*, cytosine, guanine, adenine and thymine, abbreviated as C, G, A and T, respectively) are nitrogen-containing biological compounds that form the building blocks of the DNA double helix. The linear dimension of 1 bp is approximately equal to 0.3 nm.

[16] molecules that can form more than one bond to a metal ion.

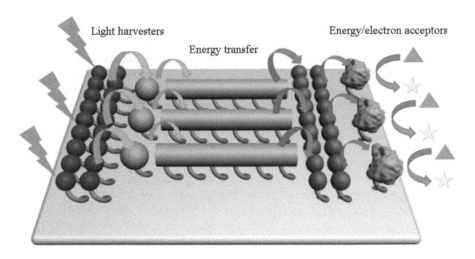

FIGURE 13.12 Schematic of a DNA-based machine for energy transfer and photonics (from Pinheiro *et al.* [67]). The shaded spheres and rods represent photonic components that can serve as light-harvesting and energy-transfer materials. The uneven spheres represent enzyme or membrane complexes that can be used as final energy or electron acceptors. These act as molecular transducer units, where light is transformed into chemical potential, represented by the transformation of substrate (triangles) into a higher-energy product (stars).

(triangles) into a higher-energy product (stars). The physical separation of photonic components creates layers of spectral separation, allowing the construction of larger and more complex photonic circuitry.

Extrapolating from the above discussion, the extension to man-detector interfacing (whilst still in the realm of science fiction) is obvious and will revolutionize radiation detection and measurement with far-reaching technological and social consequences. In theory, in addition to "direct" detection, the operator could analyse and interpret signals in real time, forming an intimate 3-D picture of his radiation environment. In fact, a multi-modal approach should be possible by biologically interfacing (or even integrating) with other sensor technologies.

13.4.2 Exploiting Other Degrees of Freedom

As we have seen in Chapter 8, section 8.5, conventional radiation photon detectors work by manipulating the electron charge by means of electric fields. This charge is generated through a process of ionization by the radiation the device is design to detect and represents the basic information carriers of the system (*i.e.*, the signal). The signal-to-noise ratio in such a system is dependent on not only the number of carriers generated, but also on the level of thermally generated noise. Both are intimately dependent on the bandgap. For high-resolution systems, the need is to generate a large number of carriers; therefore, it is necessary to resort to low bandgap materials. For materials with bandgaps < 1.4 eV, cooling is required to suppress thermally generated carriers, which will degrade the signal. For bandgaps in this region, thermoelectric coolers can be used to achieve near Fano limited operation. However, for bands gaps < 1 eV, cryogenic cooling becomes essential. For very small bandgaps, of the order of 0.1 eV, other loss mechanisms in the energy partition come into play, which effectively destroy the achievable energy resolution. In this regime, it becomes much more efficient to utilize the superconductivity properties of materials and manipulate cooper pairs to achieve very high resolution. However, very specialized cryogenic techniques are now required to reach temperatures of 300 mK or less. At these temperatures, devices are by necessity very small to minimize parasitic heat losses and are limited to niche optical or soft X-ray applications [73].

In the long term, it is clear that for semiconductors, conventional detection techniques based on manipulating the electron's charge using electric fields will reach an impasse in terms of sensitivity. A major step forward should be possible, by exploiting other obscure internal degrees of freedom of the electron in addition to its charge for non-volatile information processing, for example, using its spin, or alternately the valley degree of freedom implicit in the band structure of some semiconductors – in effect utilizing information encoded in a quantum number as opposed to charge.

13.4.2.1 Spintronics

While conventional detection systems utilize the electron's charge to produce the signal, Gary *et al.* [74] have proposed using the spintronic[17] properties of materials to generate information carriers, taking advantage of both the charge and spin properties of the electron. At present, spintronics has found application in the production of computer hard drives that rely

---

[17] a contraction of "spin transport electronics" or spin based electronics.

on giant magnetoresistance, a spintronic effect; it is actively being pursued as an enabling quantum computer technology [75]. The idea is to fabricate logic gates that operate using not just an electron's charge, but also its spin and its associated magnetic moment. In an electron, spin behaves like angular momentum but is not related to any real rotational motion. The spin of an electron can be switched more quickly than charge can be moved around; consequently, spintronic devices should operate at far higher speeds than their electronic counterparts. In addition, since moving charges requires energy, spin-based electronics should have operational advantages over conventional charge-based electronics, such as less power dissipation, which inevitably causes heat. In fact, information sent using spin remains fixed even after loss of power.

Spintronic radiation detectors are based on the concept that the electromagnetic field associated with incident nuclear radiation can interact with the spin of polarized electrons *via* the Rashba spin-orbit interaction [76] mechanism.[18] The physical realization of such a detector is schematically represented in Fig. 13.13. Spin-polarized electrons are injected into the bulk semiconductor by a ferromagnetic electrode. By suitable biasing between electrodes, the injected charge will drift through the device, during which time any interaction with incident radiation will result in a precession in electron spin, such that the current collected by a second ferromagnetic electrode (aligned parallel or anti-parallel to the first electrode) changes, providing a detectable signal. Because electron spin is not energy dependent, and spin readout does not require the precise manipulation and control of charge, the spin degree of freedom interacts less with its environment than the charge degree of freedom, offering a higher degree of noise immunity and non-volatility. Another advantage is that since the generation and collection of actual charge carriers is not actually required, wide bandgap semiconductors can be used as sensing media. This will significantly reduce thermal noise and in theory allow for reliable room-temperature operation.

The Rashba effect observed in well-established semiconductors (for example, silicon or gallium arsenide) is very small, so electrons have to travel large distances before any spin rotation is detectable. This requires materials of very high purity and very low operating temperatures. However, recent work on bismuth selenide [77] has demonstrated that the Rashba effect can be made to work at room temperature. This material is unusual in that its inner bulk structure behaves as a semiconductor while its surface behaves as a metal (it's a topological insulator). King *et al.* [77] report an amount of spin-splitting at least 10 times larger than other semiconductors for temperatures up to 100 °C.

At present, the major difficulty is to provide efficient means to inject and detect spin-polarized carriers into semiconductors. Recently, Gary *et al.* [74] investigated the flow of polarized electrons in a silicon bar and demonstrated that spin-polarized electrons could be efficiently injected, transported and later collected using ferromagnetic NiFe/MgO tunnel-barrier contacts. By measuring the change in voltage across the device, they found that it responded to both incident gamma and beta radiation with good sensitivity.

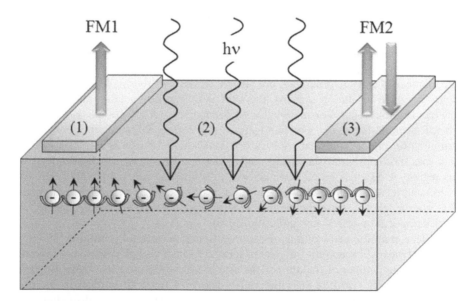

FIGURE 13.13 Conceptual schematic of a spintronics detector, with two ferromagnetic electrodes deposited onto a semiconductor. Also depicted are the three main steps required for spintronics devices: (1) injection, (2) travel/manipulation and (3) detection of spin polarized carriers. Spin is induced on the carriers by a polarized source (such as a ferromagnetic metal (*e.g.*, Ni or Fe) (FMl) before injection into the active semiconductor region. By suitable biasing between electrodes, the injected charge will drift through the device. The polarized electrons then interact with the incident radiation changing the spin of the electrons. The new polarization of the electron is then sensed at the other end of the semiconductor by a second ferromagnetic electrode (FM2). Charge can also be collected in the usual way by applying a potential across the device.

---

[18] spin splitting resulting from an applied electric rather than a magnetic field.

13.4.2.2 Valleytronics

Like spintronics, valley-based electronics, or valleytronics is a recent development that makes use of another obscure internal degree of freedom of the electron, for non-volatile information processing [78,79]. Valleytronics and spintronics are similar in that information is coded in a quantum number as opposed to charge in conventional electronics. They differ only in the quantum number used (spin or valley). Valleytronics relies on the fact that the conduction bands of some materials have two or more minima at equal energies but at different positions in momentum space. By controlling the number of electrons that occupy a particular minima (or alternatively, valley), it is possible induce a valley "polarization", which can then be used to transmit/process information without the physical movement of electrons. This new degree of freedom behaves mathematically in a way similar to the electron spin in that it acts like additional intrinsic angular momentum of the electron and arises as a direct consequence of the peculiar band structure of valleytronic materials. The effect was first observed in graphene, although a number of other bulk and monolayer materials, such as diamond [80], h-BN [81], AlAs [78,79], WS_2 and MoS_2 [82] have also shown valleytronic properties.

The valleytronic degree of freedom can be explained as follows, using graphene as an example. In graphene, the lattice consists of a planar, hexagonal ring of carbon atoms as shown in Fig. 13.14 (left). The six carbon atoms that form the ring make two families – the solid **A** atoms and open **B** atoms that occupy two distinct positions/orientations in the lattice. With reference to Fig. 13.14 (left), we can see that **A** atom can replace another **A** atom by a simple translation. Similarly, a **B** atom can replace a **B** atom by a simple translation. However, an **A** cannot replace a **B** or vice versa without breaking the symmetry of the lattice. **A** atoms are connected <u>only</u> to **B** atoms and vice versa, thus forming a bipartite lattice. The Brillouin zone also reflects this bipartite symmetry. The resulting band structure is shown in Fig. 13.7, from which we see that the conduction band and the valence band touch at six discrete points in momentum space – the so-called K points (Fig. 13.14 right). These in turn can be divided into two sets of three points, known as the K set and K` set – each set corresponding to one of the sub-lattices. The two sets have the same energy but different momenta. Points within each set are all equivalent, because they can reach each other by reciprocal lattice vectors. This difference in momenta between the K and K` points is the basis of the valley[19] pseudo-spin degree of freedom in valleytronic materials. By creating a charge carrier imbalance between the valleys it should be possible to exploit this degree of freedom. However, this is difficult to achieve in practice because a carrier requires a large momentum transfer to scatter it from K to K`. Valley polarization has been demonstrated using strain and magnetic fields, but neither of these approaches allows for dynamic control. Recently, it has been proposed that such a transfer can take place at defects that congregate along the edges of the graphene flake or due to adatoms on its surface [84]. These act as sharp scattering potentials – enough to mix K and K`. In essence, scattering off a line defect can polarize the valley degree of freedom and unlike previously proposed valley "filters", which rely on confinement structures that have proved impossible to fabricate, the line defect valley "filter" has already been observed experimentally [85]. Implementing such a filter should be easy to achieve, since these defects are naturally occurring in graphene. Thus, it should be possible to produce a valley polarized current out of an unpolarized stream of electrons, offering the possibility that valleytronic devices similar in concept to spintronic devices could be developed for radiation detection that are.

A second approach for achieving a valley imbalance has been proposed by Cao *et al.* [86], which is based on circularly polarized optical excitation. The work was carried out on MoS_2, which like graphene has a hexagonal lattice structure when viewed from above, but from the side the atoms are staggered between two planes. This leads to the creation of two energy-

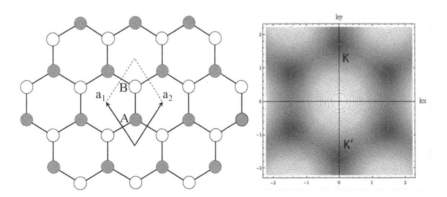

FIGURE 13.14 Left. The honeycomb lattice of graphene. The unit cell is defined by vectors a_1 and a_2 and contains two atoms belonging to sub-lattices A (filled circles) and B (open circles). Right. Contour plot of the energy in the first Brillouin zone (the zone is outlined by the hexagon) [83]. The conduction π band and the valence π band of graphene meet exactly at the corners of the hexagonal first Brillouin zone and only there. These corners are called the Dirac points.

---

[19] Valley, in this context simply refers to the similarity of the shape of these points in k-space to a valley.

degenerate valleys at the corners of the first Brillouin zone. First principle calculations showed that the two valleys absorb left- and right-handed photons differently – a phenomenon referred to as circular dichroism (CD). Therefore, it should be possible to use polarized photons to preferentially populate one of the two valleys, which under ordinary circumstances would be equally populated. The effect has been experimentally verified by Mak *et al.* [87], who measured near 100% valley polarization when MoS_2 samples were optically pumped by a polarized He-Ne laser. In a similar experiment Zeng *et al.* [88] achieved a valley polarization of 30%. It is important to note that the effect vanishes when two layers of MoS_2 are used, since the second layer overlaps the first in such a way as to eliminate the left/right asymmetry present in the single layer. While the CD approach naturally lends itself to polarized THz radiation detection [89], the practical extension to much shorter wavelengths is not so obvious.

While valley-polarized states have been created in 2-D materials such as graphene or molybdenum disulphide, it should be noted that 3-D materials such as silicon and diamond also have similar energy-degenerate valleys. In fact both have six equivalent conduction band valleys oriented along the {100} axes. Isberg *et al.* [80] demonstrated the generation, transport across macroscopic distances (0.7 mm) and detection of valley-polarized electrons in bulk diamond at 77K. The measurements were made on a high-purity CVD diamond sample equipped with semi-transparent metal contacts on two parallel (100) surfaces. The front semi-transparent contact was illuminated by short (3 ns) above-bandgap (5.8 eV) laser pulses. The light has a 3 μm absorption length in diamond and therefore creates electron hole pairs in close proximity to the front semi-transparent contact. Within 10 ps, electrons excited to the conduction band thermalize and populate the six valleys equally. With a negative bias applied to the front electrode and the back electrode grounded, the holes are extracted within 1 ns at the front electrode whereas the electrons drift to the back electrode under the influence of the applied electric field. The shape of the induced current pulse should then be a composition of contributions from the various valleys, which because of the different transport properties in the valleys will have different time signatures and polarizations. By correlating the temporal profile of the current pulse with the Hall angle, it was possible to separate electrons from different valleys. The work is particularly interesting because in diamond the electrons were found to reside in their respective valley for extremely long times, about 300 nanoseconds, as opposed to less than a nanosecond in graphene and MoS_2. As Isberg *et al.* [80] point out, this time is long enough for information processing, especially in the field of quantum computing.

REFERENCES

[1] M. Lundstrom, *"Fundamentals of Carrier Transport"*, Cambridge University Press, New York, 2nd ed. (2002) ISBN-13: 9780521631341.

[2] H.E. Ruda, "A theoretical analysis of electron transport in ZnSe", *J. Appl. Phys.*, Vol. **59** (1986), pp. 1220–1231.

[3] A. Owens, A. Peacock, "Compound semiconductor radiation detectors", *Nucl. Instr. and Meth.*, Vol. **A531** (2004), pp. 18–37.

[4] C.A. Klein, *"Bandgap* dependence and related features of radiation ionization energies in semiconductors", *J. Appl. Phys.*, Vol. **4** (1968), pp. 2029–2033.

[5] W. Que, J.A. Rowlands, "X-ray photogeneration in amorphous selenium: Geminate versus columnar recombination", *Phys. Rev. B*, Vol. **51** (1995), pp. 10500–10507.

[6] P.N. Luke, M. Amman, "Room-temperature replacement for Ge detectors – Are we there yet?", *IEEE Trans. Nucl. Sci.*, Vol. **54**, no. 4 (2007), pp. 834–842.

[7] F.J. Ryan, S.H. Shin, D.D. Edwall, J.G. Pasko, M. Khoshnevisan, C.I. Westmark, C. Fuller, "Gamma ray detectors with HgCdTe contact layers", *Appl. Phys. Lett.*, Vol. **46** (1985), pp. 274–276.

[8] A. Silenas, K. Pozela, L. Dapkus, V. Jasutis, V. Juciene, J. Pozela, K.M. Smith, "Graded-gap $Al_xGa_{1-x}As$ X-ray detector with collected charge multiplication", *Nucl. Instr. and Meth.*, Vol. **509** (2003), pp. 30–33.

[9] J.E. Lees, D.J. Bassford, J.S. Ng, C.H. Tan, J.P.R. David, "AlGaAs diodes for X-ray spectroscopy", *Nucl. Instr. and Meth.*, Vol. **594** (2003), pp. 202–205.

[10] W. Lu, C.M. Lieber, "Semiconductor nanowires", *J. Phys. D: Appl. Phys.*, Vol. **39**, no. 21 (2006), pp. R387–R406.

[11] M. Law, J. Goldberger, P. Yang, "Semiconductor nanowires and nanotubes", *Ann. Rev. Mater. Res.*, Vol. **34** (2004), pp. 83–122.

[12] Z.L. Wang, ed., *"Nanowires and Nanobelts: Materials, Properties and Devices: Volume 2: Nanowires and Nanobelts of Functional Materials"*, Springer-Verlag GmbH, Heidelberg (2005) ISBN 978-0-387-28747-8.

[13] K. Schmidt, "Nanofrontiers: Visions for the future of nanotechnology", Report of the Woodrow Wilson International Center for Scholars Project on Emerging Nanotechnologies PEN6, Washington DC (2007).

[14] N. Taniguchi, "On the basic concept of 'nano-technology'", *Proc. Intl. Conf. Prod. Eng., Tokyo, Part II, Japan Society of Precision Engineering* (1974).

[15] "U.S. Environmental Protection Agency nanotechnology whitepaper", EPA 100/B-07/001, co-chairs J. Morris, J. Willis, U.S. Environmental Protection Agency (EPA), Washington DC, February (2007) p. 5.

[16] F.D. Amaro, C.M.B. Monteiro, J.M.F. dos Santos, A. Antognini, "Novel concept for neutron detection: Proportional counter filled with ^{10}B nanoparticle aerosol", *Sci. Rep.*, Vol. **7**, Article number, 41699 (2017), pp. 1–6.

[17] E.A. McKigney, R.E. Del Sesto, L.G. Jacobsohn, P.A. Santi, R.E. Muenchausen, K.C. Ott, T.M. McCleskey, B.L. Bennett, J. F. Smith, D.W. Cooke, "Nanocomposite scintillators for radiation detection and nuclear spectroscopy", *Nucl. Instr. and Meth.*, Vol. **A579** (2007), pp. 15–18.

[18] Z. Luo, J.G. Moch, S.S. Johnson, C.C. Chen, "A review on x-ray detection using nanomaterials", *Curr. Nanosci.*, Vol. **13** (2017), pp. 364–372.

[19] G. Kim, J. Huang, M.D. Hammig, "An investigation of nanocrystalline semiconductor assemblies as a material basis for ionizing-radiation detectors", *IEEE Trans. Nucl. Sci.*, Vol. **56**, no. 3 (2009), pp. 841–848.

[20] M.D. Hammig, "Nanoscale methods to enhance the detection of ionizing radiation", in *Current Topics in Ionizing Radiation Research*, ed. M. Nenoi, InTechOpen, London, Chapter 26 (2012), pp. 557–559: ISBN 978-953-51-0196-3.

[21] M.D. Hammig, "Silicon-Based Examination of Gamma-Ray and Neutron Interactions with Solid State Materials", Technical report DTRA-TR-16-31, Defense Threat Reduction Agency, Fort Belvoir VA (2018) pp. 1–64.

[22] C. Dujardin, D. Amans, A. Belsky, F. Chaput, G. Ledoux, A. Pillonnet, "Luminescence and scintillation properties at the nanoscale", *IEEE Trans. Nucl. Sci.*, Vol. **57** (2010), pp. 1348–1354.

[23] N.V. Klassen, V.N. Kurlov, S.N. Rossolenko, O.A. Krivko, A.D. Orlov, S.Z. Shmurak, "Scintillation fibers and nanoscintillators for improving the spatial, spectrometric, and time resolution of radiation detectors", *Bull. Russ. Acad. Sci. Phys.*, Vol. **73**, no. 1 (2009), pp. 1369–1373.

[24] D.-R. Jung, J. Kim, C. Changwoo, H. Choi, S. Nam, B. Park, "Review paper: Semiconductor nanoparticles with surface passivation and surface plasmon", *Electr. Mater. Letts.*, Vol. **7**, no. 3 (2011), pp. 185–194.

[25] R.G. Chaudhuri, S. Paria, "Core/shell nanoparticles: Classes, properties, synthesis mechanisms, characterization, and applications", *Chem. Rev.*, Vol. **112**, no. 4 (2012), pp. 2373–2433.

[26] S.E. Letant, T.-F. Wang, "Semiconductor quantum dot scintillation under gamma-ray irradiation", *Nano Lett.*, Vol. **6**, no. 12 (2006), pp. 2877–2880.

[27] J.B. Plumley, N.J. Withers, A.C. Rivera, B.A. Akins, J.M. Vargas, K. Carpenter, G.A. Smolyakov, R.D. Busch, M. Osiński, "Thermal neutron detectors based on gadolinium-containing lanthanide-halide nanoscintillators", *Proc. SPIE*, Vol. **7665** (2010), pp.76651F-1–6651F-13.

[28] S.J. Koh, "Controlled placement of nanoscale building blocks: Toward large-scale fabrication of nanoscale devices", *JOM*, Vol. **59**, no. 3 (2007), pp. 22–28.

[29] C. Soci, A. Zhang, X.Y. Bao, H. Kim, Y. Lo, D. Wang, "Nanowire photodetectors", *J. Nanosci. Nanotechnol.*, Vol. **10**, no. 3 (2010), pp. 1430–1449.

[30] M. Kobayashi, J. Komori, K. Shimidzu, M. Izaki, K. Uesugi, A. Takeuchi, Y. Suzuki, "Development of vertically aligned ZnO nanowires scintillators for high spatial resolution X-ray imaging", *Appl. Phys. Lett.*, Vol. **106**, no. 8 (2015), pp. 081909-1–081909-4.

[31] M. Izaki, M. Kobayashi, T. Shinagawa, T. Koyama, K. Uesugi, A. Takeuchi, "Electrochemically grown ZnO vertical nanowire scintillator with light-guiding effect", *Phys. Status Solidi (A)*, Vol. **214**, no. 11 (2017), pp. 1700285–1700300.

[32] T. Gandhi, K.S. Raja, M. Misra, "Cadmium zinc telluride (CZT) nanowire sensors for detection of low energy gamma-ray detection", *Proc. SPIE*, Vol. **6959** (2008), pp. 695904-1–695904-13.

[33] M.F. Becchetti, M.D. Hammig, G. Kim, J. Il Park, N.A. Kotov, "Ionizing radiation detection via CdTe nanowires assembled using vacuum filtration", *Paper Presented at the IEEE NSS/MIC/RTSD Conference*, Anaheim, CA, 27th October – 3rd November (2012).

[34] A.L. Rogach, L. Katsikas, A. Kornowski, D. Su, A. Eychmüller, H. Weller, "Synthesis and characterization of thiol-stabilized CdTe nanocrystals", *Ber. Bunsenges. Phys. Chem.*, Vol. **100** (1996), pp. 1772–1778.

[35] G. Zhang, K. Tateno, H. Gotoh, T. Sogawa, "Towards new low-dimensional semiconductor nanostructures and new possibilities", *NTT Tech. Rev.*, Tokyo, Vol. **8**, no. 8 (2010), pp. 1–8.

[36] D. Bera, L. Qian, T.-K. Tseng, P.H. Holloway, "Quantum dots and their multimodal applications: A review", *Materials*, Vol. **3** (2010), pp. 2260–2345.

[37] A.M. Ealias, M.P. Saravanakumar, "A review on the classification, characterisation, synthesis of nanoparticles and their application", *IOP Conf. Ser. Mater. Sci. Eng.*, Vol. **263**, no. 032019 (2017), pp. 1–15.

[38] H.W. Kroto, J.E. Fischer, D.E. Cox, eds., "*The Fullerenes*", Pergamon, Oxford (1993).

[39] S.Z. Butler, S.M. Hollen, L. Cao, Y. Cui, J.A. Gupta, H.R. Gutiérrez, T.F. Heinz, S.S. Hong, J. Huang, A.F. Ismach, E. Johnston-Halperin, M. Kuno, V.V. Plashnitsa, R.D. Robinson, R.S. Ruoff, S. Salahuddin, J. Shan, L. Shi, M.G. Spencer, M. Terrones, W. Windl, J.E. Goldberg, "Progress, challenges, and opportunities in two-dimensional materials beyond graphene", *ACS Nano*, Vol. **7** (2013), pp. 2898–2926.

[40] B. Radisavljevic, A. Radenovic, J. Brivio, V. Giacometti, A. Kis, "Single-layer MoS$_2$ transistors", *Nat. Nanotechnol.*, Vol. **6** (2011), pp. 147–150.

[41] R.F. Frindt, "Single crystals of MoS$_2$ several molecular layers thick", *J. Appl. Phys.*, Vol. **37** (1966), pp. 1928–1929.

[42] P. Joensen, R.F. Frindt, S.R. Morrison, "Single-layer MoS$_2$", *Mater. Res. Bull.*, Vol. **21** (1986), pp. 457–461.

[43] A.H. Castro Neto, F. Guinea, N.M.R. Peres, K.S. Novoselov, A.K. Geim, "The electronic properties of graphene", *Rev. Mod. Phys.*, Vol. **81** (2009), pp. 109–162.

[44] M. Foxe, G. Lopez, I. Childres, R. Jalilian, C. Roecker, J. Boguski, I. Jovanovic, Y.P. Chen, "Graphene field-effect transistors on undoped semiconductor substrates for radiation detection", *IEEE, Trans. Nanotech.*, Vol. **11**, no. 3 (2012), pp. 581–587.

[45] A. Patil, O. Koybasi, G. Lopez, M. Foxe, I. Childres, C. Roecker, J. Boguski, J. Gu, M.L. Bolen, M.A. Capano, P. Ye, I. Jovanovic, Y.P. Chen, "Graphene field effect transistor as radiation sensor", *IEEE Nucl. Sci. Symp. Conf. Rec.*, 23rd Oct. – 29th Oct., 2011, Valencia, Spain (2011), pp. 455–459.

[46] H. Wang, IEEE, A. Hsu, J. Wu, J. Kong, Member, T. Palacios, "Graphene-based ambipolar RF mixers", *IEEE Electron Device Letts.*, Vol. **31**, no. 9 (2010), pp. 906–908.

[47] M. Foxe, G. Lopez, I. Childres, R. Jalilian, C. Roecker, J. Boguski, I. Jovanovic, Y.P. Chen, "Detection of ionizing radiation using graphene field effect transistors", *IEEE Nucl. Sci. Symp. Conf. Rec.*, Orlando, FL, Oct. 24th – Nov. 1st (2009), pp. 90–95.

[48]   O. Koybasi, I. Childres, I. Jovanovic, Y.P. Chen, "Design and simulation of a graphene DEPFET detector", *IEEE Nucl. Sci. Symp. Conf. Rec.*, 23rd Oct. – 29th Oct., 2011, Valencia, Spain (2011), pp. 4249–4254.

[49]   J. Kemmer, G. Lutz, "New detector concepts", *Nucl. Instr. and Meth.*, Vol. **A253**, no. 3 (1987), pp. 365–377.

[50]   M. Foxe, E. Cazalas, H. Lamm, A. Majcher, C. Piotrowski, I. Childres, A. Patil, Y.P. Chen, I. Jovanovi, "Graphene-based neutron detectors", *Proc. IEEE Nucl. Sci. Symp. Conf. Rec.*, 23rd Oct. – 29th Oct. (2011), pp. 352–355.

[51]   L. Esaki, R. Tsu, "Superlattice and negative differential conductivity in semi-conductors", *J. Res. Develop.*, Vol. **24** (1970), pp. 61–65.

[52]   C.C. Eugster, P.L. Hagelstein, "X-ray detection using the quantum well exciton nonlinearity", *IEEE J. Quant. Electron.*, Vol. **26**, no. 1 (1990) pp. 75–84.

[53]   B.J. Cooke, "Superlattice electroabsorption radiation detector", *USDOE Technical Report*, Contract no. W-7405-ENG-36, USDOE, Washington, DC, United States (1993).

[54]   S. Basu, "Possibility of X-ray detection using quantum wells", *IEEE J. Quant. Electron.*, Vol. **27**, no. 9 (1991), pp. 2116–2121.

[55]   Y. Chang, C.H. Grein, C.R. Becker, X.J. Wang, Q. Duan, S. Ghosh, P. Dreiske, R. Bommena, J. Zhao, M. Carmody, F. Aqariden, S. Sivananthan, "CdZnTe radiation detectors with HgTe/HgCdTe superlattice contacts for leakage current reduction", *J. Electron. Mat.*, Vol. **40**, no. 8 (2011), pp. 1854–1859.

[56]   Y. Chang, C.H. Grein, C.R. Becker, J. Huang, S. Ghosh, F. Aqariden, S. Sivananthan, "Reduced leakage currents of CdZnTe radiation detectors with HgTe/HgCdTe superlattice contacts," *Proc. SPIE*, Vol. **8507** (2012), pp. 850715-1–850715-10.

[57]   H. Li, J.D. Carter, T.H. LaBean, "Nanofabrication by DNA self-assembly", *Mater. Today*, Vol. **12**, no. 5 (2009), pp. 24–32.

[58]   B.P. Isaacoff, K.A. Brown, "Progress in top-down control of bottom-up assembly", *Nano Lett.*, Vol. **17**, no. 11 (2017), pp. 6508–6510.

[59]   S.J. Koh, "Strategies for controlled placement of nanoscale building blocks", *Nanoscale Res. Lett.*, Vol. **2** (2007), pp. 519–545.

[60]   H. Föll, J. Carstensen, S. Langa, M. Christophersen, I.M. Tiginyanu, "Porous III-V compound semiconductors: Formation, properties, and comparison to silicon", *Phys. Stat. Sol. (A)*, Vol. **197** (2003), pp. 61–70.

[61]   S.C. Bayliss, P.J. Harris, L.D. Buckberry, C. Rousseau, "Phosphate and cell growth on nanostructured semiconductors", *J. Mat. Sci. Lett.*, Vol. **17** (1997), pp. 737–740.

[62]   S.C. Bayliss, R. Heald, D.I. Fletcher, L.D. Buckberry, "The culture of mammalian cells on nanostructured silicon", *Advanced Materials*, Vol. **11**, no. 4 (1999), pp. 318–321.

[63]   Y.W. Kwon, C.H. Lee, D.-H. Choi, J.-I. Jin, "Materials science of DNA", *J. Mater. Chem.*, Vol. **19** (2009), pp. 1353–1380.

[64]   K. Matczyszy, J. Olesiak-Banska, "DNA as scaffolding for nano-photonic structures", *J. Nanophotonics*, Vol. **6**, no. 1 (2012), pp. 064505-1–064505-15.

[65]   A.V. Pinheiro, D. Han, W.M. Shih, H. Yan, "Challenges and opportunities for structural DNA nanotechnology", *Nat. Nanotechnol.*, Vol. **6** (2011), pp. 763–772.

[66]   S. Vyawahare, S. Eyal, K.D. Mathews, S.R. Quake, "Nanometer-scale fluorescence resonance optical waveguides", *Nano Lett.*, Vol. **4**, no. 6 (2004), pp. 1035–1039.

[67]   J. Muller, "Chemistry: Metals line up for DNA", *Nature*, Vol. **444** (2006), p. 698.

[68]   G.H. Clever, M. Shionoya, "Metal-base pairing in DNA", *Coord. Chem. Rev.*, Vol. **254** (2010), pp. 2391–2402.

[69]   K. Tanaka, G.H. Clever, Y. Takezawa, Y. Yamada, C. Kaul, M. Shionoya, T. Carel, "Programmable self-assembly of metal ions inside artificial DNA duplexes", *Nat. Nanotech.*, Vol. **1** (2006), pp. 190–194.

[70]   S. Liu, G.H. Clever, Y. Takezawa, M. Kaneko, K. Tanaka, X. Guo, M. Shionoya, "Direct conductance measurement of individual metallo-DNA duplexes within single-molecule break junctions", *Angew. Chem. Int. Edit.*, Vol. **50** (2011), pp. 8886–8890.

[71]   W. Su, V. Bonnard, G.A. Burley, "DNA-templated photonic arrays and assemblies: Design principles and future opportunities", *Chem.-Eur. J.*, Vol. **17** (2011), pp. 7982–7991.

[72]   R.J. Kershner, L.D. Bozano, C.M. Micheel, A.M. Hung, A.R. Fornof, J.N. Cha, C.T. Rettner, M. Bersani, J. Frommer, P.W. K. Rothemund, G.M. Wallraff, "Placement and orientation of individual DNA shapes on lithographically patterned surfaces", *Nat. Nanotech.*, Vol. **4** (2009), pp. 557–561.

[73]   P. Verhoeve, N. Rando, A. Peacock, D. Martin, R. den Hartog, "Superconducting tunnel junctions as photon counting imaging spectrometers from the optical to the x-ray band", *Opt. Eng.*, Vol. **41**, no. 6 (2002), pp. 1170–1184.

[74]   N. Gary, S. Teng, A. Tiwari, H. Yang, "Room temperature radiation detection based on spintronics", *IEEE/MIC/RTSD, 19th International Workshop on Room Temperature Semiconductor Detectors*, Anaheim, CA, 27 October – 3 November, Conference Record (2012), pp. 4152–4155.

[75]   D.D. Awschalom, M.E. Flatté, N. Samarth, "Spintronics", *Sci. Am.*, Vol. **286** (2002) pp. 66–73.

[76]   E.I. Rashba, "Properties of semiconductors with an extremum loop. 1. Cyclotron and combinational resonance in a magnetic field perpendicular to the plane of the loop", *Sov. Phys. Solid. State*, Vol. **2** (1960), pp. 1109–1122.

[77]   P.D.C. King, R.C. Hatch, M. Bianchi, R. Ovsyannikov, C. Lupulescu, G. Landolt, B. Slomski, J.H. Dil, D. Guan, J.L. Mi, E.D. L. Rienks, J. Fink, A. Lindblad, S. Svensson, S. Bao, G. Balakrishnan, B.B. Iversen, J. Osterwalder, W. Eberhardt, F. Baumberger, P. Hofmann, "Large tunable Rashba spin splitting of a two-dimensional electron gas in Bi_2Se_3", *Phys. Rev. Lett.*, Vol. **107** (2011), pp. 096802–0968025.

[78]   Y.P. Shkolnikov, E.P. De Poortere, E. Tutuc, M. Shayegan, "Valley splitting of AlAs two-dimensional electrons in a perpendicular magnetic field", *Phys. Rev. Lett.*, Vol. **89** (2002), pp. 226805-1–226805-4.

[79]   O. Gunawan, Y.P. Shkolnikov, K. Vakili, T. Gokmen, E.P. De Poortere, M. Shayegan, "Valley susceptibility of an interacting two-dimensional electron system", *Phys. Rev. Lett.*, Vol. **97** (2006), pp. 186404-1–186404-6.

[80] J. Isberg, M. Gabrysch, J. Hammersberg, S. Majdi, K.K. Kovi, D.J. Twitchen, "Generation, transport and detection of valley-polarized electrons in diamond", *Nat. Mater.*, Vol. **12** (2013), pp. 760–764.

[81] D. Pacilé, J.C. Meyer, Ç.Ö. Girit, A. Zett, "The two-dimensional phase of boron nitride: Few-atomic-layer sheets and suspended membranes", *Appl. Phys. Lett.*, Vol. **92**, no. 13 (2008) pp. 133107-1–133107-3.

[82] H.S.S. Ramakrishna Matte, A. Gomathi, A.K. Manna, D.J. Late, R. Datta, S.K. Pati, C.N.R. Rao, "MoS_2 and WS_2 analogues of graphene", *Angew. Chem.*, Vol. **122** (2010) pp. 4152–4156.

[83] L. Biró, P. Nemes-Incze, P. Lambin, "Graphene: Nanoscale processing and recent applications", *Nanoscale*, Vol. **4** (2012), pp. 1824–1839.

[84] D. Gunlycke, C.T. White, "Graphene valley filter using a line defect", *Phys. Rev. Lett.*, Vol. **106** (2011), pp. 136806-1–136806-4.

[85] D. Gunlycke, S. Vasudevan, C.T. White, "Confinement, transport gap, and valley polarization in graphene from two parallel decorated line defects", *Nano. Lett.*, Vol. **13** (2013), pp. 259–263.

[86] T. Cao, G. Wang, W. Han, H. Ye, C. Zhu, J. Shi, Q. Niu, P. Tan, E. Wang, B. Liu, J. Feng, "Valley-selective circular dichroism of monolayer molybdenum disulphide", *Nat. Commun.*, Vol. **3**, no. 887 (2012), pp. 1–5.

[87] K.F. Mak, K. He, J. Shan, T.F. Heinz, "Control of valley polarization in monolayer MoS_2 by optical helicity", *Nat. Nanotechnol.*, Vol. **7** (2012), pp. 494–498.

[88] H. Zeng, J. Dai, W. Yao, D. Xiao, X. Cui, "Valley polarization in MoS_2 monolayers by optical pumping", *Nat. Nanotechnol.*, Vol. **7** (2012), pp. 490–493.

[89] C.J. Tabert, E.J. Nicol, "Valley-spin polarization in the magneto-optical response of silicene and other similar 2D crystals", *Phys. Rev. Letts.*, Vol. **110** (2013), pp. 197402-1-1197402-5.

Appendix A: Supplementary Reference Material and Further Reading List

The following books are recommended for reference and supplementary reading:

Title/Author	Publisher	Top level review of content
Radiation Detection and Measurement, 4th edition, G. Knoll.	John Wiley & Sons, New York. (2010)	General purpose text book, covering many technologies other than semiconductors.
Semiconductor Radiation Detectors – Device Physics, G. Lutz.	Springer Berlin. (2007)	Concentrates only on silicon devices, a little on applications.
Semiconductor Devices: Physics and Technology, S.M. Sze.	John Wiley & Sons, New York (1995).	Mainly concentrates on semiconductor devices and fabrication technology.
Landolt-Börnstein: Numerical Data and Functional Relationships in Science and Technology. Group III: Crystal and Solid State Physics, Eds. O. Madelung, H. Weiss, M. Schultz.	Springer Berlin. (1987)	A systematic and extensive data collection in all areas of physical sciences and engineering. Volume 17, Subvolume A: Physics of Group IV Elements and III-V Compounds. Subvolume B: Physics of II-VI and I-VII Compounds, Semi-Magnetic Semiconductors, Volume 22, Subvolume A: Intrinsic Properties of Group IV Elements and III-V, II-VI, and I-VII Compounds.
Semiconductors: Data Handbook, 3rd edition, O. Madelung	Springer Berlin Heidelberg. (2004)	Condensed listing of basic semiconductor data compiled from the 17 volumes of the New Series of the Landolt–Börnstein data handbooks, dealing with semiconductors.
Properties of Group-IV, III-V and II-VI Semiconductors, S. Adachi	John Wiley & Sons, New York. (2005)	Covers only semiconductor properties.
Properties of Group-IV, III-V and II-VI Semiconductor Alloys, S. Adachi	John Wiley & Sons, New York. (2009)	Covers semiconductor properties of alloys.
Semiconductor Materials, L.I. Berger.	CRC Press, Boca Raton, New York. (1996)	Covers mainly semiconductor properties, some material on the principles of semiconductor device functioning.
Springer Handbook of Electronic and Photonic Materials, S. Kasap, P. Capper (Eds.),	Springer New York (2007).	Concentrates on photonic materials and devices.

The following web sites provide valuable supporting information:

http://xdb.lbl.gov/
Comprehensive X-ray database plus downloadable X-ray Data Booklet.

http://pdg.lbl.gov/2014/reviews/contents_sports.html
Particle data Group website. Comprehensive resource on the properties of particles and fundamental interactions. The Review of Particle Physics and its pocket version the Particle Physics Booklet, are available as downloads.

www.ioffe.ru/SVA/NSM/Semicond/
A comprehensive archive of information on the fundamental characteristics and properties of III-V compounds and their alloys.

www.cleanroom.byu.edu/

A Brigham Young University resource containing a large amount of reference material relevant to microfabrication and semiconductor processing.

www.semi1source.com/
An "All things" semiconductor resource site.

www.nist.gov/
A one-stop shop for reference material including web based calculators and applications.

Appendix B: Table of Physical Constants

Quantity	Symbol	Value	Units	Error ppm
Principal Constants				
π (circumference/diameter)		3.141 592 653 589 793	-	defined
e (limit $(1 + 1/n)^n$ as $n \rightarrow \infty$)		2.718 281 828 459 045	-	defined
speed of light in vacuum	c	$2.997\ 924\ 58 \times 10^8$	m s^{-1}	exact
permeability of free space	$\mu_\circ$	$4\pi \times 10^{-7}$ $= 1.2566\ 370\ 614 \times 10^{-6}$	H m^{-1} Vs A^{-1} m^{-1}	exact
permittivity of free space	$\varepsilon_\circ$	$1/m_\circ c^2$ $= 8.854\ 187\ 817 \times 10^{-12}$	F m^{-1} m^{-3}kg^{-1}s^{4}A^2	exact
Newtonian constant of gravitation	G	$6.672\ 59 \times 10^{-11}$	m^{3}kg^{-1}s^{-2}	128
Planck constant	h	$6.626\ 075\ 5 \times 10^{-34}$	J s	0.60
in electron volts		$4.135\ 669\ 2 \times 10^{-15}$	eV s	0.30
h/2π	$\hbar$	$1.054\ 572\ 66 \times 10^{-34}$	J s	0.60
in electron volts		$6.682\ 122\ 0 \times 10^{-16}$	eV s	0.30
Planck mass, $(\hbar c/G)^{1/2}$	m$_P$	$2.176\ 470 \times 10^{-8}$	kg	23
Planck length, $\hbar/m_P c$	l$_P$	$1.616\ 229 \times 10^{-35}$	m	24
Planck time, $l_{P/c}$	t$_P$	$5.391\ 16 \times 10^{-44}$	s	24
Avogadro constant	N$_A$	$6.022\ 136\ 7 \times 10^{-23}$	mol^{-1}	0.59
atomic mass unit	u	$1.660\ 539\ 040 \times 10^{-27}$	kg	0.01
Faraday constant	F	$9.648\ 530\ 9 \times 10^4$	C mol^{-1}	0.30
Molar gas constant, N$_A$e	R	8.314 46321	J mol^{-1} K^{-1}	0.91
Boltzmann constant, R/N$_A$	k	$1.380\ 658 \times 10^{-23}$	J K^{-1}	8.50
in electron volts		$8.617\ 385 \times 10^{-5}$	eV K^{-1}	8.40
kT(300K)	kT	2.585×10^{-2}	eV	8.40
Stefan-Boltzmann constant, $(\pi^2/60)k^4/h^3c^2$	σ	$5.670\ 51 \times 10^{-8}$	W m^{-2}K^{-4}	34
Wien displacement law constant, $b=\lambda_{max}T$	b	$2.897\ 756 \times 10^{-3}$	m K	8.40
standard atmosphere	atm	$1.013\ 25 \times 10^5$	N m^{-2}	exact
acceleration of free fall:				
standard (Sèvres, France)	g	9.806 65	m s^{-2}	exact
local – US datum	g (CB)	9.801 043	m s^{-2}	
local – UK datum	g (BFS)	9.811 818	m s^{-2}	
Atomic Constants				
fine-structure constant, $m_\circ ce^2/2h$	α	$7.297\ 353\ 08 \times 10^{-3}$	-	0.05
inverse fine-structure constant	α^{-1}	137.035 989 5	-	0.05
Rydberg constant, $m_e c\alpha^2/2h$	R$_\infty$	$1.097\ 373\ 153\ 4 \times 10^7$	m^{-1}	0.001
in hertz, R$_\infty$c		$3.289\ 841\ 949\ 9 \times 10^{15}$	Hz	0.001
in joules, R$_\infty$hc		$2.179\ 874\ 1 \times 10^{-18}$	J	0.60
in electron volts, R$_\infty$hc/{e}		13.605 698 1	eV	0.30
Bohr radius, $\alpha/4\pi R_\infty$	a$_\circ$	$5.291\ 772\ 49 \times 10^{-11}$	m	0.05
Bohr magneton	μ_B	$9.274\ 015\ 4 \times 10^{-24}$	J T^{-1}	0.34
nuclear magneton	μ_N	$5.050\ 786\ 6 \times 10^{-27}$	J T^{-1}	0.34
Particle Constants				
elementary charge	e	$1.602\ 176\ 62 \times 10^{-19}$	C	0.30
in e.s.u.		$4.803\ 24 \times 10^{-10}$	e.s.u.	0.30

(Continued)

(Cont.)

Quantity	Symbol	Value	Units	Error ppm
electron mass	m_e	$9.109\ 383\ 56 \times 10^{-31}$	kg	0.59
in a.m.u.		$5.485\ 799\ 090\ 70 \times 10^{-4}$	u	0,23
in energy units		$0.510\ 998\ 9461$	MeV	0.30
proton mass	m_p	$1.672\ 623\ 1 \times 10^{-27}$	kg	0.59
in a.m.u.		$1.007\ 276\ 470$	u	0.01
in energy units		$938.272\ 31$	MeV	0.30
neutron mass	m_n	$1.674\ 928\ 6 \times 10^{-27}$	kg	0.59
		$1.008\ 664\ 904$	u	0.01
		$939.565\ 63$	MeV	0.30
muon mass	m_μ	$1.883\ 531\ 6 \times 10^{-28}$	kg	0.48
in a.m.u.		$0.113\ 428\ 913$	u	0.15
in energy units		$105.658\ 389$	MeV	0.32
proton-electron mass ratio	m_p/m_e	$1.836\ 152\ 673 \times 10^3$	-	0.17
neutron-electron mass ratio	m_n/m_e	$1.838\ 683\ 662 \times 10^3$	-	0.90
electron specific charge	$-e/m_e$	$-1.758\ 820\ 024 \times 10^{11}$	$C\ kg^{-1}$	0.30
proton specific charge	e/m_p	$9.578\ 830\ 9 \times 10^7$	$C\ kg^{-1}$	0.30
Compton wavelength	λ_C	$2.426\ 310\ 58 \times 10^{-12}$	m	0.09
$\lambda_C/2\pi = \alpha a_\circ = \alpha^2/4\pi R_\infty$	D_C	$3.861\ 593\ 23 \times 10^{-13}$	m	0.09
proton Compton wavelength, $h/m_p c$	$\lambda_{C,p}$	$1.321\ 410\ 02 \times 10^{-15}$	m	0.09
$\lambda_{C,p}/2\pi$	$D_{C,p}$	$2.103\ 089\ 37 \times 10^{-16}$	m	0.09
electron magnetic moment	μ_e	$9.284\ 770\ 1 \times 10^{-24}$	$J\ T^{-1}$	0.34
proton magnetic moment	μ_p	$1.410\ 607\ 61 \times 10^{-26}$	$J\ T^{-1}$	0.34
classical electron radius, $\alpha^2 a_\circ$	r_e	$2.817\ 940\ 92 \times 10^{-15}$	m	0.13
Thomson cross-section, $(8\pi/3)(r_e)^2$	σ_e	$6.652\ 461\ 6 \times 10^{-27}$	m^2	0.27
Astronomical Constants				
astronomical unit	AU	$1.495\ 978\ 706\ 91 \times 10^{11}$	m	exact
light year	ly	$9.460\ 730\ 472\ 5808 \times 10^{15}$	m	exact
in AU		$63,240$	AU	exact
parsec (1 AU/1 arc sec)	pc	$3.085\ 677\ 581\ 28 \times 10^{16}$	m	exact
in AU		$206,265$	AU	-
Solar mass	$M_\odot$	$(1.9891 \pm 0.0002) \times 10^{30}$	kg	-
Solar radius	$R_\odot$	$(6.9551 \pm 0.001) \times 10^8$	m	-
Solar temperature	$T_\odot$	5771.8 ± 0.7	K	-
Solar Luminosity	$L_\odot$	$(3.828 \pm 0.008) \times 10^{26}$	W	-
mean Solar constant at 1 AU	sc	1360.8 ± 0.5	$W\ m^{-2}$	-
Earth mass	$M_\oplus$	$(5.9737 \pm 0.0006) \times 10^{24}$	kg	-
equatorial Earth radius	$R_\oplus$	$6\ 378136.6$	m	-
mean solar day	msd	24h 03m 56s.555	hms	-
in sidereal time	sd	1d.00273791	days	-
sidereal year (fixed star to fixed star)	syr	365.2564	days	-

A comprehensive and up-to-date listing of fundamental constants can be found in:

P.J. Mohr, D.B. Newell, B.N. Taylor, "CODATA recommended values of the fundamental physical constants: 2014", *Rev. Mod. Phys.*, Vol. **88** (2016) pp. 035009 (73 pages).

The data are also available on the WWW at http://physics.nist.gov/cuu/Constants/index.html

Appendix C: Units and Conversions[i]

Parameter	Unit	Unit
Length		
	1 meter (SI) = $1.000\ 00 \times 10^2$	centimetre (cgs)
	1 light year = $9.460\ 53 \times 10^{15}$	meter (SI)
	1 parsec = $3.085\ 68 \times 10^{16}$	meter (SI)
	1 Ångstrom = $1.000\ 00 \times 10^{-10}$	meter (SI)
	1 Ångstrom = $1.000\ 00 \times 10^{-8}$	centimetre (cgs)
	1 micron = $1.000\ 00 \times 10^{-6}$	meter (SI)
	1 XU = $1.002\ 09 \times 10^{-13}$	meter (SI)
	1 fermi = $1.000\ 00 \times 10^{-15}$	meter (SI)
	1 nautical mile = $1.852\ 00 \times 10^3$	meter (SI)
	1 statute mile = $1.609\ 34 \times 10^3$	meter (SI)
	1 astron. unit (AU) = $1.495\ 99 \times 10^{11}$	meter (SI)
	1 solar radius = $6.959\ 90 \times 10^8$	meter (SI)
	1 centimetre (cgs) = $3.240\ 78 \times 10^{-19}$	parsec
	1 centimetre (cgs) = $6.684\ 56 \times 10^{-14}$	astron. unit (AU)
	1 meter (SI) = $3.240\ 78 \times 10^{-17}$	parsec
	1 meter (SI) = $6.684\ 54 \times 10^{-12}$	astron. unit (AU)
	1 inch (Eng) = $2.540\ 00 \times 10^{-2}$	meter (SI)
Mass		
	1 kilogram (SI) = $1.000\ 00 \times 10^3$	gram (cgs)
	1 at. mass unit (amu) = $1.660\ 54 \times 10^{-24}$	gram (cgs)
	1 at. mass unit (amu) = $1.660\ 54 \times 10^{-27}$	kilogram (SI)
	1 solar mass = $1.989\ 10 \times 10^{33}$	gram (cgs)
	1 solar mass = $1.989\ 10 \times 10^{30}$	kilogram (SI)
	1 gram (cgs) = $6.022\ 14 \times 10^{23}$	at. mass unit (amu)
	1 gram (cgs) = $5.027\ 40 \times 10^{-34}$	solar mass
	1 kilogram (SI) = $6.022\ 14 \times 10^{26}$	at. mass unit (amu)
	1 kilogram (SI) = $5.027\ 40 \times 10^{-31}$	solar mass
	1 kilogram (SI) = $2.204\ 62$	pound (avdp.)
	1 kilogram (SI) = $3.527\ 40 \times 10$	ounce (avdp.)

(*Continued*)

---

[i] Adapted from M.V. Zombeck [1].

(Cont.)

Parameter	Unit	Unit
	1 pound (avdp.) = $4.535\,92 \times 10^{-1}$	kilogram (SI)
	1 pound (avdp.) = $1.600\,00 \times 10$	ounce (avdp.)
	1 ounce (avdp.) = $2.834\,95 \times 10$	gram (cgs)
	1 gram (cgs) = $3.527\,40 \times 10^{-2}$	ounce (avdp.)
Energy		
	1 joule (SI) = $1.000\,00 \times 10^{7}$	erg (cgs)
	1 joule (SI) = $6.241\,51 \times 10^{18}$	electron volt (eV)
	1 erg (cgs) = $1.000\,00 \times 10^{-7}$	joule (SI)
	1 erg (cgs) = $6.241\,51 \times 10^{11}$	electron volt
	1 electron volt = $1.602\,18 \times 10^{-12}$	erg (cgs)
	1 amu $\times c^2$ = $9.314\,95 \times 10^{8}$	electron volt
	1 gm (cgs) $\times c^2$ = $5.609\,59 \times 10^{32}$	electron volt
	1 calorie = $4.184\,00 \times 10$	joule (SI)
Force		
	1 newton (SI) = $1.000\,00 \times 10^{5}$	dyne (cgs)
	1 dyne (cgs) = $1.000\,00 \times 10^{-5}$	newton (SI)
Pressure		
	1 pascal (SI) = $1.000\,00 \times 10$	newton m^{-2} (SI)
	1 pascal (SI) = 1.019716×10^{-7}	kg force mm^{-2}
	1 Knoop = 9.80665×10^{6}	pascal
	1 bar = $1.000\,00 \times 10^{6}$	dyne cm^{-2} (cgs)
	1 bar = $9.869\,23 \times 10^{-1}$	atmosphere
	1 torr = $1.333\,22 \times 10^{-3}$	bar
Power		
	1 watt (SI) = $1.000\,00 \times 10^{7}$	erg s^{-1} (cgs)
	1 horsepower = $7.457\,00 \times 10^{2}$	watt (SI)
	1 BTU s^{-1} (Eng) = $1.055\,80 \times 10^{3}$	watt (SI)
Time		
	1 second (SI) = 1	second (cgs)
	1 minute = $6.000\,00 \times 10$	second
	1 hour = $3.600\,00 \times 10^{3}$	second
	day = $8.640\,00 \times 10^{4}$	second
	1 tropical year = $3.155\,69 \times 10^{7}$	second
	1 tropical year = $3.652\,42 \times 10^{2}$	day
	1 second = $3.168\,88 \times 10^{-8}$	tropical year
	1 sidereal second = $9.972\,70 \times 10^{-1}$	second
	1 sidereal year = $3.652\,56 \times 10^{2}$	day
Temperature		
	T Kelvin = T $-$ 273.15	Celsius
	T Kelvin = $(9/5) \times$ (T $-$ 273.15) + 32	Fahrenheit

(*Continued*)

(Cont.)

Parameter	Unit	Unit
	T Celsius = T + 273.15	Kelvin
	T Fahrenheit = (5/9) × (T – 32) + 273.15	Kelvin
	T Celsius = (9/5) × T + 32	Fahrenheit
	T Fahrenheit = (5/9) × (T – 32)	Celsius
	1 electron volt : $1.160\ 48 \times 10^4$	Kelvin
	1 Kelvin = $8.617\ 12 \times 10^{-5}$	electron volt
Electricity and magnetism		
	1 coulomb = $2.997\ 92 \times 10^9$	statcoulomb
	1 coulomb m^{-3} = $2.997\ 92 \times 10^3$	statcoul cm^{-3}
	1 ampere (coul s-1) = $2.997\ 92 \times 10^9$	statampere
	1 ampere m^{-2} = $2.997\ 92 \times 10^5$	statamp cm^{-2}
	1 volt m-1 = $3.335\ 65 \times 10^{-5}$	statvolt cm^{-1}
	1 volt = $3.335\ 65 \times 10^{-3}$	statvolt
	1 ohm = $1.112\ 65 \times 10^{-12}$	s cm^{-1}
	1 ohm m = $1.112\ 65 \times 10^{-10}$	s
	1 siemens, mho = $8.987\ 52 \times 10^{11}$	cm s^{-1}
	1 mho m^{-1} = $8.987\ 52 \times 10^9$	s^{-1}
	1 farad = $8.987\ 52 \times 10^{11}$	cm
	1 weber = $1.000\ 00 \times 10^8$	gauss cm^2 (maxwell)
	1 tesla = $1.000\ 00 \times 10^4$	gauss
	1 ampere-turn m^{-1} = $1.256\ 64 \times 10^{-2}$	oersted
	1 henry = $1.112\ 65 \times 10^{-12}$	$s^2\ cm^{-1}$
Miscellaneous		
	1 curie (SI) = $3.700\ 00 \times 10^{10}$	disintegrations s^{-1}
	1 rayleigh = $7.957\ 75 \times 10^4$	ph $cm^{-2}\ s^{-1}\ sr^{-1}$
	1 fu or jansky = $1.000\ 00 \times 10^{-26}$	watt $m^{-2}\ Hz^{-1}$
	1 jansky = $1.000\ 00 \times 10^{-8}$	erg $cm^{-2}\ s^{-1}\ Hz^{-1}$
	1 jansky = $2.420\ 00 \times 10^{-9}$	erg $cm^{-2}\ s^{-1}\ keV^{-1}$
	1 jansky = $1.509\ 00 \times 10$	keV $cm^{-2}\ s^{-1}\ keV^{-1}$
	1 eV = $1.239\ 85 \times 10^4$	Ångstrom
	1 eV = $2.417\ 97 \times 10^{14}$	Hz
	1 Ångstrom = $1.239\ 85 \times 10^4$	eV
	1 arcsec = $4.848\ 14 \times 10^{-6}$	radian
	1 arcmin = $2.908\ 88 \times 10^{-4}$	radian
	1 degree = $1.745\ 33 \times 10^{-2}$	radian
	1 $arcsec^2$ = $2.350\ 40 \times 10^{-11}$	steradian
	1 $arcmin^2$ = $8.461\ 70 \times 10^{-8}$	steradian
	1 deg^2 = $3.046\ 20 \times 10^{-4}$	steradian

REFERENCE

[1] M.V. Zombeck, "*Handbook of Space Astronomy & Astrophysics*", 2nd Edition, Cambridge University Press, Cambridge, UK (1990). ISBN-13: 9780521782425: ISBN-10: 0521782422.

Appendix D: Periodic Table of the Elements

Legend: ATOMIC NUMBER — SYMBOL — RELATIVE ATOMIC MASS (1) — ELEMENT NAME. GROUP NUMBERS: IUPAC RECOMMENDATION (1985), CHEMICAL ABSTRACT SERVICE (1986).

Group 1 (IA)	2 (IIA)	3 (IIIB)	4 (IVB)	5 (VB)	6 (VIB)	7 (VIIB)	8 (VIIIB)	9	10	11 (IB)	12 (IIB)	13 (IIIA)	14 (IVA)	15 (VA)	16 (VIA)	17 (VIIA)	18 (VIIIA)
1 H 1.008 HYDROGEN																	2 He 4.0026 HELIUM
3 Li 6.94 LITHIUM	4 Be 9.0122 BERYLLIUM											5 B 10.81 BORON	6 C 12.011 CARBON	7 N 14.007 NITROGEN	8 O 15.999 OXYGEN	9 F 18.998 FLUORINE	10 Ne 20.180 NEON
11 Na 22.990 SODIUM	12 Mg 24.305 MAGNESIUM											13 Al 26.982 ALUMINIUM	14 Si 28.085 SILICON	15 P 30.974 PHOSPHORUS	16 S 32.06 SULPHUR	17 Cl 35.45 CHLORINE	18 Ar 39.948 ARGON
19 K 39.098 POTASSIUM	20 Ca 40.078 CALCIUM	21 Sc 44.956 SCANDIUM	22 Ti 47.867 TITANIUM	23 V 50.942 VANADIUM	24 Cr 51.996 CHROMIUM	25 Mn 54.938 MANGANESE	26 Fe 55.845 IRON	27 Co 58.933 COBALT	28 Ni 58.693 NICKEL	29 Cu 63.546 COPPER	30 Zn 65.38 ZINC	31 Ga 69.723 GALLIUM	32 Ge 72.64 GERMANIUM	33 As 74.922 ARSENIC	34 Se 78.971 SELENIUM	35 Br 79.904 BROMINE	36 Kr 83.798 KRYPTON
37 Rb 85.468 RUBIDIUM	38 Sr 87.62 STRONTIUM	39 Y 88.906 YTTRIUM	40 Zr 91.224 ZIRCONIUM	41 Nb 92.906 NIOBIUM	42 Mo 95.95 MOLYBDENUM	43 Tc (98) TECHNETIUM	44 Ru 101.07 RUTHENIUM	45 Rh 102.91 RHODIUM	46 Pd 106.42 PALLADIUM	47 Ag 107.87 SILVER	48 Cd 112.41 CADMIUM	49 In 114.82 INDIUM	50 Sn 118.71 TIN	51 Sb 121.76 ANTIMONY	52 Te 127.60 TELLURIUM	53 I 126.90 IODINE	54 Xe 131.29 XENON
55 Cs 132.91 CAESIUM	56 Ba 137.33 BARIUM	57-71 La-Lu Lanthanide	72 Hf 178.49 HAFNIUM	73 Ta 180.95 TANTALUM	74 W 183.84 TUNGSTEN	75 Re 186.21 RHENIUM	76 Os 190.23 OSMIUM	77 Ir 192.22 IRIDIUM	78 Pt 195.08 PLATINUM	79 Au 196.97 GOLD	80 Hg 200.59 MERCURY	81 Tl 204.38 THALLIUM	82 Pb 207.2 LEAD	83 Bi 208.98 BISMUTH	84 Po (209) POLONIUM	85 At (210) ASTATINE	86 Rn (222) RADON
87 Fr (223) FRANCIUM	88 Ra (226) RADIUM	89-103 Ac-Lr Actinide	104 Rf (267) RUTHERFORDIUM	105 Db (268) DUBNIUM	106 Sg (271) SEABORGIUM	107 Bh (272) BOHRIUM	108 Hs (277) HASSIUM	109 Mt (276) MEITNERIUM	110 Ds (281) DARMSTADTIUM	111 Rg (280) ROENTGENIUM	112 Cn (285) COPERNICIUM	113 Nh (285) NIHONIUM	114 Fl (287) FLEROVIUM	115 Mc (289) MOSCOVIUM	116 Lv (291) LIVERMORIUM	117 Ts (294) TENNESSINE	118 Og (294) OGANESSON

LANTHANIDE

57 La 138.91 LANTHANUM	58 Ce 140.12 CERIUM	59 Pr 140.91 PRASEODYMIUM	60 Nd 144.24 NEODYMIUM	61 Pm (145) PROMETHIUM	62 Sm 150.36 SAMARIUM	63 Eu 151.96 EUROPIUM	64 Gd 157.25 GADOLINIUM	65 Tb 158.93 TERBIUM	66 Dy 162.50 DYSPROSIUM	67 Ho 164.93 HOLMIUM	68 Er 167.26 ERBIUM	69 Tm 168.93 THULIUM	70 Yb 173.05 YTTERBIUM	71 Lu 174.97 LUTETIUM

ACTINIDE

89 Ac (227) ACTINIUM	90 Th 232.04 THORIUM	91 Pa 231.04 PROTACTINIUM	92 U 238.03 URANIUM	93 Np (237) NEPTUNIUM	94 Pu (244) PLUTONIUM	95 Am (243) AMERICIUM	96 Cm (247) CURIUM	97 Bk (247) BERKELIUM	98 Cf (251) CALIFORNIUM	99 Es (252) EINSTEINIUM	100 Fm (257) FERMIUM	101 Md (258) MENDELEVIUM	102 No (259) NOBELIUM	103 Lr (262) LAWRENCIUM

(1) Atomic weights of the elements 2013, Pure Appl. Chem. **88**, 265-291 (2016)

www.periodni.com

Image courtesy: E. Generalic, https://www.periodni.com/images.html

Copyright © 2017 Eni Generalic

Appendix E: Properties of the Elements

The following table lists the atomic weights, densities, melting and boiling points, first ionization potentials, and specific heats of the elements. The table was taken from the *X-ray Data Booklet* [1]. Atomic weights apply to elements as they exist naturally on earth or, in the cases of radium, actinium, thorium, protactinium, and neptunium, to the isotopes with the longest half-lives. Values in parentheses are the mass numbers for the longest-lived isotopes. Specific heats are given for the elements at 25°C. Densities for solids and liquids are given at 20°C, unless otherwise indicated by a superscript temperature (in °C); densities for the gaseous elements are for the liquids at their boiling points.

Z	Element	Atomic weight	Density (g/cm³)	Melting point (°C)	Boiling point (°C)	Ionization potential (eV)	Specific heat (cal/g•k)
1	Hydrogen	1.00794	0.0708	−259.14	−252.87	13.598	3.41
2	Helium	4.00260	1.122	−272.2	−268.934	24.587	1.24
3	Lithium	6.941	0.533	180.54	1342	5.392	0.834
4	Beryllium	9.01218	1.845	1278	2970	9.322	0.436
5	Boron	10.81	2.34	2079	2550^c	8.298	0.245
6	Carbon	12.011	2.26	3550	3367^c	11.260	0.170
7	Nitrogen	14.0067	0.81	−209.86	−195.8	14.534	0.249
8	Oxygen	15.9994	1.14	−218.4	−182.962	13.618	0.219
9	Fluorine	18.998403	1.108	−219.62	−188.14	17.422	0.197
10	Neon	20.179	1.207	−248.67	−246.048	21.564	0.246
11	Sodium	22.98977	0.969	97.81	882.9	5.139	0.292
12	Magnesium	24.305	1.735	648.8	1090	7.646	0.245
13	Aluminum	26.98154	2.6941	660.37	2467	5.986	0.215
14	Silicon	28.0855	2.32^{25}	1410	2355	8.151	0.168
15	Phosphorus	30.97376	1.82	44.1	280	10.486	0.181
16	Sulfur	32.06	2.07	112.8	444.674	10.360	0.175
17	Chlorine	35.453	1.56	−100.98	−34.6	12.967	0.114
18	Argon	39.948	1.40	−189.2	−185.7	15.759	0.124
19	Potassium	39.0983	0.860	63.25	760	4.341	0.180
20	Calcium	40.08	1.55	839	1484	6.113	0.155
21	Scandium	44.9559	2.980^{25}	1541	2831	6.54	0.1173
22	Titanium	47.88	4.53	1660	3287	6.82	0.1248
23	Vanadium	50.9415	6.10$^{18.7}$	1890	3380	6.74	0.116
24	Chromium	51.996	7.18	1857	2672	6.766	0.107
25	Manganese	54.9380	7.43	1244	1962	7.435	0.114
26	Iron	55.847	7.860	1535	2750	7.870	0.1075
27	Cobalt	58.9332	8.9	1495	2870	7.86	0.107
28	Nickel	58.69	8.876^{25}	1453	2732	7.635	0.1061
29	Copper	63.546	8.94	1083.4	2567	7.726	0.0924
30	Zinc	65.38	7.112^{25}	419.58	907	9.394	0.0922
31	Gallium	69.72	5.877$^{29.6}$	29.78	2403	5.999	0.088
32	Germanium	72.59	5.307^{25}	937.4	2830	7.899	0.077
33	Arsenic	74.9216	5.72	817$^{28\ atm}$	613^c	9.81	0.0785
34	Selenium	78.96	4.78	217	684.9	9.752	0.0767

(Continued)

(Cont.)

Z	Element	Atomic weight	Density (g/cm³)	Melting point (°C)	Boiling point (°C)	Ionization potential (eV)	Specific heat (cal/g•k)
35	Bromine	79.904	3.11	−7.2	58.78	11.814	0.0537
36	Krypton	83.80	2.6	−156.6	−152.30	13.999	0.059
37	Rubidium	85.4678	1.529	38.89	686	4.177	0.0860
38	Strontium	87.62	2.54	769	1384	5.695	0.0719
39	Yttrium	88.9059	4.456^{25}	1522	3338	6.38	0.0713
40	Zirconium	91.22	6.494	1852	4377	6.84	0.0660
41	Niobium	92.9064	8.55	2468	4742	6.88	0.0663
42	Molybdenum	95.94	10.20	2617	4612	7.099	0.0597
43	Technetium	(98)	11.48[a]	2172	4877	7.28	0.058
44	Ruthenium	101.07	12.39	2310	3900	7.37	0.0569
45	Rhodium	102.9055	12.39	1966	3727	7.46	0.0580
46	Palladium	106.42	12.00	1554	2970	8.34	0.0583
47	Silver	107.8682	10.48	961.93	2212	7.576	0.0562
48	Cadmium	112.41	8.63	320.9	765	8.993	0.0552
49	Indium	114.82	7.30	156.61	2080	5.786	0.0556
50	Tin	118.69	7.30	231.9681	2270	7.344	0.0519
51	Antimony	121.75	6.679	630.74	1950	8.641	0.0495
52	Tellurium	127.60	6.23	449.5	989.8	9.009	0.0481
53	Iodine	126.9045	4.92	113.5	184.35	10.451	0.102
54	Xenon	131.29	3.52	−111.9	−107.1	12.130	0.0378
55	Cesium	132.9054	1.870	28.40	669.3	3.894	0.0575
56	Barium	137.33	3.5	725	1640	5.212	0.0362
57	Lanthanum	138.9055	6.127^{25}	921	3457	5.577	0.0479
58	Cerium	140.12	6.637^{25}	799	3426	5.47	0.0459
59	Praseodymium	140.9077	6.761	931	3512	5.42	0.0467
60	Neodymium	144.24	6.994	1021	3068	5.49	0.0453
61	Promethium	(145)	7.20^{25}	1168	2460	5.55	0.0442
62	Samarium	150.36	7.51	1077	1791	5.63	0.0469
63	Europium	151.96	5.228^{25}	822	1597	5.67	0.0326
64	Gadolinium	157.25	7.8772^{25}	1313	3266	6.14	0.056
65	Terbium	158.9254	8.214	1356	3123	5.85	0.0435
66	Dysprosium	162.50	8.525^{25}	1412	2562	5.93	0.0414
67	Holmium	164.9304	8.769^{25}	1474	2695	6.02	0.0394
68	Erbium	167.26	9.039^{25}	159	2863	6.10	0.0401
69	Thulium	168.9342	9.294^{25}	1545	1947	6.18	0.0382
70	Ytterbium	173.04	6.953	819	1194	6.254	0.0287
71	Lutetium	174.967	9.811^{25}	1663	3395	5.426	0.0285
72	Hafnium	178.49	13.29	2227	4602	7.0	0.028
73	Tantalum	180.9479	16.624	2996	5425	7.98	0.0334
74	Tungsten	183.85	19.3	3410	5660	7.98	0.0322
75	Rhenium	186.207	20.98	3180	5627[b]	7.88	0.0330
76	Osmium	190.2	22.53	3045	5027	8.7	0.0310
77	Iridium	192.22	22.39^{17}	2410	4130	9.1	0.0312
78	Platinum	195.08	21.41	1772	3827	9.0	0.0317
79	Gold	196.9665	18.85	1064.43	3080	9.225	0.0308
80	Mercury	200.59	13.522	−38.842	356.58	10.437	0.0333
81	Thallium	204.383	11.83	303.5	1457	6.108	0.0307
82	Lead	207.2	11.33	327.502	1740	7.416	0.0305

(Continued)

(Cont.)

Z	Element	Atomic weight	Density (g/cm³)	Melting point (°C)	Boiling point (°C)	Ionization potential (eV)	Specific heat (cal/g•k)
83	Bismuth	208.9804	9.730	271.3	1560	7.289	0.0238
84	Polonium	(209)	9.3	254	962	8.42	0.030
85	Astatine	(210)	–	302	337[b]	–	–
86	Radon	(222)	4.4	−71	−61.8	10.748	0.0224
87	Francium	(223)	–	27	677	–	–
88	Radium	226.0254	5	700	1140	5.279	0.0288
89	Actinium	227.0278	10.05[a]	1050	3200[b]	6.9	–
90	Thorium	232.0381	11.70	1750	4790	–	0.0281
91	Protactinium	231.0359	15.34[a]	<1600	–	–	0.029
92	Uranium	238.0289	18.92	1132.3	3818	–	0.0278
93	Neptunium	237.0482	20.21	640	3902[b]	–	–
94	Plutonium	(244)	19.80	641	3232	5.8	–
95	Americium	(243)	13.64	994	2607	6.0	
96	Curium	(247)	13.49[a]	1340	–	–	–
97	Berkelium	(247)	14[b]	–	–	–	–
98	Californium	(251)	–	–	–	–	–
99	Einsteinium	(252)	–	–	–	–	–
100	Fermium	(257)	–	–	–	–	–
101	Mendelevium	(258)	–	–	–	–	–
102	Nobelium	(259)	–	–	–	–	–
103	Lawrencium	(260)	–	–	–	–	–

[a] Calculated
[b] Estimated
[c] Sublimes

REFERENCE

[1] D. Vaughan, ed., "*X-Ray Data Booklet*", Lawrence Berkeley Laboratory PUB-940, University of California, Berkeley, CA (1986).

Appendix F: General Properties of Semiconducting Materials

The data compiled in these tables have been gathered from many sources. Particularly useful compilations can be found in references [1–10]. Note that published values can vary widely, particularly for transport parameters. Indeed, no values exist for many materials and so the relevant entries have been left blank. Where wide discrepancies in published data exist, average values have been used or a judicious choice has been made. Needless to say, values quoted in these tables should be used with caution.

F.1 Table Headings, Nomenclature and Explanation

Material: Semiconductors are denoted by their chemical symbols; so for example the elemental semiconductors can be described by *C*, where *C* is an element belonging to Group IV of the Periodic Table – *e.g.*, Si. Similarly, we describe binary compound semiconductors by *AB*, where *A* and *B* are elements derived from Groups I and VII (*e.g.*, AgCl), II and VI (*e.g.*, CdTe), III and V (*e.g.*, GaAs) of the Periodic Table. We consider only the stable solid form. Where more than one form exists, we consider its thermodynamically favoured allotrope at ambient temperature. Alloys are denoted by $A_xB_{(1-x)}$ where $0<x<1$ is the concentration of element *A* atoms forming the alloy.

    Crystal structure: The structure of a crystal is usually defined in terms of lattice points, which mark the positions of the atoms forming the basic unit cell of the crystal. Cullity [11] defines a lattice point as *"An array of points in space so arranged that each point has (statistically) identical surroundings"*. The word "statistically" is introduced to allow for solid solutions, where fractional atoms would otherwise result. The following abbreviations have been used. Amorphous material is denoted by **a**, crystalline material by **c**. For cubic lattices: **dia** denotes a diamond type structure (*i.e.*, two intersecting face centered cubic lattices), **NaCl** denotes a sodium chloride type lattice structure – also referred to as "rock-salt" structure (face centered cubic); **CsCl** a caesium chloride type structure (body centered cubic) and **ZB** for a zincblende structure (diamond structure for binary compounds where the two atom types form two interpenetrating face-centered cubic lattices). For hexagonal lattices – **W** denotes a wurtzite structure (an AB binary system of alternating tetrahedrally coordinated atoms stacked in an ABABAB pattern) and **H**, a pure hexagonal structure (containing six atoms per unit cell) – essentially the hexagonal analog of zincblende. **ortho** refers to orthorhombic, **rhomb** to rhombohedral, **trig** to trigonal **tetra** to tetragonal and **layered** to a layered structure (*i.e.*, cubic or hexagonal structured layers separated by van der Waal forces). Most elemental semiconductors crystallize in a diamond structure. Most compound semiconductors crystallize in either the **ZB** or **W** forms (see Fig. F.1).

    Pearson Symbols, Space Groups: Parameters which define a crystals structure and its underlying symmetry. While the Bravais lattice designations identify crystal types they cannot uniquely identify particular crystals. There are several

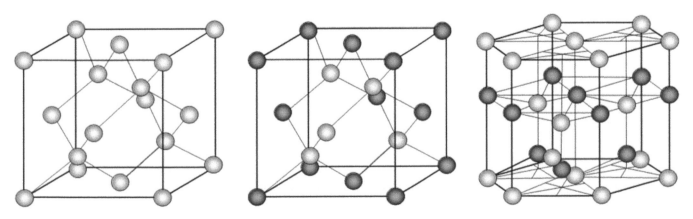

FIGURE F.1 The most common semiconductor lattice structures. Left to right – diamond, zincblende and wurtzite.

systems for classifying crystal structure, most of which are based on assigning a specific letter to each of the Bravais lattices. However, with the exception of the Pearson classification scheme, none are self-defining and even the Pearson scheme does not uniquely define a particular crystal structure. The Pearson notation [12] is a simple and convenient scheme based on the so-called Pearson symbols, of which there are three. The first symbol is a lower case letter designating the crystal type (*i.e.*, a = triclinic, m = monoclinic, o = othorhombic, t = tetragonal, h = hexagonal and rhombohedral, c = cubic). The second symbol is a capital letter which designates the lattice centering (*i.e.*, P = primitive, C = side face centered, F = all face centered, I = body centered, R = rhombohedral). Thus, the 14 unique Bravais lattices can be characterized by a two-letter mnemonic as summarized in Table 3.1. The third Pearson symbol is a number which designates the number of atoms in the conventional unit cell. Therefore, a diamond structure which is cubic, face centered and has 8 atoms in its unit cell is represented by *cF8*. To use the Pearson system effectively, however, we need to know a structure, or prototype, which is the classic example of that particular structure. For example, both GaAs and MgSe have a Pearson designation cF8, but GaAs has a classic zincblende *i.e.*, "ZnS" structure and MgSe a "NaCl" type structure. While both structures are formed by two interpenetrating face-centered cubic lattices, they differ in how the two lattices are positioned relative to one another.

While Pearson symbols categorise crystal structures into particular patterns and are conceptually simple and easy to use, not every structure is uniquely defined. The space group notation, also known as the International or Hermann-Maguin system [13], is a mathematical description of the symmetry inherent in a crystals structure and is also represented by a set of numbers and symbols. The space groups in three dimensions are made from combinations of the 32 crystallographic point groups (given in parentheses in the tables) with the 14 Bravais lattices which belong to one of the 7 basic crystal systems. This results in a space group being some combination of the translational symmetry of a unit cell including lattice centering and the point group symmetry operations of reflection, rotation and improper rotation. The combination of all these symmetry operations results in a total of 230 unique space groups describing all possible crystal symmetries. The relationship between the basic crystal systems, the Bravais lattices and the point and space groups is illustrated in Fig. F.2. The International Union of Crystallography publishes comprehensive tables [14] of all space groups and assigns each a unique number.

Lattice constant: These are the lattice parameters in units of nm, generally denoted by the letters *a*, *b* and *c*. For cubic crystal structures, the unit cell sizes are equal and we only refer to *a*. Similarly, for hexagonal crystal structures, the *a* and *b* lengths are equal and so we only give the a and c values. Orthorhombic structures are rectangular prisms so that the sides

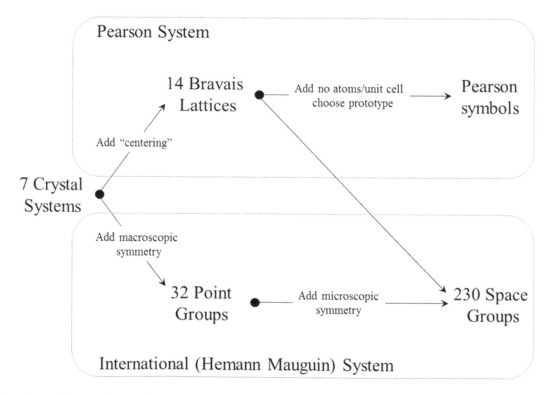

FIGURE F.2 The inter-relationship between the basic crystal systems, the Bravais lattices, Point groups and the Pearson Symbols and Space groups. The macroscopic symmetry elements are those operations (*e.g.*, reflection and translation) which take place over unit cell dimensions, whereas the microscopic symmetry elements add small translations (less than a unit cell vector) to the macroscopic symmetry operations. A point group is a representation of the ways that the macroscopic symmetry elements (operations) can be self-consistently arranged around a single geometric point. There are 32 unique ways in which this can be done.

have unequal lengths. Therefore, the *a*, *b* and *c* lengths need to be specified. As expected, the lattice constant tends to be smaller when a compound consists of smaller diameter atoms than when it consists of larger atoms.

Atomic No(s): The atomic number(s) of the constituent element(s). The atomic number corresponds to the number of protons in the nucleus of an atom of that element.

Molecular mass (molecular weight) is the mass of one molecule of a substance and is expressed in unified atomic mass units (amu). One amu is equal to 1/12 the mass of one atom of carbon-12. For the elemental semiconductors, the molecular weight is given by the atomic weight of the constituent atom. For $A^N B^{8-N}$ (N $\neq$ 4) compounds, the molecular weight is given by the sum of the atomic weights of atoms *A* and *B*.

Density: The density (mass per unit volume) of the material in units of g cm^{-3}. Data given here refer to the solid. Density is temperature dependent and different allotropes possess different densities. Values are given for the thermodynamically most favoured allotrope at ambient temperature.

Melting point: Perhaps the most important thermophysical parameter is the temperature at which a material changes from the solid to the liquid state and at which both phases exist in equilibrium. Values are given in units of K at normal pressure unless otherwise stated. Materials which sublime, that is, transit directly from the solid to the gaseous phase, are noted.

$\varepsilon_r(0)$ is the static dielectric constant, or more correctly, the relative static permittivity of a material which is a measure of its ability to concentrate electrostatic lines of flux. It is defined as the ratio of the amount of stored electrical energy in a material when a potential is applied, relative to that stored in a vacuum. In reality, if a material with a high relative permittivity is placed in an electric field, the magnitude of that field will be measurably reduced within the volume of the dielectric, by an amount proportional to the relative dielectric constant. Since a high dielectric constant is generally associated with a high dielectric strength, dielectric breakdown will occur at higher electric fields. This can have some advantage in detection systems, in that higher biases can be applied to these materials, leading to better charge collection. However, a highly dielectric medium can also lead to polarization effects, which introduce a time dependence in their detection properties. Numerically, the relative static permittivity is the same as the relative permittivity evaluated for a frequency of zero, $\varepsilon(0)$. The static dielectric constant is important in that it forms the constant of proportionality between the potential and the charge density in Poisson's equation. It is a key parameter for several scattering mechanisms, while both the zero frequency and high frequency dielectric constants are employed in the description of polar optical scattering. Note for zincblende structures, ε_r values are the same along each crystallographic axis and only one value is quoted, while for wurtzite materials, the values along the **a** and **c** axes are different and so two values are usually given: $\varepsilon(0)$ perpendicular ($\perp$) to the **c**-axis and $\varepsilon(0)$ parallel (II) to the **c**-axis. In some cases, however, only the **c**–axis value is quoted for wurtzite materials.

Ionicity f_i is an important physical concept and generally exists only in chemical bonds between different atoms. It allows us to form a quantitative description of chemical bonding and in turn structural phase stability. In essence, the ionicity of a semiconductor is a measure of the partial charges created due to the asymmetric distribution of electrons in chemical bonds. This charge is a property only of zones within the distribution and not the assemblage as a whole. For example, when an electrically neutral atom bonds chemically to another neutral atom that is more electronegative, its electrons are partially drawn away. This leaves the region about that atom's nucleus with a partial positive charge, which in turn creates a partial negative charge on the atom to which it is bonded. The ionicity of a bond can be defined as the fraction of ionic or heteropolar part of the bond, f_i, compared with the fraction of covalent or homopolar bonding, f_h. By definition, $f_i + f_h = 1$. Values are quoted for the Phillips ionicity scale [15] unless otherwise stated.[1] For the elemental semiconductors, such as Si, bonding is entirely covalent and $f_i = 0$, whereas for some of the alkali halides the bond is more than 90% ionic and $f_i \rightarrow 1$. Generally, compounds with ionicities less than 0.1 are classified as covalent, compounds with ionicities in the range 0.2–0.7 are considered partially ionic and, compounds with ionicities greater than 0.7 are considered ionic. For ionic semiconductors, electrical conductivity is due primarily to the movement of ions rather than electrons and holes.

Phillips [15] demonstrated that structural properties such as the cohesive energy and heats of formation depend linearly on f_i. Structurally, covalent bonding gives rise to structures with small coordination numbers while ionic bonding appears in high symmetry structures. A value of $f_i = 0.785$ is found to mark the transition between tetrahedral (four-fold) and octahedral (six-fold) coordination, meaning that covalent bonding is not sufficiently strong to stabilize tetrahedrally bonded structures [17] above this value.

KH denotes the hardness of the material and is defined in terms of the Knoop micro-hardness scale [18] named after Frederick Knoop of the U.S. National Bureau of Standards. The Knoop hardness (KH) number of a substance is determined by measuring the area of a "diamond-shaped" indentation made by a particular set of facets of a diamond pressed into the substance at a set pressure for a set time. Specifically, it is given by the ratio of the load applied to the indenter *P* to the unrecovered projected area *A* illustrated in Fig. F.3.

---

[1] In some cases the Tubbs values are given. The Tubbs ionicity scale [16] was devised to order the many compounds intermediate between the alkali halides and the group III-V semiconductors for which conventional ionicity definitions based on the heats of formation (*e.g.*, [15]) are of limited value.

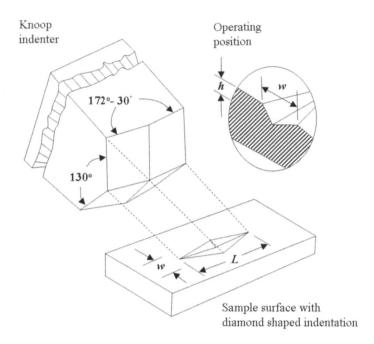

FIGURE F.3 Diagram showing the mechanics of the Knoop indenter and hardness measurement. *L* is the long diagonal of the diamond-shaped indentation, measured in microns (from [19]).

$$KH = P/A = P/CL^2, \qquad\qquad (F.1)$$

where L = measured length of the long diagonal of indentation in mm and C is a constant relating the shape of the indenter to L (= 0.07028). KH values are usually given in units of kgf·mm^{-2}, although pascals are also sometimes used (1 kgf·mm^{-2} = 9.80665 Mpa). Typical KH values lie in the range from 10 (refrigerated butter) to 10,000 (diamond) and vary with temperature – materials are softer at higher temperatures (see [20]). For example, hardness values for GaAs decrease by a factor of ~25 between room temperature and 300°C. Other measurement protocols are the Brinell harness scale [21], Vickers hardness test [22] and the Mohs[2] scale [23].

Note that Brinell and Vickers hardness measurements can be accurately related, but only over a limited range of values due to the different indenter morphologies. For example, the Brinell 3000 kgf, 10 mm steel ball test and the Vickers diamond pyramid test are essentially the same for values between 100 and 500 kg mm^{-2}. The Knoop and Vickers scales are linearly related ($HV \sim 0.2\ KH$); however, the Mohs scale is not simply related to any other test, since it is benchmarked on a set of standard minerals. The scale runs from 1 (talc), the softest, to 10 (diamond), the hardest, such that any mineral assigned a greater value can scratched a mineral of a lesser value. As a consequence, the Mohs scale is non-linear, since the procedure for assigning values cannot easily cope with non-integer values. In Fig. F.4 we show a conversion for the reference minerals used to define the Mohs scale which can be used to estimate KH values when only Mohs values are available.[3]

Type: Denotes whether the semiconductor is a direct (D) or indirect (I) bandgap material as illustrated in Fig. F.5. In a D bandgap material, the minimum energy of the conduction band lies directly above the maximum energy of the valence band in momentum (k) space, such that the electrons at the conduction-band minimum can combine directly with holes at the valence band maximum, whilst conserving momentum. In an I bandgap material, the minimum energy in the conduction band is shifted in k space relative to the valence band. An electron must therefore undergo a significant change in momentum to move from the bottom of the conduction band to the top of the valence band which can only be achieved with the mediation of a third body, such as a phonon or crystallographic defect. Examples of direct bandgap materials are GaAs, InP and CdTe; examples of indirect bandgap materials are Si, Ge and GaP. SM signifies that the material is a semi-metal. Semi-metals have indirect bandgaps, but unlike other indirect materials, the top of the valence band is at a higher energy than the bottom of the conduction band, and consequently bandgap energies are frequently negative.

Bandgap: The bandgap energy in units of eV. The bandgap is defined as the energy difference between the top of the valence band and the bottom of the conduction band (see Fig. F.5). The bandgap energy is a weak function of temperature. Values are given at 300K unless otherwise stated.

---

[2] Named after the German mineralogist Friedrich Mohs (b. 1773–d. 1839) who devised the scale in 1812.

[3] Generally given for soft materials for which the Knoop indenter may not leave a clear and undistorted impression.

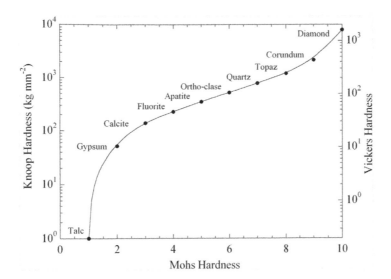

FIGURE F.4 A plot of the Knoop hardness scale versus the Mohs scale for the standard minerals used to define the scale. The right hand ordinate gives the corresponding values on the Vickers hardness (HV) scale. Note: by convention HV is usually quoted without units.

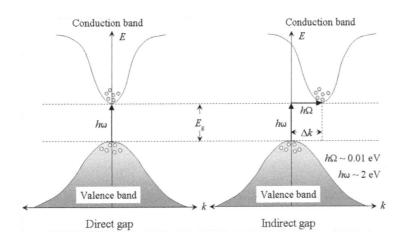

FIGURE F.5 Illustration of direct (left) and indirect (right) bandgaps in which we show the band structure in energy and momentum space, E,k. Here, E_g is the bandgap energy. Circles (o) indicate holes in the valence bands, and filled circles ($\cdot$) indicate electrons in the conduction bands. The important difference is that for a direct bandgap material, an electron can transit from the lowest potential in the conduction band to the highest potential in the valence band without a change in momentum, Δk, whereas for an indirect bandgap material it cannot without the mediation of a third body (*e.g.*, phonon) to conserve momentum. For photon induced excitation, almost all of the energy is provided by the photon ($\hbar\omega$), whereas all of the momentum ($\hbar\Omega$) is provide by the phonon.

Pair energy: The energy consumed to create an electron hole pair in units of eV. The pair energy is roughly 3 times larger than the bandgap energy – the difference being due to the energy lost to phonons and plasmons when crossing the bandgap.

Resistivity: The electrical resistivity, or specific resistance, ρ, is the resistance between the opposite faces of a metre cube of a material. It is a property of a semiconductor that depends on the free electron and hole densities and their respective mobilities. Resistivity has units of Ω-cm and is the reciprocal of conductivity. It is strongly dependent on temperature. For hexagonal or layered materials, resistivity can also depend upon the direction along which the measurement is made.

Electron mobility, μ_e: Defined as the ratio of the electron velocity (cm s^{-1}) to the electric field (V cm^{-1}) through which it is moving. Mobility has units of cm^{2}V^{-1}s^{-1} and is usually determined using one of two methods – Hall or drift. Hall mobilities, μ_H, are derived from Hall effect measurements [24], and drift or conductivity mobilities, μ_d, are derived using the Haynes-Shockley technique [25]. It is important to note that they can be different. The ratio μ_H/μ_d is usually close to unity for direct bandgap semiconductors but can be greater than unity for indirect bandgap semiconductors. Unless otherwise stated, the values quoted here are drift mobilities at a temperature of 300K. Note that for hexagonal and wurtzite structures the mobilities can be different along the **a** and **c** crystallographic axis. Whenever possible, two values are given: μ perpendicular ($\perp$) to the c-axis and μ parallel (ll) to the c-axis.

Hole mobility, μ_h: Defined as the ratio of the hole velocity (cm s^{-1}) to the electric field (V cm^{-1}) through which it is moving. Mobility has units of cm^{2}V^{-1}s^{-1}. All remarks in the previous paragraph are equally valid for hole mobilities.

$\mu_e\tau_e$: This is the electron mobility-lifetime product expressed in units of cm^2V^{-1}. Values are quoted at a temperature of 300K unless otherwise stated.

$\mu_h\tau_h$: This is the hole mobility-lifetime product expressed in units of cm^2V^{-1}. Values are quoted at a temperature of 300K unless otherwise stated.

F.2 Accompanying Notes for the Tables

The following tables list the physical and electronic properties of the various groupings of the elemental and compound semiconductors. These are groups III, IV, III-V, IV-IV, II-VI, I-VII, their binary alloys and ternary compounds. For detector applications, there are numerically approximately 16 group III-V compounds of interest, 19 group II-VI compounds, six group I-VII and a virtual infinite number of ternary and quaternary compounds. Fig. F.6 shows the section of the Periodic Table from which the bulk of the semiconductor groups are derived. The main groups and how they are formed are described in some detail in the following paragraphs.

Group VI (B) elements. These are the classical elemental semiconductors, Si, Ge, C (diamond) and gray tin (α-Sn) which crystallize in the diamond structure and are unique in the Periodic Table in that their outer shells are exactly half filled. Consequently, they bond exclusively covalently. An examination of properties of group IV elements shows that bandgap energies, hardness and melting points all decrease with increasing Z, while charge carrier mobilities, densities and lattice constants generally increase. These trends may be attributed to the progressive "metallization" of the elements with increasing Z within the group. One can also combine two different group IV semiconductors to obtain compounds such as SiC and SiGe whose physical and electronic properties are intermediate between both species.

Group III-V compounds. These are compounds which combine an anion[4] from group V (nitrogen on down) and a cation[5] from group III (usually, Al, Ga or In). Each group III atom is bound to 4 group V atoms and *vice versa* – thus each atom has a filled (8 electron) valence band, as shown in Fig. 3.7b. Although bonding would appear to be entirely covalent, the shift of valence charge from the group V atoms to the group III atoms induces a component of ionic bonding to the crystal [15]. This ionicity causes significant changes in semiconducting properties. For example, it increases both the Coulomb attraction between the ions and the forbidden bandgap energy. When grown epitaxially (MBE, MOCVD and variants), III-V materials usually assume a zincblende (ZB) structure – so in their basic electronic and crystal structures they are completely analogous to the group IV elements. The stable bulk allotrope often has a wurtzite structure. Both the zincblende and wurtzite lattices are tetrahedrally bonded. They differ only by the orientation of the nearest-neighbour tetrahedrons. The zincblende form differs from a diamond lattice only in that the two interpenetrating face-centered lattices are occupied by different atoms. Representative III-V compounds are InSb, InAs, GaAs, GaP, AlAs and AlP. In terms of bandgap, indium compounds have the smallest energy gap, followed by gallium, boron and aluminum compounds. Likewise within these groups the antimonides have the smallest bandgaps followed by the arsenides, phosphides and finally the nitrides – reflecting

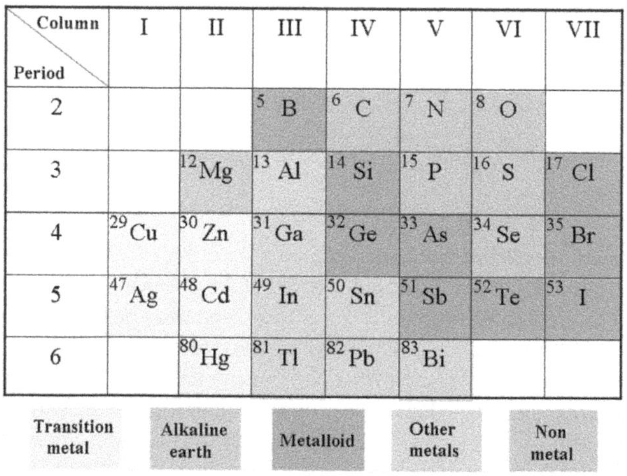

Column \ Period	I	II	III	IV	V	VI	VII
2			5 B	6 C	7 N	8 O	
3		12 Mg	13 Al	14 Si	15 P	16 S	17 Cl
4	29 Cu	30 Zn	31 Ga	32 Ge	33 As	34 Se	35 Br
5	47 Ag	48 Cd	49 In	50 Sn	51 Sb	52 Te	53 I
6		80 Hg	81 Tl	82 Pb	83 Bi		

Transition metal   Alkaline earth   Metalloid   Other metals   Non metal

FIGURE F.6   Section of the periodic table from which the bulk of the semiconductor groups are derived. Most compounds derived from this table will form tetrahedral diamond structures which satisfy the Mooser-Pearson [26] rules, a simple set of tests to determine whether a compound is likely to display semiconducting properties.

---

[4] The electronegative component of a compound, *e.g.*, As in GaAs, Te in CdTe.
[5] The electropositive component of a compound, *e.g.*, Ga in GaAs, Cd in CdTe.

the decreasing size of the atomic nuclei (r = 145 pm (Sb), 115 pm (As), 100 pm (P), 65 pm (N), where 1 pm = 10^{-12}m). Similarly, since the bond lengths are reducing with decreasing z (going from the antimonides to the nitrides), ionicity and hardness correspondingly increase. Electronically, group III-V materials tend to have larger electron mobilities at low electric fields than the elemental semiconductors, which make them attractive candidates for high speed applications. For zincblende structures there is a transition from a direct to an indirect bandgap somewhere between GaAs (1.4 eV direct) and AlSb (1.6 eV indirect).

Ternary III-V alloys have the general form $(A_{1x}, A_{21-x})B$ with two group III atoms used to fill the group III atoms in the lattice, or $A(B_{1x}, B_{21-x})$ using two group V atoms in the group V atomic positions in the lattice. Here A and B represent elements from groups III and V, respectively and x is the mole fraction in the range 0 to 1. The quaternary semiconductors use two group III and two group V elements, yielding a general form $(A_{1x}, A_{21-x})(B_{1y}, B_{21-y})$ for $0<x<1$; $0<y<1$ – e.g., $Ga_{0.12}In_{0.88}As_{0.23}P_{0.77}$. The lattice constants of ternary and quaternary compounds can be calculated with good precision using Vegard's law [27] which gives a value equal to the weighted average of all of the four possible constituent binaries. For example, the lattice constant of the quaternary compound $A_{1-x}B_xC_yD_{1-y}$, is given by

$$a(x,y) = x y\, a_{BC} + x(1-y)a_{BD} + (1-x)y\, a_{AC} + (1-x)(1-y)a_{AD} \qquad (F.2)$$

Except for binary alloys, a similar expression relating the weighted bandgap energy to the bandgap energies of the constituent elements does not really exist, due the presence of multiple minima in the conduction band which, in theory, should be taken into account in the weighting. In fact, for the higher order alloys, the bandgap can actually change type (direct to indirect and *vice-versa*), depending on composition and whether the semiconductor is strained or not. Generally, however, a so-called one valley bandgap fit can give reasonable agreement over a limited range of x and y.

Group II-VI compounds. These are compounds which combine a group IIb metal (for example, Zn, Cd and Hg, in periods 3, 4 and 5, respectively) with a group VIa cation. The latter is usually S, Se, or Te. Structurally, it forms when atomic elements from one type bonds to the four neighbours of the other type, as shown in Chapter 3, Fig. 3.7c. Three crystal structures dominate II-VI compounds. These are zincblende, wurtzite and the rocksalt (NaCl) structure. A major motivation for developing II-VI semiconductors is their broad range of bandgaps (from 0.15 eV for HgTe to 4.4 eV for MgS), high effective Z and a demonstrated capability for making MBE and MOCVD grown heterostructures, as for III-V systems. Additionally, all II-VI binaries have direct bandgaps which make them particularly attractive for optoelectronic applications. In terms of bandgap, mercury compounds have the smallest energy gap, followed by cadmium, zinc and magnesium compounds. Likewise within these groups the tellurides have the smallest bandgaps followed by the selenides and the sulphides – again reflecting the decreasing size of the atomic nuclei. In this case, the atomic radii are 140 pm (Te), 115 pm (Se) and 100 pm (S), respectively. Compounds generally crystallize naturally in a hexagonal or NaCl-structure. Representative compounds are CdTe and HgTe. Pseudo-binary alloys with Zn, Se, Mn or Cd are also common, particularly for radiation detector and optoelectronic applications (*e.g.*, $Cd_{(1-x)}Zn_xTe$, $Cd_{(1-x)}Mn_xTe$ and $Hg_{(1-x)}Cd_xTe$). Group II-VI compounds typically exhibit a larger degree of ionic bonding than III-V materials, since their constituent elements differ more in electron affinity due to their location in the Periodic Table. A major limitation of II-VI compounds is the difficulty in forming n-type and p-type material of the same compound. Also, it is difficult to control the defect state density within the bandgap due to self-compensation. Group II-VI semiconductors can be created in ternary and quaternary forms, although these are less common than III-V varieties.

Group III-VI compounds. Most of the III-VI compounds crystallize in layer type structures. The bonding is predominantly covalent within the layers and much weaker van der Waals type between layers. Because of this, the behaviour of electrons within the layers is quasi-two-dimensional. Ionicities tend to be higher than in III-V materials but lower than in II-VI compounds. Examples of this group are at present limited to the Ga-based compounds GaS, GaSe and GaTe which are being studied primarily for photoconductivity and luminescence applications.

Group I-VII compounds. In comparison to other semiconductors groups, all the I-VII group of compounds are very similar, displaying little variation in their mechanical and electrical properties (such as melting point, density, hardness and bandgap). They bond ionically and are consequently characterized by high ionicities, which in turn increases both the Coulomb attractions between the ions and the energy gaps which are considerably larger than in many III-V materials. In, fact there is a clear tendency for an increase in the energy gap and melting point with increasing ionic bonding, going from $A^{IV}A^{IV}$ to $A^{III}B^V$ to $A^{II}B^{VI}$ through to A^IB^{VII} systems. Specific examples of this group include the silver halogenides, AgCl and AgBr, which were some of the first compound semiconductors demonstrated to be sensitive to ionizing radiation. Under normal conditions, they crystallize in a rock salt (NaCl) form (primarily a consequence of high ionicity) while the other main members of the group, the copper halogenides (CuCl, CuBr and CuI) crystallize in a zincblende configuration.

Group IV-VI compounds. Stable IV-VI compounds (also known as the group IV-chalcogenides[6]) exist in various stoichiometric compositions and generally have very narrow bandgaps. In fact, most are aligned with the IR waveband.

---

[6] Derived from the Greek words *Chalcos* and *gen* meaning "ore-forming". These are elements from group VIB of the periodic table (*i.e.,* O, S, Se and Te).

They have large ionicities, 6-fold coordination, high mobilities and are electronically highly polarizable. Perhaps the most interesting are the lead chalcogenides which are unique in that their energy gaps increase with increasing temperature, as opposed to other semiconductors which have negative temperature coefficients. PbSe is a particularly interesting material in that excess Pb atoms in PbSe act as electron donors and excess Se atoms act as electron acceptors. Thus by changing the relative concentrations of these elements, it is possible to choose the nature of the material, n-type or p-type. The main application of IV-VI compounds is in the production of light emitting devices and IR detectors.

Group n-VII compounds. Group n-VII (where n = II, III, IV, V) materials generally belong to the family of layered structured, heavy metal iodides and tellurides. The group VII anions form a hexagonal close packed arrangement while the group η cations fill all the octahedral sites in alternate layers. The resultant structure is a layered lattice with the layers held in place by van der Waals forces and is typical for compounds of the form AB_2. The bonding within the layers is mainly covalent. Physically, materials in these groups tend to be mechanically soft with easy cleavage planes, have low melting points, large dielectric constants and show strong polarization effects. The fact that layered compounds are strongly bound in two directions and weakly bound in the third direction (along the c axis), leads to an anisotropy of their structural and electronic properties. Because of this, electrons can display quasi-two-dimensional behaviour within layers, which can potentially be exploited in transistor-like structures. In addition, since only van der Walls forces act between the layers, it is possible to introduce a variety of foreign atoms and organic molecules between the layers forming intercalation compounds which can significantly modify the original physical and electronic properties. The reversibility of intercalation processes can also be exploited, for example, in the production of high energy density rechargeable batteries in which the intercalation layers form the cathodes and the cell reaction powers the reverse intercalation [28].

Group I-II-V compounds. These are the filled tetrahedral semiconductors of the form $A^IB^{II}C^V$. Structurally the lattices can be viewed as "zincblende" derivatives partially filled with group I interstitials. An example would be LiZnX, where X = N, P and As. However, while complete analogs of III-V zincblende materials, the interstitials have the effect of preserving both a direct and a wide bandgap. The threshold between direct and indirect bandgaps in III-V zincblende materials lies somewhere between GaAs (1.4 eV direct) and AlSb (1.6 eV indirect). Consequently, these semiconductors are currently being investigated for optoelectronic applications.

Group I-III-VI$_2$ compounds. Chalcopyrites,[7] such as CuAlS$_2$, CuGaS$_2$ and CuInSe$_2$. Other, non-Cu based, members of the I-III-VI group are also usually referred to as chalcopyrites, since they adopt a similar crystal structure.[8] This is a ternary-compound equivalent of the diamond structure, in which every atom is bonded to four first-neighbours in a tetrahedral structure. The atomic bonds are mainly covalent. Instead of bonding to four group II elements as in a group II-VI semiconductor, the group VI element bonds to two group I and two group III elements in the I-III-VI$_2$ ternary system. Cu-In-Se systems are the most studied variants at this time, especially the alloy CuIn$_{1-x}$Ga$_x$Se$_2$ (Cu(In,Ga)Se$_2$). Group I-III-VI$_2$ systems offer direct-gap semiconductors over a broad range of lattice constants and bandgaps and have interesting non-linear optical properties. They are presently being investigated for exploitation as photovoltaic materials. In fact, the Cu-chalcopyrite family of semiconductors produces some of the best thin-film solar cell absorbers with power conversion efficiencies up to ~ 20% [29].

Ternary compounds, such as $A^{II}B^{IV}C_2^V$ or $A_2^{II}B^VC^{VII}$ and combinations of binary alloys (*e.g.*, $A_xB_{1-x}C$ where x is the fractional concentration of A) offer a much wider choice of physical parameters, such as effective atomic number and bandgap energy. Whereas the bandgap of ternary compounds is fixed, alloying gives access to a continuous range of bandgaps. Alloys of binary compounds from the same column of the periodic table can also be alloyed to arbitrary composition such that their physical properties, such as bandgap width, change smoothly with x. However, it is difficult to achieve a high degree of compositional homogeneity and crystal quality. Consequently, these materials are generally grown by MBE or MOCVD which provide better control over stoichiometry than other techniques. Most properties, such as effective mass, vary quadratically and monotonically with alloy fraction. Alloy scattering is largest near a 50% mix and transport properties tend not to vary monotonically. As a rule, ternary compounds based on the heavier metallic elements (*e.g.*, Hg and Bi) are of lower mechanical strength, poorer phase and chemical stability than $A^{III}B^V$, $A^{II}B^{VI}$ or A^IB^{VII} compounds. Also, since they do not possess four valence electrons per atom and are not full analogs of diamond, it is more difficult to synthesize these materials to the required purity and stoichiometry or even to predict their properties.

Quaternary compounds. These generally have the form $A_yB_{1-y}C_xD_{1-x}$, where A and B are group II/III elements and C and D are group VI/V elements. While ternary compositions offer a distinct advantage in that the bandgap can be adjusted within the range of the involved binary compounds, quaternary and higher compositions also allow the simultaneous adjustment of the lattice constant, which in the case of photonic applications means that radiant efficiencies can be increased over a wider range of wavelengths. As with ternary compounds it is difficult to achieve a high degree of compositional homogeneity and

---

[7] Named after the mineral chalcopyrite, CuFeS$_2$.
[8] Also adopted by a number of II-IV-V$_2$ compounds.

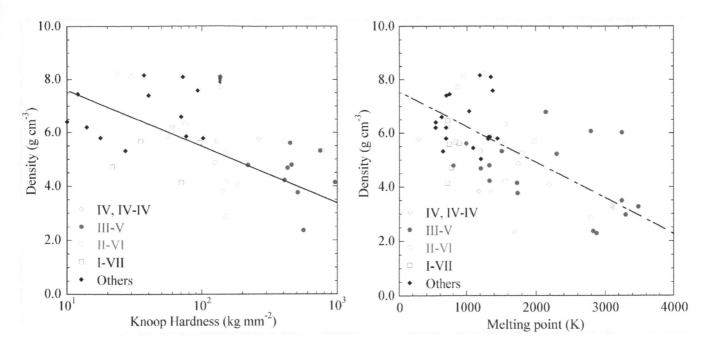

FIGURE F.7 Correlation between density and hardness (left) and density and melting point (right). The solid lines are drawn to guide the eye. Others here refer to semiconductors not in groups IV, IV-IV, II-V, II-VI and I-VII (*e.g.,* groups IV-VI).

crystal quality. For optoelectronic applications, these materials are generally grown by MBE or MOCVD to better control stoichiometry.

Organic compounds (small molecule, oligomers polymer and derivatives). These are the so-called "plastic" semiconductors. They are polymers with a de-localized π-electron system along the polymer backbone. This gives rise to the creation of alternating single and double bonds by weak *pz-pz* bonding, which in turn results in the creation of a bandgap of energy ~ 2.5 eV. Such materials offer numerous advantages in terms of easy processing (spin or dip coating as opposed to epitaxial growth) and good compatibility with a wide variety of substrates. At present, organic materials are being exploited for use in low cost flexible displays, low-end data storage media and inexpensive solar cells. For the latter, efficiencies of up to 10.5% have been achieved for single junction devices [30] and most recently up to 17% for tandem cell devices [31]. They are also of interest for medical applications because they are "tissue-equivalent" materials, in that the low atomic numbers of their constituent atoms closely match those of biological tissue.

Insulators (wide-gap materials). We have also included a separate table for wide gap materials that would normally qualify as insulators. We arbitrarily define these as having a bandgap > 3 eV, and in fact, most ceramics fall into this class. Although seldom used for their "semiconducting properties" they are technologically important for device construction – for example, in forming gate dielectrics in FETS. Although not listed or considered here, for completeness we mention Mott insulators which are a class of materials that should conduct electricity under conventional band theories but are insulators. This effect is due to electron–electron interactions, which are not considered in conventional band theory.

F.2.1 General Properties of the Groups

Based on the above summaries and the accompanying tables, we can make several generalizations about semiconductors. To begin with, hexagonal materials are generally harder than cubic materials. However, when a material can crystallize into, say, stable wurtzite and zincblende forms, the zincblende structure is generally stronger. Following on, heavy materials tend to be mechanically soft. This is borne out in Fig. F.7 (left) in which we plot density versus Knoop microhardness (KH) for a range of semiconductors. Note, "*Others*" here refer to semiconductors not in groups IV, IV-IV, II-V, II-VI and I-VII (*e.g.,* groups IV-VI). The data show a power law correlation with the heaviest materials having lower KH values than light materials. In fact, on average the KH decreases by an order of magnitude for every ~ 2 g cm^{-3} increase in density. Similarly, we might then expect heavier compounds to have lower melting points because they are likely to be softer. In this case (see Fig. F.7 right), the correlation is much weaker, but does show the expected trend and appears to be independent of the semiconductor group.

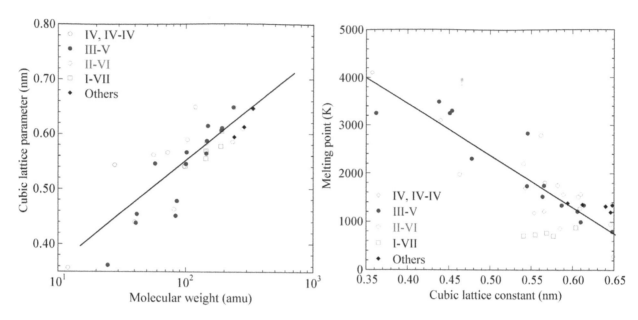

FIGURE F.8 Molecular weight (left) and melting point (right) as a function of lattice parameter for a range of semiconductors. The solid lines are drawn to guide the eye.

The most obvious material variations are those with lattice constant. For $A^N B^{8-N}$ semiconductors, a plot of lattice constant, a, versus molecular weight[9] (see Fig. F.8 left) can be approximated by a linear function of the form [6],

$$a(\text{nm}) = 0.08 + 0.23 \ln MW(\text{amu}), \tag{F.3}$$

where $MW = M_A + M_B$. In Fig. F.8 (right), we note there is a clear tendency for the melting point to increase with decreasing interatomic bond length by roughly 1,000 degrees K per Ångstom. It can be approximated by,

$$T_m(\text{K}) = 7743 - 10723a(\text{nm}). \tag{F.4}$$

Note that for I-VII materials, the melting point is independent of lattice constant. For hexagonal semiconductors, a should be replaced by an effective lattice constant, a_{eff}, given by

$$a_{\text{eff}} = \left(\sqrt{3}a^2 c\right)^{1/3}, \tag{F.5}$$

where c, is the interatomic bond length along the c crystallographic direction.

In Fig. F.9 (left) we plot Knoop microhardness as a function of lattice parameter for a number of materials, from which we can see the larger the lattice parameter, the softer or more fragile the material. Interestingly, group III-V and II-VI materials lie on different curves but with the same slope and can be approximated by,

$$KH = a_1 \exp\left(-12.13 a_o(nm)\right) \text{ kgmm}^{-2}. \tag{F.6}$$

Here $a_1 = 5.884 \times 10^5$ kg mm^{-2} for group IV, IV-IV and groups III-V materials and 1.556×10^5 for group II-VI and "other" materials.

Note that KH values also decrease with increasing temperature. For example, GaAs values decrease by a factor of ~25 between room temperature and ~ 300°C.

In Fig. F.9 (right) we plot the bandgap energy as a function of molecular weight, MW, from which we note that the different groups lie on separate curves. In fact, for groups IV, IV-IV, III-V and II-VI, the curves are very nearly parallel to each other and can be reasonably well approximated by the function

---

[9] Note: the molecular weight for an $A^N B^{8-N}$ ($N \neq 4$) compound is simply given by the sum of the atomic weights of atoms A and B. For elemental semiconductors ($N=4$), it is given the atomic weight of the elemental atom. The molecular weight M of alloy semiconductors can be obtained by linear interpolation.

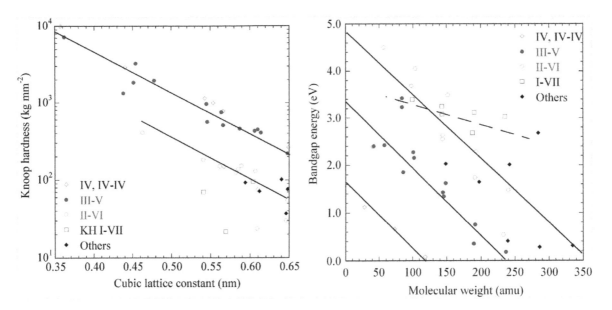

FIGURE F.9 Left: Knoop microhardness as a function of lattice parameter for a range of semiconductors. Notice that the Knoop hardness data seems to separate into two distinct distributions: groups IV, IV-IV and IIV and groups II-VI and "other" materials. Right: the variation of bandgap energy with molecular weight for group IV, III-V, II-VI and I-VII semiconductors.

$$\varepsilon_g = a_1 - 0.014\,MW\ (\mathrm{eV}), \tag{F.7}$$

where a_1 = 1.64 eV for group IV, IV-IV materials, 3.3 for group III-V materials and 4.8 for group II-VI materials. However, it is also clear from Fig. F.9 (right) that group I-VII materials lie on a quite different slope;

$$\varepsilon_g = 3.63 - 0.0032287\,MW\ (\mathrm{eV}). \tag{F.8}$$

In Fig. F.10 (left) we see there is also a clear tendency for the bandgap energy, ε_g, to decrease with increasing lattice constant. In fact, ε_g changes by roughly a factor of 1.4 per Ångstrom, except for group I-VII materials, which show a much less pronounced variation.

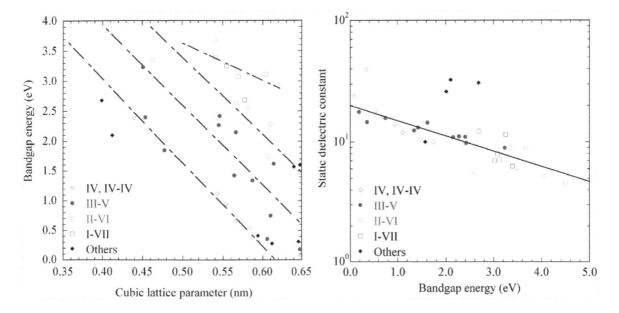

FIGURE F.10 Left: Bandgap energy as a function of lattice parameter for a range of semiconductors. Notice that the data seems to separate into distinct distributions for each semiconductor family. Right: the variation of bandgap energy with static dielectric constant for groups IV, IV-IV, III-V, II-VI, I-VII and other semiconductors.

In Fig. F.10 (right), we note that the static dielectric constant varies exponentially with bandgap energy. Smaller bandgap energies have larger dielectric constants; this is particularly well correlated for group III-V materials. A reasonably good approximation is

$$\varepsilon(0) = 14.9\exp(-0.289\varepsilon_g). \tag{F.9}$$

Thus, for room temperature operation we should expect to have a dielectric constant of ~ 12.

Electronically, we note that mobilities tend to be larger in direct bandgap materials than in indirect bandgap materials and lower in heavy compounds. In Fig. F.11 (left) we plot both electron and hole mobilities as a function of bandgap energy. The electron mobilities show a weak correlation with bandgap energy for bandgaps less than 3 eV which appear to increase with decreasing bandgap. Above 3 eV they are essentially constant. The hole mobilities show no obvious variation with bandgap energy. We also note that while alloying can result in an increase in resistivity, it is usually accompanied by a fall in mobilities, due to impurity scattering [32,33].

In Fig. F.11 (right) we plot the electron and hole effective masses as a function of the direct bandgap energy for a range of semiconductors. In the case of indirect bandgap materials, such as Ge and Si, we have used Γ energy gap (see Chapter 2, Fig. 2.6) at $k = 0$. From the figure we see there is a good correlation between the electron effective mass and the bandgap. The data can be reasonably represented by

$$m_e* = 0.0625\,\varepsilon_{g\Gamma}, \tag{F.10}$$

where $m_e* = m*/m_o$ is the effective mass in units of electron mass. The correlation between the hole effective mass and the bandgap is much weaker than that of the electrons; however, we note that the slope of the curve is similar to that for the electrons and may be represented by

$$m_h* = 0.41 + m_e*. \tag{F.11}$$

Thus,

$$m_h* = 0.0625\varepsilon_{g\Gamma} + 0.41. \tag{F.12}$$

Since effective mass and mobility are inversely related, Eq. (F.10) implies that for application where high speed is required, one should choose a semiconductor with a small bandgap.

In Fig. F.12 we plot the bandgap energy as a function of electron-hole pair creation energy for a number of compounds. As can be seen, semiconductors align along two curves, the classical curve first reported by Klein [34] (solid line) and

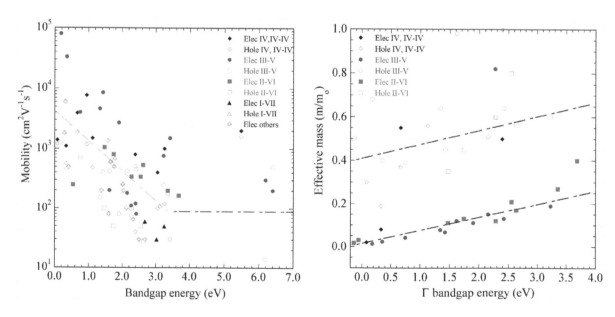

FIGURE F.11 Left: Electron and hole mobilities plotted as a function of bandgap energy for the various semiconductor groups. The lines are drawn to guide the eye. Right: variation of effective mass with direct bandgap energy. For indirect bandgap materials this corresponds to the Γ valley.

(Cont.)

Material	Crystal Structure	Pearson symbols/ Space/point group	Lattice constants *a* or *a, c* (nm)	Atomic numbers	Av. Mol. Weight (g mole^{-1})	Density (g cm^{-3})	Melting Point (K)	Dielectric Constant $\varepsilon_r(0)$	Phillips Iconicity f_i	KH (kg mm^{-2})
$CuInS_2$	tetragonal	t/16/122	0.5517, 1.1122	29, 49, 16	242.5	4.73	1298		0.41	287(V)
$CuInSe_2$	tetragonal	t/16/122	0.5751, 1.1552	29, 49, 34	336.3	5.78	1260	16.0	0.41	235(V)
CuI_2Se_6	rhomb	R3m	1.4039, 1.4153	29, 53, 34	791.1	5.29	670			
$AuGeSe_2$	tetragonal	I42d/4m2	0.5992, 1.0886	79, 32, 34	427.5	5.70	1124			
$LiInS_2$	ortho	oP16/Pna2, C_{2v}^9/mm2	0.6893,0.8058,0.6482	3, 49, 16	185.9	3.54	1153			4 (M)
$LiInSe_2$	ortho	oP16/Pna2, C_{2v}^9/mm2	0.7192,0.8412,0.6793	3, 49, 34	279.7	4.47	1188	8.5		342 (V)
$LiInTe_2$	tetragonal	t/16, 122	0.6419, 1.2486	3, 49, 52	377.0	4.87	935	7.9		
$LiZnP$	ZB	F-43m	0.576	3, 30, 15	103.3	3.80	1123	9.6		
$LiZnAs$	ZB	F-43m	0.591	3, 30, 31	147.3	4.58	1223	10		
$Pb_2P_2S_6$	monoclinic	mP20, P2$_1$/c	0.9402,0.7466,0.6612	82, 15, 16	668.7	4.79	1198			
$Pb_2P_2Se_6$	monoclinic	mP20,P2$_1$/c	0.6606,0.7446,1.1374	82, 15, 34	950.1	6.13	1057			106

Electronic Properties – Ternary Compounds

Material	Type	Bandgap (eV)	Pair energy (eV)	Resistivity (Ω-cm)	Electron mobility (cm^{2}V^{-1}s^{-1})	Hole Mobility (cm^{2}V^{-1}s^{-1})	$\mu_e\tau_e$ (cm^{2}V^{-1})	$\mu_h\tau_h$ (cm^{2}V^{-1})
$SbSeI$	I	1.66		10^8	26	7	4.4×10^{-4}	3.5×10^{-4}
$Hg_3S_2Cl_2$	I	2.56		10^{10}			1.4×10^{-4}	7.5×10^{-5}
$Hg_3Se_2Br_2$	D	2.22		10^{11}			1×10^{-4}	9×10^{-5}
$Hg_3S_2I_2$	I	2.25		10^{11}	10		2×10^{-6}	
$Hg_3Se_2I_2$	I	2.12		10^{12}	104		1×10^{-5}	
$Hg_3Te_2I_2$	D	1.93		10^{12}	10		3×10^{-6}	
Tl_6SeI_4	D	1.86		10^{11}	112	81	1×10^{-4}	6×10^{-4}
Tl_6I_4S	D	2.04		10^{10}			2×10^{-3}	2×10^{-5}
Tl_2SnS_3	I	1.44		8×10^9			1×10^{-4}	3×10^{-5}
$TlSn_2I_5$	I	2.14		10^{10}	94		1×10^{-3}	10^{-4}
$TlSbS_2$	D	1.67		2×10^{10}	13.2		3×10^{-6}	
Tl_3SbS_3	D	1.48		2×10^9			1×10^{-4}	8×10^{-5}
$TlGaS_2$	D	2.45		10^6	1766	1097		
$TlGaSe_2$	I	1.93		2×10^7	23–99	61–65	1×10^{-4}	3×10^{-5}
$TlGaTe_2$	I	1.2		3×10^3	66	96		
$TlInS_2$	I	2.2		10^7	68	170		
$TlInSe_2$	I	1.2			225	450		
$TlInTe_2$	I	0.86			420	600		
$Tl_2Au_4S_3$	D	1.63		6×10^7			1×10^{-4}	1×10^{-6}
Tl_4CdI_6	D	2.8		2×10^{10}			6×10^{-4}	1.0×10^{-4}
$TlPbI_3$	I	2.3		10^{11}				
$Tl_7Bi_3I_{16}$	D	1.74		5×10^6			4×10^{-5}	1×10^{-5}
$CsInSe_2$		2.81						
$CsInTe_2$		2.16						
CsP_bCl_3	D	2.86		3×10^8			1×10^{-4}	9×10^{-5}

(Continued)

(Cont.)

Material	Type	Bandgap (eV)	Pair energy (eV)	Resistivity (Ω-cm)	Electron mobility $(cm^2V^{-1}s^{-1})$	Hole Mobility $(cm^2V^{-1}s^{-1})$	$\mu_e\tau_e$ (cm^2V^{-1})	$\mu_h\tau_h$ (cm^2V^{-1})
CsP_bBr_3	D	2.28	5.3	10^9		52	1.7×10^{-3}	1.3×10^{-3}
CsP_bI_3	D	1.74			25			
$Cs_2Cd_3Te_4$	D	2.45		$2\times10^6(\perp)$, $1\times10^6(\parallel)$			8×10^{-4}	2×10^{-4}
$Cs_2Hg_3S_4$	D	2.81		10^8			4×10^{-4}	6×10^{-5}
$Cs_2Hg_6S_7$	D	1.64		10^7			1.9×10^{-3}	3×10^{-4}
$Cs_2Hg_3Se_4$		2.6		10^7			8×10^{-4}	
$Cs_3Bi_2I_9$	I	2.2		$1\times10^{10}[001]$ $8\times10^9[100]$	6.1		2×10^{-5}	
$CuInS_2$	D	1.55		10^6	200	15		
$CuInSe_2$	D	1.02		10^4	90–900	15–150		
CuI_2Se_6	I	1.95		10^{12}	46			
$AuGeSe_2$		1.73		10^{11}			6×10^{-6}	
$LiInS_2$	D	3.57		10^9				
$LiInSe_2$	D	2.86		10^{12}	137		9×10^{-5}	
$LiInTe_2$	D	1.50						
$LiZnP$	D	2.04		10^7	5.7	25	8.0×10^{-4}	$< 10^{-6}$
$LiZnAs$	D	1.50		10^7	4.8		9.0×10^{-4}	$< 10^{-6}$
$Pb_2P_2S_6$	I	2.15		10^{11}			3.5×10^{-5}	
$Pb_2P_2Se_6$	I	1.88		10^{11}	10		3.1×10^{-4}	4.8×10^{-5}

Ternary Alloys

Physical Properties – Ternary Alloys

Material	Crystal structure	Pearson symbols/ space/point group	Lattice constants a or a, c (nm)	Atomic numbers	Density $(g\ cm^{-3})$	Melting point (K)	Dielectric Constant $\varepsilon_r(0)$	Phillips Iconicity f_i	KH $(kg\ mm^{-2})$
$B_xGa_{1-x}N$	ZB	cF8	0.308 for $x = 0.02$	5, 31, 7	6.07 $x = 0.02$				
$In_xAs_{1-x}Sb$	ZB	cF8,F43m(T_d^2)	0.636 for $x = 0.35$	49, 13, 51	5.68 $+0.09x$		$15.15+1.65x$		
$In_xGa_{1-x}As$[10]	ZB	cF8	0.5869 for $x = 0.53$	49, 31, 33			$\sim$10		
$In_xGa_{1-x}P$[11]	ZB	cF8	0.564 for $x = 0.52$	49, 31, 15					
$In_xAl_{1-x}P$	ZB	cF8	0.5653 for $x = 0.5$	49, 13, 15					
$Ga_xIn_{1-x}Sb$	ZB	cF8	$0.6489-0.0391x$	31, 49, 51	$5.77-0.16x$		$16.8-1.1x$		265–459
$Ga_xAs_{1-x}Sb$	ZB	cF8		31, 33, 51	$5.32 + 0.29x$		$12.90 + 2.8x$		450–750
$Ga_xIn_{1-x}As$	ZB	cF8,F43m(T_d^2)	$0.60583-0.0405x$	31, 49, 33	$5.68-0.37x$	$\sim$1373	$15.1-2.87x$ $+0.67x^2$		$\sim$560
$Ga_xIn_{1-x}P$	ZB	cF8,F43m(T_d^2)	$0.5868-0.00418x$	31, 49, 15	$4.8-0.67x$		$12.5-1.4x$		1122–530
$Ga_xIn_{1-x}N$	ZB	cF8	$0.4986-0.0460x$	31, 49, 7					
$Al_xGa_{1-x}Sb$	ZB	cF8	$0.6096+0.00344x$	13, 31, 51					
$Al_xIn_{1-x}Sb$	ZB	cF8		13, 49, 33					

(Continued)

---

[10] Can be latticed matched to InP.
[11] Can be latticed matched to GaAs.

(Cont.)

Material	Crystal structure	Pearson symbols/ space/point group	Lattice constants a or a, c (nm)	Atomic numbers	Density (g cm^{-3})	Melting point (K)	Dielectric Constant $\varepsilon_r(0)$	Phillips Iconicity f_i	KH (kg mm^{-2})
Al$_x$In$_{1-x}$As[1]	ZB	cF8		13, 49, 33					
Al$_x$In$_{1-x}$P	|ZB	cF8	0.5893–0.0416x	13, 49, 15					
Al$_x$Ga$_{1-x}$As	ZB	cF8/F43m(T$_d^2$)	0.56533+0.00078x	13, 31, 33	5.32–1.56x	[12]~1687	12.90–2.84x	0.31–0.36x	730–510
Cd$_{1-x}$Zn$_x$Te	ZB	cF8	0.6477–0.03772x	48, 30, 52	5.85	1365–1568	10.4 x = 0, 9.7 x = 1		60–80; 102 (x = 4%)
CdTe$_{1-x}$Se$_x$	ZB	cF8	0.6481–0.411x	48, 52, 34	5.85–6.3	1323			
Cd$_x$Zn$_{1-x}$Se	ZB[13]	cF8/P6$_3$	0.567+0.04x	48, 30, 34	5.4-5.8	1512–1793			
Cd$_{1-x}$Mg$_x$Te	ZB[14]	CF8/F43m(T$_d^2$)	0.648–0.00499x	48, 12, 52					
Cd$_{1-x}$Mn$_x$Te	ZB[15]	cF8	0.6482–0.0150x	48, 25, 52	5.3–5.7	1343, x = 0.45			76 (x = 45%)
Hg$_{1-x}$Cd$_x$Te	ZB	cF8	0.6461+ (8.4x+ 11.68x^2-5.7x^3)10^{-4}	80, 48, 52	8.05–2.3x	1123	19.5 (x = 0.2) T = 4.2K		80 max at x = 0.75
HgBr$_x$I$_{2-x}$[16]	ortho, layered	oP, C$_{2v}^{12}$	0.772, 1.3101, 0.604	80, 35, 53	6.2	502–532			14
TlBr$_x$Cl$_{1-x}$[17]	CsCl	cP2/Pm3m	0.3835+0.015x	81, 35, 17	7.19	696, x = 0.4 (KRS-6)			29.9
TlBr$_x$I$_{1-x}$[18]	CsCl[19]	cP2	0.4125	81, 35, 53	7.4	688	[i]32.5		40[20]

---

[12] Liquidus surface.

[13] Zincblende structure observed for 0.7<x<1.0, wurtzite structure for x<0.68.

[14] Zincblende structure observed for x<0.5, wurtzite structure above.

[15] Zincblende structure observed for 0<x<0.77.

[16] Forms solid solutions for 0.2<x<1.0.

[17] Also known as KRS-6 for x=0.3.

[18] Also known as KRS-5 for x=0.4.

[19] For x<0.3. At x=0.3, bulk TlBr$_x$I$_{1-x}$ transforms from the cubic to the orthorhombic phase when cooled to liquid nitrogen temperatures.

[20] Quoted for KRS-5.

Electronic Properties – Ternary Alloys

Material	Type	Bandgap (eV)	Pair energy (eV)	Resistivity (Ω-cm)	Electron mobility (cm^{2}V^{-1}s^{-1})	Hole mobility (cm^{2}V^{-1}s^{-1})	$\mu_e\tau_e$ (cm^{2}V^{-1})	$\mu_h\tau_h$ (cm^{2}V^{-1})
$B_xGa_{1-x}N$	D ($x<0.75$)	$2.097+2.672x$		$\sim10^5$				
$In_xAs_{1-x}Sb$	D	0.1 for $x = 0.35$			5×10^5 for $x = 0.35$	< 500		
$In_xGa_{1-x}As$	D	0.75 for $x = 0.53$		2430	10,000 $x = 0.53$	150		
$In_xGa_{1-x}P$	D/I[a]	$2.46-1.72x + 0.604x^2$ (for $0.45 < x < 0.51$)	4.95 ($x = 0.5$)		3300 $x = 0.5$			
$In_xAl_{1-x}P$	D/I	2.35 for $x = 0.5$						
$Ga_xIn_{1-x}Sb$	D	$0.172+0.165x+0.43x^2$			5000–8000 $x = 0.08$–0.14	7000–80,000		
$Ga_xAs_{1-x}Sb$	D	$1.42-1.9x+1.2x^2$ (for $0<x<0.3$)						
$Ga_xIn_{1-x}As$	D	$0.36+0.63x+0.43x^2$			13,800	250–450		
$Ga_xIn_{1-x}P$	I/D	$1.34 + 0.511x + 0.604x^2$ (for $0.49 < x < 0.55$)			6500–1000	180–140		
$Ga_xIn_{1-x}N$	D	$3.21 + 1.969x$			1200–200			
$Al_xGa_{1-x}Sb$	D/I	$0.73 + 1.10x + 0.47x^2$			12,000–1000	1000–100		
$Al_xIn_{1-x}Sb$	D/I	$0.172 + 1.621x + 0.43x^2$			76,000–1000	1100–400		
$Al_xIn_{1-x}As$	D/I	$0.36+2.35x+0.24x^2$			34,000–1000	450–100		
$Al_xIn_{1-x}P$	D $x<0.4$ I $x>0.4$	$1.351+2.23x$ 2.31 ($x = 0.52$)	5.32 ($x = 0.52$)					
$Al_xGa_{1-x}As$	D ($x<0.45$) I ($x>0.45$)	$1.424 + 1.087x + 0.438x^2$ $1.9+0.125x+0.1438x^2$	$7.327-0.0077$ T ($x = 0.8$)		$8\cdot10^3$–$2.2\cdot10^4x$ $+10^4\cdot x^2$	370–$970x+740x^2$		
$Cd_{1-x}Zn_xTe$	D	$1.51+0.606x+0.139x^2$	4.6 ($x = 0.1$)	$>10^{10}$	1350	120	1×10^{-2}	2×10^{-4}
$CdTe_{1-x}Se_x$	D	$1.48(1-x)+1.74x-0.88x(1-x)$		10^9	59	33	10^{-2}	10^{-2}
$Cd_xZn_{1-x}Se$	D	$2.63x+1.74(1-x)-0.73x(1-x)$		10^{11}	~500		10^{-4}	
$Cd_{1-x}Mg_xTe$	D	$1.52+1.7x$		3×10^{10}			5×10^{-3}	
$Cd_{1-x}Mn_xTe$	D	$1.526+1.316x$	5.0 ($x = 0.1$)	3×10^{10}	718	40	1×10^{-3}	
$Hg_{1-x}Cd_xTe$	D (for $x>0.15$)	-0.3 ($x = 0$) $- 1.6$ ($x = 1$)[b]	$3.4\varepsilon_g+0.95$	100	$10^4/$ $(8.754x-1.044)^c$	10^2 $x = 0.2$, 27 $x = 0.4$		
$HgBr_xI_{2-x}$	D	2.1–3.4		5×10^{13}	30 ($x = 0.25$), 0.45 ($x = 0.5$)	0.1 ($x = 0.25$), 0.01 ($x = 0.5$)	7.1×10^{-6}	1×10^{-5}
$TlBr_xCl_{1-x}$ [d]		3.25 (KRS-6)		7.95×10^9 exp $(0.635x)$			$\sim3\times10^{-4}$ $x>0.5$	$\sim4\times10^{-5}$ $x>0.5$
$TlBr_xI_{1-x}$ [e]	I	$0.983x+1.82$		3×10^{10} $x = 0.35$, 5×10^{11} $x = 1$			[f]1×10^{-3}	

[a] Indirect for $x>0.245$

[b] ε_g(eV) $= -0.313+1787x+0.444x^2-1.237x^3+0.932x^4+(6.67\times10^{-4}-1.714\times10^{-3}x+7.6\times10^{-4}x^2).T$(K), for $T>70$K

[c] $0.18\leq x\leq0.25$.

[d] also known as KRS-6 for $x = 0.3$.

[e] also known as KRS-5 for $x = 0.4$.

[f] for $x = 0.35$.

Quaternary Compounds and Alloys

Physical Properties – Quaternary Compounds and Alloys

Material	Crystal structure	Pearson symbols/ space/point group	Lattice constants a or a, c (nm)	Atomic numbers	Av. mol. weight (g mole^{-1})	Density (g cm^{-3})	Melting Point (K)	Dielectric Constant $\varepsilon_r(0)$	Phillips Iconicity f_i	KH (kg mm^{-2})
$CsHgInS_3$	monoclinic	C2/c	1.1250, 1.1257, 2.2146	55,80,49,16	544.5	5.168				
$CsCdInSe_3$	monoclinic	C2/c	1.1708, 1.1712, 2.3051	55,48,49,34	597.0	5.059				
$CsCdInTe_3$	monoclinic	C2/c	1.2523, 1.2517, 2.4441	55,48,49,52	742.9	5.195				
$TlHgInS_3$	monoclinic	C2/c	1.3916, 0.3913, 2.1403	81,80,49,16	616.0	7.241	1073			
$Ga_xIn_{1-x}As_ySb_{1-y}$ *	ZB	Td2-F43m	0.60583 (InAs) – 0.60959 (GaSb)	32,49,33,51		5.69–0.08x	~1273	15.3+0.4x		220–750
$Ga_xIn_{1-x}As_yP_{1-y}$ **	ZB	Td2-F43m	0.5869–0.0417x +0.01896y +0.00125xy	32,49,33,15		4.81+0.552y +0.138y^2	~1373	12.5 +1.44y		380–500
$Cs_2Hg_{6-x}Cd_xS_7$	tetragonal	P42nm	1.4051–0.0249x +0.017596 x^2, 0.4104 +1.01158x-0.01185x^2	55,80,48,16		6.9962–0.5539x +0.1574x^2				

Electronic Properties – Quaternary Compounds and Alloys

Material	Type	Bandgap (eV)	Pair energy (eV)	Resistivity (Ω-cm)	Electron mobility (cm^{2}V^{-1}s^{-1})	Hole mobility (cm^{2}V^{-1}s^{-1})	$\mu_e\tau_e$ (cm^{2}V^{-1})	$\mu_h\tau_h$ (cm^{2}V^{-1})
$CsHgInS_3$	D	2.30		$10^{10}(\perp)$, 3×10^8(II)			3.6 × 10^{-5}	2.9 × 10^{-5}
$CsCdInSe_3$	D	2.40		$<4\times10^9$			1.2 × 10^{-5}	2.7 × 10^{-6}
$CsCdInTe_3$	D	1.78		$<2\times10^8$			1.1 × 10^{-4}	1.3 × 10^{-5}
$TlHgInS_3$	I	1.74		4×10^9			3.6 × 10^{-4}	2.0 × 10^{-4}
$Ga_xIn_{1-x}As_ySb_{1-y}$ *	D	0.29–0.65x+0.6x^2			<30,000			
$Ga_xIn_{1-x}As_yP_{1-y}$ **	D	1.35 +0.668x -1.068y +0.758x2 +0.078y2 -0.069xy -0.332x2y +0.03xy2			$\leq$(5400-7750y +14400y^2) x = 0.47	$\leq$200-400y+500y^2 x = 0.47		
$Cs_2Hg_{6-x}Cd_xS_7$		1.6358+0.0441x+0.07723x^2		2×10^6				

* lattice matched to GaSb
** lattice matched to InP

Organic Semiconductors

Physical and Electronic Properties – Organic Semiconductors

Material (abbreviation)	Formula	Structure (RT)	Space group	Density (g·cm^{-3})	Bandgap (eV)	Melting point (K)	Molecular mass (g. mol^{-1})	Conductivity σ (S. m^{-1})	Dielectric constant ε(0)	Mobility electron (cm^{2}s^{-1}V^{-1})	hole (cm^{2}s^{-1}V^{-1})
Anthracene	$C_{14}H_{10}$	monoclinic	P2$_1$/b	1.28	4.0	489	178.23	$10-10^2$	2.9	1.7	2.1
Napthalene	$C_{10}H_8$	monoclinic	P2$_1$/b	1.16	5.0	351	128.17		3.4	0.6	1.5
Perylene	$C_{20}H_{12}$	monoclinic	P2$_1$/a	1.30	3.1	550	252.32	10^{-3}	12	5.5@60K	87@60K
Pentacene	$C_{22}H_{14}$	triclinic	P-1	1.30	1.6	645[21]	278.36	0.01–2.5			2.0
Polyacetylene	$[C_2H_2]_n$	polymer[22]		0.4	1.5			10^3-10^7			0.1–0.5
Polythiophene (spin coated)	$[C_4H_2S]_n$	polymer		1.18	2.0	235	84.14	$10-10^3$			$10^{-4}-10^{-2}$
Polythiophene (printed)	$[C_4H_2S]_n$	polymer		1.35	2.0	235	84.14	$10-10^3$			0.1
Phenothiazine	$C_{12}H_9NS$	ortho	Pnma	1.35	3.4	458	199.27			2.5	0.02
Oligothiophene (Py-4T)		polymer			1.5			100		1.7	0.1–0.5
Polyphenylene		polymer			3.0			10^2-10^3			
Poly(p-phenylene vinylene) (PPV)	$[C_8H_6]_n$	polymer		1.28	2.5	~500		$3-5\times10^3$		10^{-4}	40
Poly(3,4-ethylenethiophene) (PEDOT)	$[C_6H_4O_2S]_n$	polymer		1.01	1.1			300			0.8–9.7
Polypyrrole (PPy)	C_4H_5N	polymer		1.46	3.1	250	67.089	10^2-10^4			10^{-2}
Polyaniline (PANI)	$[C_6H_7N]_n$	polymer		1.36	3.2	603	214.272	30–200			10^{-3}
Organic-inorganic semiconductors											
Methylammonium lead bromide (MAPbBr$_3$)	$CH_3NH_3PbBr_3$	cubic	Pm3m no. 221	3.58	2.18		478.98	5×10^{-6}	25.5	8	115
Methylammonium lead iodide (MAPbI$_3$)	$CH_3NH_3PbI_3$	cubic	Pm3m no. 221	3.95	1.55		619.98	5×10^{-6}	32	20–100	164
Methylammonium tin iodide (MASnI$_3$)	$CH_3NH_3SnI_3$	cubic	Pm3m no. 221	3.62	1.21		531.46	5	10	2320	200–300
Formonidium lead bromide (FAPbBr$_3$)	$CH(NH_2)_2PbBr_3$	cubic	Pm3m no. 221		2.15		491.98	2×10^{-8}	43.6	14	62
Formonidium lead iodide (FAPbI$_3$)	$CH(NH_2)_2PbI_3$	cubic	Pm3m no. 221		1.41		632.98	2×10^{-5}	49.3	4	35
Formonidium tin iodide (FASnI$_3$)	$CH(NH_2)_2SnI_3$	cubic	Pm3m, no. 221	3.58	1.41		544.46	10^{-5}	49.3	103	20

---

[21] Sublimation temperature.

[22] We designate the structure here simply as polymer, since in reality polymer structures are generally neither amorphous nor crystalline; they are semi-crystalline. The degree of crystallinity typically ranges between 10 and 80%.

Insulators (Wide Gap Materials)

Physical Properties – Insulators (Wide Gap Materials)

Material	Crystal structure	Pearson symbols/ space/point group	Lattice constants a or a, c (nm)	Atomic numbers	Av. mol weight	Density (g cm^{-3})	Melting Point (K)	Phillips Iconicity f_i	KH (kg mm^{-2})
TiO$_2$	tetragonal	C4, tP6,P4$_2$/mnm,136	0.45869, 0.29536	22, 8	79.9	3.78	2383	0.69	879
ZrO$_2$	mono	C43, mP12, 14	0.5150, 0.5212, 0.5315	20, 8	123.2	5.84	2988		1100
C	dia	cF8, Fd3m (O$_h^7$)	0.357	6	12.0	3.52	4100	0	7000
α-BN	H	hP4, P6$_3$/mmc(D$_{6h}$)	0.250, 0.666	5, 7	24.8	2.18	2873[a]	0.221	3400
AlN	W	hP4, P6$_3$mc(C$_{6v}^4$)	0.311, 0.498	13, 7	41.0	3.26	3487	0.449	1020–1427
SrO	cubic	cF8, Fm3m, 225	0.257	90, 8	103.6	4.70	2703	0.926	4.5 (M)[a]
Si$_3$N$_4$	trigonal	hP28, P31c, 159	0.7766, 0.5615	14, 7	140.3	3.17	2170	0.40	2200
CaO	cubic	cP2, cF12, Fm3m	0.4811	38, 8	56.1	3.34	2886	0.913	4.5 (M)[a]
MgO	NaCl	cF8, Fm3m, O$_h^5$	0.4212	12, 8	40.3	3.61	3098	0.841	816
Al$_2$O$_3$	trigonal	hR30, R3c, 167	0.4785, 1.2991	13, 8	102.0	3.99	2345	0.80	2000
SiO$_2$	H	hP9, P3121, 152	0.4916, 0.5405	14, 8	60.1	2.65	1986	0.57	710

[a] Mohs scale.

Electronic Properties – Insulators (Wide Gap Materials)

Material	Type	Bandgap (eV)	Pair energy (eV)	Dielectric constant $\varepsilon_r(0)$	Resistivity (Ω-cm)	Electron mobility (cm^{2}V^{-1}s^{-1})	Hole mobility (cm^{2}V^{-1}s^{-1})	$\mu_e\tau_e$ (cm^{2}V^{-1})	$\mu_h\tau_h$ (cm^{2}V^{-1})
TiO$_2$	I	3.2		86 (⊥), 170(∥)	10^{13}–10^{18}	19	16		
ZrO$_2$	D	5.16		19.7	10^{14}				
C (dia)	I	5.48	13.2	5.7	10^{16}	2000	1600	2×10^{-5}	$<1.6 \times 10^{-5}$
BN (h)	I	5.96		6.85(⊥), 5.06(∥)	10^{11}			8.3×10^{-7}	2.2×10^{-5}
AlN (h)	D	6.19	15.3	8.3(⊥), 8.9(∥)	10^{13}	300	14		
SrO	D	6.7		13.3	10^{14}	5			
Si$_3$N$_4$	D	5.2		7.5	10^{14}	29			
CaO	D	5.4		11.8	10^{14}	8 (700K)			
MgO	D	7.8		9.8	10^{14}	2			
Al$_2$O$_3$	D	8.3		11.5	10^{19}	3			
SiO$_2$	D	8.9		4.6	10^{17}	30			

REFERENCES

[1] O. Madelung, M. Schulz, H. Weiss, eds., *"Lundolt-Börnstein, 'Numerical Data and Functional Relationships in Science and Technology', Semiconductors"*, Springer-Verlag, New York, vol. **17** (1982).

[2] M. Levinstein, S. Rumyantsev, M. Shur, eds.,*"Handbook Series on Semiconductor Parameters"*, Vols **1, 2**, World Scientific, London (1996, 1999) ISBN-13: 978–9810229344, ISBN-13: 978–9810229351.

[3] O. Madelung, *"Semiconductors: Data Handbook"*, 3rd ed., Springer Verlag, Berlin (2004) ISBN: 978–3–540–40488–0.

[4] S. Adachi, *"Properties of Group-IV, III-V and II-VI Semiconductors"*, John Wiley & Sons, Hoboken, NJ (2005) ISBN 0470090324, 9780470090329.

[5] L.I. Berger, *"Semiconductor Materials"*, CRC Press, Boca Raton, FL, New York (1997) ISBN 0849389127, 9780849389122.

[6] S. Adachi, *"Properties of Semiconductor Alloys: Group-IV, III–V and II–VI Semiconductors"*, John Wiley & Sons, Hoboken, NJ (2009) ISBN 9780470743690.

[7] "*CRC Handbook of Chemistry & Physics, 82nd Edition, Properties of Semiconductors*", ed. D.R. Lind, CRC Press, Boca Raton, FL (2001) pp. 12–97–12–106.

[8] www.ioffe.ru/SVA/NSM/Semicond/

[9] S.M. Sze, "*Physics of Semiconductor Devices*", John Wiley & Sons, 2nd ed., Hoboken, NJ (1981) ISBN-13: 9780471056614.

[10] R.C. Alig, S. Bloom, C.W. Struck, "Scattering by ionization and phonon emission in semiconductors", *Phys. Rev. B*, Vol. **27**, no. 12 (1980) pp. 5565–5582.

[11] B.D. Cullity, "*Elements of X-Ray Diffraction*", Addison-Wesley Pub. Co., Reading, MA (1956) Revised edition: B.D. Cullity, S. R. Stock, "*Elements of X-Ray Diffraction*", Prentice Hall: New Jersey, 2nd ed. (2001) **ISBN**: 9780201610918.

[12] W.B. Pearson, "*A Handbook of Lattice Spacings and Structures of Metals and Alloys*", vol. **2**, Pergamon Press, Oxford (1967).

[13] C. Hermann, ed., "*Internationale tabellen zur bestimmung von kristallstrukturen*", Gebruder Borntraeger, Berlin, vol. **I** and **II** (1935).

[14] T. Hahn, "*International Tables for Crystallography, Volume A: Space Group Symmetry*", Springer-Verlag, Berlin, New York, 5th ed. (2002) ISBN 978-0-7923-6590-7.

[15] J.C. Phillips, "Ionicity of the chemical bond in crystals", *Rev. Mod. Phys.*, Vol. **42** (1970) pp. 317–356.

[16] M.R. Tubbs, "A spectroscopic interpretation of crystalline ionicity", *Phys. Stat. Sol.*, Vol. **41** (1970) pp. K61–K64.

[17] N.E. Christensen, S. Satpathy, Z. Pawlowska, "Bonding and ionicity in semiconductors", *Phys. Rev. B*, Vol. **36** (1987) pp. 1032–1050.

[18] F. Knoop, C.G. Peters, W.B. Emerson, "A sensitive pyramidal-diamond tool for indentation measurements", *J. Res. Natl. Bureau Stand.*, Vol. **23** (1939) pp. 39–61.

[19] A. Banerjee, M. Sherriff, E.A.M. Kidd, T.F. Watson, "A confocal microscopic study relating the autofluorescence of carious dentine to its microhardness", *Br. Dent. J.*, Vol. **187**, no. 4 (1999) pp. 206–210.

[20] I. Yonenaga, "High-temperature strength of III–V nitride crystals", *J. Phys. Condens. Mat.*, Vol. **14** (2002) pp. 12947–12951.

[21] H. Chandler, "*Hardness Testing*", 2nd ed., ASM International, Novelty, OH (1999) ISBN-13: 978–0871706409.

[22] R.L. Smith, G.E. Sandland, "An accurate method of determining the hardness of metals, with particular reference to those of a high degree of hardness", *Proc. Inst. Mech. Eng.*, Vol. **I** (1922) pp. 623–641.

[23] F. Mohs, "*Grundriss der Mineralogie*", 2 volumes, (as Prismatisches Scheel-Erz), Dresden (1822).

[24] "*Appendix A: Hall Effect Measurements*", Lake Shore 7500/9500 Series Hall System User's Manual, Lake Shore Cryotronics, Inc., Westerville, OH (1995).

[25] J.R. Haynes, W. Shockley, "The mobility and life of injected holes and electrons in germanium", *Phys. Rev.*, Vol. **81**, no. 5 (1951) pp. 835–843.

[26] E. Mooser, W.B. Pearson, "The chemical bond in semiconductors", *J. Electron.*, Vol. **1** (1956) pp. 629–645.

[27] L. Vegard, "Die Konstitution der Mischkristalle und die Raumfüllung der Atome", *Z. Phys.*, Vol. **5** (1921) pp. 17–26.

[28] C. Julien, G.A. Nazri, "*Solid State Batteries: Materials Design and Optimization*", The Springer International Series in Engineering and Computer Science, Berlin, Vol. **271** (1994) ISBN: 978–0-7923-9460-0.

[29] I. Repins, M.A. Contreras, B. Egaas, C. DeHart, J. Scharf, C.L. Perkines, B. To, R. Noufi, "19·9%-efficient ZnO/CdS/CuInGaSe$_2$ solar cell with 81·2% fill factor", *Prog. Photovolt.*, Vol. **16**, no. 3 (2008) pp. 235–239.

[30] M.C. Scharber, N.S. Sariciftci, "Efficiency of bulk-heterojunction organic solar cells", *Prog. Polym. Sci.*, Vol. **38**, no. 12 (2013) pp. 1929–1940.

[31] L. Meng, Y. Zhang, X. Wan, C. Li, X. Zhang, Y. Wang, X. Ke, Z. Xiao, L. Ding, R. Xia, H.-L. Yip, Y. Cao, Y. Chen, "Organic and solution-processed tandem solar cells with 17.3% efficiency", *Science*, Vol. **361**, no. 6407, (2018) pp. 1094–1098.

[32] H.E. Ruda, "A theoretical analysis of electron transport in ZnSe", *J. Appl. Phys.*, Vol. **59** (1986) pp. 1220–1231.

[33] M. Lundstrom, "*Fundamentals of Carrier Transport*", 2nd ed., Cambridge University Press, Cambridge (2002) ISBN-13: 9780521631341.

[34] C.A. Klein, "*Bandgap* dependence and related features of radiation ionization energies in semiconductors", *J. Appl. Phys.*, Vol. **4** (1968) pp. 2029–2033.

[35] A. Owens, A. Peacock, "Compound semiconductor radiation detectors", *Nucl. Instrum. Methods*, Vol. **A531** (2004) pp. 18–37.

[36] W. Que, J.A. Rowlands, "X-ray photogeneration in amorphous selenium: Geminate versus columnar recombination", *Phys. Rev. B*, Vol. **51** (1995) pp. 10500–10507.

[37] J. Wu, W. Walukiewicz, K.M. Yu, J.W. Ager, E.E. Haller, H. Lu, W.J. Schaff, Y. Saito, Y. Nanishi, "Unusual properties of the fundamental bandgap of InN", *Appl. Phys. Lett.*, Vol. **80**, no. 21 (2002) pp. 3967–3969.

[38] C. Wu, T. Li, L. Lei, S. Hu, Y. Liu, Y. Xie, "Indium nitride from indium iodide at low temperatures: Synthesis and their optical properties", *New J. Chem.*, Vol. **29** (2005) pp. 1610–1615.

[39] J.E. Moore, "The birth of topological insulators", *Nature*, Vol. **464** (2010) pp. 194–198.

Appendix G: Radiation Environments

G.1 Sources of Radiation

Every detection system sits in a sea of radiation, and therefore its sensitivity in detecting radiation ultimately depends on the level of the instrumental background induced by the local radiation field. This is simply given by the square root of the background within a certain bandwidth divided by the detector efficiency within that bandwidth. The environments encountered at the Earth and in space form the backdrop for all radiation measurements, and therefore a good understanding of the local radiation field is essential in detector design and operation.

G.2 Radiation in the Terrestrial Environment

At sea level, most of a detection system's background arises from a combination of cosmic ray secondaries and more than 60 naturally occurring radionuclides, whose progenitors can be grouped into three broad categories (primordial, cosmogenic and man-made – for a review see [1]), which are described below.

G.2.1 Cosmic Rays and Their Secondaries

Cosmic rays originate from outside of the solar system and consist primarily of 87% protons, 11% alpha particles, 1% heavier ions of atomic number 4–26 and about 1% electrons of very high energy. They have energies extending from a few tens of MeV per nucleon to 10^{13} GeV per nucleon – the average energy of a cosmic ray proton being around a GeV. The primary cosmic rays constantly bombard the Earth and interact with atoms in the upper atmosphere to create a cascade of secondary radiation. The flux of the incoming primaries is about 0.1 particles $cm^{-2}s^{-1}$ at the top of the atmosphere and decreases with increasing depth in the atmosphere, whereas the secondary radiation component increases, reaching a maximum in the upper troposphere at an altitude of around 20 km – the so-called Pfotzer maximum. At the maximum, in addition to particle fluxes, there is a sizable continuous bremsstrahlung component whose origin lies in cosmic ray induced electromagnetic cascades. As the altitude now decreases, particle fluxes steadily decrease as particles lose their energy by additional collisions until the majority either decay or are absorbed. Given that the absorption length of fast protons is ~120 g cm^{-2}, virtually none of the primary radiation survives the ~8 radiation lengths down to the Earth's surface. At the surface, the contribution to exposed dose (about 26 mrem per year) is therefore entirely due to the secondary radiation the primaries produce, consisting mainly of muons, electrons and 511 keV gamma rays from the annihilation of positrons. Most of the natural neutron background is also a product of cosmic rays interacting with the atmosphere. The neutron energy peaks at around 1 MeV and rapidly drops above. At the Earth's surface, the total average particle flux is around 1 $cm^{-2}s^{-1}$ and is composed of ~78% muons, 20% electrons (muon decay being the dominant source of low energy electrons), 1% protons and 1% neutrons. Secondary particle fluxes are not uniform around the earth due to the earth's magnetic field. Fluxes are lowest at the equator due to the higher geomagnetic cutoff rigidity (R~17 GV/nucleon) and highest at the poles where the field is at its weakest (R<1 GV/nucleon) and the field lines extend to the surface. The production of secondary cosmic rays is also modulated by the 11-year solar cycle, which primarily affects the number of primary particles with energies < 1 GeV/nucleon incident on the top of the atmosphere. At solar minimum, primary fluxes are about twice as high as at solar maximum. At sea level, the variation in secondary particle fluxes is much less – being < 20%. However, secondary particle fluxes can increase dramatically during solar flare events.

G.2.2 Cosmogenic Radionuclides

Cosmic rays also cause elemental transmutation in the atmosphere, in which the secondary radiation generated by the cosmic rays interacts with atomic nuclei in the atmosphere to generate different nuclides. Although many nuclides are produced, in terms of dose, the main ones are; ^{14}C, which is produced primarily in secondary interactions with atmospheric nitrogen (e.g., $^{14}N(n,p)^{14}C$), ^{3}T from secondary neutron interactions with N and O (e.g., $^{14}N(n,^{12}C)^{3}He$) and ^{7}Be and ^{10}B, produced in secondary neutron and proton interactions with N and O. The most notable reaction product is ^{14}C, which eventually reaches the Earth's surface and is absorbed into living organisms. The constant production rate coupled with the relatively short half-life of ^{14}C (*i.e.*, 5730 years) is the principle behind radiocarbon dating.

G.2.3 Primordial Radionuclides

The primordial radionuclides (also called terrestrial radionuclides) were left over from when the universe and Earth were created and have their origins in stellar nucleosynthesis. They are typically long lived, with half-lives often on the order of hundreds of millions of years. Their progeny or decay products are also usually considered as primordial. The main radionuclides are ^{238}U, 232Th and ^{40}K and their decay products. Most of these sources have been decreasing, due to radioactive decay since the formation of the Earth. Thus, the present activity on Earth from ^{238}U is only half as much as it originally was because of its 4.5 billion year half-life, and ^{40}K (half-life 1.25 billion years) is now only at ~ 8% of its original activity. Many shorter half-life and thus more intensely radioactive isotopes have not decayed out of the terrestrial environment, because of natural on-going production. Examples of these are ^{226}Ra (decay product of ^{238}U) and ^{222}Rn (a decay product of ^{226}Ra). Other primordial radionuclides include ^{50}V, ^{87}Rb, ^{113}Cd, ^{115}In, ^{123}Te, ^{138}La, ^{142}Ce, 144Nd, ^{147}Sm, ^{152}Gd, ^{174}Hf, ^{176}Lu, ^{187}Re, ^{190}Pt, ^{192}Pt and ^{209}Bi.

G.2.4 Man-made Radionuclides

Man-made radionuclides are generated primarily for the nuclear power industry, weapons production and nuclear medicine. The isotopes generated in the first two areas tend to be long lived, and are fuelled by refined isotopes, such as ^{235}U and ^{239}Pu, while nuclear medicine relies on short-lived isotopes, such as ^{99m}Tc, ^{123}I, ^{131}I, ^{201}Tl, ^{18}F and ^{111}In. In terms of human exposure, made-made radiation only accounts for about 18% of the received dose. However, the largest and most familiar source of exposure is due not to radioisotopes, but to medical X-rays.

A problem of concern is the highly variable contamination found in a lot of the materials used in fabricating detection systems. This is especially important in the design of low-level counting systems used for metrology or the assaying of low-level materials. Kreger and Mather [2] showed that reworked Al usually has more radioactive contamination than virgin material, since the Ra in instrument dials is often melted down with Al scrap. In addition, the use of tracers in modern industrial production processes tends to enhance the activity present in various materials. This is particularly true for iron and steel manufactured after 1940. In fact, Grindberg and Le Gallic [3] have pointed out that steel prepared after 1953 often contains amounts of ^{60}Co, ^{106}Ru, 232Th and rare earths. Lead is often contaminated with uranium during the purification process. The principle contaminant is ^{210}Pb, which has a half-life of 22 years. Therefore, the authors recommend that lead used for low-level shielding should be over 100 years old. Further information on the intrinsic activities found in various materials used in gamma-ray instruments may be found in Kreger and Mather [2] and Van Lieshout *et al.* [4].

Until recently, fall-out from atmospheric atomic tests was sometimes a surface contaminant. Initially it contributes many peaks to the background spectrum from heavy fission fragments, but a few years after the explosion only ^{137}Cs is significant (Aten *et al.* [5]). More recently, ^{131}I and ^{137}Cs were detected worldwide in air and food samples following both the Chernobyl and Fukushima nuclear accidents. For low-level counting systems, this form of contamination can be removed by sandblasting all exposed surfaces. For airborne contaminants, such as radon and thoron, rapid ventilation prevents equilibrium with their radioactive daughters and the subsequent deposition of radioactive dust. For ultra-low background systems, filtration of the air can be important (Grindberg and Le Gallic [3]).

G.2.5 The Radiation Environment at Sea Level

The radiation environment at sea level is comprised primarily of three components: the cosmic ray induced muon background, terrestrial radiation and airborne radionuclides. The terrestrial radiation arises from the primordial radionuclides, which exist naturally in soil and rock and all building materials. The contribution from the air arises from gaseous ^{222}Rn, which is a decay product of ^{226}Ra that occurs naturally in rocks and soils. It is in turn derived from the decay of primordial ^{238}U. On the ground, cosmic rays contribute less than 10% of the total absorbed dose rate of natural background radiation, of which radon in air is the largest contributor [6]. In Fig. G.1, we show a background energy loss spectrum measured at sea level with an 71 cm^3 intrinsically pure Ge detector. It is characterized by a strong continuum up to the ^{208}Tl line at 2.62 MeV upon which many line features are superimposed. The continuum arises from cosmic-ray muons whose energies usually peak outside the spectral region of interest. The myriad of line features emanate primarily from the primary decay chains – the uranium-radium, actinium and thorium decay series, which begin with ^{238}U, ^{235}U and ^{232}Th, respectively, and after a chained sequence of decays end up as isotopes of Pb or Bi. Some features arise from the detector itself (*e.g.*, ^{73m}Ge). The broad bump around 90 keV is mainly due to the K X-rays of the various daughter products. The only spectral line that can be attributed to cosmic rays is the annihilation line at 511 keV.

In Fig. G.2, we show diagrams of the uranium-radium, actinium and thorium decay series. The absence of the neptunium decay chain is due to the lack of sufficiently long-lived members of this chain and as such, complete decay of the parent

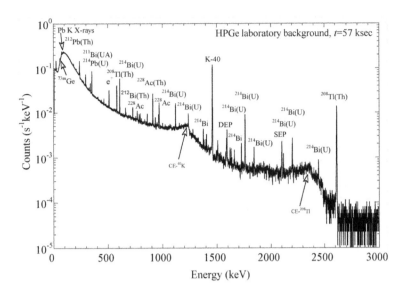

FIGURE G.1 Typical low-level background spectrum at sea level measured in Noordwijk, the Netherlands, with a 71 cm³ intrinsically pure Ge detector. Note the many radionuclide lines resulting from the primordial isotopes of the thorium (Th), uranium (U) and actinium (UA) decay chains. CE-⁴⁰K and CE-²⁰⁸Tl denote the Compton edges of ^{40}K and ^{208}Tl, respectively.

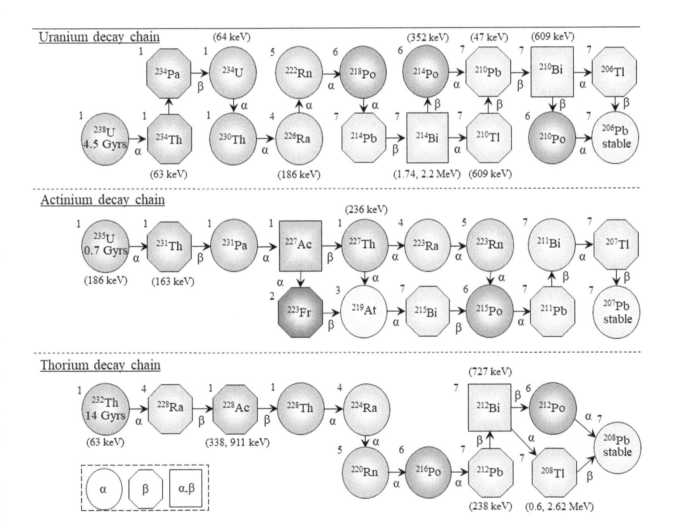

FIGURE G.2 Simplified diagram of the primordial uranium, actinium and thorium decay series. The energies of the major γ-ray lines are indicated in parentheses next to the corresponding nuclide. The numbers to the upper left sides denote different family groups, for example, 1 denotes the actinides; 2 – the alkali metals; 3 – the halogens; 4 – the alkali earth metals; 5 – the noble gasses; 6 – the metalloids and 7 – the poor metals.

radionuclides and their progeny has already occurred. Furthermore, the progenitor (^{237}Np) is transuranic[1] meaning that it does not occur naturally. In general, only the uranium-radium and thorium series dominate, since these isotopes occur in virtually all materials. The decay to stable elements for each series is accompanied by numerous emission (α, β, γ) from the parent and daughter nuclides before eventually forming stable isotopes of Pb (^{206}Pb, ^{207}Pb and ^{208}Pb). Thorium and uranium primarily undergo alpha and beta decay and are not easily directly detectable. However, many of their daughter products are strong gamma emitters; ^{232}Th is detectable *via* a 239 keV peak from ^{212}Pb; lines at 511 keV, 583 keV and 2613 keV from ^{208}Tl and lines at 911 keV and 969 keV from ^{228}Ac. ^{233}U is similar but lacks the ^{228}Ac peak, which distinguishes it from ^{232}Th. ^{238}U manifests as peaks at 609 keV, 1120 keV and 1764 keV from ^{214}Bi (*cf.* the same peak for atmospheric radon). ^{40}K is detectable directly *via* its 1461 keV gamma peak. The resulting spectrum, as illustrated in Fig. G.1, is therefore continuous with many superimposed peaks. The outstanding peak at 2.62 MeV (with pair peaks at 2.11 MeV (denoted as SEP) and 1.60 MeV, denoted as DEP) is due to 3.1 min ^{208}Tl – a daughter activity in the decay chain of ^{232}Th, which has a half-life of 1.41×10^{10} years. It is prevalent in all building materials. Above 2.62 MeV the background spectrum falls steeply by an order of magnitude. There are no radioactive isotopes in the natural environment that emit gamma rays having energies higher than 2.62 MeV.

G.3 The Space Radiation Environment

The space radiation environment is dominated by high-energy protons as they are the major component in both the Galactic Cosmic Rays and Solar Energetic Particles (GCRs and SEPs). Deuterons and alpha particles can be neglected because of their relatively low fluences. Likewise, the ambient flux of photons at MeV energies is many orders of magnitudes less than that of protons. The actual radiation environment depends on where you are in the solar system. We consider three cases: *a*) the inner heliosphere which we arbitrarily define to extend form the sun to 1 AU, *b*) Earth orbit at 1 AU and *c*) the outer heliosphere which we arbitrarily define as extending beyond 1 AU.

G.3.1 The Inner Heliosphere

For the inner heliosphere (< 1AU) one needs only to consider two components of radiation – solar particle emissions and the Galactic Cosmic Rays (GCRs). Solar particle emission can be considered to consist of two components, a more or less steady stream of low energy particles known as the solar wind and a highly energetic transient component emitted during periods of increased solar activity, known as the Solar Energetic Particles (SEPs). The intensities of both components fall off with heliospheric distance as the inverse square distance[2] from the Sun. Consequently, solar emissions are an important radiation component in the inner solar system.

G.3.1.1 The Solar Wind

The solar wind is constantly emitted from the upper atmosphere of the sun and consists of a stream of energetic particles in all directions at speeds of several hundred kms^{-1}. Its composition reflects the materials found in photosphere, consisting mainly of ionized hydrogen (electrons and protons of a few keV) with an 8% component of helium (alpha particles) and trace amounts of heavier ions and atomic nuclei: C, N, O, Ne, Mg, Si, S and Fe. Their fluxes are heavily influenced by solar activity and can vary by a factor of ~20 depending on the solar cycle. An important effect of the solar wind occurs at a distance of ~90 AU, at which it undergoes a transition from supersonic to subsonic speeds setting up a shock front. This acts as a barrier to the galactic cosmic rays, decreasing its flux at lower energies (≤ 1 GeV) by about 90%. Needless to say, this "filtering" at low energies is modulated by the solar cycle.

G.3.1.2 Solar Energetic Particles

Solar particle events (SPEs) are impulsive events characterized by copious particle emission (the so-called Solar Energetic Particles or SEPs) from the sun's chromosphere and occur mainly during periods of increased solar activity and most notably solar flares. The composition reflects the solar composition and is similar to that of the GCRs. The emissions are highly directed and propagate along interplanetary magnetic field (IMF) lines into interplanetary space. Events can last for days and fluxes can vary by as much as six orders of magnitude. Peak fluxes can stay high for periods of hours. Particle spectra tend to be flat, rolling over at about 30 MeV, but providing significant fluxes up to a few hundred MeV.

---

[1] Having an atomic number greater than 92 and so is not synthesized naturally.
[2] Controversial for SEPs.

G.3.1.3 The Galactic Cosmic Rays

The GCRs originate outside of the solar system and were produced in supernovae explosions and at compact objects such as pulsars and neutron stars. On passing through the galactic magnetic field, the particles become diffuse and on arrival in the solar system, they are seen as a low-level isotropic flux with an equal intensity from all directions but modulated by the sun's 11-year cycle. The total flux of GCRs is about 0.5 $cm^{-2}s^{-1}sr^{-1}$ near solar minimum and about half that value near solar maximum. The average annual fluence is about 2×10^8 cm^{-2}. Protons account for approximately 87% of all GCR particles, helium nuclei ~11% and heavier nuclei with atomic numbers range 4 to 26 for the remainder. They typically have energies in the range of 10 MeV per nucleon up to a TeV per nucleon. Interestingly, the shape of the energy spectrum of all species is similar, consisting of a broad peak-like structure with a maximum at about 300 MeV per nucleon at solar minimum and a few GeV per nucleon at solar maximum. The shift in the peak is due to low energy filtering caused by the increased solar wind during solar maximum.

G.3.2 Earth Orbit

Near the Earth one needs to consider three main components – the GCRs, SPE and radiation trapped in the Van Allen belts. The total flux of GCRs at the Earth ranges from about 2 $cm^{-2}s^{-1}$ to 4 $cm^{-2}s^{-1}$, depending on the position in the solar cycle and consists mainly of GeV protons (~90%). Direct effects of the solar wind can usually be neglected because of the moderating effects of Earth's magnetosphere. Only the SEPs need be considered, particularly during a solar flare. During such an event the Earth can be showered in energetic solar particles (primarily protons) released from the flare site. Some of these particles spiral down Earth's magnetic field lines, penetrating the upper layers of the atmosphere where they create additional ionization, which may produce a significant increase in the radiation environment. The occurrence of solar particle events is tied to the solar cycle and is typically characterized by a period of four years of relative inactivity, followed by seven years with increasing activity. Solar events are stochastic in nature, so the probabilities of exceeding a particular fluence can only be realistically assessed on a statistical basis. As an example, at the Earth, there is roughly a 15% chance per annum of encountering at least one solar flare of total fluence $>10^9$ protons cm^{-2} and a 6% chance of encountering at least one with a fluence $>10^{10}$ protons cm^{-2}, for fluences above 30 MeV.

G.3.2.1 Radiation Belts

The motion of Earth's molten iron core generates a magnetic field, which can collect and trap solar and low-energy GCR particles, creating two bands of increased radiation that surround the Earth, known as the Van Allen belts. Since the Earth's magnetic field is offset and tilted from its rotation axis, the radiation belts are similarly tilted and offset. The geometry and morphology of the belts is shown in Fig. G.3. The trapping of particles occurs where the magnetic field lines come close together and is dependent on the incoming particle's energy and angle of incidence. Once trapped, the particles spiral around the magnetic field lines bouncing back and forth between the magnetic poles at the so-called "mirror" points (indicated in Fig. G.3). Without this "mirroring", ions and electrons would not be trapped in the Earth's magnetosphere, but follow the field lines into the atmosphere, where they would be absorbed. The solar cycle has two main effects on the belts. The increased pressure on the magnetosphere during solar maximum causes them to compress and thereby increases the fluxes in the radiation belts. In addition, there is a net influx of solar particles from many sporadic solar flares. Physically, the belts are donut-shaped crescents centered near the equator. The inner belt extends in altitude from approximately 1,000 km to 6,000 km and is populated by high concentrations of electrons with energies of a few hundred keV and energetic protons with energies up to a few hundred MeV. The average omnidirectional proton flux above 10 MeV is ~10^5 $cm^{-2}s^{-1}$. The outer belt extends from about 13,000 to 60,000 km above the Earth's surface and is populated mainly by high-energy (~few MeV) electrons with average omnidirectional fluxes of the order of 10^6 $cm^{-2}s^{-1}$. These fluxes are far more variable than the inner belt as the field is weaker and therefore more easily influenced by solar activity. Near the poles, the trapped radiation belts extend almost down to the surface. The trapped radiation belts are not static; their altitude distribution and intensity are greatly dependent on solar activity, with hourly, daily and seasonal changes.

G.3.2.2 The South Atlantic Anomaly (SAA)

The Earth's dipole field is tipped and displaced in a region over the South Atlantic Ocean. The resultant field when combined with local magnetic perturbations, exhibits unusually low values extending over a 100° in longitude and 60° in latitude centered off the Brazilian coast (indicated in Fig. G.3). The inner Van Allen belt in this region reaches lower altitudes extending deep down into the atmosphere. As a result, very large numbers (greater than a factor of 100) of low energy charged particles congregate in this region (as illustrated in Fig. G.4), which is termed the South Atlantic Anomaly (SAA). The SAA is responsible for nearly all trapped radiation encounters in low Earth orbit and can only be avoided by satellites in low Earth orbits with inclinations <15°. The passage of a satellite through this region produces anomalously high counting rates, primarily due to the interaction of low energy (< 600 MeV) trapped protons with the instruments and

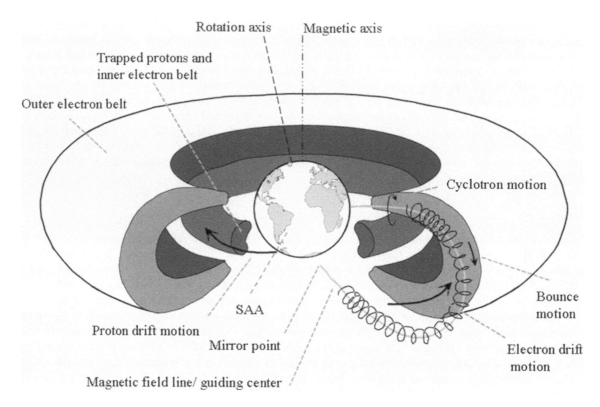

FIGURE G.3 Schematic illustrating the geometry and morphology of the Van Allen radiation belts (adapted from [7]). Charged particles encountering the belts undergo three types of motion – cyclotron motion around a geomagnetic field line, bounce motion along the field line and drift motion around the Earth. In essence, particles circle their guiding field line in a spiral pattern and are repelled from regions where the field lines converge, causing the particle to bounce back from a "mirror" point. Without this "mirroring", ions and electrons would not be trapped in the Earth's magnetosphere, but would follow their guiding field lines into the atmosphere, where they would be absorbed. The drift motion is caused by particles jumping from one field line to an adjacent one.

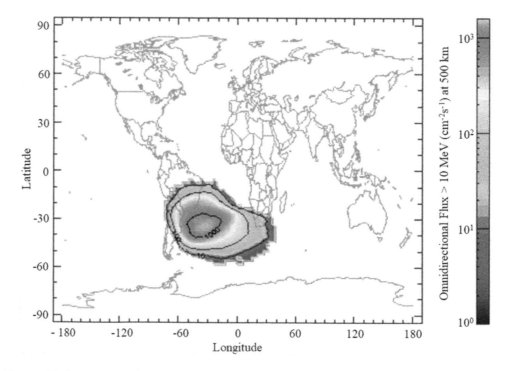

FIGURE G.4 World map of the integral proton flux >10 MeV at an altitude of 500 km. The influence of the SAA shows up clearly (Image courtesy of the European Space Agency).

spacecraft. For a typical orbit (inclination ~28° and altitude ~400 km), a satellite will pass through the SAA about 6 times a day, each transit lasting about 20 minutes. The proton spectrum encountered in this region may be expressed as ~ exp ($-E_p$ /25 MeV) for proton energies $50 < E_p < 200$ MeV and the electrons as $E^{-3.8}$ ($200 < E_e < 1500$ keV). The excess count rate is proportional to the degree of penetration of the satellite into the SAA.

G.3.3 The Outer Heliosphere

For most of the outer heliosphere (say > 1AU), only the CGRs need be considered, since the direct contribution from solar particles is diminished simply because of the inverse scale scaling of particle fluxes. However, the same is not true in the region of Jupiter, which hosts the most severe radiation environment in the solar system – a direct consequence of its strong magnetic field. Its magnetosphere is in fact the largest "object" in the solar system – the action of which is to trap and accelerate solar particles, producing intense radiation belts similar to Earth's but with particle fluences thousands of times larger and ten times more energetic. The belts themselves are dominated by high-energy electrons with energies up to ~200 MeV, although there are also significant fluxes of protons with energies up to ~90 MeV. Peak electron and proton fluxes encountered near Jupiter are typically 1,000 times that at the Earth. Lastly, it should be noted, that Saturn, Uranus and Neptune also have magnetic fields with strengths comparable to or larger than the Earth's. However, the radiation belts surrounding these planets are not expected to be much larger than the Earth's.

REFERENCES

[1] M. Eisenbud, T.F. Gesell, *"Environmental Radioactivity: From Natural, Industrial and Military Sources"*, 4th ed., Academic Press, Bozeman, MT (1997). ISBN-13: 978–0122351549.

[2] W.E. Kreger, R.L. Mather, *"Scintillation Spectroscopy of Gamma Radiation"*, ed. S.M. Shafroth, Gordon & Breach, New York (1967).

[3] B. Grindberg, Y. Le Gallic, "Basic characteristics of a laboratory designed for measuring very low activities", *Int. J. Appl. Rad. Isotope*, Vol. **12** (1961) pp. 104–117.

[4] R. Van Lieshout, A.W. Wapstra, R.A. Ricci, R.K. Girgis, *"Alpha, Beta and Gamma-Ray Spectroscopy"*, vol. **1**, ed. K. Seigbahn, North Holland Publishing Co., Amsterdam (1966) pp. 501–557.

[5] A.H.W. Aten Jr., I. Heertje, W.M.C. de Jong, "Measurements of low level environmental radiation by means of Geiger Müller counters with observations in the Amsterdam area", *Physica*, Vol. **27** (1961) pp. 809–820.

[6] E.R. Benton, E.G. Yukihara, A.S. Arena Jr., A.C. Lucas, *"Tissue Equivalent Detectors for Space Crew Dosimetry and Characterization of the Space Radiation Environment"*, A NASA Experimental Program to Stimulate Competitive Research (EPSCoR) Project, NSF, Alexandria, Virginia (2007).

[7] E.R. Benton, E.V. Benton, "Space radiation dosimetry in low-Earth orbit and beyond", *Nucl. Instrum. Methods*, Vol. **B184**, no. 1–2 (2001) pp. 255–294.

Appendix H: Table of Radioactive Calibration Sources

H.1 Introduction

The activity of a radioactive source is defined as its rate of decay and is given by

$$\frac{dn}{dt}\Big|_{decay} = -\lambda n, \qquad (H.1)$$

where n is the number of radioactive nuclei and λ is defined as the decay constant. Historically, the unit of activity has been the Curie (Ci), which is defined as exactly 3.7×10^{-10} disintegrations/sec and initially arose as the best estimate for the total activity of 1 gm of pure ^{226}Ra. For laboratory measurements involving calibrating standards, the mCi (10^{-3} Ci) and the μCi (10^{-6} Ci) are most commonly used. Other commonly used units are the Roentgen and the REM. The Roentgen is named after the discoverer of X-rays. It is defined as that amount of radiation required to produce 0.001293 grams of air ions carrying one electrostatic unit of electricity of either sign. This unit is only used with X-rays and gamma-rays. The common submultiple is the milli-roentgen or mR. For ionizing radiations we use the Roentgen Equivalent Man (REM), which includes a factor to take into account the abilities of the different radiations to produce biological injury. The common submultiple is the milli-rem or mrem. The amount of radiation received by a particular body is known as the dose and is defined as the amount of energy imparted to matter by ionizing particles, per unit mass of irradiated material, at the place of interest. Radiation doses are usually quantified in terms of the REM and the rad. The rad is a measure of the amount of the energy absorbed in a certain amount of material (100 ergs per gram). The SI unit of absorbed dose is the Gray (Gy) where 1 Gy = 100 rads. Because the rad does not specify the medium, a medium should be stated unless clearly implied. For example, the term "tissue rad" should be used in the case of exposure of soft tissue. Lastly, the REM and the rad are related by the Relative Biological Effectiveness (RBE), which is an empirical factor that expresses the effectiveness of a particular type of radiation in producing the same biological response as X− or γ−radiation. For applied radiation protection, the quality factor (QF) is the preferred term for most practical standards.

H.2 Radiation Quantities and Units[1]

Unit of activity = **Curie**:
1 Ci = 3.7×10^{10} disintegration s^{-1}.

SI unit of activity = **Becquerel**:
1 Bq = 1 disintegration/sec = 2.703×10^{11} Ci

Unit of exposure dose for X- and γ-radiation = **Roentgen**:
1 R = 1 esu cm^{-3} = 87.8 erg g^{-1} (5.49×10^{7} MeV g^{-1}) of air.
Gray (Gy) is the SI unit of absorbed dose. One gray is equal to an absorbed dose of 1 joule/kilogram (100 rads).

Unit of absorbed dose = **rad**:
1 rad = 100 erg g^{-1} (6.25×10^{-7} MeV g^{-1}) or 0.01 J/kg (0.01 gray) in any material.
1 Gy = 100 rad.

Fluxes (per cm^2) to liberate 1 rad in carbon:
3.5×10^{7} minimum ionizing singly charged particles,
1.0×10^{9} photons of 1 MeV energy
(These fluxes are correct to within a factor of 2 for all materials.)

Unit dose equivalent (for protection): **REM,** which is the special unit of any of the quantities, expressed as dose equivalent. The dose equivalent in REMs is equal to the absorbed dose in rads multiplied by the quality factor, **QF,** i.e.

1 rem = 1 rad × QF,

---

[1] This section was adapted, with permission, from the [1,2].

TABLE H.1

Radiation weighting factors, w_R, for various types of radiation. While w_R has largely superseded the QF, in the definition of equivalent dose, QF is still used in calculating the operational dose equivalent quantities used in monitoring.

Type of radiation	Radiation weighting factor, w_R	Absorbed dose[1] equal to a unit dose equivalent
X- and γ-rays, all energies	1	1
Electrons and muons, all energies	1	1
Protons > 2 MeV	5	0.1
neutrons < 10 keV	5	0.1
Fast neutrons	20	0.05
Alpha particles	20	0.05
Heavy ions, fission fragments	20	0.05

[1] Absorbed dose in rad equal to 1 REM or the absorbed dose in gray equal to 1 sievert.

The quality factor is a measure of the ability of ionizing radiation to cause biological damage, relative to a standard dose of X-rays. The SI dose equivalent unit is the **Sievert,** which is equal to the absorbed dose in Gy multiplied by the quality factor (1 Sv=100 rems).

1 Sv = 1 Gy × QF,

The QF is related functionally to the unrestricted linear energy transfer (LET) of a given radiation and is multiplied by the absorbed dose to derive the dose equivalent at the point of interaction. However, the incident radiation may not be the radiation type actually depositing the dose (as in the case of secondary interactions). In view of this limitation, the International Commission on Radiation Units & Measurements [3,4] has recommended that the QF be replaced by the radiation weighting factor, w_R, which is a dimensionless factor that takes this limitation into account by assigning a pre-defined value to radiation of a given type and energy incident on a body. It is broadly comparable with previously recommended quality factors with the exception of medium energy neutrons that are a factor of ~2 higher. Table H.1 lists radiation weighting factors for a range of radiation types and energies. The relationship between the absorbed dose and the dose equivalent illustrated in Fig. H.1. It should be noted that while the radiation weighting factor, w_R has largely

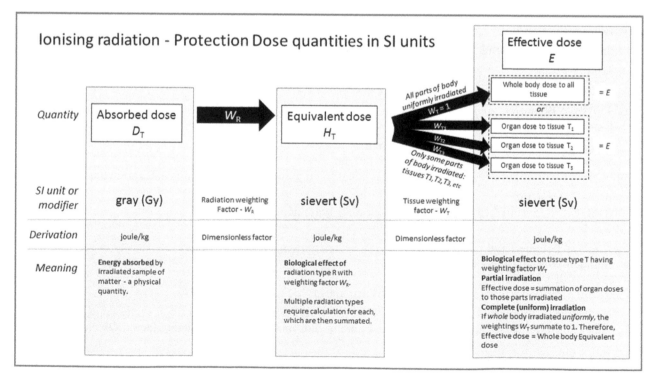

FIG. H.1 Graphic illustrating relationship between SI radiation dose units (Image attribution: By Doug Sim – Own work, CC BY-SA 3.0, https://commons.wikimedia. org/w/index.php?curid=29827386).

TABLE H.2

Radiation exposure Sv yr^{-1} (mrem yr^{-1}) of a typical person in the US (averaged over the population).

Natural sources	Exposure mSv yr^{-1} /(mrem yr^{-1})	Artificial sources	Exposure mSv yr^{-1} /(mrem yr^{-1})
Cosmic radiation[1]	0.27 (27)	Medical[3]	0.53 (53)
Cosmogenic[1]	0.01 (1)	Consumer products[4]	0.10 (10)
Terrestrial radiation[1]	0.28 (28)	Miscellaneous	0.03 (3)
Internal isotopes[2]	0.39 (39)	Nuclear power	0.003 (0.3)
Inhaled radon	2.0 (200)	Occupational	0.01 (1)
Total natural	2.95 (295)	Total man made	0.67 (67)
Total dosage 3.62 (362)			
Dosage range (allowing for different geographical and other factors) 1.98–10.3 (198–1032)			

[1] See Appendix G.
[2] Ingested in food, mainly ^{40}K and ^{210}Pb.
[3] Mainly diagnostic X-rays, nuclear medicine.
[4] Primarily drinking water, building materials, smoke alarms.

superseded the quality factor in the definition of equivalent dose, QF is still used in calculating the operational dose equivalent quantities used in monitoring.

The accepted value for the average background radiation dose from natural and man-made sources to people living in the Unites States is 362 mrem yr^{-1} effective dose equivalent (EDE) [5]. However, it should be noted that across the US there is a substantial geographic variability in exposures to radon, terrestrial and cosmic radiation. In fact, total doses can range between 1.98–10.3 mSv yr^{-1} (198–1032 mrem yr^{-1}).

Table H.2 presents a breakdown of the sources of background radiation and the average annual effective dose equivalents[2] associated with those sources, as described by the Committee on the Biological Effects of Ionizing Radiations in their 1990 publication [5]. This figure illustrates that the dose from exposure to indoor radon (200 mrem/year EDE) constitutes over 50% of the total dose.

H.3 Exposure Levels and Limits

Recommended exposure limits for radiation workers (whole body dose):
ICRP[3]: 20 mSv yr^{-1} (2 REM yr^{-1}) averaged over 5 years with the dose in any one year not exceeding 50 mSv (5 REM).
Maximum *permissible occupational dose* for the whole body:
50 mSv yr^{-1} (5 REM yr^{-1}) or ~ 1 mSv wk^{-1} (100 mrem wk^{-1})
Lethal dose: Whole-body dose from penetrating ionizing radiation resulting in 50% mortality in 30 days (LD 50/30), assuming no medical treatment. Typically, the LD 50/30 is about 5 Gy (500 rads) received over a very short period.

REFERENCES

[1]. *"Particle Properties Data Booklet"*, Lawrence Berkeley Laboratory, Berkeley, CA (1987).
[2]. A.C. Upton, "The biological effects of low-level ionizing radiation", *Sci. Am.*, Vol. **246** (1982) pp. 41–49.
[3]. ICRU Chairman H.-G. Menzel, "ICRU Report 90, key data for ionizing-radiation dosimetry: Measurement standards and applications", *J. ICRU Meas.*, Vol. **14**, no. 1 (2014) Pages NP, https://doi.org/10.1093/jicru/ndw043.
[4]. "1990 recommendations of the international commission on radiological protection", *Ann. ICRP*, ICRP publication 60, C. Clement, ICRP Scientific Secretary and Editor-in-Chief, Pergamon Press, NY, Vol. **21**, no. 1–3 (1991). ISBN 978-0-08-041144-6.
[5]. Committee on the Biological Effects of Ionizing Radiations (BEIR), *"Health Effects of Exposure to Low Levels of Ionizing Radiation, BEIR V"*, National Academy Press, Washington, DC (1990).

---

[2] The effective dose equivalent is the radiation dose to any organ or by any type of radiation (*i.e.*, alpha, beta, gamma, neutron) that is equivalent in terms of health risk to a uniform whole-body exposure to external gamma radiation.
[3] International Commission on Radiological Protection.

H.4 Table of Radionuclides

In the following table, we list over 50 radionuclides commonly used as α, β, X- and γ-ray calibration standards. Sources may be obtained from the following vendors: Eckert & Ziegler Isotope Products Inc., 24937 Avenue Tibbitts, Valencia, CA 91355, USA; The Institute of Isotopes Co., Ltd. (IZOTOP), 1121 Budapest, Konkoly Thege Miklós út 29–33, Hungary or Standard Reference Materials, 100 Bureau Drive, Stop 2302, National Institute of Standards and Technology Gaithersburg, MD 20899–2302, USA and are generally available with strengths ranging from μCi to mCi and calibrated activities from 1 to 20%. The first column of the table lists the parent nuclide. Some daughter nuclides may be in equilibrium with the parent nuclide when source is supplied. In cases where this may occur, the transition probabilities for the daughters relate to the disintegrations of each daughter. This is stated in the table. Daughters with half-lives greater than the parent nuclide have not been listed since they would be present only in insignificant amounts. The second column gives the type of decay (*i.e.*, α, β, X or γ) and the third lists the corresponding decay energies in units of MeV. For β-emission, the end-point energy is also quoted. The fourth column gives the transition probabilities for each mode of the primary decay. They are expressed as percentages of the total number of nuclear transformations of the relevant nuclides. The last two columns give the photon energies and branching ratios for electromagnetic transitions.

Abbreviations:

Half-lives
  y-years
  d-days
  h-hours
  m-minutes
  s-seconds
  ms-milliseconds
  μs-microseconds

Type of decay
  e.c.-electron capture
  i.t.-isometric transition
  s.f.-spontaneous fission

Photons emitted
  IC-indicates that photons of the stated energy are ~100% internally converted.

NUCLIDE INDEX

Nuclide and half-life	Type of decay	Particle energies and transition probabilities		Electromagnetic transitions	
		energy MeV	Transition probability	photon energy Mev	photons emitted
Americium-241	α	5.387	1.6%	0.026	2.5%
(433y)		5.442	12.5%	0.033	0.1%
		5.484	85.2%	0.043	0.1%
		5.511	0.20%	0.0595	35.9%
		5.543	0.34%	0.099	0.02%
		others	low	0.103	0.02%
				0.125	0.004%
				others	low
				Np LX-rays	~40%
				(0.012–0.022)	
Antimony-124	β−	0.21	9%	0.603	98.0%
(60.2d)		0.61	52%	0.646	7.2%
		0.86	4%	0.709	1.4%
		0.94	2%	0.714	2.3%

(*Continued*)

(Cont.)

Nuclide and half-life	Type of decay	Particle energies and transition probabilities		Electromagnetic transitions	
		energy MeV	Transition probability	photon energy Mev	photons emitted
		1.57	5%	0.723	11.2%
		1.65	3%	0.791	0.7%
		2.30	23%	0.968	1.9%
		others	low	1.045	1.9%
				1.325	1.5%
				1.355	0.9%
				1.368	2.5%
				1.437	1.1%
				1.691	50.4%
				2.091	6.1%
				others	<0.5% each
Antimony-125 (2.77y)	β−	0.094	13.5%	0.035	4.5%
		0.124	5.7%	0.176	6.8%
		0.130	18.1%	0.321	0.5%
		0.241	1.5%	0.381	1.5%
		0.302	40.2%	0.428	29.8%
		0.332	0.3%	0.463	10.4%
		0.445	7.2%	0.601	17.8%
		0.621	13.5%	0.607	4.9%
				0.636	11.4%
				0.671	1.7%
				others	<0.5% each
				0.027 - 0.031	~50%(Te KX-rays)
	Daughter ¹²⁵ᵐTe(58d)			(23% of ¹²⁵Sb decays via ¹²⁵ᵐTe)	
	i.t. 100%			0.035	7%
				0.109	0.3%
				0.027 - 0.032	~110%(Te KX-rays)
Barium-133 (10.8y)	e.c.		100%	0.053	2.2%
				0.080	2.6%
				0.081	33.9%
				0.161	0.7%
				0.223	0.4%
				0.276	7.1%
				0.303	18.4%
				0.356	62.2%
				0.384	8.9%
				Cs KX-rays (0.030–0.036)	~120%
Barium-140 (12.80d)	β-	0.468	24%	0.014	1.3%
		0.582	10%	0.030	14%
		0.886	2.6%	0.163	6.2%
		1.005	46%	0.305	4.5%
		1.019	17%	0.424	3.2%

(*Continued*)

(Cont.)

Nuclide and half-life	Type of decay	Particle energies and transition probabilities		Electromagnetic transitions	
		energy MeV	Transition probability	photon energy Mev	photons emitted
		others	0.4%	0.438	2.1%
				0.537	23.8%
				0.602	0.6%
				0.661	0.7%
				others	low intensity
	Daughter ¹⁴⁰La				
Beryllium-7 (53.3d)	e.c.		100%	0.478	10.4%
Bromine-82 (35.3h)	β-	0.263	1.7%	0.221	2.3%
		0.444	98.3%	0.554	72%
				0.606	1.0%
				0.619	43%
				0.698	27%
				0.777	83%
				0.828	24%
				1.008	1.7%
				1.044	29%
				1.317	28%
				1.475	17%
				1.651	0.9%
				others up to 1.96	<1% each
Cadmium-109 (462d)	e.c		100%	Ag KX-rays (0.022–0.026)	67.7%
	via ¹⁰⁹ᵐAg(40s)			0.088	3.8%
				Ag KX-rays (0.022–0.026)	~34.5%
Calcium-45 (164d)	β−	0.257	100%		
Calcium-47 (4.54d)	β-	0.69	82%	0.489	6.8%
		1.22	0.1%	0.530	0.1%
		1.99	17.9%	0.767	0.2%
				0.808	6.8%
				1.297	75.1%
				others	low intensity
	Daughter ⁴⁷Sc(3.48d)				
		0.44	70%	0.159	69.7%
		0.60	30%		
Carbon-14 (5730y)	β−	0.156	100%		
Cerium-139 (137.5d)	e.c.		100%	0.166	79.9%
				0.033–0.039	~90%(La KX-rays)
Cerium-141 (32.5d)	β−	0.436	70%	0.145	48%
		0.581	30%	0.035–0.042	~17%(Pr KX-rays)

(Continued)

(Cont.)

Nuclide and half-life	Type of decay	Particle energies and transition probabilities		Electromagnetic transitions	
		energy MeV	Transition probability	photon energy Mev	photons emitted
Cerium-144	β–	0.182	19.1%	0.034	0.1%
(284.3d)		0.216	0.2%	0.040	0.4%
		0.236	4.4%	0.053	0.1%
		0.316	76.3%	0.080	1.5%
				0.100	0.03%
				0.134	10.8%
		via 7.2m^{144}Pr			
	β–	1.534	0.05%	0.059	0%
	i.t.		1.15%	0.696	0.05%
				0.814	0.05%
		via 7.2m^{144}Pr			
	β–	0.808	1.0%	0.696	1.53%
		2.298	1.2%	1.489	0.28%
		2.994	97.75%	2.186	0.72%
Cesium-134	β–	0.09	26%	0.475	1.5%
(2.06y)		0.42	2.5%	0.563	8.1%
		0.66	71.5%	0.569	14.0%
				0.605	97.5%
				0.796	85.4%
				0.802	8.6%
				1.038	1.0%
				1.168	2.0%
				1.365	3.3%
Cesium-137	β–	0.512	94.6%		
(30.17y)		1.174	5.4%		
	via ^{137m}Ba(2.6m):			0.662	85.1%
				Ba KX-rays (0.032–0.038)	~7%
Chlorine-36	β–	0.709	98.1%		
(3.01 × 10^{5}y)	e.c.		1.9%		
Chromium-51	e.c.		100%	0.320	9.83%
(27.7d)				0.005–0.006	~22%(V KX-rays)
Cobalt-56	β+	0.4	1%	0.511	from β+
(78.0d)		1.5	18%	0.847	99.97%
	e.c.		81%	0.977	1.4%
				1.038	14.0%
				1.175	2.3%
				1.238	67.6%
				1.360	4.3%
				1.771	15.7%
				2.015	3.1%
				2.035	7.9%
				2.599	16.9%
				3.010	1.0%
				3.202	3.0%
				3.254	7.4%
				3.273	1.8%

(Continued)

(Cont.)

Nuclide and half-life	Type of decay	Particle energies and transition probabilities		Electromagnetic transitions	
		energy MeV	Transition probability	photon energy Mev	photons emitted
				3.452	0.9%
				others	<1% each
Cobalt-57	e.c.		100%	0.014	9.5%
(271.7d)				0.122	85.5%
				0.136	10.8%
				0.570	0.01%
				0.692	0.16%
				others	low
				Fe KX-rays (0.006–0.007)	~55%
Cobalt-58	β+	0.475	15.0%	0.511	from β+
(70.8d)	e.c.		85.0%	0.811	99.4%
				0.864	0.7%
				1.675	0.5%
				0.006–0.007	~26%(Fe KX-rays)
Cobalt-60	β−	0.318	99.9%	1.173	99.86%
(5.27y)		1.491	0.1%	1.333	99.98%
				others	<0.01%
Curium-244	α	5.763	23.6%	0.043	0.02%
(17.8y)		5.806	76.4%	0.099	0.0013%
		others	low	0.152	0.0014%
				others (up to ~0.8)	low
				Pu LX-rays (0.012–0.023)	~8%
Europium-152	β−	0.185	1.8%	0.122	28.2%
(13.3y)		0.394	2.4%	0.245	7.4%
		0.705	13.8%	0.344	26.3%
		1.484	8.0%	0.411	2.2%
		others	1.7%	0.444	3.1%
	β+		~0.02%	0.779	12.8%
	e.c.		72.3%	0.867	4.1%
				0.964	14.4%
				01.086	10.0%
				1.090	1.7%
				1.112	13.6%
				1.213	1.4%
				1.299	1.6%
				1.408	20.6%
				~75 others	(<1%each)
Gallium-67	e.c.		100%	0.091	3.6%
(78.26h)				0.185	23.5%
				0.209	2.6%
				0.300	16.7%
				0.394	4.4%
				0.494	0.1%
				0.704	0.02%
				0.795	0.06%

(Continued)

(Cont.)

Nuclide and half-life	Type of decay	Particle energies and transition probabilities		Electromagnetic transitions	
		energy MeV	Transition probability	photon energy Mev	photons emitted
				0.888	0.17%
				0.008–0.010	43%(Zn KX-rays)
				via 9.2×s ^{67m}Zn	
				0.093	37.6%
				0.008–0.010	13%(Zn KX-rays)
Gold-198 (2.696d)	β–	0.285	1.32%	0.412	95.45%
		0.961	98.66%	0.676	1.06%
		1.373	0.02%	1.088	0.23%
Gold-199 (3.13d)	β–	0.25	21%	0.050	0.3%
		0.29	72%	0.158	39.6%
		0.45	7%	0.208	8.8%
				0.069–0.083	~18%(Hg KX-rays)
Hydrogen-3 (12.43y)	β–	0.0186	100%		
Indium-111 (2.804d)	e.c.		100%	0.171	90.9%
				0.245	94.2%
				0.023–0.027 ~84%(Cd	~84%(Cd KX-rays)
Iodine-125 (60.0d)	e.c.		100%	0.035	7%
				Te KX-rays (0.027–0.032)	138%
Iodine-129 (1.57 × 10^{7}y)	β–	0.150	100%	0.040	7.5%
				Xe KX-rays (0.030–0.035)	~69%
Iodine-131 (8.04d)	β–	0.247	1.8%	0.080	2.4%
		0.304	0.6%	0.284	5.9%
		0.334	7.2%	0.364	81.8%
		0.606	89.7%	0.637	7.2%
		0.806	0.7%	0.723	1.8%
	1.3% of ^{131}I decays via ^{131m}Xe (12d)				
	i.t.		100%	0.164	2%
			(percentages related to disintegrations of ^{131m}Xe)		
Iron-55 (2.69y)	e.c.		100%	Mn KX-rays (0.0059–0.0065)	~28%
Iron-59 (44.6d)	β–	0.084	0.1%	0.143	0.8%
		0.132	1.1%	0.192	2.8%
		0.274	45.8%	0.335	0.3%
		0.467	52.7%	0.383	0.02%
		1.566	0.3%	1.099	55.8%
				1.292	43.8%
				1.482	0.06%
Krypton-85 (10.73y)	β–	0.158	0.43%		
		0.672	99.57%		
		via ^{85m}Rb(0.96μs)			
				0.514	0.43%

(*Continued*)

(Cont.)

Nuclide and half-life	Type of decay	Particle energies and transition probabilities		Electromagnetic transitions	
		energy MeV	Transition probability	photon energy Mev	photons emitted
Lanthanum-140	β−	1.247	11%	0.131	0.8%
(40.27h)		1.253	6%	0.242	0.6%
		1.288	1%	0.266	0.7%
		1.305	5%	0.329	21%
		1.357	45%	0.432	3.3%
		1.421	5%	0.487	45%
		1.685	18%	0.752	4.4%
		2.172	7%	0.816	23%
		others	low	0.868	5.5%
				0.920	2.5%
				0.925	6.9%
				0.950	0.6%
				1.597	95.6%
				2.348	0.9%
				2.522	3.3%
				others	<0.5% each
Lead-210	α		2×10^{-6}%	0.046	~4%
(22.3y)	β−	0.015	~80%	0.009 -	
		0.061	~20%	0.017	~21%
					(Bi LX-rays)
	Daughter ^{210}Bi				
		(5.01d)			
	α	~4.67	$~1.3 \times 10^{-4}$%		
	β−	1.161	~100%		
	Daughter ^{210}Po				
		(138.38d)			
	α	5.305	100%		
Manganese-54	e.c.		100%	0.835	100%
(312.5d)				0.0055	~25%
					(Cr KX-rays)
Mercury-203	β−	0.212	100%	0.279	81.5%
(46.6d)				0.071–0.085	12.8%
					(Tl KX-rays)
Molybdenum-99	β−	0.454	18.3%	0.041	1.2%
(66.2h)		0.866	1.4%	0.141	5.4%
		1.232	80%	0.181	6.6%
		others	0.03%	0.366	1.4%
				0.412	0.02%
				0.529	0.05%
				0.621	0.02%
				0.740	13.6%
				0.778	4.7%
				0.823	0.13%
				0.961	0.1%
				via 6.02h ^{99m}Tc in equilibrium	
				0.002	~0%
				0.141	83.9%
				0.143	0.03%
Neptunium-237	α	4.638	6%	0.029	12%

(Cont.)

Nuclide and half-life	Type of decay	Particle energies and transition probabilities		Electromagnetic transitions	
		energy MeV	Transition probability	photon energy Mev	photons emitted
$(2.14 \times 10^6 y)$		4.663	3.3%	0.087	13%
		4.765	8%	0.106	0.08%
		4.770	25%	0.118	0.18%
		4.787	47%	0.131	0.09%
		4.802	~3%	0.134	0.07%
		4.816	2.5%	0.143	0.44%
		4.872	2.6%	0.151	0.25%
		others	<2%	0.155	0.10%
				0.169	0.08%
				0.193	0.06%
				0.195	0.21%
				0.202	0.05%
				0.212	0.17%
				0.214	0.05%
				0.238	0.07%
				others	low intensity
	Daughter ^{233}Pa				
		(27.0d)			
		0.15	27%	0.075	1.3%
		0.17	15%	0.087	2.0%
		0.23	36%	0.104	0.7%
		0.26	17%	0.300	6.5%
		0.53	2%	0.312	38%
		0.57	3%	0.340	4.3%
				0.375	0.7%
				0.398	1.3%
				0.416	1.7%
Nickel-63 (100y)	β−	0.066	100%		
Niobium-95 (35.0d)	β−	0.160	>99.9%	0.766	99.8%
Phosphorus-32 (14.3d)	β−	1.709	100%		
Plutonium-238 (87.7y)	α	5.445	28.7%	0.043	IC
		5.499	71.1%	U LX-rays (0.011–0.022)	~13%
		others	0.2%	U KX-rays (0.094–0.115)	~2.1 × 10^{-4}%
Polonium-210 (138.38d)	α	4.5	0.001%	0.802	0.0012%
		5.305	100%		
Potassium-42 (12.36h)	β−	1.683	0.3%	0.312	0.3%
		1.995	17.6%	0.900	0.05%
		3.520	82%	1.021	0.02%
		others	0.1%	1.525	17.9%
				1.921	0.04%
				2.424	0.02%

(*Continued*)

(Cont.)

Nuclide and half-life	Type of decay	Particle energies and transition probabilities		Electromagnetic transitions	
		energy MeV	Transition probability	photon energy Mev	photons emitted
				others	<0.01% each
Promethium-147 (2.623y)	β−	0.103	low	0.121	2.85×10^{-3}%
		0.225	~100%		
Radium-226 (1600y) via daughters in equilibrium:	α	4.598	5.5%	0.186	3.4%
		4.781	94.5%		
^{222}Rn(3.824d)	α	5.486	100%		
^{218}Po(3.05m)	α	6.000	~100%		
	β−	0.277	~0.02%		
^{218}At(~2s)	α,β−		very low		
^{218}Rn(3.0 x10^{-2}s)	α		very low		
^{214}Pb(26.8m)	β−	0.21	0.5%	0.053	IC
		0.51	15.5%	0.242	6.7%
		0.69	42%	0.295	16.9%
		0.74	36%	0.352	32.0%
		1.03	6%		
^{214}Bi(19.8m)	α	4.9–5.5	0.02%	0.273	5.3%
	β−	0.42	11%	0.609	41.7%
		1.02	23%	0.769	5.3%
		1.51	18%	1.120	14.3%
		1.55	15%	1.238	5.0%
		1.88	9%	1.378	4.8%
		2.6	4%	1.764	15.9%
		3.27	20%	2.204	5.3%
^{214}Po(1.62 ×10^{-4}s)	α	7.688	100%		
^{210}Tl(1.30m)	β−		very low		
^{210}Pb and daughters (not necessarily in equilibrium)					
^{210}Pb(22.3y)	α		2×10^{-6}%	0.046	~4%
	β−	0.015	~80%	Bi LX-rays	~21%
		0.61	~20%	(0.009–0.017)	
daughters:†					
^{210}Bi(5.01d)	α	4.67	1.3×10^{-4}%		
	β−	1.161	100%		
^{210}Po(138.38d)	α	4.5	0.001%	0.802	0.0012%
		5.305	100%		
^{206}Tl(4.20m)	β−	present in very low abundance			
		† percentages relate to disintegration of each daughter			
Rubidium-86 (18.7d)	β−	0.69	8.8%	1.077	8.8%
		1.77	91.2%		
Ruthenium-103 (39.26d)	β−	0.101	6.3%	0.053	0.4%
		0.214	89.0%	0.113	~0.01%
		0.456	0.3%	0.242	~0.01%
		0.711	4.4%	0.295	0.3%
				0.444	0.4%
				0.497	88.2%
				0.557	0.8%

(Continued)

(Cont.)

Nuclide and half-life	Type of decay	Particle energies and transition probabilities		Electromagnetic transitions	
		energy MeV	Transition probability	photon energy Mev	photons emitted
				0.610	5.5%
				0.020–0.023	~0.9(Rh KX-rays)
				via ^{103m}Rh(56m)	
				0.040	0.1%
				0.020–0.023	~8%(Rh KX-rays)
Ruthenium-106 (369d)	β−	0.039	100%		
	via ^{106}Rh(30.4s)				
	β−	1.98	1.7%	0.512	20.6%
		2.41	10.5%	0.616	0.7%
		3.03	8.4%	0.622	9.9%
		3.54	78.9%	0.874	0.4%
		others	0.5%	1.050	1.5%
				1.128	0.4%
				1.562	0.2%
				others	<0.1% each
Scandium-46 (83.3d)	β−	0.357	~100%	0.889	100%
		1.48	0.004%	1.121	100%
Selenium-75 (119.8d)	e.c.		100%	0.066	1.1%
				0.097	3.5%
				0.121	17.3%
				0.136	59.0%
				0.199	1.5%
				0.265	59.1%
				0.280	25.2%
				0.401	11.6%
				others	<0.05% each
				As KX-rays (0.010–0.012)	~50%
	via ^{75m}As(16.4ms):				
				0.024	0.03%
				0.280	5.4%
				0.304	1.2%
				As KX-rays (0.010–0.012)	~2.6%
Silver-110m (249.8d)	β−	0.084	67.6%	0.116	0%
		0.531	31%	0.447	3.4%
		others	low	0.620	2.7%
	i.t.		1.4%	0.658	94.2%
				0.678	11.1%
				0.687	6.9%
				0.707	16.3%
				0.744	4.5%
				0.764	22.5%
				0.818	7.2%
				0.885	71.7%
				0.937	34.4%

(Continued)

(Cont.)

Nuclide and half-life	Type of decay	Particle energies and transition probabilities		Electromagnetic transitions	
		energy MeV	Transition probability	photon energy Mev	photons emitted
				1.384	25.7%
				1.476	4.1%
				1.505	13.7%
				1.562	1.2%
	via ^{110}Ag(24.5s)				
	β–	2.23	0.1%	0.658	0.1%
		2.89	1.3%	others	low intensity
Sodium-22	β+	0.546	90.49%	0.511	from β+
(2.60y)		1.820	0.05%	1.275	99.95%
	e.c.		9.46%		
Sodium-24	β–	0.284	0.08%	1.369	100%
(15.02h)		1.392	99.92%	2.754	99.85%
				3.861	0.08%
Strontium-85	e.c.		100%	0.36	0.002%
(64.8d)				0.88	0.01%
				0.013–0.015	~60%
					(Rb KX-rays)
				via ^{85m}Rb	
				(0.96×s)	
				0.514	99.2%
Strontium-89	β–	0.554	~0.01%		
(50.5d)		1.463	~100%		
				via ^{89m}Y (16s)	
				0.909	~0.01%
Strontium-90	β–	0.546	100%		
(28.6y)					
	via ^{90}Y(64.1h):				
		0.513	~0.02%	1.761	IC
		2.274	~99.98%		
Sulfur-35	β–	0.167	100%		
(87.4d)					
Technetium-99	β–	0.204	low	0.089	6×10^{-4}%
(2.13×10^5y)		0.293	~100%		
Technetium-99m	i.t.		100%	0.002	~0%
(6.02h)				0.141	88.5%
				0.143	0.03%
	Daughter ^{99}Tc				
Terbium-160	β–	0.441	4.4%	0.087	13.8%
(72.3d)		0.481	10.0%	0.197	5.2%
		0.553	3.3%	0.216	3.9%
		0.575	46.4%	0.299	26.9%
		0.791	6.8%	0.765	2.0%
		0.874	26.8%	0.879	29.8%
		others	2.3%	0.962	9.9%
				0.966	24.9%
				1.178	15.1%
				1.200	2.3%

(*Continued*)

(Cont.)

Nuclide and half-life	Type of decay	Particle energies and transition probabilities		Electromagnetic transitions	
		energy MeV	Transition probability	photon energy Mev	photons emitted
				1.272	7.6%
				1.312	2.9%
				others	<2% each
Thallium-201	e.c.		100%	0.031	0.22%
(73.1h)				0.032	0.22%
				0.135	2.65%
				0.166	0.16%
				0.167	10.0%
				0.068–0.082	~95%(Hg KX-rays)
Thallium-204	β−	0.763	97.4%	0.069–0.083	~1.5%
(3.78y)	e.c.		2.6%		(Hg KX-rays)
Thorium-228	α	5.140	0.03%	0.085	1.6%
(1.913y)		5.176	0.2%	0.132	0.19%
		5.211	0.4%	0.167	0.12%
		5.341	28%	0.216	0.29%
		5.424	71%		
		others	low		
	via daughters in equilibrium (percentages refer to disintegrations of each daughter)				
Radium-224	α	5.447	5.2%	0.241	4.2%
(3.64d)		5.684	94.8%		
Radon-220	α	5.747	0.07%	0.542	0.07%
(55.3s)		6.288	99.93%		
Polonium-216	α	5.984	~0.002%	0.808	0.002%
(0.15s)		6.777	~100%		
Lead-212	β−	0.155	5%	0.115	0.6%
(10.6h)		0.332	82%	0.239	44.8%
		0.571	13%	0.300	3.4%
Bismuth-212	α	5.607	0.4%	0.040	1%
(60.6m)		5.768	0.6%	0.288	0.3%
		6.051	25.2%	0.328	0.1%
		6.090	9.6%	0.453	0.4%
		others	low	0.727	6.6%
				0.785	1.1%
	β−	0.445	0.7%	0.893	0.4%
		0.572	0.3%	0.952	0.2%
		0.630	1.9%	1.079	0.5%
		0.738	1.5%	1.513	0.3%
		1.524	4.5%	1.621	1.5%
		2.251	55.2%	1.680	0.1%
		others	low	1.806	0.1%
Polonium-212					
(3.05 × 10⁻⁷s)	α	8.785	100%		
Thulium-170	β−	0.884	22.8%	0.084	3.4%
(128d)		0.968	77%		
	e.c.		0.2%		
Tin-113	e.c.		100%	0.255	2.1%

(*Continued*)

(Cont.)

Nuclide and half-life	Type of decay	Particle energies and transition probabilities		Electromagnetic transitions	
		energy MeV	Transition probability	photon energy Mev	photons emitted
(115.1d)				0.024–0.028	73% (In KX-rays)
	Daughter ^{113m}In		99.5%	0.392	64.9%
	i.t.		100%	0.024–0.028	24%(In KX-rays)
Tritium		see Hydrogen-3			
Tungsten-185	β−	0.304	low	0.125	~0.005%
(75.1d)		0.429	>99.9%		
Uranium-238	α	4.145	23%	0.048	IC
(4.49 × 10^{9}y)		4.195	77%		
daughters in equlibrium:					
^{234}Th	β−	0.100	12%	0.030	IC
(24.1d)		0.101	21%	0.063	5.7%
		0.193	67%	0.092	3.2%
				0.093	3.6%
^{234m}Pa	β−	2.29	98%	0.043	IC
(1.17m)		others	0.13%	0.767	0.2%
	i.t.		low	0.810	0.5%
^{234}Pa	β−			1.001	0.6%
(6.70h)			very low		
^{234}U	α	4.723	27.5%	0.053	0.1%
(2.48 × 10^{5}y)		4.773	72.5%		
Xenon-133	β−	0.266	0.9%	0.080	0.4%
(5.25d)		0.346	99.1%	0.081	36.6%
				0.160	0.05%
				0.030–0.036	~46%(Cs KX-rays)
Yttrium-88	β+	0.763	~0.2%	0.511	from β+
(106.6d)	e.c.		99.8%	0.898	93.2%
				1.383	0.04%
				1.836	99.36%
				2.734	0.6%
				3.219	0.009%
				3.52	0.007%
				Sr KX-rays (0.014–0.016)	~60%
Yttrium-90	β−	0.513	~0.02%	1.761	0.01%
(64.1h)		2.274	~99.98%		
Yttrium-91	β−	0.340	0.3%	1.205	0.3%
(58.5d)		1.545	99.7%		
Zinc-65	β+	0.325	1.46%	0.345	~0.003%
(243.8d)	e.c.		98.54%	0.511	from β+
				0.770	~0.003%
				1.115	50.7%
				0.008–0.009	~38%(Cu KX-rays)

(*Continued*)

(Cont.)

Nuclide and half-life	Type of decay	Particle energies and transition probabilities		Electromagnetic transitions	
		energy MeV	Transition probability	photon energy Mev	photons emitted
Zirconium-95	β−	0.365	54.7%	0.724	44.5%
(64.0d)		0.398	44.6%	0.757	54.6%
		0.887	0.7%		
		1.12	low		
				via 86.6h ^{95m}Nb in equilibrium	
				0.235	0.2%
	Daughter ^{95}Nb				

Index

9 780367 779689